MW01492271

NOISE AND VIBRATION CONTROL ENGINEERING

NOISE AND VIBRATION CONTROL ENGINEERING

PRINCIPLES AND APPLICATIONS

Edited by
Leo L. Beranek
István L. Vér

A WILEY-INTERSCIENCE PUBLICATION

JOHN WILEY & SONS, INC.

New York • Chichester • Brisbane • Toronto • Singapore

The statistics, data and other information in this book have been obtained from many sources, including government organizations and professional engineering organizations. The authors and publisher have made every reasonable effort to make this book accurate and authoritative, but do not warrant and assume no liability for, the accuracy or completeness of the text or its fitness for any particular purpose. The authors and publisher do not intend for this book to be used as a substitute for professional engineering advice.

In recognition of the importance of preserving what has been written, it is a policy of John Wiley & Sons, Inc., to have books of enduring value published in the United States printed on acid-free paper, and we exert our best efforts to that end.

Copyright © 1992 by John Wiley & Sons, Inc.

All rights reserved. Published simultaneously in Canada.

Reproduction or translation of any part of this work beyond that permitted by Section 107 or 108 of the 1976 United States Copyright Act without the permission of the copyright owner is unlawful. Requests for permission or further information should be addressed to the Permissions Department, John Wiley & Sons, Inc.

Library of Congress Cataloging in Publication Data:

Noise and vibration control engineering: principles and applications / edited by L. L. Beranek, I. L. Vér.

p. cm.

Includes index.

ISBN 0-471-61751-2

1. Noise control. 2. Vibration. 3. Soundproofing. I. Beranek, Leo Leroy, 1914– . II. Vér, I. L. (István L.), 1934–

TD892.N6512 1992

620.2'3—dc20 92-11347

CIP

Printed in the United States of America

10 9 8 7 6 5 4 3

CONTRIBUTORS

Grant S. Anderson, Harris Miller Miller and Hanson Inc., Lexington, Massachusetts

Hans D. Baumann, H. D. Baumann Associates, Portsmouth, New Hampshire

Leo L. Beranek, Consultant, Cambridge, Massachusetts

Larry J. Eriksson, Nelson Industries, Stoughton, Wisconsin

Anthony G. Galaitsis, Bolt Beranek and Newman, Inc., Cambridge, Massachusetts

Henning E. von Gierke, Harry G. Armstrong Aerospace Medical Research Laboratory, Wright-Patterson Airforce Base, Ohio

Murray Hodgson, National Research Council, Ottawa, Ontario, Canada

Michael S. Howe, Bolt Beranek and Newman, Inc., Cambridge, Massachusetts

Ulrich J. Kurze, Müller BBM GmbH, München, Germany

William W. Lang, IBM Corporation, Poughkeepsie, New York

George C. Maling, Jr., IBM Corporation, Poughkeepsie, New York

William D. Mark, Bolt Beranek and Newman, Inc., Cambridge, Massachusetts Currently: Pennsylvania State University, University Park, Pennsylvania

Fridolin P. Mechel, Fraunhofer Institut Für Bauphysik, Stuttgart, Germany

Donald J. Nefske, General Motors Corporation, Warren, Michigan

Charles W. Nixon, Harry G. Armstrong Aerospace Medical Research Laboratory, Wright-Patterson Airforce Base, Ohio

Allan G. Piersol, Piersol Engineering Company, Woodland Hills, California

Theo Priede, University of Capetown, Capetown, South Africa

Shung H. Sung, General Motors Corporation, Warren, Michigan

László Timár-Peregrin, Technical University Budapest, Budapest, Hungary

Eric E. Ungar, Bolt Beranek and Newman, Inc., Cambridge, Massachusetts

István L. Vér, Bolt Beranek and Newman, Inc., Cambridge, Massachusetts

Eric W. Wood, Acentech, Inc., Cambridge, Massachusetts

Alfred C. C. Warnock, National Research Council, Ottawa, Ontario, Canada

PREFACE

The control of noise and vibration has commanded the attention of increasing numbers of manufacturers and mechanical engineers as the public's demand for quiet has expanded. Concert hall music is played with satisfaction in today's better automobiles. Factory and outdoor machinery, motors for automobiles and boats, power plants, air moving devices, and home appliances can be made significantly quieter with today's techniques. The aim of this book is to provide design engineers with specific technical information needed for solutions of the most frequently encountered noise and vibration problems. The intent is to convey physical insights into the processes of noise generation and attenuation so that a chosen solution is not only acoustically sufficient but also cost effective.

This text is new; only the introductory material and a few figures resemble material that appeared in the 1971 book, *Noise and Vibration Control*, edited by one of the present authors. Only four of the present twenty-three chapter authors participated in the writing of the earlier text. Completely new are chapters on the control of noise by active devices, and the quieting of internal combustion engines, electrical machinery, and gears. Also, the book draws on the experience of foreign experts where their publications revealed important advances.

We are particularly indebted to R. B. Coleman, M. Heckl, J. B. Moreland, C. L. Morfey, D. J. Nefske, M. G. Prasad, P. J. Remington, and J. H. Rindel who rendered help and criticism during the preparation of this book, and the many individuals and organizations who gave us permission to use copywrited technical information. We are also indebted to Bolt Beranek and Newman, Inc. for encouraging members of its senior technical staff to contribute chapters and for assistance in many ways, and to the Alexander von Humboldt Foundation for enabling one of the editors to reside in Germany while studying recent results of European acoustic research. It gives us special pleasure to acknowledge the help of Mrs. Judy A. Derle who assisted in the preparation of many of the chapters and to Frank Cerra and Edward Cantillon, our editors at John Wiley & Sons, for their effective help and guidance in the production of this book.

LEO L. BERANEK
ISTVÁN L. VÉR

Cambridge, Massachusetts
May 1992

CONTENTS

NOISE AND VIBRATION CONTROL ENGINEERING

CHAPTER 1

Basic Acoustical Quantities: Levels and Decibels

LEO L. BERANEK
Consultant
Cambridge, Massachusetts

1.1 BASIC QUANTITIES OF SOUND WAVES

Sound Waves and Noise

In the broadest sense, a *sound wave* is any disturbance that is propagated in an elastic medium, which may be a gas, a liquid, or a solid. Ultrasonic, sonic, and infrasonic waves are included in this definition. Most of this text deals with sonic waves, those sound waves that can be perceived by the hearing sense of a human being. *Noise* is defined as any perceived sound that is objectionable to a human being. The concepts basic to this chapter can be found in references 1–5. Portions are further expanded in Chapter 2.

Sound Pressure

A person who is not deaf perceives as sound any vibration of the eardrum in the audible frequency range that results from an incremental variation in air pressure at the ear. A variation in pressure above and below atmospheric pressure is called *sound pressure*, in units of pascals (Pa). A young person with normal hearing can

Noise and Vibration Control Engineering: Principles and Applications, Edited by Leo L. Beranek and István L. Vér.
ISBN 0-471-61751-2 © 1992 John Wiley & Sons, Inc.

perceive sound in the frequency range of roughly 15–16,000 Hz (hertz), defined as the normal audible frequency range.

Because the hearing mechanism responds to sound pressure, it is one of two quantities that is usually measured in engineering acoustics. The normal ear is most sensitive at frequencies between 3000 and 6000 Hz, and a young person can detect pressures as low as about 20 μPa, compared to the normal atmospheric pressure (101.3×10^3 Pa) around which it varies, a fractional variation of 2×10^{-10}.

Sound Spectra

A sound wave may be comprised of a pure tone (single frequency, e.g., 1000 Hz), a combination of single frequencies harmonically related, or a combination of single frequencies not harmonically related, either finite or infinite in number. A combination of a finite number of tones is said to have a *line spectrum*. A combination of an infinite (large) number of tones has a *continuous spectrum*. A continuous-spectrum noise for which the amplitudes versus time occur with a normal (Gaussian) distribution is called *random noise*. Three of these types of noise are shown by the frequency spectra in Figs. 1.1*a–c*. A combination of a line and a continuous spectrum, called a complex spectrum, is shown in Fig. 1.1*d*.

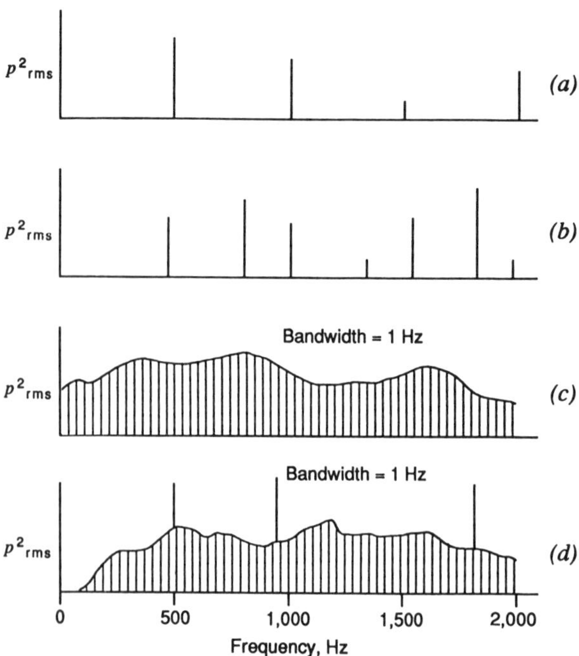

Fig. 1.1 Mean-square sound pressure spectra: (*a*) Harmonically related line spectrum; (*b*) inharmonically related line spectrum; (*c*) continuous power spectral density spectrum; (*d*) combination line and continuous power spectral density spectrum (complex spectrum).

Regardless of which type of sound wave is considered, when propagating at normal sound pressure amplitudes (to avoid nonlinearity) in air over reasonably short distances (so that sound attenuation in the air itself, which becomes significant at frequencies above 1000 Hz, can be neglected), the waveform is unchanged. Thus, a violin heard at a distance of 30 m sounds the same as at 5 m, although it is less loud.

Sound Intensity

The second quantity commonly measured in engineering acoustics is *sound intensity*, defined as the continuous flow of power carried by a sound wave through an incrementally small area at a point in space. The units are watts per square meter (W/m^2). This quantity is important for two reasons. First, at a point in free space, it is related to the total power radiated into the air by a sound source, and second, it bears at that point a fixed relation to the sound pressure.

Sound intensity at a point is directional (a vector) in the sense that the position of the plane of the incrementally small area can vary from being perpendicular to the direction in which the wave is traveling to being parallel to that direction. It has its maximum value, I_{max}, when its plane is perpendicular to the direction of travel. When parallel, the sound intensity is zero. In between, the component of I_{max} varies as the cosine of the angle formed by the direction of travel and a line perpendicular to the incremental area.

Another equation, which we shall develop in the next chapter, relates sound pressure to sound intensity. In an environment in which there are no reflecting surfaces, the sound pressure *at any point in any type of freely traveling* (plane, cylindrical, spherical, etc.) wave is related to the maximum intensity I_{max} by

$$p_{rms}^2 = I_{max} \cdot \rho c \qquad \text{Pa}^2 \tag{1.1}$$

where p_{rms} = root-mean-square (rms) sound pressure, Pa (N/m^2)
ρ = density of air, kg/m^3
c = speed of sound in air, m/s [see Eq. (1.4)]
N = force, N

Sound Power

A sound source radiates a measurable amount of power into the surrounding air, called *sound power*, in watts. If the source is nondirectional, it is is said to be a *spherical sound source* (see Fig. 1.2). For such a sound source the measured (maximum) sound intensities at all points on an imaginary spherical surface centered on the acoustic center of the source are equal. Mathematically,

$$W_s = (4\pi r^2) I_s(r) \qquad \text{W} \tag{1.2}$$

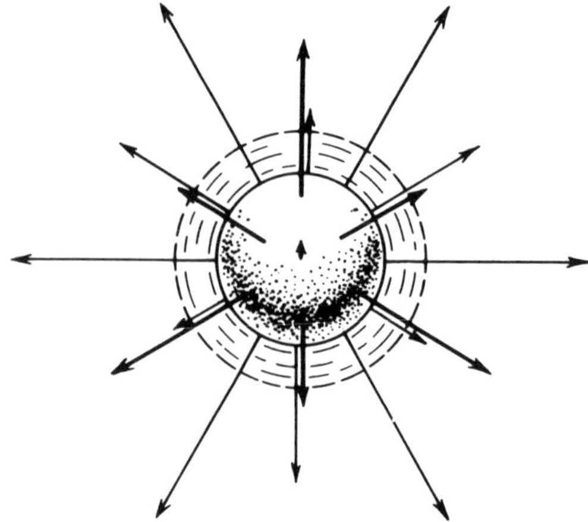

Fig. 1.2 Generation of a one-dimensional spherical wave. A balloonlike surface pulsates uniformly about some equilibrium position and produces a sound wave that propagates radially away from the balloon at the speed of sound.

where $I_s(r)$ = maximum sound intensity at radius r at surface of an imaginary sphere surrounding source, W/m^2.
W_s = total sound power radiated by source, W (N · m/s)
r = distance from acoustical center of source to surface of imaginary sphere, m

A similar statement can be made about a line source, that is, the maximum sound intensities at all points on an imaginary cylindrical surface around a *cylindrical sound source*, $I_c(r)$, are equal:

$$W_c = (2\pi rl)I_c(r) \qquad \text{W} \tag{1.3}$$

where W_c = total sound power radiated by cylinder of length l, W
r = distance from acoustical centerline of cylindrical source to imaginary cylindrical surface surrounding source,

Particle Velocity

Consider that the surface of the spherical source of Fig. 1.2 is expanding and contracting sinusoidally. During the first half of its sinusoidal motion, it pushes the air particles near its surface outward. Because air is elastic and compresses, the pressure near the surface will increase. This increased pressure overcomes the inertia of the air particles a short distance away and they move outward. That outward movement causes the pressure to build up at this new distance and, in turn, pushes more removed particles outward. This outward movement of the disturbance takes place at the speed of sound.

In the next half of the sinusoid, the sinusoidally vibrating spherical source reverses direction, creating a drop of pressure near its surface that pulls nearby air particles toward it. This reverse disturbance also propagates outward with the speed of sound. Thus, at any one point in space, there will be sinusoidal to-and-fro movement of the particles, called the *particle velocity*. Also at any point, there will be sinusoidal rise and fall of sound pressure.

From the basic equations governing the propagation of sound, as is shown in the next chapter, we can say the following:

1. In a *plane wave* (approximated at a large distance from a point source) propagated in free space (no reflecting surfaces) the sound pressure and the particle velocity reach their maximum and minimum values at the same instant and are said to be in phase.
2. In such a wave the particles move back and forth along the line in which the wave is traveling. In reference to the spherical radiation discussed above, this means that the particle velocity is always perpendicular to the imaginary spherical surface (the wave front) in space. This type of wave is called a longitudinal, or compressional, wave. By contrast, a transverse wave is illustrated by a surface wave in water where the particle velocity is perpendicular to the water surface while the wave propagates in a direction parallel to the surface.

Speed of Sound

A sound wave travels outward at a rate dependent on the elasticity and density of the air. Mathematically, the *speed of sound* in air is calculated as

$$c = \sqrt{\frac{1.4P_s}{\rho}} \quad \text{m/s} \tag{1.4}$$

where P_s = atmospheric (ambient) pressure, Pa
ρ = density of air, kg/m^3

For all practical purposes, the speed of sound is dependent only on the absolute temperature of the air. The equations for the speed of sound are

$$c = 20.05\sqrt{T} \quad \text{m/s} \tag{1.5}$$

$$c = 49.03\sqrt{R} \quad \text{ft/s} \tag{1.6}$$

where T = absolute temperature of air in degrees Kelvin, equal to 273.2 plus the temperature in degrees Celsius
R = absolute temperature in degrees Rankine, equal to 459.7 plus the temperature in degrees Fahrenheit

For temperatures near 20°C (68°F), the speed of sound is

$$c = 331.5 + 0.58°\text{C} \quad \text{m/s} \tag{1.7}$$

$$c = 1054 + 1.07°\text{F} \quad \text{ft/s} \tag{1.8}$$

Period

The reciprocal of the frequency f of a pure tone is the *period* T in seconds. It is the time required for one complete cycle of the sinusoidal tone. Thus the period of a 500-Hz wave is 0.002 s.

Wavelength

Wavelength is defined as the distance the pure-tone wave travels during a full period. It is denoted by the Greek letter λ and is equal to the speed of sound divided by the frequency of the pure tone:

$$\lambda = cT = \frac{c}{f} \quad \text{m} \tag{1.9}$$

Sound Energy Density

In standing-wave situations, such as sound waves in closed, rigid-wall tubes, rooms containing little sound-absorbing material, or reverberation chambers, the quantity desired is not sound intensity, but rather the *sound energy density*, namely, the energy (kinetic and potential) stored in a small volume of air in the room owing to the presence of the standing-wave field. The relation between the *space-averaged sound energy density* D and the *space-averaged squared sound pressure* is

$$D = \frac{p_{\text{av}}^2}{\rho c^2} = \frac{p_{\text{av}}^2}{1.4P_s} \quad \text{W} \cdot \text{s/m}^3$$

$$(\text{J/m}^3 \text{ or simply N/m}^2) \tag{1.10}$$

where p_{av}^2 = space average of mean-square sound pressure in space, determined from data obtained by moving a microphone along a tube or around a room or from samples at various points, Pa^2

P_s = atmospheric pressure, Pa; under normal atmospheric conditions, at sea level, $P_s = 1.013 \times 10^5$ Pa

1.2 SOUND SPECTRA

In the previous section we described sound waves with line and continuous spectra. Here we shall discuss how to quantify such spectra.

Continuous Spectra

As stated before, a continuous spectrum can be represented by a large number of pure tones between two frequency limits, whether those limits are apart 1 Hz or thousands of hertz (see Fig. 1.3). Because the hearing system extends over a large frequency range and is not equally sensitive to all frequencies, it is customary to measure a continuous spectrum sound in a series of contiguous frequency bands using a sound analyzer.

Customary bandwidths are one-third octave and one octave (see Fig. 1.4 and Table 1.1). The rms value of such a filtered sound pressure is called the one-third-octave-band or the octave-band sound pressure, respectively. If the filter bandwidth is 1 Hz, a plot of the filtered *mean-square* pressure of a *continuous-spectrum* sound versus frequency is called the *power spectral (or spectrum) density spec-*

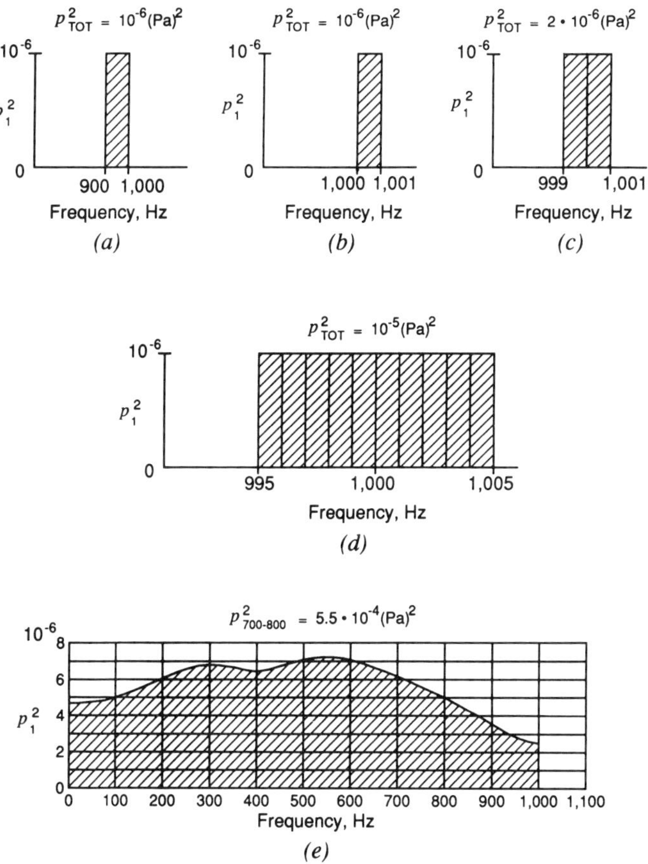

Fig. 1.3 Power spectral density specta showing the linear growth of total mean-square sound pressure when the bandwidth of noise is increased. In each case, p_1^2 is the mean-square sound pressure in a band 1 Hz wide.

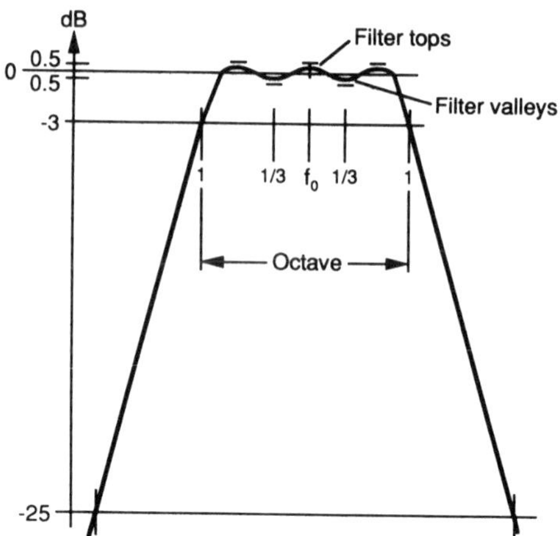

Fig. 1.4 Frequency response of one manufacturer's octave-band filter. The 3-dB down points for the octave band and for a one-third-octave-band filter are indicated.

trum. Narrow bandwidths are commonly used in analyses of machinery noise and vibration, but the term "spectral density" has no meaning in cases where pure tones are being measured.

The mean-square (or rms) sound pressure can be determined for each of the contiguous frequency bands and the result plotted as a function of frequency (Fig. 1.3*e*).

Bandwidth Conversion. It is frequently necessary to convert sounds measured with one set of bandwidths to a different set of bandwidths or to reduce both sets of measurements to a third set of bandwidths. Let us imagine that we have a machine that at a point in space produces a mean-square sound pressure of $p_1^2 = 10^{-6}$ Pa^2 in a 1-Hz bandwidth between 999 and 1000 Hz (Fig. 1.3*a*). Now imagine that we have a second machine the same distance away that radiates the same power but is confined to a bandwidth between 1000 and 1001 Hz (Fig. 1.3*b*). The total spectrum now becomes that shown in Fig. 1.3*c* and the total mean-square pressure is twice that in either band. Similarly, 10 machines would produce 10 times the mean-square sound pressure of any one (Fig. 1.3*d*).

In other words, if the power spectral density spectrum in a frequency band of width Δf is flat (the mean-square sound pressures in all the 1-Hz-wide bands within the band are equal), the total mean-square sound pressure for the band is given by

$$p_{\mathrm{tot}}^2 = p_1^2 \frac{\Delta f}{\Delta f_0} \qquad \mathrm{Pa}^2 \tag{1.11}$$

where $\Delta f_0 = 1$ Hz.

As an example, assume that we wish to convert the power spectral density

TABLE 1.1 Center and Approximate Cutoff Frequencies (Hz) for Standard Set of Contiguous-Octave and One-Third-Octave Bands Covering Audio frequency Range[a]

	Octave			One-Third Octave		
Band	Lower Band Limit	Center	Upper Band Limit	Lower Band Limit	Center	Upper Band Limit
12	11	16	22	14.1	16	17.8
13				17.8	20	22.4
14				22.4	25	28.2
15	22	31.5	44	28.2	31.5	35.5
16				35.5	40	44.7
17				44.7	50	56.2
18	44	63	88	56.2	63	70.8
19				70.8	80	89.1
20				89.1	100	112
21	88	125	177	112	125	141
22				141	160	178
23				178	200	224
24	177	250	355	224	250	282
25				282	315	355
26				355	400	447
27	355	500	710	447	500	562
28				562	630	708
29				708	800	891
30	710	1,000	1,420	891	1,000	1,122
31				1,122	1,250	1,413
32				1,413	1,600	1,778
33	1,420	2,000	2,840	1,778	2,000	2,239
34				2,239	2,500	2,818
35				2,818	3,150	3,548
36	2,840	4,000	5,680	3,548	4,000	4,467
37				4,467	5,000	5,623
38				5,623	6,300	7,079
39	5,680	8,000	11,360	7,079	8,000	8,913
40				8,913	10,000	11,220
41				11,220	12,500	14,130
42	11,360	16,000	22,720	14,130	16,000	17,780
43				17,780	20,000	22,390

[a]From ref. 6.

spectrum of Fig. 1.3*e*, which is a plot of $p_1^2(f)$, the mean-square sound pressure in 1-Hz bands, to a spectrum for which the mean-square sound pressure in 100-Hz bands, p_{tot}^2, is plotted versus frequency. Let us consider only the 700–800-Hz band. Because $p_1^2(f)$ is not equal throughout this band, we could painstakingly determine and add together the actual p_1^2's or, as is more usual, simply take the average value for the p_1^2's in that band and multiply by the bandwidth. Thus, for each 100-Hz band, the total mean-square sound pressure is given by Eq. (1.11), where p_1^2 is the average 1-Hz band quantity throughout the band. For the 700–800-Hz band, the average p_1^2 is 5.5×10^{-6} and the total is $5.5 \times 10^{-6} \times 100\text{ Hz} = 5.5 \times 10^{-4}\text{ Pa}^2$.

If mean-square sound pressure levels have been measured in a specific set of bandwidths such as one-third-octave bands, it is possible to present accurately the data in a set of wider bandwidths such as octave bands by simply adding together the mean-square sound pressures for the component bands. Obviously, it is not possible to reconstruct a narrower bandwith spectrum accurately (e.g., one-third-octave bands) from a wider bandwidth spectrum (e.g., octave bands.) However, it is sometimes necessary to make such a conversion in order to compare sets of data measured differently. Then the implicit assumption has to be made that the narrower band spectrum is continuous and monotonic within the larger band. In either direction, the conversion factor for each band is

$$p_B^2 = p_A^2 \frac{\Delta f_B}{\Delta f_A} \tag{1.12}$$

where p_A^2 is the measured mean-square sound pressure in a bandwidth Δf_A and p_B^2 is the desired mean-square sound pressure in the desired bandwidth Δf_B.

As an example, assume that we have measured a sound with a continuous spectrum using a one-octave-band filterset and have plotted the intensity for the contiguous bands versus the midfrequency of each band (see the upper three circles in Fig. 1.5). Assume that we wish to convert to an approximate one-third-octave-band spectrum to make comparisons with other data possible. Not knowing how the mean-square sound pressure varies throughout each octave band, we assume it to be continuous and monotonic and apply Eq. (1.12). In this example, approximately, $\Delta f_B = \frac{1}{3}\Delta f_A$ for all the bands (see Table 1.1), and the mean-square sound pressure in each one-third-octave band with the same midfrequency will be one-third that in the corresponding one-octave band. The data would be plotted versus the midfrequency of each one-third-octave band (see Table 1.1 and Fig. 1.5). Then the straight, sloping line is added under the assumption that the true one-third-octave spectrum is monotonic.

Complex Spectra

The mean-square sound pressure resulting from the combination of two or more pure tones of different amplitudes p_1, p_2, p_3 and *different frequencies* f_1, f_2, f_3 is given by

$$p_{\text{rms}}^2(\text{total}) = p_1^2 + p_2^2 + p_3^2 + \cdots \tag{1.13}$$

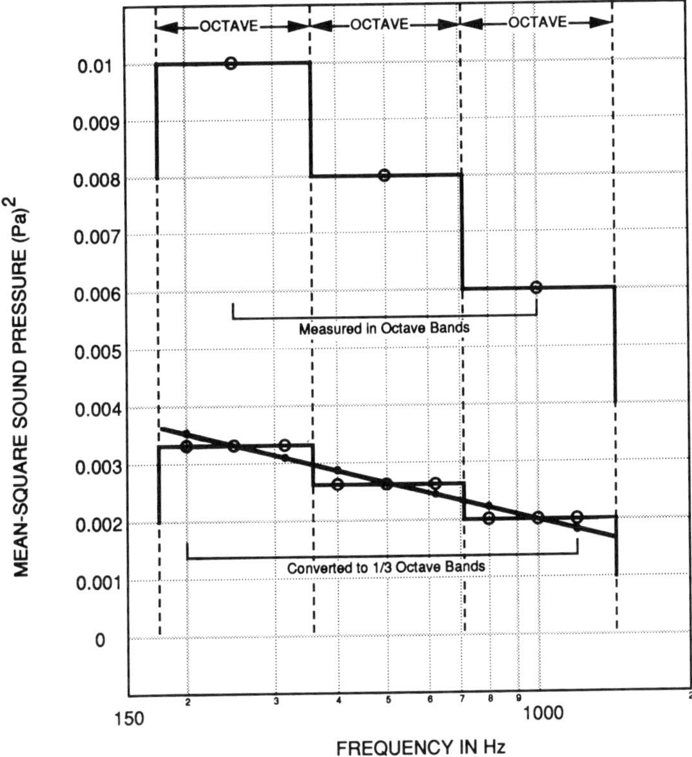

Fig. 1.5 Conversion of octave-band mean-square sound pressures into third-octave-band mean-square sound pressures. Such conversion should be made only where necessary under the assumption that the third-octave-band mean-square pressure levels decrease monotonically with band midfrequency. The sloping solid curve is the assumed-correct converted spectrum.

The mean-square sound pressure of two pure tones of the *same frequency* but different amplitudes and phases is found from

$$p_{\text{rms}}^2(\text{total}) = p_1^2 + p_2^2 + 2p_1p_2\cos(\theta_1 - \theta_2) \tag{1.14}$$

where the phase angle of each wave is represented by θ_1 or θ_2.

Comparison of Eqs. (1.13) and (1.14) reveals the importance of phase when combining two sine waves of the same frequency. If the phase difference $\theta_1 - \theta_2$ is zero, the two waves are in phase and the combination is at its maximum value. If $\theta_1 - \theta_2 = 180°$, the third term becomes $-2p_1p_2$ and the sum is at its minimum value. If the two waves are equal in amplitude, the minimum value is zero.

If one wishes to find the mean-square sound pressure of a number of waves all of which have different frequencies except, say, two, these two are added together according to Eq. (1.14) to obtain a new mean-square pressure. Then this mean-square pressure and the mean-square pressures of the remainder of the components are summed according to Eq. (1.13).

1.3 LEVELS[6]

Because of the wide range of sound pressures to which the ear responds (a ratio of 10^5 or more for a normal person), sound pressure is an inconvenient quantity to use in graphs and tables. This is also true for the other acoustical quantities listed above. Early in the history of the telephone it was decided to adopt logarithmic scales for representing acoustical quantities and the voltages encountered in associated electrical equipment.

As a result of that decision, sound powers, intensities, pressures, velocities, energy densities, and voltages from electroacoustic transducers are commonly stated in terms of the logarithm of the ratio of the measured quantity to an appropriate reference quantity. Because the sound pressure at the threshold of hearing at 1000 Hz is about 20 μPa, this was chosen as the fundamental reference quantity around which the other acoustical references have been chosen.

Whenever the magnitude of an acoustical quantity is given in this logarithmic form, it is said to be a *level* in *decibels* (dB) *above* or *below* a zero *reference level* that is determined by a *reference quantity*. The argument of the logarithm is always a ratio and, hence, is dimensionless. The level for a very large ratio, for example the power produced by a very powerful sound source, might be given with the unit *bel*, which equals 10 dB.

Power and Intensity Levels

Sound Power Level. Sound power level is defined as

$$L_W = 10 \log_{10} W/W_0 \qquad \text{dB } re\ W_0 \tag{1.15}$$

and conversely

$$W = W_0 \text{antilog}_{10} \frac{L_W}{10} = W_0 \times 10^{L_W/10} \qquad \text{W} \tag{1.16}$$

where W = sound power, W
W_0 = reference sound power, standardized at 10^{-12} W

As seen in Table 1.2, a *ratio* of 10 in the power W corresponds to a *level difference* of 10 dB regardless of the reference power W_0. Similarly, a ratio of 100 corresponds to a level difference of 20 dB. Power ratios of less than 1 are allowable: they simply lead to negative levels. For example (see Table 1.2), a power ratio of 0.1 corresponds to a level difference of -10 dB.

TABLE 1.2 Sound Powers and Sound Power Levels

Radiated Sound Power W, W		Sound Power Level L_W, dB	
Usual Notation	Equivalent Exponential Notation	Relative to 1 W	Relative to 10^{-12} W (standard)
100,000	10^5	50	170
10,000	10^4	40	160
1,000	10^3	30	150
100	10^2	20	140
10	10^1	10	130
1	1	0	120
0.1	10^{-1}	−10	110
0.01	10^{-2}	−20	100
0.001	10^{-3}	−30	90
0.000,1	10^{-4}	−40	80
0.000,01	10^{-5}	−50	70
0.000,001	10^{-6}	−60	60
0.000,000,1	10^{-7}	−70	50
0.000,000,01	10^{-8}	−80	40
0.000,000,001	10^{-9}	−90	30

Column 4 of Table 1.2 gives sound power levels relative to the standard reference power level $W_0 = 10^{-12}$ W.

Some sound power ratios and the corresponding sound-power-level differences are given in Table 1.3. We note from the last line that the sound power level for the *product of two ratios* is equal to the *sum of the levels* for the two ratios. For example, determine L_W for the quantity 2×4. From Table 1.3, $L_W = 3.0 + 6.0 = 9.0$ dB, which is the sound power level for the ratio 8. Similarly, L_W for a ratio of 8000 equals the sum of the levels for 8 and 1000, that is, $L_W = 9 + 30 = 39$ dB.

Sound Intensity Level. Sound intensity level, in decibels, is defined as

$$\text{Intensity level} = L_I = 10 \log \frac{I}{I_{\text{ref}}} \qquad \text{dB } re\ I_{\text{ref}} \tag{1.17}$$

where I = sound intensity whose level is being specified, W/m^2
I_{ref} = reference intensity standardized as $10^{-12}\ \text{W/m}^2$

Sound power levels should not be confused with intensity levels (or with sound pressure levels, which are defined next), which also are expressed in decibels. Sound power is a measure of the *total* acoustic power radiated by a source. Sound intensity and sound pressure specify the acoustical "disturbance" produced at a

TABLE 1.3 Selected Sound Power Ratios and Corresponding Power-Level Differences

Sound Power Ratio W/W_0, R	Sound-Power-Level Difference[a] 10 log W/W_0, L_W (dB)
1,000	30
100	20
10	10
9	9.5
8	9.0
7	8.5
6	7.8
5	7.0
4	6.0
3	4.8
2	3.0
1	0.0
0.9	−0.5
0.8	−1.0
0.7	−1.5
0.6	−2.2
0.5	−3.0
0.4	−4.0
0.3	−5.2
0.2	−7.0
0.1	−10
0.01	−20
0.001	−30
$R_1 \times R_2$	$L_{W_1} + L_{W_2}$

[a]To the nearest 0.1 dB.

point removed from the source. For example their levels depend on the distance from the source, losses in the intervening air path, and room effects (if indoors). A helpful analogy is to imagine that sound power level is related to the total rate of heat production of a furnace, while either of the other two levels is analogous to the temperature produced at a given point in a dwelling.

Sound Pressure Level

Almost all microphones used today respond to sound pressure, and in the public mind, the word *decibel* is commonly associated with sound pressure level or A-weighted sound pressure level (see Table 1.4). Strictly speaking, sound pressure level is analogous to intensity level, because, in calculating it, pressure is first

TABLE 1.4 A and C Electrical Weighting Networks for Sound-Level Meter[a]

Frequency, Hz	A-weighting Relative Response, dB	C-weighting Relative Response, dB
10	−70.4	−14.3
12.5	−63.4	−11.2
16	−56.7	−8.5
20	−50.5	−6.2
25	−44.7	−4.4
31.5	−39.4	−3.0
40	−34.6	−2.0
50	−30.2	−1.3
63	−26.2	−0.8
80	−22.5	−0.5
100	−19.1	−0.3
125	−16.1	−0.2
160	−13.4	−0.1
200	−10.9	0
250	−8.6	0
315	−6.6	0
400	−4.8	0
500	−3.2	0
630	−1.9	0
800	−0.8	0
1,000	0	0
1,250	+0.6	0
1,600	+1.0	−0.1
2,000	+1.2	−0.2
2,500	+1.3	−0.3
3,150	+1.2	−0.5
4,000	+1.0	−0.8
5,000	+0.5	−1.3
6,300	−0.1	−2.0
8,000	−1.1	−3.0
10,000	−2.5	−4.4
12,500	−4.3	−6.2
16,000	−6.6	−8.5
20,000	−9.3	−11.2

[a]These numbers assume a flat, diffuse-field (random-incidence) response for the sound-level meter and microphone.

squared, which makes it proportional to intensity (power per unit area):

$$\text{Sound pressure level} = L_p = 10 \log \left[\frac{p(t)}{p_{\text{ref}}}\right]^2$$

$$= 20 \log \frac{p(t)}{p_{\text{ref}}} \qquad \text{dB } re\ p_{\text{ref}} \tag{1.18}$$

where p_{ref} = reference sound pressure, standardized at 2×10^{-5} N/m² (20 μPa) for airborne sound; for other media, references may be 0.1 N/m² (1 dyn/cm²) or 1 μN/m² (1 μPa)

$p(t)$ = instantaneous sound pressure, Pa

Note that L_p *re* 20 μPa is 94 dB greater than L_p *re* 1 Pa.

As we shall show shortly, $p(t)^2$ is only proportional to sound intensity if its mean-square value is taken. Thus, in Eq. (1.18), $p(t)$ would be replaced by p_{rms}.

The relations among sound pressure levels (*re* 20 μPa) for pressures in the meter-kilogram-second (mks), centimeter-gram-second (cgs), and English systems of units are shown by the four nomograms of Fig. 1.6.

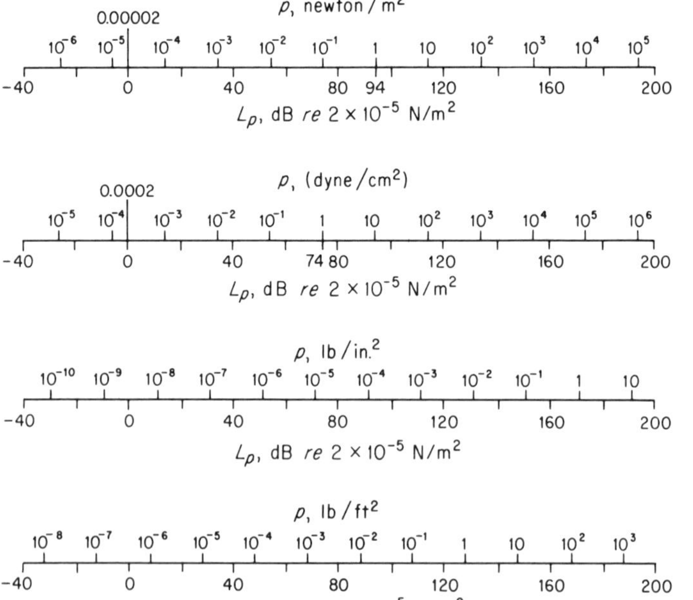

Fig. 1.6 Chart relating L_p (dB *re* 20 μPa) to p in N/m², dyn/cm², lb/in.², and lb/ft². For example, 1.0 Pa equals 94 dB *re* 20 μPa.

1.4 DEFINITIONS OF OTHER COMMONLY USED LEVELS AND QUANTITIES IN ACOUSTICS

Analogous to sound pressure level given in Eq. (1.18), *A-weighted Sound Pressure Level* L_A is given by

$$L_A = 10 \log \left[\frac{p_A(t)}{p_{\text{ref}}}\right]^2 \quad \text{dB} \tag{1.19}$$

where $p_A(t)$ is the instantaneous sound pressure measured using the standard frequency-weighting A (see Table 1.4).

Average sound level $L_{\text{av},T}$, is given by

$$L_{\text{av},T} = 10 \log \frac{\dfrac{1}{T}\displaystyle\int_0^T p^2(t)\, dt}{p_{\text{ref}}^2} \quad \text{dB} \tag{1.20}$$

where T is the (long) time over which the averaging takes place.

Average A-weighted sound level $L_{A,T}$ (also called L_{eq}, *equivalent continuous A-weighted noise level*) is given by

$$L_{A,T} = 10 \log \frac{\dfrac{1}{T}\displaystyle\int_0^T p_A^2(t)\, dt}{p_{\text{ref}}^2} \quad \text{dB} \tag{1.21}$$

The time T must be specified. In noise evaluations, its length is usually one to several hours, or 8 h (working day), or 24 h (full day).

Day–night sound (noise) level L_{dn} is given by

$$L_{\text{dn}} = 10 \log \frac{1}{24}\left[\frac{\displaystyle\int_{07:00}^{22:00} p_A^2(t)\, dt}{p_{\text{ref}}^2} + \frac{\displaystyle\int_{22:00}^{07:00} 10\, p_A^2(t)\, dt}{p_{\text{ref}}^2}\right] \quad \text{dB} \tag{1.22}$$

where the first term covers the "daytime" hours from 07:00 to 22:00 and the second term covers the nighttime hours from 22:00 to 07:00. Here, the nighttime noise levels are considered to be 10 dB greater than they actually measure. The A-weighted sound pressure p_A is sampled frequently during measurement.

A-weighted sound exposure $E_{A,T}$ is given by

$$E_{A,T} = \int_{t_1}^{t_2} p_A^2(t)\, dt \quad \text{Pa}^2 \cdot \text{s} \tag{1.23}$$

This equation is not a level. The term $E_{A,T}$ is proportional to the energy flow (intensity times time) in a sound wave in the time period T. The period T starts and stops at t_1 and t_2, respectively.

A-weighted noise exposure level L_{EAT} is given by

$$L_{EA,T} = 10 \log \left(\frac{E_{A,T}}{E_0} \right) \quad \text{dB} \tag{1.24}$$

where E_0 is a reference quantity, standardized at $(20\ \mu\text{Pa})^2 \cdot \text{s} = (4 \times 10^{-10}\ \text{Pa})^2 \cdot \text{s}$. However, international standard ISO 1999:1990-01-5, on occupational noise level, uses $E_0 = (1.15 \times 10^{-5}\ \text{Pa})^2 \cdot \text{s}$, because, for an 8-h day, L_{EAT}, with that reference, equals the average A-weighted sound pressure level $L_{A,T}$. The two reference quantities yield levels that differ by 44.6 dB. For a single impulse, the time period T is of no consequence provided T is longer than the impulse length and the background noise is low.

Hearing threshold for setting "zero" at each frequency on a pure-tone audiometer is the standardized, average, pure-tone threshold of hearing for a population of young persons with no otological irregularities. The standardized threshold sound pressure levels at the frequencies 250, 500, 1000, 2000, 3000, 4000, 6000, and 8000 Hz are, respectively, 24.5, 11.0, 6.5, 8.5, 7.5, 9.0, 8.0, and 9.5 dB measured under an earphone. An audiometer is used to determine the difference at these frequencies between the threshold values of a person (the lowest sound pressure level of a pure tone the person can detect consistently) and the standardized threshold values. Measurements are sometimes also made at 125 and 1500 Hz.[7]

Hearing impairment (hearing loss) is the number of decibels that the permanent hearing threshold of an individual at each measured frequency is above the zero setting on an audiometer, in other words, a change for the worse of the person's threshold of hearing compared to the normal for young persons.

Hearing threshold levels associated with age are the standardized pure-tone thresholds of hearing associated solely with age. They were determined from tests made on the hearing of persons in a certain age group in a population with no otological irregularities and no appreciable exposure to noise during their lives.

Hearing threshold levels associated with age and noise are the standardized pure-tone thresholds determined from tests made on the hearing of individuals who had histories of higher than normal noise exposure during their lives. The average noise levels and years of exposure were determined by questioning and measurement of the exposure levels.

Noise-induced permanent threshold shift (NIPTS) is the shift in the hearing threshold level caused solely by exposure to noise.

1.5 REFERENCE QUANTITIES USED IN NOISE AND VIBRATION

American National Standard

The American National Standards Institute has issued a standard (ANSI S1.8-1989) on "Reference Quantities for Acoustical Levels." This standard is a revision of

ANSI S1.8-1969. The authors of this book have been surveyed for their opinions on preferred reference quantities. Table 1.5 is a combination of the standard references and of references preferred by the authors. The two references are clearly distinguished. All quantities are stated in terms of the International System of units (SI) and in British units.

Relations among Sound Power Levels, Intensity Levels, and Sound Pressure Levels

As a practical matter, the reference quantities for sound power, intensity, and sound pressure (in air) have been chosen so that their corresponding levels are interrelated in a convenient way under certain circumstances.

The threshold of hearing at 1000 Hz for a young listener with acute hearing, measured under laboratory conditions, was determined some years ago as a sound pressure of 2×10^{-5} Pa. This value was then selected as the reference pressure for sound pressure level.

Intensity at a point is related to sound pressure at that point in a free field by Eq. (1.1). A combination of Eqs. (1.1), (1.17), and (1.18) yields the sound intensity level

$$\begin{aligned} L_I &= 10 \log \frac{I}{I_{\text{ref}}} = 10 \log \frac{p^2}{\rho c I_{\text{ref}}} \\ &= 10 \log \frac{p^2}{p_{\text{ref}}^2} + 10 \log \frac{p_{\text{ref}}^2}{\rho c I_{\text{ref}}} \end{aligned}$$

$$L_I = L_p - 10 \log K \qquad \text{dB } re \ 10^{-12} \ \text{W/m}^2 \tag{1.25}$$

where K = const = $I_{\text{ref}}\rho c / p_{\text{ref}}^2$, which is dependent upon ambient pressure and temperature; quantity $10 \log K$ may be found from Fig. 1.7
= $\rho c/400$

The quantity $10 \log K$ will equal zero, that is, $K = 1$, when

$$\rho c = \frac{p_{\text{ref}}^2}{I_{\text{ref}}} = \frac{4 \times 10^{-10}}{10^{-12}} = 400 \text{ mks rayls} \tag{1.26}$$

We may also rearrange Eq. (1.25) to give the sound pressure level

$$L_p = L_I + 10 \log K \qquad \text{dB } re \ 2 \times 10^{-5} \ \text{Pa} \tag{1.27}$$

In Table 1.6, we show a range of ambient pressures and temperatures for which ρc = 400 mks rayls. We see that for average atmospheric pressure, namely, 1.013×10^5 Pa, the temperature must equal 38.9°C (102°F) for ρc = 400 mks rayls. However, if T = 22°C and $p_s = 1.013 \times 10^5$ Pa^2, $\rho c \approx 412$. This yields a value

TABLE 1.5 Reference Quantities for Acoustical Levels from American National Standard ANSI S1.8-1989 and as Preferred by Authors

Name	Definition	Preferred Reference Quantities: SI	Preferred Reference Quantities: British
Sound pressure level, dB (gases)	$L_p = 20 \log_{10}(p/p_0)$	$p_0 = 20\ \mu\text{Pa} = 2 \times 10^{-5}\ \text{N/m}^2$	$2.90 \times 10^{-9}\ \text{lb/in.}^2$
Sound pressure level, dB (other than gases)	$L_p = 20 \log_{10}(p/p_0)$	$p_0 = 1\ \mu\text{Pa} = 10^{-6}\ \text{N/m}^2$	$1.45 \times 10^{-10}\ \text{lb/in.}^2$
Sound power level, dB	$L_W = 10 \log_{10}(W/W_0)$	$W_0 = 1\ \text{pW} = 10^{-12}\ \text{N} \cdot \text{m/s}$	$8.85 \times 10^{-12}\ \text{in.} \cdot \text{lb/s}$
Sound power level, bel	$L_W = \log_{10}(W/W_0)$ bel	$W_0 = 1\ \text{pW} = 10^{-12}\ \text{N} \cdot \text{m/s}$	$8.85 \times 10^{-12}\ \text{in.} \cdot \text{lb/s}$
Sound intensity level, dB	$L_I = 10 \log_{10}(I/I_0)$	$I_0 = 1\ \text{pW/m}^2 = 10^{-12}\ \text{N/m} \cdot \text{s}$	$5.71 \times 10^{-15}\ \text{lb/in.} \cdot \text{s}$
Vibratory force level, dB	$L_{F_0} = 20 \log_{10}(F/F_0)$	$F_0 = 1\ \mu\text{N} = 10^{-6}\ \text{N}$	$2.25 \times 10^{-7}\ \text{lb}$
Frequency level, dB	$N = \log_{10}(f/f_0)$	$f_0 = 1\ \text{Hz}$	1.00 Hz
Sound exposure level, dB	$L_E = 10 \log_{10}(E/E_0)$	$E_0 = (20\ \mu\text{Pa})^2 \cdot \text{s} = (2 \times 10^{-5}\ \text{Pa})^2 \cdot \text{s}$	$8.41 \times 10^{-18}\ \text{lb}^2/\text{in.}^4$
The quantities listed below are not officially part of ANSI S1.8. They either are listed there for information or are included here as the authors' choice.			
Sound energy level given in ISO 1683:1983	$L_e = 10 \log_{10}(e/e_0)$	$e_0 = 1\ \text{pJ} = 10^{-12}\ \text{N} \cdot \text{m}$	$8.85 \times 10^{-12}\ \text{lb} \cdot \text{in.}$
Sound energy density level given in ISO 1683:1983	$L_D = 10 \log_{10}(D/D_0)$	$D_0 = 1\ \text{pJ/m}^3 = 10^{-12}\ \text{N/m}^2$	$1.45 \times 10^{-16}\ \text{lb/in.}^2$
Vibration acceleration level	$L_a = 20 \log_{10}(a/a_0)$	$a_0 = 10\ \mu\text{m/s}^2 = 10^{-5}\ \text{m/s}^2$	$3.94 \times 10^{-4}\ \text{in./s}^2$
Vibration acceleration level in ISO 16831983	$L_a = 20 \log_{10}(a/a_0)$	$a_0 = 1\ \mu\text{m/s}^2 = 10^{-6}\ \text{m/s}^2$	$3.94 \times 10^{-5}\ \text{in./s}^2$
Vibration velocity level	$L_v = 20 \log_{10}(v/v_0)$	$v_0 = 10\ \text{nm/s} = 10^{-8}\ \text{m/s}$	$3.94 \times 10^{-7}\ \text{in./s}$
Vibration velocity level in ISO 1683:1983	$L_v = 20 \log_{10}(v/v_0)$	$v_0 = 1\ \text{nm/s} = 10^{-9}\ \text{m/s}$	$3.94 \times 10^{-8}\ \text{in./s}$
Vibration displacement level	$L_d = 20 \log_{10}(d/d_0)$	$d_0 = 10\ \text{pm} = 10^{-11}\ \text{m}$	$3.94 \times 10^{-10}\ \text{in.}$

Notes: Decimal multiples and submultiples of SI units are formed as follows: 10^{-1} = deci (d), 10^{-2} = centi (c), 10^{-3} = milli (m), 10^{-6} = micro (μ), 10^{-9} = nano (n) and 10^{-12} = pico (p). Also J = joule = W · s (N · m), N = newton, and Pa = pascal = 1 N/m². Note that 1 lb = 4.448 N.

Although some international standards differ, in this text, to avoid confusion between power and pressure, we have chosen to use *W* instead of *P* for power; and to avoid confusion between energy density and voltage, we have chosen *D* instead of *E* for energy density. The symbol lb means pound force.

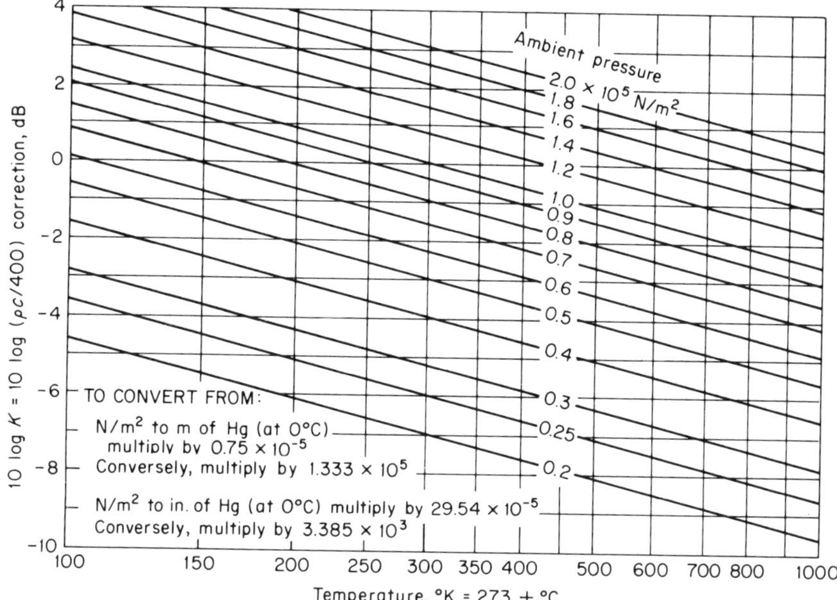

Fig. 1.7 Chart determining the value of $10 \log(\rho c/400) = 10 \log K$ as a function of ambient temperature and ambient pressure. Values for which $\rho c = 400$ are also given in Table 1.6.

of $10 \log(\rho c/400) = 10 \log 1.03 = 0.13$ dB, an amount that is usually not significant in acoustics.

Thus, for most noise measurements, we neglect $10 \log K$ and in a free progressive wave let

$$L_p \approx L_I \tag{1.28}$$

TABLE 1.6 Ambient Pressures and Temperatures for which ρc (Air) = 400 mks rayls

Ambient Pressure			Ambient Temperature T	
p_s Pa	m of Hg, 0°C	in. of Hg, 0°C	°C	°F
0.7×10^5	0.525	20.68	−124.3	−192
0.8×10^5	0.600	23.63	−78.7	−110
0.9×10^5	0.675	26.58	−27.0	−17
1.0×10^5	0.750	29.54	+30.7	+87
1.013×10^5	0.760	29.9	38.9	102
1.1×10^5	0.825	32.5	94.5	202
1.2×10^5	0.900	35.4	164.4	328
1.3×10^5	0.975	38.4	240.4	465
1.4×10^5	1.050	41.3	322.4	613

Otherwise, the value of 10 log K is determined from Fig. 1.7 and used in Eq. (1.25) or (1.27).

Under the condition that the *intensity is uniform over an area S*, the sound power and the intensity are related by $W = IS$. Hence, the sound power level is related to the intensity level as follows:

$$10 \log \frac{W}{10^{-12}} = 10 \log \frac{I}{10^{-12}} + 10 \log \frac{S}{S_0}$$

$$L_W = L_I + 10 \log S \qquad \text{dB } re\ 10^{-12}\ \text{W} \tag{1.29}$$

where S = area of surface, m^2
$S_0 = 1\ \text{m}^2$

Obviously, only if the area $S = 1.0\ \text{m}^2$ will $L_W = L_I$. Also, observe that the relation of Eq. (1.29) is not dependent on temperature or pressure.

1.6 DETERMINATION OF OVERALL LEVELS FROM BAND LEVELS

It is necessary often to convert sound pressure levels measured in a series of contiguous bands into a single-band level encompassing the same frequency range. The level in the all-inclusive band is called the *overall level* L(OA) given by

$$L_p(\text{OA}) = 20 \log \sum_{i=1}^{n} 10^{L_{pi}/20} \qquad \text{dB} \tag{1.30}$$

$$L_p(\text{OA}) = 10 \log \sum_{i=1}^{n} 10^{L_{Ii}/10} \qquad \text{dB} \tag{1.31}$$

The conversion can also be accomplished with the aid of Fig. 1.8. Assume that the contiguous-band levels are given by the eight numbers across the top of Fig. 1.9. The frequency limits of the bands are not important to the method of calculation as long as the bands are contiguous and cover the frequency range of the overall band. To combine these eight levels into an overall level, start with any two, say, the seventh and eighth bands. From Fig. 1.8 we see that whenever the difference between two band levels is zero, the combined level is 3 dB higher. If the difference is 2 dB (the sixth band level minus the new level of 73 dB), the sum is 2.1 dB greater than the larger (75 + 2.1 dB). This procedure is followed until the overall band level is obtained, here, 102.1 dB.

It is instructive to combine the bands in a different way, as is shown in Fig. 1.10. The first four bands are combined first; then the second four bands are combined. The levels of the two wider band levels are then combined. It is seen that the overall level is determined by the first four bands alone. This example points

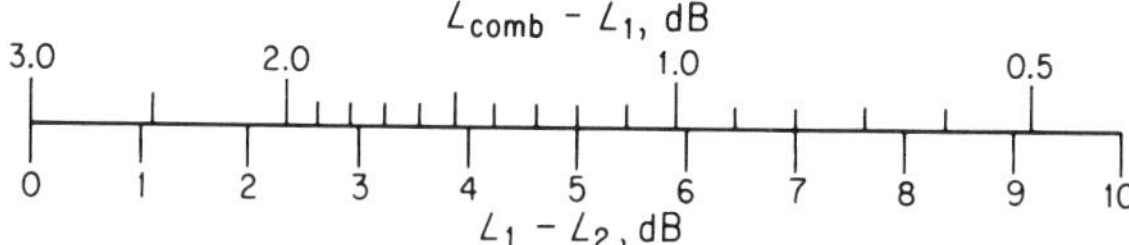

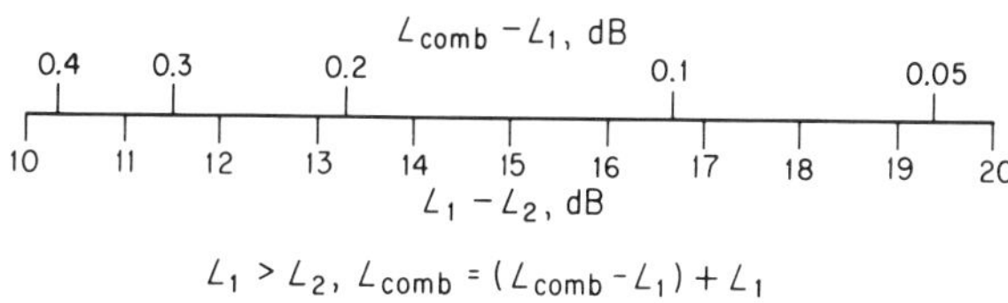

Fig. 1.8 Nanogram for combining two sound levels L_1 and L_2 (dB). Levels may be power levels, sound pressure levels, or intensity levels. Example: $L_1 = 88$ dB, $L_2 = 85$ dB, $L_1 - L_2 = 3$ dB. Solution: $L_{comb} = 88 + 1.8 = 89.8$ dB.

up the fact that characterization of a noise by its overall level may be completely inadequate for some noise control purposes because it may ignore a large portion of the frequency spectrum. If the data of Fig. 1.10 represented a genuine noise control situation, the 102.1 overall level might be meaningless for some applications. For example, the sound pressure levels in the four highest bands might be the cause of annoyance or interference with speech communication, as is discussed in Chapter 17.

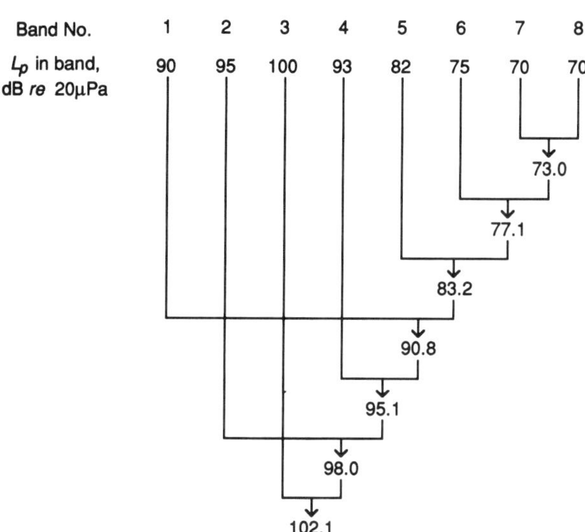

Fig. 1.9 Determination of an overall sound pressure level from levels in frequency bands (see also Fig. 1.10).

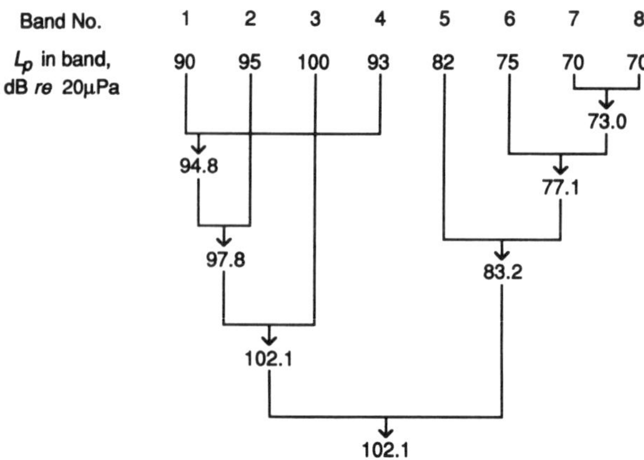

Fig. 1.10 Alternate determination of an overall sound pressure level from levels in frequency bands (see also Fig. 1.9).

Finally, it should be remembered that in almost all noise control problems, it makes no sense to deal with small fractions of decibels. Rarely does one need a precision of 0.2 dB in measurements, and quite often it is adequate to quote levels to the nearest decibel.

REFERENCES

1. L. L. Beranek, *Acoustics*, Acoustical Society of America, Woodbury, NY, 1986.
2. K. U. Ingard, *Fundamentals of Waves and Vibrations*, Cambridge University Press, New York, 1988.
3. M. Rossi, *Acoustics and Electroacoustics*, Artech House, Norwood, MA, 1988.
4. P. M. Morse and K. U. Ingard, *Theoretical Acoustics*, Princeton University Press, Princeton, NJ, 1987.
5. L. E. Kinsler, A. R. Frey, A. B. Coppens, and J. B. Sanders, *Fundamentals of Acoustics*, 3rd ed., Wiley, New York, 1982.
6. L. L. Beranek, *Noise and Vibration Control*, rev. ed., Institute of Noise Control Engineering, Poughkeepsie, NY, 1988.
7. L. L. Beranek, *Acoustical Measurements*, rev. ed., Acoustical Society of America, Woodbury, NY, 1988.

CHAPTER 2

Waves and Impedances

LEO L. BERANEK
Consultant
Cambridge, Massachusetts

2.1 THE WAVE EQUATION[1]

Sound waves must obey the laws of physics. For gases these include Newton's second law of motion, the gas law, and the laws of conservation of mass. Combined, these equations produce the wave equation that governs the behavior of sound waves regardless of the surroundings in which they occur.

Equation of Motion

The equation of motion (also called the force equation) is obtained by the application of Newton's second law to a small volume of gas in a homogeneous medium. Imagine the small volume of gas to be enclosed in a packet with weightless flexible sides and assume that there is negligible drag (friction) between particles inside and outside the packet. Suppose that this small volume exists in a part of the medium where the sound pressure p [actually $p(t)$] increases at the space rate of

$$\mathbf{grad}\, p = \mathbf{i}\frac{\partial p}{\partial x} + \mathbf{j}\frac{\partial p}{\partial y} + \mathbf{k}\frac{\partial p}{\partial z} \tag{2.1}$$

Noise and Vibration Control Engineering: Principles and Applications, Edited by Leo L. Beranek and István L. Vér.
ISBN 0-471-61751-2 © 1992 John Wiley & Sons, Inc.

where **i**, **j**, and **k**, are unit vectors in the x, y, and z directions, respectively. Obviously, **grad** p is a vector quantity.

The difference between the forces acting on the sides of the packet is a force **f** equal to the rate at which the force changes with distance times the incremental dimensions of the box:

$$\mathbf{f} = -\left[\mathbf{i}\left(\frac{\partial p}{\partial x}\Delta x\right)\Delta y\,\Delta z + \mathbf{j}\left(\frac{\partial p}{\partial y}\Delta y\right)\Delta x\,\Delta z + \mathbf{k}\left(\frac{\partial p}{\partial z}\Delta z\right)\Delta x\,\Delta y\right] \tag{2.2}$$

Note that a positive gradient causes the packet to accelerate in the negative direction.

Division of both sides of the equation by $\Delta x\,\Delta y\,\Delta z = V$ gives the force per unit volume acting to accelerate the box,

$$\frac{\mathbf{f}}{V} = -\mathbf{grad}\,p \tag{2.3}$$

By Newton's law, the force per unit volume of Eq. (2.3) equals the time derivative of the momentum per unit volume of the box. Because the box is a deformable packet, the mass inside is constant. Hence,

$$\frac{\mathbf{f}}{V} = -\mathbf{grad}\,p = \frac{M}{V}\frac{D\mathbf{q}}{Dt} = \rho'\frac{D\mathbf{q}}{Dt} \tag{2.4}$$

where $\mathbf{q}$ is the average vector velocity of the gas in the packet, ρ' is the average density of the gas in the packet, and $M = \rho' V$ is the total mass of the gas in the packet.

The partial derivative D/Dt is not a simple one but represents the total rate of the change of velocity of the particular bit of gas in the packet regardless of its position:

$$\frac{D\mathbf{q}}{Dt} = \frac{\partial \mathbf{q}}{\partial t} + q_x\frac{\partial \mathbf{q}}{\partial x} + q_y\frac{\partial \mathbf{q}}{\partial y} + q_z\frac{\partial \mathbf{q}}{\partial z} \tag{2.5}$$

where q_x, q_y, and q_z are the components of the vector particle velocity $\mathbf{q}$.

If $\mathbf{q}$ is small enough, the rate of change of momentum of the particles in the box can be approximated by the rate of change of momentum at a fixed point $Dq/Dt \doteq \partial\mathbf{q}/\partial t$, and the instantaneous density ρ' can be approximated by the average density ρ. Then

$$-\mathbf{grad}\,p = \rho\frac{\partial \mathbf{q}}{\partial t} \tag{2.6}$$

Gas Law

At audible frequencies, the wavelength of a sound wave is long compared to the spacing between air molecules, so that expansions and contractions at two different parts of the medium occur so rapidly that there is no time for heat exchange between points of differing instantaneous pressures. Hence, the compressions and expansions are adiabatic. From elementary thermodynamics,

$$PV^{\gamma} = \text{const} \tag{2.7}$$

where γ for air, hydrogen, oxygen, and nitrogen equals 1.4. If we let $P = P_s + p$ and $V = V_s + \tau$, where P_s and V_s are the undisturbed pressure and volume of the packet, we get, for small values of incremental pressure p and incremental volume τ,

$$\frac{p}{P_s} = -\frac{\gamma\tau}{V_s} \tag{2.8}$$

The time derivative of Eq. (2.8) yields

$$\frac{1}{P_s}\frac{\partial p}{\partial t} = -\frac{\gamma}{V_s}\frac{\partial \tau}{\partial t} \tag{2.9}$$

Continuity Equation

The continuity equation is a statement that the mass of the gas in the deformable packet is constant. Thus the change in the incremental volume τ depends only on the divergence of the vector displacement ξ:

$$\tau = V_s \operatorname{div} \xi \tag{2.10}$$

or

$$\frac{\partial \tau}{\partial t} = V_s \operatorname{div} \mathbf{q} \tag{2.11}$$

where $\mathbf{q}$ is the instantaneous (vector) particle velocity.

Wave Equation in Rectangular Coordinates

The three-dimensional wave equation is given by combining Eqs. (2.6), (2.9), and (2.11) and setting

$$c^2 = \frac{\gamma P_s}{\rho} \tag{2.12}$$

which yields

$$\nabla^2 p = \frac{1}{c^2} \frac{\partial^2 p}{\partial t^2} \tag{2.13}$$

where

$$\nabla^2 p = \frac{\partial^2 p}{\partial x^2} + \frac{\partial^2 p}{\partial y^2} + \frac{\partial^2 p}{\partial z^2} \tag{2.14}$$

The one-dimensional wave equation is simply

$$\frac{\partial^2 p}{\partial x^2} = \frac{1}{c^2} \frac{\partial^2 p}{\partial t^2} \tag{2.15}$$

We could also have eliminated p in the combination of the three equations and retained $\mathbf{q}$, in which case we would have had

$$\nabla^2 \mathbf{q} = \frac{1}{c^2} \frac{\partial^2 \mathbf{q}}{\partial t^2} \tag{2.16}$$

2.2 SOLUTIONS TO THE ONE-DIMENSIONAL WAVE EQUATION

General Solution

The general solution to Eq. (2.15) is the sum of two terms,

$$p(x, t) = f_1 \left(t - \frac{x}{c}\right) + f_2 \left(t + \frac{x}{c}\right) \quad \text{Pa} \tag{2.17}$$

where f_1 and f_2 are arbitrary functions. As we shall illustrate shortly, the first term represents an outgoing wave and the second term a backward-traveling wave. The functions f_1 and f_2 represent the shapes of the two sound waves being propagated. Examples of typical time histories and spectra of $p(t)$ at a fixed location are given in Fig. 2.1. We also recognize c as the speed of sound in air.

Outwardly Traveling Plane Wave

An apparatus for producing an outward-traveling plane wave is shown in Fig. 2.2. A piston at the left moving sinusoidally generates a sound wave that travels outward in the positive x direction and becomes absorbed in the anechoic termination

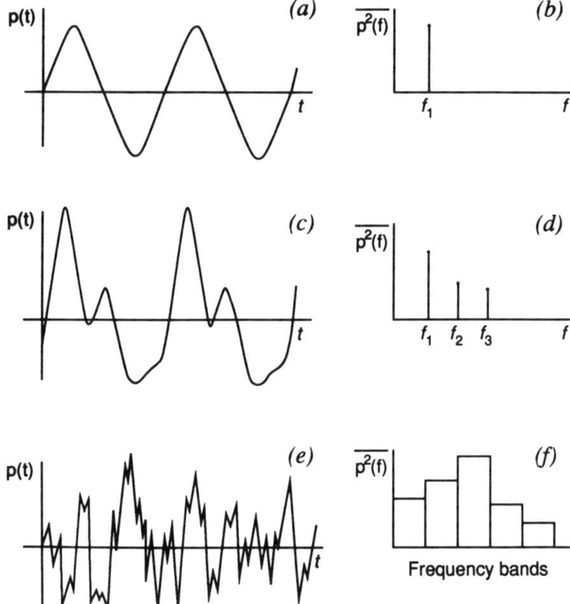

Fig. 2.1 (*a*, *c*, *e*) Forms of time function f_1, or f_2 of Eq. (2.17). Corresponding spectra are shown on the right. (*b*, *d*) Line spectra. (*f*) Complex spectrum (very large number of tones in each band).

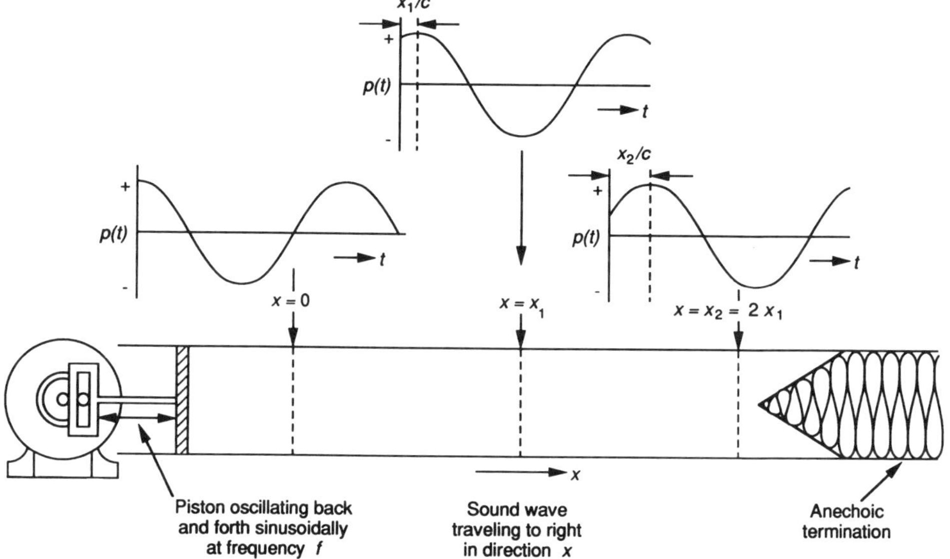

Fig. 2.2 Apparatus for producing a plane forward-traveling sound wave. A plane wave generated by the piston at the left travels to the right and is absorbed by the anechoic termination. The three waves at the top give the variation in sound pressure with time at the three points indicated, $x = 0$, $x = x_1$, and $x = x_2 = 2x_1$.

so that no reflected wave exists. Equation (2.17) becomes

$$p(x, t) = f_1\left(t - \frac{x}{c}\right) = P_R \cos k(x - ct) \qquad \text{Pa} \tag{2.18}$$

where P_R is the peak amplitude of the sound pressure.

Let us choose the space and the time origins, as shown by the left-hand sine wave in Fig. 2.2, so that P_R has its maximum value at $x = 0$ and $t = 0$. After a time t_1, the wave will have traveled a distance $x_1 = ct_1$. Similarly for $x_2 = 2x_1 = 2ct_1$.

Figure 2.3 shows a set of four spatial timeshots taken at $t = 0, \frac{1}{4}T, \frac{1}{2}T$, and $\frac{3}{4}T$, where T is the time period of the piston machine. Each shows the sound pressure over a spatial extent of one wavelength, $\lambda = c/f = cT$. The 20 vertical lines along each snapshot enable one to observe the spatial variation of sound pressure at a given point at the four different times.

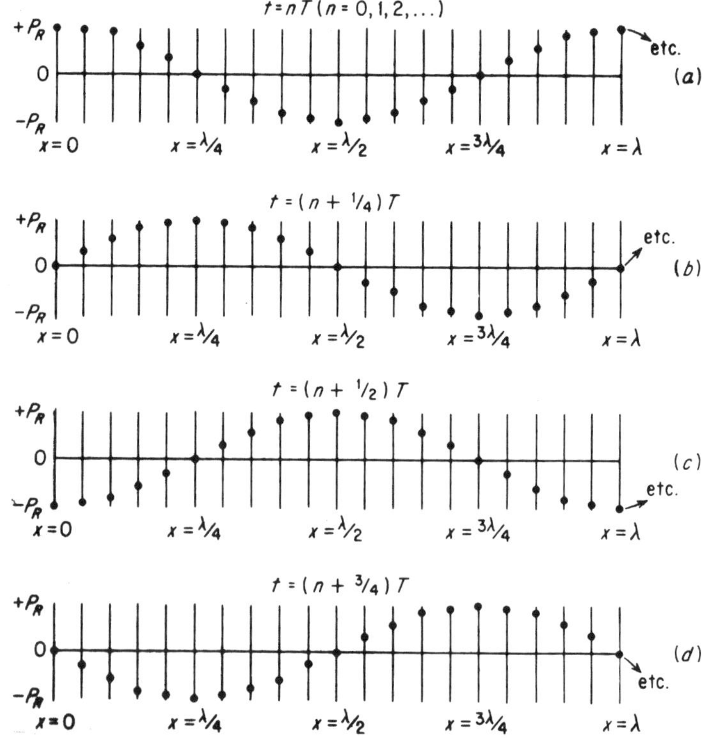

Fig. 2.3 Graphs showing sound pressure in a plane free-progressive wave traveling from left to right at 20 equally spaced axial locations at four instants of time t. The wave is produced by a source at the left and travels to the right with the speed c. The length of time it takes a wave to travel a distance equal to a wavelength is called the period T. Forward-traveling wave: $p(x, t) = P_R \cos k(x - ct)$; $k = 2\pi/\lambda = 2\pi/(cT) = 2\pi f/c = \omega/c$.

Snapshot (*a*) represents the pressure-versus-distance relation for times $t = 0$, $T, 2T, 3T, \ldots, nT$. The maximum value $+P_R$ exists at $x = 0$. Because the wave is periodic in space, the maximum value must also occur at $x = \lambda, 2\lambda, 3\lambda, \ldots$.

Snapshot (*b*) shows the sound pressure a quarter of a period, $\frac{1}{4}T$, later, that is, the wave in (*a*) has moved to the right a distance equal to $\frac{1}{4}\lambda$ to become the wave in (*b*). Similarly for (*c*) and (*d*). To convince yourself that the wave is traveling to the right, allow your eyes to jump successively from (*a*) to (*b*) to (*c*) to (*d*) and note that the peak, $+P_R$, moves successively to the right.

Wavenumber. The cosine function of Eq. (2.18) repeats its value every time the argument increases 2π radians (360°). From the definition of wavelength, $\lambda = c/f = cT$, we can write this periodicity condition as

$$\cos[k(x + \lambda - ct)] = \cos[k(x - ct) + 2\pi] \tag{2.19}$$

so that $k\lambda = 2\pi$ and $k = 2\pi/\lambda$ radians per meter. We see that the meaning of the parameter k, called the *wavenumber*, is a kind of "spatial frequency." The argument of the cosine in Eq. (2.18) may be written in any one of the following ways:

$$\begin{aligned} k(x - ct) &= \frac{2\pi}{\lambda}(x - ct) = 2\pi f\left(\frac{x}{c} - t\right) = 2\pi\left(\frac{x}{\lambda} - \frac{t}{T}\right) \\ &= \frac{2\pi x}{\lambda} - 2\pi f t = kx - \omega t \end{aligned}$$

From Eq. (2.19) we can write the equations for the snapshots of Fig. 2.3 as

$$\begin{array}{lll} (a) & t = nT & p = P_R \cos \dfrac{2\pi x}{\lambda} \\[2ex] (b) & t = (n + \frac{1}{4})T & p = P_R \cos\left(\dfrac{2\pi x}{\lambda} - \dfrac{\pi}{2}\right) \\[2ex] (c) & t = (n + \frac{1}{2})T & p = P_R \cos\left(\dfrac{2\pi x}{\lambda} - \pi\right) \\[2ex] (d) & t = (n + \frac{3}{4})T & p = P_R \cos\left(\dfrac{2\pi x}{\lambda} - \dfrac{3\pi}{2}\right) \end{array}$$

where $n = 0, 1, 2, 3, \ldots$.

Root-Mean-Square Sound Pressure

Most common sounds consist of a rapid, irregular series of positive-pressure disturbances (compressions) and negative-pressure disturbances (rarefactions) measured from the equilibrium pressure value. If we were to measure the mean value of the sound pressure disturbance, we would find that it would be zero because there are as many positive compressions as negative rarefactions. Thus, the mean value of sound pressure is not a useful measure. We must look for a measure that permits the effects of the rarefactions to be added to (rather than subtracted from) the effects of the compressions.

One such measure is the rms sound pressure p_{rms}. The rms sound pressure is obtained, first, by squaring the value of the sound pressure disturbance at each instant of time. Next the squared values are added and averaged over the sample time. The rms sound pressure is the square root of this time average, $p_{\text{rms}} = \sqrt{\langle p^2 \rangle_t}$. Since the squaring operation converts all the negative sound pressures to positive squared values, the rms sound pressure is a useful nonzero measure of the magnitude of the sound wave. (The rms value is also called the *effective value*.)

Particle Velocity

From Eq. (2.6) we may derive a relation between sound pressure p and particle velocity u, where u is the component of $\mathbf{q}$ in the x direction. In one dimension,

$$-\frac{\partial p}{\partial x} = \rho \frac{\partial u}{\partial t} \tag{2.20}$$

Substitution of Eq. (2.18) into Eq. (2.20) gives

$$-u = \frac{1}{\rho} \int \frac{\partial p}{\partial x}\, dt = \frac{-P_R}{\rho c} \cos k(x - ct) \tag{2.20a}$$

or

$$u = \frac{p}{\rho c} \tag{2.21}$$

where ρ = time-averaged density of air, = 1.18 kg/m^3 for normal room temperature T = 22°C (71.6°F) and atmospheric pressure P_s = 0.751 m (29.6 in.) Hg

c = speed of sound [see Eqs. (1.5)–(1.8)], which at normal temperatures of 22°C equals 344 m/s (1129 ft/s)

ρc = 406 mks rayls (N · s/m^3) at normal room temperature and pressure at other T's and P_s's, ρc is found from Fig. 1.7

Intensity

A freely traveling progressive sound wave transmits energy. We define this energy transfer as the *intensity* **I**, the energy that flows through a unit area in unit time. The units are watts per square meter (N/m · s). It has its maximum value, I_{max}, when the plane of the unit area is perpendicular to the direction in which the wave is traveling. Intensity, analogous to electrical power, equals the time average of the product of sound pressure and particle velocity,

$$I_{max} = \overline{p \cdot u} \qquad \text{W/m}^2 \text{ (N/m} \cdot \text{s)} \tag{2.22}$$

For the wave of Figs. 2.2 and 2.3,

$$I_{max} = \lim T \to \infty \frac{1}{T} \int_0^T \frac{P_R^2}{\rho c} \cos^2 k(x - ct)\, dt \tag{2.23}$$

where $I(\theta) = I_{max} \cos \theta$, θ being the angle between the direction of travel of the wave and a line perpendicular to the plane of the unit area through which the flow of sound power is being determined.

∞ = a time long enough that I_{max} has reached its asymptotic value within experimental error.

Because the time average of the cosine is zero,

$$I_{max} = \frac{P_R^2}{2\rho c} = \frac{p_{rms}}{\rho c} \qquad \text{W/m}^2 \tag{2.24}$$

where p_{rms} is the square root of the mean (time) square value of $p(t)$, as can be demonstrated by finding $I\rho c$ from Eq. (2.24).

Backward-traveling Plane Wave

A backward-traveling plane wave may be produced by interchanging the source and termination of Fig. 2.2. The wave now travels in the $-x$ direction and is described by

$$p(x, t) = P_L \cos k(x + ct) \qquad \text{Pa} \tag{2.25}$$

Comparison of Eqs. (2.18) and (2.25) show that if the two variables x and ct are separated by a negative sign, the wave travels in the positive direction, and that if the two variables are separated by a positive sign, the direction reverses.

The four "snapshots" of Fig. 2.4 illustrate the backward-traveling wave. Allowing your eyes to jump from a to b to c to d and following the movement of $+P_L$ from right to left convinces one that this is true.

One-Dimensional Spherical Wave

Sound Pressure. The equation for the sound pressure associated with a free-progressive, spherically traveling sound wave, produced as shown in Fig. 1.1. in Chapter 1, is

$$p(r, t) = \frac{A}{r} \cos k(r - ct) \qquad \text{Pa} \tag{2.26}$$

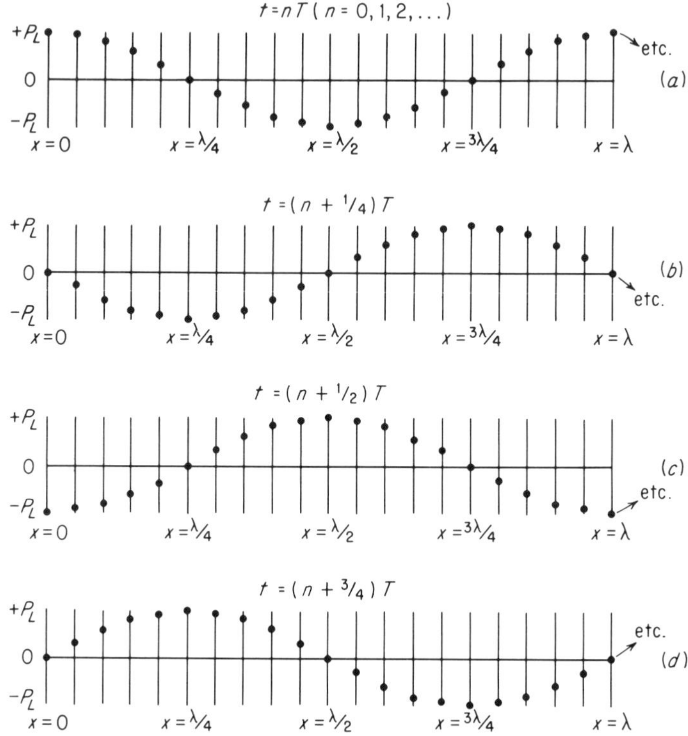

Fig. 2.4 Graphs showing sound pressure in a plane free-progressive wave traveling from right to left at 20 equally spaced axial locations at four instants of time t. The wave is produced by a source at the right (or by being reflected from a boundary at the right) and travels to the left with a speed c. The period T is defined as for Fig. 2.3. Backward-traveling wave: $p(x, t) = P_L \cos[k(x + ct)]$; $k = 2\pi/\lambda = 2\pi/(cT) = 2\pi f/c = \omega/c$.

where A is an amplitude factor with dimension newtons per meter. Because the sign between r and ct is negative, the wave is traveling outward in the positive r direction. In a spherical wave, the pressure amplitude is inversely proportional to the radial distance r.

Particle Velocity. The particle velocity for a spherical wave, using Eq. (2.20a) with r substituted for x, is

$$u(r, t) = -\frac{1}{\rho} \int \left[\frac{A}{r} k \sin k(r - ct) + \frac{A}{r^2} \cos k(r - ct) \right] dt$$

$$= -\frac{1}{\rho} \left[\frac{-kA}{kcr} \cos k(r - ct) - \frac{A}{r^2 kc} \sin k(r - ct) \right]$$

or

$$u(r, t) = \frac{A}{\rho c r} \cos k(r - ct) \left[1 + \frac{1}{kr} \tan k(r - ct) \right] \tag{2.27}$$

For large values of kr,

$$u(r, t) \doteq \frac{p(r, t)}{\rho c} \qquad k^2 r^2 \gg 1 \tag{2.28}$$

For very small values of kr, $k^2 r^2 \ll 1$,

$$u(r, t) \doteq \frac{A}{k \rho c r^2} \sin k(r - ct)$$

$$\doteq \frac{p(r, t)}{\rho c k r} \underline{/90^\circ \; re \text{ phase of } p(r, t)} \tag{2.29}$$

Equations (2.27) and (2.29) show that as one approaches the center of a spherical source, the sound pressure and particle velocity become progressively more out of phase, approaching 90° in the limit.

Intensity. For a freely traveling spherical wave, Eq. (2.26) states that for *all values of* r, the sound pressure varies as $1/r$. Because $u = p/\rho c$ [Eq. (2.28)] for the case $k^2 r^2 \gg 1$, we can write for all values of r that

$$I_{\max} \text{ at radius } r = \overline{p \cdot u} = \frac{p_{\text{rms}}^2 \text{ at } r}{\rho c} \qquad \text{W/m}^2 \tag{2.30}$$

2.3 SOUND POWER OUTPUT OF ELEMENTARY RADIATORS

Monopole (Radiating Sphere)[2]

The total power W_M radiated from a simple source, a pulsating sphere, called a monopole, is given in Table 2.1, where

$\hat{Q}_s$ = source strength, $= 4\pi a^2 \hat{v}_r$, m³/s
$\hat{v}_r$ = peak value of velocity of sinusoidally pulsating surface, m/s
a = radius of pulsating sphere, m
k = $2\pi f/c$, m^{-1}
ρc = characteristic impedance of gas, 406 mks rayls ($\text{N} \cdot \text{s/m}^3$) for air at normal room temperature and atmospheric pressure

TABLE 2.1 Sound Power Output of Elementary Radiators

Source Type	Source Behavior 180° Phase Difference	Sound Power Output, W	Auxiliary Expressions
Monopole	$+\hat{V}_r$, $-\hat{V}_r$, a	$W_M = \dfrac{\rho c k^2}{8\pi[1 + k^2a^2]} \hat{Q}_s^2$	$\hat{Q}_s = 4\pi a^2 \hat{v}_r$
Dipole	$+\hat{V}_r$, $-\hat{V}_r$, a	$W_D = \dfrac{\rho c k^4 d^2}{12\pi} \hat{Q}_s^2$	d = distance between monopoles
Oscillating sphere	$-\hat{V}_x$, $+\hat{V}_x$, a	$W_{OS} = \dfrac{2\pi\rho c k^4 a^6}{3[4 + k^4a^4]} \hat{v}^2$ $W_{BP} = \dfrac{\rho c k^2}{4\pi} \hat{Q}^2$ $ka \ll 1$	$\hat{v}_x$ = peak sinusuoidal vibration velocity
Baffled piston	WALL, PISTON, RIGID, $-\hat{V}_x$, $+\hat{V}_x$	$W_{BP} = \dfrac{\rho c}{2\pi a^2} \hat{Q}^2$ $ka \gg 1$	$\hat{Q} = \pi a^2 \hat{v}_x$ where $\hat{v}_x$ *is as above*

Dipole (Two Closely Spaced Monopoles)[2]

By definition, two monopoles constitute a dipole when $(kd)^2 \ll 1$ and when they vibrate 180° out of phase. A dipole has a figure-eight radiation pattern, with minimum radiation in the direction perpendicular to a line connecting the two monopoles. The total sound power, W_D, radiated is given in the second row of Table 2.1, where $\hat{Q}_s$, $\hat{v}_r$, a, k, and ρc are as given above and

d = separation between monopoles, m
r = distance from dipole to spherical surface over which power radiated is measured, m

Oscillating Sphere[3,4]

An oscillating sphere is defined as a rigid sphere moving axially, back and forth, around its rest position. The total power, W_{OS}, radiated is given in the third row of Table 2.1, where r is as defined for a dipole, ρc and k are as defined for a monopole, and

$\hat{v}_x$ = peak back-and-forth velocity, m/s
a = radius of oscillating sphere, m

Baffled Piston[5]

An axially vibrating diaphragm in an infinite plate is called a baffled piston. The total power radiated to one side of the rigid wall, W_{BP}, is given in the fourth row of Table 2.1, both for the case of a piston whose radius is small compared to a wavelength, $ka \ll 1$, and vice versa, $ka \gg 1$, where

$\hat{v}_x$ = peak axial velocity of piston, m/s
a = radius of piston, m

The other quantities are as defined above.

2.4 INTERFERENCE AND RESONANCE

The sound pressure and incremental density in a sound wave are generally very small in comparison with the equilibrium values on which they are superposed. This is certainly true for speech and music waves. As a result, it is possible in such acoustical situations to determine the effect of two sound waves in the same space by simple linear addition of the effects of each sound wave separately. This is a statement of the principle of *superposition*.

In the previous section we presented a series of spatial snapshots for plane sound waves traveling to the right (Fig. 2.3) and to the left (Fig. 2.4). According to the principle of superposition, the effect of the sum of these two waves will be the sum of their effects, which we can see graphically by adding Figs. 2.3 and 2.4. The result is shown in Fig. 2.5, where we have set the amplitude of the forward-traveling wave P_R equal to the amplitude of the backward-traveling wave P_L.

The interference of the two waves has produced a surprising change. No longer does the sound pressure at one place occur to the right or to the left of that place at the next instant. The wave no longer travels; it is a standing wave. We see that at each point in space the sound pressure varies sinusoidally with time, except at the points $x = \frac{1}{4}\lambda$ and $\frac{3}{4}\lambda$, where the presure is always zero. The maximum value of the pressure variation at different points is different, being greatest at $x = 0$, $x = \frac{1}{2}\lambda$, and $x = \lambda$. The sound pressures at the points between the points $x = \frac{1}{4}\lambda$ and $x = \frac{3}{4}\lambda$ always vary together, that is, increase or decrease in phase. At the same times the sound pressures for the points to the left of $x = \frac{1}{4}\lambda$ and to the right of $x = \frac{3}{4}\lambda$ decrease or increase together (in phase). Thus all pressures are in time phase in the standing wave, but there is a space difference of phase of 180° between the sound pressures at the points at $x = 0$ and $x = \frac{1}{2}\lambda$.

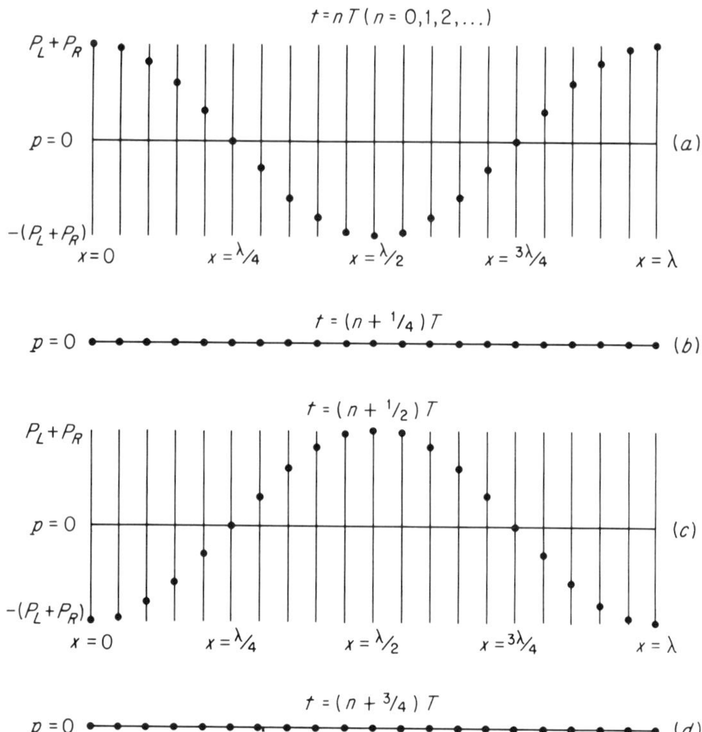

Fig. 2.5 Graphs giving sound pressure in a plane standing wave at 20 places in space at four instants of time t. The wave is produced by two sources equal in strength at the right and left of the graphs (or the right source is a perfectly reflecting boundary that sends back a wave of equal amplitude to that produced by the source on the left). Standing wave: $p(x, t) = 2P(\cos kx)(\cos 2\pi ft)$, where k is the wavenumber.

Remembering that $P_R = P_L = P$, that is, that the amplitude of the wave traveling to the right is equal to the amplitude of the wave traveling to the left, we find that the sum of the two waves is

$$p(x, t) = P\cos[k(x - ct)] + P\cos[k(x + ct)]$$
$$= 2P(\cos kx)(\cos 2\pi ft) \qquad \text{Pa} \tag{2.31}$$

From Eqs. (2.18), (2.20), and (2.31) we see very clearly the differences between a standing and a traveling wave. In a traveling wave distance x and time t occur as a sum or difference in the argument of the cosine. Hence, for the traveling wave, by adjusting both time and distance (according to the speed of sound) in the argument of the cosine, we can always keep the argument and thus the magnitude of the cosine the same. In Eq. (2.31) distance and time no longer appear together in the argument of a single cosine. So the same sound pressure cannot occur at an adjacent point in the space at a later time.

Standing waves will exist in any regular enclosure. In a rectangular room, for example, three classes of standing waves may exist (see Chapter 6). One class includes all waves that are perpendicular to one pair of opposing walls, that is, that travel at grazing incidence to two pairs of walls, the $(n_x, 0, 0)$, $(0, n_y, 0)$, $(0, 0, n_z)$ modes of vibration. A second class travels at grazing incidence to only one pair of walls, the $(n_x, n_y, 0)$, $(n_x, 0, n_z)$, $(0, n_y, n_z)$ modes of vibration. A third class involves all walls at oblique angles of incidence, the (n_x, n_y, n_z) modes of vibration. Each free-standing wave in an acoustical space is called a normal mode of vibration, or simply a resonance. The frequencies at which resonant standing waves can exist are related to the separation between the reflecting surfaces. For example, the lowest frequency for a resonant standing wave in a one-dimensional system consisting of two rigid parallel walls is given by

$$f = \frac{c}{2d} \quad \text{Hz} \tag{2.32}$$

where f = lowest frequency for resonant standing wave, Hz
c = speed of sound, m/s
d = distance separating two reflecting surfaces, m

Resonant standing waves can also exist at every integral multiple of this frequency. That is to say,

$$f = \frac{nc}{2d} \quad \text{Hz} \tag{2.33}$$

where n is an integer 1, 2, 3, ...

2.5 IMPEDANCE AND ADMITTANCE

Reference to Eqs. (2.18) and (2.21) reveals that the magnitudes of sound pressure and particle velocity are directly proportional to each other. Also, in the special case of plane-wave sound propagation, the time dependence of sound pressure is exactly the same as the time dependence of particle velocity, and at any point in the wave there is no phase difference between the two quantities. Thus, in a plane sound wave the ratio of sound pressure to particle velocity at all instants of time is a constant equal to ρc.

In general, however, for linear (small-signal) acoustical phenomena in the steady state, there is a difference in the time functions of sound pressure and particle velocity, leading to a phase difference of one relative to the other. Thus, at any point the particle velocity may lead or lag the sound pressure. In many situations, both the ratio of the magnitudes and the relative phase may be functions of frequency.

In several of the chapters that follow, it is convenient in acoustical design to avoid separate consideration of steady-state sound pressure and steady-state particle velocity (or other quantities that may be derived from them, such as force and volume velocity) and instead to deal with either one and with their complex ratio, as defined below.

Complex Notation

The foundations for the designation of steady-state signals with the same frequency but different phases by complex notation are expressed in the identities

$$|A|\cos(\omega t + \theta_1) \equiv \text{Re}\,\overline{A}e^{j\omega t} \tag{2.34}$$

where $|A| \equiv$ amplitude of cosine function
$\theta_1 \equiv$ phase shift at time $t = 0$
Re $\equiv$ means "real part of"
$j \equiv \sqrt{-1}$

$$\overline{A} \equiv A_{\text{Re}} + jA_{\text{Im}} = |A|\,e^{j\theta_1} \tag{2.35}$$

$$|A| \equiv \sqrt{A_{\text{Re}}^2 + A_{\text{Im}}^2} \tag{2.36}$$

$$\theta_1 \equiv \tan^{-1}(A_{\text{Im}}/A_{\text{Re}}) \tag{2.37}$$

We note also that

$$e^{j\theta} \equiv \cos\theta + j\sin\theta \tag{2.38}$$

so that

$$A_{\text{Re}} \equiv |A|\cos\theta \tag{2.39}$$

$$A_{\text{Im}} \equiv |A|\sin\theta \tag{2.40}$$

These equations say that a cosinusoidal time-varying function, given by the left-hand side of Eq. (2.34), can be represented by the real-axis projection of a vector of magnitude $|A|$ given by Eq. (2.36), rotating at a rate ω radians per second. The angle θ_1 is the angle of the vector (in radians) relative to the positive real axis at the instant of time $t = 0$.

We might therefore express a time-varying steady-state sound pressure or force by

$$\overline{A}e^{j\omega t} = |A|\,e^{j\omega t}e^{j\theta_1} \tag{2.41}$$

Also, a time-varying steady-state velocity or volume velocity might be expressed by

$$\overline{B}e^{j\omega t} = |B|\,e^{j\omega t}e^{j\theta_2} \tag{2.42}$$

Definitions of Complex Impedance

Complex Impedance Z. In general, complex impedance is defined as

$$\bar{Z} \equiv \frac{\bar{A}}{\bar{B}} = \frac{|A|\,e^{j(\omega t + \theta_1)}}{|B|\,e^{j(\omega t + \theta_2)}} = \frac{|A|\,e^{j\theta_1}}{|B|\,e^{j\theta_2}} = |Z|\,e^{j\theta} \tag{2.43}$$

and

$$\bar{Z} \equiv R + jX = \sqrt{R^2 + X^2}\,e^{j\theta} = |Z|\,e^{j\theta} \tag{2.44}$$

where $\bar{Z}$ = *complex impedance as given above*
A = steady-state pressure or force
B = steady-state velocity or volume velocity
$|Z|$ = magnitude of complex impedance
θ = phase angle between the time functions A and B, $= \theta_1 - \theta_2$
R, X = real and imaginary parts, respectively, of the complex impedance $\bar{Z}$
R = resistance, Re Z
X = reactance, Im $\bar{Z}$

Often, $|A|$ and $|B|$ are taken to be the rms values of the phenomena they represent, although, if so taken, a factor of $\sqrt{2}$ must be added to both sides of Eq. (2.34) to make them correct in a physical sense. Whether amplitudes or rms values are used makes no difference in the impedance ratio.

Complex impedances are of several types, according to the quantities involved in the ratios. Types common in acoustics are given below.

Acoustic Impedance Z_A. The acoustic impedance at a given surface is defined as the complex ratio of (a) sound pressure averaged over the surface to (b) volume velocity through it. The surface may be either a hypothetical surface in an acoustic medium or the moving surface of a mechanical device. The unit is $\mathrm{N \cdot s/m^5}$, also called the mks acoustical ohm. That is,

$$Z_A = \frac{p}{U} \quad \mathrm{N \cdot s/m^5} \text{ (mks acoustical ohms)} \tag{2.45}$$

Specific Acoustic Impedance Z_s. The specific acoustic impedance is the complex ratio of the sound pressure at a point of an acoustic medium or mechanical device to the particle velocity at that point. The unit is $\mathrm{N \cdot s/m^3}$, also called the mks rayl. That is,

$$Z_s = \frac{p}{u} \quad \mathrm{N \cdot s/m^3} \text{ (mks rayls)} \tag{2.46}$$

Mechanical Impedance Z_M. The mechanical impedance is the complex ratio of the force acting on a specific area of an acoustic medium or mechanical device to the resulting linear velocity through or of that area, respectively. The unit is the N · s/m, also called the mks mechanical ohm. That is,

$$Z_M = \frac{f}{u} \quad \text{N} \cdot \text{s/m (mks mechanical ohms)} \tag{2.47}$$

Characteristic Resistance ρc. The characteristic resistance is the ratio of the sound pressure at a given point to the particle velocity at that point in a free, plane, progressive sound wave. It is equal to the product of the density of the medium and the speed of sound in the medium (ρc). It is analogous to the characteristic impedance of an infinitely long, dissipationless transmission line. The unit is the N · s/m^3, also called the mks rayl. In the solution of problems in this book we shall assume for air that $\rho c = 406$ mks rayls, which is valid for a temperature of 22°C (71.6°F) and a barometric presure of 0.751 m (29.6 in.) Hg.

Normal Specific Acoustic Impedance Z_{sn}. At the boundary between air and a denser medium (such as a porous acoustic material) we find a further definition necessary, as follows: When an alternating sound pressure p is produced at the surface of an acoustic material, an alternating velocity u of the air particles is produced through the surface. The to-and-fro motions of the air particles may be at any angle relative to the surface. The angle depends both on the angle of incidence of the sound wave and on the nature of the acoustic material. For example, if the material is porous and has very low density, the particle velocity at the surface is nearly in the same direction as that in which the wave is propagating. By contrast, if the surface were a large number of small-diameter tubes packed side by side and oriented perpendicular to the surface, the particle velocity would necessarily be only perpendicular to the surface. In general, the direction of the particle velocity at the surface has both a normal (perpendicular) component and a tangential component.

The normal specific acoustic impedance (sometimes called the unit area acoustic impedance) is defined as the complex ratio of the sound pressure p to the *normal component* of the particle velocity u_n at a plane, in this example at the surface of the acoustic material. Thus

$$Z_{sn} = \frac{p}{u_n} \quad \text{N} \cdot \text{s/m}^3 \text{ (mks rayls)} \tag{2.48}$$

Definition of Complex Admittance

Complex admittance is the reciprocal of complex impedance. In all ways, it is handled by the same set of rules as given by Eqs. (2.34)–(2.44). Thus, the complex admittance corresponding to the complex impedance of Eq. (2.43) is

$$\bar{Y} \equiv \frac{\bar{B}}{\bar{A}} = \frac{|B|\, e^{j\theta_2}}{|A|\, e^{j\theta_1}} = |Y|\, e^{j\phi} \tag{2.49}$$

where $|Y| = 1/|Z|$
$\phi = -\theta$

The choice between impedance and admittance is sometimes made according to whether $|A|$ or $|B|$ is held constant during a measurement. Thus, if $|B|$ is held constant, $|Z|$ is directly proportional to $|A|$ and is used. If $|A|$ is held constant, $|Y|$ is directly proportional to $|B|$ and is used.

REFERENCES

1. W. J. Cunningham, ''Application of Vector Analysis to the Wave Equation,'' *J. Acoust. Soc. Am. 22,* 61 (1950); R. V. L. Hartley, ''Note on 'Application of Vector Analysis to the Wave Equation,''' *J. Acoust. Soc. Am. 22,* 511 (1950).
2. K. U. Ingard, *Fundamentals of Waves and Oscillations*, Cambridge University Press, New York, 1988, Chapter 17.
3. M. Rossi, *Acoustics and Electroacoustics*, Artech House, Norwood, MA, 1988, Section 2.2.30.
4. L. Cremer, *Vorlesungen uber Technische Akustik*, Springer-Verlag, New York, 1971, Section 2.5.2.
5. L. L. Beranek, *Acoustics*, Acoustical Society of America, Woodbury, NY, 1986, Sections 4.3 and 7.5.

CHAPTER 3

Data Analysis

ALLAN G. PIERSOL
Piersol Engineering Company
Woodland Hills, California

For studies related to noise and vibration control engineering, the analysis of measured acoustical noise and/or vibration data may be accomplished with a number of goals in mind. The most important of these goals can be divided into four broad categories: (a) an assessment of the severity of an environment, (b) the identification of system response properties, (c) the identification of sources, and (d) the identification of transmission paths. The first goal is commonly accomplished using one-third-octave-band-level calculations, or perhaps frequency-weighted overall level measurements, as described in Chapter 1. The other goals often require more advanced data analysis procedures. Following a brief discussion of the general types of acoustical and vibration data of common interest, the most important data analysis procedures for accomplishing goals (b)–(d) are outlined, and important applications of the results from such analyses are summarized. Because of the broad and intricate nature of the subject, heavy use of references is employed to cover details.

3.1 TYPES OF DATA SIGNALS

Acoustical and vibration data are commonly acquired in the form of analog time history signals produced by appropriate transducers (details of acoustical and vi-

Noise and Vibration Control Engineering: Principles and Applications, Edited by Leo L. Beranek and István L. Vér.
ISBN 0-471-61751-2 © 1992 John Wiley & Sons, Inc.

bration transducers and signal conditioning equipment are available from the data acquisition documents listed in the Bibliography, and the literature published by acoustical and vibration measurement system manufacturers). The signals are generally produced with the units of volts but can be calibrated into appropriate engineering units (g, m/s, Pa, etc.) as required. From a data analysis viewpoint, it is convenient to divide these time history signals into two broad categories, each with two subcategories, as follows:

1. Deterministic data signals: (a) steady-state signals; (b) transient signals.
2. Random data signals: (a) stationary signals; (b) nonstationary signals.

Deterministic Data

Deterministic data signals are those for which it is theoretically feasible to determine a mathematical equation that would predict future time history values of the signal (within reasonable experimental error), based upon a knowledge of the applicable physics or past observations of the signal. The most common type of deterministic signal, called a periodic signal, has a time history $x(t)$ that exactly repeats itself after a constant time interval T_p, called the period of the signal; that is,

$$x(t) = x(t \pm T_p) \tag{3.1}$$

The most common sources of periodic acoustical and vibration signals are constant-speed rotating machines, including propellers and fans. Ideally, such signals would have only one dominant frequency, allowing them to be represented by a simple sine wave. However, it is more likely that the periodic source will produce a complex signal that must be described, using a Fourier series representation,[1] by a collection of harmonically related sine waves, as illustrated in Fig. 3.1. In any case, periodic signals are called steady-state because their average properties (mean value, mean-square value, and spectrum) do not vary with time.

There are steady-state acoustical and vibration signals that are not rigorously periodic, for example, the data produced by a collection of independent (unsynchronized) periodic sources, such as the propellers on a multiengine propeller airplane. Such nonperiodic steady-state signals are referred to as almost periodic. Most nonperiodic deterministic signals, however, are also not steady state; that is, their average properties change with time. An important type of time-varying signal is one that begins and ends within a reasonable measurement time interval. Such signals are called transient signals. Examples of deterministic transient signals include well-controlled impacts, sonic booms, and aircraft landing loads. Such data can be described, using a Fourier integral representation,[1] by a continuous spectrum, as illustrated for an exponentially decaying oscillation in Fig. 3.2.

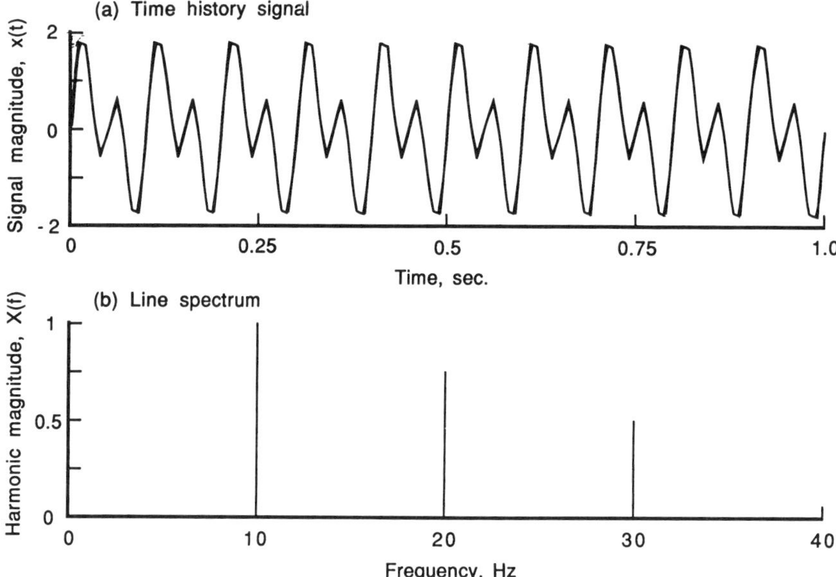

Fig. 3.1 Time history and line spectrum for periodic signal.

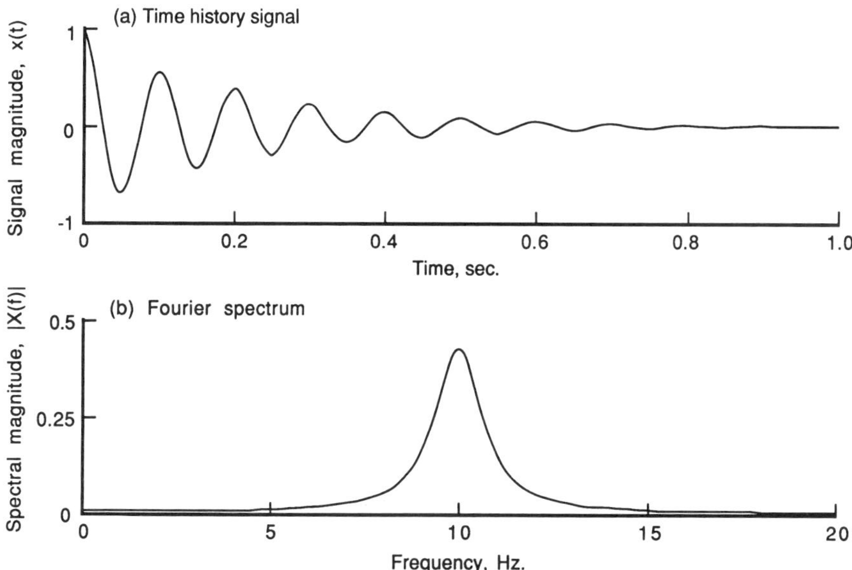

Fig. 3.2 Time history and Fourier spectrum for deterministic transient signal.

Random Data

Random acoustical and vibration data signals may be broadly defined as all signals that are not deterministic, that is, where it is not theoretically feasible to predict future time history values based upon a knowledge of the applicable physics or past observations. In some cases, the border between deterministic and random signals may be blurred. For example, the pressure field produced by a high-speed fan at its blade passage rate with a uniform inflow would be deterministic, but turbulence in the inflow would introduce a random property to the pressure signal. In other cases, the data will be more fully random in character (sometimes called "strongly mixed"). Examples include the pressure fields generated by fluid dynamic boundary layers (flow noise), the loads produced by atmospheric turbulence, and the acoustical noise caused by the exhaust gas mixing from an air blower. These sources of acoustical and vibration signals cover a wide frequency range and have totally haphazard time histories, as illustrated in Fig. 3.3.

When the mechanisms producing random acoustical or vibration data are time invariant, the average properties of the resulting signals will also be time invariant. Such random data are said to be stationary. Unlike steady-state deterministic data, stationary random data must be described by a continuous spectrum. Furthermore, because of the probabilistic character of the data, the measurement of the spectrum and all other signal properties of interest will involve statistical sampling errors that do not occur in the analysis of deterministic signals. These statistical errors will be summarized later.

If the average properties of random signals vary with translations in time, the signals are said to be nonstationary. An illustration of nonstationary data is shown in Fig. 3.4. Although there is a well-developed theoretical methodology for the analysis of arbitrary nonstationary signals,[1] the analysis procedures often require more data than are commonly available and further involve extensive and complex computer calculations. An exception is a special class of nonstationary signals, called stochastic transients, that begin and end within a reasonable measurement time interval. Common sources of stochastic transients are hard impact loads and pyrotechnic devices. An illustration of a stochastic transient signal is shown in Fig. 3.5. Such data can be analyzed by procedures similar to those used to describe

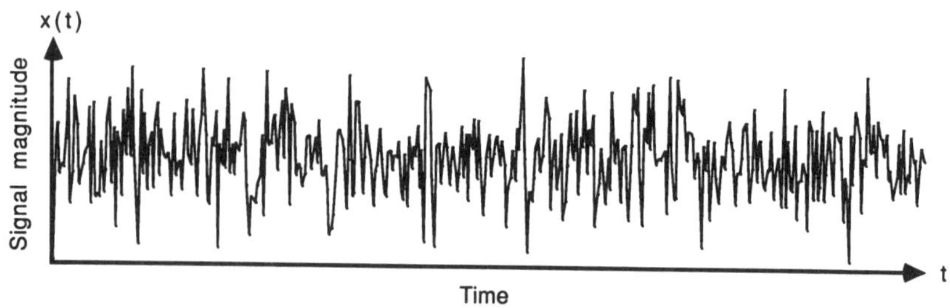

Fig. 3.3 Time history for stationary random signal.

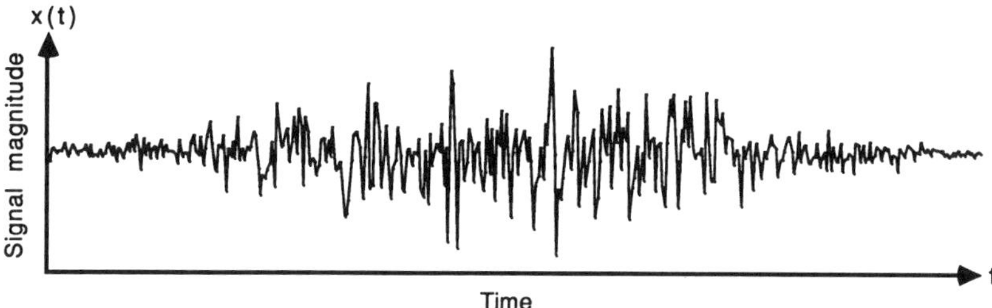

Fig. 3.4 Time history for nonstationary random signal.

deterministic transients discussed earlier, except now there will be a statistical sampling error problem that must be addressed.

3.2 MEAN AND MEAN-SQUARE VALUES

The most rudimentary measures of any steady-state or stationary data signal are the mean value and the mean-square value (or variance), which provide single-valued descriptions of the central tendency and dispersion of the signal. Mean and mean-square values of signals can be measured using either digital computations[1] or analog instruments.[2] Hence, appropriate algorithms are presented for both digital and analog analysis procedures.

Mean Values

For acoustical and vibration signals, the mean value is commonly zero, simply because most transducers used for acoustic pressure and vibration acceleration measurements do not sense static values. However, if a transducer is used that senses static values and the central tendency of the signal is of interest, the mean value for a time history measurement, $x(t)$, of duration T_r is computed by

$$m_x = \frac{1}{T_r} \int_0^{T_r} x(t)\, dt \tag{3.2a}$$

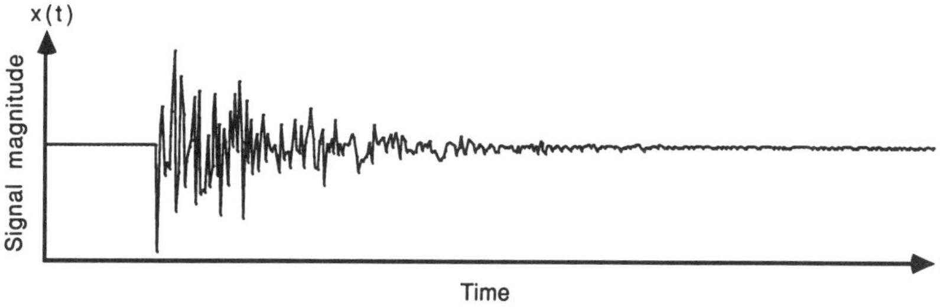

Fig. 3.5 Time history for stochastic transient signal.

For digital data with a sampling interval of Δt, $x(t) = x(n\,\Delta t)$, $n = 1, 2, \ldots, N$, and the mean value is computed by

$$m_x = \frac{1}{N} \sum_{n=1}^{N} x(n\,\Delta t) \tag{3.2b}$$

The mean value m_x in Eq. (3.2) has units of volts and is essentially the quantity computed by a DC voltmeter. If the data signal is periodic, the calculation in Eq. (3.2) will be accurate as long as T_r (or $N\,\Delta t$) is an integer multiple of the period T_p. For random data, the calculation will involve a statistical sampling error that is a function of T_r (or N) and the spectral characteristics of the signal,[1] to be discussed later.

Mean-Square Values

The mean-square (ms) value of a steady state or stationary acoustical or vibration signal $x(t)$ [or a digitized signal $x(n\,\Delta t)$] is defined by

$$w_x = \frac{1}{T_r} \int_0^{T_r} x^2(t)\,dt = \frac{1}{N} \sum_{n=1}^{N} x^2(n\,\Delta t) \tag{3.3}$$

The ms value w_x in Eq. (3.3) has the units of volts squared, which is proportional to power or power per unit area, and hence w_x is often referred to as the overall "power" or "intensity" of the acoustical or vibration signal. If the mean value of the signal is zero, the ms value in Eq. (3.3) is equal to the variance of $x(t)$, defined as

$$s_x^2 = \frac{1}{T_r} \int_0^{T_r} [x(t) - \mu_x]^2\,dt = \frac{1}{N} \sum_{n=1}^{N} [x(n\,\Delta t) - \mu_x]^2 \tag{3.4}$$

where μ_x is the true mean value of the signal (m_x as $T_r \to \infty$). The positive square roots of the quantities defined in Eqs. (3.3) and (3.4), $w_x^{1/2}$ and s_x, are called the rms value and the standard deviation, respectively, of the signal $x(t)$. Again, $w_x^{1/2} = s_x$ if the mean value of the signal is zero, as will be assumed henceforth. The value s_x in Eq. (3.4), or $w_x^{1/2}$ in Eq. (3.3) when $\mu_x = 0$, is essentially the quantity measured by a true rms voltmeter (not to be confused with an AC voltmeter that employs a linear rectifier calibrated to measure the correct rms value for a sine wave and reads about 1 dB low when measuring random noise). As for mean-value calculations, the ms value calculation in Eq. (3.3) will be precise for periodic data if T_r is an integer multiple of the period T_p. For random data, however, there will be a statistical sampling error that is a function of T_r (or N) and the spectral characteristics of the signal,[1] to be discussed later.

Weighted Averages

The averaging operation indicated in Eqs. (3.2)–(3.4) is a simple linear sum of prior values over a specific time interval. This type of average, referred to as an unweighted or linear average, is the natural way one would average any set of discrete data values and is the simplest way for a digital computer to calculate an average value. However, some acoustical and vibration data analyses are still performed using analog instruments, where a relatively expensive operational amplifier is required to accomplish an unweighted average. Hence, the averaging operation in analog instruments is commonly weighted so that it can be accomplished using inexpensive passive circuit elements.[2] The most common averaging circuit used by analog instruments is a simple low-pass filter consisting of a series resistor and shunt capacitor (commonly called an *RC* filter) that produces an exponentially weighted average. For a ms value estimate (assuming the mean value is zero), the exponentially weighted average estimate is given by

$$w_x(t) = \frac{1}{K} \int_0^t x^2(\tau) \exp[-(t - \tau)/K] \, d\tau \tag{3.5}$$

where $K = RC$ is the product of the numerical values of the resistance in ohms and capacitance in farads used in the averaging circuit. The term K has units of time (seconds) and is referred to as the *time constant* of the averaging circuit. An exponentially weighted averaging circuit provides a continuous average value estimate versus time, which is based on all past values of the signal. It follows that after starting the average calculation, a period of time must elapse before the indicated average value is accurate. As a rule of thumb, when averaging a steady-state or stationary signal, at least four time constants ($t > 4K$) must elapse to obtain an average value estimate with an error of less than 2%. The error in question is a bias error. For the analysis of random data signals, there will also be a statistical sampling error, to be discussed later.

Running Averages

When the acoustical or vibration data of interest have average properties that are time varying (nonstationary data), "running" time averages are often used to describe the data. For the case of a ms value estimate, this could be accomplished by executing Eq. (3.3) repeatedly over short, contiguous time segments of duration $T \ll T_r$. Exponentially weighted averaging is particularly convenient for the computation of running averages because it produces a continuous average value estimate versus time. However, a near-continuous estimate can also be generated using an unweighted average (as is more desirable for digital data analysis) by simply recomputing an average value every data sampling interval Δt rather than only at the end of the averaging time T. An illustration of a running average for the rms value of typical nonstationary acoustical data measured near an airport

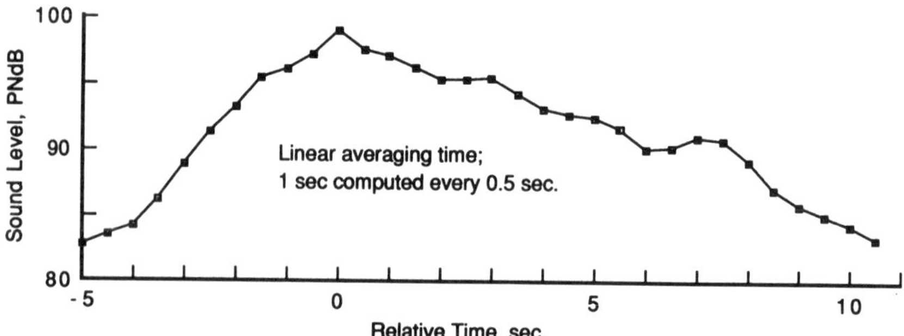

Fig. 3.6 Running average for weighted rms value of aircraft flyover noise. (Courtesy of Acoustic Analysis Associates, Inc., Canoga, Park, CA.)

during an aircraft flyover is shown in Fig. 3.6 (the measurement in this illustration is a frequency-weighted rms value called *perceived noise level.*

The basic requirement in the computation of a running average is to select an averaging time T (or averaging time constant K) that is short enough not to smooth out the time variations of the data property being measured but long enough to suppress statistical sampling errors in the average value estimate at any time (assuming the data are at least partially random in character). Analytical procedures for selecting the averaging time T that provides an optimum compromise between the smoothing and statistical sampling errors have been formulated,[1] but trial-and-error procedures coupled with experience will usually provide adequate results. Also, many of the simpler acoustical and vibration measurement instruments used in the field have fixed averaging times ("fast" and "slow" averaging circuits) built into the instrument.[2]

Statistical Sampling Errors

As mentioned earlier, there are no fundamental errors (other than instrument and calibration errors) associated with the calculation of mean and ms values of periodic data signals, assuming the averaging time is an integer multiple of the period of the signal. For random data signals, however, there will be a statistical sampling error due to the fact that the record duration and averaging time can never be long enough to cover all unique signal values. It is convenient to describe this statistical sampling error in terms of a normalized standard deviation of the resulting estimate, called the normalized random error (also called the coefficient of variation). The normalized random error of an estimate, $\hat{\theta}$, for a signal property θ is defined as

$$\epsilon_r[\hat{\theta}] = \sigma[\hat{\theta}]/\theta \tag{3.6}$$

where σ [ ] denotes the standard deviation as defined in Eq. (3.4), and the hat denotes an estimate. The interpretation of the normalized random error is as follows. If a signal property θ is repeatedly estimated with a normalized random error of, say, 0.1, then about two-thirds of the estimates, $\hat{\theta}$, will be within $\pm 10\%$ of the true value of θ.

The normalized random errors of the mean, ms, and rms value estimates are summarized in Table 3.1.[1] In these error equations, B_s is a measure of the spectral bandwidth of the data signal, defined as[3]

$$B_s = \frac{w^2}{\int_0^\infty G^2(f)\, df} \tag{3.7}$$

where w is the ms value and $G(f)$ is the auto (power) spectrum of the data signal, to be addressed later. The error formulas in Table 3.1 are approximations that are valid for $\epsilon_r \leq 0.20$. Note in Table 3.1 that the random error for mean value estimates is dependent on the true mean (μ_x) and standard deviation (σ_x) of the signal as well as the bandwidth (B_s) and averaging time (T or K). However, the random error for ms and rms value estimates is a function only of the bandwidth and averaging time.

Synchronous Averaging

Periodic vibration and acoustical data signals produced by rotating machinery (including propellers and fans) are sometimes contaminated by additive, extraneous noise such that the measured signal is $x(t) = p(t) + n(t)$, where $p(t)$ is the periodic signal of interest and $n(t)$ is the noise. In such cases, the signal-to-noise ratio of the periodic signal can be strongly enhanced by the procedure of synchronous averaging, where the data record is divided into a collection of segments $x_i(t)$, $i = 1, 2, \ldots, q$, each starting at exactly the same phase angle during a period of $p(t)$. The collection of segments can then be ensemble averaged to extract $p(t)$ from the extraneous noise as well as other periodic components that are not harmonically related to $p(t)$ as follows:

$$p(t) \approx \frac{1}{q} \sum_{i=1}^{q} x_i(t) \tag{3.8}$$

Table 3.1 Normalized Random Errors for Mean, ms, and rms Value Estimates

	Normalized Random Error, ϵ_r	
Signal Property	Linear Averaging with Averaging Time T	*RC* Weighted Averaging with Time Constant K
Mean value, m_x	$\sigma_x/[\mu_x(2B_sT)^{1/2}]$	$\sigma_x/[\mu_x(4B_sK)^{1/2}]$
ms value, w_x	$1/(B_sT)^{1/2}$	$1/(2B_sK)^{1/2}$
rms value, $w_x^{1/2}$	$1/(4B_sT)^{1/2}$	$1/(8B_sK)^{1/2}$

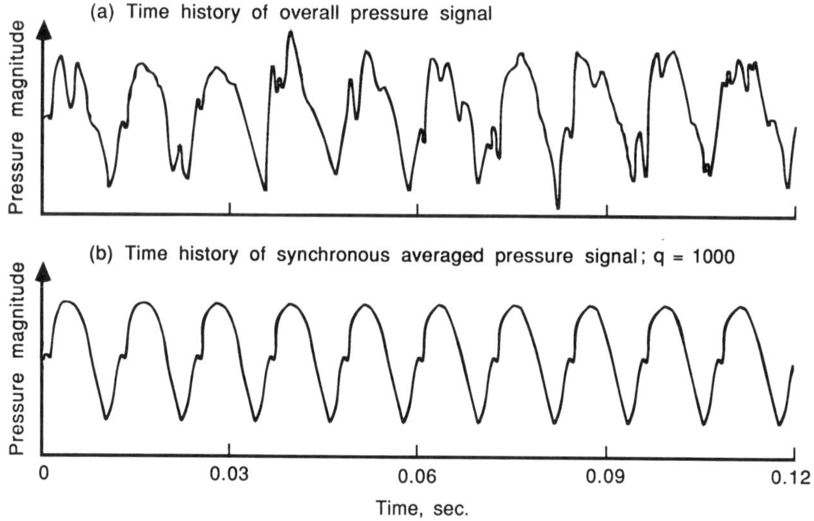

Fig. 3.7 Original and synchronous averaged time histories for pressure measurement on sidewall of propeller airplane.

Synchronous averaging is illustrated in Fig. 3.7 for the pressure field generated in the plane of the propeller on the sidewall of a propeller airplane powered by a reciprocating engine. With 1000 ensemble averages (accomplished in less than 2 min), the procedure extracts the propeller pressure signal cleanly from the engine and boundary layer turbulence noise.

The primary requirement for synchronous averaging is the ability to initiate new records at a desired instant during a period of $p(t)$. This is most effectively accomplished using a trigger signal that is a noise-free indicator of the phase during each period of $p(t)$. For rotating machines, a noise-free trigger signal is commonly obtained using either an optical detector or a magnetic pulse generator on the rotating element of the machine. The time base accuracy of the trigger signal determines the accuracy of the magnitude of the resulting synchronous averaged signal; that is, time base errors in the trigger signal cause a reduction in the indicated signal amplitude with increasing frequency. The signal-to-noise ratio enhancement for the synchronous averaged signal is given by $10 \log_{10} q$, where q is the number of segments used in the ensemble-averaging operation.

3.3 SPECTRAL FUNCTIONS

The mean square (ms) value of a digital signal constitutes an overall measure of the "power" or "intensity" represented by the measured quantity, but much more useful information is provided by a frequency decomposition of the signal values. As noted in Chapter 1, the computation of rms values in one-third-octave bands is widely used for the frequency analysis of acoustical data and sometimes for vibration data as well. However, the more advanced signal-processing techniques needed for system, source, and path identification problems require the computation of frequency spectra with a much finer resolution than one-third-octave bands. Furthermore, if the data are random in character, a frequency analysis in terms of power quantities per unit frequency (Hz) greatly facilitates the desired evaluations of the data signals.

Prior to 1965, most analysis of acoustical and vibration data signals, including the calculation of frequency spectra, was accomplished by analog instruments. Highly resolved frequency spectra were generally computed using narrow-bandwidth analog filters, often employing mechanical elements such as resonant crystals or magnetostrictive devices.[2] Of course, narrow-bandwidth frequency spectra could also be computed, even at that time, on a digital computer using Fourier transform software, but the computations were time consuming and expensive because of the large number of data values needed to represent wide-bandwidth acoustical and vibration signals. In 1965, this situation changed dramatically with the introduction of an algorithm for the fast computation of Fourier series coefficients[4] that reduced the required computer calculations by up to two orders of magnitude. Various versions of this algorithm have since come into wide use and are generally referred to as fast Fourier transform (FFT) algorithms. The vast majority of all current narrow-bandwidth spectral analysis of acoustical and vibration data signals are performed using FFT algorithms.

The FFT Algorithm

Using lowercase letters for functions of time and uppercase letters for functions of frequency, the Fourier transform of a time history signal, $x(t)$, which is measured over the time interval $0 \leq t \leq T$, is defined for all frequencies (both positive and negative) by

$$X(f, T) = \int_0^T x(t)e^{-j2\pi ft}\,dt \tag{3.9a}$$

In terms of a digital time series of N data values where $x(t) = x(n\,\Delta t)$, $n = 0, 1, \ldots, N - 1$ (starting the indexing at $n = 0$ is helpful in maintaining consistent relationships between time and frequency functions), the Fourier transform may be written as

$$X(f, T) = X(k\,\Delta f, N) = \Delta t \sum_{n=0}^{N-1} x(n\,\Delta t)\exp(-j2\pi fn\,\Delta t) \tag{3.9b}$$

where the spectral components are generally complex valued and are defined only at N discrete frequencies given by

$$f_k = k\,\Delta f = \frac{k}{N\,\Delta t} \qquad k = 0, 1, 2, \ldots, N - 1 \tag{3.9c}$$

The finite Fourier transform defined in Eq. (3.9), when divided by T (or $N\,\Delta t$), essentially yields the conventional Fourier series coefficients for a periodic function under the assumption that the time history record $x(t)$ is one period (or an integer multiple of one period) of the periodic function being analyzed. The Fourier components are unique only out to $k = \frac{1}{2}N$, that is, out to the frequency $f_k = 1/(2\,\Delta t)$, commonly called the Nyquist frequency, f_N, of the digitized signal. The Nyquist frequency is that frequency for which there are only two sample values per cycle and, hence, aliasing will initiate.[1] Comparing the digital result in Eq. (3.9b) with the analog formulation in Eq. (3.9a), the first $\frac{1}{2}N + 1$ Fourier coefficients, from $k = 0$ to $k = \frac{1}{2}N$, define the spectral components at nonnegative frequencies, while the last $\frac{1}{2}N - 1$ Fourier coefficients, from $k = \frac{1}{2}N + 1$ to $k = N - 1$, essentially define the spectral components at negative frequencies.

The details of the various FFT algorithms are fully documented in the literature (see refs. 1 and 4 and the signal analysis documents listed in the Bibliography). It is necessary here only to note a few basic characteristics of the most common algorithm used for acoustical and vibration work, which is commonly referred to as the *Cooley–Tukey algorithm* in recognition of the authors of the 1965 paper[4] that initiated the wide use of such algorithms:

1. It is convenient to restrict the number of data values for each FFT to a power of 2, that is, $N = 2^p$, where values of $p = 8$ to 12 are commonly used.
2. The fundamental frequency resolution of the Fourier components will be $\Delta f = 1/(N\,\Delta t)$.
3. The Nyquist frequency where aliasing will initiate, denoted by f_N, occurs at the $k = \frac{1}{2}N$ Fourier component, that is, $f_N = 1/(2\,\Delta t)$.
4. The first $\frac{1}{2}N + 1$ Fourier components up to the Nyquist frequency are related to the last $\frac{1}{2}N - 1$ components above the Nyquist frequency by $X(k) = X^*(N - k)$, $k = 0, 1, 2, \ldots, N - 1$, where the asterisk denotes complex conjugate.
5. The Fourier components defined only for positive frequencies (called a one-sided spectrum) are given by $X(0)$, $X(N/2)$, and $2X(k)$; $k = 1, 2, \ldots, (N/2 - 1)$.

Line and Fourier Spectral Functions

For deterministic signals that are periodic, a frequency decomposition or spectrum of the signal is directly obtained by computing the Fourier series coefficients of the signal over at least one period of the signal using an FFT algorithm. Assuming the mean value equals zero, the one-sided spectrum of Fourier components for a periodic signal, $p(t)$, are given by

$$P(f) = \frac{2X(f, T)}{T} \qquad f > 0 \text{ or } k = 1, 2, \ldots, \frac{N}{2} - 1 \tag{3.10}$$

where $X(f, T)$ is as defined in Eq. (3.9). The Fourier component magnitudes, $|P(f)|$, are usually plotted versus frequency in the form of a discrete frequency spectrum (often called a *line spectrum*), as illustrated earlier in Fig. 3.1*b*. Of course, each Fourier component is a complex number that defines a phase as well as a magnitude (the phases for the Fourier components in Fig. 3.1 are all zero). However, the phase information is generally retained only in those applications where there may be a need to reconstruct the signal time history or determine peak values.

Aliasing. Because of the aliasing problem inherent in digital spectral analysis,[1] it is important to assure there are no spectral components in the data signal being analyzed above the Nyquist frequency $[f_N = 1/(2\,\Delta t)]$ for the analysis. This can be guaranteed only when the analog signal is low-pass filtered prior to the digitization to remove any spectral components that may exist in the signal above f_N. The low-pass filters employed to accomplish this task are commonly referred to as *antialiasing* filters and should always be used for all spectral analysis.

Leakage Errors and Tapering. Ideally, in the anaysis of periodic data, the spectral computations should be performed over an exact integer multiple of one period to avoid truncation errors in the Fourier series calculation. In practice, however, the computations are often terminated at a time that is convenient for the FFT algorithm and not related to the exact period of the data. The resulting truncation error leads to a phenomenon called side-lobe leakage[1,5] that can severely distort the desired results. To suppress this leakage, it is common to taper the measured time history signal in a manner that forces the values at the start and finish of the measurement to be zero, so as to eliminate the discontinuity between the beginning and ending data values. Numerous tapering functions (often called "windows") have been proposed over the years,[5] but one of the earliest and still most widely used is the *cosine-squared* taper (commonly called the *Hanning* window) given by

$$u_h(t) = 1 - \cos^2 \frac{\pi t}{T} \qquad 0 \le t \le T \tag{3.11}$$

The FFT is then performed on the signal, $y(t) = x(t)\,u_h(t)$, rather than directly on the original measured signal, $x(t)$. Leakage suppression can also be accomplished by equivalent operations in the frequency domain.

For deterministic signals that are not periodic and further have a well-defined beginning and end (transients), a spectrum of the signal is directly obtained by computing the Fourier transform of the signal over the entire duration of the signal, again using an FFT algorithm (to obtain a one-sided spectrum, the actual computation is $2X(k\,\Delta f)$, $k = 1, 2, \ldots, N/2 - 1$. The Fourier spectrum magnitude is plotted as a continuous function of frequency, as illustrated previously in Fig. 3.2*b*. Similar to the spectra for periodic signals, there is also a phase function associated with Fourier spectra, but it is generally retained only if the signal time history is to be reconstructed or peak values are of interest. As long as the FFT computation is performed over a measurement duration that covers the entire duration of the transient event, there is no side-lobe leakage problem in the analysis.

As a final point on transient signal analysis, it should be mentioned that transient data signals, particularly those produced by short-duration mechanical shocks, are often analyzed by a technique called the *shock response spectrum*,[6] which essentially defines the peak response of a hypothetical collection of single-degree-of-freedom mechanical systems to the transient input. The shock response spectrum can be a valuable tool for assessing the damaging potential of mechanical shock loads on equipment but is not particularly useful for noise and vibration reduction applications.

Auto (Power) Spectral Density Functions

The autospectral density function (also called the ''power'' spectral density function) provides a convenient and consistent measure of the frequency composition of random data signals. The autospectrum, denoted by $G_{xx}(f)$, is most easily visualized as the mean square (ms) value of the signal passed through a narrow-bandpass filter divided by the filter bandwidth, as illustrated in Fig. 3.8. In equation form,

$$\hat{G}_{xx}(f) = \frac{1}{T\,\Delta f}\int_0^T x^2(f, \Delta f, t)\,dt \tag{3.12}$$

where $x(f, \Delta f, t)$ denotes the signal passed by the narrow-bandpass filter with a center frequency f and a bandwidth Δf. To obtain the exact autospectral density function, the operations in Fig. 3.8 would theoretically be carried out in the limit as $T \to \infty$ and $\Delta f \to 0$ such that $T\,\Delta f \to \infty$. It is clear from Fig. 3.8 and Eq. (3.12) that the units of the autospectral density function are volts squared per hertz.

The operations shown in Fig. 3.8 represent the way autospectra were computed by analog instruments[2] prior to the introduction of FFT algorithms and the transition to digital data analysis procedures. Today, with the ready availability of FFT hardware and software, the autospectral density function (at positive frequencies only) is estimated directly by[1]

$$G_{xx}(f) = \frac{2}{n_d T}\sum_{i=1}^{n_d} |X_i(f, T)|^2 \qquad f > 0 \tag{3.13}$$

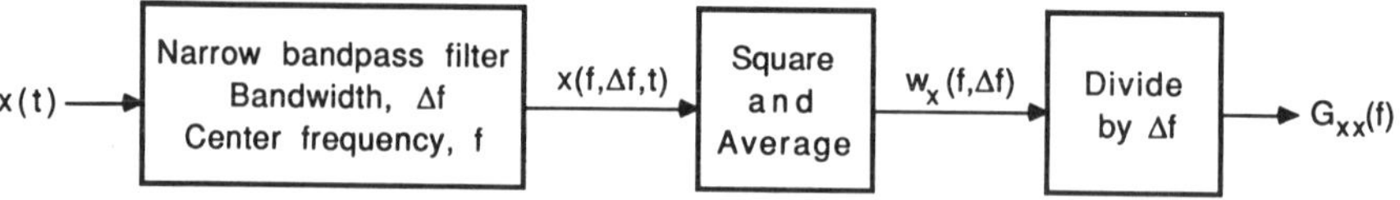

Fig. 3.8 Autospectral density function measurement by analog filtering operations.

where $X_i(f, T)$ is the FFT of $x(t)$ computed over the ith data segment of duration T, as defined in Eq. (3.9), and n_d is the number of disjoint (statistically independent) data segments used in the calculation. To obtain the exact autospectral density function, the operations in Eq. (3.13) would theoretically be carried out in the limit as $T \to \infty$ and $n_d \to \infty$. As will be seen later, the number of averages n_d determines the random error in the estimate, while the segment duration T for each FFT computation determines the resolution and, hence, a potential bias error in the estimate. The collection of disjoint data segments needed to estimate a statistically reliable autospectrum is usually created by dividing the total available measurement duration T_r into a sequence of contiguous segments of duration T, as illustrated in Fig. 3.9. It follows that $n_d = T_r/T = \Delta f\, T_r$, often referred to as the *BT product* of the estimate. It can be shown[1] that Eq. (3.13) is equal to the result in Eq. (3.12) when the appropriate limits are imposed, that is, when $T \to \infty$ and $\Delta f \to 0$ in Eq. (3.12) and $T \to \infty$ and $n_d \to \infty$ in Eq. (3.13). Note that the autospectral density function is always a real number (there is no phase information associated with autospectra).

There are a number of grooming operations that are commonly employed to enhance the quality of autospectral density estimates. A few of the more important ones are as follows (see the noted references for details).

Antialiasing Filters. As for periodic data analysis, to avoid aliasing, it is important that the random signal being analyzed have no spectral values above the Nyquist frequency, $f_N = 1/(2\,\Delta t)$. Hence, the analog signal must always be lowpass filtered prior to digitization to suppress any spectral content that may exist above f_N.[1]

Tapering Windows. Since random data signals essentially have an infinite period, there will always be a truncation error associated with the selected segment duration T. Hence, tapering operations (windows) are commonly used in random signal analysis to suppress the side-lobe leakage problem. Of the numerous available tapering functions,[5] the cosine-squared (Hanning) window defined in Eq. (3.11) is the most widely used.

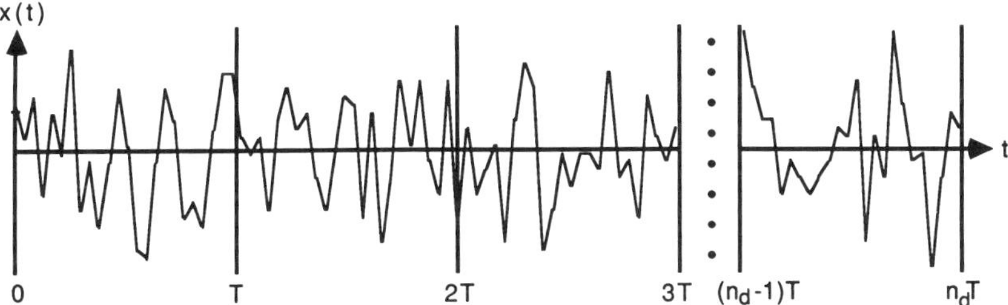

Fig. 3.9 Subdivision of measured time history into n_d contiguous segments.

Overlapped Processing. Although tapering operations on segments of the measured signal are desirable to suppress leakage, they also increase the bandwidth of the effective spectral window associated with the analysis.[5] If it is desired to maintain the same spectral window bandwidth with tapering that would have been achieved without tapering, the segment duration, T, for the analysis must be increased. However, assuming the total duration of the measurement, T_r, is fixed, this will reduce the number of disjoint averages, n_d, and increase the random error of the spectral estimates. This increase in random error can be counteracted by computing the spectrum with overlapped segments, rather than contiguous segments.[7,8] A 50% overlap is commonly used in such cases.

Zoom Transforms. As discussed earlier, FFT algorithms are usually implemented with a fixed number of data points. Hence once a desired upper frequency limit for an analysis (the Nyquist frequency f_N) has been chosen, the resolution of the analysis, Δf, as defined in Eq. (3.9c), is also fixed. Situations often arise when the desired upper frequency limit and frequency resolution are not compatible with the number of data points used by the FFT computation. In these cases, a finer resolution for a given value of f_N can be achieved using computation techniques referred to as *zoom transform* procedures.[1] The most common zoom transform techniques employ a complex demodulation calculation that essentially segments the frequency range of the signal into contiguous bands that are then analyzed separately.

Cross-Spectral Density Functions

The solution of acoustical and vibration reduction problems involving random processes is often facilitated by the identification of a linear dependence (correlation) between two measurements at different locations. The basic parameter that defines the linear dependence between two measured random signals, $x(t)$ and $y(t)$, as a function of frequency is the cross-spectral density function, which is estimated (at positive frequencies only) by[1]

$$G_{xy}(f) = \frac{2}{n_d T} \sum_{i=1}^{n_d} X_i^*(f, T) Y_i(f, T) \qquad f > 0 \tag{3.14a}$$

where $X_i(f, T)$ and $Y_i(f, T)$ are the FFTs of $x(t)$ and $y(t)$, respectively, computed over the ith simultaneous data segments of duration T, n_d is the number of disjoint records used in the calculation, and the asterisk denotes complex conjugate. As for autospectra, the exact cross-spectral density function would be obtained in the limit as $T \to \infty$ and $n_d \to \infty$. All of the computational considerations and grooming procedures discussed for autospectra apply to cross-spectra as well. Unlike the autospectrum, however, the cross-spectrum is generally a complex number that includes both magnitude and phase information and, hence, may be denoted in complex polar notation as

$$G_{xy}(f) = |G_{xy}(f)| \, e^{j\theta_{xy}(f)} \tag{3.14b}$$

Coherence Functions

For many applications, it is more convenient to work with a normalized version of the cross-spectral density function, called the coherence function (sometimes called coherency squared), which is defined as[1]

$$\gamma_{xy}^2(f) = \frac{|G_{xy}(f)|^2}{G_{xx}(f)G_{yy}(f)} \tag{3.15}$$

The coherence function is a real-valued quantity bounded by zero and unity, that is,

$$0 \leq \gamma_{xy}^2(f) \leq 1 \tag{3.16}$$

where a value of zero means there is no linear dependence and a value of unity means there is a perfect linear dependence between the signals $x(t)$ and $y(t)$ at the frequency f. A coherence value that is less than unity at one or more frequencies is usually indicative of one of the following situations[9]:

1. Extraneous noise is present in the measurements.
2. The frequency resolution of the spectral estimates is too wide.
3. The system relating $y(t)$ to $x(t)$ has time-dependent parameters.
4. The system relating $y(t)$ to $x(t)$ is not linear.
5. The output $y(t)$ is due to other inputs besides $x(t)$.

By carefully designing an experiment to minimize the first four possible reasons for a low coherence, the fifth reason provides the basis for a powerful procedure to identify acoustical noise and/or vibration sources. Specifically, if it is known that a constant-parameter linear system exists between a source and a receiver location, the source signal is measured with an adequate signal-to-noise ratio, and the spectra of the source and receiver signals are estimated with an adequate frequency resolution, then the coherence function defines the fractional portion of the receiver signal autospectral density that is due to the measured source signal. This is the basis for the coherent output power relationship, which is discussed and illustrated in Section 3.5.

Statistical Sampling Errors

There are no statistical sampling errors associated with the calculation of spectra for periodic signals, assuming the averaging time is an integer multiple of the period of the signal. The same is true of the calculation of Fourier spectra for deterministic transient signals, assuming the averging time is longer than the transient. The calculation of spectral density quantities for random signals, however, will involve a random sampling error, as discussed previously in Section 3.2. First-order approximations for these random errors in autospectra, cross-spectra, and coherence function estimates are summarized in Table 3.2.[1] The random errors

Table 3.2 Normalized Random Errors for Autospectra, Cross Spectra, and Coherence Function Estimates

Signal Property	Normalized Random Error ϵ_r or Standard Deviation σ_r
Autospectral density function, $G_{xx}(f)$	$\epsilon_r = 1/n_d^{1/2}$
Cross-spectral density magnitude, $\lvert G_{xy}(f)\rvert$	$\epsilon_r = 1/[n_d \gamma_{xy}^2(f)]^{1/2}$
Cross-spectral density phase, $\theta_{xy}(f)$	$\sigma_r = [1 - \gamma_{xy}^2(f)]^{1/2}/[2n_d \gamma_{xy}^2(f)]^{1/2}$
Coherence function, $\gamma_{xy}^2(f)$	$\epsilon_r = [1 - \gamma_{xy}^2(f)]/[0.5 n_d \gamma_{xy}^2(f)]^{1/2}$

are presented in terms of the normalized random error (coefficient of variation) defined in Eq. (3.6), except for estimates of the cross-spectrum phase where the random error is given in terms of the standard deviation of the estimated phase angle in radians.

Beyond the random errors, there is also a bias error problem in the estimation of spectral density functions that occurs at peaks and valleys in the estimates. This bias error is caused by the finite-resolution bandwidth used for the calculations. For auto- and cross-spectral density magnitude estimates, the bias error is approximated in normalized terms by[1,9]

$$\epsilon_b[\hat{G}(f)] = b[\hat{G}(f)]/G(f) = -\tfrac{1}{3}(\Delta f/B_r)^2 \tag{3.17}$$

where $b[\ ]$ denotes the bias error incurred by estimating $G(f)$ by its biased value $\hat{G}(f)$, Δf is the frequency resolution of the analysis, and B_r is the half-power-point bandwidth of a spectral peak in either $G_{xx}(f)$ or $|G_{xy}(f)|$ at that frequency. There is no general bias error equation for coherence function estimates, but error relationships have been formulated for special cases.[10]

3.4 CORRELATION FUNCTIONS

Certain noise and vibration control problems that involve relatively wide bandwidth random data signals are best addressed using time domain signal-processing procedures, as opposed to the frequency domain spectral analysis techniques discussed in the previous section. The basic calculation of interest is the correlation function between two random data signals, $x(t)$ and $y(t)$, which is estimated by

$$R_{xy}(\tau) = \frac{1}{T - \tau} \int_0^{T-\tau} x(t)\,y(t + \tau)\,dt \tag{3.18a}$$

where τ is a time delay. In digital notation,

$$R_{xy}(r\,\Delta t) = \frac{1}{N - r} \sum_{n=1}^{N-r} x[n\,\Delta t]\,y[(n + r)\,\Delta t] \tag{3.18b}$$

where r is a lag number corresponding to a time delay of $r\Delta t$. The general quantity estimated in Eq. (3.18) is called the cross-correlation function between the signals $x(t)$ and $y(t)$. For the special case where $x(t) = y(t)$,

$$R_{xx}(\tau) = \frac{1}{T - \tau} \int_0^{T-\tau} x(t)x(t + \tau)\, dt \tag{3.19}$$

is called the autocorrelation function of $x(t)$. Note that for $\tau = 0$, the autocorrelation function is simply w_x, the ms value of the signal. In both Eqs. (3.18) and (3.19), the estimated quantities will become exact in the limit as the averaging time $T \to \infty$. For finite values of T, there will be a random sampling error in the estimates, to be discussed later.

The correlation function is related to the spectral density function through a Fourier transform,[1]

$$G_{xy}(f) = 2 \int_{-\infty}^{\infty} R_{xy}(\tau) e^{-j2\pi f\tau}\, d\tau \tag{3.20}$$

Equation (3.20), often called the *Wiener-Khinchine relationship*, is the basis for computing correlation functions in practice. Specifically, the spectral density function is first computed by the FFT procedures outlined in Section 3.3. An inverse Fourier transform of the spectral density function is then computed to obtain the correlation function. Due to the remarkable efficiency of the FFT algorithm, this approach requires about one-tenth the calculations needed to compute Eq. (3.18b) directly. However, due to the *circular effects* associated with the FFT algorithm, a number of special operations are needed to obtain correct results, as detailed in reference 1.

Correlation Coefficient Function

For many applications, it is more convenient to work with the normalized cross-correlation function between $x(t)$ and $y(t)$, called the correlation coefficient function, which is given by (assuming the mean value is zero)

$$\rho^2(\tau) = \frac{R_{xy}^2(\tau)}{R_{xx}(0)R_{yy}(0)} = \frac{R_{xy}^2(\tau)}{w_x w_y} \tag{3.21}$$

The correlation coefficient function (sometimes called the squared correlation coefficient function) is similar to the coherence function, defined in Section 3.3, in that it is a real-valued quantity bounded by zero and unity, that is,

$$0 \le \rho_{xy}^2(\tau) \le 1 \tag{3.22}$$

where a value of zero means there is no linear dependence and a value of unity means there is a perfect linear dependence[1] between $x(t)$ and $y(t)$ at the time displacement τ. Hence, the correlation coefficient function is interpreted much like the frequency domain coherence function discussed in Section 3.3, except the correlation coefficient function applies to the entire frequency range of the two signals while the coherence function applies to specific frequencies. Also, from Eq. (3.14b), time delay information in the correlation coefficient function is related to the phase information in the cross-spectral density function by

$$\theta(f) = 2\pi f \tau \tag{3.23}$$

Hence, the phase of the cross spectrum can be valuable for extracting time delay information when the time delay is a function of frequency.[11]

Statistical Sampling Errors

When applied to random data signals, the computation of correlation functions will involve a statistical sampling error. In terms of a normalized random error defined in Eq. (3.6), the error in a cross-correlation estimate can be approximated by[1]

$$\epsilon_r[\hat{R}_{xy}(\tau)] = \left(\frac{1 + 1/\rho_{xy}^2(\tau)}{2B_s T_r} \right)^{1/2} \tag{3.24}$$

where $\hat{R}_{xx}(\tau)$ is an estimate of $R_{xx}(\tau)$, T_r is the total measurement duration over which the computations are performed, and B_s is the smallest statistical bandwidth for the two data signals, as defined in Eq. (3.7).

3.5 DATA ANALYSIS APPLICATIONS

The applications for signal analysis in noise and vibration studies are extensive and can become quite elaborate.[9] However, as mentioned in the introduction to this chapter, there are three specific application areas of special interest for noise and vibration control problems: (a) the identification of system response properties, (b) the identification of excitation sources, and (c) the identification of transmission paths.

Identification of System Response Properties

The control of noise and vibration is often facilitated by the determination of gain factors between excitation sources and receiver location responses. The fundamental measurement of interest here is the frequency response function (sometimes called the transfer function) between the two points of interest. Given an excitation source signal $x(t)$ and a simultaneously measured response signal $y(t)$, the frequency response function between the source and receiver signal is given by[1]

$$H_{xy}(f) = \frac{G_{xy}(f)}{G_{xx}(f)} \tag{3.25a}$$

where the auto- and cross-spectral density functions are as defined in Eqs. (3.13) and (3.14), respectively. The frequency response function is generally a complex-valued quantity that is more conveniently expressed in complex polar notation as

$$H_{xy}(f) = |H_{xy}(f)| e^{j\phi_{xy}(f)} \tag{3.25b}$$

where the magnitude function $|H_{xy}(f)|$ is the gain factor and the argument $\phi_{xy}(f)$ is the phase factor between $x(t)$ and $y(t)$. In the more advanced applications, such as normal-mode analysis,[12] both the gain and phase factor are needed. In many elementary applications, however, only the gain factor may be of interest.

The normalized random error in frequency response magnitude (gain factor) estimates is approximated by[1]

$$\epsilon_r[|\hat{H}_{xy}(f)|] \approx \frac{[1 - \gamma_{xy}^2(f)]^{1/2}}{[2n_d \gamma_{xy}^2(f)]^{1/2}} \tag{3.26}$$

where $\hat{H}_{xy}(f)$ is an estimate of $H_{xy}(f)$, $\gamma_{xy}^2(f)$ is the coherence function between the source and receiver signals, and n_d is the number of disjoint averages used to compute the autospectra and cross spectra from which the gain factor is calculated. The random error in frequency response phase estimates is the same as given for the phase of cross-spectral density estimates in Table 3.2.

Like coherence function estimates, the random error in a gain factor estimate approaches zero as the coherence function approaches unity, even for a small number of averages in the spectral density estimates. Hence, if the coherence function is large, the gain factor can be estimated with greater accuracy than the spectral density estimates used in its computation.

There are several sources of bias errors in gain factors estimates,[1,9] but the most significant is due to frequency resolution bias errors in the spectral density functions used to compute the gain factor, as given by Eq. (3.17). As a rule of thumb, if there are at least four spectral components between the half-power points of peaks in the spectral data, that is, if $\epsilon_b[G(f)] < 0.02$ in Eq. (3.17), then the bias errors in the gain factor estimate should be negligible.

As an illustration of the application of gain factor estimates, consider the experiment illustrated in Fig. 3.10, involving two vibration measurements made on a simulated spacecraft payload during a vibration test. One of the measurements is near the mounting point of the payload, and the other is on a critical payload element where vibration may adversely affect the payload performance. The gain factor clearly reveals a frequency region (around 110 Hz) where vibration at the mounting point is greatly magnified at the critical element of concern, due to a strong normal-mode response (resonance) of the payload at this frequency. It follows that efforts to reduce the vibration should be concentrated in this frequency region.

Identification of Periodic Excitation Sources

The identification of periodic acoustical and vibration excitations can usually be accomplished by a straightforward narrow-bandwidth spectral analysis plus a

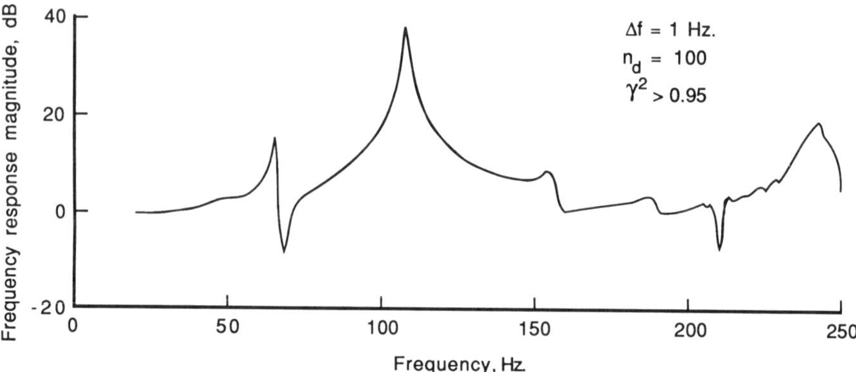

Fig. 3.10 Gain factor estimate for component in space vehicle payload.

knowledge of the rpm of all rotating machinery producing the acoustical noise and/or vibration. This is illustrated in Fig. 3.11, which shows the spectrum for the vibration on the floor of a microelectronics manufacturing facility with extensive air conditioning equipment. It is seen that the more intense spectral peaks in the vibration data can be directly identified with specific rotating machines, simply through a knowledge of the rotational frequencies. Of course, this approach cannot separate the vibration contributions of two rotating machines that operate at exactly the same frequency. However, as long as there is some difference in the rotational speed between two machines, say, Δg Hertz, the contributions of the machines can theoretically be separated by a spectral analysis with a frequency resolution Δf that is less than Δg. In practice, it is desirable to have

$$\Delta f = \frac{1}{T} < \frac{\Delta g}{3} \tag{3.27}$$

where T is the segment duration for the spectral computation. Since the signals are periodic or almost periodic, there is no random error associated with the resulting spectral estimates.

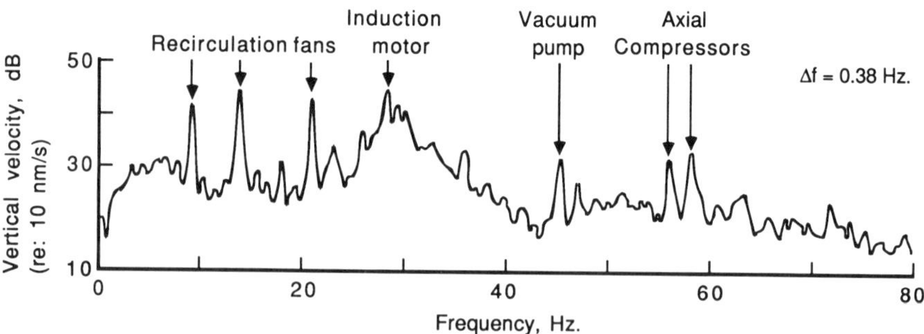

Fig. 3.11 Fourier spectrum of floor vibration in microelectronics facility. (Coutesy of BBN Laboratories, Inc., Canoga Park, CA.)

Identification of Random Excitation Sources

When two or more sources of an acoustical and/or vibration environment are random in character and further cover essentially the same frequency range, the identification of the contributions of the individual sources is more difficult. If possible, the contribution of each source of excitation should be identified by turning off all but one of the sources so that the effects of each source can be measured individually. If this is not possible, then the principle of coherent output power might be applied. Specifically, measure the acoustical or vibration signal produced at the source location by each of a collection of q suspected sources, denoted by $x_i(t)$; $i = 1, 2, \ldots, q$. For each individual source signal, simultaneously measure the receiver signal $y(t)$ and compute the coherence function between the ith source and receiver signals using Eq. (3.15). Under proper conditions, the autospectrum of the receiver signal due solely to the contribution of the ith source signal is given by

$$G_{y:i}(f) = \gamma_{iy}^2(f) G_{yy}(f) \tag{3.28}$$

where $G_{y:i}(f)$ reads "the autospectrum of the receiver signal that is due only to the source signal $x_i(t)$." Equation (3.28) is called the coherent output power relationship. If used properly, it can be a powerful tool for separating the contributions of various possible sources of random excitation in acoustical noise and vibration problems. The primary requirements for the proper application of the coherent output power relationship are as follows[9]:

1. The candidate sources of excitation (or the responses in the immediate vicinity of the sources) must be measured accurately and with negligible measurement noise. However, since the coherence function is dimensionless, any type of transducer (pressure, velocity, acceleration, or displacement) can be used for the source measurements as long as it generates a signal that has a linear relationship with the excitation phenomenon.
2. The candidate sources must be statistically independent, and there must be no interference (crosstalk) in the measurement of any one source due to energy propagating from the other sources; that is, the coherence functions among the measured source signals must all be zero.
3. There must be no significant feedback or nonlinear effects between the candidate source and receiver signals.

The normalized random error for coherent output power measurements is approximated by[1]

$$\epsilon_r[\hat{G}_{y:i}(f)] \approx \frac{[2 - \gamma_{iy}^2(f)]^{1/2}}{[n_d \gamma_{iy}^2(f)]^{1/2}} \tag{3.29}$$

where $\hat{G}_{y:i}(f)$ is an estimate of $G_{y:i}(f)$, $\gamma_{iy}^2(f)$ is the coherence function between the ith source signal and the receiver signal, and n_d is the number of disjoint averages used to compute the autospectra and cross spectra from which the coherent output is calculated. It is clear from Eq. (3.29) that the number of disjoint records (and hence the total measurement duration, $T_r = n_d T$) required for an accurate coherent output power calculation will be substantial when $\gamma_{iy}^2(f) << 1$, as will commonly occur if there are numerous independent sources contributing to $y(t)$.

A serious bias error can occur in coherent output power calculations due to time delays between the source and receiver signals that may arise when there is a substantial distance between the source and receiver measurement positions[1,9,10,13] or the measurements are made in a reverberant environment.[9,14] Since the two measurements are usually recorded and analyzed on a common time base, time delays between the source and receiver signals will cause a portion of the received signal to be uncorrelated with the source signal. These time-delay-induced bias errors can be suppressed by the use of precomputation delays or the selection of an appropriately long block duration T in the data analysis, as detailed in references 9, 13, and 14.

To illustrate the coherent output power calculation, consider the experiment outlined in Fig. 3.12, where a panel section excited by a broadband random vibration source radiates acoustical noise to a receiver microphone. The autospectrum of the radiated noise, $y(t)$, as seen by the receiver microphone with no other noise sources present, is shown by the thin solid line in Fig. 3.13. Statistically independent background noise is now introduced by a loudspeaker to produce acoustical energy with an overall value that is about 500 times more intense (27 dB higher) than the panel-radiated noise. The autospectrum of the total received microphone signal due to both the panel radiation plus the background noise is also shown in Fig. 3.13, by the heavy solid line. Finally, the coherent output power between the microphone and an accelerometer mounted on the panel is computed with the background noise present. This result is shown by the dashed line in Fig. 3.13. It is seen that the coherent output power calculation extracts the autospectrum of the radiated panel noise, as measured by an accelerometer, from the intense background noise with reasonable accuracy at most frequencies. The reason the procedure works in this example is that the radiated noise from the panel has a linear relationship with the panel motion measured by the accelerometer.

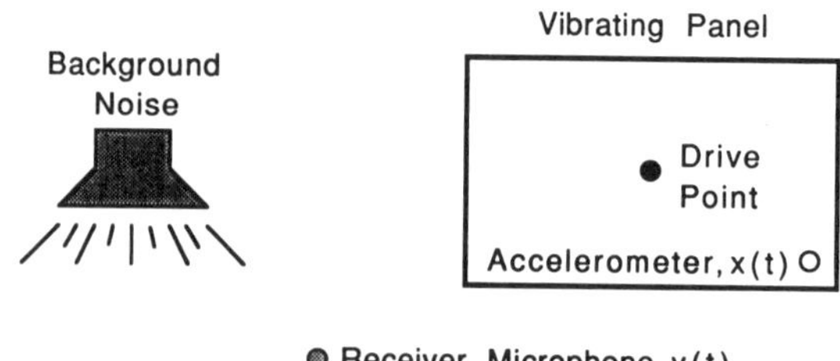

Fig. 3.12 Acoustical source identification experiment with radiating panel in background noise.

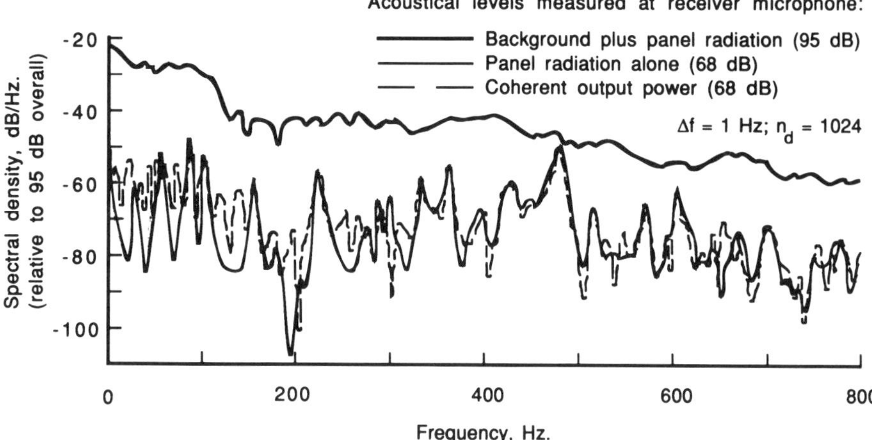

Fig. 3.13 Overall and coherent output power spectra for radiating panel in background noise. (From ref. 9 with the permission of the authors.)

Furthermore, since the panel is driven from only one point, the vibration response at any one point on the panel is representative of the vibration at all points.

Identification of Propagation Paths

Another analysis of great importance in acoustical noise and vibration reduction problems is the identification of the physical path or paths by which energy from a source of excitation travels to a receiver location. For those cases involving broadband random energy that propagates in a nondispersive manner (with a frequency-independent propagation speed), such as airborne noise, the identification of propagation paths can often be accomplished by a cross-correlation analysis. Specifically, assume a source signal $x(t)$ propagates in a nondispersive manner through r paths to produce a receiver signal $y(t)$. For simplicity, further assume the propagation paths have uniform (frequency independent) gain factors denoted by H_i, $i = 1, 2, \ldots, r$. It follows that

$$y(t) = H_1 x(t - \tau_1) + H_2 x(t - \tau_2) + \cdots + H_r x(t - \tau_r) \tag{3.30}$$

where τ_i, $i = 1, 2, \ldots, r$, are the propagation times through each of the paths. Then, from Eqs. (3.18) and (3.19),

$$R_{xy}(\tau) = H_1 R_{xx}(\tau - \tau_1) + H_2 R_{xx}(\tau - \tau_2) + \cdots + H_r R_{xx}(\tau - \tau_r) \tag{3.31}$$

In words, the cross-correlation function between the source and receiver signals will be a series of superimposed autocorrelation functions, each associated with a nondispersive propagation path and centered on a time delay equal to the propagation time along that path. Referring to Eq. (3.20), if the source signal has a wide bandwidth, these autocorrelation peaks will decay rapidly when τ deviates from τ_i, and will be sharply defined in the cross-correlation estimate, as illustrated in Fig. 3.14. Noting that the propagation time for each path is the ratio of distance to propagation speed, the physical path associated with each correlation peak can usually be identified from a knowledge of the length of the path and the propagation speed of the nondispersive waves in the medium forming the path. Finally, the portion of the receiver signal ms value that propagated through a specific path is proportional to the square of the magnitude of the correlation peak associated with that path. The normalized random error associated with the estimate of $R_{xy}(\tau)$ in Eq. (3.31) is given by Eq. (3.24).

For a cross-correlation analysis to be effective in identifying different nondispersive propagation paths, several requirements must be met, as follows.

1. The source $x(t)$ must be a broadband random signal.
2. The propagation paths must have reasonably uniform gain factors.
3. The propagation time through each path must be different from all other paths.

As a rule of thumb,[9] the difference in the propagation times through any pair of paths must be $\Delta t > 1/B_s$, where B_s is the spectral bandwidth of the receiver signal, as defined in Eq. (3.7). As the spectral bandwidth becomes small, the peaks in the cross-correlation function spread, and the ability to identify peaks representing individual propagation paths diminishes. There are a number of other signal-processing operations that will enhance the ability to detect individual propagation paths between two signals when the bandwidth is not wide.[9,15] Also, the use of envelope functions generated by Hilbert transforms can further enhance such detections.[1] Nevertheless, in the limiting narrow-band case where the source signal is a sine wave (or any periodic function), the identification of individual propagation paths cannot be achieved by any signal-processing procedure no matter how large a difference there is between the propagation times through the various paths.

To illustrate this application of cross-correlation analysis to a propagation path identification problem, consider the experiment shown in Fig. 3.15, which in-

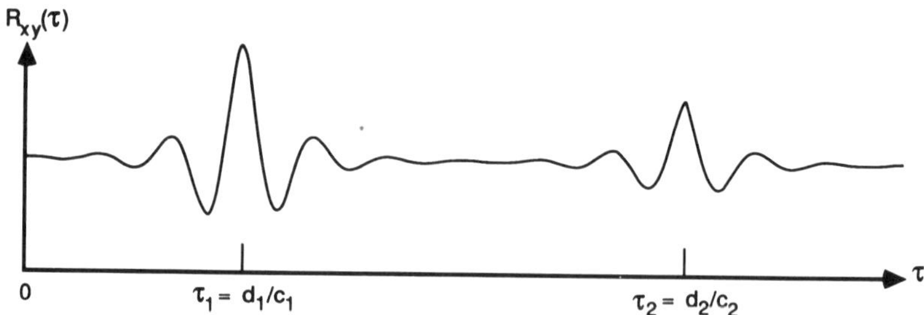

Fig. 3.14 Cross-correlation function between source and receiver signals with two nondispersive propagation paths.

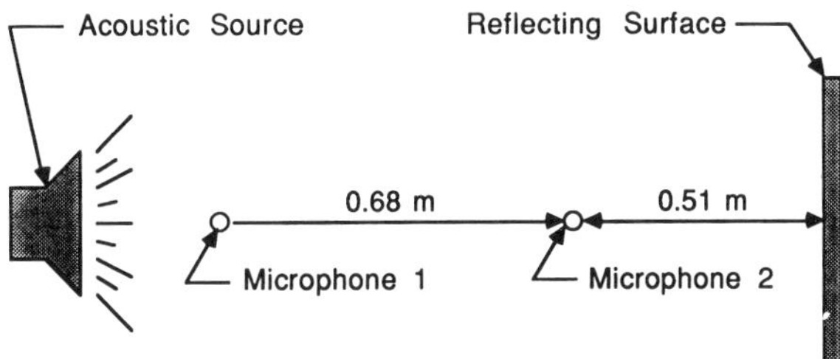

Fig. 3.15 Acoustical propagation path experiment with back reflection.

volves a speaker that produces acoustical noise with a bandwidth of approximately 8 kHz. Two microphones are used to measure the noise, one located in front of the speaker and another located 0.68 m from the speaker. There is a wall behind the receiver microphone that causes a back reflection producing a second path between the source and receiver microphones with a length of 1.7 m. The computed cross-correlation function between the source and receiver signals is shown in Fig. 3.16. It is seen that two maxima appear in the cross-correlation estimate at 2 and 5 ms. Noting that the speed of sound in air at room temperature is about 340 m/s, these two peaks clearly identify the direct path and the back reflection. The magnitude of the correlation peak corresponding to the back reflection is only about 40% of the magnitude of peak corresponding to the direct path, as would be expected due to spherical spreading loss; that is, the reflected path is 2.5 times longer than the direct path and, hence, should be about 8 dB lower in level.

The cross-correlation analysis procedure often works well in multipath acoustical problems where the propagation is in the form of longitudinal waves that are nondispersive. In structural vibration problems, there may also be some nondispersive longitudinal wave propagation, but most vibratory energy in structures propagates in the form of flexural waves, which are dispersive,[16] that is, the propagation velocity is a function of frequency. Also, flexural waves strongly reflect and/or scatter at locations where there are changes in either the material properties or the geometry of the structural path. These facts greatly complicate the detection

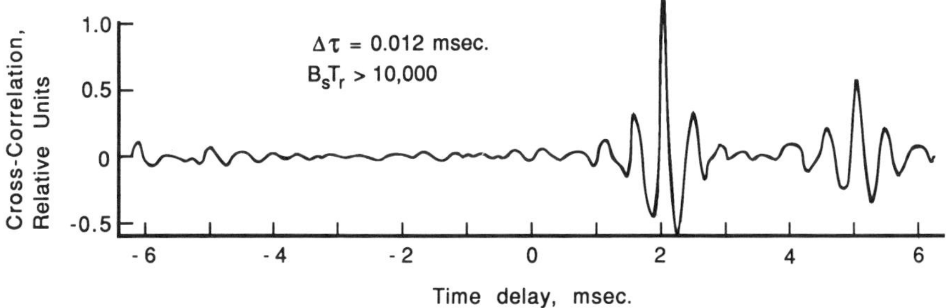

Fig. 3.16 Cross-correlation function between two microphone signals with back reflection. (From ref. 9 with the permission of the authors.)

of individual propagation paths in multipath structural vibration problems. Nevertheless, meaningful results can sometimes be obtained through the judicious use of bandwidth-limited cross-correlation analyses[1,9,17,18] if the structural paths are reasonably homogeneous.

REFERENCES

1. J. S. Bendat and A. G. Piersol, *RANDOM DATA: Analysis and Measurement Procedures*, 2nd ed., Wiley, New York, 1986.
2. L. L. Beranek, *Acoustical Measurements*, rev. ed., Acoustical Society of America, Woodbury, NY, 1988.
3. R. B. Blackman and J. W. Tukey, *The Measurement of Power Spectra*, Dover, New York, 1958, p. 19.
4. J. W. Cooley and J. W. Tukey, "An Algorithm for the Machine Calculation of Complex Fourier Series," *Math. Computat.*, **19,** 297–301 (1965).
5. F. J. Harris, "On the Use of Windows for Harmonic Analysis with the Discrete Fourier Transform," *Proc. IEEE*, **66,** No. 1, 51–83, (1978).
6. S. Rubin, "Concepts of Shock Data Analysis," C. M. Harris and C. E. Crede (eds.), in *Shock and Vibration Handbook*, 3rd ed., McGraw-Hill, New York, 1988.
7. P. D. Welch, "The Use of Fast Fourier Transforms for the Estimation of Power Spectra: A Method Based on Time Averaging Over Short, Modified Periodograms," *IEEE Trans. Audio and Electroacoustics*, **AU-15,** (2) 70–73 (1967).
8. A. H. Nuttall, "Spectral Estimates by Means of Overlapped Fast Fourier Transformed Processing of Windowed Data," NUSC TR-4169, Naval Underwater Systems Center, New London, CT, October 1971.
9. J. S. Bendat and A. G. Piersol, *Engineering Applications of Correlation and Spectral Analysis*, Wiley, New York, 1980.
10 H. Schmidt, "Resolution Bias Errors in Spectral Density, Frequency Response and Coherence Function Estimates," *J. Sound Vib.*, **101** (3), 347–427 (1985).
11. A. G. Piersol, "Time Delay Estimation Using Phase Data," *IEEE Trans. Acoust. Speech Signal Proc.*, **ASSP-29** (3) 471–477 (1981).
12. D. J. Ewins, *Modal Testing: Theory and Practice*, Research Studies Press, Letchworth, Hertfordshire, England, 1984.
13. M. W. Trethewey and H. A. Evensen, "Time-Delay Bias Errors in Estimating Frequency Response and Coherence Functions From Windowed Samples of Continuous and Transient Signals," *J. Sound Vib.*, **97** (4) 531–540 (1984).
14. K. Verhulst and J. W. Verheij, "Coherence Measurements in Multi-Delay Systems," *J. Sound Vib.*, **62** (3) 460–463 (1979).
15. C. H. Knapp and G. C. Carter, "The Generalized Correlation Method for Estimation of Time Delay," *IEEE Trans. Acoust. Speech Signal Proc.*, ASSP-24 (4) 320–327 (1976).
16. L. Cremer, M. Heckl, and E. E. Ungar, *Structure-Borne Sound*, Springer-Verlag, New York, 1973.

17. P.H.White, "Cross-Correlation in Structural Systems: Dispersive and Nondispersive Waves," *J. Acoust. Soc. Am.*, **45** (5) 1118–1128 (1969).

18. J. E. Barger, "Noise Path Diagnostics in Dispersive Structural Systems Using Cross-correlation Analysis," *Noise Control Eng.*, **6** (3) 122–129 (1976).

BIBLIOGRAPHY

Transducers and Data Acquisition

Beauchamp, K. and C. Yuen, *Data Acquisition for Signal Analysis*, George Allen & Unwin, Boston, 1980.

Beranek, L. L., *Acoustical Measurements*, rev. ed., Acoustical Society of America, Woodbury, NY, 1988.

Broch, J. T., *Mechanical Vibration and Shock Measurements*, 2nd ed., Bruel & Kjaer, Naerum, Denmark, 1984.

Doebelin, E. O., *Measurement Systems Application and Design*, 4th ed., McGraw-Hill, New York, 1990.

Fitzgerald, J. and T. S. Eason, *Fundamentals of Data Communication*, Wiley, New York, 1978.

Gregg, W. D., *Analog and Digital Communication*, Wiley, New York, 1978.

Harris, C. M. and C. E. Crede, *Shock and Vibration Handbook*, 3rd ed., McGraw-Hill, New York, 1988.

Hassall, J. R. and K. Zaveri, *Acoustic Noise Measurements*, 5th ed., Bruel & Kjaer, Naerum, Denmark, 1989.

Hnatek, E. R., *A Users Handbook of D/A and A/D Converters*, Wiley, New York, 1976.

Peterson, A. P. G. and E. E. Gross, Jr., *Handbook of Noise Measurements*, General Radio, Concord, MA, 1972.

Serridge, M. and T. R. Licht, *Piezoelectric Accelerometer and Vibration Preamplifier Handbook*, Bruel & Kjaer, Naerum, Denmark, 1986.

Signal Processing

Beauchamp, K. and C. Yuen, *Digital Methods for Signal Analysis*, George Allen & Unwin, Boston, 1979.

Bracewell, R. N., *The Fast Fourier Transform and Its Applications*, McGraw-Hill, New York, 1978.

Brigham, E. O., *The Fast Fourier Transform and Its Applications*, Prentice-Hall, Englewood Cliffs, NJ, 1988.

Broch, J. T., *Principles of Experimental Frequency Analysis*, Elsevier, New York, 1990.

DeFatta, D. J., J. G. Lucas, and W. S. Hodgkiss, *Digital Signal Processing*, Wiley, New York, 1988.

Newland, D. E., *Random Vibrations and Spectral Analysis*, Longman, New York, 1975.

Oppenheim, A. V. and R. W. Schafer, *Discrete-time Signal Processing*, Prentice-Hall, Englewood Cliffs, NJ, 1989.

Otnes, R. K. and L. D. Enochson, *Applied Time Series Analysis*, Wiley, New York, 1978.

Rabiner, L. R. and B. Gold, *Theory and Applications of Digital Signal Processing*, Prentice-Hall, Englewood Cliffs, NJ, 1975.

Randall, R. B., *Frequency Analysis*, 3rd ed., Bruel & Kjaer, Naerum, Denmark, 1987.

Roth, O., *Digital Signal Analysis Using Digital Filters and FFT Techniques*, Bruel & Kjaer, Naerum, Denmark, 1981.

CHAPTER 4

Determination of Sound Power Levels and Directivity of Noise Sources

GEORGE C. MALING, JR. and WILLIAM W. LANG

IBM Acoustics Laboratory
Poughkeepsie, New York

LEO L. BERANEK

Consultant
Cambridge, Massachusetts

4.1 INTRODUCTION

Noise control can be considered as a system problem, the system containing three major parts: the source, the path, and the receiver.[1]

In any noise control problem, sound energy from the source or sources travels over a multiplicity of paths, both in solid structures and in air, to reach the receiver—an individual, a group of people, a piece of equipment, or a structure that is affected by the noise. Three action words are associated with the source–path–receiver model: *emission*, *transmission*, and *immission*. Sound energy that is emitted by a noise source is transmitted to a receiver where it is immitted. (*Note:* the verb *immit*, meaning the opposite of *emit*, is now rare in the English language; but both the verb and the noun, immission, are used in European languages to distinguish the acts of sending out and receiving.)

Noise and Vibration Control Engineering: Principles and Applications, Edited by Leo L. Beranek and István L. Vér.
ISBN 0-471-61751-2 © 1992 John Wiley & Sons, Inc.

Sound pressure level is the physical quantity usually used to describe a sound field quantitatively because the ear responds to sound pressure. A sound-level meter may be used to measure the sound pressure level at the location in the sound field occupied by the receiver. The preferred descriptor of immission is the sound pressure level in decibels. However, the sound pressure level by itself is not a satisfactory quantity to describe the strength of a noise source because the strength varies with distance from the source and with the acoustical environment in which the source operates.

Two quantities are needed to describe completely the strength of a noise source, its *sound power level* and its *directivity*. The sound power level is a measure of the total sound power radiated by the source in all directions and is usually stated as a function of frequency, for example, in octave bands. The sound power level is then the preferred descriptor for the emission of energy by noise sources. To clearly distinguish between emission and immission, the sound power level is expressed in bels and the sound pressure level in decibels. One bel equals 10 dB. When A-weighting is applied to the overall sound power level, the result is the noise power emission level (NPEL) in bels.

The directivity of a source is a measure of the variation in its radiation with direction. Directivity is usually stated as a function of angular position around the acoustical center of the source and also as a function of frequency. Some sources radiate sound energy nearly uniformly in all directions. These are called nondirectional sources (see Fig. 4.1). Generally, such sources are small in size compared to the wavelength of the sound radiated. Most practical sources are somewhat directional (see Fig. 4.2); that is, they radiate more sound in some directions than in others.

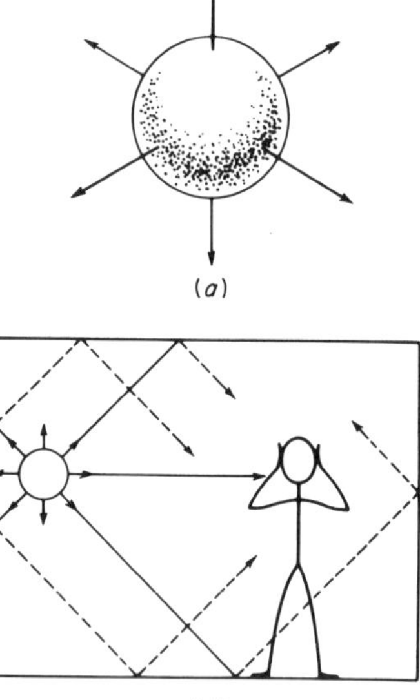

Fig. 4.1 Sound source that radiates uniformly in all directions: (*a*) sound source in free space; (*b*) same source in enclosure showing reflections from interior surfaces. Solid lines show the direct sound; dashed lines show the reflected (reverberant) sound.

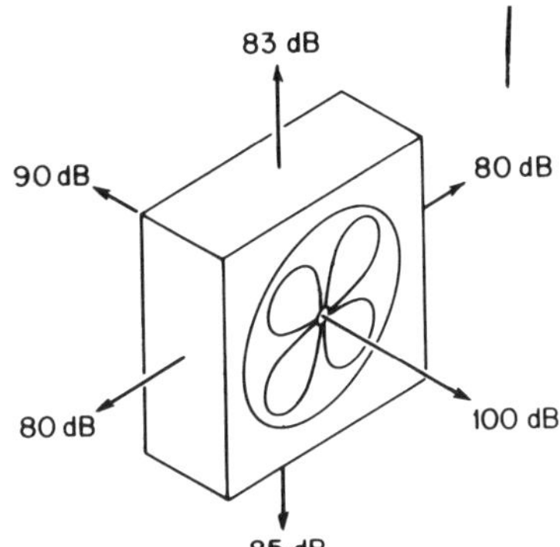

Fig. 4.2 Sound source with directive radiation into free space. This behavior is typical of equipment noise.

From the sound power level and directivity, it is possible to calculate the sound pressure levels produced by the source in the acoustical environment in which it operates. However, this method for rating a noise source is not unique because both the radiated sound power and the directivity are influenced by nearby reflecting surfaces, such as floors and walls. For most sources of practical interest, the presence of a nearby reflecting surface has a greater influence on the directivity of a source than it does on the sound power emitted by the source. Measures of source directivity are presented in Section 4.12.

A source may set a nearby surface into vibration if it is rigidly attached to that surface, causing more sound power to be radiated than if the source were vibration isolated. Both the operating and mounting conditions of the source therefore influence the amount of sound power radiated as well as the directivity of the source. Nonetheless, the sound power level alone is useful for comparing the noise radiated by machines of the same type and size as well as by machines of different types and sizes; for determining whether a machine complies with a specified upper limit of noise emission; for planning in order to determine the amount of transmission loss or noise control required; and for engineering work to assist in developing quiet machinery and equipment.

4.2 A-WEIGHTED SOUND POWER

A single-number descriptor for noise source emission is obtained when the sound power as a function of frequency is weighted using the A-weighting curve. The result is the A-weighted sound power, Chapter 1, Table 1.4.

Even though the sound power produced by a noisy machine is only a tiny fraction of the total mechanical power that the machine produces, the range of sound powers produced by sources of practical interest is enormous—from less than a microwatt to megawatts. Shaw[2] has estimated the radiation efficiency of a wide variety of noise sources, that is, the ratio of the A-weighted sound power produced by the source to the mechanical power. The estimates are shown in Fig. 4.3. The A-weighted sound powers produced by a number of small and large noise sources

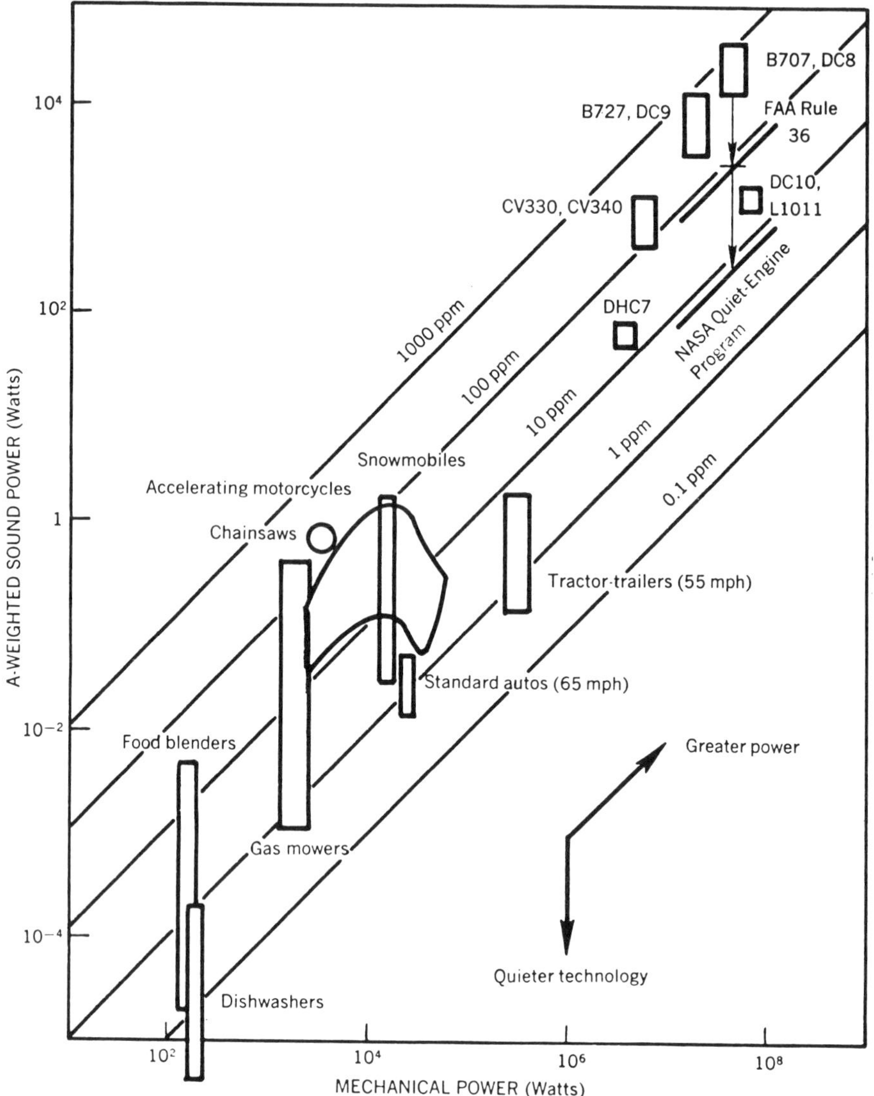

Fig. 4.3 Estimated values of A-weighted sound power versus mechanical power for various machines. The diagonal lines are lines of constant mechanoacoustical efficiency (sound power per mechanical power) in parts per million. The line labeled "FAA Rule 36" approximates 1975 noise levels for new aircraft designs, while the goal of a research program in progress at the time the figure was published is labeled "NASA Quiet-Engine Program."

are compared to their mechanical powers. The diagonal lines in the figure give conversion efficiencies ranging from 10^{-3} (1000 ppm or 0.1%) down to 10^{-7}.

The sound powers of sources of practical interest cover a range of more than 12 orders of magnitude. Hence, it is convenient to express a sound power on a logarithmic scale using an internationally agreed upon power, 10^{-12} W, as the reference for the logarithm (see Chapter 2). The A-weighted sound power level in bels is defined as

$$L_{WA} = \log \frac{W_A}{W_0} \tag{4.1}$$

where W_A = A-weighted sound power,
L_{WA} = Noise Power Emission Level (NPEL), bels
W_0 = reference sound power, internationally agreed upon as 10^{-12} W

Consequently

$$L_{WA} = \log W_A + 12 \text{ bels } re\ 10^{-12} \text{ W} \tag{4.2}$$

For convenience, the A-weighted sound power level (L_{WA}) is the NPEL, expressed in bels. Since there is an order-of-magnitude difference between NPEL values (emission) and sound pressure levels in decibels (immission), the ambiguity of expressing both emission and immission values in decibels is avoided. Use of the NPEL terminology is particularly important in dealing with the public. In many countries, people who are unfamiliar with the technical details of acoustics are unable to distinguish between different quantities that are expressed in decibels. Nonetheless, acoustical engineers often find it convenient to express sound power levels (unweighted) in decibels:

$$L_W = 10 \log \frac{W}{W_0} \quad \text{dB} \tag{4.3}$$

Example 4.1 A sound source radiates an A-weighted acoustical power of 3 W. Find the NPEL (A-weighted sound power level) in bels.

SOLUTION

$$\begin{aligned} L_{WA} &= \log 3 + 12 \\ &= 0.48 + 12 \doteq 12.5 \text{ bels } re\ 10^{-12} \end{aligned}$$

A level is a dimensionless quantity, and it is imperative that the reference be stated to avoid confusion.

4.3 RADIATION FIELD OF A SOUND SOURCE

Near Field, Far Field, and Reverberant Field

Since the sound power emitted by a source must be determined by measurement of a field quantity such as sound pressure or sound intensity (the sound energy flowing through a unit area in a unit time), it is important to understand the radiation field of a sound source when it is placed in various acoustical environments. The character of the radiation field of a typical noise source usually varies with distance from the source. In the vicinity of the source, the particle velocity is not necessarily in the direction of propagation of the sound wave, and an appreciable tangential velocity component may exist at any point. This is the near field. It is characterized by appreciable variations of the sound pressure with distance from the source along a given radius, even when the source is in a free unbounded space commonly referred to as a free field. Moreover, in the near field, the sound intensity is not simply related to the mean-square value of the sound pressure.

The distance from the source to which the near field extends is dependent on the frequency, on a characteristic source dimension, and on the phases of the radiating parts of the surface of the source. The characteristic dimension may vary with frequency and angular orientation. It is difficult, therefore, to establish limits for the near field of an arbitrary source with any degree of accuracy. It is often necessary to explore the sound field experimentally.

In the far field, the sound pressure level decreases by 6 dB for each doubling of the distance from the source, provided that either the source is in free space or the absorption in the space surrounding the source is great enough that the reverberant field has not yet been reached (see Fig. 4.4). In this free-field part of the far field, the particle velocity is primarily in the direction of propagation of the sound wave and the sound intensity is related to the mean-square sound pressure by

$$I = \frac{p_{\text{rms}}^2}{\rho c} \tag{4.4}$$

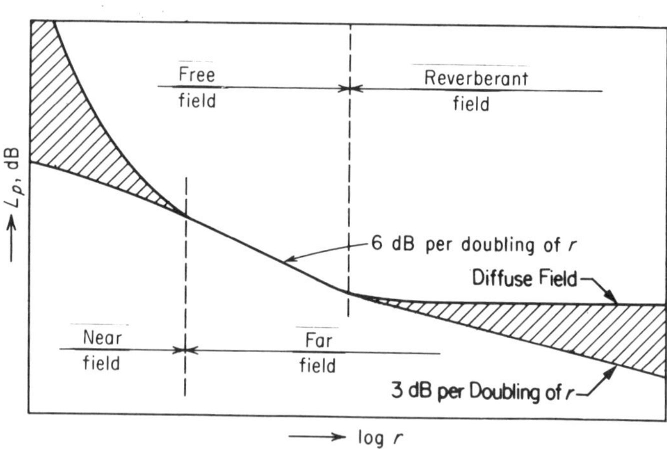

Fig. 4.4 Variation of sound pressure level in an enclosure along a radius r from a typical noise source. The free field, reverberant field, near field, and far field are shown. The near field indicates the region where the sound pressure level L_p decreases at the rate of 6 dB for each doubling of distance from the acoustical center of the source, although this region is often very short. In the far field, the sound pressure level in a highly reverberant room is constant. The lower edge of the shaded region is typical of sound fields in furnished rooms of dwellings and offices.

where I = sound intensity, W/m^2
p_{rms}^2 = mean-square sound pressure
ρ = density of air
c = speed of sound

Note that I is determined at the same point p_{rms} is measured. If the source is radiating inside an enclosure, fluctuations of sound pressure with position are observed in the reverberant part of the far field, that is, in the region where the waves reflected from the boundaries of the enclosure are superimposed upon the incident field (see Fig. 4.4). In a highly reverberant room in the region where the contribution of the sound pressure from the source is considerably smaller than the contribution from the reflected sound, the sound pressure level reaches a value essentially independent of the distance from the source. This is the diffuse field where many reflective wave trains cross from all directions and the sound energy density is nearly uniform throughout this portion of the field.

In furnished rooms in dwellings and offices where the sound field is neither a free field nor a diffuse field, the sound pressure level decreases by about 3 dB for each doubling of distance from the source.

4.4 SOUND POWER, SOUND INTENSITY, AND SOUND PRESSURE

The sound energy radiated by a source spreads out over a larger and larger area as the sound wave travels outward from the source. If a closed surface that surrounds the source is selected, the sound power radiated by the source can be calculated from the following integral:

$$W = \int_S \mathbf{I} \cdot \mathbf{dS} \tag{4.5}$$

where W = sound power, W
$\mathbf{I}$ = sound intensity, W/m^2
$\mathbf{dS}$ = element of surface area
S = surface area that surrounds source

The surface integral of Eq. (4.5) is the integral of the component of sound intensity *normal* to an element of surface area, **dS**. The integral may be carried out over a spherical or hemispherical surface that surrounds the source. Other surfaces, such as a rectangular box, are often used in practice. If the source is nondirectional and the integration is carried out over a spherical surface having a radius r, sound intensity and sound power are related by

$$I \text{ at radius } r = \frac{W}{S} = \frac{W}{4\pi r^2} \tag{4.6}$$

where W = sound power, W
S = area of spherical surface, $=4\pi r^2$, m^2
r = radius of sphere, m

In general, a source is directional, and the sound intensity is not the same at all points on the surface. Consequently, an approximation must be made to evaluate the integral of Eq. (4.5). It is customary to divide the measurement surface into a number of subsegments, each having an area S_i, and to approximate the sound intensity on each surface subsegment (in a direction normal to the surface element). The sound power of the source may then be calculated by a summation over all of the surface subsegments:

$$W = \sum_i I_i S_i \tag{4.7}$$

where I_i = sound intensity averaged over ith segment of area, W/m^2
S_i = ith segment of area, m^2

When each of the subareas S_i on the measurement surface has the same area, Eq. (4.7) reduces to

$$W = IS \tag{4.8}$$

where I = average sound intensity on measurement surface, W/m^2
S = total area of measurement surface, m^2

We may express Eq. (4.8) logarithmically as

$$L_W = 10 \log(I/I_0) + 10 \log(S/S_0) \tag{4.9}$$

$$L_W = L_I + 10 \log(S/S_0) \tag{4.10}$$

where L_W = sound power level, dB *re* 10^{-12} W
L_I = sound intensity level, dB *re* 10^{-12} W/m^2
S = area of measurement surface, m^2
S_0 = 1 m^2
I_0 = reference sound intensity, internationally agreed upon as 10^{-12} W/m^2

Equation (4.10) is usually used to determine the sound power level of a source from the sound intensity level except when the source is highly directional. For directional sources, the subareas of the measurement surface may be selected to be unequal and the logarithmic form of Eq. (4.7) [See Eq. (4.13)] should be used.

From Eq. (2.24), for the far field of a source radiating into free space,

$$I \text{ at } r = \frac{p_{\text{rms}}^2}{\rho c} \tag{4.11}$$

where ρc = characteristic resistance of air (see Section 2.5 in Chapter 2), $N \cdot s/m^3$, equal to about 406 mks rayls at normal room conditions

p_{rms} = root-mean-square sound pressure, N/m^2

Combining Eqs. (4.7) and (4.11) yields

$$W = \frac{1}{\rho c} \sum_i p_i^2 S_i \tag{4.12}$$

where p_i = average rms sound pressure over area segment S_i, N/m^2.

We may express Eq. (4.12) logarithmically as

$$L_W = 10 \log \sum_i S_i \times 10^{L_{p_i}/10} - 10 \log K \tag{4.13}$$

where L_W = sound power level, dB *re* 10^{-12} W

L_{p_i} = sound pressure level over ith area segment, dB *re* 2×10^{-5} N/m^2

S_i = area of ith segment, m^2

$10^{L_{p_i}/10} = (p_i/p_{ref})^2$

$p_{ref} = 2 \times 10^{-5}$ N/m^2

$K = \rho c/400$; values for 10 log K may be found from Fig. 1.7; at normal temperatures and atmospheric pressures, this term is negligible

Note that in Eq. (4.13) if the A-weighted sound power level in bels were being computed, the constant 10 in front of the logarithm would disappear and the result (with $K = 1$) would be

$$L_{WA} = \log \sum_i S_i \times 10^{L_{p,A_i}/10} \tag{4.14}$$

When each subsegment S_i has an equal area and $10 \log K = 0$, Eq. (4.13) can be expressed as

$$L_W = L_P + 10 \log(S/S_0) \tag{4.15}$$

where L_W = sound power level, dB *re* 10^{-12} W

L_p = sound pressure level averaged on a mean-square basis over measurement surface (surface sound pressure level), dB *re* 2×10^{-5} N/m^2

S = area of measurement surface, m^2

$S_0 = 1$ m^2

Equation (4.15) is usually used to determine the sound power level of a source from the sound pressure level except when the source is highly directional. For directional sources, the subareas of the measurement surface may be selected to be unequal; Eq. (4.13) should be used (usually with 10 log K = 0).

Hence, the sound power level of a source can be computed from sound pressure level measurements made in a free field. Equations (4.13) and (4.15) are widely used in standardized methods for determination of sound power levels in a free field or in a free field over a reflecting plane.

The sound power level of a source can also be computed from sound-pressure-level measurements made in an enclosure with a diffuse sound field because in such a field the sound energy density is constant; it is directly related to the mean-square sound pressure and, therefore, to the sound power radiated by the source. The sound pressure level in the reverberant room builds up until the total sound power absorbed by the walls of the room is equal to the sound power generated by the source. The sound power is determined by measuring the mean-square sound pressure in the reverberant field. This value is either compared with the mean-square pressure of a source of known sound power output (*comparison method*) or calculated directly from the mean-square pressure produced by the source and a knowledge of the sound-absorptive properties of the reverberant room (*direct method*). Equation (4.16) or (4.18) in Section 4.7 is used to determine the sound power level of the source.

Good approximations to free-field conditions can be achieved in properly designed anechoic rooms, or outdoors. Diffuse sound fields can be obtained in laboratory reverberation rooms. Sufficiently close engineering approximations to diffuse-field conditions can be obtained in rooms that are fairly reverberant and irregularly shaped. When these environments are not available or when it is not possible to move the noise source under test, other techniques valid for in situ determination of sound power level may be used and are described later in this chapter.

All procedures described in this chapter apply to the determination of sound power levels in octave or one-third-octave bands. The techniques are independent of bandwidth. The A-weighted sound power level (NPEL) is obtained by summing the octave-band or one-third-octave-band data after applying the appropriate A-weighting corrections. The corrections given in Table 4.1 may be applied to the unweighted octave- or one-third-octave-band levels, and the levels then summed on a mean-square basis to obtain NPEL.

Nonsteady and impulsive noises are difficult to measure under reverberant-field conditions. Measurements on such noise sources should be made either under free-field conditions or using one of the techniques described in Chapter 20. It has, however, been shown that quasi-steady noise, such as noise from a typewriter, can be measured in a diffuse-field environment.[3]

4.5 MEASUREMENT ENVIRONMENTS

Two different types of laboratory environments in which noise sources are measured are found in modern laboratories. In an anechoic room, where all of the

Table 4.1 A-weighting Corrections for Unweighted Octave and One-Third-Octave Bands

Band Center Frequency (Hz)	One-Third-Octave-Band Weightings (dB)	Octave-Band Weightings (dB)
50	−30.2	—
63	−26.2	−26.2
80	−22.5	—
100	−19.1	—
125	−16.1	−16.1
160	−13.4	—
200	−10.9	—
250	−8.6	−8.6
315	−6.6	—
400	−4.8	—
500	−3.2	−3.2
630	−1.9	—
800	−0.8	—
1,000	0.	0.
1,250	0.6	—
1,600	1.0	—
2,000	1.2	1.2
2,500	1.3	—
3,150	1.2	—
4,000	1.0	1.0
5,000	0.5	—
6,300	−0.1	—
8,000	−1.1	−1.1
10,000	−2.5	—

boundaries are highly absorbent, the free-field region extends very nearly to the boundaries of the room. A hemianechoic room has a reflective floor, but all other boundaries are highly absorbent. Both the anechoic and hemianechoic environments are used to determine the sound power level of a source, which is calculated by using Eq. (4.13) or (4.15). The surface sound pressure level is measured on an imaginary surface surrounding the source and located in the far field. In a reverberation room, where all boundaries are acoustically hard and reflective, the reverberant field extends throughout the volume of the room except for a small region in the vicinity of the source. The sound power level is determined from an estimate of the average sound pressure level in the diffuse field of the room.

The sound pressure field in an ordinary room, that is, in furnished rooms of dwellings and offices, is neither a free field nor a diffuse field. Here the relationship between the sound intensity and the mean-square pressure is more complicated. Instead of measuring the mean-square pressure, it is usually more advantageous to use a sound intensity analyzer that measures the sound intensity directly (see Section 4.11). By sampling the sound intensity at defined locations in the vicinity of the source, the sound power level of the source can be determined. Equation (4.10) is used to determine the sound power levels from the sound intensity levels. If the subareas of the measurement surface are unequal, the logarithmic form of Eq. (4.7) may be used.

4.6 INTERNATIONAL STANDARDS FOR DETERMINATION OF SOUND POWER

ISO Standards

The International Organization for Standardization (ISO) has published a series of international standards, the ISO 3740 series,[4-11] which describes several methods for determining the sound power levels of noise sources. The fundamental standards in this series are as follows:

ISO 3740: Guidelines for the Use of Basic Standards and for the Preparation of Noise Test Codes.

ISO 3741: Precision Methods for Broad-Band Sources in Reverberation Rooms.

ISO 3742: Precision Methods for Discrete-Frequency and Narrow-Band Sources in Reverberation Rooms.

ISO 3743: Engineering Methods for Special Reverberation Test Rooms.

ISO 3744: Engineering Methods for Free-Field Conditions over a Reflecting Plane.

ISO 3745: Precision Methods for Anechoic and Semi-Anechoic Rooms.

ISO 3746: Survey Method.

ISO 3747: Survey Method Using a Reference Sound Source.

The most important factor in selecting an appropriate noise measurement method is the ultimate use of the sound-power-level data that are to be obtained. Table 4.2 summarizes the applicability of each of the basic standards of the ISO 3740 series. The principal uses of acoustical data include the development of quieter machines and equipment, the noise testing of machines and equipment in production, and comparisons of several machines that may be of the same or different type and size.

In making a decision on the appropriate measurement method to be used, the experimenter should take into consideration several factors: (a) the size of the noise source, (b) the moveability of the noise source, (c) the test environments available for the measurements, (d) the character of the noise emitted by the noise source, and (e) the grade (classification) of accuracy required for the measurements. The methods described in this chapter are consistent with those of the ISO 3740 series and with the corresponding set of the American National Standards Institute (ANSI).[12-18]

Table 4.2 Description of ANSI and ISO 3740 Series of Standards for Determination of Sound Power

ANSI Standard	ISO Standard	Classification of Method	Test Environment	Size of Source	Character of Noise	Sound Power Levels Obtainable	Optional Information Obtainable
S1.31	ISO 3741	Precision (Grade 1)	Reverberation room meeting specified requirements	Steady; volume of source preferably less than 1% of test room volume	Steady broadband	In one-third-octave or octave bands	A-weighted sound level
S1.32	ISO 3742	Precision (Grade 1)			Steady, discrete frequency or narrow band	In one-third-octave or octave bands	A-weighted sound level
S1.33	ISO 3743	Engineering (Grade 2)	Special reverberant test room		Steady, broadband, narrow band, or discrete frequency	A-weighted and in octave bands	Other weighted sound power levels
S12.34	ISO 3744	Engineering (Grade 2)	Outdoors or in large room	Greatest linear dimension less than 15 m; otherwise limited only by available test environment	Any	A-weighted and in one-third-octave or octave bands	Directivity information and sound pressure levels as a function of time; other weighted sound power levels
S1.35	ISO 3745	Precision (Grade 1)	Anechoic or hemianechoic room	Volume of source preferably less than 0.5% of test room volume	Any	A-weighted and in one-third-octave or octave bands	Directivity information and sound pressure levels as a function of time; other weighted sound power levels
S1.36	ISO 3746	Survey (Grade 3)	No special test environment	No restrictions; limited only by available test environment	Any	A-weighted	Sound pressure levels as a function of time; other weighted sound power levels
None	ISO 3747	Survey (Grade 3)	No special test environment; source under test not moveable	No restrictions	Steady, broadband, narrow band, or discrete frequency	A-weighted	Sound power levels in octave bands

Source: Reprinted from ANSI S12.34 by permission of the Acoustical Society of America.

4.7 DETERMINATION OF SOUND POWER IN A DIFFUSE FIELD

Characteristics of Reverberation Rooms

Measurements of the sound power level of a device or machine may be performed in a laboratory reverberation room.[5,6] The determination of the sound power level of a noise source in such a room is based on the premise that the measurements are performed entirely in the diffuse (reverberant) sound field (see Fig. 4.4). In the reverberant field, the average sound pressure is essentially uniform, although there are fluctuations from point to point, and it is simply related to the sound power radiated by the source through Eq. (4.18). Information concerning the directivity of the source cannot be obtained in a diffuse sound field.

If the 125-Hz octave band is the lowest frequency band of interest, the reverberant room should have a volume of at least 200 m^3. The equipment being tested should have a volume no greater than 1% of the room volume. The surfaces of the room should be highly reflective; each surface should have a Sabine absorption coefficient between 0.5 and 1.5 times the mean value of the absorption coefficient of all surfaces. The surface that is closest to the equipment being evaluated should have a Sabine absorption coefficient that does not exceed 0.06. If the room is to be used to measure equipment that has discrete frequency components in its spectrum, it will usually be necessary to add low-frequency sound absorption to the room or to use rotating diffusers.[6] Procedures that may be used to qualify the room for both broadband and narrow-band measurements are outlined below.

Room Qualification

The ISO standards 3741 and 3742[5,6] contain detailed procedures for the qualification of reverberation rooms for the determination of sound power levels. For sources of broadband sound, the room is qualified by selecting eight source positions in the room and determining the standard deviation of the average sound pressure level in the room for these positions. The room is qualified according to ISO 3741 if the standard deviation does not exceed the values given in Table 4.3.

The qualification of reverberation rooms for the measurement of noise that contains discrete-frequency components in the spectrum is complicated. Standard ISO

Table 4.3 Qualification Requirements for Reverberation Room Used for Measurement of Broadband Noise Sources

Octave-Band Center Frequencies (Hz)	One-Third-Octave-Band Center Frequencies (Hz)	Maximum Allowable Standard Deviation (dB)
125	100–160	1.5
250, 500	200–630	1.0
1000, 2000	800–2500	0.5
4000, 8000	3150–10,000	1.0

Source: Reprinted from ANSI S1.31 by permission of the Acoustical Society of America.

3742 contains detailed procedures for room qualification. A loudspeaker is placed in the room and is driven by a series of discrete-frequency tones in each one-third-octave band. For example, there are 22 frequencies spaced 1 Hz apart in the 100-Hz one-third-octave band and 23 frequencies spaced 5 Hz apart in the 500-Hz octave band. The average sound pressure level in the room is determined at each frequency and corrected for the loudspeaker response. The room is qualified according to ISO 3742 if the standard deviation of the level in each one-third-octave band does not exceed the values given in Table 4.4.

Experimental Setup

An array of microphones or a single microphone that traverses a path (often circular) in the reverberant room may be used to determine the average sound pressure level in the reverberant field. If λ is the wavelength of the lowest frequency of interest, the sound pressure in the reverberant field should be determined using a path length of $\frac{3}{2}\lambda$ or 3 m, whichever is greater. If an array of microphones is used, the array should consist of at least three microphones for the measurement of broadband sound. If discrete-frequency components are present in the spectrum, a larger number of microphones is usually required for adequate sampling of the sound field, and the traversing microphone path is usually preferred. The microphone path or array should be positioned in the room so that no microphone position is within one major source dimension of the equipment being evaluated. The minimum distance between the microphone(s) and the equipment being evaluated is determined differently in the comparison method and the direct method for determination of sound power. These requirements are discussed below. If the source is highly directional, microphone positions that fall within the region of highest directivity should be avoided.

If the noise source under test is normally associated with a hard floor, wall, edge, or corner, it should be placed in a corresponding position in the reverberation room. The source should not be placed near the geometric center of the room since in that location many of the resonant modes of the room would not be excited.

Table 4.4 Qualification Requirements for Reverberation Room Used for Measurement of Narrowband Noise Sources

Octave-Band Center Frequency (Hz)	One-Third-Octave-Band Center Frequency (Hz)	Maximum Allowable Standard Deviation (dB)
125	100–160	3.0
250	200–315	2.0
500	400–630	1.5
1000, 2000	800–2500	1.0
4000	3150–5000	1.5
8000	6300–10,000	2.0

Source: Reprinted from ANSI S1.32 by permission of the Acoustical Society of America.

Near the boundaries of the room, the sound field will depart from the ideal state of diffusion.[19] Assuming a source to be radiating continuous-spectrum noise and the analyzing filter to be one-third octave wide, it is good engineering practice to place none of the microphones in an array closer than $\frac{1}{2}\lambda$ or 1 m, whichever is smaller, from any surface of the room. Here λ is the wavelength corresponding to the lowest frequency of interest.

Comparison Method

The procedure for determining the sound power level of a noise source by the comparison method[5] requires the use of a reference sound source (see Fig. 4.5 and ref. 20) of known sound power output. The calibration of a reference sound source is given in Table 4.5. Being careful to locate the measuring microphones in the reverberant field, the procedure is as follows:

1. With the equipment being evaluated at a suitable location in the room, determine, in each frequency band, the average sound pressure level (on a mean-square basis) in the reverberant field using the microphone array or traverse described above.
2. Replace the source under test with the reference sound source and repeat the measurement to obtain the average level for the reference sound source.

Fig. 4.5 Reference sound source, Bruel and Kjaer Type 4204. (Courtesy of Bruel and Kjaer Instruments, Inc.)

Table 4.5 Calibration Table for Bruel and Kjaer Reference Sound Source, Type 4204[a]

Sound power is measured in one-third-octave bands.
Octave-band values and A-weighted value are calculated.
The sound source is placed in a free field above a reflecting plane.

Calibration accuracy : 100–160 Hz: ±1.0 dB
: 200–4000 Hz: ±0.5 dB
: 5000–10,000 Hz: ±1.0 dB

Serial no., 1114279 Air temperature, 19°C
Date, April 3, 1984 Barometric pressure, 1010 mbar
Signature, Hans Christensen Relative humidity, 32%

	Sound Power, dB *re* 1 pW			
	A		B	
Center Frequency (Hz)	One-Third-Octave (dB)	One Octave (dB)	One-Third-Octave (dB)	One Octave (dB)
100	74.5		79.3	
125	74.7	79.8	78.7	83.9
160	75.7		79.4	
200	76.3		79.8	
250	76.1	81.0	80.0	84.8
315	76.4		80.2	
400	76.2		79.8	
500	76.1	80.9	79.7	84.5
630	76.2		79.8	
800	78.4		81.9	
1000	79.7	84.9	83.5	88.5
1250	81.6		85.1	
1600	81.4		85.2	
2000	80.3	85.1	84.5	89.2
2500	78.9		83.3	
3150	78.7		82.6	
4000	77.8	82.7	82.0	86.9
5000	77.4		81.7	
6300	76.1		80.6	
8000	74.1	79.2	78.8	83.8
10000	72.3		77.1	
Curve *A*	90.5		94.5	

[a]A: line frequency, 50.0 Hz; frequency of rotation, 48.4 Hz. B: line frequency, 60.0 Hz; frequency of rotation, 55.9 Hz.

The sound power level of the souce under test L_W for a given frequency band is calculated as

$$L_W = L_p + (L_{Wr} - L_{pr}) \tag{4.16}$$

where L_W = sound power level for source being evaluated, dB *re* 10^{-12} W
L_p = space-averaged sound pressure level of source being evaluated, dB *re* 2×10^{-5} N/m^2.
L_{Wr} = calibrated sound power level of reference source, dB *re* 10^{-12} W
L_{pr} = space-averaged sound pressure level of reference sound source, dB *re* 2×10^{-5} N/m^2

To make certain that the reverberant sound field predominates in the determination of the average sound pressure level, the minimum distance $d_{\min}$ between the microphone(s) and the equipment being evaluated should be at least[5]

$$d_{\min} = 0.8 \times 10^{(L_{Wr} - L_{pr})/20} \tag{4.17}$$

where L_{Wr} and L_{pr} are as defined under Eq. (4.16).

Direct Method

Instead of using a reference sound source, the direct method requires that the sound-absorptive properties of the room be determined by measuring the reverberation time in the room for each frequency band (see Chapter 20).

With this method, the space-averaged sound pressure level for each frequency band of the source being evaluated is determined as described above for the comparison method. The sound power level of the source is found from[5]

$$L_W = L_p - 10 \log \frac{T}{T_0} + 10 \log \frac{V}{V_0} + 10 \log \left(1 + \frac{S\lambda}{8V}\right) - 10 \log \left(\frac{B}{1000}\right) - 14 \text{ dB } re\ 10^{-12} \text{ Watt} \tag{4.18}$$

where L_p = space-averaged sound pressure level of equipment being evaluated, dB *re* 2×10^{-5} N/m^2
T = reverberation time of room for that frequency band, s
V = room volume, m^3
T_0 = 1 s
V_0 = 1 m^3
S = area of all boundary surfaces of room, m^2
λ = wavelength corresponding to center frequency in octave or one-third-octave band being measured
B = barometric pressure, mbar
1000 = reference barometric pressure, mbar

To make certain that the reverberant sound field predominates in the determination of the average sound pressure level, the minimum distance $d_{\min}$ between the microphone(s) and the equipment being evaluated should be at least[5]

$$d_{\min} = 0.16 \left(\frac{V}{T}\right)^{1/2} \tag{4.19}$$

where V and T are as defined above.

Example 4.2 Assume a room with a volume of 200 m^3, a surface area S = 210 m^2, and a reverberation time at 100 Hz of 3 s. The space-averaged sound pressure level L_p, in the diffuse field with a given machine operating is 100 dB. Find the sound power level for this machine. Assume a discrete-frequency spectrum and an atmospheric pressure of 1000 mbar.

SOLUTION Use Eq. (4.18) to determine the sound power. The wavelength corresponding to 100 Hz is 3.44 m. The sound power level is

$$\begin{aligned} L_W &= 100 - 10 \log(3) + 10 \log(200) \\ &\quad + 10 \log\left(1 + \frac{210 \times 3.44}{8 \times 200}\right) - 10 \log(1) - 14 \\ &= 100 - 4.8 + 23 + 1.6 - 0 - 14 \\ &= 105.8 \text{ dB } re\ 10^{-12} \text{ W (100 Hz mean frequency)} \end{aligned}$$

4.8 DETERMINATION OF SOUND POWER IN A FREE FIELD

Measurements of the sound power level produced by a device or a machine are performed in a laboratory anechoic or hemianechoic room. Alternatively, a hemianechoic environment may be provided at an open-air site above a paved area, distant from reflecting surfaces such as buildings and with a low background noise. In such an outdoor environment, only the floor or ground surface is rigid, and there are no other reflecting objects in the vicinity. This environment approximates a large room with sound-absorptive treatment on ceilings and walls with the equipment under test mounted on the concrete floor.

The determination of the sound power level radiated in an anechoic or a hemianechoic room is based on the premise that the reverberant field is negligible at the positions of measurement and that the total radiated sound power is obtained by a space integration of the sound intensity over a hypothetical surface that surrounds the noise source. The measurement surface must be in the far field of the source. Then the space average of the mean-square sound pressure over the measurement surface is determined. This may be done by choosing an array of microphone positions over the surface. The number of microphone positions that are needed will depend on the accuracy required and the directivity of the source. The more directional the source, the greater the number of points required.

Measurement in Hemianechoic Space

The sound power determination in a hemianechoic space may be performed according to ISO 3744[8] for engineering-grade accuracy or according to ISO 3745[9] for precision-grade accuracy. If a hemisphere is used to determine the location of the microphone positions, it has its center on the reflecting plane beneath the acoustical center of the sound source. To ensure that the measurements are carried out in the far field, the radius of the hemisphere should be equal to at least two major source dimensions or four times the average source height above the reflecting plane, whichever is larger. It is good engineering practice for the radius of the test hemisphere always to be greater than about 1.0 m. No microphone position should be closer to the room boundaries than $\frac{1}{4}\lambda$, where λ is the wavelength of sound at the midband frequency of the lowest frequency band of interest. Outdoors, atmospheric effects are likely to influence the measurements if the radius of the test hemisphere is much greater than about 15 m, even in favorable weather.

Figure 4.6[8] gives an array of 10 key microphone positions and 10 additional microphone positions that may be used for measurements according to ISO 3744. A hemispherical array of 10 microphone positions according to ISO 3745[9] is shown in Fig. 4.7. Symmetry in the radiation pattern could reduce the number of measurement positions, but there is danger in selecting too few measurement positions. Reflections of the sound from the hard floor may cause a far-field interference pattern. If several of the microphones are located in regions of low sound pressure in this pattern, the average mean-square sound pressure may be in error. Thus, a large number of points at various heights above the floor enhances the accuracy of the results.

A rectangular array of microphone positions is frequently used for the determination of sound power levels. Detailed requirements for the selection of a rectangular array are given in ISO 3744.[8] At least nine positions on the measurement surface as shown in Fig. 4.8 should be used.

To obtain the sound power level of the source, the background sound pressure level at each point on the measurement surface is measured. The sound pressure level produced by the source is then measured and corrected for background noise using one of the tables in the international standards.[8,9] Alternatively, the measured sound pressure level and the background sound pressure level may be converted to mean-square pressures. The difference between the two mean-square pressures is taken and converted back to a sound pressure level to obtain the corrected sound pressure level of the source.

Equation (4.15) is used to determine the sound power level of the source. For highly directional sources, it is desirable to select subareas of the measurement surface in the region of high directivity and to measure the sound pressure level associated with each subarea. The sound power level can then be determined by using Eq. (4.13).

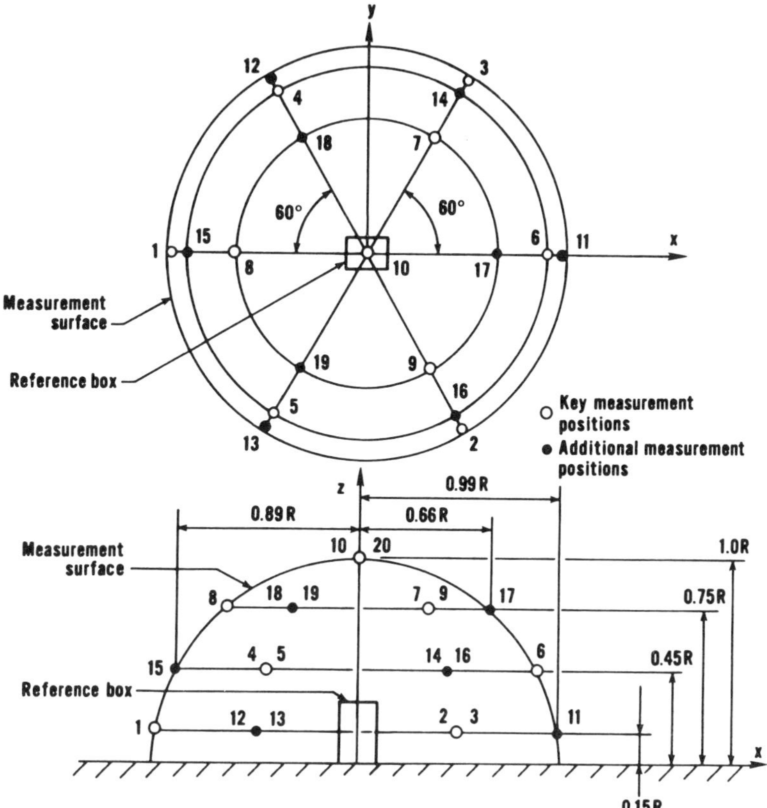

Fig. 4.6 Microphone array for a hemispherical measurement surface according to ISO 3744. Key microphone positions are numbered 1–10. Additional microphone positions are numbered 11–20 and may be used to improve the accuracy of the measurements. (Reprinted from ANSI S12.34 by permission of the Acoustical Society of America.)

Measurement in Anechoic Space

When the noise source in normal use is not mounted over a hard surface or is small enough to be placed near the center of an anechoic room, the space-averaged mean-square sound pressure level may be determined over a test sphere. Measurements may be made according to ISO 3745.[9] An array of 20 microphone positions specified in ISO 3745 is shown in Fig. 4.9. The methods described above for the hemianechoic case apply, except that all 20 of the microphone positions shown in Fig. 4.9 are to be used, and the test area is the surface of a full sphere rather than a hemisphere.

When measurements are made in an anechoic space, the sound power level of each frequency band is computed according to Eq. (4.15) except that K is included

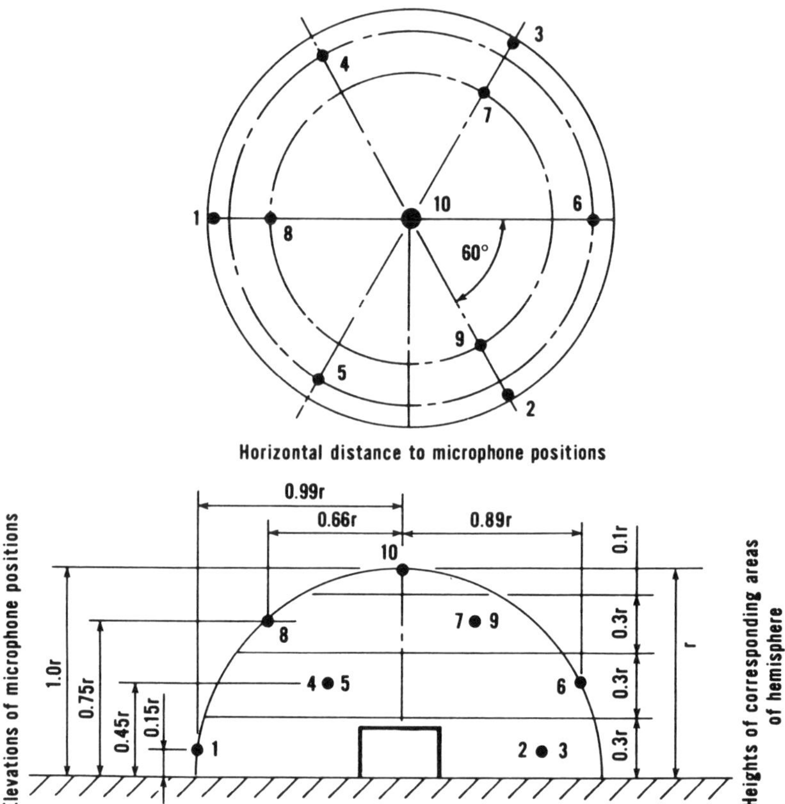

Fig. 4.7 Microphone array for a hemispherical measurement surface according to ISO 3745. (Reprinted from ANSI S1.35 by permission of the Acoustical Society of America.)

in the calculation for precision measurements. When the source is highly directional, the sound power level may be caluculated using Eq. (4.13).

4.9 ENVIRONMENTAL CORRECTIONS AND DETERMINATION OF SOUND POWER LEVEL

In the previous sections, the only corrections applied for the determination of the average sound pressure level of the measurement surface were for the presence of background noise. If reflections from the room surfaces affect the measured surface sound pressure level, it is important to apply a correction to the measured data to account for these reflections. The environmental correction is subtracted directly from the measured sound pressure level (after corrections for background noise are applied). The ISO 3744 standard[8] allows a maximum environmental correction

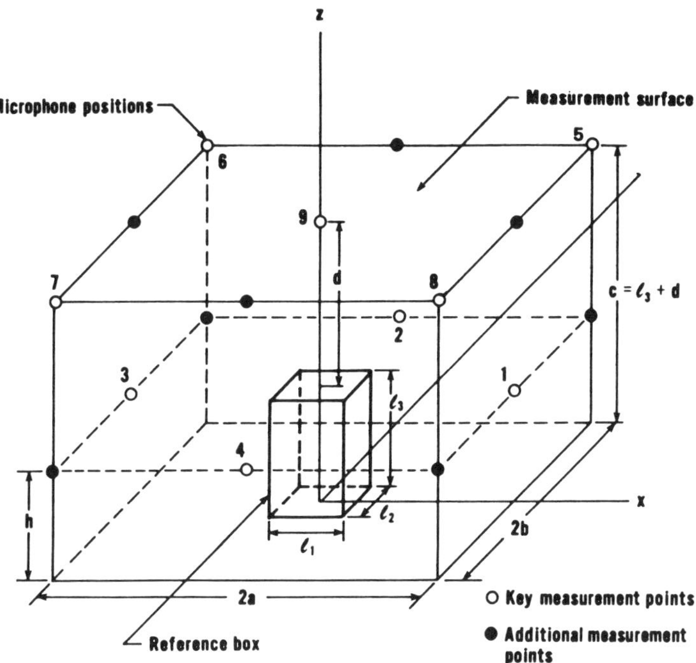

Fig. 4.8 Microphone array for a rectangular measurement surface according to ISO 3744. The distance d from the rectangular parallepiped that surrounds the source to the measurement surface is often taken as 1 m. The height h given in the figure is usually 0.5 ($l_3 = d$) but may be the shaft height for rotating machines, the center of the machine, or 1.2 m. The nine key microphone positions are usually adequate, but the additional microphone positions shown should be used if the major source dimension exceeds twice the distance d. (Reprinted from ANSI S12.34 by permission of the Acoustical Society of America.)

of 2 dB; the correction may be determined in several ways:

1. By comparing the calibrated sound power level of a reference souce, L_{Wr}, with the measured sound power level of the same source in the room, L_W. The environmental correction is $K_1 = L_W - L_{Wr}$.
2. By measuring the reverberation time in the room and determining the total Sabine absorption A according to $A = 0.16\ (V/T)$, where T is the reverberation time in seconds and V is the volume of the test room in cubic meters. The environmental correction K_1 is determined using Fig. 4.10. In the figure, S is the area of the measurement surface in square meters.

The American counterpart to ISO 3744, ANSI S12.34[16] allows the environmental correction to be determined using a two-surface method.

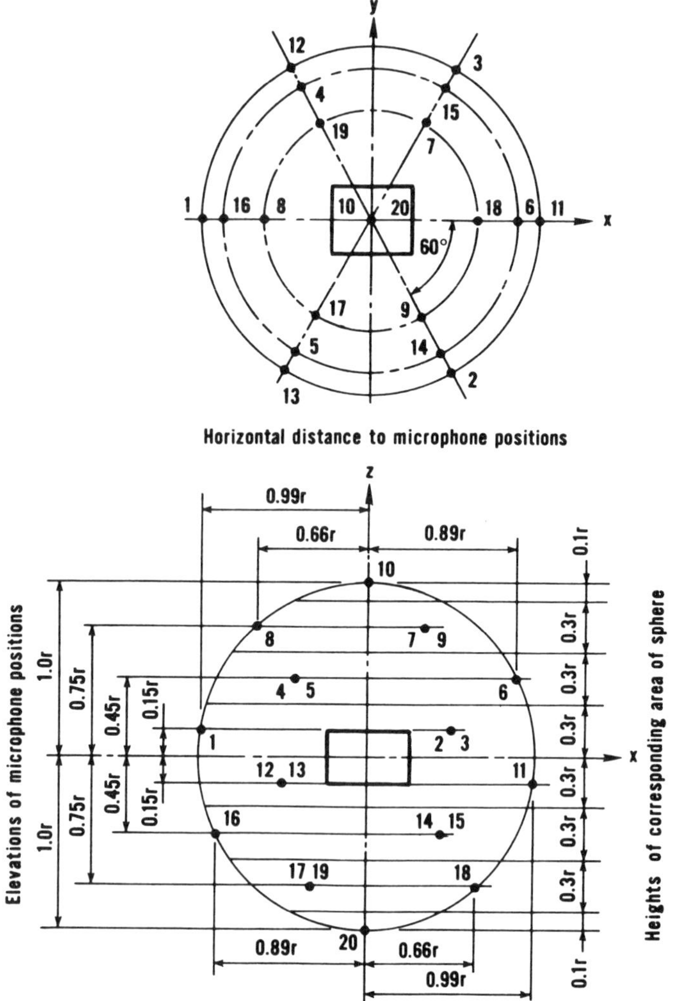

Fig. 4.9 Microphone array for a spherical measurement surface according to ISO 3745. (Reprinted from ANSI S1.35 by permission of the Acoustical Society of America.)

Reference should be made to ISO 3746[10] for determination of sound power in situ, that is, when the source cannot be moved into a laboratory environment and when the environmental correction is greater than 2 dB. The uncertainty of the sound power level determined according to ISO 3746 is greater than that obtained when ISO 3744 or ISO 3745 is used. The average sound pressure level is determined on the rectangular measurement surface of Fig. 4.11 or the hemispherical measurement surface of Fig. 4.12 (see also Table 4.6). The measured levels are corrected using the procedure described below, and the sound power level is calculated according to Eq. (4.15).

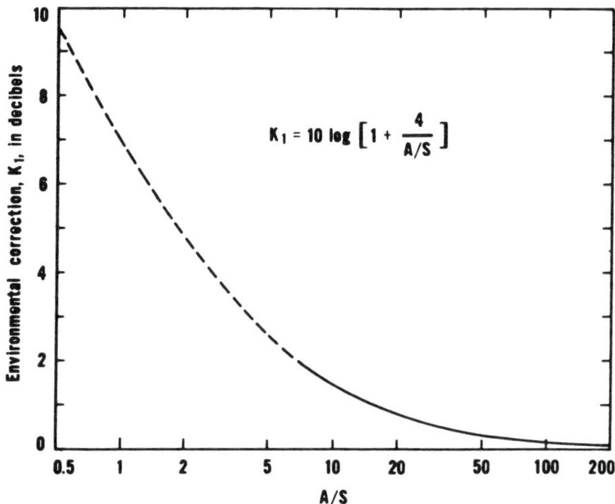

Fig. 4.10 Environmental correction K_1 in decibels. (Reprinted from ANSI S1.36 by permission of the Acoustical Society of America.)

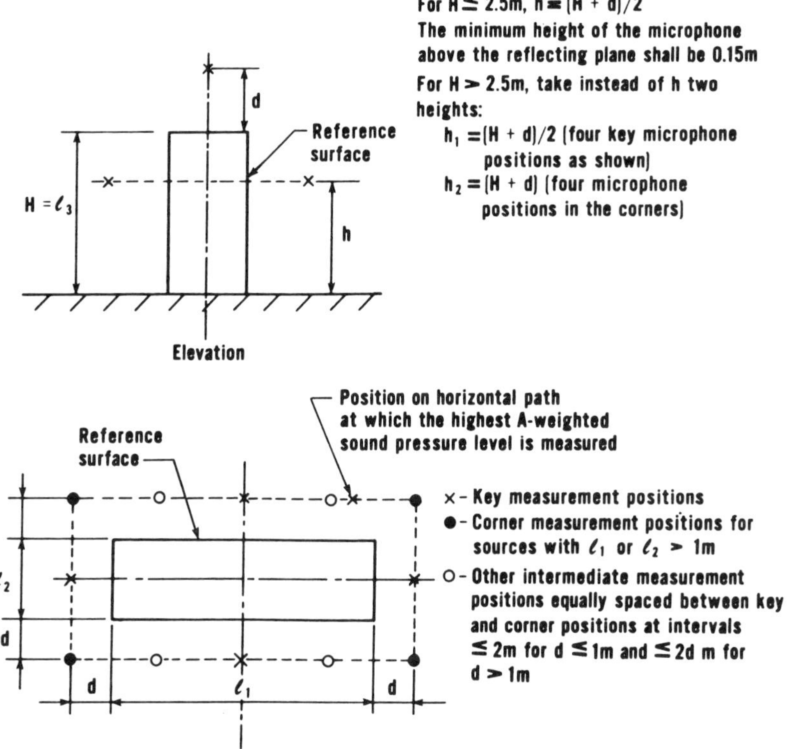

Fig. 4.11 Location of measurement points on a rectangular measurement surface according to ISO 3746. The source dimensions are l_1, l_2, l_3. (Reprinted from ANSI S1.36 by permission of the Acoustical Society of America.)

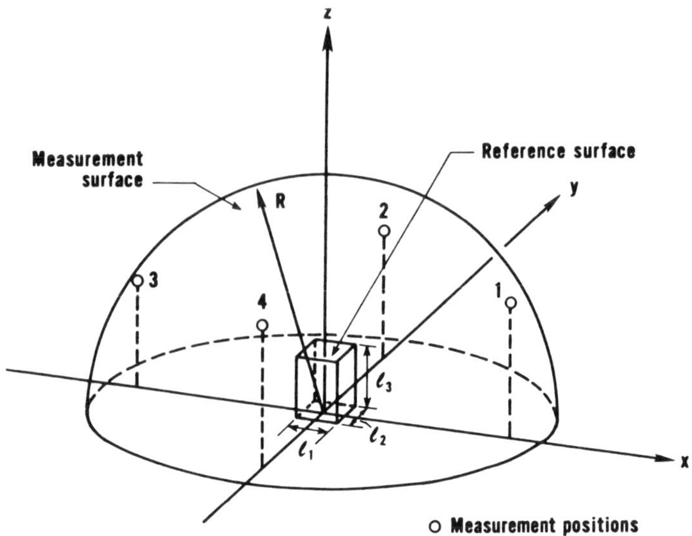

Fig. 4.12 Location of measurement points on a hemispherical measurement surface according to ISO 3746. (Reprinted from ANSI S1.36 by permission of the Acoustical Society of America.)

The environmental correction K_1 may be determined using either of the procedures described above. Alternatively, the average Sabine absorption coefficient $\overline{a}$ in metric sabines per square meters may be estimated from Table 4.7.[10] The total Sabine absorption A is calculated from

$$A = \overline{a} S_R \tag{4.20}$$

where S_R is the total area of the test room in square meters. The value of $\overline{a}$ should not exceed 0.25 and the environmental correction K_1 should not exceed 7 dB (which corresponds to a ratio $A/S \leq 1$).

Table 4.6 Microphone Positions on Hemisphere for Survey Measurements According to ANSI S1.36-1979

Microphone Number	Microphone Positions (All Heights, $z = 0.6R$)		
	x/R	y/R	z/R
1	0.8	0.0	0.6
2	0.0	0.8	0.6
3	−0.8	0.0	0.6
4	0.0	−0.8	0.6

Source: ANSI S1.36-1979.

Table 4.7 Approximate Values of Average Sabine Absorption Coefficient $\bar{a}$

Average Sabine Absorption Coefficient $\bar{a}$	Description of Room
0.05	Nearly empty room with smooth hard walls made of concrete, brick, plaster, or tile
0.1	Partly empty room, room with smooth walls
0.15	Room with furniture, rectangular machinery room, rectangular industrial room
0.2	Irregularly shaped room with furniture, irregularly shaped machinery room, or industrial room
0.25	Room with upholstered furniture, machinery or industrial room with a small amount of acoustical material (e.g., partially absorptive ceiling) or ceiling or walls

The ISO 3747 standard[11] provides an alternative method for determination of sound power level in situ using a reference sound source.

4.10 DETERMINATION OF SOUND POWER USING SOUND INTENSITY

The previously described methods for determination of the sound power level of a source depend on the fact that the sound intensity can be determined (at least approximately) from the mean-square sound pressure. When the measurement environment is a laboratory-quality room, no environmental corrections are required. As described in Section 4.9, the surface sound pressure level in other environments may be environmentally corrected to account for reflections of sound from the room surfaces.

The availability of sound intensity analyzers and probes capable of a direct measurement of sound intensity has made it possible to determine the sound power of a source without using an approximate relation between sound pressure and sound intensity. It is, however, still necessary to approximate the surface integral of Eq. (4.5) by selection of an appropriate number of surface segments, S_i, on which the component of sound intensity normal to the surface is measured. The sound intensity associated with each segment may be determined by making a measurement at a single point or by moving the intensity probe slowly over the segment (scanning).

Sound intensity is the product of sound pressure and particle velocity, normally averaged over time. In order to determine sound intensity, it is necessary to determine the particle velocity at a particular point in space. Two methods are currently available for determination of particle velocity: direct measurement using ultrasonic transmitters and receivers and measurement of the sound pressure at two closely spaced microphones. The latter method uses a finite-difference approximation to estimate the pressure gradient and the particle velocity at a given point in space. An intensity probe that uses the finite-difference approximation is shown

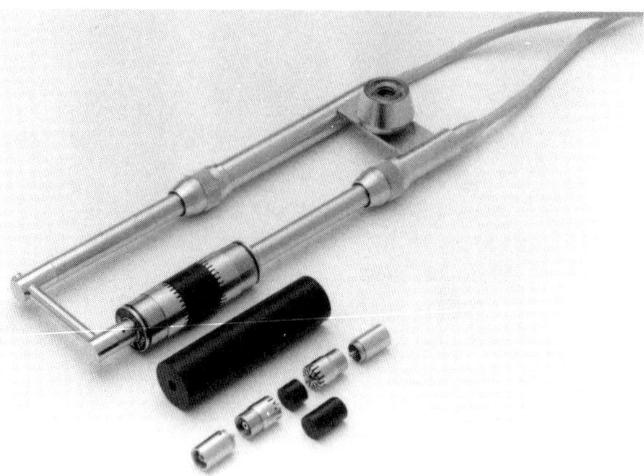

Fig. 4.13 Sound intensity probe showing two $\frac{1}{2}$-in.-diameter microphones separated precisely by a 1.2-cm spacer. Just beneath is a 5-cm spacer. Below are two $\frac{1}{4}$-in. microphones separated by a 0.6-cm spacer. (Courtesy of Bruel and Kjaer Instruments, Inc.)

in Fig. 4.13. The most commonly used sound intensity probe has two microphones spaced a distance 1.2 cm apart. The useful frequency range for this spacing is about 125 Hz to 5 kHz. As shown, other spacers are available for other frequency ranges.

There are no national or intenational standards for the calibration of sound intensity probes nor are there any standards that describe the characteristics of sound intensity analyzers; it is necessary to depend on the manufacturer for calibration of the probe and for the accuracy of the analyzer. One calibration apparatus is shown in Fig. 4.14.

By Euler's equation [see Eq. (2.6)], basic to the acoustical wave equation, the sound intensity $I = \overline{p \cdot q}$ in a given direction can be determined from measurements of the sound pressure at two points separated by a small distance along that direction. If the pressures so measured are designated P_A and P_B and the separation of the points A and B is Δr, the intensity at the midpoint between the two measurement locations and on the line joining A and B is approximately

$$I = \tfrac{1}{2}(P_A + P_B) \int \frac{P_B - P_A}{\rho \, \Delta r} \, dt \tag{4.21}$$

where ρ is the density of air in kilograms per cubic meter. The pressures are measured in newtons per square meter.

The intensity vector along Δr is proportional to the product of the sound pressure and the pressure gradient averaged over time. The measurement of P_A and P_B is accomplished with a sound intensity probe of the type shown in Fig. 4.13. A spacer between two microphones determines Δr, which may be chosen as 0.6 cm for frequencies between 250 Hz and 12 kHz, 1.2 cm for those between 125 Hz and 5 kHz, and 5 cm between 31.5 Hz and 1.25 kHz.

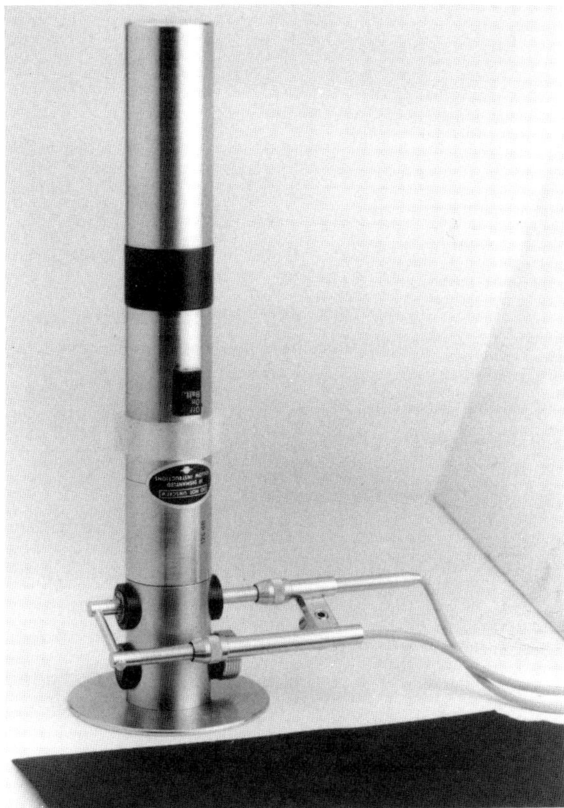

Fig. 4.14 Calibration apparatus for a sound intensity probe. (Courtesy of Bruel and Kjaer Instruments, Inc.)

Because the pressure gradient in a sound wave is maximum in the direction of travel and zero perpendicular to that direction, the output of the intensity analysis system depends on the orientation of the probe.

When determining the sound power radiated by a machine, the measurement surface specified in Eq. (4.5) must be defined. For determination of sound power using the intenational standards described in Sections 4.8 and 4.9, the measurement surface is clearly defined. No corresponding standards exist for determination of sound power via sound intensity; draft documents are, however, available.[21,22] For sound power determination via sound intensity, the measurement surface can be a rectangular box, a hemisphere, or a shape that approximates the shape of the machine. The measurement surface is divided into segments having an area S_i square meters. During each measurement the sound intensity probe **must be held perpendicular** to each segment. Thus, if a rectangular box is used as the surface, the probe must always be held perpendicular to a side of the box at each position. The intensity may be determined at a single point associated with the segment S_i or the probe may be moved slowly over the surface (scanning). The latter method has, in some cases, proved to be more accurate in practice. The sound power is determined using Eq. (4.7) and the sound power level is

$$L_W = 10 \log\left(\frac{W}{W_0}\right) \quad \text{dB } re\ 10^{-12}\ \text{W} \tag{4.22}$$

All quantities have been defined previously. The sound power level may be determined in octave or one-third-octave bands or the A-weighted sound power level (NPEL) may be determined.

The main advantage of using sound intensity to determine sound power is that, in theory, noise from sources outside the measurement surface does not affect the determination of the source sound power. Any sound entering one side of the enclosing surface leaves by other sides. (There must not be any absorbing material within the surface because all of the energy entering one side will not leave by another side.) Sound pressure measurements are not sensitive to the direction of sound propagation and cannot distinguish between sound energy flowing into the measurement surface and sound energy existing from the surface. There is, however, a direction associated with sound intensity, and the algebraic sign of the vector normal to the surface identifies the direction of sound propagation. Given this advantage of the sound intensity method, environmental corrections to account for reflections from room surfaces are not used, and the sound power of a machine can often be measured in situ, even with other machines running.

Field Indicators

It is difficult to know the accuracy of a determination of sound power level when the sound intensity method is used because the measurement conditions are often poorly defined. A series of *field indicators*[21,22] have been proposed that can be used to quantify the conditions under which the measurements are made. The values of the field indicators are believed to be related to the accuracy with which the sound power level can be determined. Selected field indicators are described below.

A useful field indicator has been called D22 in reference 22. Two deteminations of sound power are made following the procedures described at the beginning of this section: one with the source on and one with the source off. The indicator is calculated in each octave or one-third-octave band from

$$\text{D22} = 10 \log \frac{W_{\text{on}}}{W_{\text{off}}} \tag{4.23}$$

where W_{on} = sound power determined with source on, W
W_{off} = sound power determined with source off, W; known as apparent sound power of source

The D22 indicator is related to the level of the background noise relative to the level of the source and the cancellation of the background sound intensity, which can be achieved by the selected intensity probe positions on the measurement surface. It has been suggested[22] that the value of this indicator should exceed 10 dB over frequency range of interest if engineering-grade accuracy in the determination of the sound power level is desired.

A second useful indicator, called D23 in reference 22, is the difference between the sound power level of the source determined using sound pressure and using sound intensity. Reflections of sound from the room surfaces usually have a greater influence on the former than on the latter, which leads to positive values of D23. The average sound pressure level on the measurement surface is determined and the sound power level of the source is calculated using Eq. (4.15). The sound power level of the source is then determined using the intensity methods described in this section. The difference between the two results, in octave or one-third-octave bands, is D23. When the sound power is determined using sound intensity, one must include the sign of the intensity in the calculations (i.e., the sound intensity is negative if the vector points toward the source). If the sound power level calculated using sound pressure is more than about 6 dB higher than the value obtained using the sound intensity method, the measurement conditions are considerably less than ideal, and the sound power determination via sound intensity may be inaccurate.

If the same calculation is repeated using the *magnitude* of the intensity vector to determine sound power via sound intensity, a different result, called D24 in reference 22, is obtained. The difference D24 − D23 is also a useful field indicator; it is related to the fraction of the intensity vectors that point toward the source. The most accurate sound power determination is usually obtained when the sound intensity is directed outward from the source at all measurement points and the difference D24 − D23 = 0. A fourth useful field indicator is related to the normalized variance of sound intensity on the measurement surface. The indicator is called F_4 in reference 21 and D25 in reference 22. These indicators are related by

$$\mathrm{D25} = 10 \log(F_4^2) \tag{4.24}$$

$$\mathrm{D25} = 10 \log \frac{(I_i - \bar{I})^2}{(M - 1)\bar{I}^2} \tag{4.25}$$

where M = number of intensity probe positions (number of segments on measurement surface)

I_i = sound intensity at ith intensity probe position

$\bar{I}$ = average sound intensity on measurement surface (taking into account sign of sound intensity)

If D25 is negative, it should be possible to determine the sound power with engineering accuracy using about 10–15 intensity probe positions on the measurement surface. A more precise determination of the number of positions required for a given grade of accuracy may be used,[21] but the proposed methods frequently lead to a requirement that the number of intensity probe positions used for the measurement must be very large.

Table 4.8 Range of Acceptable Differences between Sound Power Level Determined using N and $2N$ Intensity Probe Positions

Octave-Band Center Frequencies (Hz)	One-Third-Octave-Band Center Frequencies (Hz)	Tolerance Limits (dB)
125	100–160	1.5
250–500	200–630	1.0
1000–4000	800–5000	0.75
8000	6300–10,000	1.25

An alternative method for determination of the required number of intensity probe positions on the measurement surface has been proposed.[22] The sound power level is determined by using both N and $2N$ positions. If the magnitude of the difference in sound power level between the two determinations (in octave or one-third-octave bands) is less than the values given in Table 4.8, the accuracy of the sound power determination obtained using $2N$ positions is of engineering grade (assuming, of course, that the sound intensity analyzer is accurate and the intensity probe is properly calibrated).

In summary, the direct determination of sound intensity has made it possible to determine sound power in environments that are not reverberation rooms, anechoic rooms, or hemianechoic rooms. The accuracy with which sound power can be determined using sound intensity analyzers and intensity probes is probably greater than can be achieved using the in situ methods of ISO 3746 because the effects of room reflections and background noise in the environment are minimized. When the characteristics of the sound intensity analyzers and probes are standardized and when standard methods, similar to those in the ISO 3740 series, are available for determination of sound power via sound intensity, intensity methods will become common for the engineering determination of sound power level. References 23–30 contain useful information on the use of sound intensity for determination of sound power and for other applications in noise control engineering.

4.11 SOUND POWER DETERMINATION IN A DUCT

The sound power level of a source in a duct can be computed easily from sound-pressure-level measurements, provided that the sound field in the duct is essentially a plane progressive wave, using the equation

$$L_W = L_p + 10 \log \frac{S}{S_0} \tag{4.26}$$

where L_W = level of total sound power traveling down duct, dB *re* 10^{-12} W

L_p = sound pressure level measured just off centerline of duct, dB *re* 2×10^{-5} $\mathrm{N/m^2}$

S = cross-sectional area of duct, m^2
$S_0 = 1\ m^2$

The above relation assumes not only a nonreflecting termination for the end of the duct opposite the source but also a uniform sound intensity across the duct. At frequencies above the first cross resonance of the duct the latter assumption is no longer satisfied. Equation (4.26) can still be used provided L_p is replaced by a suitable space average L_{pD} over the cross-sectional area S obtained from sound-pressure-level measurements at several points across the duct. The number of measurement positions across the cross section used to determine L_{pD} will depend on the accuracy desired and the frequency.

In practical situations, reflections occur at the open end of the duct, and the effect of branches and bends must be considered.[31,32] When there is flow in the duct, it is also necessary to surround the microphone by a suitable windscreen (see Chapter 14) to reduce the aerodynamic noise that might interfere with the measurements.[33]

4.12. DETERMINATION OF SOURCE DIRECTIVITY[33,34]

Most sources of sound of practical interest are directional to some degree. If one measures the sound pressure level in a given frequency band a fixed distance away from the source, different levels will generally be found for different directions. A plot of these levels in polar fashion at the angles for which they were obtained is called the *directivity pattern* of the source. A directivity pattern forms a three-dimensional surface, a hypothetical example of which is sketched in Fig. 4.15. The particular pattern shown exhibits rotational symmetry about the direction of

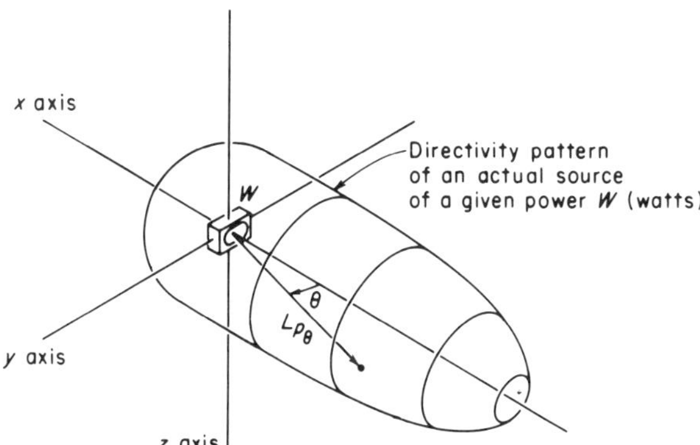

Fig. 4.15 Directivity pattern of a noise source radiating sound power W into free space. A particular sound pressure level L_p is shown as the length of a vector terminating on the surface of the directivity pattern at angle θ. Sound pressure levels were measured at various angles θ and at a fixed distance r from the actual source in free space.

maximum radiation, which is typical of many noise sources. At low frequencies many sources of noise are nondirectional, or nearly so. As the frequency increases, directivity also increases. The directivity pattern is usually determined in the far (anechoic) field (see Fig. 4.4), and in the absence of obstacles and reflecting surfaces other than those associated with the source itself, L_p decreases at the rate of 6 dB per doubling of distance.

Directivity Factor Q

A numerical measurement of the directivity of a sound source is the directivity factor Q, a dimensionless quantity. To understand the meaning of the directivity factor, we must first compare Figs. 4.15 and 4.16. We see in Fig. 4.16 the directivity pattern of a nondirectional source. It is a sphere with a radius equal in length to L_{pS}, the sound pressure level in decibels measured at distance r from a source radiating a total sound power W. The sources of Figs. 4.15 and 4.16 both radiate the same total sound power W, but because the source of Fig. 4.15 is directional, it radiates more sound than that of Fig. 4.15 in some directions and less in others.

In order to derive a directivity factor Q, we must assume that the directivity pattern does not change shape regardless of the radius r at which it is measured. For example, if L_p at a particular angle is 3 dB greater than at a second angle, the 3-dB difference should be the same whether r is 1, 2, 10, or 100 m. This can only be determined in the far field of a source located in anechoic space.

The directivity factor Q_θ is defined as the ratio of (1) the mean-square sound pressure $[(\mathrm{N/m^2})^2]$ at angle θ and distance r from an actual source radiating W watts to (2) the mean-square sound pressure at the same distance from a nondirectional source radiating the same acoustic power W. Alternatively, Q_θ is defined as the ratio of the intensity $(\mathrm{W/m^2})$ at angle θ and distance r from an actual source to the intensity at the same distance from a nondirectional source, both sources radiating the same sound power W. Thus

$$Q_\theta = \frac{p_\theta^2}{p_s^2} = \frac{I_\theta}{I_s} = \frac{10^{L_{p\theta}/10}}{10^{L_{pS}/10}} \quad \text{(dimensionless)} \tag{4.27}$$

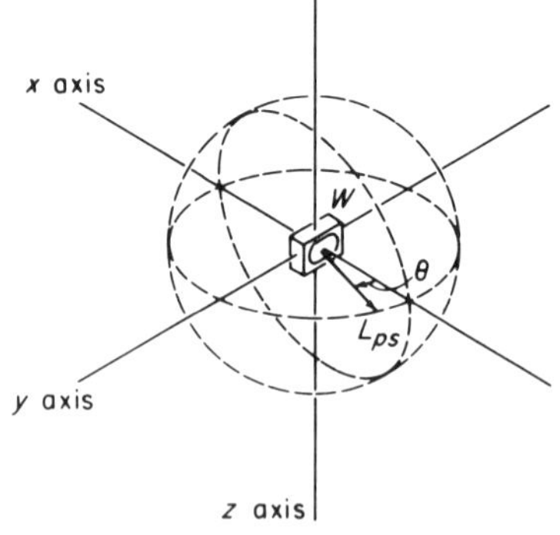

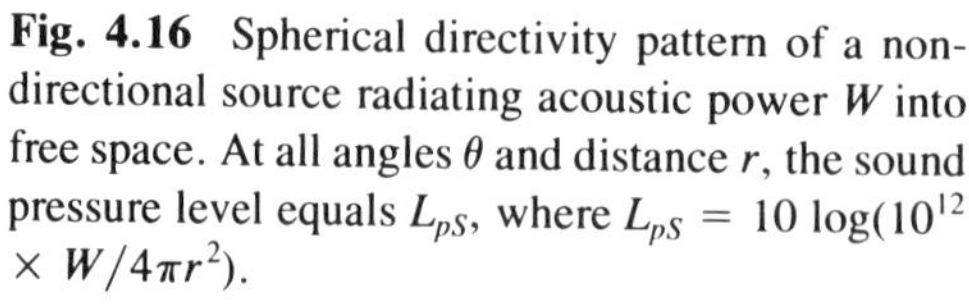
Fig. 4.16 Spherical directivity pattern of a nondirectional source radiating acoustic power W into free space. At all angles θ and distance r, the sound pressure level equals L_{pS}, where $L_{pS} = 10 \log(10^{12} \times W/4\pi r^2)$.

or

$$Q_\theta = 10^{(L_{p\theta} - L_{pS})/10} \tag{4.28}$$

where $L_{p\theta}$ = sound pressure level measured a distance r and an angle θ from a source radiating power W into an anechoic space (see Fig. 4.15)

L_{pS} = sound pressure level measured at a distance r from a nondirectional source of power W radiating into anechoic space (see Fig. 4.16)

Note that Q_θ is for the angle θ at which $L_{p\theta}$ was measured and that $L_{p\theta}$ and L_{pS} are for the same distance r.

Directivity Index

The directivity index (DI) is simply defined as

$$\mathrm{DI}_\theta = 10 \log Q_\theta \quad \text{dB} \tag{4.29}$$

or

$$\mathrm{DI}_\theta = L_{p\theta} - L_{pS} \tag{4.30}$$

Obviously, a nondirectional source radiating into spherical space has $Q_\theta = 1$ and DI = 0 at all angles θ. Conversely,

$$Q = 10^{\mathrm{DI}_\theta/10} \tag{4.31}$$

Relations between $L_{p\theta}$, Directivity Factor, and Directivity Index

The sound pressure level for the nondirectional source of Fig. 4.16 is

$$L_{pS} = 10 \log \frac{p^2 \text{ (at distance } r)}{4 \times 10^{-10}} \quad \text{dB} \tag{4.32}$$

From Eq. (4.4) assuming $\rho c = 400$, the area of a sphere $= 4\pi r^2$ and $I = W/\text{area}$, we get

$$L_{pS} = 10 \log \frac{W \times 10^{12}}{4\pi r^2} \quad \text{dB} \tag{4.33}$$

From Eqs. (4.28) and (4.33), we find that

$$L_{p\theta} = 10 \log \frac{WQ_\theta \times 10^{12}}{4\pi r^2} \tag{4.34}$$

Where W is in watts and r is in meters. In logarithmic form

$$L_{p\theta} = L_W + \mathrm{DI}_\theta - 20 \log r - 11 \text{ dB} \tag{4.35}$$

Determination of Directivity Index in Spherical Space

The directivity index DI_θ of a sound source in free space at angle θ and for a given frequency band is computed from

$$\mathrm{DI}_\theta = L_{p\theta} - \overline{L}_{pS} \quad \mathrm{dB} \tag{4.36}$$

where $L_{p\theta}$ = sound pressure level measured a distance r and an angle θ from source, dB

$\overline{L}_{pS}$ = level of space-averaged mean-square pressure detemined over test sphere of area $4\pi r^2$ surrounding source

Determination of Directivity Index in Hemispherical Space

The directivity index DI_θ of a sound source on a rigid plane at angle θ and for a given frequency band is computed from

$$\mathrm{DI}_\theta = L_{p\theta} - \overline{L}_{\mathrm{pH}} + 3\ \mathrm{dB} \tag{4.37}$$

where $L_{p\theta}$ = sound pressure level measured a distance r and an angle θ from source, dB

L_{pH} = level of space-averaged mean-square pressure determined over test hemisphere of area $2\pi r^2$ surrounding source

The 3 dB in this equation is added to the $\overline{L}_{\mathrm{pH}}$ because the measurement was made over a hemisphere instead of a full sphere, as defined in Eq. (4.36). The reason for this is that the intensity at radius r is twice as large if a source radiates into a hemisphere as compared to a sphere. That is, if a source were to radiate uniformly into hemispherical space, $\mathrm{DI}_\theta = \mathrm{DI} = 3$ dB.

Determination of Directivity Index in Quarter-Spherical Space

Some pieces of equipment are normally associated with more than one reflecting surface, for example, an air conditioner standing on the floor against a wall. The power level of noise sources of this type may be measured with those surfaces in place. This is done best in a test room with anechoic walls but with one hard wall forming an "edge" with the hard floor. The general considerations of the preceding paragraphs apply here as well. One determines the space-averaged mean-square sound pressure level for the quadrant $\overline{L}_{pQ}$ and also determines $L_{p\theta}$ as before. The directivity index is given by

$$\mathrm{DI}_\theta = L_{p\theta} - \overline{L}_{pQ} + 6\ \mathrm{dB} \tag{4.38}$$

REFERENCES

1. R. H. Bolt and K. U. Ingard, "System Considerations in Noise-Control Problems," in C. M. Harris (ed.), *Handbook of Noise Control*, 1st ed., McGraw-Hill, New York, 1957, Chapter 22.

2. E. A. G. Shaw, "Noise Pollution—What Can Be Done?" *Physics Today* **28** (1), 46 (1975).
3. K. K. Woehrle, "Impulsive Noise: Determination of Sound Power Using Reverberation Room Methods," in *Proc. INTER-NOISE 80*, Noise Control Foundation, New York, (1980), pp. 1087-1092.
4. ISO 3740, "Acoustics—Determination of Sound Power Levels of Noise Sources—Basic Standards and for the Preparation of Noise Test Codes," International Organization for Standardization, Geneva, Switzerland.
5. ISO 3741, "Acoustics—Determination of Sound Power Levels of Noise Sources—Broad-Band Sources in Reverberation Rooms," International Organization for Standardization, Geneva, Switzerland.
6. ISO 3742, "Acoustics—Determination of Sound Power Levels of Noise Sources—Discrete-Frequency and Narrow-Band Sources in Reverberation Rooms," International Organization for Standardization, Geneva, Switzerland.
7. ISO 3743, "Acoustics—Determination of Sound Power Levels of Noise Sources—Special Reverberation Test Rooms," International Organization for Standardization, Geneva, Switzerland.
8. ISO 3744, "Acoustics—Determination of Sound Power Levels of Noise Sources—Free-Field Conditions Over a Reflecting Plane," International Organization for Standardization, Geneva, Switzerland.
9. ISO 3745, "Acoustics—Determination of Sound Power Levels of Noise Sources—Anechoic and Semi-Anechoic Rooms," International Organization for Standardization, Geneva, Switzerland.
10. ISO 3746, "Acoustics—Determination of the Sound Power Levels of Noise Sources—Survey Method," International Organization for Standardization, Geneva, Switzerland.
11. ISO 3747, "Acoustics—Determination of Sound Power Levels of Noise Sources: Survey Method Using a Reference Sound Source," International Organization for Standardization, Geneva, Switzerland.
12. ANSI S1.30-1979, "Guidelines for the Use of Sound Power Standards and for the Preparation of Noise Test Codes," Acoustical Society of America, Woodbury, NY.
13. ANSI S1.31-1980 (R1986), "Precision Methods for the Determination of Sound Power Levels of Broad-Band Noise Sources in Reverberation Rooms," Acoustical Society of America, Woodbury, NY.
14. ANSI S1.32-1980 (R1986), "Precision Methods for the Determination of Sound Power Levels of Discrete-Frequency and Narrow-Band Noise Sources in Reverbation Rooms," Acoustical Society of America, Woodbury, NY.
15. ANSI S1.33-1982 (R1986), "Engineering Methods for the Determination of Sound Power Levels of Noise Sources in a Special Reverberation Test Room," Acoustical Society of America, Woodbury, NY.
16. ANSI S12.34-1988, "Engineering Methods for the Determination of Sound Power Levels of Noise Sources for Essentially Free-Field Conditions over a Reflecting Plane," Acoustical Society of America, Woodbury, NY.
17. ANSI S1.35-1979, "Precision Methods for the Determination of Sound Power Levels of Noise Sources in Anechoic and Hemi-Anechoic Rooms," Acoustical Society of America, Woodbury, NY.
18. ANSI S1.36-1979, "Survey Methods for the Determination of Sound Power Levels of Noise Sources," Acoustical Society of America, Woodbury, NY.

19. R. D. Waterhouse, ''Output of a Sound Source in a Reverberation Chamber and Other Reflecting Environments,'' *J. Acoust. Soc. Am.* **30,** 4 (1958).
20. ISO 6926, ''Acoustics—Determination of Sound Power Levels of Noise Sources; Requirements on the Performance and Calibration of Reference Sound Sources,'' International Organization for Standardization, Geneva, Switzerland. The corresponding American National Standard is ANSI S12.5-1985.
21. ISO DP 9614, ''Acoustics—Determination of the Sound Power Levels of Noise Sources Using Sound Intensity Measurement at Discrete Points,'' International Organization for Standardization, Geneva, Switzerland, September 12, 1988.
22. Proposed American National Standard S12.12, ''Engineering Method for the Determination of Sound Power Levels of Noise Sources Using Sound Intensity,'' Acoustical Society of America, New York, October 1988.
23. ''Recent Developments in Sound Intensity Measurement,'' Proceedings of a conference on sound intensity measurement, September 30–October 2, 1981, CETIM, Senlis, France.
24. S. Gade, ''Sound Intensity: Theory,'' Technical Notes (Part I, 1982), Bruel and Kjaer, Naerum, Denmark.
25. M. J. Crocker, ''Direct Measurement of Sound Intensity and Practical Applications in Noise Control Engineering,'' in *Proc. INTER-NOISE 84*, Noise Control Foundation, New York, 1984, pp. 21–36.
26. Second International Congress on Sound Intensity, Proceedings of an international conference on sound intensity, September 23–26, 1985, CETIM, Senlis, France.
27. G. Rasmussen, ''Intensity Measurement,'' Publication BA-7196-11, Bruel and Kjaer, Naerum, Denmark, 1985.
28. C. C. Maling, Jr., ''Progress in the Application of Sound Intensity Techniques to Noise Control Engineering,'' in *Proc. INTER-NOISE 86*, Noise Control Foundation, New York, 1986, pp. 39–74.
29. F. J. Fahy, *Sound Intensity*, Elsevier Science Publishers, Barking, Essex, United Kingdom, 1989.
30. J. Tichy, ''Noise Control Applications of Sound Intensity,'' in *Proc. INTER-NOISE 89*, Noise Control Foundation, New York, 1989, pp. 45–68.
31. ASHRAE Standard 68-1986, ''Laboratory Method of Testing In-Duct Sound Power Measurement Procedure for Fans,'' American Society of Heating, Ventilating and Air-conditioning Engineers, Atlanta, GA.
32. P. K. Baade, ''Effects of Acoustic Loading on Axial Flow Fan Noise Generation,'' *Noise Control Eng. J.* **8** (1), 5–15 (1977).
33. L. L. Beranek, *Acoustical Measurements*, Acoustical Society of America, Woodbury, NY, 1988.
34. L. L. Beranek, *Acoustics*, Acoustical Society of America, Woodbury, NY. 1986.

CHAPTER 5

Outdoor Sound Propagation

GRANT S. ANDERSON
Harris Miller Miller & Hanson Inc.
Lexington, Massachusetts

ULRICH J. KURZE
Müller-BBM GmbH
München, Germany

5.1 ELEMENTS OF OUTDOOR SOUND PROPAGATION

Outdoor sound propagation is commonly discussed in three components: source, path, and receiver. First, the source emits sound power, resulting in sound levels that can be measured in the vicinity of the source. Next, these sound levels diminish as the sound propagates outward along the path from source to receiver. Finally, sound levels from all sources combine at the receiver. Chapter 4 discusses sound levels measured in the vicinity of the source. Chapter 17 discusses assessment criteria at the receiver. This chapter discusses outdoor sound propagation along the path from source to receiver.

Early investigations of outdoor sound propagation have been reviewed previously.[1–4] This chapter draws upon these early investigations, as well as upon more recent investigations for which selected references are cited here. Figure 5.1 illustrates the most significant mechanisms of outdoor sound propagation.

The equations in this chapter are based mostly on a mixture of experimental measurements and physical scale models, guided by theory. Primary emphasis is

Noise and Vibration Control Engineering: Principles and Applications, Edited by Leo L. Beranek and István L. Vér.
ISBN 0-471-61751-2 © 1992 John Wiley & Sons, Inc.

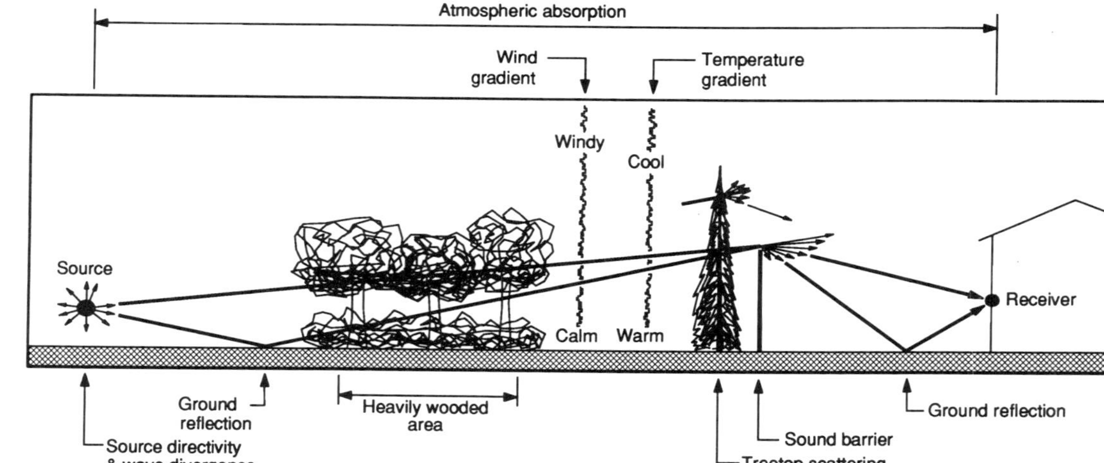

Fig. 5.1 The most significant mechanisms of outdoor sound propagation. Sound diminishes with distance as it diverges from its source, which may be "directive." Atmospheric absorption attenuates the sound along its path. Ground reflections interfere with the direct sound, causing either attenuation or (less often) amplification. Heavily wooded areas provide attenuation, as do man-made and natural sound barriers. Treetop scattering may reduce barrier effectiveness. Vertical gradients of both wind and temperature refract (bend) sound paths up or down, causing sound shadow zones, altering ground interference, and modifying the effectiveness of sound barriers.

put on (overall) A-weighted sound pressure levels L_A for spectra without prominent pure tones. For A-weighted levels an accuracy of ± 5 dB is generally achievable out to source–receiver distances of approximately 500 m. Although this chapter does discuss sound in octave bands and narrower, prediction accuracy at 500 m is sometimes as poor as ± 15–20 dB in individual bands.

The attenuation of sound as it propagates outward from its source depends on frequency. For this reason the reduction of A-weighted level depends on spectral composition. The equations in this chapter that relate to A-weighted levels pertain best to source spectra similar (± 5 dB) to the following:

Octave-band center frequency, Hz	63	125	250	500	1000	2000	4000	8000
L_p (octave) minus L_A (overall), dB	−2	+1	−1	−3	−5	−8	−12	−23

This spectrum is typical of muffled diesel engines, roadway and rail traffic, aircraft traffic, impulsive sounds such as rifle fire, and many industrial sources of outdoor sound. It is labeled "typical source spectrum" throughout this chapter.

Caution: This chapter is not intended to substitute for official sound propagation prediction methods, which vary from country to country and often from agency to agency within each country. Such official methods generally place a high value on consistency and simplicity, and a lesser value on accuracy. Instead, this chapter attempts to facilitate understanding and engineering judgment, as a supplement to official methods.

5.2 THE BASIC EQUATION OF OUTDOOR SOUND PROPAGATION

The basic equation of outdoor sound propagation explicitly contains the effects of wave divergence, source directivity, and large surfaces near the source. For a "point source" (explained later) with sound power level L_W, the basic equation for the sound pressure level $L_p(r)$ at source–receiver distance r is[5]

$$L_p(r) = L_W - 20 \log \frac{r}{1 \text{ m}} + \text{DI}_{\text{rcvr}} - 10 \log \frac{\Omega}{4\pi} - 11 - A_{\text{combined, rcvr}} \quad (5.1)$$

where DI_{rcvr} is the source directivity index in the receiver direction, Ω is the solid angle at the source that is available for sound propagation, and $A_{\text{combined, rcvr}}$ is the combined attenuation (Section 5.11) from all significant propagation mechanisms between source and receiver.

For use in this equation, L_W is most often obtained from the sound pressure level $[L_p]_{\text{ref}}$ measured at a reference position near the source, using the inverse equation

$$L_W = [L_p]_{\text{ref}} + 20 \log \frac{r_{\text{ref}}}{1 \text{ m}} - \text{DI}_{\text{ref}} + 10 \log \frac{\Omega}{4\pi} + 11 + A_{\text{combined, ref}} \quad (5.2)$$

where r_{ref} is the distance from source to reference position, DI_{ref} is the source directivity index in the reference direction, Ω is the solid angle at the source that is available for sound propagation, and $A_{\text{combined, ref}}$ is the combined attenuation from all significant propagation mechanisms between source and reference position.

Equations (5.1) and (5.2) are valid in the acoustical and geometric "far field" of the source of sound, shown in Fig. 4.4. In addition, they are valid whether the source is moving or not, as long as the effects of speed upon L_W, or upon $[L_p]_{\text{ref}}$, are properly incorporated.

Wave Divergence, Source Directivity, and Large Surfaces Near the Source

Concerning wave divergence, Eq. (5.1) states that the sound level from a point source decreases by 6 dB for every doubling of distance (6 dB/dd), and correspondingly by 20 dB for every 10-fold increase in distance.

Concerning source directivity, when the reference position is chosen so that $\text{DI}_{\text{ref}} \approx \text{DI}_{\text{rcvr}}$, the DI terms vanish in the algebraic combination of Eqs. (5.1) and (5.2). In addition, the horizontal directivity of roadway traffic sources (and many others) can often be ignored for two additional reasons. First, directivity of such sources is often "blurred" by (1) multiple reflections and scattering from nearby objects and surfaces, (2) scattering by atmospheric turbulence, and (3) multiple sources facing in different directions—for example, as happens when roadway vehicles progress along their route. Second, a positive DI_{rcvr} in one frequency band can be offset by a negative DI_{rcvr} in another band, when the bands are combined into an A-weighted overall sound pressure level.

Two transportation sources of sound for which source directivity cannot be ignored are jet aircraft and wheel–rail sources on trains. A-weighted levels caused by jet aircraft are highly directive toward the rear quarter. Wheel–rail sources on trains exhibit a nearly dipolar directivity, aimed perpendicular to the track. Calculations for such vehicles should include proper source directivity toward receivers of interest, relative to the vehicle's direction of travel.

For sound sources on roofs or building facades, Fig. 5.2 shows approximate source directivities as influenced by the building surfaces.[5]

With large surfaces near the source, the solid-angle term $10 \log(\Omega/4\pi)$ in Eqs. (5.1) and (5.2) is a broadband approximation for the complex interaction between direct and reflected waves from the nearby surfaces. In this approximation the surfaces prevent sound energy from propagating in certain directions, thereby increasing sound levels in other directions of propagation. For example, when a source is located close to the ground, sound energy originally headed downward

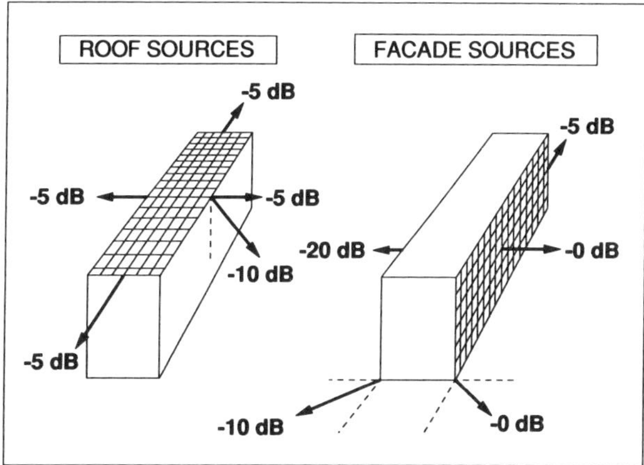

Fig. 5.2 Approximate directivities for sources on roofs and building facades or for sound emanating from roof/facade openings in a building. For sources of sound located on the hatched surfaces, this figure shows approximate values of DI for use in Eqs. (5.1) and (5.2) but not the effects of the nearby surface terms Ω. These DI values also incorporate approximate shielding due to the building itself.

is reflected upward by the ground, thereby doubling the sound energy radiated upward. In this case the sound energy is constrained within a solid angle of 2π, which is one-half its unobstructed solid angle, 4π. Therefore, the solid-angle term in Eq. (5.1) becomes $-10 \log (2\pi/4\pi)$, which equals +3 dB. This is the "baseline" case for soft-ground attenuation in Section 5.5. For a source located near both the ground and a vertical building facade, the solid-angle term becomes $-10 \log (\pi/4\pi)$, which equals +6 dB.

Approximation of all Sources of Sound as Point Sources

All sources of sound—localized and large, stationary and moving—must be approximated by "point sources" in the use of Eqs. (5.1) and (5.2). A point source is one for which: (1) all source dimensions are smaller than one-half the source–receiver distance, and (2) essentially the same propagation conditions exist to the receiver from all portions of the source. Sources of sound too large to satisfy these two point source requirements must be approximated by several smaller sources, each of which does satisfy the two requirements. Once Eq. (5.1) is applied separately to each of these several smaller sources S to obtain their respective sound pressure levels, $[L_p(r)]_S$, then the combined sound pressure level is obtained by logarithmic summation, in the normal manner:

$$[L_p]_{\text{combined}} = 10 \log\left(\sum_{\text{all sources}, S} 10^{[L_p(r)]_S/10} \right) \tag{5.3}$$

Typical point sources of sound include an individual piece of mechanical equipment, a siren, a front-end loader confined to a designated operating area, a localized group of well-defined industrial or construction sources of sound, and a portion of sound-radiating piping. Point sources are almost always incoherent with one another. When they are not (as with electrical transformers), the mutual interference of sound from individual sources can cause propagation complexities beyond the scope of this chapter.

Time-integrated Measures of Cumulative Exposure

Time-integrated measures of cumulative exposure are commonly used to assess impact of time-varying (fluctuating) sound. Several of these time-integrated measures are defined in Section 1.4. The most commonly used measure is the average A-weighted sound level, L_{AT} (also called L_{eq}, the equivalent continuous A-weighted noise level). Combination of Eqs. (1.19) and (1.21) from Chapter 1 yields this alternative expression for L_{AT}:

$$L_{AT} = L_{eq} = 10 \log \left(\frac{1}{t_2 - t_1} \int_{t_1}^{t_2} 10^{L_A(t)/10} \, dt \right) \tag{5.4}$$

In this equation $L_A(t)$ is the instantaneous A-weighted sound pressure level, as measured on a sound level meter with frequency weighting A. The time-averaging extends from t_1 to t_2 in seconds.

For use in this equation, Eq. (5.1) must be generalized to be a function of time:

$$L_A(t) = L_{WA}(t) - 20 \log \frac{r(t)}{1 \text{ m}} + \text{DI}_{\text{rcvr}}(t) - 10 \log \frac{\Omega(t)}{4\pi} - 11 - A_{\text{combined, rcvr}}(t) \tag{5.5}$$

For stationary (nonmoving) sources, all quantities in Eq. (5.5) are constant in time except the source's sound power level, L_{WA} (and perhaps also the source's directivity index, DI_{rcvr}, if the source is rotating). Combination of Eqs. (5.4) and (5.5) for a stationary, nonrotating source therefore yields

$$L_{AT,\text{stat}} = L_{\text{eq,stat}} = 10 \log \left(\frac{1}{t_2 - t_1} \int_{t_1}^{t_2} 10^{L_{WA}(t)/10} \, dt \right) - 20 \log \frac{r}{1 \text{ m}}$$
$$+ \text{DI}_{\text{rcvr}} - 10 \log \frac{\Omega}{4\pi} - 11 - A_{\text{combined, rcvr}}$$

The term with the integral in this equation is the source's so-called "energy average" A-weighted sound power level. The resulting L_{AT} (i.e., L_{eq}) can be measured directly with an integrating sound level meter.

For moving sources such as transportation vehicles, the source–receiver distance r is also a function of time; often DI_{rcvr} and $A_{\text{combined, rcvr}}$ are, as well. An integrating sound level meter still yields L_{AT} (i.e., L_{eq}) directly; the complexity arises in calculating this cumulative measure.

A common method of calculation is illustrated in Fig. 5.3. In this figure the route of a single, moving vehicle is approximated by a string of straight-line segments (route elements), each short enough to satisfy both point source requirements above. Farther from the receiver, these point source requirements allow longer route elements, as shown in the figure. The point source requirements ensure that the receiver's $L_A(t)$ is essentially constant over each individual route element, as the vehicle traverses the element. Therefore, after some algebraic manipulation, Eq. (5.4) approximates to

$$\begin{aligned} L_{AT,\text{move}} &= L_{\text{eq,move}} \\ &\approx 10 \log \left(\sum_{\text{each element},\, n} 10^{[L_{A,n} + 10 \log(\Delta t_n / 1\text{ s})]/10} \right) - 10 \log \frac{t_2 - t_1}{1 \text{ s}} \end{aligned} \tag{5.6}$$

In this equation $L_{A,n}$ is the receiver's $L_A(t)$ when the vehicle is in element n, as calculated with Eq. (5.5). Note that this approximation includes the effects of source directivity, through the individual DI_{rcvr} terms for all the $L_{A,n}$. In this equation Δt_n is the amount of time (duration) the vehicle is in element n (equal to the element length, divided by vehicle speed). A changing vehicle speed along its route will affect the duration adjustment, $10 \log (\Delta t_n / 1 \text{ s})$.

Comparison of Eq. (5.6) with Eq. (5.3) shows the equivalence between this single-moving-vehicle, average calculation and a multivehicle, static calculation. By this equivalence, the time integral for the single, moving vehicle has been

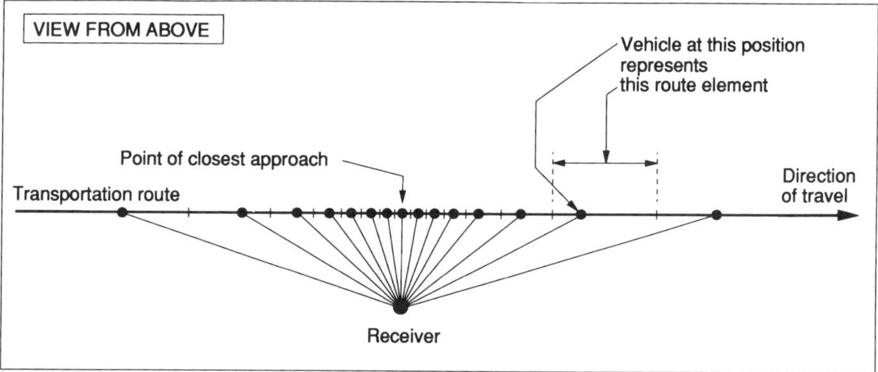

Fig. 5.3 Approximation of the route of a single, moving vehicle by a string of straight-line route elements, each short enough to satisfy the two point source requirements. After adjustment by the length of time the moving vehicle spends in each route element, logarithmic summation over a series of vehicles, one per route element, approximates several time-integrated measures for the passage of a single, moving vehicle.

approximated by a static calculation in which a series of vehicles (one per route element, each adjusted for duration) substitutes for the moving vehicle.

By this same type of calculation method, other time-integrated measures from Chapter 1 approximate to

$$L_{E,AT,\text{move}} = \text{SEL}_{\text{move}} \approx 10 \log \left(\sum_{\text{each element},\, n} 10^{[L_{A,n} + 10\log(\Delta t_n / 1\,\text{s})]/10} \right)$$

and

$$L_{\text{dn, move}} \approx 10 \log \left[\frac{1}{24} \left(\sum_{\text{each daytime hour},\, h} 10^{(L_{A,1\text{h}})_h/10} + \sum_{\text{each nighttime hour},\, h} 10^{[(L_{A,1\text{h}})_h + 10]/10} \right) \right]$$

where $L_{E,AT}$ is the A-weighted noise exposure level (also called SEL, the sound exposure level), and L_{dn} is the day–night sound level. In the equation for L_{dn}, the first summation covers the ''daytime'' hours from 07:00 to 22:00, and the second summation covers the ''nighttime'' hours from 22:00 to 07:00 the next morning.

The exact instant of passage is unimportant to the integrals (or summations) in the equations for these time-integrated measures. For this reason time-integrated measures for extended sources such as trains can be computed in one application of these equations, even though the full train is too long to qualify as a point source. In essence, each of the train's ''subsources'' of sound qualifies as a point source and contributes its share to the integral (or summation) as it passes the receiver.

Note that the transportation route in Fig. 5.3 is straight. This is not necessary for the validity of these approximations. For a curved route the individual source–receiver distances for each route element incorporate the effect of the curve.

Also in Fig. 5.3, note that the route elements are drawn to subtend equal angles at the receiver. This is not necessary, either. When elements do subtend equal angles, they contribute nearly equally to the level at the receiver. More distant elements are longer in length; their sound pressure level at the receiver, decreased by distance, is compensated by their larger duration adjustment. Exact compensation occurs for nondirective sources with no distance-dependent, excess attenuation (such as soft-ground attenuation) between source and receiver.

Also in Fig. 5.3, note that a doubling of the receiver's distance to the point of closest approach does not result in a distance doubling to other route elements. To more distant elements, the increase is less. Calculation with an infinite string of route elements and no excess attenuation, $A_{\text{combined, rcvr}}$, shows that the total sound pressure level at the receiver decreases by 3 dB/dd (10 dB for every 10-fold increase in perpendicular distance) for this ''line'' source.

Similar approximations cannot be used to compute measures, such as L_{10} (the A-weighted sound pressure level exceeded 10% of the time), that are based on temporal percentages. Such measures do not obey "conservation of energy." Their proper calculation for a traffic stream requires a statistically complex method that spaces vehicles realistically along their route.

5.3 OVERVIEW OF SIGNIFICANT MECHANISMS OF OUTDOOR SOUND PROPAGATION

The basic equation in the preceding section accounts for the effects of wave divergence, of source directivity, and of large surfaces near the source. Figure 5.4 summarizes all other significant mechanisms of outdoor sound propagation, indicating under what conditions each is significant. This figure does not include the effects of fog, precipitation, and atmospheric turbulence, which are generally negligible.[6]

5.4 ATMOSPHERIC ABSORPTION

Atmospheric absorption is the attenuation of sound during its passage through air. It is caused mostly by the vibrational relaxation of oxygen and nitrogen molecules, which depends on humidity, temperature, atmospheric pressure, and strongly on the frequency of sound. Because outdoor temperature and humidity vary considerably from moment to moment and place to place along typical propagation paths, calculations with laboratory-derived equations are not practicable for outdoor predictions, for lack of adequate input. Fortunately, atmospheric absorption is generally small compared to attenuations caused by other propagation mechanisms, and therefore approximate equations are generally sufficient.

A-weighted Sound Pressure Levels

For the A-weighted level of the "typical source spectrum," Fig. 5.5 shows source–receiver distances for the first 3-dB reduction caused by atmospheric absorption. After the first 3 dB of atmospheric absorption, the A-weighted sound pressure level drops off more slowly with distance, because high-frequency energy has been partially depleted from the spectrum. Therefore, succeeding 3-dB reduction distances are greater than shown in the figure.

Pure Tones and Octave-Band Sound Pressure Levels

The exact calculation of pure-tone atmospheric absorption has been standardized recently.[7] At 1 atm pressure, 10°C, and 70% relative humidity, the standardized calculation of pure tones is approximated at a source–receiver distance r by

$$A_{\text{atm},f} = \frac{r}{1000\ \text{m}}\left[0.6 + 1.6\left(\frac{f}{1\ \text{kHz}}\right) + 1.4\left(\frac{f}{1\ \text{kHz}}\right)^2\right] \tag{5.7}$$

		ATTENUATION EQUALS APPROXIMATELY 5 dB		
MECHANISM	BRIEF DESCRIPTION	UNDER THESE CONDITIONS	AT THESE DISTANCES	
ATM ABSORP Section 5.4 	Absorption of sound directly by the atmosphere	At 10 deg C and 70 % relative humidity	800 m	A
			1500 m at 500 Hz 250 m at 4000 Hz	Oct
SOFT GROUND Section 5.5 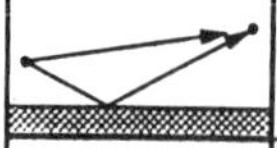	Interference (mostly destructive) between direct and reflected sound rays, over acoustically "soft" ground	For source and receiver heights approximately 1.2 m	85 m	A
			10 m at 250 and 500 Hz 50 m at 125 and 1000 Hz Never at 63 and 2000 Hz	Oct
BARRIER Section 5.6 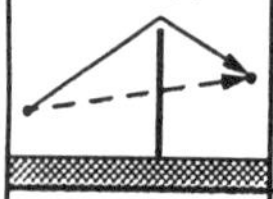	Attenuation due to an intervening sound barrier, combined with partial loss of ground attenuation over acoustically "soft" ground, resulting in barrier insertion loss, IL	When receiver is just inside geometrical shadow of barrier, with neutral temperature conditions and no wind	All	----
BUILDINGS Section 5.7 	Partial shielding by row(s) of intervening buildings	With one intervening row of buildings approximately 25 % open	All	----
HEAVY WOODS Section 5.8 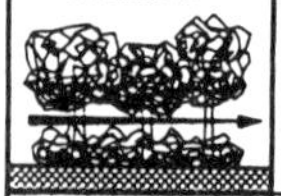	Partial shielding by intervening areas of heavy woods	With dense trees and underbrush	30 m	A
			100 m at 500 Hz 50 m at 4000 Hz	Oct
URBAN REVERB Section 5.9 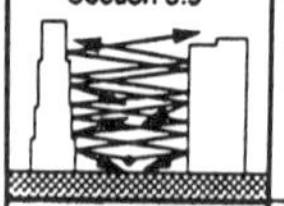	Amplification due to multiple reflections in urban canyons	With buildings at least 10 m tall on both sides of street	All	----
WIND/TEMP Section 5.10 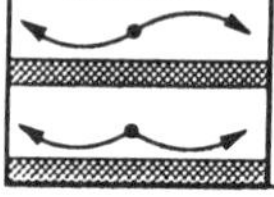	Modification of soft-ground attenuation and/or barrier insertion loss, or creation of shadow zones -- all caused by vertical wind and temperature gradients	On sunny day, for source and receiver heights approximately 1.2 m	150 m	A
			150 m at 500 Hz 50 m at 4000 Hz	Oct

Fig. 5.4 Summary of all significant mechanisms of outdoor sound propagation. Omitted are the effects of wave divergence, source directivity, and large surfaces near the source (which are discussed in Section 5.2) as well as fog, precipitation, and air turbulence (which are generally negligible). Tabulations labeled A relate to A-weighted levels for the typical source spectrum.

The following octave-band equation predicts less atmospheric absorption at high frequencies, because it takes into account the lesser attenuation at the lower end of each octave band.[8]

$$A_{\text{atm, band}} = \frac{r}{1000 \text{ m}} \left[0.2 + 3.6 \left(\frac{f_{\text{band}}}{1 \text{ kHz}} \right) + 0.36 \left(\frac{f_{\text{band}}}{1 \text{ kHz}} \right)^2 \right]$$

where f_{band} is the band's center frequency.

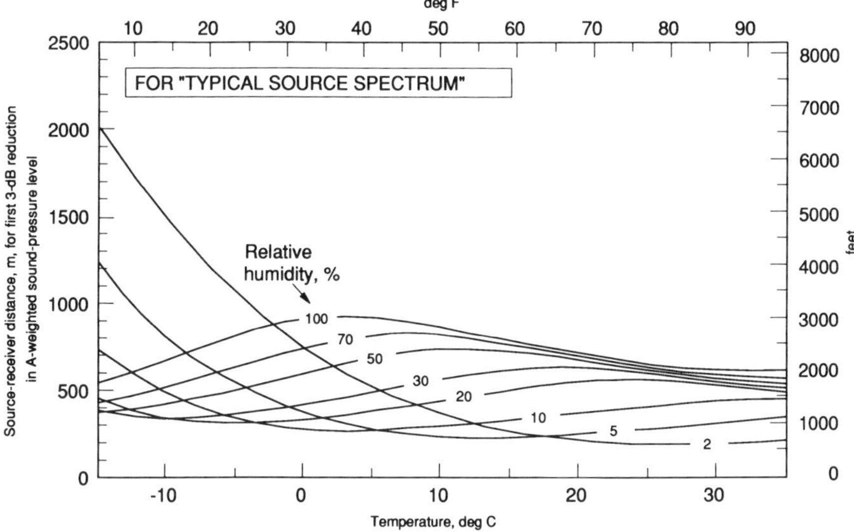

Fig. 5.5 Atmospheric absorption for the typical source spectrum. Source–receiver distance for the first 3-dB reduction in A-weighted sound pressure level, caused by atmospheric absorption, at 1 atm. These curves are computed from the typical source spectrum and equations in ref. 7.

5.5 GROUND ATTENUATION

As shown in the top frame of Fig. 5.6, reflection from the ground produces a second sound ray from source to receiver. This ground-reflected ray can interfere with the direct, nonreflected ray to produce either net attenuation or net amplification, depending on mutual phase.[9] Such interference is a function of frequency,

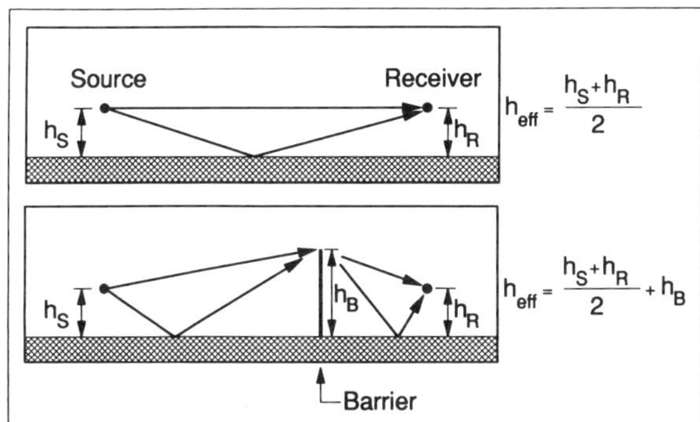

Fig. 5.6 Parameters for use in calculating ground attenuation over grassland with Eqs. (5.8) and (5.9). Top frame: without intervening barrier. Bottom frame: with intervening barrier.

and varies between an attenuation of 20–30 dB (destructive interference at frequencies for which the two rays are 180° out of phase) and an amplification of 6 dB (pressure doubling at frequencies for which they are in phase).

For nongrazing propagation over acoustically "hard" ground (asphalt, hard-packed earth, water, water-soaked earth), several frequency regions of attenuation and amplification often occur within an octave of each other. For this reason the effects of acoustically hard ground often average to an approximate 3-dB increase over that of the direct sound, for both octave-band and A-weighted levels. In this chapter this +3-dB correction for hard ground is included as the solid-angle term $-10 \log (2\pi/4\pi)$ of the basic equation.

In contrast, near-grazing propagation over acoustically "soft" ground (grassland or other ground containing root structure, plowed or aerated earth, snow, other "fissured" ground) introduces a phase reversal upon reflection, except at very low frequencies. It thereby produces significant broadband attenuation. This section concentrates upon near-grazing propagation over acoustically soft ground (mostly over grassland or plowed earth), relative to the hard-ground baseline of +3 dB. Propagation over snow and other fissured surfaces beside grassland and plowed earth is discussed in references 10 and 11.

A-weighted Sound Pressure Levels

Many experimentally based equations for soft-ground attenuation of A-weighted level appear in the literature and in standards documents. Most of these equations are specialized to propagation over grassland or plowed earth. These equations differ from one another in their simplifying assumptions and in their inherent conservatism.

For sound from individual roadway vehicles (point sources), the equations of the Ontario Ministry of Environment incorporate a constant grassland attenuation per distance doubling, depending on the "effective" height h_{eff} of the source–receiver path. Relative to acoustically hard ground ($\Omega = 2\pi$ in the basic equation), the attenuation at source–receiver distance r is[12]

$$A_{\text{grass, Ontario}} = (10G)\log \frac{r}{15\text{ m}} \geq 0 \tag{5.8}$$

$$0 \leq G = 0.75 \left(1 - \frac{h_{\text{eff}}}{12.5\text{ m}}\right) \leq 0.66$$

$$h_{\text{eff}} = \begin{cases} \frac{1}{2}(h_S + h_R) & \text{without intervening barrier} \\ \frac{1}{2}(h_S + h_R) + h_B & \text{with intervening barrier} \end{cases}$$

The geometric parameters for this equation appear in Fig. 5.6. Note that uneven terrain must be approximated by flat ground that duplicates the location and reflection angles of the actual ground-reflected rays.

In this equation no ground effects are considered closer than 15 m from the source, nor above an h_{eff} of 12.5 m. In addition, the 0.66 upper limit on G causes the computed ground attenuation to be constant for effective heights between 0 and 1.5 m.

Note that the insertion of a sound barrier between source and receiver increases the effective path height h_{eff}, which reduces the ground attenuation. In physical terms an intervening barrier destroys part of the soft-ground attenuation because it increases the grazing angles of incidence with the ground and (often) elevates the sound path further above the ground. This partial loss of soft-ground attenuation is discussed further in Section 5.6, where it is incorporated into the definition of barrier insertion loss. Equation (5.8) is compared to experimental data in Section 5.11.

When Eq. (5.8) is integrated over an infinitely long, straight transportation route (line source), the equation's normalization distance of 15 m changes to 8.8 m for monopole sources and 12.8 m for dipole sources.[13,14]

Soft-ground attenuation currently prescribed by the U.S. Federal Highway Administration is independent of path height but becomes abruptly zero at an average path height of 3 m. It is obtained from Eq. (5.8) by setting G equal to 0.5.[13] Equation (5.9) of the Society of German Engineers (VDI) results in less soft-ground attenuation. It applies, however, only to receivers downwind from the source and more than 4 m above the ground, where less soft-ground attenuation is expected.[5,8]

$$A_{\text{grass, VDI}} = 4.8 - \frac{h_S + h_R}{r}\left(17 + \frac{300\ \text{m}}{r}\right) \geq 0 \tag{5.9}$$

Pure Tones and Octave-Band Sound Pressure Levels

Ground attenuation in quiet, isotropic air over flat, homogeneous, acoustically soft ground can be calculated theoretically from the acoustic wave equation with proper boundary conditions.[3,15,16] The solution consists of three sound wave terms: (1) a direct wave, (2) a reflected wave with a phase shift upon reflection, and (3) the remaining terms, collectively called the surface wave. The resulting sound pressure levels follow from an addition of these three terms, taking phase into account.

Figure 5.7 contains an approximation to the theoretical pure-tone results over grassland or plowed earth, relative to acoustically hard ground. This approximation is a function of two easily calculated geometric quantities. At midfrequencies and for small-enough grazing angles and small-enough pathlength differences, the reflected wave's 180° phase shift upon reflection results in destructive interference between the direct and reflected waves. The surface wave dominates in this central region of destructive interference. Attenuations here can approach 30–40 dB, but recommended are the horizontal dotted-line limits in the figure. At higher frequencies the path length difference is no longer negligible compared to wavelength, and so the destructive interference is lost. At lower frequencies the ground is not soft enough and therefore the 180° phase shift upon reflection does not occur. Consequently, there is no destructive interference.

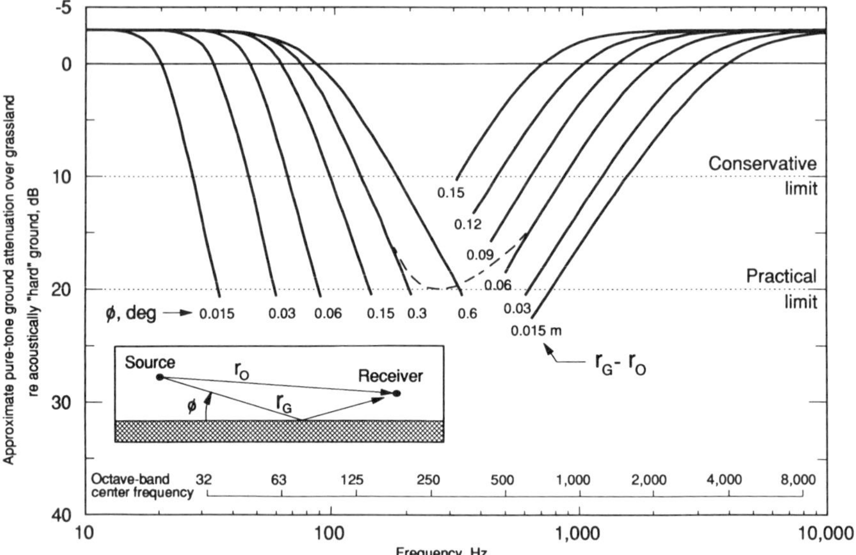

Fig. 5.7 Highly approximate pure-tone ground attenuation over grassland or plowed earth, relative to acoustically "hard" ground ($\Omega = 2\pi$ in the basic equation). Use the grazing angle ϕ to match a curve at low frequencies; then use the pathlength difference $r_G - r_0$ to match a curve at high frequencies; and then fare the curves together smoothly, as shown by the dashed example. These curves were computed from equations in ref. 16 (after minor corrections) for limited ranges of relative source–receiver positions, then averaged and smoothed.

In Fig. 5.7 note that soft-ground attenuation is greatest for small grazing angles (left side) and small pathlength differences (right side). As the source–receiver distance increases, both ϕ and $r_G - r_0$ get smaller, which means that the soft-ground attenuation increases as the distance increases. Both ϕ and $r_G - r_0$ also get smaller as either the source or receiver approaches the ground; thus soft-ground attenuation increases as heights decrease.

More accurate than Fig. 5.7 are the following empirical equations for octave-band ground attenuation in quiet isotropic air.[5] These equations are also relative to acoustically hard ground ($\Omega = 2\pi$ in the basic equation):

$$A_{\text{ground, oct}} = \begin{cases} -3M & \text{at 63 Hz} \\ G_S A_S + G_R A_R - 3M(1 - G_P) & \text{at 125 Hz} \\ G_S B_S + G_R B_R - 3M(1 - G_P) & \text{at 250 Hz} \\ G_S C_S + G_R C_R - 3M(1 - G_P) & \text{at 500 Hz} \\ G_S D_S + G_R D_R - 3M(1 - G_P) & \text{at 1000 Hz} \\ 1.5G_S + 1.5G_R - 3M(1 - G_P) & \text{at 2000, 4000, and 8000 Hz} \end{cases} \tag{5.10}$$

In these octave-band equations the terms G depend on the acoustical flow resistivity (see Chapter 8) of the ground close to the source (G_S), close to the receiver (G_R), and along the propagation path (G_P). Separately for each of these three regions: G equals zero for acoustically hard ground, unity for acoustically soft ground, and an intermediate value for intermediate conditions. For a source–receiver distance r_{SR}, the height and distance terms in Eq. (5.10) are

$$M = 1 - \frac{30(h_S + h_R)}{r_{SR}} \geq 0$$

$$A = 1.5 + 3.0\left[1 - \exp\left(\frac{-r}{50\text{ m}}\right)\right]\exp\left[-0.12\left(\frac{h - 5\text{ m}}{1\text{ m}}\right)^2\right] + 5.7\left[1 - \exp\left(\frac{-2.8 \times 10^{-6} r^2}{1\text{ m}^2}\right)\right]\exp\left[-0.09\left(\frac{h}{1\text{ m}}\right)^2\right]$$

$$B = 1.5 + 8.6\left[1 - \exp\left(\frac{-r}{50\text{ m}}\right)\right]\exp\left[-0.09\left(\frac{h}{1\text{ m}}\right)^2\right]$$

$$C = 1.5 + 14.0\left[1 - \exp\left(\frac{-r}{50\text{ m}}\right)\right]\exp\left[-0.46\left(\frac{h}{1\text{ m}}\right)^2\right]$$

$$D = 1.5 + 5.0\left[1 - \exp\left(\frac{-r}{50\text{ m}}\right)\right]\exp\left[-0.9\left(\frac{h}{1\text{ m}}\right)^2\right]$$

where exp(x) means e (the natural-logarithm base) raised to the x power.

To correspond to the subscripts of A, B, C, and D in Eq. (5.10), the variables r and h must be subscripted with either S or R, as appropriate, in these equations for A, B, C and D. The subscripted variables r are defined and limited by $r_S = 30h_S \leq r_{SR}$ and $r_R = 30h_R \leq r_{SR}$.

Some Complications

Reference Microphone above Acoustically Soft Ground. When ground between the source and the reference position is acoustically soft, the term $A_{\text{combined, ref}}$ in Eq. (5.2) may contain ground attenuation. Any attenuation between the source and reference position is highly undesirable. To avoid soft-ground attenuation at the reference microphone, select a reference position sufficiently far above the ground and sufficiently close to the source to minimize interference complexities between direct and reflected sound rays. A commonly used 1.2-m reference microphone height at a 15-m source–reference distance is often not free of such complexities, especially in octave bands. Recommended instead is a 15-m source–reference distance and a reference microphone height of 3–4 m above acoustically soft ground.

Undulating Ground. Undulating acoustically soft ground can greatly complicate the interference between direct and reflected sound rays. Physical scale modeling with appropriate soft-ground materials can approximate this interference,[17] but should be done within specially designed, quiet wind tunnels to account for the effects of wind on ray curvature. For maximum accuracy field measurements are recommended to determine propagation over undulating ground.

Atmospheric Turbulence. Attenuation over acoustically soft ground often varies in time because of atmospheric turbulence. Turbulence tends to reduce ground attenuation because it "randomizes" the mutual phase between direct and reflected rays. Where the direct and reflected sound rays pass through "different" turbulence, this randomization partially destroys their coherence.[6, 18] In general, the coherence between adjacent rays is inversely proportional to the separation between their two paths, the atmosphere's turbulence strength (which increases with wind speed and is affected by the atmosphere's thermal structure), and the square of the frequency. If turbulence is ignored, both calculations and physical scale modeling generally overestimate attenuation over acoustically soft ground.

5.6 BARRIER INSERTION LOSS

A sound barrier is any large object that blocks the line of sight between source and receiver—including the ground itself if it protrudes upward through the line of sight.[3, 19] As shown in the insert of Fig. 5.8, for long barriers where diffraction of

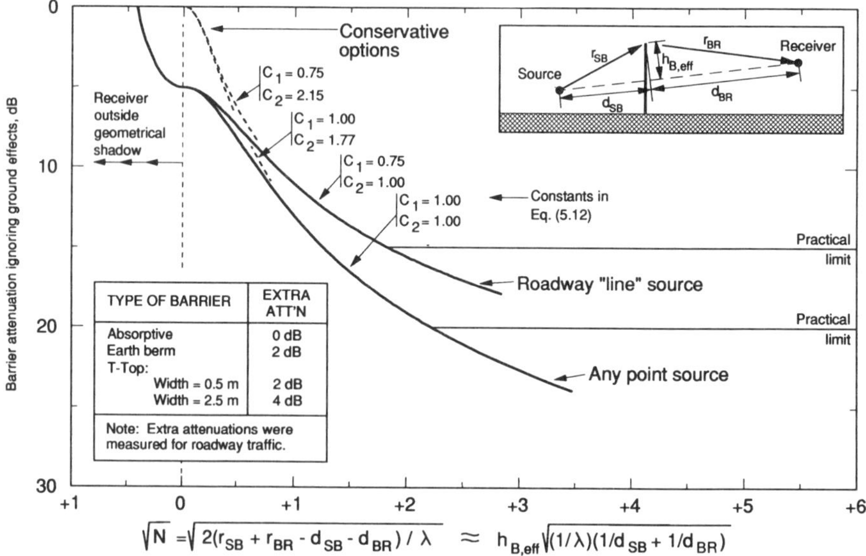

Fig. 5.8 Barrier attenuation for point and roadway line sources, with adjustments for different types of barriers. To be conservative, use the dashed rather than the solid curves. The barrier Fresnel number N is unitless; λ is wavelength in meters. The constants C_1 and C_2 differ for line/point sources and for conservative/regular options, as shown.

sound by the side edges of the barrier is negligible, the sound that reaches the receiver arrives across the barrier top, diffracted (bent) downward out of the so-called Fresnel zone above the top edge of the barrier and into the barrier "shadow."

The sound that enters the shadow is reduced in level by the diffraction. This reduction is called the barrier attenuation A_{barrier}. In addition, the barrier causes partial loss of soft-ground attenuation, as discussed in Section 5.5. The net effect of barrier diffraction, combined with this partial loss of soft-ground attenuation, is called the barrier insertion loss IL.

$$\begin{aligned} \text{IL}_{\text{barrier}} &= A_{\text{barrier}} - (\text{lost soft-ground attenuation}) \\ &= A_{\text{barrier}} - [(A_{\text{ground}})_{\text{no barrier}} - (A_{\text{ground}})_{\text{barrier}}] \end{aligned} \tag{5.11}$$

This section contains equations for barrier attenuation, A_{barrier}, for use in Eq. (5.11).

Pure Tones

In common use for predicting barrier attenuation are the following approximations. For single-edge diffraction, barrier attenuation A_{barrier} is approximated by[13,14]

$$A_{\text{barrier}} = \begin{cases} 20 \log \dfrac{\sqrt{2\pi N}}{\tan\sqrt{2\pi N}} + 5 \geq 0 & \text{outside shadow} \\[2ex] (20C_1) \log \dfrac{\sqrt{2\pi N}}{\tanh\,(C_2\sqrt{2\pi N})} + 5 \leq 20 & \text{inside shadow} \end{cases} \tag{5.12}$$

Alternatively,[20]

$$\begin{aligned} A_{\text{barrier}} &= 10 \log \left[3 + (C)(N) \exp\left(\frac{-1}{2000 \text{ m}} \sqrt{\frac{r_{SB} r_{BR} (d_{SB} + d_{BR})}{2(r_{SB} + r_{BR} - d_{SB} - d_{BR})}} \right) \right] \\ &\leq 20 \end{aligned} \tag{5.13}$$

where the square-root term in Eq. (5.13) accounts for average downwind propagation. The parameters in these two equations, including the Fresnel number N and the constants C_1 and C_2, are defined in Fig. 5.8. The constant C in Eq. (5.13) equals 20 for receivers high above the ground and 10 for receivers close to a reflecting plane. The barrier does not need to be perpendicular to the source–receiver line for Eqs. (5.12) and (5.13) to be valid.

Equation (5.12) appears in Fig. 5.8, along with several adjustments for different types of barriers along roadways.[21,22] In this figure barrier attenuation is plotted against the square root of Fresnel number N. This square root is approximately proportional to the parameter $h_{B,\text{eff}}$, the "effective" barrier height (above the line of sight). In this figure note that attenuation increases as (1) the effective barrier height increases and (2) as the barrier approaches either the source or the receiver.

Figure 5.8 also shows the net barrier attenuation for a roadway "line" source that is shielded by a barrier parallel to the roadway. For this line source attenuation, the geometric parameters must be measured perpendicular to the roadway. This line source curve in the figure follows from direct integration of Eq. (5.12) along the line, assuming nondirective sources.

Sometimes two barriers intervene in succession between source and receiver. The increased attenuation caused by such "double barriers" can be estimated by multiplying the more effective barrier's Fresnel number N in Eqs. (5.12) or (5.13) by the factor[20] $[1 + (5\lambda/s)^2]/[\frac{1}{3} + (5\lambda/s)^2]$, where λ is the wavelength and s is the separation between barriers. This factor limits the computed increase in attenuation to 5 dB.

A-weighted Sound Pressure Levels: Simplification

Generally the pure-tone equations for barrier attenuation also suffice for A-weighted levels, through use of an "effective" source frequency. The A-weighted attenuation of the "typical source spectrum," calculated band by band, has been found from experience to equal the pure-tone attenuation at 500–1000 Hz. Accordingly, in the Fresnel number definition in Fig. 5.8, λ equals 0.34–0.67 m for the typical source spectrum.

Some Complications

Flanking through, under, and around the Barrier. The barrier attenuation equations assume that diffracted sound dominates at the receiver. This means that the sound passing through the barrier must be negligible (down 5 dB or more) compared to the diffracted sound. This asumption requires that the barrier be relatively free of holes in its face and free of any continuous gap between it and the ground. More sound energy gets through such holes/gaps than intuition suggests. Also required is a sufficiently large transmission loss through the surface material of the barrier. A mass per unit surface area (excluding framing) of 20 kg/m^2 is usually sufficient to provide a transmission loss of 25 dB or more at 500 Hz. Such a mass per unit area is easily achievable with thicknesses and materials needed for mechanical stability of the barrier.[23]

Flanking around Barrier Ends and through Large Penetrations. When shielding a transportation route, barriers will often leave unshielded a portion of the route far down the line in both directions. As a result, the sound flanks the ends of the barrier. Figure 5.9 illustrates the approximate limits in barrier attenuation caused by this end flanking. Less end flanking occurs over acoustically soft ground, as shown in the left frame of the figure, compared to the right frame, because the end-flanking sound is attenuated more by the ground, compared to that arriving perpendicular to the route (over the barrier). Similarly, for dipole sources such as wheel–rail sources on trains, less end flanking occurs because the source

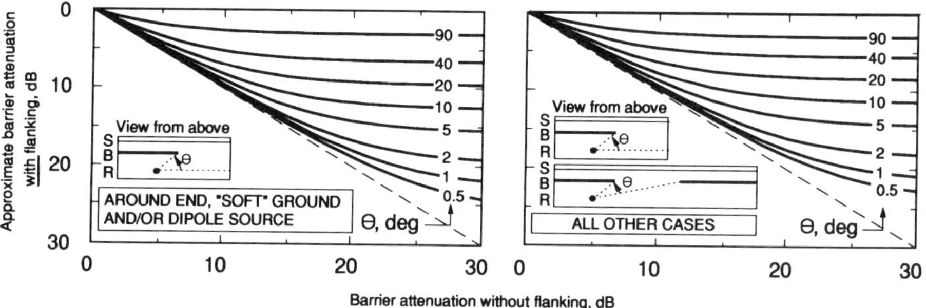

Fig. 5.9 Approximate decrease in barrier attenuation for a line source, caused by flanking through a large penetration of the barrier or around the end of the barrier. In the views from above, *S* denotes the line source, *B* the barrier, and *R* the receiver.

directs relatively less sound up and down the track, compared to that directed perpendicularly.

Flanking over Barrier Top. Three objects commonly scatter sound over the tops of barriers, thereby reducing their effectiveness: (1) nearby trees that protrude above the barrier top, (2) scattering elements on nearby buildings, and (3) atmospheric turbulence. The reduction in barrier effectiveness caused by trees is confined to relatively high frequencies, 2000 Hz and higher.[24] It is one of the factors that can limit barrier attenuation of point sources to 20 dB, as shown in Fig. 5.8. The reduction in barrier effectiveness caused by scattering from urban buildings is more serious, since it happens also at lower frequencies. Conservatively, urban scattering limits barrier attenuation to 5–10 dB whenever a clearly identifiable scattering path exists off a nearby building. Detailed physical scale modeling can predict such limits.[17] Atmospheric turbulence in the presence of wind generally provides an attenuation limit of 15–25 dB, no matter how free the area is from other flanking.[25]

Flanking Caused by Wind. As discussed further in Section 5.10, wind blowing from source to receiver can reduce barrier effectiveness, especially for barriers located midway between source and receiver. The wind-induced reduction in effectiveness is most important when the barrier just interrupts the line of sight between source and receiver. For this geometry the barrier provides 5 dB of attenuation without wind, as shown by the solid lines in Fig. 5.8, at $N = 0$. The slightest wind from source to receiver can eliminate this attenuation.

Three methods exist to estimate this wind-induced flanking. The first method approximates each source–receiver line of sight as a downward-bending circular arc with a radius of 3–5 km. Such refraction radii are typical with wind and temperature gradients of moderate magnitude. The arched, no-barrier line of sight will lie further above the ground than the unarched one, thereby reducing the effective barrier protrusion through the line of sight, $h_{B,\text{eff}}$, and thereby also reducing the barrier attenuation and insertion loss. In Eq. (5.13) the square-root expression takes this downwind flanking into account.[20]

The second method to estimate the effect of wind-induced flanking involves modifying the parameter C_2 in Eq. (5.12). The modified value (either 2.15 or 1.77) depends on whether the barrier is shielding a line or a point source, as shown in Fig. 5.8. This modification results in the dashed curves in Fig. 5.8, which pass through 0 dB for a grazing line of sight ($N = 0$).

The third estimation method consists of using specific adjustments to account for wind, as described in Section 5.10.

Reduced Effectiveness Caused by Reflections from Parallel Barrier on Far Side of Sound Source. Sometimes a second barrier exists on the far side of the sound source. When this second barrier is parallel to the one intervening between source and receiver, then multiple sound reflections between the two parallel barriers will produce reverberation between them. In turn this reverberation will reduce the effectiveness of the intervening barrier.[23,26] The reduction in barrier effectiveness is worst when the receiver can see the opposite barrier over the top of the intervening barrier, in which case barrier effectiveness can be reduced by 5 dB or more. Reverberation is also possible between the side of a rail-transit vehicle and a close-in sound barrier meant to shield wheel–rail sources on trains.

Common solutions to counter the effects of reverberation include the following: (1) applying a sound-absorbing material, with noise reduction coefficient NRC (arithmetic average of the Sabine absorption coefficients in the 250-, 500-, 1000-, and 2000-Hz octave bands) of 0.6 or greater, to the source side of one or both barriers; (2) tilting one or both barriers away from each other, which ''spills'' the reverbation by aiming direct reflections up to the sky;[27] or (3) increasing the barrier heights. Generally, a 10° tilt of both barriers is sufficient to prevent direct reflections from passing too close to opposite diffracting edges, although each geometry should be investigated individually.

More Precise Pure-Tone Insertion Loss near Ground. The barrier attenuation equations, Eqs. (5.12) and (5.13), apply directly to pure tones. However, the insertion loss equation, Eq. (5.11), does not, because the lost ground attenuation in this equation is often simplified for A-weighted levels.

Pure-tone calculation of barrier insertion loss can be approximated (for a quiet, isotropic atmosphere) through a detailed consideration of four paths: a fully airborne path, a second path with one ground reflection on the source side of the barrier, a third path with one ground reflection on the receiver side, and a fourth path with two ground reflections, one on each side. Each of these four paths will be differently affected by diffraction and by reflection from the ground. The sound level behind the barrier may then be obtained by combining the contributions of these four paths, taking phase into account. A second calculation yields the level without the barrier, again taking phase into account for the direct wave, the reflected wave, and the surface wave. Finally, the insertion loss is the difference between these two levels.[3] Note that phase effects in this calculation become uncertain in the presence of wind.

5.7 ATTENUATION THROUGH/OVER ROWS OF BUILDINGS AND OTHER OBSTACLES

A row of buildings provides attenuation when it intervenes between the source and receiver. The sound that reaches the receiver arrives both across the building tops and through the gaps between the buildings. For this reason the row of buildings provides less attenuation than does a solid barrier (one without gaps) of the same height. An approximation to the attenuation, $A_{1\,\text{row}}$, from a single intervening row of buildings is[8,28,29]

$$A_{1\,\text{row}} = -10 \log \left[1 - \min(F_l, F_\theta) + 10^{-\text{IL}_\text{barrier}/10}\right] \leq 10 \qquad (5.14)$$

In this equation F_l is the fractional linear blockage caused by intervening buildings, computed as $\Sigma(l_i/L)$ in Fig. 5.10. Also, F_θ is the fractional angular blockage caused by intervening buildings, computed as $\Sigma(\theta_i/\Theta)$ in the same figure. In this equation IL_barrier is the barrier insertion loss of the intervening buildings, ignoring the gaps between them. Use of the minimum of F_l or F_θ in Eq. (5.14) accounts for geometries where the receiver has a direct view of a portion of the source, through gaps between buildings. The recommended 10-dB upper limit is based on field experience.

This equation applies only to the first row of intervening buildings. Subsequent rows provide less attenuation. A conservative estimate is 1.5 dB for each additional row, up to a limit of 10–15 dB total attenuation.[13] For sound propagation through areas of industrial buildings that are not arranged in regular rows, an attenuation of $(0.05\ \text{dB/m})(r_\text{buildings})$ is typical, caused by scattering of sound energy skyward.[8] In this expression $r_\text{buildings}$ is the length of the source–receiver path that lies within the industrial-building zone.

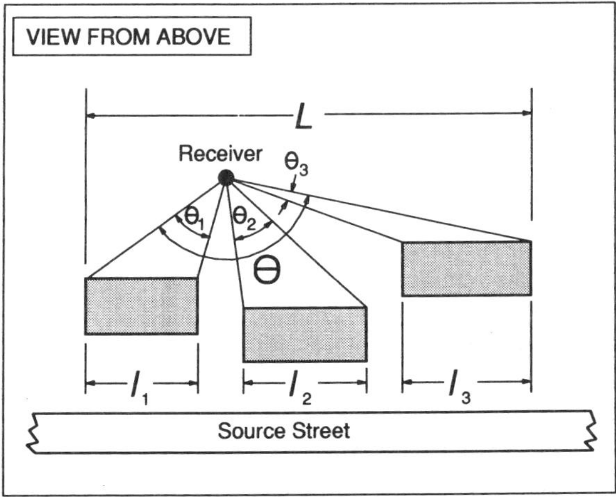

Fig. 5.10 Parameters for use in calculating attenuation through/over a single row of buildings with Eq. (5.14).

5.8 ATTENUATION THROUGH HEAVILY WOODED AREAS

Sufficiently dense and wide wooded areas provide attenuation when they intervene between source and receiver.[30–32] Such attenuation is caused by sound scattering into the sky from trunks and limbs (middle frequencies) and leaves (very high frequencies). Sound absorption by leaves is generally not significant. For some types of trees loss of leaves during the winter reduces wooded-area attenuation somewhat; for others it does not. In addition, some low-frequency attenuation results from ground attenuation within the wooded area, where the roots of underbrush produce acoustically soft ground. The attenuation caused by heavy woods can be estimated from[8]

$$A_{\text{woods}} = 6 \ldots 10 \left(\frac{f}{1\ \text{kHz}}\right)^{1/3} \left(\frac{r_{\text{woods}}}{100\ \text{m}}\right) \leq 10$$

In this equation r_{woods} is the length of the source–receiver path that lies within the heavily wooded area.

To be certain of this attenuation, (1) the wooded area must be dense with trees and have sufficient underbrush to block direct view of the source from the receiver and to produce acoustically soft ground, and (2) the trees must generally protrude above the line of sight by 5 m or more.

Some Complications

Variability Caused by Different Tree Types and Different Times of Year. Data vary widely for different tree types and different times of year. For this reason conservative estimates of environmental impact often ascribe no attenuation to wooded areas.

Possible Benefits that A-weighted Sound Pressure Level Misses. Even when measurements show no significant attenuation from intervening trees, many people believe strongly that such trees do quiet their environment. Perhaps the reason is purely psychological: people like trees and trees "soften" the environment. In addition, perhaps people also hear distinctions missed by sound level meters. Trees in leaf do significantly scatter the very high frequency sounds that can convey "mechanical harshness," and thereby may reduce harshness without significantly lowering the A-weighted level. In addition, scattering within wooded areas can add a type of "forest reverberation" to the sound, thereby blurring both harshness and impulsive transients that also annoy. Both these mechanisms "soften" the sound and perhaps make it less annoying; neither will significantly reduce the A-weighted level. In addition, wind motion through leaves produces a pleasant sound, which can partially mask more annoying sounds.

5.9 AMPLIFICATION CAUSED BY URBAN REVERBERATION

Urban canyons can amplify roadway sounds, through multiple reflections from the parallel building facades that flank urban streets. In effect, these building facades restrict the sound's divergence outward from the roadway, cause urban reverberation, and thereby increase sound levels. Within urban canyons, urban amplification AMP_{reverb} can be calculated for a single lane of traffic with the following equation:[8]

$$AMP_{reverb} = 10 \log \left[1 + \left(\frac{r}{r + 2d_{facade}} \right)^2 (1 - \alpha) \right] + R$$

$$R = 4 \left(\frac{h_{building}}{w_{street}} \right) \leq 3$$

In this equation r is the perpendicular distance from traffic to receiver, d_{facade} is the distance between receiver and the nearest building facade, α is the absorption coefficient of the building facades, $h_{building}$ is the building height, and w_{street} is the width of the street. Urban reverberation depends on the structure and roughness of the building facades and is only approximated by this equation.[33] More precise prediction is possible with physical scale models that incorporate realistic scattering elements[17,34] or with computer programs that use ray-tracing or imaging, preferably with scattering algorithms.

5.10 EFFECTS OF VERTICAL WIND/TEMPERATURE GRADIENTS (REFRACTION)

Vertical gradients of wind speed and temperature can both affect sound propagation, by refracting (bending) sound waves either upward or downward. Figure 5.11 shows several possibilities.[35,36] In the two frames on the left, refraction alters the apparent barrier protrusion above the line of sight. In the two frames on the right, refraction produces sound shadows where the ground "prevents" refracted rays from entering. This figure shows the qualitative relationship between wind/temperature conditions and upward/downward refraction. In general, upward refraction results in increased attenuation, while downward refraction may cause the loss or reduction of soft-ground attenuation, wooded-area attenuation, or barrier insertion loss.

Upwind, a sound shadow is typically caused by vertical wind gradients. Downwind, the thermal shadow is often eliminated by wind-induced downward refraction. Sideways to the wind direction, the wind has no effect and therefore the less significant temperature lapse shadow generally predominates (during daytime).

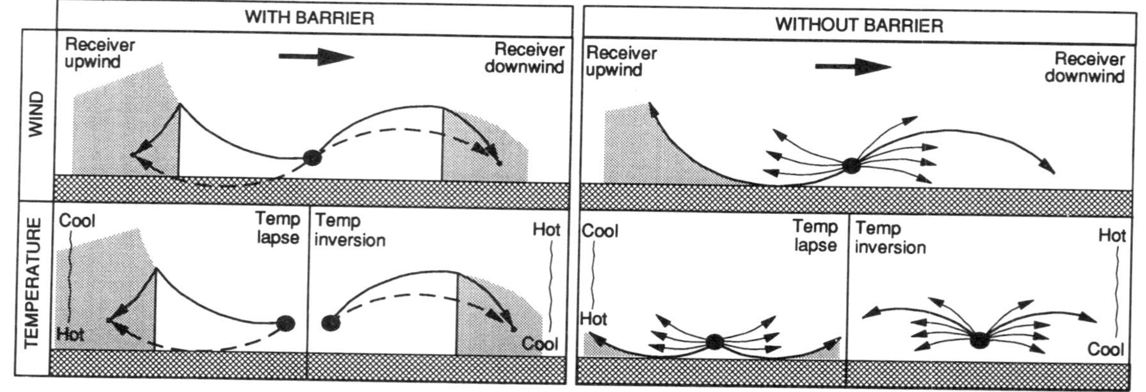

Fig. 5.11 General effects of vertical wind/temperature gradients with and without barriers. Top frames: wind, with (solid line) and without (dotted line) a noise barrier and ground. Bottom frames: temperature, with and without a noise barrier.

A-weighted Sound Pressure Levels

Wind and temperature refraction affects both soft-ground attenuation and barrier insertion loss, the latter because it is partly a function of soft-ground attenuation. Without an intervening barrier, when sound propagates a distance r at an average (unrefracted) height h over acoustically soft ground, the attenuation caused by wind and temperature gradients can be estimated by[37]

$$A_{\text{wind/temp, no barrier}} = \begin{cases} -3.0 \log \dfrac{r}{15 \text{ m}} & \text{for case 1} \\ 0 & \text{for case 2} \\ \left(10 - \dfrac{6.2h}{1 \text{ m}} + \dfrac{0.03h^2}{1 \text{ m}^2}\right) \log \dfrac{r}{15 \text{ m}} \geq 0 & \text{for case 3} \\ \left(14 - \dfrac{7.9h}{1 \text{ m}} + \dfrac{0.3h^2}{1 \text{ m}^2}\right) \log \dfrac{r}{15 \text{ m}} \geq 0 & \text{for case 4} \end{cases} \tag{5.15}$$

With an intervening barrier, the attenuation caused by wind and temperature gradients can be estimated by[37]

$$A_{\text{wind/temp, barrier}} = \begin{cases} -3 & \text{for case 1} \\ 0 & \text{for case 2} \\ +3 & \text{for case 3} \end{cases} \tag{5.16}$$

In these equations case 1 (moderate downward refraction) occurs where one mechanism (either wind or temperature) contributes to downward refraction and the other is neutral. Case 2 (no refraction) occurs either when both mechanisms are neutral or when one produces upward and the other downward refraction. Case 3 (moderate upward refraction) occurs when one mechanism contributes to upward refraction and the other is neutral. Finally, case 4 (strong upward refraction) occurs when both mechanisms produce upward refraction.

These equations apply to a wind component in the source–receiver direction of approximately 5 m/s, measured 3 m above the ground. Greater wind components will produce somewhat greater effects; lesser wind components, lesser effects. Conservatively, the full speed of the wind can be used, rather than just its component, if the wind direction fluctuates $\pm 45°$ or less from the source–receiver direction. Equations (5.15) and (5.16) are compared to long-term outdoor measurements in Section 5.11.

Wind and Temperature Refraction Paths

For vertical wind and temperature gradients independent of height, the refracted path between any two points consists of a circular arc with the following radius of curvature r_c:

$$\frac{r_c}{1\ \text{km}} = \frac{T_{10\ \text{m}} - T_{0.5\ \text{m}}}{1.9\ \text{K}} + \frac{S_{10\ \text{m}} - S_{0.5\ \text{m}}}{3.2\ \text{m/s}} \tag{5.17}$$

In this equation T is the air temperature in degrees Kelvin, and S is the wind speed in meters per second—at the two heights shown.

Some Complications

Great Variability. The effects of wind and temperature gradients are highly variable from moment to moment because of atmospheric turbulence. Because of this variability, reliance upon long-term sound reduction from upward refraction is risky. Some official prediction methods require the assumption of downwind propagation (downward refraction) under all circumstances, as a conservative approximation for long-term-averaged wind conditions.[8]

The average attenuation caused by wind and temperature, A_{ave}, during mixed upwind and downwind conditions, can be approximated at a source–receiver distance r and for receiver height greater than 4 m by

$$A_{\text{ave}} = \frac{3}{(10^5\ \text{m}^2)/r^2 + 1.6} \tag{5.18}$$

Channeling over Water. Downwind over water in the daytime, sound first refracts downward, then reflects without attenuation from the water's surface, and then continues on to refract/reflect many times again. As a result, sound is channeled into a moderately thick layer of air above the water, and levels can be 10–20 dB higher downwind than would otherwise be expected. In effect, the vertical divergence of the sound has been partially eliminated by the downward refraction and the unattenuated reflection from the water. Because of this effect, cannons fired in France have been heard in England.

5.11 INTERACTION AMONG PROPAGATION MECHANISMS

Figure 5.12 summarizes the major interactions among all significant propagation mechanisms. These interactions significantly affect resulting sound levels. Unfortunately, they are often difficult to predict because they are inherently complex.

Crucial interactions include those (1) between barrier and soft-ground attenuation, which is incorporated into the barrier insertion loss equation of Section 5.6, (2) between wind/temperature gradients and the ground, which are incorporated into the equations of Section 5.10, and (3) between wind/temperature gradients and barrier attenuation, which are similarly incorporated in Section 5.10.

Figures 5.13 and 5.14 summarize a classic series of measurements in Eng-

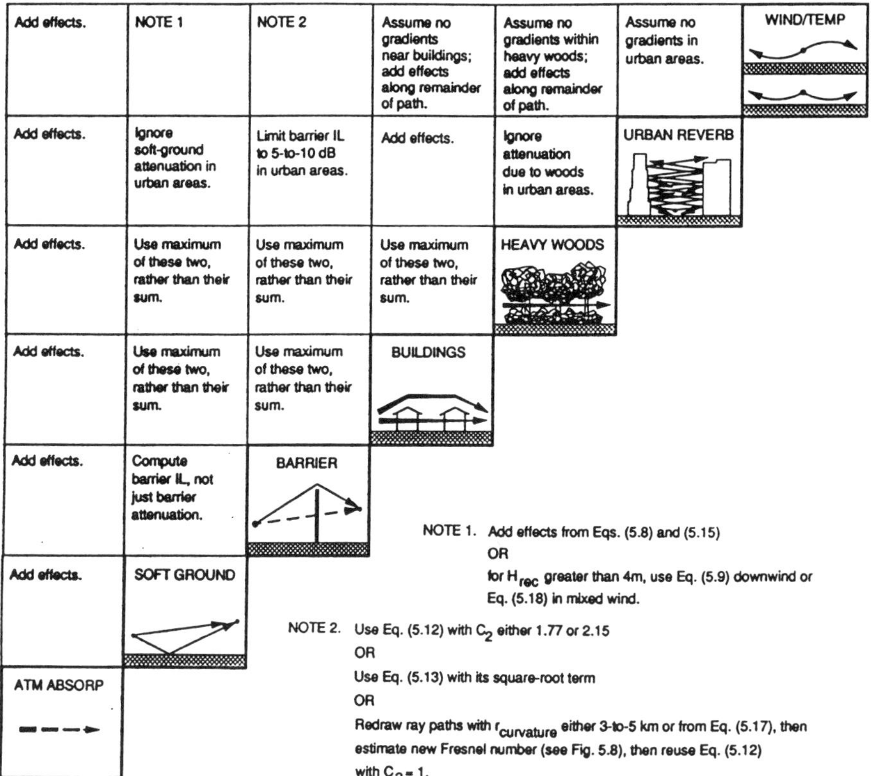

Fig. 5.12 Summary of major interactions among all significant mechanisms of outdoor sound propagation.

land,[38-40] where wind and temperature were monitored and resulting levels measured over a year, both with and without an intervening sound barrier. Without intervening barrier, these measurements utilized an aircraft engine at a height of 2 m as the source and microphones at a height of 1.2 m. With barrier, an array of loudspeakers served as source at a height of approximately 0.7 m, with microphone heights between 1.5 and 12 m. The intervening barrier was positioned 10–15 m from the source and ranged in height between 1.8 and 4.9 m.

The top frames of Figs. 5.13 and 5.14 show the data, after the expected effects of wave divergence and atmospheric absorption are subtracted out. If no other mechanisms were appreciable, these data would center around 0 dB at all distances, with some experimental scatter. The actual biases and scatter are the combined effects of all other propagation mechanisms. Lower frames in the figure subtract out succeedingly the effect of more propagation mechanisms. As expected, they show less and less bias and scatter.

Figures 5.13 and 5.14 use the following equations for subtraction of each propagation mechanism: Equation (5.1) combined with Eq. (5.2) for divergence, Eq.

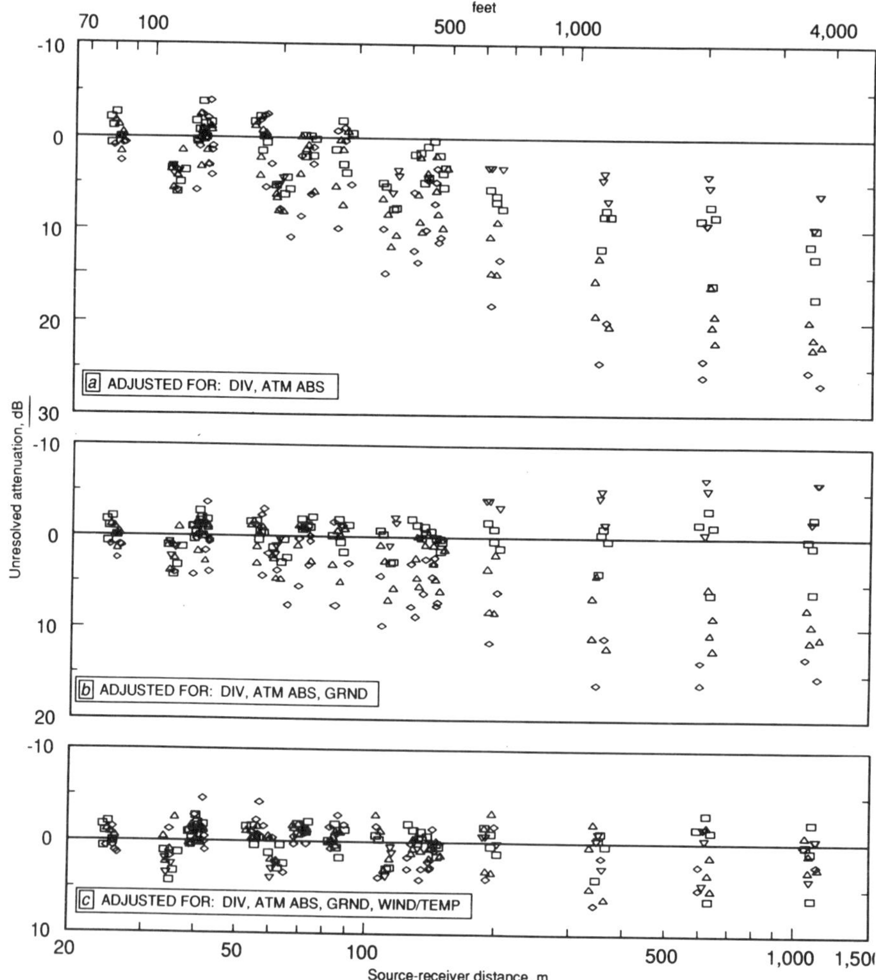

Fig. 5.13 Propagation measurements over grassland, showing scatter relative to selected equations: (*a*) Data adjusted to subtract the effects of wave divergence and atmospheric absorption. (*b*) Data further adjusted for soft-ground attenuation. (*c*) Data further adjusted for wind/temperature gradients. (∇) Moderate downward refraction; (□) no refraction; (△) moderate upward refraction; (◇) strong upward refraction. (From refs. 38 and 39.)

(5.7) for atmospheric absorption, Eq. (5.8) for soft-ground attenuation, Eqs. (5.12) and (5.8) for barrier insertion loss, and Eqs. (5.15) and (5.16) for wind and temperature effects.

The bottom frames of these two figures show the unresolvable bias and scatter, after all subtractions have been made. As shown, the bias is not appreciable and the scatter corresponds to a standard deviation of approximately 2.5 dB. These data show the accuracy and precision that can be expected from a selected combination of equations in this chapter.

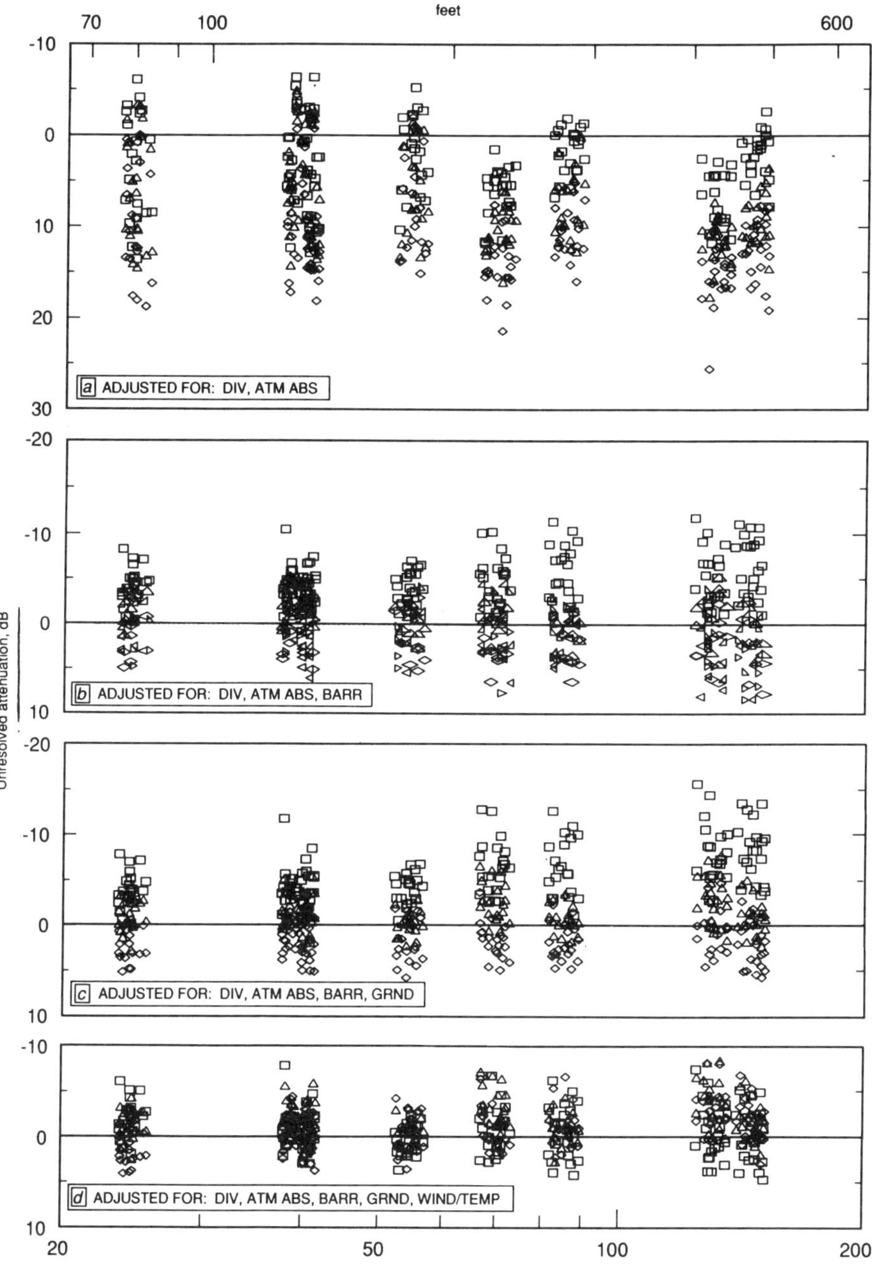

Fig. 5.14 Propagation measurements over a barrier on grassland, showing scatter relative to selected equations. (*a*) Data adjusted to subtract the effects of wave divergence and atmospheric absorption. (*b*, *c*) Data further adjusted, respectively, for barrier insertion loss and for soft-ground attenuation. (*d*) Data further adjusted for wind/temperature gradients. (∇) Moderate downward refraction; (□) no refraction; (△) moderate upward refraction; (◇) strong upward refraction (From ref. 40.)

REFERENCES

1. J. E. Piercy, T. F. W. Embleton, and L. C. Sutherland, "Review of Noise Propagation in the Atmosphere," *J. Acoust. Soc. Am.* **61**, 1403–1418 (1977).
2. M. E. Delany, "Sound Propagation in the Atmosphere: A Historical Review," *Proc. Inst. Acoust.* **1**, 32–72 (1978).
3. A. D. Pierce, *Acoustics: An Introduction to Its Physical Principles and Applications,* McGraw-Hill, New York, 1981.
4. T. F. W. Embleton, "Sound Propagation Outdoors: Improved Prediction Schemes for the 80's," *Noise Control Eng.* **18**, 30–39 (1982).
5. International Standards Organization, "Attenuation of Sound During Propagation Outdoors—Part 2: A General Method of Calculation" first draft proposal ISO-DP 9613-2 for Acoustics, International Standards Organization, Geneva, 1990.
6. S. F. Clifford and R. J. Lataitis, "Turbulence Effects on Acoustic Wave Propagation Over a Smooth Surface," *J. Acoust. Soc. Am.* **73**, 1545–1550 (1983).
7. American National Standards Institute, "Method for the Calculation of the Absorption of Sound by the Atmosphere," American National Standard ANSI S1.26-1978, Acoustical Society of America, New York, 1978. See L. L. Beranek, *Acoustical Measurements*, Acoustical Society of America, Woodbury, NY, 1988, Chapter 2, for graphs of tables in ANSI S1.26-1978.
8. Verein Deutscher Ingenieure, "Schallausbreitung im Freien," (Outdoor Sound Propagation), Report No. VDI 2714, VDI-Verlag GmbH, Düsseldorf, 1988.
9. K. Attenborough, "Review of Ground Effects on Outdoor Sound Propagation from Continuous Broadband Sources," *Appl. Acoust.* **24**, 289–319 (1988).
10. T. F. W. Embleton, J. E. Piercy, and G. A. Daigle, "Effective Flow Resistivity of Ground Surfaces Determined by Acoustical Measurements," *J. Acoust. Soc. Am.* **74**, 1239–1244 (1983).
11. J. Nicolas, J.-L. Berry, and G. A. Daigle, "Propagation of Sound above a Finite Layer of Snow," *J. Acoust. Soc. Am.* **77**, 67–73 (1985).
12. H. Gidamy, C. T. Blaney, C. Chiu, J. E. Coulter, M. DeLint, L. G. Kende, A. D. Lightstone, J. D. Quirt, and V. Schroter, "ORNAMENT: Ontario Road Noise Analysis Method for Environment and Transportation," Environment Ontario, Noise Assessment and Systems Support Unit, Advisory Committee on Road Traffic Noise, 1988.
13. T. M. Barry and J. A. Reagan, "FHWA Highway Traffic Noise Prediction Model," Report No. FHWA-RD-77-108, U.S. Federal Highway Administration, Washington, DC, 1978.
14. C. E. Hanson, H. J. Saurenman, G. S. Anderson, and D. A. Towers, "Guidance Manual for Transit Noise and Vibration Impact Assessment," Report No. UMTA-DC-08-9091-90-1, U.S. Urban Mass Transportation Administration, Washington, DC, 1990.
15. S.-I. Thomasson, "On the Concepts of Hard, Soft and Surface Wave Boundaries," *Acustica* **48**, 209–217 (1981).
16. C. I. Chessell, "Propagation of Noise Along a Finite Impedance Boundary," *J. Acoust. Soc. Am.* **62**, 825–834 (1977).

17. G. S. Anderson, ''Acoustical Scale Modeling of Roadway Traffic Noise: A Literature Review,'' Report No. 3630, Bolt Beranek and Newman, Cambridge, MA, 1978.

18. G. A. Daigle, J. E. Piercy, and T. F. W. Embleton, ''Line-of-sight Propagation through Atmospheric Turbulence Near the Ground,'' *J. Acoust. Soc. Am.* **74,** 1505–1513 (1983).

19. K. B. Rasmussen, ''On the Effect of Terrain Profile on Sound Propagation Outdoors,'' *J. Sound Vib.* **98**, 35–44 (1985).

20. Verein Deutscher Ingenieure, ''Schallschutz durch Abschirmung im Freien,'' (Noise Control by Means of Shielding Outdoors), Report No. VDI 2720, VDI-Verlag GmbH, Düsseldorf, 1987.

21. D. N. May and M. M. Osman, ''The Performance of Sound Absorptive, Reflective, and T-Profile Noise Barriers in Toronto,'' *J. Sound Vib.* **71**, 65–71 (1980).

22. D. N. May and M. M. Osman, ''Highway Noise Barriers: New Shapes,'' *J. Sound Vib.* **71**, 73–101 (1980).

23. M. A. Simpson, ''Noise Barrier Handbook,'' Report No. FHWA-RD-76-58, U.S. Federal Highway Administration, Washington, DC, 1976.

24. W. M. Schuller and J. H. de Zeeuw, ''Acoustic Effect of Trees on Barriers,'' *Proc. Inter-Noise '81*, pp. 253–256, 1981.

25. G. A. Daigle, ''Diffraction of Sound by a Noise Barrier in the Presence of Atmospheric Turbulence,'' *J. Acoust. Soc. Am.* **71**, 847–854 (1982).

26. W. Bowlby and L. F. Cohn, ''A Model for Insertion Loss Degradation for Parallel Highway Noise Barriers,'' *J. Acoust. Soc. Am.* **80**, 855–868 (1986).

27. C. W. Menge, ''Highway Noise: Sloped Barriers as an Alternative to Absorptive Barriers,'' *Noise Control Eng.* **14**, 74–78 (1980).

28. DB-Informationsschriften, ''Richtlinie zur Berechnung der Schallimmissionen von Schienenwegen'' (Guidelines for the Calculation of Sound Immission from Railways), Schall 03, Deutche Bundesbahn BZA, Munich, 1990.

29. DB-Informationsschriften, ''Richtlinie für schalltechnische Untersuchungen bei der Planung von Rangier- und Umschlagbahnhöfen'' (Guidelines for Acoustical Investigations during Planning of Railway Switch Yards), Akustik 04, Deutche Bundesbahn BZA, Munich, 1986.

30. M. A. Price, K. Attenborough, and N. W. Heap, ''Sound Attenuation through Trees: Measurements and Models,'' *J. Acoust. Soc. Am.* **84**, 1836–1844 (1988).

31. F. Fricke, ''Sound Attenuation in Forests,'' *J. Sound Vib.* **92**, 149–158 (1984).

32. M. J. M. Martens, F. G. P. Corten, and W. H. T. Huisman, ''A New Type of Noise Screen Constructed from and by Living Plants,'' *Proc. Inter-Noise '85*, pp. 499–502, 1985.

33. U. J. Kurze, ''Traffic Noise Propagation in Open and Built-up Areas,'' *Proc. FASE '84*, pp. 1–11, 1984.

34. G. S. Anderson, ''Acoustical Scale Modeling of Roadway Traffic Noise: Final Report, Vols. 1 and 2,'' Report No. 3939, Bolt Beranek and Newman, Cambridge, MA, 1979.

35. B. A. de Jong, *The Influence of Wind and Temperature Gradients on Outdoor Sound Propagation*, Delft University Press, Delft, 1983.

36. W. K. Van Moorhem and G. K. Landheim, ''The Propagation of Plane Waves in a Thermally Stratified Atmosphere,'' *J. Acoust. Soc. Am.* **76**, 867–870 (1984).

37. G. S. Anderson, to be published.

38. P. H. Parkin and W. E. Scholes, "The Horizontal Propagation of Sound from a Jet Engine Close to the Ground, at Radlett," *J. Sound Vib.* **1**, 1–13 (1964).

39. P. H. Parkin and W. E. Scholes, "The Horizontal Propagation of Sound from a Jet Engine Close to the Ground, at Hatfield," *J. Sound Vib.* **2**, 353–374 (1965).

40. W. E. Scholes, A. C. Salvidge, and J. W. Sargent, "Field Performance of a Noise Barrier," *J. Sound Vib.* **16**, 627–642 (1971).

CHAPTER 6

Sound in Small Enclosures

DONALD J. NEFSKE and SHUNG H. SUNG

Engineering Mechanics Department
General Motors Research Laboratories
Warren, Michigan

6.1 INTRODUCTION

Only in an anechoic room may sound waves travel outwardly in any direction without encountering reflecting surfaces. In practice, one must deal with every shape and size of enclosure containing an infinite variety of sound-diffusing, -reflecting, and -absorbing objects and surfaces. However, in small rooms of fairly regular shape, with smooth walls, the sound is not statistically diffuse and will depend on the acoustical modal response in the room. Typical examples where this is the case are the passenger compartments of transportation vehicles, ductwork and small rooms in buildings, enclosures used to enhance the response of audio equipment, and enclosures designed for sound isolation. While the modal theory for rectangular enclosures can be found in most standard reference books on acoustics,[1-5] the extended approach for irregular geometries with arbitrary wall impedances and the use of the finite-element method are more amenable to practical application and are described here.

Noise and Vibration Control Engineering: Principles and Applications, Edited by Leo L. Beranek and István L. Vér.
ISBN 0-471-61751-2 © 1992 John Wiley & Sons, Inc.

6.2 SOUND PRESSURE IN A VERY SMALL ENCLOSURE

Before considering the general case, it will be instructive to consider the sound pressure response in a very small enclosure. Frequently, in noise control problems, a noise source is enclosed in a very small box to prevent it from radiating noise to the exterior, as conceptually represented in Fig. 6.1*a*. When the noise source has a frequency low enough so that the wavelength of the sound is long compared to the largest dimension of the box, the sound pressure produced by the source will be uniform throughout the interior cavity. A uniform sound pressure field will also be created in an enclosure when the walls of the enclosure are forced to vibrate at low frequency by the pressures of an exterior sound source, as conceptually rep-

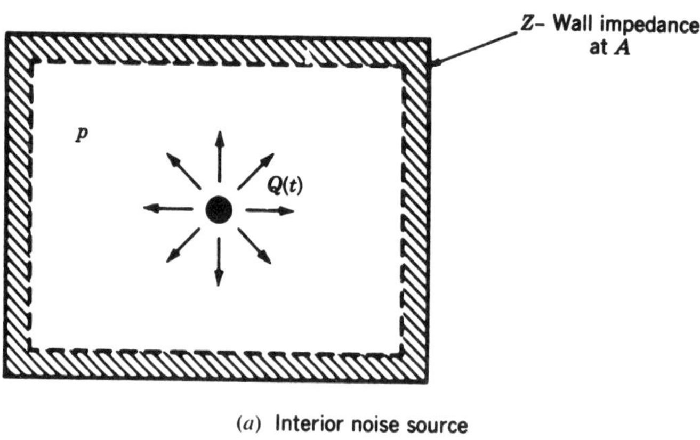

(*a*) Interior noise source

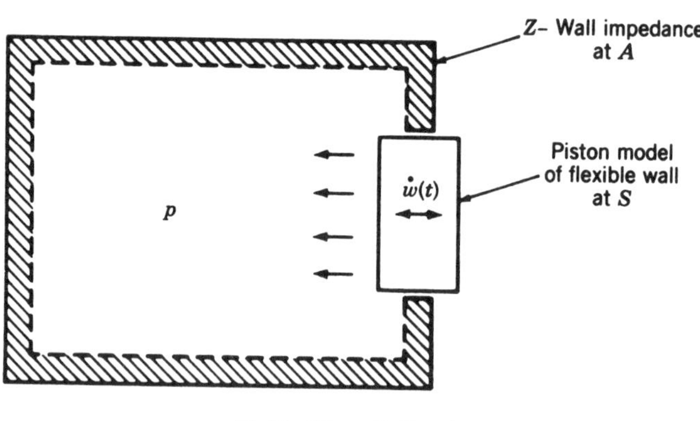

(*b*) Flexible wall vibration

Fig. 6.1 Sound in small enclosure with impedance boundaries generated by (*a*) interior noise source and (*b*) piston model of flexible-wall vibration. Symbols: $Q(t)$, volume velocity; Z, wall impedance; A, lined wall area; $\dot{w}(t)$, piston velocity; S, piston area.

resented by the vibrating piston in Fig. 6.1*b*. Frequently, the walls of the enclosure will be damped by applying panel-damping treatment or absorption material over the walls.

The uniform, steady-state sound pressure in such enclosures can be shown to be $p(t) = \hat{p}\cos(\omega t + \theta)$, where[6]

$$\hat{p} = \frac{\rho c^2 \hat{Q}}{V[(\omega + 2\delta_i)^2 + (2\delta_r)^2]^{1/2}} \quad \text{N/m}^2$$

$$\theta = \phi - \tan^{-1}\frac{\omega + 2\delta_i}{2\delta_r} \quad \text{rad} \tag{6.1}$$

Here ρ is the air density, c is the speed of sound, V is the enclosure volume, $Q(t) = \hat{Q}\cos(\omega t + \phi)$ is the *volume velocity* of the noise source operating at the forcing frequency $f = \omega/2\pi$ (Hz) with the amplitude $\hat{Q}$ and phase ϕ, and $\delta = \delta_r + i\delta_i$ is a complex damping factor ($i = \sqrt{-1}$) that accounts for the acoustical impedance $Z = R + iX$ of the wall area A, where

$$\delta_r = \frac{cA}{2V}\,\text{Re}\,\frac{\rho c}{Z} \qquad \delta_i = \frac{cA}{2V}\,\text{Im}\,\frac{\rho c}{Z} \quad \text{s}^{-1} \tag{6.2}$$

For the interior noise source in Fig. 6.1*a*, $Q(t)$ is taken as positive for outward volume flow (m^3/s), while for the flexible-wall vibration in Fig. 6.1*b*, $Q(t) = -S\dot{w}(t)$ is an *equivalent* volume velocity with outward piston velocity $\dot{w}(t)$ taken as positive.

From Eq. (6.1), we see that the magnitude $\hat{p}$ of the sound pressure in the enclosure depends not only on the noise source magnitude $\hat{Q}$ and forcing frequency $f = \omega/2\pi$ but also on the enclosure volume V and total wall impedance in δ. For a rigid-wall enclosure, $|Z| \to \infty$, so that $\delta_r, \delta_i = 0$, and

$$\hat{p} = \frac{\rho c^2 \hat{Q}}{\omega V} \quad \text{N/m}^2$$

$$\theta = \phi - \tfrac{1}{2}\pi \quad \text{rad} \tag{6.3}$$

In this case, the sound pressure depends only on the magnitude and frequency of the volume velocity of the source and on the enclosure volume. The sound pressure lags the volume velocity by exactly 90° indicating that the source does not radiate any acoustic power.

Example 6.1 The piston of Fig. 6.1*b* has an area of 1 cm^2 and is driven harmonically with a peak-to-peak displacement of 4 mm at a frequency of 100 Hz. The volume of the cavity is 0.0125 m^3 and it has rigid walls. What is the sound pressure level in the cavity?

SOLUTION For harmonic excitation $w = \hat{w} \sin \omega t$, and we obtain $\dot{w} = (\omega\hat{w})\cos \omega t = \hat{\dot{w}} \cos \omega t$, so that $\hat{\dot{w}} = \omega\hat{w}$. Therefore $\hat{Q} = S\hat{\dot{w}} = S\omega\hat{w}$, and from Eq. (6.3),

$$\hat{p} = \frac{\rho c^2 S\hat{w}}{V} = \rho c^2 \frac{\Delta \hat{V}}{V} \quad \mathrm{N/m^2} \tag{6.4}$$

where $\Delta \hat{V} = S\hat{w}$ is the volume change of the enclosure due to the piston displacement. Inserting the appropriate numerical values, we have $\Delta \hat{V} = 10^{-4} \times 2 \times 10^{-3} = 2 \times 10^{-7}\ \mathrm{m^3}$, so that

$$\hat{p} = 1.21 \times 343^2 \times \frac{2 \times 10^{-7}}{1.25 \times 10^{-2}} = 2.28\ \mathrm{N/m^2}$$

The sound pressure level is then

$$L_p = 20 \log_{10} \frac{p_{\mathrm{rms}}}{p_{\mathrm{ref}}} = 20 \log_{10} \frac{0.707 \times 2.28}{2. \times 10^{-5}} = 98\ \mathrm{dB}$$

Example 6.2 A noise source is enclosed in a very small box, as in Fig. 6.1*a*, that has flexible but very stiff walls. Determine the formula for the interior sound pressure.

SOLUTION For a mass–spring–dashpot model of the box walls, the acoustical impedance is $Z = A^{-1}[C + i\omega(M - K/\omega^2)]$, where A is the wall surface area. For very stiff walls, $Z \approx -iK/A\omega$ so that, from Eq. (6.2),

$$\delta_r = 0 \qquad \delta_i = \frac{\omega}{2} \frac{\rho c^2 A^2/V}{K} = \frac{\omega}{2} \frac{K_{\mathrm{air}}}{K} \quad \mathrm{s^{-1}}$$

where $K_{\mathrm{air}} = \rho c^2 A^2/V$ (N/m) is the *stiffness* of the air. Substituting δ_r and δ_i into Eq. (6.1) then gives

$$\hat{p} = \frac{1}{1 + K_{\mathrm{air}}/K} \frac{\rho c^2 \hat{Q}}{\omega V} \quad \mathrm{N/m^2}$$
$$\theta = \phi - \tfrac{1}{2}\pi \quad \mathrm{rad} \tag{6.5}$$

Note that the sound pressure in an enclosure with compliant walls is lower than the sound pressure in an equivalent rigid-wall enclosure given by Eq. (6.3), and the addition of damping to the walls can be shown to produce a similar effect.

Example 6.3 A *Helmholtz resonator* is a rigid-wall enclosure with a small aperture of cross-sectional area S that connects the enclosure to a column of air of length L that oscillates as the piston in Fig. 6.1*b*. Determine the natural frequency.

SOLUTION The mass of the column of air is $M = \rho SL$ and the force on it is pS. Hence we must have $M\hat{\ddot{w}} = pS$ where $\dot{w}$ is the velocity of the air column. Since $\hat{\dot{w}} = \omega\hat{w}$, we obtain $\hat{Q} = S\hat{\dot{w}} = S^2\hat{p}/\omega M$. Substituting this for $\hat{Q}$ in Eq. (6.3) gives $(1 - \rho c^2 S^2/\omega^2 MV)\hat{p} = 0$ or $(1 - K_{\text{air}}/\omega^2 M)\hat{p} = 0$, where $K_{\text{air}} = \rho c^2 S^2/V$ is the air stiffness. The natural frequency is then $\omega_0 = \sqrt{K_{\text{air}}/M} = c\sqrt{S/LV}$.

6.3 GOVERNING EQUATIONS FOR ACOUSTICAL MODAL RESPONSE

For larger enclosures or for higher frequencies, the sound pressure field in the enclosure is no longer uniform but depends on the acoustical modal response in the enclosure. More importantly, the sound pressure can be amplified considerably near discrete frequencies corresponding to the acoustical cavity resonances. This modal nature of sound field in an enclosure results from the superposition of sound waves that propagate according to the well-known *acoustical wave equation*

$$\nabla^2 p - \frac{\ddot{p}}{c^2} = 0 \qquad \text{N/m}^4 \tag{6.6}$$

where $\ddot{p}$ denotes the second partial derivative with respect to time t and ∇^2 is the Laplacian operator, which involves spatial coordinates such as (x, y, z).

Noise sources interior to an enclosed cavity can be included as forcing terms in the wave equation. For the example of the monopole source (e.g., a loudspeaker in a cabinet) as in Fig. 6.1*a*, the time-varying mass flow rate is $\dot{m}(x, y, z, t) = \rho Q(x, y, z, t)$ (kg/s), so that

$$\nabla^2 p - \frac{\ddot{p}}{c^2} = -\frac{\rho\dot{Q}}{V} \qquad \text{N/m}^4 \tag{6.7}$$

where $\dot{Q} = \partial Q/\partial t$. Other interior noise sources can be represented as combinations of monopole sources or else can be included directly in the wave equation in a similar manner. For simple harmonic motion, $p(x, y, z, t) = \text{Re}[p_0(x, y, z)\exp(i\omega t)]$ and $Q(x, y, z, t) = \text{Re}[(Q_0(x, y, z)\exp(i\omega t)]$, and we obtain the inhomogeneous *Helmholtz equation* for the steady-state sound pressure response,

$$\nabla^2 p_0 + \left(\frac{\omega}{c}\right)^2 p_0 = -\frac{i\omega\rho Q_0}{V} \qquad \text{N/m}^4 \tag{6.8}$$

where $f = \omega/2\pi$ (Hz) is the forcing frequency of the vibration and $\lambda = c/f$ (m) is the wavelength of the sound produced by the source.

The boundary conditions for p determine the reflection, absorption, and transmission of the sound waves at the enclosure's surfaces and are derived from fluid mechanical considerations. For small-amplitude motions, a momentum balance at the boundary requires that the air particle velocity u normal to a boundary surface to be related to p through

$$\frac{1}{\rho}\frac{\partial p}{\partial n} = -\dot{u} \qquad \text{m/s}^2 \tag{6.9}$$

where $\partial/\partial n$ is the outward surface-normal derivative. For an impervious wall surface, u is the normal-velocity component of the surface itself, while for a porous wall surface, u is the normal-velocity component of the air into the pores of the surface. Table 6.1 lists the boundary conditions for different wall surfaces that are characterized by their acoustical impedance Z and surface-normal vibration velocity $\dot{w}$. In what follows, we will investigate the steady-state and transient sound pressure response in enclosures with flexible, absorbent boundaries where Z and $\dot{w}$ are defined as in Table 6.1.

TABLE 6.1 Acoustical Boundary Conditions

WALL TYPE	BOUNDARY CONDITION	AIR PARTICLE VELOCITY
1. RIGID air $\|Z\| \to \infty$	$\frac{\partial p}{\partial n} = 0$	$u = 0$
2. FLEXIBLE $\dot{w}$ - wall panel surface-normal velocity wall air	$\frac{1}{\rho}\frac{\partial p}{\partial n} = -\ddot{w}$	$u = \dot{w}$
3. ABSORBENT air Z_a-acoustic impedance of layer (see chapt. 8)	$\frac{1}{\rho}\frac{\partial p}{\partial n} = -\frac{1}{Z_a}\dot{p}$	$u = \frac{p}{Z_a}$
4. FLEXIBLE AND ABSORBENT Z_w - wall panel impedance wall Z_a air	$\frac{1}{\rho}\frac{\partial p}{\partial n} = -\frac{1}{Z_a}\dot{p} - \ddot{w}$ $= -(\frac{1}{Z_a} + \frac{1}{Z_w})\dot{p}$	$u = \frac{p}{Z_a} + \dot{w}$ $= \frac{p}{Z_a} + \frac{p}{Z_w}$
5. OPEN $Z = 0$ air	$p = 0$ (pressure release)	determine from analysis

6.4 NATURAL FREQUENCIES AND MODE SHAPES

Acoustical resonances that result in high sound pressure occur at discrete natural frequencies in an enclosed cavity. The acoustical resonances are found from the free-vibration solution of the wave equation by substituting $p_n = p_{n0} \Psi_n(x, y, z)\exp(i\omega_n t)$, which gives

$$\nabla^2\Psi_n + \left(\frac{\omega_n}{c}\right)^2 \Psi_n = 0 \qquad \text{m}^{-2} \tag{6.10}$$

with free-vibration boundary conditions for the walls. The nondimensional pressure distributions $\Psi_n(x, y, z)$ for $n = 0, 1, 2, \ldots$ are the *mode shapes*, with $f_n = \omega_n/2\pi$ hertz being the corresponding *natural frequencies*. For absorbent boundaries, the mode shapes and natural frequencies are complex, and the acoustical modes are damped. However, for rigid boundaries ($|Z| \to \infty$) or for fully reflective open boundaries ($|Z| = 0$), one obtains real, undamped modes, or *standing waves*, that depend only on the geometric shape of the cavity.

Table 6.2 gives formulas for the natural frequencies and mode shapes of the undamped acoustical modes of a few regular-shape enclosures with rigid walls as well as those of a tube with fully reflective open boundaries. (A more complete table can be found in ref. 7.) One can show that the acoustical modes are orthogonal over the cavity volume such that

$$\int_V \Psi_m\Psi_n \, dV = \begin{cases} 0, & m \neq n \\ V_n, & m = n \end{cases} \qquad \text{m}^3 \tag{6.11}$$

and constitute *normal modes* of the enclosure. In Table 6.2, the number of indices used to identify each mode is based on the dimensionality of the enclosure, whereas we have used a single index to identify a mode, so the equivalence for a three-dimensional enclosure would be $f_n \equiv f_{ijk}$ and $\Psi_n \equiv \Psi_{ijk}$.

It is noteworthy that the first natural frequency of a cavity that is completely enclosed by rigid walls is zero, $f_0 = 0$, and is a *uniform pressure mode* (sometimes called the *Helmholtz mode*) with $\Psi_0 = 1$ and

$$V_0 = \int_V \Psi_0^2 \, dV = V \qquad \text{m}^3 \tag{6.12}$$

where V is the volume of the enclosure. Also, the fundamental frequency of the first spatially varying mode in an enclosed cavity is approximately $f_1 = c/2L$ hertz, where L is the maximum linear dimension of the cavity.

For a cavity with an open boundary, the $n = 0$ mode is absent because a (nonzero) uniform pressure cannot be sustained. Also, the natural frequencies and mode shapes in Table 6.2 for a tube with an open boundary are approximate because the boundary condition ($p = 0$) does not fully model the physics of the fluid behavior

TABLE 6.2 Acoustical Modes and Natural Frequencies[a]

DESCRIPTION	FIGURE	NATURAL FREQUENCY f_{ijk} (Hz)	MODE SHAPE Ψ_{ijk}
1. SLENDER TUBE BOTH ENDS CLOSED	$D \ll L$; D; x; L	$\frac{ic}{2L}$ $D \ll \lambda$ where $\lambda = c/f$	$\cos\frac{i\pi x}{L}$ $\quad i = 0, 1, 2, \ldots$
2. SLENDER TUBE ONE END CLOSED ONE END OPEN	$D \ll L$; D; x; L	$\frac{ic}{4L}$ $D \ll \lambda$ where $\lambda = c/f$	$\cos\frac{i\pi x}{2L}$ $\quad i = 1, 3, 5, \ldots$
3. SLENDER TUBE BOTH ENDS OPEN	$D \ll L$; D; x; L	$\frac{ic}{2L}$ $D \ll \lambda$ where $\lambda = c/f$	$\sin\frac{i\pi x}{L}$ $\quad i = 1, 2, 3, \ldots$
4. CLOSED RECTANGULAR VOLUME	y; x; z; L_y; L_x; L_z	$\frac{c}{2}\left(\frac{i^2}{L_x^2} + \frac{j^2}{L_y^2} + \frac{k^2}{L_z^2}\right)^{1/2}$	$\cos\frac{i\pi x}{L_x}\cos\frac{j\pi y}{L_y}\cos\frac{k\pi z}{L_z}$ $\quad i = 0, 1, 2, \ldots$; $j = 0, 1, 2, \ldots$; $k = 0, 1, 2, \ldots$
5. CLOSED CYLINDRICAL VOLUME	x; θ; r; $2R$; L	$\frac{c}{2\pi}\left(\frac{\lambda_{jk}^2}{R^2} + \frac{i^2\pi^2}{L^2}\right)^{1/2}$ λ_{jk} from Table 6.2a below	$J_j(\lambda_{jk}\frac{r}{R})\cos\frac{i\pi x}{L}\begin{cases}\sin j\theta; \\ or; \\ \cos j\theta\end{cases}$ $\quad i = 0, 1, 2, \ldots$; $j = 0, 1, 2, \ldots$; $k = 0, 1, 2, \ldots$ $J_j = j$-th order Bessel Function
6. CLOSED SPHERICAL VOLUME	r; R	Modes Symmetric about Center $\frac{c\lambda_i}{2\pi R}$ λ_i from Table 6.2b below	$\frac{R}{\lambda_i r}\sin\frac{\lambda_i r}{R}$ $\quad i = 0, 1, 2, \ldots$
7. ARBITRARY CLOSED VOLUME	L	L - Maximum Linear Dimension Fundamental Natural Frequency (approximate): $\frac{c}{2L}$	FINITE ELEMENT ANALYSIS

Table 6.2a

λ_{jk}	j						
k	0	1	2	3	4	5	6
0	0.	1.8412	3.0542	4.2012	5.3176	6.4156	7.5013
1	3.8317	5.3314	6.7061	8.0152	9.2824	10.5199	11.7349
2	7.0156	8.5363	9.9695	11.3459	12.6819	13.9872	15.2682
3	10.173	11.7060	13.1704	14.5859	15.9641	17.3128	18.6374

$\lambda_{j=0,k} = \pi(k + 1/4)$ for $k \geq 3$ $\quad (J_j'(\lambda_{jk}) = 0)$

Table 6.2b

i	0	1	2	3	4
λ_i	0.	4.4934	7.7253	10.9041	14.0662

$\lambda_i = \pi(i + 1/2)$ for $i \geq 4$ $\quad (\tan\lambda_i = \lambda_i)$

[a]From ref. 7.

at an open boundary. The accuracy of the formulas in Table 6.2 for an open tube increases with increasing slenderness of the tube. In general, the dimension of the open boundary must be small compared with the acoustical wavelength so that sound is fully reflected from the open boundary.

Example 6.4 A closed rectangular cavity has dimensions $0.41 \times 0.51 \times 0.61$ m. From Table 6.2, frame 4, the formulas for the normal-mode frequencies and mode shapes are

$$f_{ijk} = \frac{c}{2}\sqrt{\left(\frac{i}{L_x}\right)^2 + \left(\frac{j}{L_y}\right)^2 + \left(\frac{k}{L_z}\right)^2} \quad \text{(Hz)}$$
$$\Psi_{ijk} = \cos\frac{i\pi x}{L_x}\cos\frac{j\pi y}{L_y}\cos\frac{k\pi z}{L_z} \tag{6.13}$$

Let $L_x = 0.61$ m, $L_y = 0.51$ m, $L_z = 0.41$ m. (1) Find the natural frequencies of the $i = 2, j = 0, k = 0$; the $i = 1, j = 1, k = 0$; and the $i = 2, j = 1, k = 0$ normal modes of vibration. (2) Plot the sound pressure distribution for these three normal modes of vibration. The speed of sound, adjusted for temperature, is 347.3 m/s.

SOLUTION

1. From Eq. (6.13) we have

$$f_{2,0,0} = \frac{347.3}{2}\sqrt{\left(\frac{2}{0.61}\right)^2} = 569.3 \text{ Hz}$$

$$f_{1,1,0} = \frac{347.3}{2}\sqrt{\left(\frac{1}{0.61}\right)^2 + \left(\frac{1}{0.51}\right)^2} = 443.8 \text{ Hz}$$

$$f_{2,1,0} = \frac{347.3}{2}\sqrt{\left(\frac{2}{0.61}\right)^2 + \left(\frac{1}{0.51}\right)^2} = 663.4 \text{ Hz}$$

2. The pressure distributions $|\Psi_{ijk}|$ of the modes are shown in Fig. 6.2, where the normalization is such that the maximum pressure at the corners is $|\Psi_{ijk}| = 1$. Since $k = 0$ for all three modes, the pressure distributions are uniform in the z direction.

6.5 ACOUSTICAL FINITE-ELEMENT ANALYSIS

For simple geometries and boundary conditions, the acoustical modes can be expressed analytically as in Table 6.2, but for more complicated geometries and boundary conditions, numerical approaches such as the finite-element method[8,9] are required. The finite-element method provides a means of analyzing an arbitrarily shaped continuum based upon the representation of the continuum by a finite number of interconnected elements. While originally developed for structural analysis, the finite-element method is applicable to numerous nonstructural problems. The acoustical cavity problem falls into this category, where an acoustical finite-

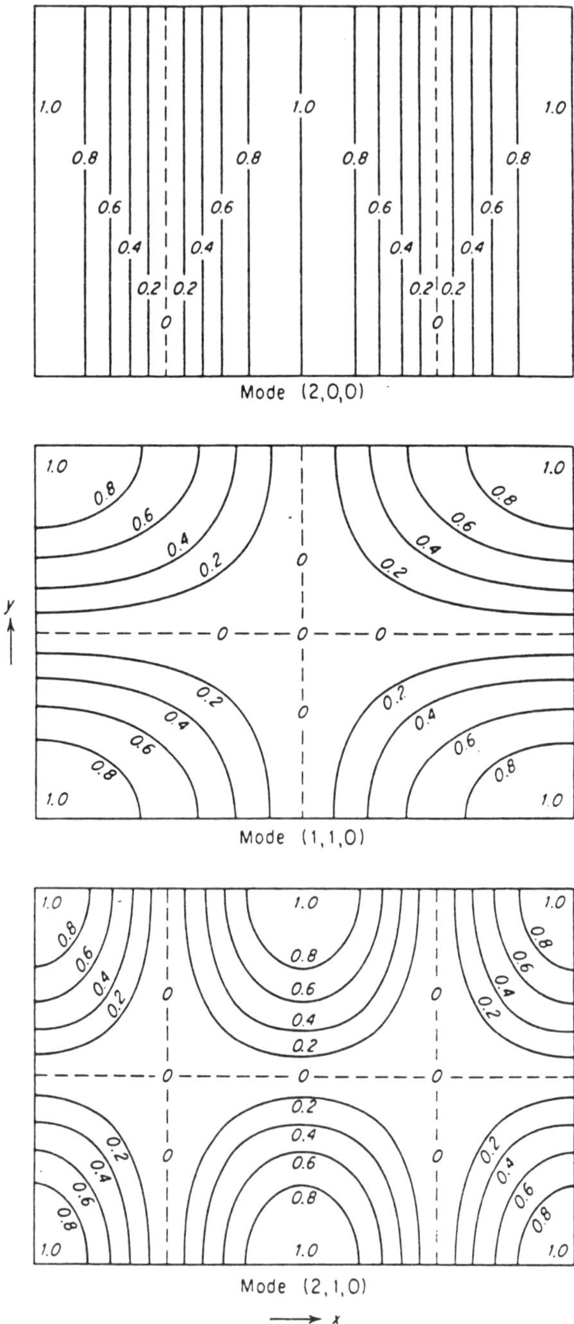

Fig. 6.2 Sound pressure contour plots on a section through a rectangular enclosure for three different modes of vibration. The numbers on the plots indicate the relative sound pressure amplitude. (Reprinted with permission from ref. 2, p. 210.)

element model of the cavity can be developed and solved within most structural finite-element computer codes by employing the so-called *structural–acoustical analogy*[10] outlined in Table 6.3. By specifying the structural material properties as shown in the table, the structural equations of motion reduce to the acoustical wave equation and the structural boundary conditions reduce to the required acoustical boundary conditions.

The accuracy of linear finite elements to solve for the acoustical resonances is illustrated in Fig. 6.3, where the computed natural frequencies for a tube with closed ends are compared with the exact frequencies given by the formula from Table 6.2. From these results, the percentage of error in the computed frequency can be seen to be proportional to $(n/N)^2$, where n is the mode number and N is the number of rectangular elements. These results can also be applied to an irregularly shaped acoustical cavity to estimate the number of elements required in a particular direction to develop a finite-element model that will give a desired degree of accuracy of a particular mode. For example, one can estimate that accuracy to within 10% in frequency can be obtained for the first four modes in a particular direction by using about 10 elements in that direction.

To illustrate the acoustical modal analysis of a complex shaped enclosure, Fig. 6.4 shows the first two modes of an automobile passenger compartment. The acoustical finite-element model in Fig. 6.4*a* was developed from linear hexahedral and pentahedral elements, where only a half-model is required because of symmetry.[11] The first acoustical mode of the passenger compartment in Fig. 6.4*b* is a longitudinal mode with the nodal surface passing vertically through the compartment. The nodal surface on which $|\Psi_n| = 0$ and the antinodal surfaces on which $|\Psi_n|$ is a local maximum are of practical importance because they indicate low- and high-noise regions, respectively, which result from excitation of the mode. For the example in Fig. 6.4*b*, excitation of the mode may go unnoticed by a front-seat occupant because of the proximity of the nodal surface, but it may be heard by a rear-seat occupant. The situation is just the opposite for the second mode illustrated in Fig. 6.4*c*. Because of the irregular geometric shape of the compartment, these modes could not be predicted by simple formulas like those presented in Table 6.2.

6.6 FORCED SOUND PRESSURE RESPONSE IN ENCLOSURE

Often the sound pressure response in an enclosure for known sound sources is of interest. The enclosure may be complicated in shape, and it may have walls that are both flexible and absorbent as well as multiple interior surfaces. If the undamped modes $\Psi_n(x, y, z)$ and natural frequencies $f_n = \omega_n/2\pi$ for $n = 0, 1, 2, \ldots$ are known, the forced acoustical response in such an enclosure can be directly expressed using the *modal analysis* technique as the *normal-mode expansion*,[12,13]

$$p(x, y, z, t) = \sum_n P_n(t)\Psi_n(x, y, z) \qquad \mathrm{N/m^2} \tag{6.14}$$

TABLE 6.3 Structural–Acoustical Analogy

DESCRIPTION	STRUCTURE	ANALOGY	ACOUSTIC CAVITY
FINITE ELEMENT	$w_k = \sum_{l=1}^{N} (w_k)_l N_l^k \quad k = x, y, z$	$w_z = p$ $N_l^z = N_l$	$p = \sum_{l=1}^{N} p_l N_l$
CONSTITUTIVE EQUATIONS	$\begin{pmatrix} \sigma_{zz} \\ \tau_{zy} \\ \tau_{zz} \end{pmatrix} = \begin{pmatrix} G_{11} & G_{14} & G_{16} \\ G_{14} & G_{44} & G_{46} \\ G_{16} & G_{46} & G_{66} \end{pmatrix} \begin{pmatrix} \varepsilon_{zz} \\ \gamma_{zy} \\ \gamma_{zz} \end{pmatrix}$	$\sigma_{zz} = u_z \quad \frac{\partial w_z}{\partial x} = \frac{\partial p}{\partial x}$ $\tau_{zy} = u_y \quad \frac{\partial w_z}{\partial y} = \frac{\partial p}{\partial y}$ $\tau_{zz} = u_z \quad \frac{\partial w_z}{\partial z} = \frac{\partial p}{\partial z}$	$\begin{pmatrix} u_z \\ u_y \\ u_z \end{pmatrix} = \begin{pmatrix} -1/\rho & 0 & 0 \\ 0 & -1/\rho & 0 \\ 0 & 0 & -1/\rho \end{pmatrix} \begin{pmatrix} \partial p/\partial x \\ \partial p/\partial y \\ \partial p/\partial z \end{pmatrix}$
EQUILIBRIUM EQUATIONS	$\frac{\partial \sigma_{zz}}{\partial x} + \frac{\partial \tau_{zy}}{\partial y} + \frac{\partial \tau_{zz}}{\partial z} = -\rho_s \ddot{w}_z$	MATERIAL PROPERTIES $G_{11} = 1/\rho$ $G_{44} = 1/\rho$ $G_{66} = 1/\rho$ $G_{14} = 0$ $G_{16} = 0$ $G_{46} = 0$ $\rho_s = \frac{1}{\rho c^2}$	$\frac{\partial u_z}{\partial x} + \frac{\partial u_y}{\partial y} + \frac{\partial u_z}{\partial z} = -\frac{1}{\rho c^2}\ddot{p}$
FINITE ELEMENT EQUATIONS	$[K]\{w\} - \omega^2[M]\{w\} = \{0\}$ $K_{ij} = \int_V \{\nabla N_i^z\}^T [G] \{\nabla N_j^z\} dV$ $M_{ij} = \int_V \rho_s N_i N_j dV$		$[K]\{p\} - \omega^2[M]\{p\} = \{0\}$ $K_{ij} = \int_V \frac{1}{\rho} \{\nabla N_i\}^T \{\nabla N_j\} dV$ $M_{ij} = \int_V \frac{1}{\rho c^2} N_i N_j dV$

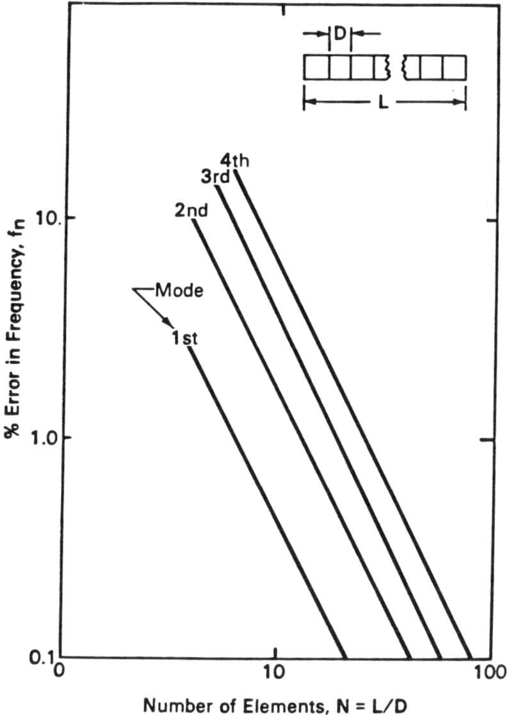

Fig. 6.3 Rate of convergence of linear acoustical finite elements. Exact solution: $f_n = nc/2L$. (Reprinted with permission from ref. 11, p. 393.)

where the P_n are time-varying coefficients that must be determined to satisfy the acoustical wave equation, the boundary conditions, and the initial conditions.

For flexible and absorbent boundaries described by the boundary conditions in Table 6.1, the modal coefficients $P_n(t)$ are determined from

$$\ddot{P}_n + 2\delta_n \dot{P}_n + \omega_n^2 P_n = \frac{\rho c^2}{V_n} F_n(t) \qquad \mathrm{N/m^2 \cdot s^2} \tag{6.15}$$

where V_n is given by Eq. (6.11) and $F_n(t)$ is the *modal force*

$$F_n(t) = \int_V \frac{\dot{Q}}{V} \Psi_n \, dV - \int_S \ddot{w} \Psi_n \, dS \qquad \mathrm{m^3/s^2} \tag{6.16}$$

Q is the volume velocity of interior noise sources, and $\dot{w}$ is the vibration velocity of the wall panels at the boundary surface S. The integrations are carried out over the enclosure volume V and the wall panel surface S, respectively. In Eq. (6.15), δ_n is the *modal damping constant* or *modal damping decrement*, which as defined

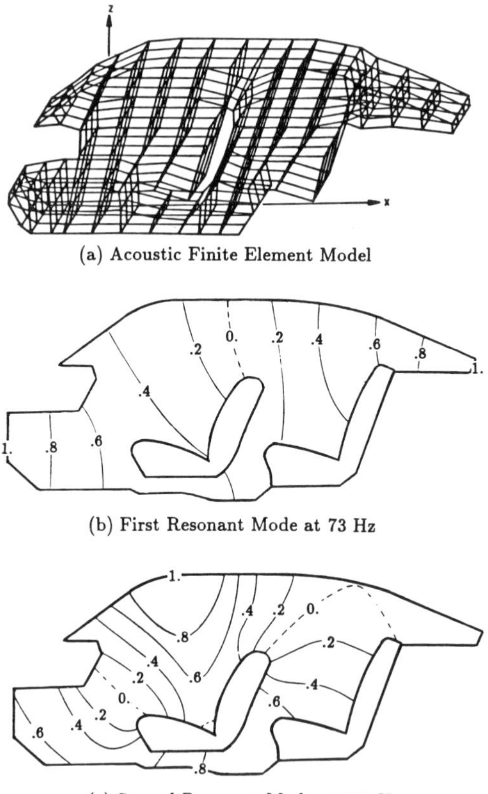

(a) Acoustic Finite Element Model

(b) First Resonant Mode at 73 Hz

(c) Second Resonant Mode at 130 Hz

Fig. 6.4 Acoustical finite-element analysis of automobile passenger compartment to determine the acoustical mode shapes and natural frequencies. The mode shapes become increasingly more complex as the natural frequency increases and as the complexity of the enclosure geometry increases.

here is complex and given by

$$\delta_n = \delta_n^r + i\delta_n^i = \frac{\rho c^2}{2V_n} \int_A \frac{\Psi_n^2}{Z}\, dA \qquad \mathrm{s}^{-1} \tag{6.17}$$

where $Z = R + iX$ is the complex wall impedance and the integration is carried out over the impedance wall area A. Table 6.1 determines the particular interpretation of $\dot{w}$ and Z in Eqs. (6.16) and (6.17) for different boundaries. The undamped modes Ψ_n are assumed to satisfy the rigid-wall boundary condition ($\partial\Psi_n/\partial n = 0$) on both the wall panel surface S and the impedance boundary A.

From Eq. (6.17), the real and imaginary parts of δ_n can be expressed as [cf. Eq. (6.2)]

$$\delta_n^r = \frac{cA_n}{2V_n} \operatorname{Re} \frac{\rho c}{Z_n} \qquad \delta_n^i = \frac{cA_n}{2V_n} \operatorname{Im} \frac{\rho c}{Z_n} \qquad \mathrm{s}^{-1} \tag{6.18}$$

where Z_n is an average impedance for the nth mode and

$$A_n = \int_A \Psi_n^2 \, dA \qquad \text{m}^2 \tag{6.19}$$

Recalling the interpretation of the acoustical impedance, we see that δ_n^r relates to the wall resistivity (~ damping) and δ_n^i relates to the wall reactance (~ flexibility). In the above formulation, we have assumed that light "damping" exists such that $|\delta_n^r| << \omega_n$ and $|\delta_n^i| << \omega_n$. For more heavily "damped" enclosures that often occur in practice, the modal equations (6.15) are coupled through the boundary impedance and the wall flexibility, and they must be solved simultaneously. A complete discussion of the procedure is given in references 13 and 14.

Example 6.5 The acoustical modes Ψ_n and natural frequencies ω_n are given by the finite-element analysis in Table 6.3. Determine the normal-mode expansion for the forced sound pressure response.

SOLUTION From the finite-element analysis, the acoustical modes are given as the $M \times N$ matrix

$$\{\Psi_1 \Psi_2 \cdots \Psi_n \cdots \Psi_N\} = \begin{bmatrix} \psi_{11} & \psi_{12} & \cdots & \psi_{1n} & \cdots & \psi_{1N} \\ \psi_{21} & \psi_{22} & \cdots & \psi_{2n} & \cdots & \psi_{2N} \\ \vdots & \vdots & \ddots & \vdots & \ddots & \vdots \\ \psi_{m1} & \psi_{m2} & \cdots & \psi_{mn} & \cdots & \psi_{mN} \\ \vdots & \vdots & \ddots & \vdots & \ddots & \vdots \\ \psi_{M1} & \psi_{M2} & \cdots & \psi_{Mn} & \cdots & \psi_{MN} \end{bmatrix} \tag{6.20}$$

where each row m corresponds to the response at a particular grid point of the finite-element model and each column n corresponds to the response of a particular mode. In matrix form, the normal-mode expansion in Eq. (6.14) for the forced sound pressure response becomes

$$\begin{bmatrix} p_1(t) \\ p_2(t) \\ \vdots \\ p_m(t) \\ \vdots \\ p_M(t) \end{bmatrix} = \begin{bmatrix} \psi_{11} & \psi_{12} & \cdots & \psi_{1n} & \cdots & \psi_{1N} \\ \psi_{21} & \psi_{22} & \cdots & \psi_{2n} & \cdots & \psi_{2N} \\ \vdots & \vdots & \ddots & \vdots & \ddots & \vdots \\ \psi_{m1} & \psi_{m2} & \cdots & \psi_{mn} & \cdots & \psi_{mN} \\ \vdots & \vdots & \ddots & \vdots & \ddots & \vdots \\ \psi_{M1} & \psi_{M2} & \cdots & \psi_{Mn} & \cdots & \psi_{MN} \end{bmatrix} \begin{bmatrix} P_1(t) \\ P_2(t) \\ \vdots \\ P_n(t) \\ \vdots \\ P_N(t) \end{bmatrix} \tag{6.21}$$

Each modal coefficient $P_n(t)$ is determined from Eq. (6.15) with ω_n known and $V_n = \{\Psi_n\}^T[M]\{\Psi_n\}$ with $[M]$ having been developed in the finite-element analysis. The modal forcing in Eq. (6.15) is

$$\{F_n(t)\} = [\psi]^T(\{\dot{Q}\} - [S]\{\ddot{w}\}) \tag{6.22}$$

where $\{\dot{Q}\}$ and $\{\ddot{w}\}$ are vectors of the grid point volume acceleration and wall panel acceleration and $[S]$ is a matrix of surface areas associated with each wall panel grid. Similarly, the damping constant in Eq. (6.15) is

$$\{\delta_n\} = \frac{\rho c^2}{2V_n}[\psi]^T\left[\frac{A}{Z}\right][\psi] \tag{6.23}$$

where $[A/Z]$ is a matrix of grid point impedance values weighted by the surface area A_m associated with each grid point m. Most large structural finite-element computer codes, when adapted for acoustical normal-mode analysis according to Table 6.3, also have capabilities for modal frequency response and modal transient response to give the forced acoustical response in the form of Eq. (6.21).

6.7 STEADY-STATE SOUND PRESSURE RESPONSE

The steady-state sound pressure in an enclosure is often of interest where the noise emanates from a point source, as in Fig. 6.1*a*. For a point source located at (x_0, y_0, z_0) and having a volume velocity $Q(t)$, the mass flow from the source function can be mathematically represented using the delta function as $\dot{m}(t) = \rho Q(t) \delta(x - x_0)\delta(y - y_0)\delta(z - z_0)$. Letting $Q(t) = \hat{Q}\cos(\omega t + \phi)$ be the steady-state excitation at a particular forcing frequency $f = \omega/2\pi$, the sound pressure response in the enclosure from Eqs. (6.14)–(6.16) can then be expressed as

$$p(x, y, z, t) = \sum_n p_n(x, y, z)\cos(\omega t + \theta_n) \quad \text{N/m}^2 \tag{6.24}$$

where

$$p_n(x, y, z) = \frac{\rho c^2\omega\hat{Q}\Psi_n(x, y, z)\Psi_n(x_0, y_0, z_0)}{V_n[(\omega^2 - \omega_n^2 + 2\delta_n^i\omega)^2 + (2\delta_n^r\omega)^2]^{1/2}} \quad \text{N/m}^2 \tag{6.25}$$

and

$$\theta_n = \phi - \tan^{-1}\frac{\omega^2 - \omega_n^2 + 2\delta_n^i\omega}{2\delta_n^r\omega} \quad \text{rad} \tag{6.26}$$

The amplitude of the modal sound pressure is $\hat{p}_n = |p_n|$.

Equation (6.24) shows that the steady-state sound pressure at one point in the enclosure can be considered as the superposition of numerous components of the same frequency ω but with different amplitudes $|p_n|$ and phase angles θ_n. If the source location (x_0, y_0, z_0) or the observer location (x, y, z) in Eq. (6.25) is on the nodal surface of a particular mode n, then the minimum participation of that mode will be observed in the response. It is also noteworthy that the solution is symmetric in the coordinates (x_0, y_0, z_0) of the sound source and the point of observation (x, y, z). If we put the sound source at (x, y, z), we observe at point (x_0, y_0, z_0) the same sound pressure as we did at (x, y, z) when the source was at (x_0, y_0, z_0). This is the famous reciprocity theorem (see Chapt. 9.9) that can sometimes be applied with advantage to measurements in room acoustics.

When only the uniform pressure mode ($\omega_0 = 0$) is excited, Eqs. (6.24)–(6.26) reduce to $p = \hat{p}_0 \cos(\omega t + \theta_0)$, where

$$\hat{p}_0 = \frac{\rho c^2 \hat{Q}}{V[(\omega + 2\delta_0^i)^2 + (2\delta_0^r)^2]^{1/2}} \qquad \text{N/m}^2$$

$$\theta_0 = \phi - \tan^{-1} \frac{\omega + 2\delta_0^i}{2\delta_0^r} \qquad \text{rad} \tag{6.27}$$

which are equivalent to Eqs. (6.1). The uniform pressure mode applies to the case of sound sources operating at very low frequencies, well below the first ($n = 1$) resonance of the enclosure. This requirement is generally met if $L < \frac{1}{10}\lambda$, where L is the maximum linear dimension of the cavity and where $\lambda = c/f$ is the wavelength of the sound generated by the source operating at the forcing frequency $f = \omega/2\pi$.

Example 6.6 For the rectangular cavity of Example 6.4, determine the steady-state sound pressure response in the cavity for loudspeaker excitation (a) when the enclosure walls are rigid and there is no damping; (b) when one wall is covered uniformly with absorption material of known impedance Z; and (c) when the damping is expressed in terms of the *critical damping ratio* ζ_n. Assume the loudspeaker acts as a simple monopole source.

SOLUTION

(a) For an undamped ($\delta_n^r = 0$) and rigid-wall ($\delta_n^i = 0$) enclosure, we obtain, from Eqs. (6.24)–(6.26),

$$p(x, y, z, t) = \rho c^2 \hat{Q} \sum_{i=0}^{I} \sum_{j=0}^{J} \sum_{k=0}^{K} \frac{\omega \Psi_{ijk}(x, y, z) \Psi_{ijk}(x_0, y_0, z_0)}{V_{ijk}(\omega^2 - \omega_{ijk}^2)}$$

$$\cdot \cos(\omega t + \phi - \tfrac{1}{2}\pi) \qquad \text{N/m}^2 \tag{6.28}$$

where Ψ_{ijk} and $f_{ijk} = \omega_{ijk}/2\pi$ are obtained from Table 6.2, V_{ijk} is determined from Eq. (6.11), and I, J, K are the number of modes we include to obtain a converged solution. For a closed rectangular enclosure,

$$\frac{V_{ijk}}{V} = \epsilon_i \epsilon_j \epsilon_k \quad \text{where } \epsilon_n = \begin{cases} 1 & \text{for } n = 0 \\ \frac{1}{2} & \text{for } n \geq 1 \end{cases} \tag{6.29}$$

with $V = L_x L_y L_z$. Note from Eq. (6.13) that for every mode of vibration, the sound pressure is a maximum at the corners of the rectangular enclosure. Also, for every mode of vibration for which one of the indexes i, j, or k is *odd*, the sound pressure is zero at the *center* of the enclosure; hence at the geometric center of the enclosure only one-eighth of the modes of vibration produce a finite sound pressure. Extending this further, at the center of any one wall, the modes for which two of the indexes (i, j, k) are odd will have zero pressure, so that only one-fourth of them will participate. Finally, at the center of one edge of the enclosure, the modes for which one index is odd will have zero pressure, so that only one-half of them participate there.

(b) Assume the uniform absorption material is on the $z = 0$ wall. Then, from Eq. (6.18),

$$\delta^r_{ijk} = \frac{cA_{ijk}}{2V_{ijk}} \operatorname{Re} \frac{\rho c}{Z_{ijk}} = \frac{c}{2L_z \epsilon_k} \operatorname{Re} \frac{\rho c}{Z}$$

$$\delta^i_{ijk} = \frac{cA_{ijk}}{2V_{ijk}} \operatorname{Im} \frac{\rho c}{Z_{ijk}} = \frac{c}{2L_z \epsilon_k} \operatorname{Im} \frac{\rho c}{Z} \tag{6.30}$$

where $Z_{ijk} = Z$ and $A_{ijk} = \epsilon_i \epsilon_j A_z$ with $A_z = L_x L_y$. Substituting Eq. (6.30) into Eqs. (6.24)–(6.26), one can evaluate the series solution for the sound pressure when Z is known.

(c) From the theory of vibration (see Chapt. 12), the critical damping ratio ζ_n is related to the damping constant through $\delta_n = \zeta_n \omega_n$. In complex form, $\zeta_n = \zeta^r_n + i\zeta^i_n$, so that $\delta^r_n = \zeta^r_n \omega_n$ and $\delta^i_n = \zeta^i_n \omega_n$, which can be substituted into Eqs. (6.25) and (6.26) to evaluate the sound pressure for given ζ_n.

Figure 6.5 shows the predicted versus measured sound pressure level in a rectangular enclosure for volume-velocity (loudspeaker) excitation. The response was predicted by Eqs. (6.24)–(6.26). Figure 6.5*a* is the response in the undamped, rigid-wall enclosure, where the resonance peaks resulting from excitation of the rigid-wall cavity modes are identified. Theoretically, the response at these resonances in an undamped enclosure should be infinite. In practice, however, damping is present even in an enclosure with very rigid walls since viscous and thermal losses occur in the air and at the boundaries. Modal damping provides a convenient method of accounting for these losses, and it is easily included in the solution as described in Example 6.6(c). Since the losses are primarily resistive, a real value of modal damping is used. Figure 6.5*b* shows the sound pressure response when sound absorptive material of known impedance covers the bottom wall of the enclosure. The predicted response is obtained using the complex "damping" for-

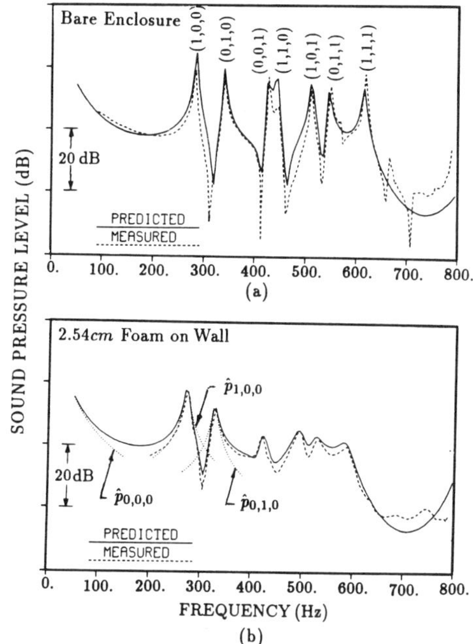

Fig. 6.5 Predicted versus measured comparison of sound-pressure-level curves for a constant volume–velocity source in a rectangular enclosure of dimensions 0.61 × 0.51 × 0.41 m. The microphone was at (0.51 m, 0.10 m, 0.30 m) and the source was at (0.0 m, 0.10 m, 0.10 m). (*a*) Bare enclosure with 0.5% modal damping. (*b*) Foam 2.54 cm thick with measured impedance $\rho c/Z = (0.12 + 0.47i)(f/800)$ for $0 \leq f < 800$ Hz covering the 0.61 × 0.51-m wall.

mulas given in Eq. (6.30) and by using acoustical modes based on the reduced height of the enclosure due to the absorption material thickness. The resonant responses of the $|p_{0,0,0}|$, $|p_{1,0,0}|$, and $|p_{0,1,0}|$ terms in the series in Eq. (6.24) are also shown. The attenuation and frequency shift of the resonant peaks are evident by comparing Figs. 6.5*a* and 6.5*b*, and these are due to the resistive and reactive nature of the absorption material.

6.8 ENCLOSURE DRIVEN AT RESONANCE

If the point sound source of volume velocity Q is driven by a steady-state generator of frequency ω and if that frequency equals a normal-mode frequency ω_n, then the modal sound pressure from Eq. (6.25) is

$$p_n(x, y, z) = \frac{\rho c^2 \hat{Q} \Psi_n(x, y, z) \Psi_n(x_0, y_0, z_0)}{V_n[(2\delta_n^i)^2 + (2\delta_n^r)^2]^{1/2}} \quad \text{N/m}^2 \qquad (6.31)$$

From this we conclude that $|p_n| \to \infty$ when both $\delta_n^r = 0$ and $\delta_n^i = 0$, so that ω_n is the *resonance frequency* when we have an enclosure with undamped and rigid-walls.

For an enclosure with reactive-walls ($\delta_n^i \neq 0$, $\delta_n^r \equiv 0$), the frequency dependence in the denominator of Eq. (6.25) can be approximated by

$$\omega^2 - \omega_n^2 + 2\delta_n^i \omega \approx \omega^2 - \omega_n^2 \left(1 - \frac{2\delta_n^i}{\omega_n}\right) = \omega^2 - \omega_{n0}^2 \qquad (6.32)$$

where

$$\omega_{n0} = \omega_n \sqrt{1 - \frac{2\delta_n^i}{\omega_n}} \quad \text{rad/s} \qquad (6.33)$$

Since now $|p_n| \to \infty$ when $\omega \to \omega_{n0}$, we see that the new resonance frequency of an enclosure with undamped, reactive-walls is ω_{n0}. The resonance frequency *shifts* from the rigid-wall resonance value ω_n due to the reactance δ_n^i of the walls, and ω_{n0} can be either higher or lower than the rigid-wall frequency ω_n, depending on whether $\delta_n^i < 0$ or $\delta_n^i > 0$, respectively. From Eq. (6.26), the phase angle between the modal pressure and the volume velocity source at resonance (i.e., when $\omega = \omega_{n0}$) is either 0° or 180°, and this provides a means of experimentally identifying the resonance frequency ω_{n0} and the enclosure reactance δ_n^i.

The enclosure dissipation δ_n^r controls the amplitude of the resonant peak when $\omega = \omega_{n0}$ since

$$p_n(x, y, z) = \frac{\rho c^2 \hat{Q} \Psi_n(x, y, z) \Psi_n(x_0, y_0, z_0)}{2\delta_n^r V_n} \quad \text{N/m}^2 \qquad (6.34)$$

The dissipation also shifts the resonance frequency an additional amount. The actual resonance frequency ω_{res} at which $|p_n|$ becomes a maximum can be shown to be

$$\omega_{\text{res}} = \sqrt{\omega_{n0}^2 - 2(\delta_n^r)^2} = \sqrt{\omega_n^2 - 2\delta_n^i \omega_n - 2(\delta_n^r)^2} \quad \text{rad/s} \qquad (6.35)$$

The dissipative effect (δ_n^r) is always to reduce the resonance frequency, but this effect is of second order and generally less important than the reactive effect (δ_n^i), which is of first order, unless δ_n^r is large. In general, therefore, to first order, the resonance frequency of an enclosure is $\omega_{\text{res}} = \omega_{n0}$ with ω_{n0} given by Eq. (6.33), and the peak sound pressure is given by Eq. (6.34). The *half-power bandwidth* of the resonant peak is the width of the resonance curve at 3 dB below the peak power (6 dB below the peak pressure) and is

$$\Delta\omega_{\text{res}} = 2\delta_n^r \quad \text{rad/s} \qquad (6.36)$$

This provides a means of experimentally determining the enclosure dissipation δ_n^r.

Finally, for light damping we can introduce from diffuse field theory (Chapter 7) the *random-incidence absorption coefficient* $\alpha = 8\ \mathrm{Re}(\rho c/Z)$, which is valid when both $|\mathrm{Re}(\rho c/Z)| << 1$ and $|\mathrm{Im}(\rho c/Z)| << 1$. Inserting this into Eq. (6.18) and substituting for δ_n^r in Eq. (6.34), we obtain

$$p_n(x, y, z) = \frac{8\rho c\hat{Q}\Psi_n(x, y, z)\,\Psi_n(x_0, y_0, z_0)}{\alpha_n A_n} \quad \mathrm{N/m^2} \qquad (6.34a)$$

where $\alpha_n = 8\ \mathrm{Re}(\rho c/Z_n)$ is the modal absorption coefficient. From this result we see that doubling the modal absorption halves the modal peak pressure, which reduces the modal sound pressure level at the resonance frequency by 6 dB.

Example 6.7 Determine the resonance frequency of the first mode in Fig. 6.5*b*. The measured impedance is $\rho c/Z = (0.12 + 0.47i)(f/800)$.

SOLUTION For the (1, 0, 0) mode, we have $f_{1,0,0} = 284.8$ Hz from Eq. (6.13). For this frequency, $\rho c/Z = (0.12 + 0.47i)(284.8/800) = 0.043 + 0.167i$. Noting that the thickness of the absorption material is 2.54 cm, we have $L_z = 0.41 - 0.0254 = 0.3846$ m, and from Eq. (6.30),

$$\delta_{1,0,0}^r = \frac{347.3}{2 \times 0.3846 \times 1} \times 0.043 = 19.6$$

$$\delta_{1,0,0}^i \frac{347.3}{2 \times 0.3846 \times 1} \times 0.167 = 76.2$$

Substituting these into Eq. (6.35) gives the resonance frequency

$$f_{\mathrm{res}} = \frac{\omega_{\mathrm{res}}}{2\pi} = f_n \sqrt{1 - \frac{2\delta_n^i}{\omega_n} - 2\left(\frac{\delta_n^r}{\omega_n}\right)^2}$$

$$= 284.8(1. - 0.085 - 0.00024)^{1/2} = 272.4 \text{ Hz}$$

Actually, the calculated frequency should be slightly greater than this because the impedance is frequency dependent and should be evaluated at f_{res}. An iterative calculation gives $f_{\mathrm{res}} = 273$ Hz, while the measured resonance frequency in Fig. 6.5*b* is 274 Hz.

6.9 FLEXIBLE-WALL EFFECT ON SOUND PRESSURE

A flexible wall may exhibit structural resonances and affect the sound pressure in two ways: (1) by acting as a noise source when exterior structural or pressure loads excite vibrations of the wall and (2) by acting as a boundary impedance that can alter the cavity sound pressure. Both of these effects may occur simultaneously in a room with flexible walls, in which case a coupled *structural–acoustical analysis* is required to predict the sound pressure response. We shall first consider each of the two effects of a flexible wall separately and then consider the coupled structural–acoustical response.

Flexible Wall as Noise Source

When the vibration velocity $\dot{w}(x, y, z, t)$ of a flexible wall is known, boundary condition 4 from Table 6.1 can be applied to Eqs. (6.15) and (6.16), which can then be solved for the steady-state sound pressure response:

$$p_n(x, y, z) = \frac{\rho c^2 \omega \hat{Q}_n \Psi_n(x, y, z)}{V_n[(\omega^2 - \omega_n^2 + 2\delta_n^i \omega)^2 + (2\delta_n^r \omega)^2]^{1/2}} \quad \text{N/m}^2 \tag{6.37}$$

$$\theta_n = \phi_n - \tan^{-1} \frac{\omega^2 - \omega_n^2 + 2\delta_n^i \omega}{2\delta_n^2 \omega} \quad \text{rad} \tag{6.38}$$

where δ_n^r, δ_n^i relate to the layer of absorption material of acoustical impedance Z_a covering the impervious walls. The forcing is through the equivalent modal volume-velocity of the wall vibration.

$$Q_n = \hat{Q}_n cos(\omega t + \phi_n) = -\int_S \dot{w}\Psi_n \, dS \quad \text{m}^3/\text{s} \tag{6.39}$$

where the vibration velocity of the wall $\dot{w} = \hat{\dot{w}} \cos(\omega t + \phi_w)$ is assumed to have a known amplitude $\hat{\dot{w}}(x, y, z)$ and phase $\phi_w(x, y, z)$.

Figure 6.6*a* shows a calculation of the forced acoustical response in the automobile passenger compartment using the above method with measured panel acceleration amplitude and phase data to represent the wall vibration and with measured wall impedance data to represent the absorption materials. The solution is expressed in matrix form as in Eq. (6.21) with the acoustical modes obtained from a finite-element analysis (Fig. 6.4). Figure 6.6*a* shows that a large spatial variation in sound pressure level occurs in the passenger compartment because of the modal nature of the acoustical response. Figure 6.6*b* shows the sound pressure computed separately for the vibration of each individual wall panel. The individual sound pressures can be combined considering the magnitude and phase to obtain the resultant sound pressure, which is shown. This provides a method to identify the major noise sources and the extent to which they must be controlled to yield a specified noise reduction. Figure 6.6*b* illustrates the large amplitude of the sound pressure due to the back-window vibration. Also note that the elimination of the roof vibration would increase the noise at the driver's ear.

Flexible Wall as Reactive Impedance

Wall panel vibration can generally be expressed as the normal-mode expansion

$$w = \sum_m W_m(t)\Phi_m(x, y, z) \quad \text{m} \tag{6.40}$$

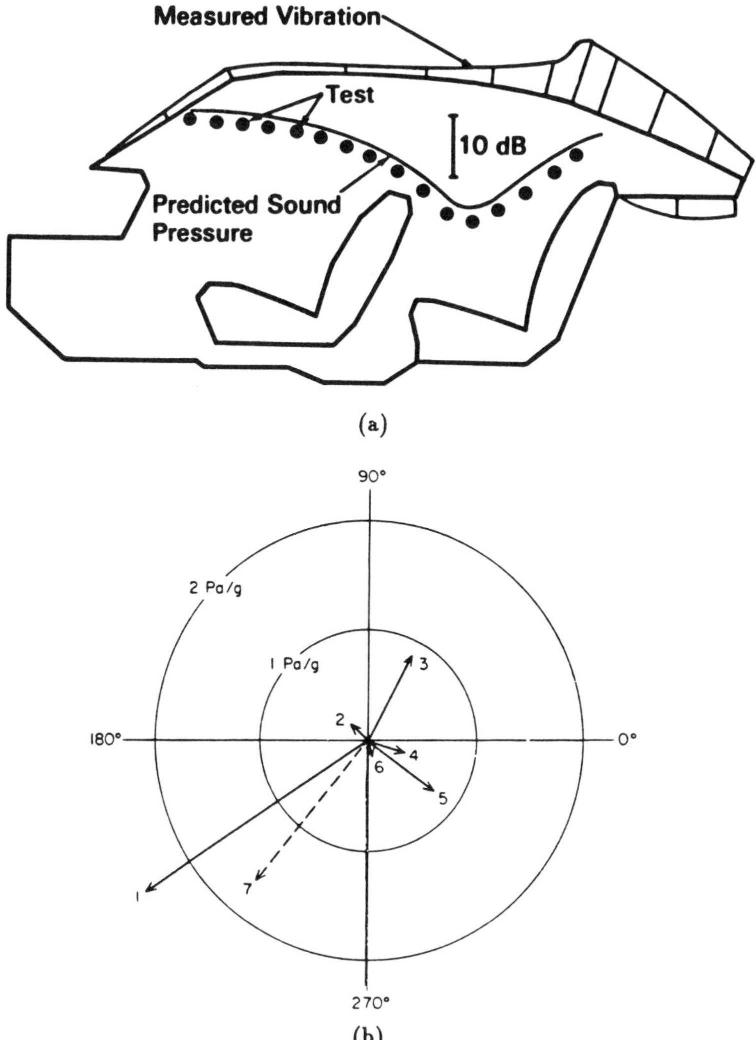

Fig. 6.6 (*a*) Measured versus predicted comparison of sound-pressure-level spatial variation in automobile passenger compartment for 40-Hz structural excitation. (*b*) Polar amplitude–phase diagram indicating panel contributions to resultant sound pressure at driver's ear (1, back window; 2, rear floor; 3, roof; 4, windshield; 5, rear shelf; 6, front floor; 7, resultant). (Reprinted with permission from ref. 16.)

where the Φ_m are the structural vibration mode shapes and W_m are the modal amplitudes. If we consider the coupling between the mth structural mode and the nth acoustical mode, the modal acoustical impedance of the wall is

$$Z_{mn} = (S_{mn})^{-1}\left[C_m + i\omega\left(M_m - \frac{K_m}{\omega^2}\right)\right] \qquad \text{N} \cdot \text{s/m}^3 \tag{6.41}$$

where M_m, C_m, and K_m are the modal mass, damping, and stiffness of the mth wall mode, and

$$S_{mn} = \int_S \Phi_m \Psi_n \, dS \qquad \text{m}^2 \tag{6.42}$$

is the *structural–acoustical coupling*. [Note that when $S_{mn} = 0$, there is no coupling between the structural mode m and the acoustical mode n, and Eq. (6.41) does not apply. In practice, when S_{mn} is sufficiently small, the coupling can be neglected to simplify the solution.]

For an undamped enclosure, the structural–acoustical coupling modifies Eq. (6.25) to

$$p_n(x, y, z) = \frac{\rho c^2 \omega \hat{Q} \Psi_n(x, y, z) \Psi_n(x_0, y_0, z_0)}{V_n |\omega^2 - \omega_{n0}^2|} \left| \frac{\omega^2 - \Omega_m^2}{\omega^2 - \Omega_{m0}^2} \right| \qquad \text{N/m}^2 \tag{6.43}$$

where $\Omega_m = \sqrt{K_m/M_m}$ (rad/s) is the natural frequency of the mth wall mode and where ω_{n0}^2 and Ω_{m0}^2 are solutions of

$$(\omega^2 - \omega_n^2)(\omega^2 - \Omega_m^2) - D^2\omega^2 = 0 \tag{6.44}$$

where $D = \sqrt{K_{\text{air}}/M_m}$ with $K_{\text{air}} = \rho c^2 S_{mn}^2 / V_n$ (N/m) being the *stiffness* of the nth acoustical mode relative to the mth structural mode.

Equation (6.43) shows that the flexible wall introduces an additional resonance at $\omega = \Omega_{m0}$, which will appear as a resonance peak in the sound pressure response and corresponds to the wall resonance. One can show from Eq. (6.44) that, with wall modes at frequencies below the rigid-wall cavity resonance ($\Omega_m < \omega_n$, the case of *mass-controlled* boundaries), the cavity resonance is raised ($\omega_{n0} > \omega_n$) and the wall resonance is lowered ($\Omega_{m0} < \Omega_m$). With wall modes at frequencies above the rigid-wall cavity resonance ($\Omega_m > \omega_n$, the case of *stiffness-controlled* boundaries), the cavity resonance is lowered ($\omega_{n0} < \omega_n$) and the wall resonance is raised ($\Omega_{m0} > \Omega_m$). In addition, the wall acts as a vibration absorber when $\omega = \Omega_m$, so that the modal sound pressure $p_n = 0$. However, the total sound pressure p may not be zero at this frequency because of the participation of other acoustical modes.

Coupled Structural–Acoustical Response

We have considered above the coupling of one acoustical mode n with one structural mode m, which is the case when a single coupling coefficient S_{mn} dominates for the mode pair (m, n). In practice, however, each acoustical mode n may couple with several structural modes, and the general case can be treated by substituting Eq. (6.40) into the second integral of Eq. (6.16) to obtain, from Eq. (6.15),

$$\ddot{P}_n + 2\delta_n \dot{P}_n + \omega_n^2 P_n + \frac{\rho c^2}{V_n} \sum_m S_{mn} \ddot{W}_m = \frac{\rho c^2}{V_n} F_n(t) \qquad \text{N/m}^2 \cdot \text{s}^2 \tag{6.45}$$

where the summation is carried out over all coupled structural modes m. Similarly, each structural mode m may in general couple with several acoustical modes, and the corresponding equation for the structural modal amplitude $W_m(t)$ in Eq. (6.40) is

$$\ddot{W}_m + 2\Delta_m \dot{W}_m + \Omega_m^2 W_m - \frac{1}{M_m} \sum_n S_{mn} P_n = \frac{R_m(t)}{M_m} \quad \text{m/s}^2 \tag{6.46}$$

where the summation is carried out over all coupled acoustical modes n. In this latter equation, Δ_m is the modal damping constant, Ω_m is the natural frequency, M_m is the modal mass, and $R_m(t)$ is the modal force. The coupling between the acoustical and structural systems makes it necessary to solve Eqs. (6.45) and (6.46) simultaneously for all coupled P_n and W_m. A complete derivation of the equations and discussion of the procedure can be found in reference 13, and its finite-element implementation is described in references 16 and 17.

To illustrate a typical coupled structural–acoustical response, Fig. 6.7 shows an application to a small metallic box used to isolate instruments from an external noise field (cf. Fig. 6.1*b*). The analysis of such enclosures to predict their noise attenuation is discussed in detail in Chapter 13. The finite-element method can be used to model the box wall panels and the acoustical cavity (Figs. 6.7*b*, *c*) in order to compute the uncoupled structural and acoustical modes. Equations (6.45) and (6.46) are then solved for the coupled frequency response, where the exterior noise field in Fig. 6.7*a* is specified as an oscillating pressure $\hat{p}_E \cos(\omega t)$ applied uniformly to the wall panels. The noise attenuation inside the enclosure is characterized by its *insertion loss* defined in Chapter 13 as $20 \log_{10}(\hat{p}_I/\hat{p}_E)$, where $\hat{p}_I$ is the interior sound pressure. The predicted versus measured insertion loss is shown in Fig. 6.7*d*. The structural–acoustical coupling effect is particularly evident in the rms panel vibration shown in Fig. 6.7*e*, which compares the coupled versus the uncoupled (i.e., in vacuo) panel response.

6.10 TRANSIENT SOUND PRESSURE RESPONSE

When the source of sound in a room is turned off, the sound dies out, or decays, at a rate that depends on the dissipation, or damping in the room. The acoustical response that exists in the room can be found from the transient solution of Eq. (6.15), which is

$$P_n(t) = \frac{\rho c^2}{V_n \omega_{nD}} \int_0^t F_n(\tau) e^{-\delta_n^r (t-\tau)} \sin \omega_{nD}(t - \tau)\, d\tau \quad \text{N/m}^2 \tag{6.47}$$

where $\omega_{nD} = \sqrt{\omega_n^2 - (\delta_n^r)^2 + (\delta_n^i)^2}$ is the "damped" modal frequency of the acoustical response in the enclosure. Equation (6.47) is a particular solution that satisfies zero initial conditions. For general initial conditions, the following free-

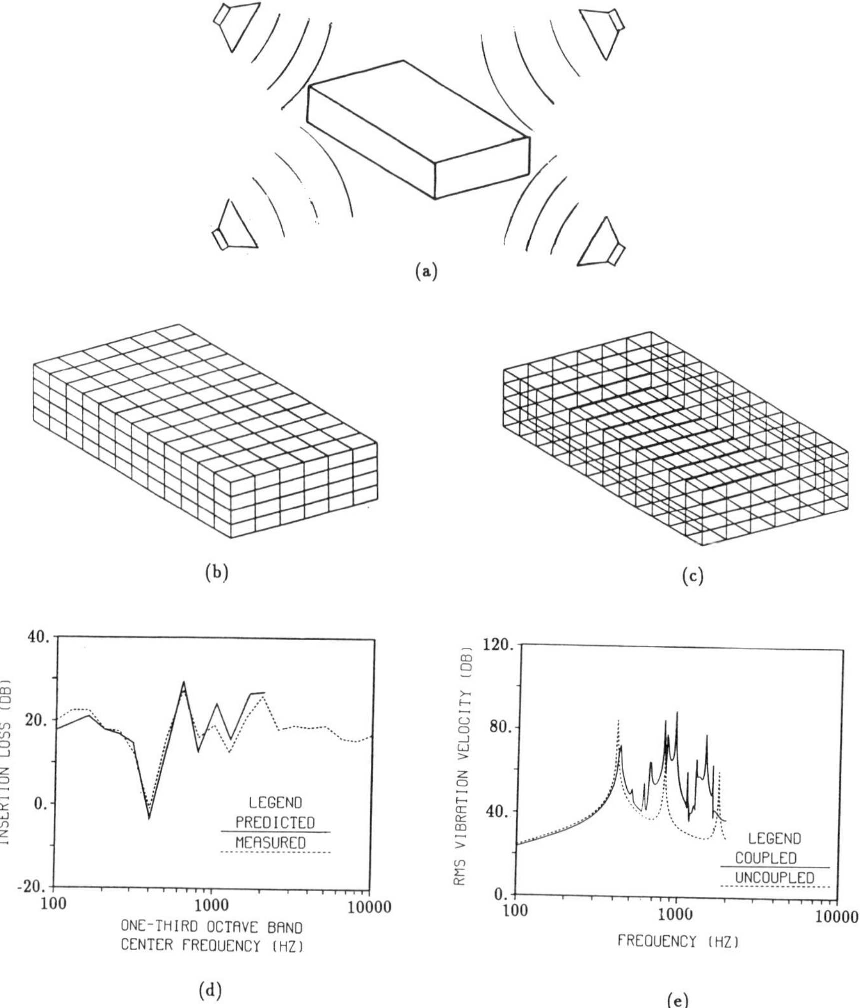

Fig. 6.7 Structural–acoustical analysis to predict insertion loss of a 30 × 15 × 5-cm unlined aluminum box with 0.16-cm-thick walls. (*a*) Box in uniform noise environment. (*b*) Structural finite-element model of box wall panels. (*c*) Acoustical finite element of box cavity. (*d*) Predicted versus measured insertion loss at box center. (*e*) Coupled versus uncoupled surface-averaged vibration of 30 × 15-cm wall panel. [Measured data in (d) from Chapter 13, ref. 9.]

vibration response must be added to the above equation,

$$P_n(t) = e^{-\delta_n^r t} \left[\frac{\dot{P}_n(0) + P_n(0)\delta_n^r}{\omega_{nD}} \sin \omega_{nD} t + P_n(0) \cos \omega_{nD} t \right] \qquad \text{N/m}^2 \qquad (6.48)$$

Equation (6.48) can also be used to determine the decay of the modes after the sound source is turned off at $t = 0$, when the initial conditions $P_n(0)$ and $\dot{P}_n(0)$ are known. For light damping, each mode of vibration behaves independently of the others, and the total process of sound decay is the summation in Eq. (6.14) of the sound pressures associated with all of the individual modes of vibration that fall within the frequency band of interest. The long-time sound decay may be different than the short-time sound decay because of the different damping factors and initial conditions for the modes.

The *reverberation time* T is defined as the time in seconds required for the level of sound to drop by 60 dB or for the pressure to drop to $\frac{1}{1000}$ of its initial value (see Chapter 7). One can similarly define a *modal* reverberation time T_n as that for which the pressure decays in that mode by 60 dB or $\frac{1}{1000}$ of its initial value. Since the reverberation time is that associated with the decaying part of the solution in Eq. (6.48),

$$T_n = \frac{6.91}{\delta_n^r} = \frac{13.82 V_n}{c A_n \operatorname{Re}(\rho c / Z_n)} \qquad \text{s} \tag{6.49}$$

where we have substituted from Eq. (6.18). With $c = 343$ m/s (20°C) and expressed in terms of the random incidence absorption coefficient, $\alpha_n = 8 \operatorname{Re}(\rho c / Z_n)$, with V in cubic meters and A in square meters, one obtains the modal reverberation time ($n > 0$)

$$T_n = 0.322 \frac{V_n}{\alpha_n A_n} \qquad \text{s} \tag{6.50}$$

which can be evaluated for a given Z by using the formulas in Table 6.2. The uniform pressure mode ($n = 0$) must be treated separately. The general transient solution when $n = 0$ reduces to

$$P_0(t) = P_0(0) e^{-2\delta_0^r t} + \frac{\rho c^2}{V} \int_0^t \int_0^\tau F_0(\sigma)\, d\sigma\, e^{-2\delta_0^r (t-\tau)}\, d\tau \qquad \text{N/m}^2 \tag{6.51}$$

and the reverberation time for the decay of a uniform sound pressure is $T_0 = 6.91/2\delta_0^r = 6.91V/cA \operatorname{Re}(\rho c/Z)$ so that, for V in cubic meters and A in square meters,

$$T_0 = 0.161 \frac{V}{\alpha A} \qquad \text{s} \tag{6.52}$$

In this case, the formula for the reverberation time of a uniform sound pressure is identical to the formula of Sabine for the reverberation time of a diffuse sound field.

If the pressure–time history is dominated by a single mode, then the reverberation time is equal to the appropriate modal reverberation time. If the sound source is turned off at $t = 0$, the magnitude of the sound pressure associated with a particular mode at a response location (x, y, z) is given by Eq. (6.48). The decay of the response from its maximum amplitude $[\dot{P}_n(0) = 0]$ can then be written as

$$p_n(x, y, z, t) = P_n(0) \Psi_n(x, y, z) e^{-\delta_n^r t} \cos(\omega_n t + \theta_n) \qquad \text{N/m}^2 \tag{6.53}$$

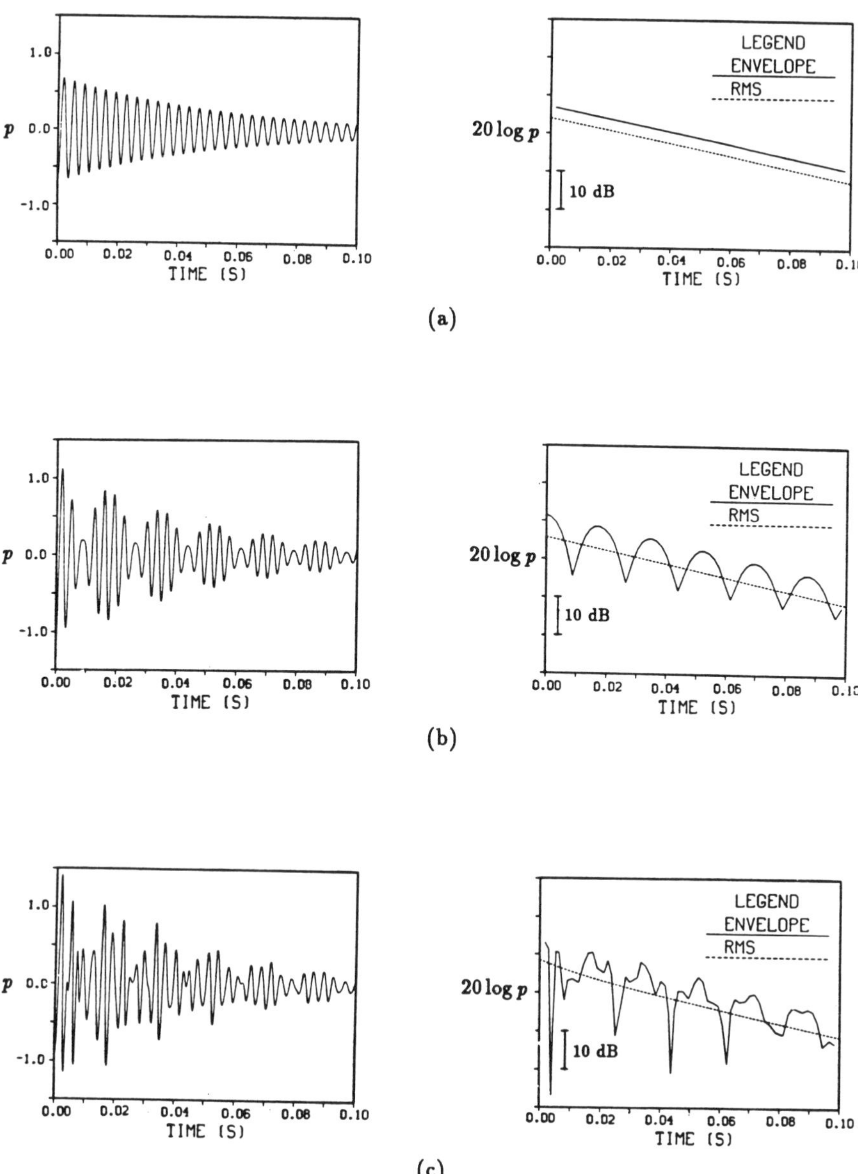

Fig. 6.8 Sound pressure decay curves for rectangular enclosure of Fig. 6.5*b*: (*a*) for the (1, 0, 0) mode of vibration; (*b*) for the (1, 0, 0) and (0, 1, 0) modes of vibration together; (*c*) for all modes of vibration up to 800 Hz. The graphs on the left show the course of the instantaneous sound pressure at the microphone location, and those on the right show the curve of the envelope of the left graphs and the rms pressure of Eq. (6.54) plotted on a log p versus t coordinate system.

where $\theta_n = \tan^{-1}(\delta_n^r/\omega_{nD})$ is the modal phase. If we take the rms time average of $\cos(\omega_n t + \theta_n)$ and the rms spatial average of $\Psi_n(x, y, z)$ and we designate the resultant as $\bar{p}_n(t)$, then Eq. (6.53) indicates that on a plot of $\log \bar{p}_n$ versus time, as in Fig. 6.8*a*, both the envelope of the sound pressure and the rms sound pressure decay linearly with time and have the constant reverberation time T_n.

When many normal modes of vibration (each with its own amplitude, phase, resonance frequency, and damping constant) decay simultaneously, the total rms (spatial and temporal) sound pressure is obtained from Eq. (6.14) as

$$\bar{p}(t) = \sqrt{\bar{p}_I^2 + \bar{p}_{I+1}^2 + \cdots + \bar{p}_{I+N}^2} \qquad \text{N/m}^2 \tag{6.54}$$

where I is the first mode in the frequency band being considered and $I + N$ is the last mode in the decay. In this case, the decay envelope is generally not linear as above, even when the modal reverberation times do not vary greatly from one mode to another so that all modes in the frequency band have a similar damping constant. The decay envelope is irregular, as in Figs. 6.8*b*, *c*, because the modes of vibration have different frequencies and beat with each other during the decay. However, the rms decay from Eq. (6.54) will be nearly linear if the modal reverberation times are similar (Fig. 6.8*b*) or if the long-time decay is governed by a single least-damped mode, which may occur when the modal reverberation times or initial conditions differ (Fig. 6.8*c*).

This brief description of sound decay in small rooms is applicable to rooms of any size and shape, and it can be extended to include the effects of structural–acoustical coupling. The fact that sound fields in large enclosures involve too many modes for the calculations to be practical does not mean that there are not distinct normal modes of vibration, each with its own normal frequency and damping constant. Alternate and more practical methods for handling large enclosures involving large numbers of modes are discussed in the next chapter.

REFERENCES

1. P. M. Morse and K. U. Ingard, *Theoretical Acoustics*, McGraw-Hill, New York, 1968.
2. L. L. Beranek (Ed.), *Noise and Vibration Control*, rev. ed., Institute of Noise Control Engineers, Poughkeepsie, NY, 1988.
3. H. Kuttruff, *Room Acoustics*, Applied Science, London, 1979.
4. L. E. Kinsler, A. R. Frey, A. B. Coppens, and J. V. Sanders, *Fundamentals of Acoustics*, 3rd ed., Wiley, New York, 1982.
5. L. Cremer, H. A. Muller, and T. J. Schultz, "Wave Theoretical Room Acoustics," in *Principles and Applications of Room Acoustics*, Vol. 2, Applied Science, London, 1982.
6. D. J. Nefske and S. H. Sung, "Sound in Small Enclosures," Research Publication GMR-7069, General Motors Research Laboratories, Warren, MI, June 1990.
7. R. D. Blevins, "Fluid Systems," in *Formulas for Natural Frequency and Mode Shape*, Van Nostrand Reinhold, New York, 1984, Chapter 13.

8. O. C. Zienkiewicz, *The Finite Element Method*, McGraw-Hill, London, 1977.
9. K. H. Huebner and E. A. Thornton, *The Finite Element Method for Engineers*, Wiley, New York, 1982.
10. G. C. Everstine, "Structural Analogies for Scalar Field Problems," *Int. J. Numer. Methods Eng.*, **17,** 471–476 (1981).
11. M. M. Kamal and J. A. Wolf, Jr. (Eds.), *Modern Automotive Structural Analysis*, Van Nostrand Reinhold, New York, 1982.
12. P. M. Morse and R. H. Bolt, "Sound Waves in Rooms," *Rev. Modern Phys.*, **16**(2), 69–150 (1944).
13. E. H. Dowell, G. F. Gorman III, and D. A. Smith, "Acoustoelasticity: General Theory, Acoustic Natural Modes and Forced Response to Sinusoidal Excitation, Including Comparisons with Experiment," *J. Sound Vib.*, **52**(4), 519–542 (1977).
14. E. H. Dowell, "Reverberation Time, Absorption, and Impedance," *J. Acoust. Soc. Am.*, **64**(1), 181–191 (1978).
15. E. H. Dowell, C. F. Chao, and D. B. Bliss, "Absorption Material Mounted on a Moving Wall–Fluid/Wall Boundary Condition," *J. Acoust. Soc. Am.*, **70**(1), 244–245 (1981).
16. D. J. Nefske, J. A. Wolf, Jr., and L. J. Howell, "Structural–Acoustic Finite Element Analysis of the Automobile Passenger Compartment: A Review of Current Practice," *J. Sound Vib.*, **80**(2), 247–266 (1982).
17. S. H. Sung and D. J. Nefske, "A Coupled Structural–Acoustic Finite Element Model for Vehicle Interior Noise Analysis," *ASME J. Vib. Acoust. Stress Reliabil. Design* **106,** 314–318 (1984).

CHAPTER 7

Noise in Rooms

MURRAY HODGSON and ALFRED C. C. WARNOCK

Institute for Research in Construction
National Research Council of Canada
Ottawa, Canada

Noise problems experienced by room users include temporary or permanent hearing loss in a factory, poor speech communication in a conference room, an inability to concentrate in a classroom or library, and inadequate speech privacy in an open-plan office. To modify sound fields and improve acoustical conditions, the acoustician must understand the relationship between sound sources, the room and its contents, and the characteristics of room sound fields. Models for predicting room sound fields are of primary importance; they allow acoustical conditions to be optimized during design and noise-control measures to be evaluated for cost effectiveness in new or existing rooms. In this chapter, we take an energy-based approach, ignoring the phase or modal effects discussed in Chapter 6. This is justified, except perhaps at low frequencies, since we are interested in large rooms of complex shape, which may not be empty. Further, we are interested in results in octave or one-third-octave bands. In this chapter, the sound field is quantified by the time-averaged mean-square sound pressure p^2 in pascals squared or the associated sound pressure level L_p in decibels ($L_p = 10 \log_{10} p^2/p_0$, $p_0 = 2 \times 10^{-5}$ Pa). Readers wishing to do further reading on more fundamental aspects of room acoustics should consult reference 1.

Noise and Vibration Control Engineering: Principles and Applications, Edited by Leo L. Beranek and István L. Vér.
ISBN 0-471-61751-2 © 1992 John Wiley & Sons, Inc.

7.1 NOISE SOURCES AND FREE-FIELD RADIATION

Sound sources radiate energy over some frequency range and, at each frequency, with a certain intensity. The rate of emission of energy is described by the sound power W in watts or the associated sound power level L_w in decibels ($L_w = 10 \log_{10} (W/W_0)$, $W_0 = 10^{-12}$ watts) usually measured in octave or one-third octave frequency bands. Depending on its physical characteristics and frequency, a source may radiate uniformly in all directions (omnidirectional source) or preferentially in certain directions (directional source) (see also Chapter 4).

Point Spherical Sources

In free space the sound pressure p due to a point spherical source of power W decreases with the distance between the source and the receiver, r in meters, due to spherical divergence according to

$$p^2 = \frac{W\rho c Q}{4\pi r^2} \quad (\text{Pa})^2 \tag{7.1a}$$

where ρ = density of air, kg/m^3
c = speed of sound in air, m/s
Q = source directivity factor (ratio of sound power radiated in receiver direction to average power radiated in all directions)

In terms of sound pressure level, this may be written as

$$\begin{aligned} L_p &= L_w + 10 \log_{10} \frac{Q}{4\pi r^2} \quad \text{dB } re \text{ } 20 \ \mu\text{Pa} \\ &= L_w + 10 \log_{10}(Q) - 20 \log_{10}(r) - 11 \quad \text{dB} \end{aligned} \tag{7.1b}$$

Note that sound pressure level decreases by 6 dB for each doubling of distance (dd) from the source.

Real sources are usually more complex than the idealized point source. Most real sources (large machines, conveyors, building walls) are extended in space in one or more dimensions. It is instructive to consider the behavior of some more complex sources.

Infinite Linear Array of Point Sources

If we have an infinitely long row of equal, incoherent, omnidirectional point sources of sound power level L_w on a straight line with a constant distance b meters between adjacent sources, then the sound pressure at a perpendicular distance r meters from the line is given by[2]

$$r < \frac{b}{\pi}: \quad L_p = L_w - 20 \log_{10}(r) - 11 \quad \text{dB} \tag{7.2a}$$

$$r > \frac{b}{\pi}: \quad L_p = L_w - 10 \log_{10}(r) - 10 \log_{10}(4b) \quad \text{dB} \tag{7.2b}$$

For small values of r, close to the source array, the sound pressure level decreases at 6 dB/dd. For large values of r, far from the array, the sound pressure level decreases at only 3 dB/dd.

Finite Extended Sources

Finite extended sources can be approximated by a finite array of point sources. The mean-square pressure at distance r from incoherent linear and planar sources of total sound power W and sound power level L_w are approximately as follows[2]:

Finite straight-line source of length d meters:

$$r < \frac{d}{\pi}: \quad L_p = L_w - 10 \log_{10}(r) - 10 \log_{10}(4d) \quad \text{dB} \tag{7.3a}$$

$$r > \frac{d}{\pi}: \quad L_p = L_w - 20 \log_{10}(r) - 11 \quad \text{dB} \tag{7.3b}$$

Close to the line source, the sound pressure level decreases at 3 dB/dd. Far from the source, the rate of decrease is 6 dB/dd as for a point source.

Finite planar source (dimensions bc with c > b, b and c in meters):

$$r < \frac{b}{\pi}: \quad L_p = L_w + 10 \log_{10} \frac{\pi}{4bc} \quad \text{dB} \tag{7.4a}$$

$$\frac{b}{\pi} < r < \frac{c}{\pi}: \quad L_p = L_w - 10 \log_{10}(r) - 10 \log_{10}(4c) \quad \text{dB} \tag{7.4b}$$

$$r > \frac{c}{\pi}: \quad L_p = L_w - 20 \log_{10}(r) - 11 \quad \text{dB} \tag{7.4c}$$

Near the source, the sound pressure does not decrease with increasing source–receiver distance. At greater distances, the planar source begins to look like a line source and the sound pressure level decreases at 3 dB/dd. Still further away, the planar source looks like a point source and the sound pressure level decreases at 6 dB/dd.

Multiple Sources

Where a room contains more than one incoherent source, the mean-square sound pressure at a point is the sum of the contributions p_i^2 of the individual sources; that is, $p_{\text{tot}}^2 = \Sigma\, p_i^2$. The relation between the total sound pressure level and the individual source sound pressure level contributions L_{pi} is $L_{p,\text{tot}} = 10 \log_{10}(\Sigma\, 10^{L_{pi}/10})$ in decibels.

Reaction of Nearby Surfaces on Source

If an omnidirectional source of constant volume–velocity (velocity times source area) is located near one or more reflecting surfaces, the source radiates into a smaller space and the sound power it radiates appears to increase.[3] The effect increases with decreasing source–surface distance and is significant when the distance from the surface to the source is less than half a wavelength at the frequency of interest. For a source located near a hard, plane surface, the sound radiates only into a halfspace and the sound pressure level in the halfspace increases by 3 dB (the directivity factor Q doubles). Similarly, for a source next to a two- and three-dimensional corner, the sound pressure level increases by 6 and 9 dB, respectively. The corresponding increases in directivity factor Q are 4 and 8. Practical sources may not behave like constant volume-velocity sources; while there may be increases in sound power due to this effect, the increases may be less than the values above. This effect must, in any case, be taken into account for equipment, fans, ventilation outlets, and so on, located near room surfaces.

7.2 FACTORS CONTROLLING ROOM SOUND FIELDS

Propagation of Sound

Sound waves propagate away from a source, being reflected (or scattered) by room boundaries, barriers, and furnishings. The resulting sound field at the receiver is composed of two parts. Sound that propagates directly from the source to the receiver is direct sound. Its amplitude decreases with increasing distance from the source as in a free field, as previously discussed. Sound that reaches the receiver after reflection from room surfaces or contents is reverberant sound. Its amplitude decreases with increasing source-receiver distance at a rate that depends on the acoustical properties of the room and on the type of source, as previously discussed. To examine sound propagation in rooms independent of source sound power, we define the sound propagation function, $\mathrm{SPF}(r) = L_p(r) - L_w$ in decibels, as the variable characterizing the effect of the room alone on the sound field. Sound pressure levels may then be obtained using $L_p(r) = \mathrm{SPF}(r) + L_w$. Thus, for a point source in a free field with $Q = 1$, Eq. (7.1b) yields $\mathrm{SPF}(r) = -20 \log_{10}(r) - 11$ dB. In the case of multiple sources, the individual source contributions are obtained from the sound power level of each source and the sound propagation functions for the room at the appropriate source–receiver distances for each source; the total equals the mean-square sum of the individual contributions.

Decay of Sound

Shortly after a source begins to radiate, an equilibrium (or steady state) is established between the rate of energy absorption in the room and the rate of energy emission by the source. If the source then ceases to radiate, the energy in the room decreases with time at a rate determined by the rate of energy absorption. This sound decay is usually characterized by the reverberation time T_{60}, the time in seconds required for the sound pressure level to decrease by 60 dB. Calculations of T_{60} are based on the average rate of decay over some part of the decay curve, preferably the first 20–30 dB.

Air Absorption

Energy is absorbed continuously as sound propagates in air. This process follows an exponential law $E(r) = E_0\exp(-2mr)$, where the constant $2m$ in the exponent is called the energy air absorption exponent, expressed in nepers per meter (1 Np = 8.69 dB). Air absorption depends on air temperature, relative humidity, ambient pressure and frequency. Table 7.1 presents some typical values calculated using reference 4 (see also Chapter 5).

Surface Absorption

The ability of a surface to absorb incident sound energy is characterized by the energy absorption coefficient α. This is the fraction of the energy striking the surface that is absorbed on reflection, usually measured in octave or one-third-octave bands. If E_i is the energy incident on a surface with absorption coefficient α, the reflected energy is $E_r = E_i(1 - \alpha)$. Chapter 8 discusses sound-absorptive materials and sound-absorber design in detail.

Surface Reflection

Acoustical energy not absorbed by a surface it strikes is reflected. If the surface is flat and homogeneous and large compared to the wavelength, reflection is specular—the angle of reflection is equal to the incident angle. In practice, because of

Table 7.1 Energy Air Absorption Exponents (2 m in Terms of 10^{-3} Np/m) Predicted Using Ref. 4, Assuming an Ambient Pressure of 101.3 kPa.

Temperature (°C)	Relative Humidity (%)	125 (Hz)	250 (Hz)	500 (Hz)	1000 (Hz)	2000 (Hz)	4000 (Hz)	8000 (Hz)
10	25	0.1	0.2	0.6	1.7	5.8	18.7	43.0
	50	0.1	0.2	0.5	1.0	2.8	9.8	33.6
	75	0.1	0.2	0.5	0.9	2.0	6.5	23.6
20	25	0.1	0.3	0.6	1.2	3.5	12.4	41.6
	50	0.1	0.3	0.7	1.2	2.3	6.5	22.4
	75	0.1	0.2	0.6	1.3	2.2	5.0	15.8
30	25	0.1	0.4	0.9	1.5	3.0	8.3	28.8
	50	0.1	0.2	0.8	1.7	3.0	5.8	16.4
	75	0.0	0.2	0.6	1.7	3.3	5.8	13.4

surface roughness as well as physical or impedance discontinuities, the energy is reflected into a range of angles.

Furnishings

The density and distribution of furnishings vary from room to room. Often furnishings occupy the lower region of a room (factory machines, classroom desks) and there are fewer objects at higher elevations. Of course, the horizontal distribution of room furnishings may vary considerably. Acoustical energy propagating in furnished regions is scattered, as well as partially absorbed, significantly modifying the sound field.

7.3 DIFFUSE-FIELD THEORY

By far the best known theoretical models for predicting room sound fields are based on diffuse-field theory. Diffuse-field theory is widely applied because of its simplicity. Often forgotten is the fact that it is of limited applicability because of its restrictive assumptions. The sound field in a room is diffuse if it has the following attributes: (1) at any position in the room, energy is incident from all directions with equal intensities (and with random phase, though we are ignoring phase in this chapter) and (2) the intensity of the sound field does not vary with receiver position.

Average Diffuse-Field Surface Absorption Coefficient

Diffuse-field theory uses the average rate of random-incidence surface sound absorption averaged over all the room surfaces. Thus, we define the average diffuse-field surface sound absorption coefficient as

$$\overline{\alpha}_d = \frac{\sum_i S_i \overline{\alpha}_{di}}{\sum_i S_i} \tag{7.5}$$

where S_i and $\overline{\alpha}_{di}$ are the surface area and diffuse-field absorption coefficient, respectively, of the ith surface.

If $\overline{\alpha}_d$ is to be useful, it is necessary that no part of the room be strongly absorbing, since in this case a diffuse sound field cannot exist. Absorptive objects such as seats, tables, and even people must be included when calculating $\overline{\alpha}_d$ despite the fact that such objects have ill-defined surface areas. It is common practice to assign a diffuse-field absorption, A_i in square meters, to each object, where $A_i = \overline{\alpha}_{di} S_i$ with S_i the surface area. The absorptions of all objects are summed and the total absorption is included in the calculation of $\overline{\alpha}_d$ with no modification made to the total area. In other words, the total area $\Sigma\, S_i$ in Eq. (7.5) is taken to be that of the room boundaries, excluding objects and people. In the case of closely spaced absorbers, such as auditorium seats or suspended ceiling baffles, caution is necessary since the total absorption may depend on the total area covered and may not be the sum of the absorptions of the individual objects. Also, the absorption coefficient of small patches of absorptive material will be higher than that of larger areas of the same material due to effects of diffraction at the material's edges.

Sabine and Eyring Theory

Several diffuse-field models exist. The most generally applicable is the Eyring theory, expressed by the following equations applicable for a *point source:*

Steady-state conditions:

$$L_p(r) = L_w + 10 \log_{10}\left(\frac{Q}{4\pi r^2} + \frac{4}{R}\right) \quad \text{dB} \tag{7.6}$$

Sound decay:

$$T_{60} = \frac{55.3V}{c[4mV - S \cdot \ln(1 - \overline{\alpha}_d)]} \quad \text{s} \tag{7.7}$$

where L_p = sound pressure level, dB *re* 2×10^{-5} Pa
L_w = source sound power level, dB *re* 10^{-12} W
T_{60} = reverberation time, s
c = speed of sound in air, m/s
R = room constant, = $[4mV - S \cdot \ln(1 - \overline{\alpha}_d)]/(1 - \overline{\alpha}_d)$, m^2
r = source-receiver distance, m
Q = directivity factor
$2m$ = energy air absorption exponent, Np/m
V = room volume, m^3
$\overline{\alpha}_d$ = average diffuse-field surface absorption coefficient
S = total room (surface and barrier) surface area, m^2

The Eyring theory is applicable to rooms with arbitrary average surface sound absorption coefficient. It should be used, for example, to determine absorption coefficients of room surfaces from measured reverberation times. If, however, the average surface absorption coefficient is sufficiently low (say <0.25) it is possible to use the Sabine theory; that is, Eqs. 7.6 and 7.7 with $-\ln(1 - \overline{\alpha}_d)$ replaced by $\overline{\alpha}_d$. Thus, the Sabine theory can be used when determining the sound absorption coefficient of materials in reverberation rooms. Of course, the theory applied in a particular sound-field prediction must also be consistent with that used to obtain the sound absorption coefficient data used in the prediction. Refer to reference 1 for further discussion of the relative merits of the Sabine and Eyring theories.

Diffuse-field theory predicts the following characteristics of the sound field:

1. *Steady State.* The total sound field is the sum of direct and reverberant components described, respectively, by the first and second terms in paren-

theses of Eq. (7.6). The direct field, which dominates near the source, is independent of the room's properties. Its sound pressure level decreases at 6 dB/dd. The reverberant field, which dominates far from the source, does not vary with source–receiver distance. Its sound pressure level is, to a first approximation (when both $\overline{\alpha}_d$ and m are small), inversely proportional to $\overline{\alpha}_d$ and to S. Under these conditions, the sound propagation function is as shown in Fig. 7.1.

2. *Sound Decay/Reverberation Time.* After the sound source ceases to radiate, the mean-square sound pressure at the receiver decays exponentially with time; the corresponding sound pressure level decreases linearly with time. To a first approximation, the reverberation time is directly proportional to the ratio $V/S\,\overline{\alpha}_d$.

Diffuse-field theory can also be used in the case of extended sources. The direct-field terms for a point source in Eq. (7.6) are simply replaced with the corresponding extended-source expressions from Eq. (7.2), (7.3), or (7.4).

Note that diffuse-field theory accounts for only some of the relevant room-acoustical parameters and accounts for some of these in an approximate manner. In particular, room geometry and source directivity are modeled only approximately. Neither the distribution of the surface absorption nor the presence of barriers or furnishings is modeled. This and the related fact that, as mentioned previously, the theory is based on restrictive hypotheses seriously limit its applicability. To give one concrete example, the constant sound pressure level predicted by diffuse-field theory at sufficiently large distances from sources is seldom found in practice; as a rule, noise levels due to a single sound source in a room decrease monotonically with distance from the source.

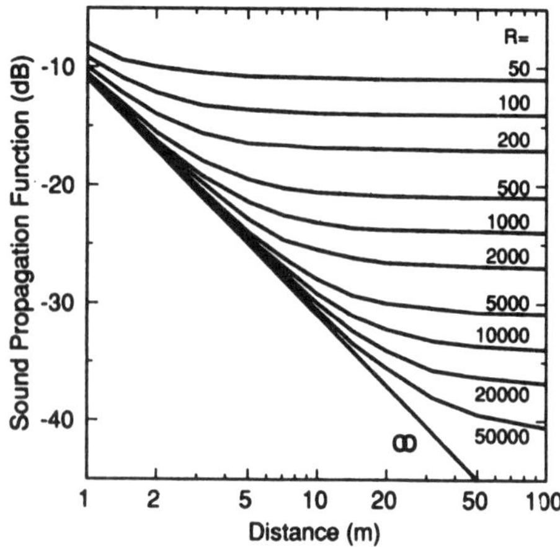

Fig. 7.1 Sound propagation function predicted by diffuse-field theory for various values of the room constant R (m^2) (assuming low surface and air absorption).

Diffuse-field theory, when applicable, can be used to estimate the reductions of noise level or of reverberation time that will occur when absorptive material is added to a room. Equation (7.6) is used to calculate the difference in noise levels before and after treatment. At positions near sources, where the direct field dominates, the reduction in sound pressure level is small. At large distances from all sources, where the reverberant field dominates, the reduction approaches that of the reverberant sound pressure level. This, to a first approximation (when both $\overline{\alpha}_d$ and m are small), is given by $10 \log_{10} (A'/A)$, where A and A' are the total room absorptions ($A = \overline{\alpha}_d S$) before and after treatment. Doubling the room absorption reduces the sound pressure level of the reverberant field by 3 dB. Similarly, using Eq. (7.7), the reverberation time after treatment is $T'_{60} = T_{60}(A/A')$. Doubling the acoustical absorption halves the reverberation time.

As an example, consider a room with dimensions $10 \times 5 \times 2.4$ m (volume $V = 120$ m^3, surface area $S = 172$ m^2) and all concrete surfaces with $\alpha(1 \text{ kHz}) = 0.05$. Thus, ignoring air absorption, we have $A(1 \text{ kHz}) = \alpha(1 \text{ kHz})S = 0.05 \times 172 = 8.6$ m^2 and reverberation time $T_{60} = 55.3 \times 120/(340 \times 8.6) = 2.3$ s. A suspended acoustical ceiling with $\alpha_c(1 \text{ kHz}) = 0.9$ and surface area of 50 m^2 is installed to improve the acoustical environment. After treatment $A' = [0.9 \times 50 + (172 - 50) \times 0.05] = 51.1$ m^2 and, to a first approximation, $T'_{60} = (A/A')T_{60} = 2.3(8.6/51.1) = 0.4$ s.

7.4 OTHER PREDICTION MODELS

When diffuse-field theory cannot be applied, more comprehensive models may be used to predict the steady-state sound pressure level and the sound decay in a room. These models apply only to point sources. Extended sources must be approximated by an array of point sources and the prediction model applied to each source. Two main approaches are available to accomplish this objective. These are computer calculations (method of images and ray tracing) and empirical models. The different approaches make assumptions that define and limit their applicability and accuracy. Further, they take the various room-acoustical parameters into account to a greater or lesser extent. Empirical prediction models are usually limited in application to specific building types and are discussed in subsequent sections.

Method of Images

The method-of-images approach is based on the assumption that reflections from surfaces, assumed specular, can be replaced by the direct sound rays of image sources. Figure 7.2 shows the simple example of a two-dimensional array of image rooms and sources.

The simplest implementation of the method of images applies to empty, rectangular-parallelepiped rooms with no barriers.[5] In this case, the image sources corresponding to the infinite number of reflections are located on a three-dimensional grid. The total mean-square pressure is the sum of the contributions of all

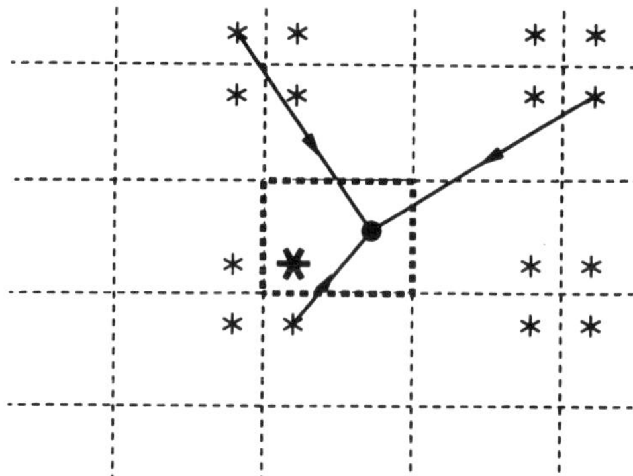

Fig. 7.2 Image space of an empty two-dimensional room, showing the source (*) and receiver (●), the image sources (*), the image surface planes (-----), and the propagation paths of sound from three of the image sources (—▶—).

the images, allowing for spherical divergence and energy losses due to absorption at each surface encountered by the ray from the image. For decaying sound, the sound pressure level at time t after the cessation of the sound source is calculated by summing the contribution of image sources located at distances greater than ct from the receiver.

The method of images can be extended to empty rooms of arbitrary shape bounded by planes.[6] However, calculation times are greatly increased and may become impracticable. The method of images can also account for the presence of furnishings under the assumption that they are isotropically distributed in random fashion throughout the room.[7,8] In this case, the furnishings, as well as the sources, are mirrored in the room surfaces. The steady-state and decaying sound fields are formed as before. However, the contribution of each image source is modified due to scattering and absorption by the furnishings; absorption at surfaces is also increased due to scattering.

Ray Tracing

Ray-tracing techniques[9] can be used to predict sound fields in empty or furnished rooms of arbitrary shape with arbitrary surface absorption distributions, different surface reflection properties, and variable furnishing density. Source directivity can also be modeled. A computer program simulates the emission of a large number of rays or cones from each source in either random or deterministic fashion. Each ray is followed as it propagates in the room, being reflected and scattered by surfaces, barriers, and furnishings until it reaches a receiver position. The energy of a ray is attenuated according to spherical divergence, as well as surface, furnishing, and air absorption. In principle, any surface reflection law (specular, diffuse) can be modeled. Ondet and Barbry[10] proposed an algorithm for accounting for quasi-arbitrary furnishing distributions in a ray-tracing model.

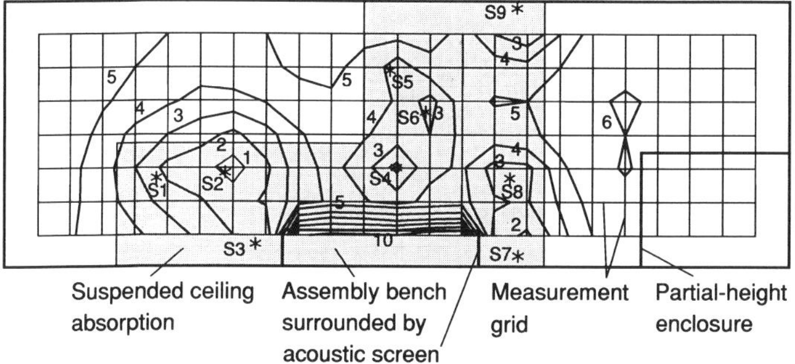

Fig. 7.3 Contour map of the predicted reduction of A-weighted sound pressure level in a furnished workshop (dimensions 48.0 m × 16.0 × 7.2 m) that results from the introduction of an acoustical barrier around a noise-sensitive assembly bench and of a ceiling-absorption treatment suspended over the region of the sources. The map was produced from noise levels predicted before and after treatment for positions at the interception points of an imaginary horizontal 2 × 2-m grid 1.5 m above the floor as shown. Also shown are the noise sources (e.g., S4*) and the contour lines and values (e.g., −5). (Reproduced with permission from ref. 11.)

Figure 7.3 illustrates the potential of ray-tracing techniques. It shows the contour map of the predicted reduction of A-weighted sound pressure level over the floor of a furnished workshop that results from the introduction of an acoustical barrier around a noise-sensitive assembly bench and of a ceiling-absorption treatment suspended over the region of the sources. This kind of detailed analysis is not possible using diffuse-field theory. It would only be possible to estimate noise reductions from the increase of absorption. These reductions could then be corrected as a function of the proximity of the receiver to the acoustical treatment and of the estimated insertion loss of the barrier.

7.5 DOMESTIC ROOMS AND CLOSED OFFICES

Schultz[12] studied the prediction of sound pressure levels in domestic rooms and closed offices. He measured the variation of sound pressure level with distance from a source in a variety of small rooms, observing that levels never become constant at larger distances as predicted by diffuse-field theory. In fact, he found that the curves always had a slope of about −3 dB/dd, though the absolute level of the curves varied considerably. He further found that existing models developed for predicting sound propagation in factories and corridors, for example, could not predict his experimental results. Schultz therefore proposed the following empirical formula:

$$L_p(r) = L_w - 10 \log_{10}(r) - 5 \log_{10}(V) - 3 \log_{10}(f) + 12 \quad \text{dB} \qquad (7.8)$$

where f is the frequency in hertz. Note that this formula does not explicitly contain a room-absorption term. Such behavior contrasts markedly with that predicted by diffuse-field theories and is an excellent example of their limitations.

7.6 INDUSTRIAL BUILDINGS

Industrial buildings represent rooms that often have serious acoustical problems due to excessive noise levels and reverberation. Most countries have regulations aimed at limiting the risk of hearing damage by limiting the noise exposure of factory workers. Excessive reverberation can lead to poor verbal communication or a reduced ability to identify warning signals, and thus to danger, as well as to stress and fatigue.

Acoustical Properties

Factory buildings come in every shape and size. However, many are rectangular in plan (with widely varying dimension ratios and plan sizes) with flat or nonflat (e.g., pitched or sawtooth) roofs. The floors of most factories are made of concrete. Their walls are often of brick or blockwork, sometimes of metal cladding. The cladding may be a single profiled-metal layer or, if providing thermal insulation, a double-panel (insulation or air between the impervious faces). Factory roofs are usually of suspended-panel construction consisting of metal or other panels supported by metal trusswork or portal frames. A common newer construction is the steel deck, consisting of, for example, profiled metal inside, a vapor barrier, several centimeters of thermal insulation, tar paper, and ballast outside, again supported by metal trusswork. Acoustical steel decks exist; the inner metal layer is perforated and its profiles filled with absorptive material, providing high acoustical absorption over a broad range of frequencies. Acoustical decks can support loads almost as great as normal decks and cost only about 10% more.

Hodgson[13] has estimated average sound absorption coefficients of the surfaces of untreated factory buildings. He found the coefficients varied little between buildings of similar construction and little at all at mid- and high frequencies. Typical values for $\overline{\alpha}_d$ are 0.06–0.08 for the 500–4000-Hz octave bands. The variations with construction, and of sound absorption coefficients, are greatest at low frequency due to the vibration and transmission characteristics of the suspended-panel roofs. Typical values for $\overline{\alpha}_d$ are 0.08–0.16 for the 125- and 250-Hz octave bands. Note that these are average sound absorption coefficients for all of the building surfaces including the concrete floor; if the high values at low frequency are attributable to the roof alone, then the ceiling absorption coefficient attains 0.3 in some cases.

Figures 7.4*a* and 7.4*b* show, respectively, reverberation times and 1000-Hz sound propagation functions measured in an empty, untreated factory with average dimensions 45 × 42.5 × 4 m high and a double-panel roof.[14] As is often the case, the sound propagation function at most source-receiver distances, and the rever-

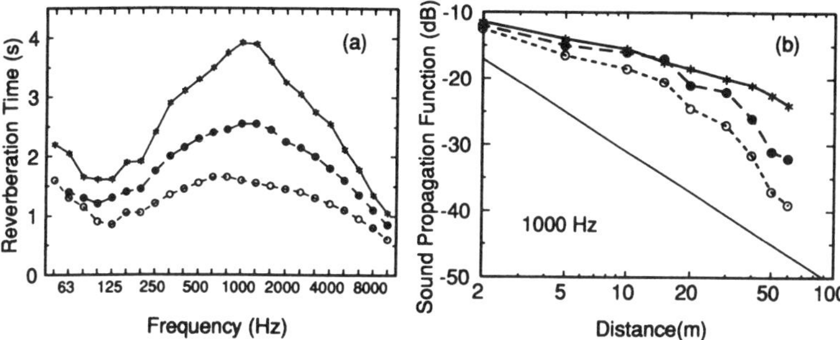

Fig. 7.4 Measured (*a*) third-octave-band reverberation times; (b) 1000-Hz sound propagation curves in a factory when empty (*) and containing 25 (●) and 50 (○) metal machines and, for reference, for a point source in a free field (——). (Reproduced with permission from ref. 14.)

beration time, are found to be highest at midfrequencies. Both variables decrease at low and high frequency due, respectively, to increased roof and air absorption.

The effect of factory furnishings is also illustrated in Figs. 7.4*a*,*b*, showing the sound propagation functions and reverberation times after first 25 and then an additional 25 printing machines were introduced. These metal machines had average dimensions of 3 × 3 × 2 m high. Introduction of the furnishings significantly decreases reverberation time and the sound propagation function. According to diffuse-field theory, the decreases of reverberation time for this particular building correspond to the introduction of as much as 400–900 m^2 of acoustical absorption.[14] The percentage change of reverberation time and the magnitudes of the changes of sound propagation function with increasing furnishing density vary little with frequency. These results are supported by similar measurements in other factories.

Simplified Models for Factory Noise Prediction

While ray tracing is probably the most generally accurate method for predicting factory noise levels, it has the disadvantage of involving long calculation times for factories of complex shape and/or containing many noise sources. There is considerable practical need for simplified prediction methods. Several such methods have been developed. We present three of them and emphasize that *they have not been generally validated.*

Wilson[15] proposed a simple method for predicting the slope of the dBA sound-propagation-function curve in factories with width and length at least four times the height. It allows approximate initial estimates to be made quickly. It is based on his experimental observations of the extent to which the slope of the sound-propagation-function curve changes with increased absorption and/or furnishings. The slope is assumed to be constant with the following values: −3 dB/dd in acoustically hard, empty factories; −4 dB/dd in acoustically hard, furnished factories or in absorbent-lined, empty factories; and −5 dB/dd in absorbent-lined factories with furnishings.

Friberg[16] developed an empirical formula, based on measurements in many factories, for predicting the slope, assumed constant, of the A-weighted sound-propagation-function curve. It applies only to sources with fairly flat sound power spectra. It can be applied to long factories of any shape. The factory furnishings are assumed to be located on the floor and to have heights equal to the average furnishing height. The surface absorption is quantified by the ceiling absorption coefficient at "around 1000 Hz" (α'). The slope D (in dB/dd) is given by $D = a\alpha' + b$, where a and b are tabulated constants whose values depend on the furnishing density, as determined by qualitative descriptors, and on the room shape, as shown in Table 7.2. The Friberg model involves more calculation effort than the Wilson model but would be expected to be more accurate.

Zetterling[17] proposed an even more comprehensive and potentially more accurate model to predict the reduction of A-weighted sound pressure level relative to the level at 1 m from a single source in factories of any shape. First, the *acoustical quality of the room* is quantified by a *total score* using Figs. 7.5*a–c*, which relate this quantity to (1) the room volume; (2) the ceiling height and average midfrequency absorption coefficient; and (3) the room width and average midfrequency wall absorption coefficient, respectively. The attenuation is then determined from

Table 7.2 Constants Required for Predictions According to Friberg Model[a]

Room category	a	b
BH	−3.0	−4.0
BM	−2.5	−3.75
BL	−2.0	−3.5
NH	−3.0	−3.0
NM	−2.75	−2.75
NL	−2.5	−2.5

[a]From ref. 16. The room shape and contents are categorized as follows (h = average furnishing height, H = room height, W = room width):

Room shape (left-hand letter):

N = rooms with $W < 4H$
M = rooms with $4H < W < 6H$
B = rooms with $W > 6H$

Room contents (right-hand letter):

L = rooms with zero or low furnishing density, or densely furnished with $h < \frac{1}{8}H$
M = rooms with medium density of high furnishings, or high density of furnishings with $\frac{1}{8}H < h < \frac{1}{4}H$
H = rooms with high density of furnishings with $h > \frac{1}{4}H$

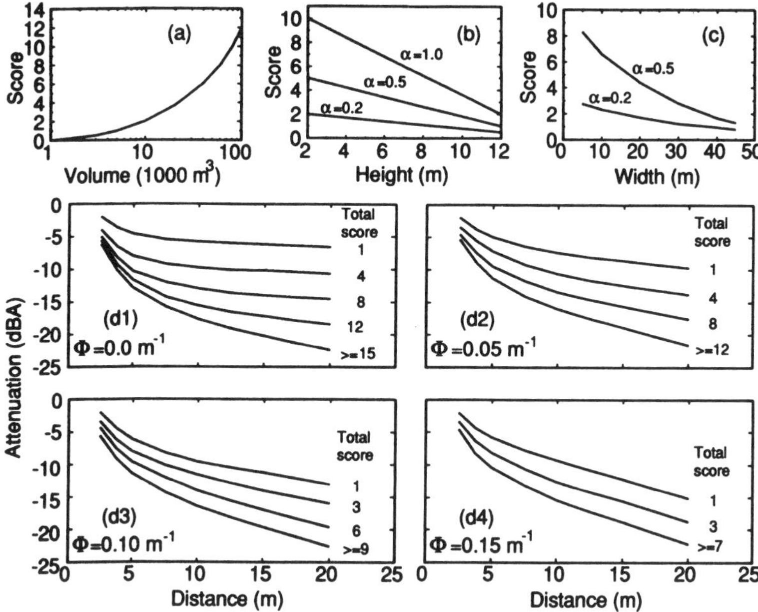

Fig. 7.5 Graphs used to predict factory noise levels by the Zetterling empirical model[17]. A total score is determined from (*a*)–(*c*). This, the estimated furnishing density Φ, and the source–receiver distance are used to determine the attenuation (of level relative to that at 1 m from a source) from the appropriate version of (*d*) (or by extrapolating between them). (Reproduced with permission.)

the total score, the source-receiver distance, and the estimated furnishing density using the appropriate version of Fig. 7.5*d* (or extrapolating between them).

As an example, let us use the Zetterling model to calculate the A-weighted sound pressure levels at a height of 1.5 m at the center of the workshop illustrated in Fig. 7.3 before and after introduction of an absorptive ceiling treatment. The workshop dimensions are length 48.0 m, width 16.0 m, and height 7.2 m. According to Fig. 7.5*a*, the room volume of 5000 m^3 gives a volume score contribution of 1. Assume the average midfrequency absorption coefficients of the surfaces are 0.07 for walls, 0.07 for untreated ceiling, and 0.9 for treated ceiling. Figure 7.5*b* gives ceiling score contributions of 0.5 (untreated) and 5.5 (treated). Figure 7.5*c* gives a wall score contribution of 1. Thus the total scores are 2.5 (untreated) and 7.5 (treated). The workshop contains nine noise sources located as shown by asterisks in Fig. 7.3 and with A-weighted sound power levels in Table 7.3. The further assumption that the sources are compact and omnidirectional and thus that $L_p(1 \text{ m}) = L_w - 20 \log_{10}(1 \text{ m}) - 11 = L_w - 11$ dB leads to levels at 1 m from each source (with only that source running), as shown in Table 7.3. From these levels, the source-room center distances shown in Table 7.3, the total scores, and the attenuations, the A-weighted levels at the room center due to each source are calculated. In Table 7.3, these levels were determined using Fig. 7.5*d*3,

Table 7.3 Details of Calculation of Noise Levels in Workshop before and after Acoustical Treatment Using the Zetterling Prediction Model

				Untreated		Treated	
Source Number	L_w (dBA)	L_p (1 m) (dBA)	Distance d (m)	Attenuation (dBA)	L_p (d) (dBA)	Attenuation (dBA)	L_p (d) (dBA)
1	84	73	15	−13	60	−18	55
2	91	80	11	−11	69	−15	65
3	82	71	11	−11	60	−15	56
4	84	73	2	−2	71	−5	68
5	76	65	4	−5	60	−9	56
6	80	69	2	−2	67	−5	64
7	81	70	10	−11	59	−15	55
8	79	68	7	−8	60	−12	56
9	83	72	10	−11	61	−15	57
Total A-weighted sound pressure level					75		72

assuming the workshop contained a moderate density of furnishings ($\Phi = 0.1$ m^{-1}). Total levels are determined by energy-based addition of the individual source contributions. The result is 75 dBA before treatment and 72 dBA after treatment indicating that introduction of the ceiling treatment is expected to reduce the A-weighted sound level at the room center by 3 dB.

Several comments concerning the above methods must be made. First, the Wilson and Friberg methods predict only the slope of the sound-propagation-function curve, providing no information as to how its absolute level, necessary for prediction, should be determined. It is necessary to set the level so as to predict levels equal to those in the free field at some distance [e.g., 1 m, for which SPF(1 m) = −11 dB] from a source. Second, these methods assume the slope to be constant. A glance at most of the predicted or measured sound-propagation-function curves presented in this chapter will convince the reader that this is not generally an accurate assumption; in fact, a double-slope curve would be more accurate. However, if the models accurately predict the short-distance or initial slope, they should be fairly accurate for factories in which the average source–receiver distance is small. Third, the models rely on either a quantitative or a qualitative description of the factory furnishing density. It is not yet known how to determine this quantity accurately. It can be estimated as the total surface area of the furnishings (or, more practically, of imaginary boxes that would fit around the individual objects) divided by four times the volume of the furnished region.[8] However, there is some evidence that this procedure underestimates the correct value.[11] Alternatively, since the furnishing density is the inverse of the mean free path between the objects, it can be estimated from the inverse of the mean distance between all pairs of objects, including those that are the (infinite number of) images of the real objects in the building's bounding surfaces. Experience has shown that average factory furnishing densities may attain 0.15 m^{-1}; those of a factory's furnished regions may attain 0.3 m^{-1}. In any case, no simplified model will ever predict noise levels in factories of complex shape, furnishing distributions, containing barriers, and so on, with acceptable accuracy.

Factory Noise Control

To obtain maximum cost efficiency and to comply with specific industrial noise regulations, the reduction of factory noise levels should be achieved by the application of noise-control principles in the following order of priority:

1. *Control at Source.* Reduce the sound power output of the equipment by design or acoustical treatment.
2. *Control of Direct Field.* Isolate receiver positions from noisy sources by increasing the distance between them and by the use of source enclosures (see Chapter 13), control rooms, and screens. Often separation can be achieved by appropriate planning of the factory layout, taking full advantage of the building geometry, natural barriers such as stock piles, and furnishings.
3. *Control of Reverberant Field by Applying Sound-absorptive Materials to Room Surfaces.* Such treatments should be accorded lower priority since they tend to be expensive and not very effective near noise sources, such as at typical operator positions. In practice, reductions of 0–6 dBA are possible. The best surfaces to treat are those closest to noise sources and/or receiver positions. In low-height industrial rooms, this usually implies the ceiling; treatments consist of sound-absorptive materials applied to or suspended from the roof. For example, acoustical baffles (rectangular pieces of sound-absorptive material) can be hung as flags in appropriate patterns at appropriate densities. Difficulties may arise due to interference with lighting and sprinkler systems and with respect to fire regulations and cleanliness requirements. The treatment of walls may be warranted in enclosures with three dimensions fairly equal and when noise sources are located close to walls, where it is important to absorb the strong wall reflections. Note that in factories of any shape, surface absorption, even when it has little effect on noise levels, may significantly reduce reverberation.

When dealing with multiple sources in factories, the effect of noise-control measures may be determined to a first approximation by considering what happens at a distance corresponding to the *average* source-receiver distance. Changes in total noise levels due to noise-control measures are approximated by changes in the sound propagation function for a single source at this distance. For machine operator positions, the average source-receiver distance is small—typically 1–2 m. For a room with a more-or-less-square floor plan, and with sound sources uniformly distributed over the floor, the average distance between all possible source-receiver positions is about one-half the average horizontal dimension of the room.

7.7 OPEN-PLAN OFFICES

An open-plan office is a large space that accommodates many seated workers separated by low barriers that provide visual separation and acoustical isolation between work stations. The barriers may be free-standing or integrated into the furniture. The ceiling and the carpeted floor form two large, extended, sound-absorptive planes; their horizontal dimensions are much greater than the height of the ceiling. The space between these surfaces is filled with office furniture and barriers, usually sound absorptive. Diffuse-field theory certainly does not apply in such a space; the sound pressure level decreases continuously with distance from sources. The principal problem in an open-plan office is not propagation to large distances but the provision of privacy between neighboring work stations. Short-range sound paths between neighboring work stations are most important. Sound emanating from one work position reflects from extended surfaces (ceiling, walls, and windows) and bends over or around the edges of barriers, as illustrated in Fig. 7.6. These sound paths must be controlled to provide reasonable privacy for workers.

Open-plan offices can provide a reasonable degree of acoustical privacy if they are carefully designed as a system and if adjacent work functions are compatible and not too close together. Special protection must be provided, however, against noisier devices such as printers and copiers, for example, by locating them in a shielded area. Conference rooms with full-height partitions providing good sound insulation should be available for activities that require low background noise levels and high privacy.

Open-Plan Office Barriers

Barriers (also called screens, partial-height partitions, and office dividers) provide sound attenuation and visual privacy between work stations for seated persons. Systems furniture combines the functions of the barrier and amenities such as stor-

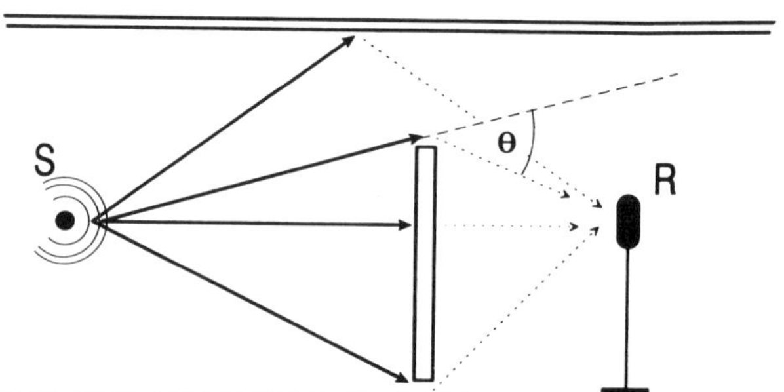

Fig. 7.6 Possible sound paths between work stations in an open-plan office cross section. Reflection from the ceiling is reduced by making it very absorptive, transmission through the barrier is reduced by selecting a large enough weight, and diffraction losses over the barrier are increased by increasing barrier height.

age compartments, lighting, power, communications, and desk or work surfaces into a single unit. A barrier should attenuate sound that passes through it so the transmitted sound is negligible.

Sound diffracts over the top of the barrier to reach the next work station. For an infinitely wide barrier between a source and a receiver position, the insertion loss IL in decibels relative to the level at the receiver in the absence of the barrier may be approximated by[18]

$$IL = 13.9 + 7 \log (N) + 1.4 (\log N)^2 \qquad N \geq 0.001 \tag{7.9}$$

where $N = 2f(A + B - d)/c$, d is the straight-line distance from the source to the receiver in meters, A is the distance from the source to the top of barrier in meters, and B is the distance from the top of the barrier to the receiver in meters. Barrier attenuation is treated in more detail in Chapter 5. The greater the angle through which sound has to bend to reach the receiving position on the other side of the barrier (θ in Fig. 7.6), the greater will be the insertion loss due to the barrier. Thus, high barriers are more effective than low ones, and barriers placed close to the speaker or listener are more effective than those equidistant from each.

The effects of diffraction around and transmission through the barrier must be combined to determine the total insertion loss of a barrier. Sound transmitted through the screen should be negligible relative to the sound diffracted around it—especially at those frequencies important for speech intelligibility. Specifying that the normal-incidence sound transmission loss at 1000 Hz should be 6 dB greater than the theoretical insertion loss due to diffraction is a satisfactory criterion. This criterion leads to the requirement that the minimum mass per unit area of the barrier ρ_s (in kg/m^2) should be $\rho_s \geq 2.7(A + B - d)$. The total effect of the barrier can be calculated by applying Eq. (7.9) to each edge in turn and then summing the acoustical energies.

The maximum barrier dimensions are usually limited by physical convenience, possible interference with air flow, and preservation of the open look. Recommended minimum dimensions are height 1.7 m and width 1.8 m. Barriers can be joined to increase the effective length. If the attenuation of a high barrier is desired but visual openness must be maintained, a plate of glass or transparent plastic may be fitted to the top of a low barrier to increase its height. The gap between the bottom edge of the barrier and the floor should be small, otherwise sound is transmitted under the barrier to the opposite side. This is not critical, however, as sound following this path tends to be diffused or absorbed by furniture and carpets; a gap of 100 mm or so may be left to allow floor cleaning. A standard test method for evaluating barriers for use in open-plan offices is available.[19]

Open-Plan Office Ceilings

To ensure that the full benefit of the barriers between work stations is obtained, the ceiling assembly must be highly sound absorptive, especially at the frequency bands important for determining speech intelligibility (500 Hz to 4 kHz). A specific test procedure is available for the evaluation of ceiling systems for use in open-plan offices.[20] The results are summarized by a single-number rating called articulation class *AC*.[21] The rating is designed to correlate with transmitted speech intelligibility between office spaces and is based on articulation index[22] (see also Chapter 16). Articulation class information may not always be available, in which case sound absorption coefficients have to be used to evaluate or specify ceilings.

Figure 7.7 shows calculated articulation-class values for sound propagating between work positions for different heights of barriers and for ceilings having different sound absorption coefficients. An open-plan office ceiling should have an articulation class of 200 or more. This corresponds to an attenuation of 20 dBA for pink noise propagating between work positions. In Fig. 7.7, this value is only reached for average ceiling sound absorption coefficients greater than 0.9. The average absorption coefficient may be taken as approximately equal to the noise reduction coefficient *NRC* (the average of the absorption coefficients at 250, 500, 1000, and 2000 Hz, rounded to the nearest 0.05). Increasing the height of a barrier increases sound attenuation significantly only when the sound absorption of the ceiling is high.

Flat plastic lighting fixtures in the ceiling can reflect considerable amounts of sound and can seriously reduce the sound isolation between adjacent work positions; they should be avoided or their number kept to a minimum. Open-grid fixtures or those that scatter sound are preferred. Small lamps can provide additional illumination at individual desks as necessary.

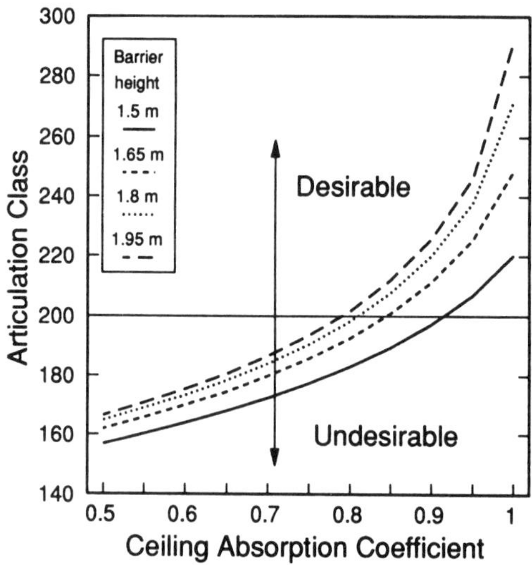

Fig. 7.7 Calculated articulation class for two work stations 4 m apart separated by a barrier midway between them. The ceiling is assumed to have the same absorption coefficient at all frequencies in the range 200–5000 Hz. The benefits of higher barriers are obtained only when the ceiling is highly absorptive. The effects of attenuation due to distance were included in the calculation.

Application of Sound-absorptive Materials to Vertical Surfaces

Sound propagating in the horizontal plane and bypassing barriers by reflection from vertical surfaces can reduce the sound attenuation between work positions. To prevent this, surfaces such as walls, office barriers, square columns, backs of cabinets, and systems furniture, as well as bookcases, should be covered with sound-absorptive material having a noise reduction coefficient of 0.7 or higher. A thickness of 2.5 cm or more of glass fiber or mineral wool with a porous fabric cover will satisfy this requirement. The application of carpets directly to hard surfaces is not an effective solution, since typical carpets have low sound absorption coefficients. Round columns with diameters of 0.5 m or less may be left uncovered. A simple way of avoiding wall reflections is to avoid gaps between the wall and barriers. To prevent reflections from their surfaces, office barriers are usually covered on both faces with sound-absorptive material (such as glass fiber) covered with fabric. Such barriers have a noise reduction coefficient of about 0.8; reflections from them are substantially reduced. A standard test method for evaluating the sound-reflective properties of wall finishes and panels is available.[23]

Reflections from Windows

It is difficult to prevent sound from propagating between work positions by reflecting from nearby closed windows. This problem cannot easily be solved using drapes, which would need to be heavy and closed; also, most slatted blinds do not reduce reflections. Using the area next to the windows as a corridor reduces this problem; barriers positioned so as to define the corridor block the reflected sound. If work positions must be located next to windows, additional panels blocking the gap between the edge of the barrier and the glass can be used to reduce reflections.

Masking Sound

Open offices are highly absorptive environments and typically have very low background sound levels and, hence, reduced speech privacy. A masking-sound system should be used to generate a uniform background sound field that does not fall below a predetermined value throughout the open office area. An array of loudspeakers above the ceiling radiates a random noise with an appropriate spectrum to the office below. The increased background sound makes it more difficult to hear speech from adjacent work positions.

The level, spectrum, and spatial variation of the masking sound require careful adjustment to ensure occupant acceptance. A masking spectrum is usually an approximately straight line with slope of about −5 dB/octave. Sound levels at high and low frequencies may be reduced relative to this straight line to increase occupant acceptance of the sound. The overall A-weighted sound level is usually in the range of 45–50 dB. Higher levels may induce complaints and tend to defeat the purpose of the masking since the occupants compensate by speaking in a louder voice—even at close range.

Spatial and temporal variations[24] of masking noise should be less than 3 dBA. Since sound from the loudspeakers must pass through the ceiling panels, the transmission loss of the ceiling affects the uniformity of the sound field in the office below. Transmission through lighting fixtures and other ceiling elements can lead to localized areas of high sound level below and to occupant complaints. The installation of masking-noise systems is best left to experienced professionals.

Speech Privacy in Existing Open-Plan Offices

For the evaluation of the degree of speech privacy in an existing office or a mock-up of a proposed office, a standard test procedure[25] has been written. Sound propagation between work positions is measured using sound generated by a standard loudspeaker and speech privacy is calculated using the articulation index.[21]

The example that follows illustrates the calculation of speech privacy between two work positions separated by a screen and shows how reflections from the ceiling and from a nearby wall can drastically reduce speech privacy. Speech privacy is expressed in terms of articulation index *AI*. An *AI* of 0.05 or less may be considered to represent confidential privacy; $0.05 \leq AI \leq 0.15$ represents acceptable privacy in the open office. In this example, the following physical parameters were used: The speaker and listener are both seated 2 m from the screen, with head heights of 1.2 m; the screen height is 1.65 m and the ceiling height is 2.4 m. A wall runs parallel to the line joining the speaker and the listener at a distance of 1.25 m. The absorption coefficient of the ceiling is assumed to be 0.85, constant with frequency; that for the wall is taken as 0.5. These values correspond to reductions of reflected energy of 8.2 and 3 dB, respectively. Attenuations due to distance by way of the ceiling and wall-reflection paths are both equal to 13.5 dB. The details of the geometric calculations are left to the reader. Table 7.4 gives the details of the rest of the calculation. Voice levels and *AI* weighting factors are taken from reference 25. The masking-noise spectrum has an A-weighted level of 45 dB. To simplify the calculations, diffraction at other screen edges is ignored. The calculations show that, in this case, confidential privacy is only achieved in the absence of the ceiling and wall. Further efforts are needed to reduce the strength of reflections from these surfaces.

7.8 REVERBERATION ROOMS

Reverberation rooms are designed and equipped to give a close approximation to a diffuse sound field. Measurements in these rooms are used to characterize the sound-absorptive properties of materials,[26,27] the sound power of sources,[28,29] and sound transmission through building elements,[30–32] among other things. A typical reverberation room has a volume of about 200 m^3 or more; some are constructed with nonparallel walls. The walls and all surfaces in the room are made highly reflective so that reverberation times are long and the region dominated by the direct field of sources is as small as possible.

Table 7.4 Details of Calculation of Articulation Index in Idealized Open-Office Situation[a]

One-third-Octave	Masking		AI Screen		Screen only		Ceiling + Screen			Ceiling + Wall + Screen		
Frequency (Hz)	V_0	45 dBA	Weight	IL	V_1	dAI	V_2	$V_1 + V_2$	dAI	V_3	$V_1 + V_2 + V_3$	dAI
200	60	44	0.0004	8	39	0.000	38	41	0.000	43	45	0.001
250	64	43	0.0010	9	42	0.000	42	45	0.002	47	49	0.006
315	63	41	0.0010	9	41	0.000	41	44	0.003	46	48	0.007
400	65	40	0.0014	10	42	0.004	43	46	0.009	48	50	0.015
500	66	38	0.0014	10	43	0.007	44	46	0.012	49	51	0.018
630	64	36	0.0020	11	40	0.008	42	44	0.016	47	49	0.025
800	58	35	0.0020	11	34	0.000	36	38	0.007	41	43	0.016
1000	58	33	0.0024	12	33	0.001	36	38	0.012	41	43	0.024
1250	59	31	0.0030	12	34	0.008	37	38	0.022	42	44	0.038
1600	56	29	0.0037	13	30	0.003	34	35	0.023	39	41	0.043
2000	52	27	0.0038	14	25	0.000	30	31	0.016	35	36	0.036
2500	53	25	0.0034	15	25	0.003	31	32	0.025	36	37	0.044
3150	53	22	0.0034	15	25	0.010	31	32	0.034	36	37	0.053
4000	50	19	0.0024	16	21	0.005	28	29	0.024	33	34	0.038
5000	46	16	0.0020	17	16	0.001	24	24	0.018	29	30	0.029
Total AI						0.05			0.22			0.39
Privacy					Confidential		Higher than acceptable			Higher than acceptable		

[a]*Note:* V_0 is voice level at 1 m, V_1 is level due to diffraction over screen, V_2 is level due to reflection from ceiling, V_3 is level due to reflection from wall. These levels are combined by summing energies. Screen insertion loss is calculated using Eq. (7.9).

At low frequencies, the frequency response of reverberation and other rooms to wide-band noise shows several peaks corresponding to room modes. As frequency increases, the spacing between modes becomes less, the modes begin to overlap and the individual modes are less obvious. At some transition frequency the room response for bands of noise becomes approximately constant, the properties of the sound field become more uniform, and the room response may be described in statistical terms. This transition frequency is usually defined as the Schroeder frequency,[33] $f_S = 2000(T_{60}/V)^{1/2}$ hertz. For a 250-m^3 room with $T_{60} = 5$ s, $f_S = 282$ Hz.

To make the response of the room more uniform at low frequencies, it is usually advisable to add low-frequency sound-absorptive elements. Even in rooms with a large enough volume, correctly chosen dimensions and the recommended amount of sound absorption, the spatial variances of pressure and sound decay rate are often too large to satisfy precision requirements in standards. It is common to add rotating diffusers as well as fixed panels suspended at random positions and orientations throughout the room to perturb the room modes. Rotating diffusers are very effective in decreasing the time-averaged spatial variance of measured data whether it be of sound pressure level or sound decay rate. Fixed diffusing panels do little to change the variance of sound pressure level or of decay rate but are usually necessary for sound absorption measurements. Despite such measures, it is still necessary to sample the sound pressure at many statistically independent locations within the room to measure mean-square sound pressure and decay rate correctly.

It has been shown[3] that close to one, two, or three intersecting infinite reflecting planes (edges and one- and two-dimensional corners) the sound pressure and energy increase. The mean-square pressure p^2 can be expressed as

$$p^2 = 1 + \sum_{n=1}^{N_{\text{im}}} \frac{\sin(kr_n)}{kr_n} \quad \text{Pa}^2 \tag{7.10}$$

where $k = 2\pi/\lambda$ (λ = wavelength)
p^2 = normalized to unity in absence of images or far from reflecting surfaces
r_n = distances from the images of the source point to measurement point
N_{im} = number of images: 1 for measurement near a plane surface, 3 near an edge, and 7 near a corner

Closer than $\lambda/2$ to highly reflecting surfaces, the sound pressure increases significantly because of these interference effects. To allow for the increase in sound energy close to plane surfaces, an adjustment term $1 + S\lambda/8V$ is included in the calculation of the sound power W from the mean-square sound pressure. Thus, the relationship used to determine the sound power W of a source from the space-averaged mean-square sound pressure p^2 it creates in the room is[3]

$$W = \frac{55.3p^2}{\rho c^2 T_{60}(1 + S\lambda/8V)} \quad \text{W} \tag{7.11}$$

where ρ = density of air, kg/m^3
c = speed of sound in air, m/s
V = room volume, m^3
T_{60} = reverberation time, s

It is assumed that sampling of the sound field in the room is confined to the central regions away from room surfaces and from the sound source.

The Sabine formula (see section 7.3) relating reverberation time to room absorption, is assumed to hold and is the basis for measurements of the sound absorption coefficient in reverberation rooms.[26,27]

7.9 ANECHOIC ROOMS

An anechoic room is a room with highly absorptive surfaces such that a source in the room radiates as in a free field. The resulting sound field has only a direct component. Tests are made in an anechoic room when it is necessary to measure accurately the unperturbed sound radiated by a source, for example when measuring its directivity pattern or sound power. A variant on this concept is the hemi-anechoic room, which has a flat, reflecting floor and sound-absorptive walls and ceiling. It is used to test sources that normally are mounted on or operate in the presence of a reflecting surface. The surfaces of anechoic rooms are made highly absorptive by lining them with deep sound-absorptive materials. The lining typically consists of wedges made of mineral wool or glass fiber. All anechoic rooms are more anechoic at high than at low frequencies. The lowest frequency at which an anechoic room can be used depends on the room volume and the depth of the wedges. Rooms with dimensions of the order of several meters and 0.6-m-deep wedges are anechoic down to a few hundred hertz. Very large rooms with 1–2-m-deep wedges may be effective down to about 100 Hz. To provide a floor working surface in an anechoic room, a wire grid may be attached to the walls. Such a grid may degrade the high-frequency performance of the room.

REFERENCES

1. L. Cremer, H. A. Muller, and T. J. Schultz, *Principles and Applications of Room Acoustics*, Applied Science Publishers, New York, 1982.
2. E. J. Rathe, "Note on Two Common Problems of Sound Propagation," *J. Sound Vib.*, **10**(3), 472–476 (1969).
3. R. V. Waterhouse, "Output of a Sound Source in a Reverberation Chamber and in Other Reflecting Environments," *J. Acoust. Soc. Am.*, **30**(1), 4–13 (1958).
4. ANSI S1.26-1978, "Method for the Calculation of the Absorption of Sound by the Atmosphere," Acoustical Society of America, New York.
5. J. B. Allen and D. A. Berkeley, "Image Method for Efficiently Simulating Small-Room Acoustics," *J. Acoust. Soc. Am.*, **65**(4), 943–950 (1979).

6. J. Borish, "Extension of the Image Model to Arbitrary Polyhedra," *J. Acoust. Soc. Am.*, **75,** 1827–1836 (1984).
7. E. A. Lindqvist, "Noise Attenuation in Factories," *Appl. Acoust.*, **16,** 183–214 (1983).
8. S. Jovicic, "Anleitung zur Vorausbestimmung des Schallpegels in Betriebsgebäuden," Report to the Ministerium für Arbeit, Gesundheit und Soziales des Landes Nordrhein/Westfalen, 1979.
9. A. Krokstad, S. Strom, and S. Sorsdal, "Calculating the Acoustical Room Response by the Use of a Ray-tracing Technique," *J. Sound Vib.*, **8**(1), 118–125 (1968).
10. A. M. Ondet and J.L. Barbry, "Modelling of Sound Propagation in Fitted Workshops Using Ray Tracing," *J. Acoust. Soc. Am.*, **85**(2), 787–796 (1989).
11. M. R. Hodgson, "Case History: Factory Noise Prediction Using Ray Tracing—Experimental Validation and the Effectiveness of Noise Control Measures," *Noise Control Eng. J.*, **33**(3), 97–104 (1989).
12. T. J. Schultz, "Improved Relationship between Sound Power Level and Sound Pressure Level in Domestic and Office Spaces," Report No. 5290, BBN Inc., Cambridge, MA, 1983.
13. M. R. Hodgson, "Towards a Proven Method for Predicting Factory Sound Propagation," in *Proc. Inter-Noise 86*, Noise Control Foundation, New York, 1986, pp. 1319–1322.
14. M. R. Hodgson, "Measurement of the Influence of Fittings and Roof Pitch on the Sound Field in Panel-Roof Factories," *Appl. Acoust.*, **16,** 369–391 (1983).
15. P. M. Wilson, "A Pragmatic Look at Sound Propagation in Real Factory Spaces," *Proc. IOA Conference on Noise Control in Factory Buildings*, 1982.
16. R. Friberg, "Noise Reduction in Industrial Halls Obtained by Acoustical Treatment of Ceilings and Walls," *Noise Control Vib. Red.*, pp. 75–79, March 1975.
17. T. Zetterling, "Simplified Calculation Model for Noise Propagation in Large Factories," in *Proc. Inter-Noise 86*, Noise Control Foundation, New York, 1986, pp. 767–770.
18. Z. Maekawa, "Noise Reduction by Screens," *Applied Acoustics*, Vol. 1, p. 157, 1968.
19. ASTM E1375, "Measuring Interzone Attenuation of Furniture Panels Used as Acoustical Barriers," Philadelphia, PA.
20. ASTM E1111, "Standard Test Method for Measuring Interzone Attenuation of Ceiling Systems," Philadelphia, PA.
21. ASTM E1110, "Standard Classification for Determination of Articulation Class," Philadelphia, PA.
22. ANSI S3.5, "American National Standard Methods for the Calculation of the Articulation Index," Acoustical Society of America, New York.
23. ASTM E1376, "Measuring Interzone Attenuation of Sound Reflected by Wall Finishes and Furniture Panels," Philadelphia, PA.
24. ASTM E1041, "Standard Guide for Measurement of Masking Sound in Open Offices," Philadelphia, PA.
25. ASTM E1130, "Standard Test Method for Objective Measurement of Speech Privacy in Open Offices Using Articulation Index," Philadelphia, PA.
26. ASTM C423, "Standard Test Method for Sound Absorption and Sound Absorption Coefficients by the Reverberation Room Method," Philadelphia, PA.

27. ISO 354, "Measurement of Sound Absorption in a Reverberation Room," Geneva, Switzerland.

28. ANSI S1.31, "Precision Methods for the Determination of Sound Power Levels of Broad-Band Noise Sources in Reverberation Rooms," and ANSI S1.32, "Precision Methods for the Determination of Sound Power Levels of Discrete-Frequency and Narrow-Band Noise Sources in Reverberation Rooms," Acoustical Society of America, New York.

29. ISO 3740, 3741, 3742, "Determination of Sound Power Levels of Noise Sources," Geneva, Switzerland.

30. ISO 140/III, "Laboratory Measurements of Airborne Sound Insulation of Building Elements," Geneva, Switzerland.

31. ASTM E90, "Standard Test Method for Laboratory Measurement of Airborne Sound Transmission Loss of Building Partitions," Philadelphia, PA.

32. ASTM E492, "Standard Method of Laboratory Measurement of Impact Sound Transmission through Floor–Ceiling Assemblies using the Tapping Machine," Philadelphia, PA.

33. M. R. Schroeder, "Frequency-Correlation Functions of Frequency Responses in Rooms," *J. Acoust. Soc. Am.*, **34,** 1819 (1962).

ASTM standards are available from ASTM HQ, 1916 Race Street, Philadelphia, PA 19103-1187.

ANSI standards are available from Standards Secretariat, c/o Acoustical Society of America, 500 Sunnyside Blvd., Woodbury, NY, USA 11797.

ISO standards are available from ISO Central Secretariat, Case Postale 56, CH-1211, Geneve 20, Switzerland.

CHAPTER 8

Sound-absorbing Materials and Sound Absorbers

FRIDOLIN P. MECHEL
Fraunhofer Institute of Building Physics
Stuttgart, Germany

ISTVÁN L. VÉR
Bolt Beranek and Newman, Inc.
Cambridge, Massachusetts

8.1 INTRODUCTION

Sound-absorbing materials are utilized in almost all areas of noise control engineering. This chapter deals with the following aspects:

1. Description of the key physical attributes and parameters that cause a material to absorb sound.
2. Description of the acoustical performance of sound absorbers used to perform specific noise control functions.
3. Compilation of acoustical parameters that allow the quantitative design of sound-absorbing configurations on the bases of material and geometric parameters.
4. Experimental methods to measure the acoustical parameters of sound-absorbing materials and the acoustical performance of sound absorbers.

Noise and Vibration Control Engineering: Principles and Applications, Edited by Leo L. Beranek and István L. Vér.
ISBN 0-471-61751-2 © 1992 John Wiley & Sons, Inc.

8.2 POROUS SOUND-ABSORBING MATERIALS

Porous sound-absorbing materials are available in the form of mats, boards, or preformed elements manufactured of glass, mineral or organic fibers, wood chips, coco fibers, or felted textile or open cell foam (usually polyurethane). They have open pores with typical dimensions below 1 mm that are very much smaller than the wavelength of sound. Hence, each can be treated as a lossy homogeneous medium. The goal of this acoustical characterization is the prediction of the complex characteristic impedance Z_a and propagation constant Γ_a. (See Section 2.5.)

How Fibrous Materials Absorb Sound

Owing to the acting sound pressure, the air molecules (in addition to their random thermal motion) oscillate in the interstices of a porous material with the frequency of the exciting sound wave. The oscillations result in frictional losses. Changes in flow direction and expansions and contractions of the flow through irregular pores result in loss of momentum in the direction of wave propagation. These two phenomena account for most of the energy losses at high frequencies.

At low frequencies heat conduction is another source for the energy loss. Owing to the exciting sound, the air in the pores undergoes periodic compression and decompression and an accompanying change of temperature. Because of the long time during each half-period of oscillation, the large surface-to-volume ratio, and the relatively high heat conduction of the fibers, the efficient exchange of heat means that the compressions are essentially isothermal. At high frequencies the compression process is adiabatic. In the frequency range between isothermal and adiabatic compression the heat exchange process results in further loss of sound energy. In a fibrous material this loss is especially high if the sound propagates parallel to the plane of the fibers and may account for up to 40% of sound attenuation (energy lost per meter of propagation).

Finally, the losses owing to forced mechanical oscillations of the skeleton of a porous material are generally so low that it is reasonable to neglect them.

Physical Characteristics of Fibrous Acoustical Materials

Glass and mineral wool fibers are manufactured by melting the material and forcing it through a series of nozzles to form liquid fibers that are then split into many small-diameter fibers. These small fibers are then impregnated with a binder, compressed to desirable density and thickness, and heat treated to solidify the binder. The preferred orientation of the fibers is parallel to the plane of surface of the finished product. The angular orientation of the fibers within this plane is random.

Most glass or mineral wools for nonaircraft use are constituted of fibers of a statistically distributed fiber diameter between 1 and 10 μm with the maximum of the distribution in the range of 2–5 μm. The diameter of fibers in aircraft blankets are half as large. The density, including any droplets, ρ_A, is usually in the range of 20–200 kg/m^3 (1.3–13 lb/ft^3). In mineral wools (not used in aircraft) the droplets (or shot with diameter over 100 μm) constitute about 30% of the weight. Frequently used fibrous acoustical materials fall in the density range of 30–100 kg/m^3 (2–6 lb/ft^3) (in aircraft, 7–20 kg/m^3).

Porosity. Frequenctly, the designer only knows the thickness d and the density ρ_A of the porous absorber material. For glass fiber and mineral wool products the density of the fiber material is $\rho_M = 2450$ kg/m^3. This yields the porosity h (defined as the ratio of pore volume to total volume):

$$h = 1 - \frac{\rho_A}{\rho_M} \tag{8.1}$$

The droplets are too large to influence viscous and thermal losses. Therefore, it is more appropriate to use the fiber-based density $\rho_A' \simeq \frac{2}{3}\rho_A$ in Eq. (8.1).

Fiber Diameter. Because fibrous sound-absorbing materials do not have uniform fiber diameter, microscopic examinations leads to describing the fiber diameter statistics by a Poisson distribution.[1]

Structure Factor. The dimensionless structure factor χ, which takes into account the effect of the pores and cavities that are perpendicular to the propagation direction of the sound wave, is determined from acoustical measurements as

$$\chi = h^2 \mathrm{Re}\left\{\left(\frac{Z_a}{Z_0}\right)^2\right\} \tag{8.2}$$

where $Z_0 = \rho_0 c_0$ is the characteristic impedance of the gas filling the voids between fibers. The structure factor decreases with increasing frequency and ranges from extreme high values of $\chi = 6$ down to $\chi = 1$ but generally falls in the range of $\chi = 1.3$. Most numerical calculations use $\chi = 1$.

Flow Resistivity. Flow resistivity (specific flow resistance per unit thickness) R_1 is the most important physical characteristic of a porous material. It is defined as

$$R_1 = -\frac{1}{v}\frac{\Delta p}{\Delta x} = -\frac{tS}{V}\frac{\Delta p}{\Delta x} \quad \mathrm{N \cdot s/m^4 \ (mks\ rayls/m)} \tag{8.3}$$

where Δp is the static pressure differential across a homogeneous layer of thickness Δx, v is face velocity of the flow through the material, V is the volume of air passing through the test sample during the time period t, and S is face area (one side) of the sample.

Since R_1 generally depends on the velocity v, it is customary to extrapolate measured R_1 versus v to R_1 ($v = 0.05$ cm/s) because below this particle velocity the flow resistivity of most fibrous materials does not depend any more on the velocity.

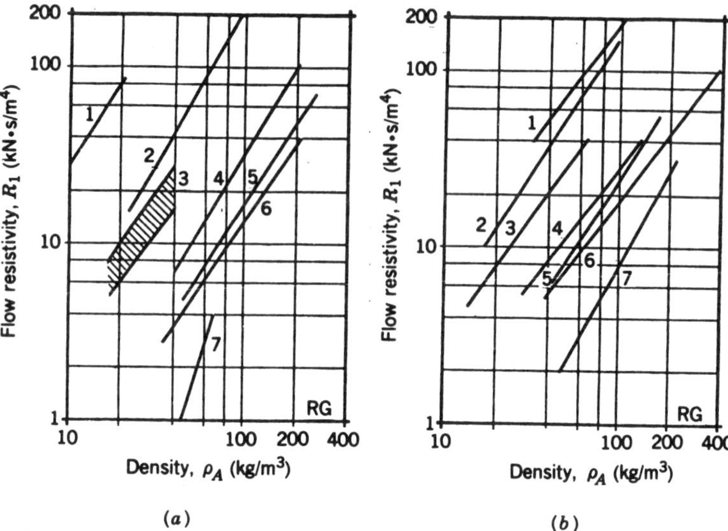

Fig. 8.1 Flow resistivity R_l of fibrous sound-absorbing materials as a function of density ρ_A. (*a*) 1, ISOVER glass fiber "hyperfine"; 2, cotton; 3, polyurethane foam; 4, KLIMALIT, mineral wool; 5, fiberglass, not weaveable; 6, fiberglass, textile fiber; 7, aluminum wool. (*b*) 1, Kaoline wool; 2, glass fiber "superfine"; 3, ISOVER glass fiber; 4, ISOVER basalt wool; 5, basalt wool; 6, SILLAN, mineral wool; 7, glass fiber, thick.

Flow Resistivity versus Density and Fiber Diameter. Figures 8.1*a* and 8.1*b* show the flow resistivity R_l as a function of density for various fibrous sound-absorbing materials manufactured in Europe. The data in Fig. 8.1 is presented in the form

$$R_l = A\rho_A^B \tag{8.4}$$

where ρ_A is in kilograms per cubic meter and the coefficients A and B are those listed in Table 8.1. Some of the fiber diameters listed in Table 8.1 are estimates. The differences in regression parameters A and B obtained for different materials results mostly from the different statistical distribution of the fiber diameters and, to a smaller extent, from the different percentages of droplets (shot) included in the material density but that make no contribution to the flow resistivity.

Figure 8.2 shows the approximate flow resistivity as function of density with fiber diameter as parameter for a family of glass fiber products of a U.S. manufacturer. These idealized curves correspond to a regression exponent $B = 1.16$ in Eq. (8.4). They also show that the flow resistivity increases very strongly with decreasing fiber diameter d. For the same density, a 50% reduction of the fiber diameter results in an eightfold increase in flow resistivity.

The effects of fiber diameter on flow resistivity have been studied both theoretically[1] and experimentally.[2] The regression analyses of the measured data

TABLE 8.1 Regression Parameters for Fig. 8.1 and Eq. (8.4)

Figure	Curve No.	Fiber	A	B	Approximate Fiber Diameter (μm)[a]
8.1*a*	1	ISOVER glass fiber "Hyperfine"	0.656	1.621	1–2
	2	Cotton	0.601	1.789	4–7
	4	KLIMALIT mineral wool	0.0148	1.667	~ 10
	5	FIBERGLASS, not textile	0.0118	1.567	10–15
	6	Fiberglass, textile	0.0134	1.491	10–15
	7	Aluminum wool	7.3×10^{-6}	3.123	30–50
8.1*b*	1	Kaoline wool	0.367	1.359	1–3
	2	ISOVER glass fiber "Superfine"	0.122	1.564	2–4
	3	ISOVER glass fiber	0.121	1.401	3–7
	4	ISOVER basalt wool	0.080	1.271	4–7
	5	basalt wool	0.0192	1.551	4–10
	6	SILLAN mineral wool	0.044	1.304	7–12
	7	glass fiber, not textile	0.00166	1.831	15–20

[a] $1\ \mu\text{m} \approx 4 \times 10^{-5}$ in.

yield the following empirical relationship between fiber radius a, density, and flow resistivity R_l:

$$R_l \frac{\langle a_m^2 \rangle}{\eta} = \begin{cases} 3.2\mu^{1.42} & \text{for glass fiber} \\ 4.4\mu^{1.59} & \text{for mineral and basalt wool} \end{cases} \tag{8.5}$$

where $\langle a_m^2 \rangle$ is the mean square average fiber radius, η is the dynamic viscosity of air, and μ is the massivity of the absorber material defined as $\mu = \rho_A/\rho_M$.

Effect of Compression on Flow Resistivity. Manufacturers of fibrous sound-absorbing materials prefer to use the same fibers to produce an entire product line of boards of different thickness and density. The desired density ρ_A is achieved by baking in the binder at appropriate compression of the bulk fiber material. Frequently, the user installs the sound-absorbing layers under further compression (either to achieve a higher flow resistance or, at high-temperature applications, to assure that the material does not fall apart after the binder evaporates when the temperature first reaches 150°C (300°F).

Figure 8.3*a* shows the effect of compression of the basalt wool board on flow resistivity R_l, indicating that for all three samples R_l increases with increasing compression. Figure 8.3*b* is a plot of the same data but here R_l is plotted on the horizontal scale, as a function of the density ρ_A. The data points in Fig. 8.3*b* fall on a straight line, indicating that $R_l \sim \rho_A^{1.6}$, which is in good agreement with Eq. (8.5).

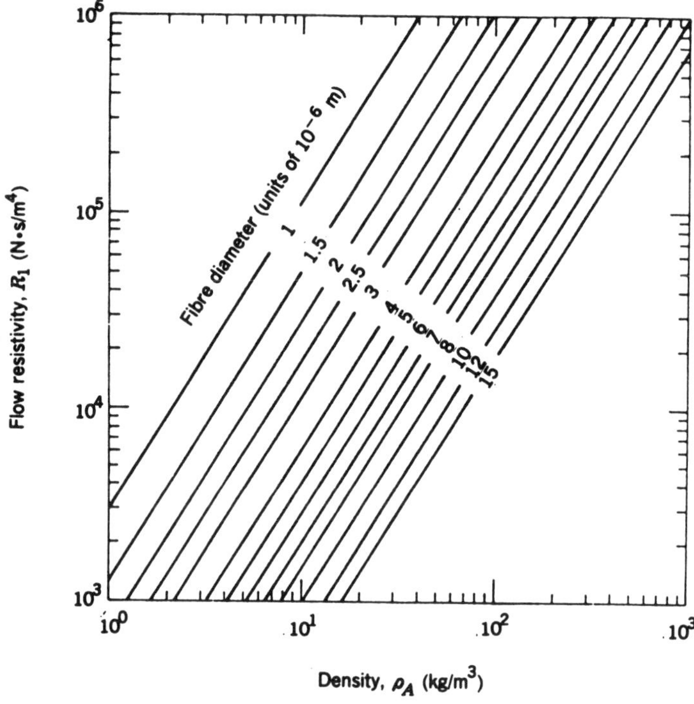

Fig. 8.2 Approximate flow resistivity R_1 as a function of density ρ_A with average fiber diameter as parameter for Fiberglass products. (Courtesy, Owens Corning Fiberglass Corp.)

Nonlinearity. Nonlinearity of the acoustical characteristics of fibrous sound-absorbing materials is observed[3] only at very high sound pressure levels, SPL $\geq$ 165–170 dB (*re* 20 μPa).

The flow resistivity R_1 of fibrous materials depends on flow velocity as described in the approximate empirical relationship

$$R_1(V) = R_{10} + FV \tag{8.6}$$

where $R_1(V)$ and R_{10} are in kNs/m^4 and V in m/s and the empirically determined nonlinearity factor F is

$$F = 0.75 R_{10}^{0.57} \tag{8.7}$$

and is subject to considerable scatter.

Acoustical Characterization of Fibrous Sound-absorbing Materials

Fibrous sound-absorbing materials can be fully characterized by their complex propagation constant Γ_a and complex characteristic impedance Z_a, which are de-

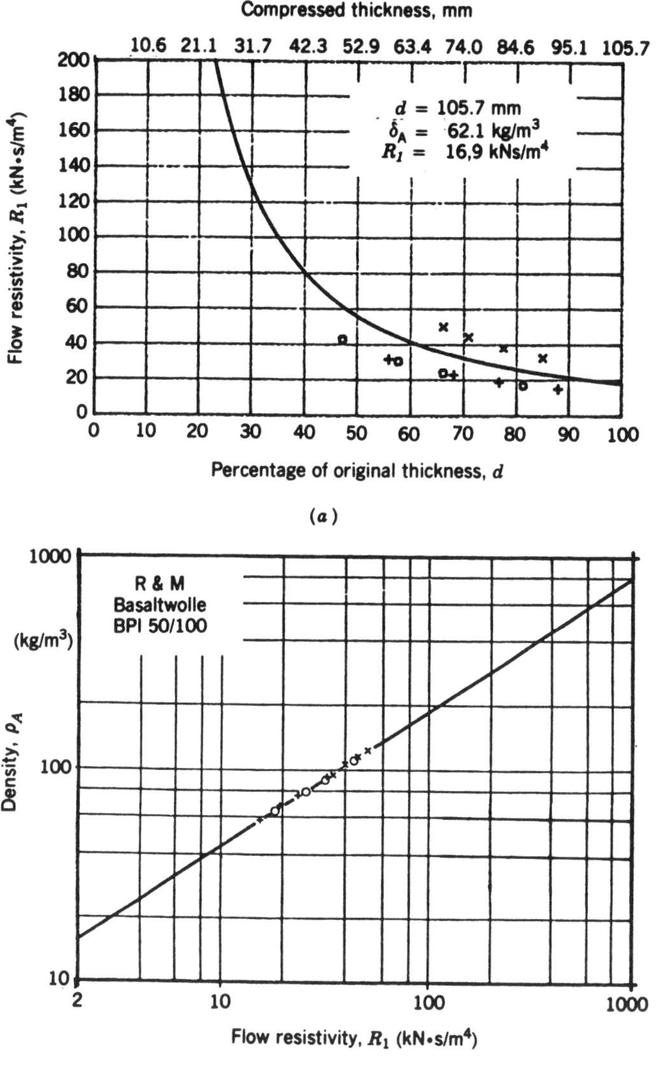

Fig. 8.3 Change in flow resistivity R_1 (in kN · s/m^4) owing to compression of the test samples as a function of (*a*) compression; (*b*) density of compressed layer, ρ_A: (+) sample 1; (×) sample 2; (○) sample 3.

fined by

$$p(x, t) = \hat{p}e^{-x\Gamma_a}e^{j\omega t} \qquad \text{N/m}^2 \tag{8.8}$$

$$\langle p(x, t)\rangle_y = Z_a\langle v_x(x, t)\rangle_y \qquad \text{N/m}^2 \tag{8.9}$$

where $\langle \cdots \rangle$ means taking an average perpendicular to the propagation direction over an area that is small compared with the wavelength but large compared with the size of pores. From the four acoustical characteristics

$$\Gamma_a = \Gamma'_a + j\Gamma''_a \quad \text{and} \quad Z_a = Z'_a + jZ''_a$$

the attenuation constant Γ'_a is the easiest to determine experimentally by measuring with a probe tube microphone the decrease of the sound pressure level (in nepers per meter) of a plane sound wave propagating in a very thick layer of material. The phase exponent Γ''_a is obtained by measuring the change of phase with distance. The characteristic impedance Z_a is obtained by measuring the input impedance of a sufficiently deep (reflection from the end is not any more noticeable) layer of the absorber material placed in an impedance tube.

This section provides information on how to predict the key acoustical characteristics Γ_a and Z_a of fibrous sound-absorbing materials on the bases of their physical characteristics. Empirical prediction, on the bases of regression analyses of measured data, is emphasized because it yields the most accurate result.

Empirical Prediction on the Bases of Regression Analyses of Measured Data. There are two families of parameters that determine the sound absorption coefficient. The characterization would be simplest if both Γ_a and Z_a would depend on a single parameter.

Figure 8.4*a* shows the measured[4] normal incidence sound absorption coefficient α_0 of different Rockwool materials of practically infinite thickness (between 0.5 and 1 m, depending on the bulk density of the absorber material) as a function of frequency with bulk density of the material as parameter. Figure 8.4*a* clearly indicates that the bulk density is not the sought single parameter that would collapse the measured data points into a single curve. Figure 8.4*b* shows the same data points as Fig. 8.4*a* but now as a function of the nondimensional variable $E = \rho_0 f/R_1$ on the horizontal scale, where ρ_0 is the density of air, f is the frequency, and R_1 is the flow resistivity of the bulk material at the density at which α_0 was measured. Observing Fig. 8.4*b*, note that $E = \rho_0 f/R_1$ is the sought single parameter that collapses all measured data. The normalized, nondimensional frequency variable $E = \rho_0 f/R_1$ is useful in describing not only the sound-absorbing capability of the semi-infinite layer of fibrous material but also the propagation constant and characteristic impedance of the bulk material.

To facilitate uniform presentation, it is useful to present the propagation constant and characteristic impedance in a dimensionless manner as

$$\Gamma_{an} = \Gamma_a/k_0 = \Gamma'_{an} + j\Gamma''_{an} \tag{8.10}$$

$$Z_{an} = Z_a/Z_0 = Z'_{an} + jZ''_{an} \tag{8.11}$$

where $k_0 = \omega/c_0$ is the wavenumber in air and $Z_0 = \rho_0 c_0$ is the characteristic impedance of the gas filling the voids between the fibers for plane waves.

Figures 8.5 and 8.6 show plots of the real and imaginary parts of the normalized propagation constant Γ_{an} and that of the normalized characteristic impedance Z_{an}

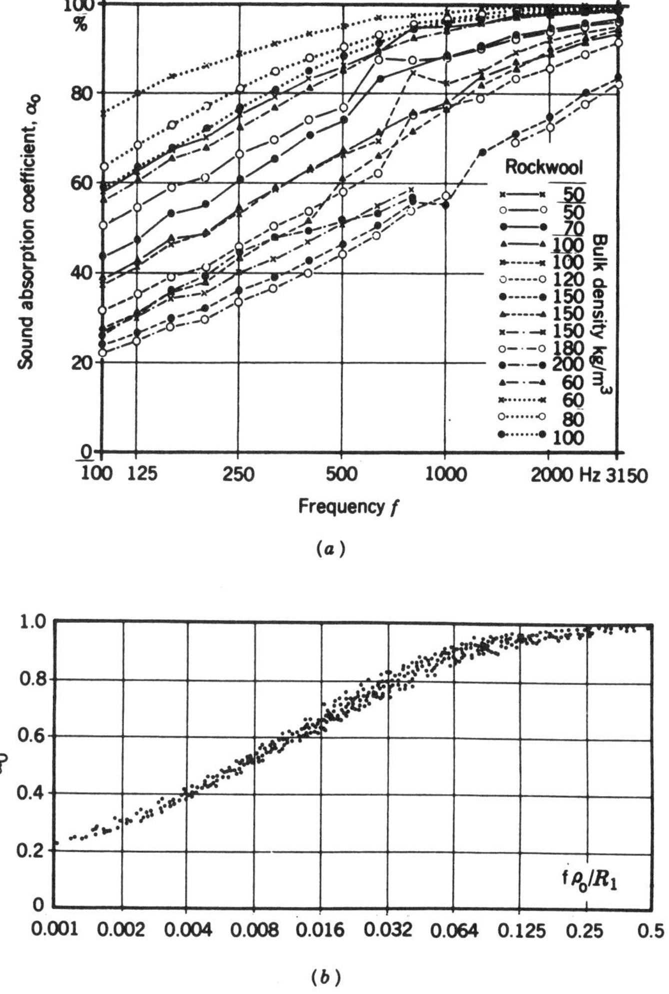

Fig. 8.4 Normal-incidence sound absorption coefficients of different rockwool materials of practically infinite thickness (0.5–1 m) measured in an impedance tube, plotted as a function of (*a*) frequency with bulk density ρ_A and (*b*) nondimensional variable $\rho_0 f/R_1$.

for a large variety of minearl wool sound-absorbing materials plotted as a function of the normalized frequency parameter, indicating that indeed the normalized frequency parameter $E = \rho_0 f/R_1$ is a universal descriptor of fibrous porous sound-absorbing materials. The data presented in Figs. 8.5 and 8.6 and similar curves for glass fiber materials were obtained by careful measurements of the acoustical and material characteristics (Γ'_{an}, Γ''_{an}, Z'_{an}, Z''_{an}, and R_1) of over 70 different types of materials.[5] The solid lines in Figs. 8.5 and 8.6 resulted from the regression

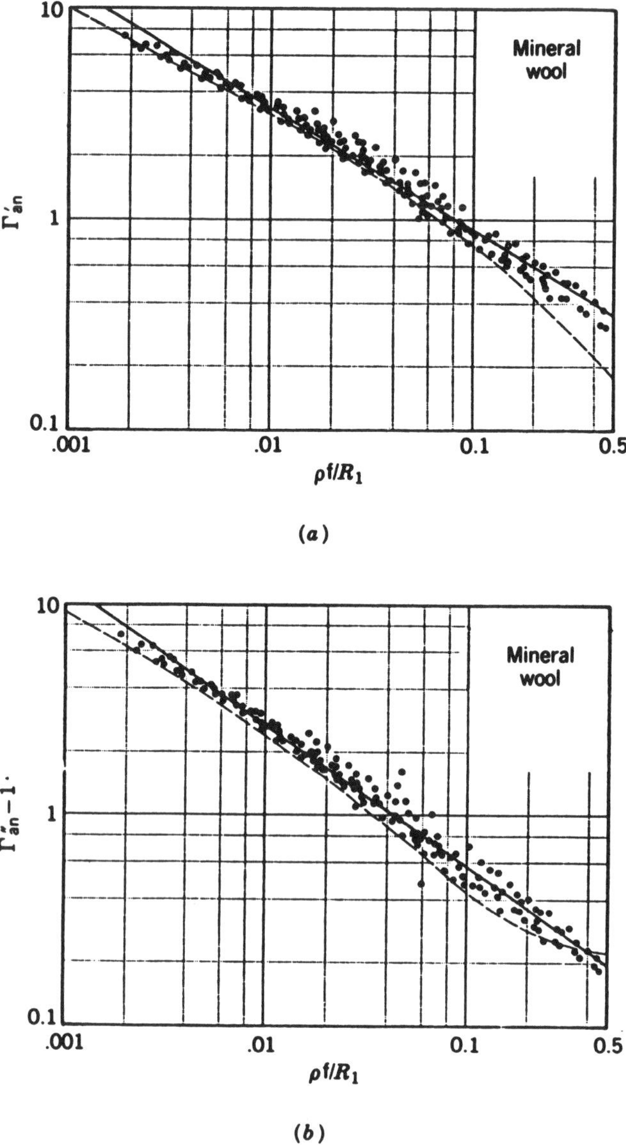

Fig. 8.5 Measured normalized propagation constant $\Gamma_{an} = \Gamma_a/k_0 = \Gamma'_{an} + j\Gamma'_{an}$ for mineral wool as a function of normalized frequency parameter $E = \rho_0 f/R_1$: (*a*) Γ'_{an}; (*b*) $\Gamma''_{an} - 1$. (——) Regression line; Eq. (8.12a). (-----) Rayleigh approximation.

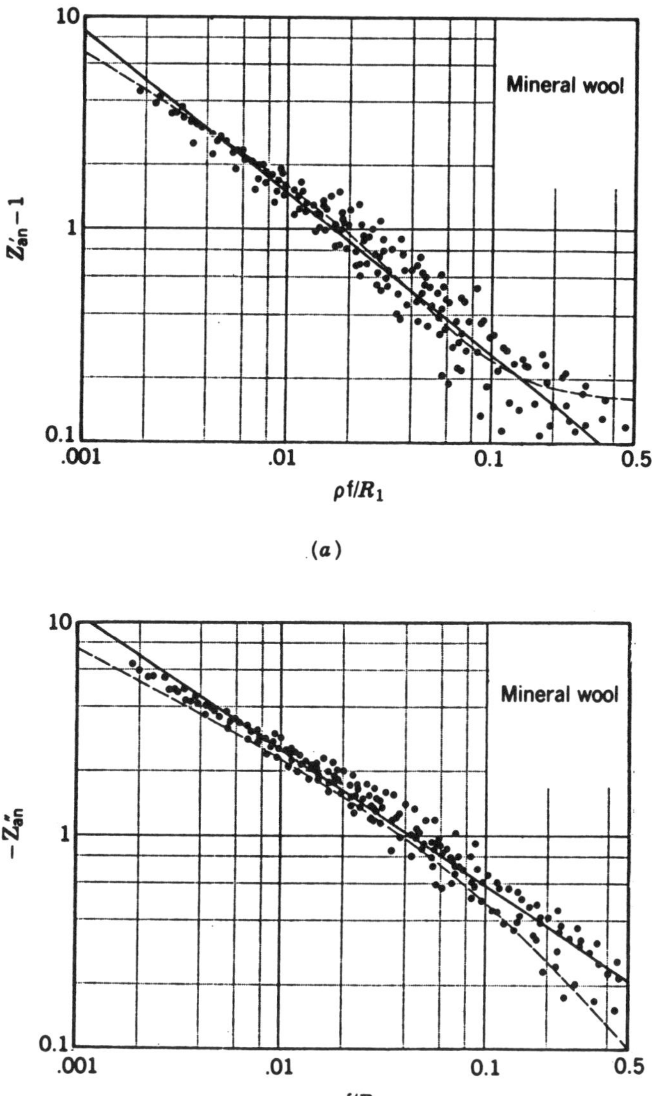

Fig. 8.6 Measured normalized characteristic impedance $Z_{an} = Z_a/Z_0 = Z'_{an} + jZ''_{an}$ for mineral wool as a function of normalized frequency parameter $E = \rho_0 f/R_1$: (*a*) $Z'_{an} - 1$; (*b*) $-Z''_{an}$. (——) Regression line; Eq. (8.12b). (-----) Rayleigh approximation.

analyses of the data and have the form

$$\Gamma_{an} = \frac{\Gamma_a}{k_0} = a'E^{-\alpha'} + j(1 + a''E^{-\alpha''}) \tag{8.12a}$$

$$Z_{an} = \frac{Z_a}{Z_0} = (1 + b'E^{-\beta'}) - jb''E^{-\beta''} \tag{8.12b}$$

The regression parameters a', a'', b', b'', α', α'', β', and β'' in Eq. (8.12) are compiled in Table 8.2. There are different regression parametes for the normalized frequency regions below and above $E = 0.025$. It was found that all fibrous materials measured could be divided into two categories, namely (1) mineral wool and basalt wool and (2) glass fiber.

Effect of Temperature. In sound absorbers or silencers designed to operate at high temperatures it is important to know the acoustical characteristics of the fibrous porous sound-absorbing materials at design temperature. Fortunately, it is not necessary to measure the propagation constant Γ_a and characteristic impedance Z_a at design temperature T because the values of these acoustical characteristics measured at room temperature, T_0, can be scaled. The dimensionless frequency variable $E = \rho_0 f/R_1$ on which both Γ_{an} and Z_{an} depend must be evaluated at design temperature. Considering that the influence of temperature on ρ and η is

$$\rho(T) = \rho(T_0)\frac{T_0}{T} \tag{8.13}$$

$$\eta(T) = \eta(T_0)\left(\frac{T}{T_0}\right)^{0.65} \tag{8.14}$$

yields

$$E(T) = E(T_0)\left(\frac{T}{T_0}\right)^{-1.65} = \frac{\rho_0 f}{R_1(T_0)}\left(\frac{T}{T_0}\right)^{-1.65} \tag{8.15}$$

TABLE 8.2 Regression Coefficients[a] for Predicting Propagation Constant and Characteristic Impedance of Fibrous Sound-absorbing Materials

Material	E Region	a'	α'	a''	α''	b'	β'	b''	β''
Mineral and basalt wool	$E \leq 0.025$	0.322	0.502	0.136	0.641	0.081	0.699	0.191	0.556
	$E > 0.025$	0.179	0.663	0.103	0.716	0.0563	0.725	0.127	0.655
Glass fiber	$E \leq 0.025$	0.396	0.458	0.135	0.646	0.0668	0.707	0.196	0.549
	$E > 0.025$	0.179	0.674	0.102	0.705	0.0235	0.887	0.0875	0.770

[a]According to Eq. (8.12).

where T and T_0 are the design temperature and room temperature, respectively, measured on an absolute scale (i.e., Kelvin and Rankine). Considering further that $\Gamma_{an} = \Gamma_a/k = \Gamma_a c/\omega$ and $Z_{an} = Z_a/\rho c$ and that $\rho = \rho_0(T/T_0)^{-1}$ and $c = c_0(T/T_0)^{1/2}$, the acoustical characteristics at design temperature T (degrees Kelvin) are determined as follows:

$$\Gamma_a(T) = \Gamma_{an}[E(T)]\,\frac{\omega}{c_0}\left(\frac{T}{T_0}\right)^{-1/2} \tag{8.16a}$$

$$Z_a(T) = Z_{an}[E(T)]\,(\rho_0 c_0)\left(\frac{T}{T_0}\right)^{-1/2} \tag{8.16b}$$

where $\Gamma_{an}[E(T)]$ and $Z_{an}[E(T)]$ are the normalized attenuation constant and normalized characteristic impedance according to Eqs. (8.12a) and (8.12b) computed for $E = E(T)$ determined from Eqs. (8.13) and (8.14).

Example Compute the Γ'_{an}, Γ''_{an}, Z'_{an}, and Z''_{an} of a very (practically infinite) thick layer of glass fiber at a frequency of 100 Hz for 20°C and for 500°C and determine α_N. The flow resistivity of the fibrous material at 20°C, $R_1(T_0) = 16{,}000$ $\text{N}\cdot\text{s/m}^4$ and $\rho_0 = 1.2\ \text{kg/m}^3$:

$$T_0 = 273 + 20 = 293\ \text{K} \qquad T = 273 + 500 = 773\ \text{K}$$

$$E(T_0) = \frac{\rho_0 f}{R_1} = 1.2 \times \frac{100}{16{,}000} = 0.0075 < 0.025$$

$$E(T) = E(T_0)\left(\frac{T}{T_0}\right)^{-1.65} = 0.0075\left(\frac{773}{293}\right)^{-1.65} = 1.5 \times 10^{-3} < 0.025$$

The regression parameters from Table 8.2 are:

$$a' = 0.396 \qquad a'' = 0.135 \qquad \alpha' = 0.458 \qquad \alpha'' = 0.646$$

$$b' = 0.0668 \qquad b'' = 0.196 \qquad \beta' = 0.707 \qquad \beta'' = 0.549$$

According to Eq. (8.12), we obtain

	Temperature	
Parameters	20°C	500°C
$\Gamma'_{an} = a'E^{-\alpha'}$	1.3	7.78
$\Gamma''_{an} = 1 + a''E^{-\alpha''}$	1.72	10.0
$Z'_{an} = 1 + b'E^{-\beta'}$	1.41	7.62
$Z''_{an} = b''E^{-\beta''}$	0.81	6.9

The normal-incidence sound absorption coefficient according to Eq. (8.18) is

$$\alpha_N = 4\,\frac{Z'_{an}}{Z'^2_{an} + 2Z'_{an} + 1 + Z''^2_{an}}$$

yielding

$$\alpha_N = \begin{cases} 0.88 & \text{for } 20°\text{C} \\ 0.25 & \text{for } 500°\text{C} \end{cases}$$

8.3 SOUND ABSORPTION BY LARGE FLAT ABSORBERS

In dealing with sound absorption the goal is usually to determine the absorbed or reflected portion of a sound wave incident on the absorber. This is easiest when the surface of the absorber is flat and is sufficiently large so that sound waves scattered at the edges of the absorber can be neglected. Then, for the special case of a plane incident sound wave, it is possible to assign a sound energy absorption coefficient α for each point on the absorber surface given by

$$\alpha = \frac{\text{absorbed energy}}{\text{incident energy}} = 1 - |R^2| \tag{8.17}$$

where R is the reflection factor, which is defined as the ratio of the reflected and incident sound pressure at the interface. A high sound absorption coefficient ($\alpha \to 1$) requires that $|R| \to 0$. Note that $|R| = 0.1$ corresponds to $\alpha = 0.99$. In this chapter we shall deal only with infinitely large, flat, homogeneous absorbers. Edge effects manifest themselves in increased sound absorption with increasing perimeter-surface area ratio of the absorber.[6]

Perpendicular Sound Incidence

For perpendicular sound incidence it is sufficient to know the complex normal specific wall impedance $Z_l = Z'_l + jZ''_j$ of the absorber, which is the ratio of sound pressure and the *normal* component of the particle velocity at the interface (see Appendix, Chapter 1). The reflection factor and absorption coefficient are related by

$$R = \frac{Z_l - Z_0}{Z_l + Z_0} \qquad \alpha = \frac{4Z'_l Z_0}{(Z_l + Z_0)^2 + Z''^2_l} \tag{8.18}$$

where $Z_0 = \rho_0 c_0$ is the characteristic impedance of air for plane waves and ρ_0 is the density and c_0 the speed of sound in air.

For sound absorbers of limited porosity h of the surface (e.g., those with perforated or cloth facings) one must use in Eq. (8.18) the so-called air-side wall impedance. This is obtained by taking the air-side volume velocity averaged over the face area of the absorber. If the absorber is covered by a perforated facing and the specific acoustical impedance of an orifice (with the absorber behind) is Z_i, then $Z_1 = Z_i/h$, where h is the porosity of the perforated plate.

Equation (8.18) indicates $|R| \rightarrow 0$ requires that $Z_1 \rightarrow Z_0$. This means that the ideal absorber should not resist the sound more than the unbounded air does. Because the characteristic impedance of porous sound-absorbing materials, Z_a, should be only slightly above that for air, Z_0, it is imperative to keep the porosity high: for fibrous sound-absorbing materials in the range of 0.95–0.99 and for perforated facings above 0.25.

Porous Absorbers of Infinite Thickness. Homogeneous porous, sound absorbing materials (glass fiber, mineral wool, open cell foam, etc.) are best characterized by their complex propagation constant $\Gamma_a = \Gamma'_a + j\Gamma''_a$ and complex characteristic impedance $Z_a = Z'_a + jZ''_a$ (plane sound wave assumed). Thus,

$$R = \frac{Z_a - Z_0}{Z_a + Z_0} \tag{8.19}$$

For practical purposes a layer thickness d can be considered "infinite" if $\Gamma'_a d > 2$.

We will see that the higher the flow resistivity of the porous material, the more Z_a exceeds Z_0. Even in the extreme case of practically infinite layer thickness (which at low frequencies and low flow resistivity becomes very large) the sound absorption coefficient will be small. At mid and high frequencies the general tendency is $Z_a \rightarrow Z_0/h$, and consequently, the good impedance matching results in a high absorption coefficient.

Porous Sound Absorbers of Finite Thickness in Front of a Rigid Wall. In this case the wall impedance is controlled by the combination of the incident and (multiple) reflected sound waves in the layer, yielding

$$Z_1 = Z_a \coth(\Gamma_a d) \qquad \text{N} \cdot \text{s/m}^3 \tag{8.20}$$

With this wall impedance Eq. (8.20) exhibits the following behavior:

(a) In case $d \ll \frac{1}{4}\lambda_a$ ($\lambda_a = 2\pi/\Gamma''$ is the wavelength in the absorbing material), when the layer is thin or the frequency is low, the magnitude of $\coth(\Gamma_a d)$ is always large and the lack of impedance matching between Z_1 and Z_0 leads to a small sound absorption coefficient. This is the reason that there is no "sound-absorbing paint" and that rugs absorb sound only modestly.

(b) For large $\Gamma_a d$ in Eq. (8.20), when $\Gamma''_a d = 2\pi d/\lambda_a > 2$, $\coth(\Gamma_a d) \rightarrow 1$, provided the attenuation Γ'_a is not too small. In this case the acoustical behavior of the layer approximates that of an "infinitely thick" layer.

(c) For low internal attenuation Γ_a' and for increasing frequency, $\coth(\Gamma_a d)$ has its first minimum at $d \simeq \frac{1}{4}\lambda_a$, and the absorption coefficient-versus-frequency curve exhibits its first maximum. The following minima at $d \simeq \frac{1}{2}\lambda$ and further maxima at $d \simeq \frac{3}{4}\lambda_a$ and $\frac{5}{4}\lambda_a$ are usually not perceivable any more for most commercially available porous material because of the high attenuation coefficient Γ_a' of such materials.

In designing porous sound absorbers one must strike a compromise between the requirements $Z_a \rightarrow Z_0$ (important for obtaining high absorption coefficient at low frequencies) that requires low flow resistivity and usually a low density and the requirement to keep Γ_a' as high as possible to keep layer thickness small. This, however, requires high flow resistivity and usually high density. As we will see in Eqs. (8.27) and (8.28), the angle of sound incidence also will play a role in reaching the right compromise. Consequently, this issue will be covered in more detail in Section 8.3 in the discussion of random incidence (under "Oblique Sound Incidence").

Sound-absorbing Layer with Airspace Behind and in Front of Rigid Wall. The sound-absorbing configuration depicted in Fig. 8.7 is frequently used in practical applications (e.g., hung acoustical ceilings). The analytical model is derived by assuming in the airspace an incident and a reflected sound wave that yields an impedance at the back surface of the porous layer

$$Z_2 = -jZ_0 \cot(k_0 t) \qquad \text{N} \cdot \text{s/m}^3 \tag{8.21}$$

where $k_0 = \omega / c_0$ is the acoustical wavenumber for air. The internal reflection coefficient at the back side of the layer is $R_B = (Z_2 - Z_a)/(Z_2 + Z_a)$ and the wall

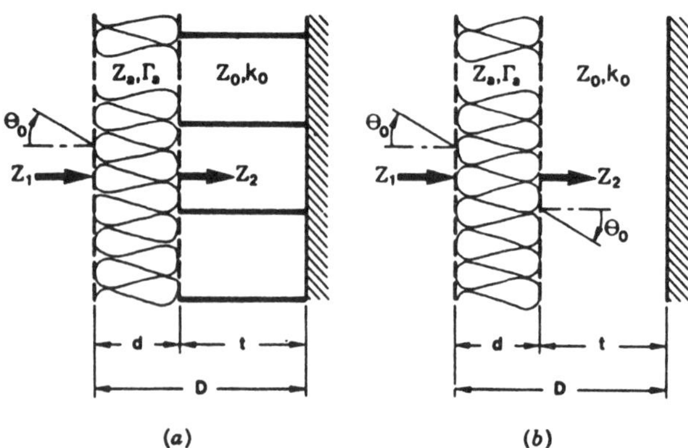

Fig. 8.7 Combination of a bulk-reacting absorber layer with a (*a*) locally reacting air gap and (*b*) bulk-reacting air gap.

impedance at the front face of the absorbing layer is obtained as

$$Z_1 = Z_a \frac{Z_2 \cosh(\Gamma_a d) + Z_a \sinh(\Gamma_a d)}{Z_a \cosh(\Gamma_a d) + Z_2 \sinh(\Gamma_a d)} \quad \mathrm{N \cdot s/m^3} \tag{8.22a}$$

$$Z_1 = \frac{Z_h(Z_2 + Z_w)}{Z_2 + Z_h} = Z_h \frac{1 + Z_w/Z_2}{1 + Z_h/Z_2} \quad \mathrm{N \cdot s/m^3} \tag{8.22b}$$

In Eq. (8.22b) the symbol Z_h is the input impedance of the sound-absorbing layer backed by a hard wall (i.e., $Z_2 \to \infty$) and Z_w is the input impedance of the layer backed by zero impedance ($Z_2 = 0$), which can be realized by placing the layer at a distance $t = \frac{1}{4}\lambda_0$ in front of a rigid wall (λ_0 is acoustical wavelength in air). The relationship $Z_a = \sqrt{Z_h Z_w}$ used in Eq. (8.22b) follows from a different definition of the characteristic impedance. The reflection factor and absorption coefficient α are computed by using Z_1 in Eq. (8.19).

For small airspace thickness, $t/\lambda_0 < \frac{1}{8}$, and $|Z_h| >> |Z_w|$ give the following approximation:

$$Z_a \simeq \frac{Z_h}{1 + Z_h/Z_2} \quad \mathrm{N \cdot s/m^3} \tag{8.23}$$

When the thickness of the absorbing layer is not too small, Eq. (8.23) results in $Z_1 \simeq Z_h$. Consequently at low frequencies a very thin airspace between the absorbing layer and rigid wall is ineffective. When the thickness of the airspace becomes $t = \frac{1}{4}\lambda_0$, $Z_2 = 0$ and Eq. (8.22a) yields

$$Z_1 = Z_w = Z_a \tanh(\Gamma_a d) \quad \mathrm{N \cdot s/m^3} \tag{8.24}$$

As long as the layer thickness is small compared with the wavelength ($d < \frac{1}{8}\lambda_a$), the magnitude of $\tanh(\Gamma_a d)$ is smaller than unity. Consequently, the airspace behind the absorbing layer results in a reduction of the wall impedance, and at low frequencies the magnitude of the wall impedance shifts toward the characteristic impedance of air Z_0. This results in a decrease of the reflection factor R and in a corresponding increase of the normal incidence sound absorption coefficient. The largest improvement is observed when the thickness of the airspace is $t = \frac{1}{4}\lambda_0$.

For frequencies where the airspace thickness corresponds to a multiple of $\frac{1}{2}\lambda_0$, Z_2 is very large and the airspace becomes totally ineffective.

Oblique Sound Incidence

For oblique sound incidence one must distinguish between locally reacting and bulk-reacting absorbers. Locally reacting absorbers are those where sound propagation parallel to the absorber surface is prohibited (e.g., partitioned porous layer, Helmholtz resonators with partitioned volume, small plate absorbers). In bulk-reacting absorbers, such as a porous layer possibly with unpartitioned airspace behind, sound propagation in the sound-absorbing layer or in the airspace behind it is possible in the direction parallel to the absorber surface. The reason for calling an absorber locally reacting is that the particle velocity at the interface depends only on the local sound pressure. For bulk-reacting absorbers, the particle velocity at the interface depends not only on the local sound pressure but also on the particular distribution of the sound pressure in the entire absorber volume.

Oblique Incidence Sound on Locally Reacting Absorber. Locally reacting absorbers are characterized by a wall impedance Z_l that is independent of the angle of incidence θ_0.

For locally reacting absorbers the reflection factor $R(\theta_0)$ is determined by how well the wall impedance Z_l matches the field impedance of $Z_0/\cos\theta_0$ and is given by

$$R(\theta_0) = \frac{Z_l \cos\theta_0 - Z_0}{Z_l \cos\theta_0 + Z_0} \tag{8.25}$$

Equation (8.25) indicates that best impedance match, and correspondingly the lowest reflection factor for a given incidence angle θ_0, is achieved by $|Z_l| > Z_0$, and that overmatched wall impedances, $|Z_l| > Z_0$, will yield absorption maxima at a specific angle of incidence.

Oblique Incidence Sound on Bulk-reacting Absorber. Mineral fiber and open-pore foams are the most frequently used bulk-reacting sound-absorbing materials. The acoustical behavior of these materials can be fully characterized by their propagation constant Γ_a and characteristic acoustical impedance Z_a. Methods for measuring and predicting these key acoustical parameters are given in Section 8.2 under "Acoustical Characterization of Fibrous Sound-absorbing Materials."

Semi-infinite Layer. Figure 8.8 shows the incident, reflected, and transmitted waves at the interface. The combination of these waves must satisfy the following boundary conditions: (1) equal normal components of the impedances and (2) equality of the wavenumber components parallel to the interface. The second of these requirements yields the diffraction law

$$\frac{\sin\theta_l}{\sin\theta_0} = \frac{jk_0}{\Gamma_a} \tag{8.26}$$

where θ_l is the complex propagation angle of the sound inside the absorber. The complex reflection factor is given by

$$R(\theta_0) = \frac{1 - (Z_0/Z_l)/\cos\theta_0}{1 + (Z_0/Z_l)/\cos\theta_0} \tag{8.27a}$$

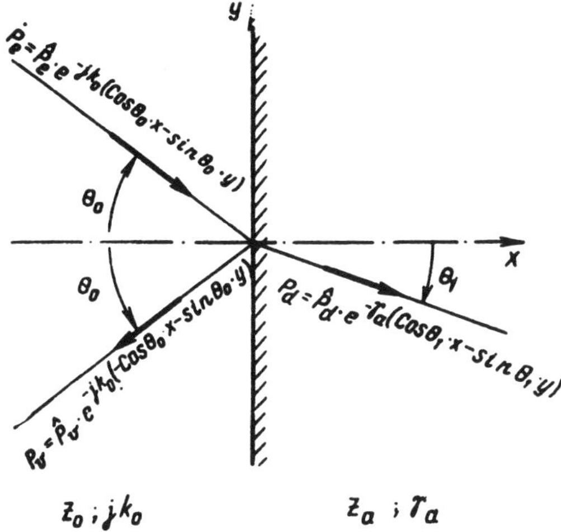

Fig. 8.8 Reflection and transmission of an oblique-incidence sound wave by a semi-infinite, bulk-reacting absorber.

and

$$Z_1 = \frac{Z_a}{[1 + (k_0/\Gamma_a)^2 \sin \theta_0^2]^{1/2}} \quad \text{N} \cdot \text{s/m}^3 \tag{8.27b}$$

Finite-Layer Thickness. For a finite layer of bulk-absorbing material of thickness d in front of a hard wall, the reflection factor is

$$R(\theta_0) = \frac{1 - (Z_0/Z_{1d})/\cos \theta_0}{1 + (Z_0/Z_{1d})/\cos \theta_0} \tag{8.28}$$

where $Z_{1d} = (Z_a/\cos \theta_1)\coth(\Gamma_a d \cos \theta_1)$ and θ_1 is defined in Eq. 8.26.

Multilayers. Multilayer absorbers may consist of a porous sound-absorbing layer with an airspace behind, as illustrated in Fig. 8.7, or a number of porous layers of different thicknesses and different acoustical characteristics, as shown schematically in Fig. 8.9. Analytical expressions describing the absorber functions and design charts to predict the sound absorption coefficients for those multilayer absorbers are provided in reference 4.

Random Incidence

For a diffuse sound field where the intensity $I(\theta) = I$ is independent of the incident angle θ, the sound power incident on a small surface area, dS, of the absorber is

$$dW_{\text{inc}} = I\, dS \int_0^{2\pi} d\phi \int_0^{\pi/2} \cos \theta \sin \theta \, d\theta = \pi I \, dS \tag{8.29}$$

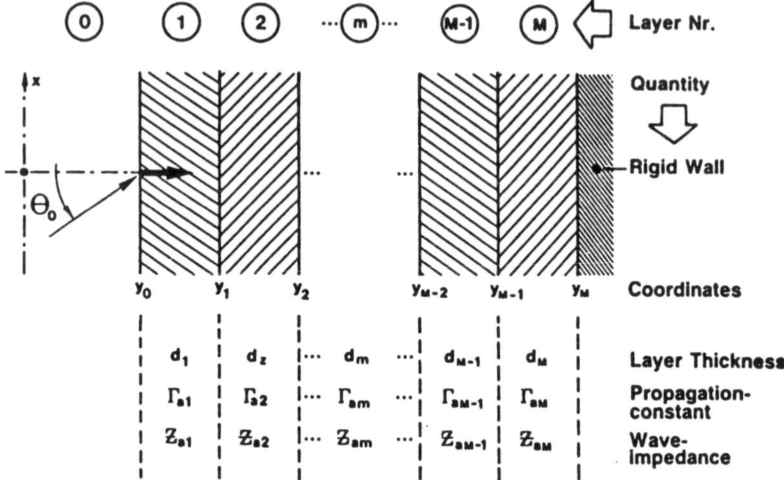

Fig. 8.9 Scheme of a multilayer absorber configuration.

and the absorbed portion is

$$dW_a = I\, ds \int_0^{2\pi} d\phi \int_0^{\pi/2} \alpha(\theta) \cos\theta \sin\theta\, d\theta$$

$$= 2\pi I\, dS \int_0^{\pi/2} \alpha(\theta) \cos\theta \sin\theta\, d\theta \qquad (8.30)$$

and the random-incidence sound absorption coefficient

$$\alpha_R = \frac{dW_a}{dW_{\text{inc}}} = 2 \int_0^{\pi/2} \alpha(\theta) \cos\theta \sin\theta\, d\theta \qquad (8.31)$$

Random-Incidence Sound Absorption Coefficient of Locally Reacting Absorbers. The combination of Eqs. (8.17), (8.25), and (8.31) will yield the random-incidence sound absorption coefficient.

However, the resulting equation is of limited use because the wall impedance, which is necessary as input data, is difficult to obtain. Experimentally, it is easier to measure α_R than Z_1. In addition, it gives no information how α_R depends on the primary geometric and physical parameters, such as layer thickness, frequency, and flow resistivity.

Random-Incidence Sound Absorption Coefficient of Bulk-reacting Absorbers. For bulk-reacting absorbers the integration in Eq. (8.31) can only be evaluated numerically. It is more appropriate to use the geometric and acoustical characteristics of the porous liner than its impedance to generate design charts for general use.

8.4 DESIGN CHARTS FOR FIBROUS SOUND-ABSORBING LAYERS

For design use it is essential to have charts where the sound absorption coefficient is plotted as a function of the first-order parameters of the absorber, such as thickness d, flow resistivity R_1, and frequency f. This section contains such design charts. Design charts for a multiple-layer absorber, such as shown in Fig. 8.9, are given in reference 4.

Monolayer Absorbers

As shown in Section 8.2, the key acoustical parameters Γ_a and Z_a of fibrous sound-absorbing materials depend only on a single material parameter, namely the flow resistivity R_1 (specific flow resistance per unit thickness; see Chapter 1). Consequently, the absorption coefficient can be computed and plotted as a function of two dimensionless variables:

$$F = \frac{fd}{c_0} \qquad R = \frac{R_1 d}{Z_0} \tag{8.32}$$

Then $k_0 d = 2\pi F$ and $E = \rho_0 f / R_1 = F/R$. At oblique incidence, the angle θ is the third input variable. Design charts in the sense mentioned in the introduction are obtained when contour lines of constant values of α are plotted on a log-log chart of R versus F. A variation of frequency f would produce a horizontal path, an increase of the flow resistivity R_1 would generate an upward move, and an increase in layer thickness d would lead to a diagonal shift up to the right. Frequency curves for an absorber layer may be derived by a horizontal intersection, with a starting point determined by the individual parameter values.

Figure 8.10 shows lines of constant absorption of a monolayer absorber for normal incidence (the absorber may be bulk or locally reacting). The maximum absorption is reached at about $R = 1.2$ for $F = 0.25$, that is, at a thickness d equal to a quarter of a free-field wavelength and at the flow resistance of the layer of $1.2Z_0$. Higher resonances at odd multiples of a quarter wavelength are visible for a small flow resistance. The "summit line" of the first maximum, which goes through the points of relative maxima of α at the first resonance, is inclined so that the first relative maximum will occur below $F = 0.25$ for overmatched flow resistances, $R > 1$, and the first relative maximum of the frequency curve will occur at somewhat higher frequencies than $F = 0.25$ for undermatched flow resistances, $R < 1$. The higher orders of resonances are relatively stationary in frequency when R varies.

At *oblique incidence* the sound absorption for small angles, below about $\theta = 30°$, is only weakly modified. The modifications in $\alpha(\theta)$ are pronounced for larger angles with locally reacting absorbers. The curves on the left in Figs. 8.11*a–c* were computed for a locally reacting absorber for $\theta = 30°$, $45°$, and $60°$. The resonance structure of the contour plots become very distinct at low flow

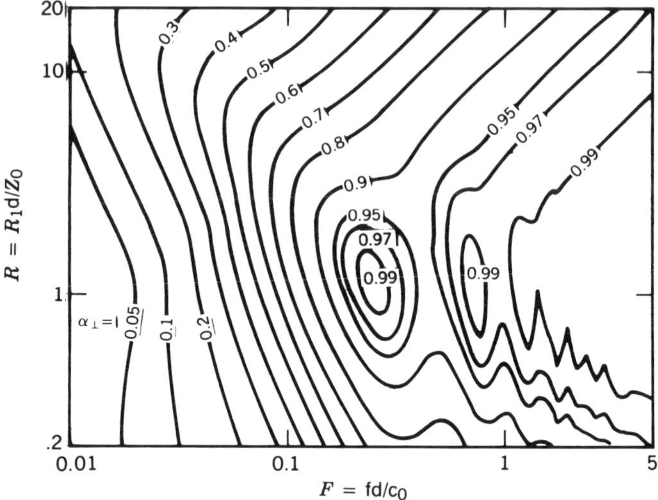

Fig. 8.10 Lines of constant absorption at normal-incidence sound; absorption coefficient for fibrous absorbers of thickness d and flow resistivity R_1; no airspace.

resistances and high frequencies. Curves on the right in Figs. 8.11*a–c* show the analogous contour plots of $\alpha(\theta)$ for the same values of θ but now for bulk-reacting (homogeneous) absorber layers. The contours are relatively continuous, as compared to the local absorber. The maxima of absorption are shifted downward toward smaller R values for larger angles.

The *random-incidence* absorption coefficient α_R for locally reacting and for bulk-reacting absorbers are plotted on the left and right sides of Fig. 8.12, respectively. For locally reacting absorbers the absolute maximum of $\alpha_R = 0.95$ is attained for a matched flow resistance $R = 1$ at $F = 0.367$, that is, at a thickness $d = 0.367\lambda_0$, which is larger than a free-field quarter wavelength, $\frac{1}{4}\lambda_0$. With homogeneous absorbers, the absorption coefficient α_R become larger than with locally reacting absorbers. Only a weak resonance maximum (belonging to the second resonance) is observed with homogeneous absorbers. The contour lines of absorption are smooth and steady.

Key Information Contained in Design Charts. Many of the general conclusions that were known qualitatively can be answered now in a quantitative manner by observing Figs. 8.10–8.12.

Determining the Thickness of an Absorber. What is the thickness d_∞ from which an absorber layer starts to behave acoustically as infinitely thick and yields no further increase in absorption? This thickness starts where the contour lines become 45° diagonally straight lines. The exact position of the curve for thickness d_∞ will depend on the criterion chosen and on the tolerance allowed. If, at this limit, the final value of α is supposed to be reached with a deviation between 1

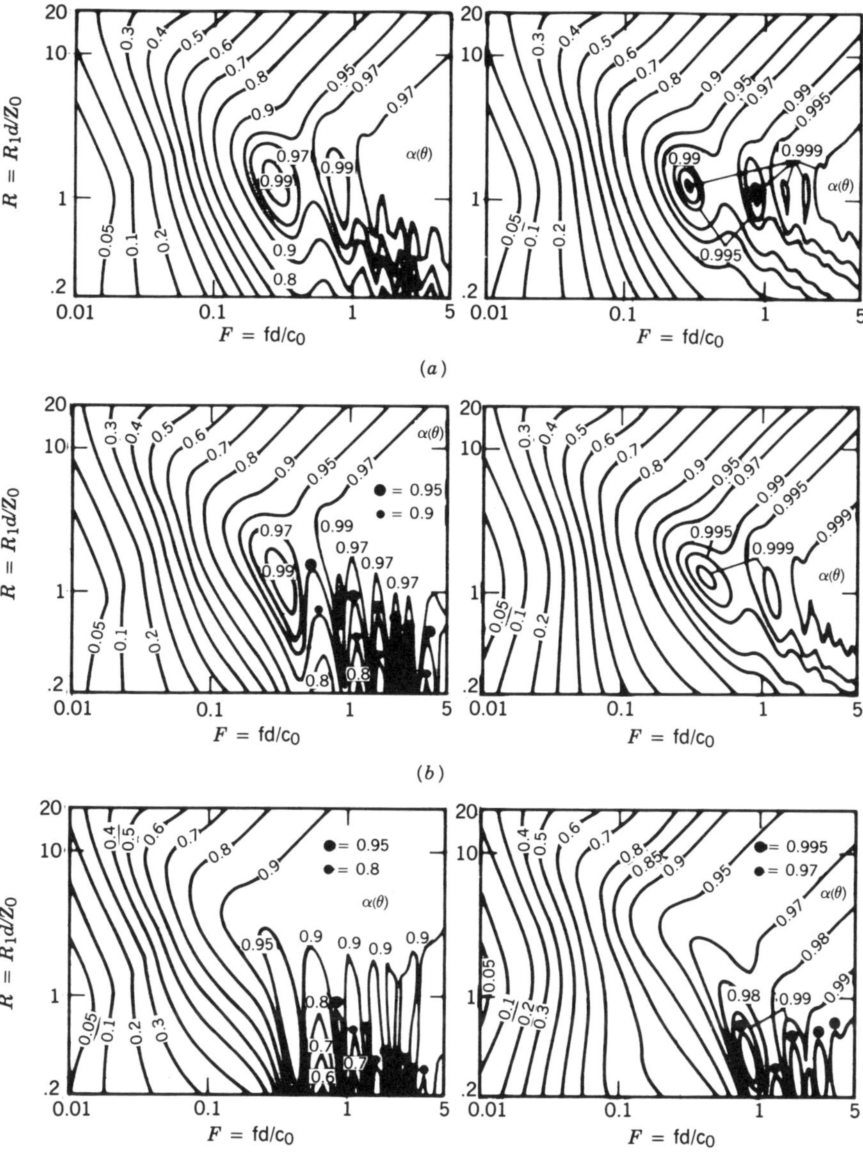

Fig. 8.11 Lines of constant absorption $\alpha(\theta)$ for fibrous absorbers of thickness d and flow resistivity R_1 for discrete angles of incidences θ; no airspace. Left: locally reacting. Right: bulk reacting. (*a*) $\theta = 30°$. (*b*) $\theta = 45°$. (*c*) $\theta = 60°$.

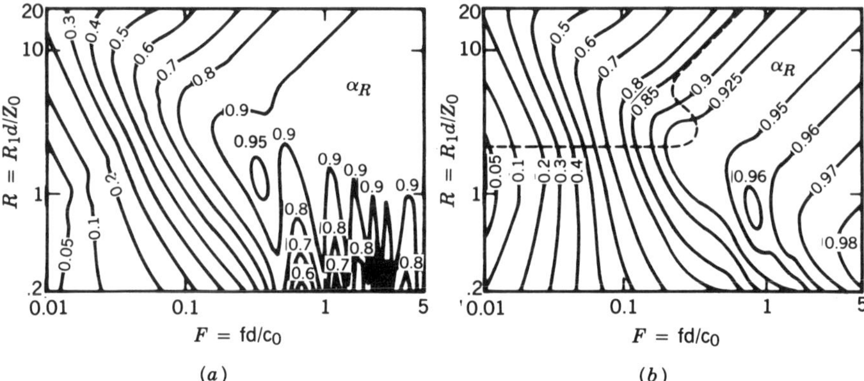

Fig. 8.12 Lines of constant random-incidence absorption coefficient α_R for fibrous absorbers of thickness d; no airspace: (a) locally reacting liner; (b) bulk reacting liner.

and 3%, then the limit curve for normal incidence (Fig. 8.10) is defined by

$$F = 7.45R^{-1.67} \tag{8.33}$$

The layer becomes practically infinite when the sound attenuation during propagation of the sound wave through the layer from the surface to the rigid rear wall is $8.68\Gamma'_a d_\infty = 24$ dB. This is a much higher attenuation than usually assumed as sufficient (about 6–10 dB) for neglecting the influence of the reflection from the rigid wall on the sound absorption at the front side of the absorber layer.

A lower number can be obtained if, from Fig. 8.10, for the limit of clearly visible relative maxima, in the third-octave-band frequency curves of α, we take the relation $F = 1.0R^{-1.63}$. This value would correspond to an attenuation of $\Delta L(d) = 8.68\Gamma'_a d_\infty \simeq 11$ dB. The relationship for "practically infinite" layer thickness d_∞ for which $\Delta L(d_\infty) = 24$ dB can be formulated as

$$fd_\infty^{2.67} R_1^{1.67} = 59 \times 10^6 \quad \text{for } \theta = 0° \tag{8.34a}$$

$$fd_\infty^{2.56} R_1^{1.56} = 34.3 \times 10^6 \quad \begin{array}{l}\text{for } \theta = 45°, \text{ locally reacting} \\ \text{and for random incidence}\end{array} \tag{8.34b}$$

$$fd_\infty^{2.4} R_1^{1.4} = 7.7 \times 10^6 \quad \text{for } \theta = 45°, \text{ bulk reacting} \tag{8.34c}$$

$$fd_\infty^{2.2} R_1^{1.2} = 1.1 \times 10^6 \quad \begin{array}{l}\text{for random incidence} \\ \text{bulk reacting}\end{array} \tag{8.34d}$$

where f is the frequency in hertz, d_∞ is the layer thickness in meters, and R_1 is the flow resistivity in N · s/m^4.

Optimal Choice of Flow Resistance. Optimal values for the normalized specific flow resistance of the layer $R = R_1 d/Z_0$, plotted on the vertical scale of Figs. 8.10–8.12, depend on what the designer wants to accomplish.

If the maximum sound absorption coefficient is the aim, then the optimal choice of R is in the range from 1 to 2 for both locally reacting and bulk-reacting absorbers and for normal, oblique, or random angle of sound incidence with the exception of bulk-reacting absorbers for $\theta > 45°$ incidence where $R = 0.7$ yields the best results.

The absorption coefficient depends strongly on flow resistivity R_l, where the curves in Figs. 8.10–8.12 have nearly horizontal contours. Observing the design charts reveals that the typical orientation of the curves of constant α is nearly vertical, indicating only a slight dependence on R_l. This is fortunate because our ability to determine R_l and our accurate control over its value in the manufacturing process is limited.

Bulk-reacting versus Locally Reacting Absorbers. A comparison of two graphs in Fig. 8.12 indicates that the random-incidence sound absorption coefficient of locally reacting and bulk reacting absorbers of the same thickness is practically identical (within 10%) in the left upper quadrant where $R > 2$ and $F = fd/c_0 < 0.25$, above the dotted line in the graph on the right in Fig. 8.12.

A comparison of $\theta = 45°$ curves in Fig. 8.11*b* with the random incident curves presented in Fig. 8.12 indicates that the $\theta = 45°$ curves for both locally reacting and bulk reacting absorbers agree with the random-incidence curves within an error of 5%, except in the range where $\alpha > 0.9$, which is of little practical interest. This close agreement between $\alpha(45°)$ and α_R indicates that it is permissible to compute $\alpha(45°)$ instead of the much more difficult α_R.

Two-Layer Absorbers

The two-layer absorbers shown in Fig. 8.7 have a layer of fibrous sound-absorbing material of thickness d, backed by an airspace of thickness t, with total thickness $D = d + t$. Figure 8.7*a* shows a partitioned (e.g., by a honeycomb) airspace resulting in a locally reacting absorber, and Fig. 8.7*b* shows an unpartitioned airspace resulting in a bulk reacting absorber. The sound absorber types shown in Fig. 8.7 are usually employed when it is impractical to fill the entire absorber thickness with porous material because no fibrous material of sufficiently low flow resistivity is available to keep the normalized flow resistance $R = R_l d/Z_0 < 2$. Based on the analyses described in reference 4, design charts similar to those presented in a preceding section for monolayer absorbers were computed.

Figure 8.13 shows contour maps of the random-incidence sound absorption coefficient α_R in a diffuse sound field for a bulk absorber layer of thickness d in front of a locally reacting air gap of thickness t (see Fig. 8.7*a*) for three fractions $d/D = 0.25, 0.5, 0.75$ of the absorber thickness d, and the total thickness $D = d + t$. Because the two-layer design curves in Fig. 8.13 are not plotted as functions of the total layer thickness D, they are not directly comparable with the monolayer design curves presented in Fig. 8.12. Some differences may be noticed by comparing these figures to Fig. 8.12 applicable for the bulk monolayer absorber.

First, the variable $F = fd/c_0$ for maximum absorption in the first resonance maximum is shifted toward smaller values with decreasing d/D, which results in

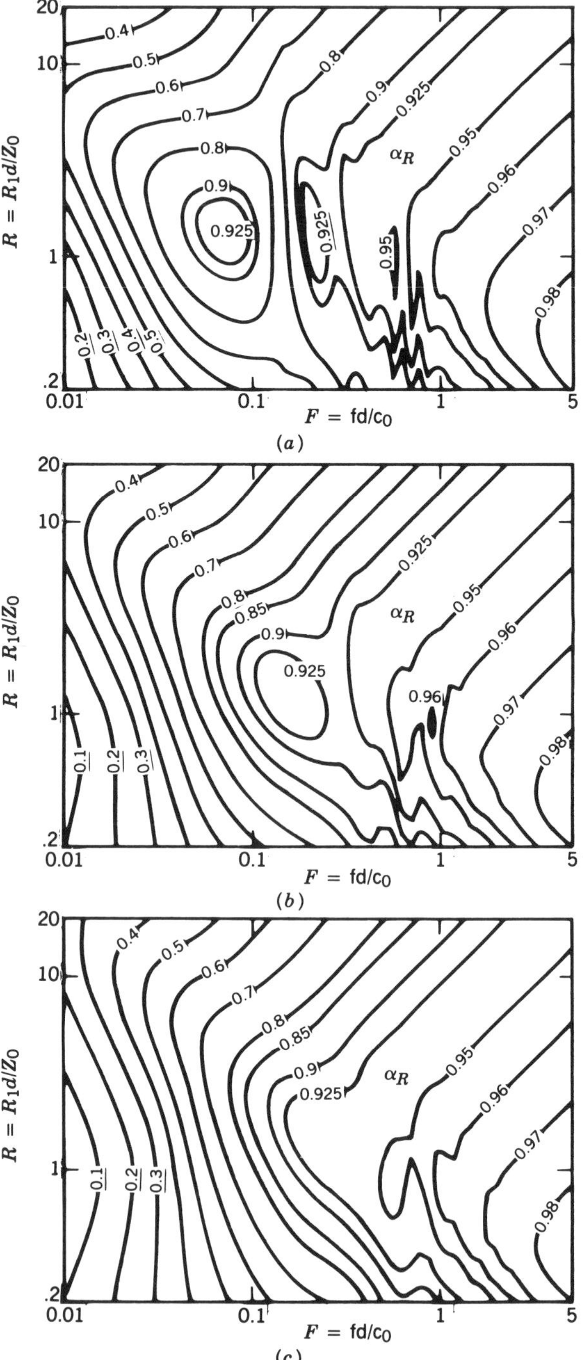

Fig. 8.13 Lines of constant random-incidence absorption coefficient α_R for a bulk-reacting absorber layer of thickness d in front of a locally reacting air gap of thickness t for thickness ratios $d/D = d/(d + t)$ of (*a*) 0.25; (*b*) 0.5; (*c*) 0.75.

a higher absorption at low frequencies for less absorber material thickness. Generally, however, the reduction in weight of the material is not as large. The horizontal shift of the absorption maximum toward lower F values, when realized by a reduction in d, makes necessary a compensation for the smaller thickness by an inversely larger flow resistivity necessary on the ordinate in order to hold R on a constant value.

The increase in R_1 is done mostly by an increase of the material bulk density ρ_A. For most fibrous absorber materials R_1 is proportional to about $\rho_A^{1.5}$. Hence, only a small net reduction in absorber material is possible by the addition of an air gap.

For thick absorber layers, that is, d large, the air gap has no effect, as expected. The limit for the onset of an "infinite" thickness is the same as with the monolayer absorber.

The resonance structure of the layered absorber with small d/D resembles more that of the locally reacting monolayer absorber (see Fig. 8.12*a*). The character of the bulk-reacting monolayer absorber plot (see Fig. 8.12*b*), which must be the asymptotic limit for increasing d/D, becomes dominant for $d/D \geq 0.5$.

The sound absorption coefficient of a porous layer of thickness d with an air gap behind is generally higher than that of a monolayer with equal d. However, the absorption of the layered absorber is smaller than that of a monolayer absorber with equal total thickness $d' = D$. In the range $R > 1$ of the absorption of the layered absorber with $d/D = 0.75$, there is a correspondence to a monolayer absorber with thickness $d' = 1.25d$ and with $d/D = 0.5$ to a monolayer having a thickness $d' = 1.67d$. The quite different character of the plot for $d/D = 0.25$ excludes a similar equivalence.

Finally, the lines of constant α_R of the layered absorber have strong deflections toward the left for values of α_R between about 0.7 and 0.9. As a consequence, the optimum flow resistance $R_1 d$ (which is defined by the leftmost point of a curve) has a more distinct meaning than with monolayer absorbers.

Design graphs for layered absorbers with bulk-reacting absorber layers in front of a bulk-reacting air gap (see Fig. 8.7*b*) are given in reference 4.

Multilayer Absorbers

Multilayer absorbers, such as shown in Fig. 8.9, have been treated in reference 4. Best results are obtained if the flow resistivity of the layers, R_1, increases from the interface toward the rigid wall. These results obtained with such multilayer absorbers are somewhat better than those obtained with a monolayer absorber of equal thickness. However, the improvements seldom justify the added cost and complexity.

Even the most elaborate multiple-layer absorber cannot, by far, match the random-incidence sound absorption performance of anechoic wedges.

Thin Porous Surface Layers

Thin porous surface layers such as mineral wool felt sprayed on plastic, steel wool, mineral wool, or glass fiber cloth; wire mesh cloth; and thin perforated metal are frequently used to provide mechanical protection. They also reduce the loss of fibers. They are acoustically characterized by their flow resistance $R_s = \Delta p / v$ and their mas per unit area ρ_s. The impedance of such surface layers is

$$Z_s = \begin{cases} R_s & (8.35a) \\ \dfrac{(j\omega\rho_s)R_s}{j\omega\rho_s + R_s} & (8.35b) \end{cases}$$

which must be added to the wall impedance of the absorber. If the porous surface layer is not free to move, use Eq. (8.35a). For surface layers that are free to move due to the pressure differential produced by their flow resistance, use Eq. (8.35b).

Tables 8.3 and 8.4 provide design information on wire mesh cloth and glass fiber cloth, respectively. The flow resistance values listed in these tables represent the linear part of the flow resistance, which is appropriate for design use only if the particle velocity is low. At high sound pressure levels (above 140 dB), the nonlinear behavior of the flow resistance must be determined by measuring the flow resistance as a function of the face velocity.

Sintered porous metals have been developed for silencers in jet engine inlets. If backed by a honeycomb-partitioned airspace, they provide good sound absorption and remain linear up to high sound pressure levels and for high Mach number grazing flow. Their additional advantage is that they do not require any surface protection. Table 8.5 gives the flow resistance R_s and mass per unit area ρ_s for some commercially available porous metal sheets.

Steel wool mats typically have a thickness from 10 to 45 mm, mass per unit area ρ_s from 1 to 3.8 kg/m^4, and a specific flow resistance R_s from 100 N · s/m^3 (0.006ρc) to 500 N · s/m^3 (0.03ρc). Recently, such steel wool mats needled on mineral wool of specified thickness and density have become available on special order.

Perforated metal facings, when they rest directly on a porous sound-absorbing layer, can be accounted for by a series impedance Z_s given by

$$Z_s = \frac{j\omega\rho_0}{\epsilon}\left\{l + \Delta l\left[H(1 - |v_h|) - j\frac{\Gamma_a}{k_0}\frac{Z_a}{Z_0}\right]\right\} \qquad (8.36)$$

TABLE 8.3 Mechanical Characteristics and Flow Resistance R_s of Wire Mesh Cloths

		Wire Diameter		Mass per Unit Area		Flow Resistance R_s	
Wires/cm	Wires/in.	μm (10^{-6} m)	mils (10^{-3} in.)	kg/m^2	lb/ft^2	N · s/m^3	$\rho_0 c_0$
12	30	330	13.0	1.6	0.32	5.7	0.014
20	50	220	8.7	1.2	0.25	5.9	0.014
40	100	115	4.5	0.63	0.13	9.0	0.022
47	120	90	3.6	0.48	0.1	13.5	0.033
80	200	57	2.25	0.31	0.63	24.6	0.06

TABLE 8.4 Mechanical Characteristics and Flow Resistance R_s of Glass Fiber Cloths[a]

Manufacturer[b]	Cloth No.	Surface Density oz/yd^2	Surface Density g/m^2	Construction, Ends × Picks	Flow Resistance, mks rayls (N · s/m^3)
1, 2, 3	120	3.16	96	60 × 58	300
1, 2, 3	126	5.37	164	34 × 32	45
1, 2, 3	138	6.70	204	64 × 60	2,200
1, 2, 3	181	8.90	272	57 × 54	380
3	1044	19.2	585	14 × 14	36
2	1544	17.7	535	14 × 14	19
3	3862	12.3	375	20 × 38	350
1	1658	1.87	57	24 × 24	10
1	1562	1.94	59	30 × 16	<5
1	1500	9.60	293	16 × 14	13
1	1582	14.5	442	60 × 56	400
1	1584	24.6	750	42× 36	200
1	1589	12.0	366	13 × 12	11

[a]Averaged over a large sample.
[b]Code numbers for manufacturers are as follows: (1) Burlington Glass Fabrics Company, (2) J. P. Schwebel and Company, and (3) United Merchants Industrial Fabrics.

TABLE 8.5 Specific (Unit-Area) Flow Resistance R_s, Thickness, and Mass Per Unit Area of Sintered Porous Metals Manufactured by the Brunswick Corporation[a]

Specific Flow Resistance Air, 70°F $\rho_0 c_0$	Specific Flow Resistance Air, 70°F N · s/m^3	NLF[b] 500/20	Designation	Thickness mm	Thickness in.	Mass Per Unit Area kg/m^2	Mass Per Unit Area lb/ft^2
0.25	100	3.6	FM 125	1.0	0.04	3.9	0.79
		5.0	FM 127	0.76	0.03	3.3	0.67
		2.6	FM 185	0.5	0.02	2.0	0.4
		2.0	347-10-20-AC3A-A	0.5	0.02	1.32	0.27
		2.0	347-10-30-AC3A-A	0.76	0.03	1.1	0.23
		2.0	802	0.5	0.02	1.3	0.27
0.88	350	4.7	FM 134	0.89	0.035	3.8	0.77
1.25	500	1.8	FM 122	0.76	0.03	1.4	0.28
		3.6	FM 126	0.66	0.026	3.7	0.76
		3.3	FM 190	0.41	0.016	2.0	0.4
		2.0	347-50-30-AC3A-A	0.76	0.03	1.4	0.29

[a]Except FM 802, which is made of Hastelloy x, all materials are type 347 stainless steel. (Courtesy of Brunswick Corporation.)
[b]Nonlinearity factor, calculated as the ratio of flow resistances obtained at flow velocities of 500 and 20 cm/s, respectively.

where ϵ is the fractional open area of the perforated facing, l is the plate thickness, Δl is the end correction length of the perforations given in Section 8.5, H is a step function that is 1 for $|v_h| \leq 1$ m/s and zero for $|v_h| > 1$ m/s, and v_h is the particle velocity in the holes of the perforations. For thin, unrestrained perforated surface protection plates, it is necessary to take the parallel combination of Z_s and $j\omega\rho_s$ according to Eq. (8.35b).

8.5 RESONANCE ABSORBERS

In building acoustics the most frequently used type of resonant absorber is the Helmholtz resonator consisting of the mass of an air volume in a cross-sectional area restriction (such as holes or slits in a covering plate) and the compliance of the air volume behind the covering plate. Such a Helmholtz resonator is shown in Fig. 8.14. In the following treatment it is assumed that all dimensions of a single resonator are small compared with the acoustical wavelength (except in the case of two-dimensional resonators with slits) and that the skeleton of the resonator is rigid.

Acoustical Impedance of Resonators

The specific acoustical impedance of the resonator opening Z_R is the sum of the impedance of the enclosed air volume Z_v and that of the air volume that oscillates in and around the resonator mouth Z_m; namely,

$$Z_R = Z_v + Z_m = (Z'_v + jZ''_v) + (Z'_m + jZ''_m) \quad \mathrm{N \cdot s/m^3} \tag{8.37}$$

The volume impedance $Z_v = jZ''_v$ is purely imaginary and predominantly of spring character while the mouth impedance has a real part Z'_m and an imaginary part Z''_m that is predominantly of mass character.

The impedance of a rectangular resonator volume is

$$Z_v = jZ''_v = -j\rho_0 c_0 \cot(k_0 t) \frac{S_a}{S_b} \quad \mathrm{N \cdot s/m^3} \tag{8.38a}$$

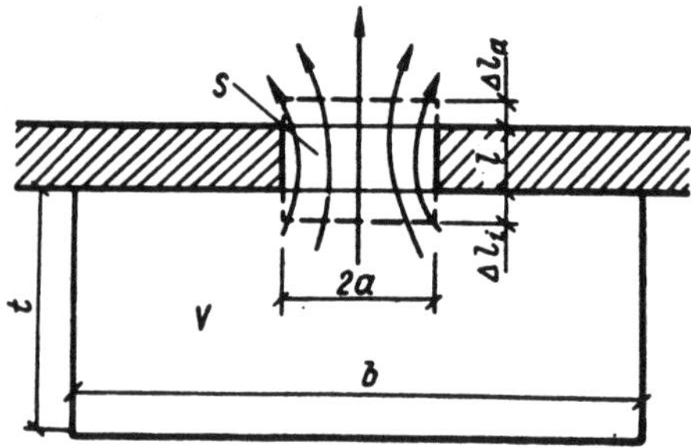

Fig. 8.14 Key geometric parameters of a Helmholtz resonator.

where S_b is the surface area of the resonator cover plate, $S_a = \pi a^2$ is the area of the resonator mouth, t is the depth of resonator cavity, and $V = S_b t$ is the resonator volume. If the resonator dimensions are small compared with the wavelength ($k_0 t \ll 1$), Eq. (8.38a) yields

$$Z_v = jZ_v'' = -j \frac{\rho_0 c_0^2}{\omega} \frac{S_a}{V} \quad \text{N} \cdot \text{s/m}^3 \tag{8.38b}$$

The impedance of the resonator mouth Z_m consists of

$$Z_m = Z_m' + jZ_m'' = Z_{mi} + Z_{m0} + Z_{me} + Z_s \quad \text{N} \cdot \text{s/m}^3 \tag{8.39}$$

where Z_{mi}, Z_{m0}, and Z_{me} are the components attributable to the oscillations of air internal to the mouth, within the mouth, and external to the mouth of the resonator, respectively and Z_s is the impedance of the screen that may be placed across the resonator mouth to provide resistance. For round resonator openings of radius a and orifice plate thickness l, these impedance components are predicted as

$$Z_{m0} = \rho_0 \frac{l}{2a} \sqrt{8\nu\omega} + j\omega\rho_0 l \left(1 + \frac{\sqrt{8\nu/\omega}}{2a}\right) \quad \text{N} \cdot \text{s/m} \tag{8.40a}$$

$$Z_{me} = Z_{\text{rad}} + j\omega\rho_0 \sqrt{\frac{8\nu}{\omega}} \quad \text{N} \cdot \text{s/m} \tag{8.40b}$$

$$Z_{mi} = j\omega\rho_0 \left(\frac{8}{3\pi} a + \sqrt{\frac{8\nu}{\omega}}\right) \quad \text{N} \cdot \text{s/m} \tag{8.40c}$$

where Z_{rad} is the radiation impedance of a baffled piston of radius a, which for $k_0 a < 1$ is given by

$$Z_{\text{rad}} = \tfrac{1}{4}\rho_0 c_0 (k_0 a)^2 + j\omega\rho_0 \frac{8}{3\pi} a \quad \text{N} \cdot \text{s/m} \tag{8.41}$$

where ν is the kinematic viscosity ($\nu = 15 \times 10^{-6}$ m^2/s for air at room temperature). The quantity $\delta_\nu = 0.5(8\nu/\omega)^{1/2}$ in Eqs. (8.40a) and (8.40b) is the so-called viscous boundary layer thickness. The combinations of Eqs. (8.40) and (8.41) yields

$$Z_m' = \rho_0 \left[\sqrt{8\nu\omega}\left(1 + \frac{l}{2a}\right) + \frac{(2\omega a)^2}{16 c_0}\right] + Z_s \quad \text{N} \cdot \text{s/m}^3 \tag{8.42a}$$

$$Z_m'' = \omega\rho_0 \left[l + \left(\frac{8}{3\pi}\right) 2a\right] + \left(1 + \frac{l}{2a}\right)\sqrt{8\nu\omega} \quad \text{N} \cdot \text{s/m}^3 \tag{8.42b}$$

In general, the first term in Eq. (8.42a) applies only for well-rounded orifice edges. In case of sharp orifice edges with burrs it can be many times higher. Also note that the second term in Eq. (8.42b) is usually small compared with the first term and can be neglected.

The results of Eq. (8.42), which are valid only for round orifices and for a single resonator, can be generalized for other orifice shapes and for situations where the orifice is located on a large wall or at a two-dimensional or three-dimensional corner. This generalized form is

$$Z'_m = \frac{P(l + P/2\pi)}{4S_a} \rho_0 \sqrt{8\nu\omega} + \rho_0 c_0 \frac{k_0^2 S_a/\Omega}{1 + k_0^2 S_a/\Omega} + Z_s \quad \mathrm{N \cdot s/m} \qquad (8.43a)$$

$$Z''_m = j\omega\rho_0 \left[l + 2\,\Delta l + \frac{P(l + P/2\pi)}{4S_a} \sqrt{8\nu/\omega} \right] \quad \mathrm{N \cdot s/m} \qquad (8.43b)$$

where P is the perimeter of the orifice, S_a is its surface area, and Ω is the spatial angle the resonator "looks into":

$$\Omega = \begin{cases} 4\pi & \text{for resonator away from all walls} \\ 2\pi & \text{flush mounted on wall far from corners} \\ \pi & \text{flush mounted on a wall at two-dimensional corner} \\ \frac{1}{2}\pi & \text{flush mounted on a wall at three-dimensional corner} \end{cases}$$

Resonance Frequency

The resonance frequency f_0 of the Helmholtz resonator occurs where Z''_m and Z''_v are equal in magnitude. Combination of Eqs. (8.38b) and (8.42b) yields

$$f_0 = \frac{c_0}{2\pi} \sqrt{\frac{S_a}{V\langle \cdots \rangle}} \simeq \frac{c_0}{2\pi} \sqrt{\frac{S_a}{V(l + 16a/3\pi)}} \qquad (8.44)$$

where $\langle \cdots \rangle = l + \Delta l$ represents the quantity in angular brackets in Eq. (8.42b). The quantity $\Delta l = 16a/3\pi$ represents the combined internal and external end corrections. For resonators where the area of the mouth, S_a, is not much smaller than the area of the top plate of the resonator, the vicinity of the side walls of the cavity influences the flow in the resonator mouth. In this case it is necessary to determine the combined end correction from $\Delta l = 8a/3\pi + \Delta l_{\text{int}}$, where the internal end correction Δl_{int} can be found in reference 7.

Absorption Cross Section of Individual Resonators

For individual resonators (or groups of resonators where the distance between the individual resonators is large enough that interaction is negligible), the absorption coefficient α is meaningless. In this case, the sound-absorbing performance must be characterized by the absorption cross section A of the indiviudal resonator. The absorption cross section is defined as that surface area (perpendicular to the direction of sound incidence) through which, in the undisturbed sound wave (resonator not present), the same sound power would flow through as the sound power absorbed by the resonator.

The power dissipated in the mouth of a resonator is

$$W_m = 0.5\ |v_a|^2 S_a R_T = 0.5\ 2^n |p_{\text{inc}}|^2 S_a R_T / |Z_R|^2 \quad \text{W} \tag{8.45}$$

Accordingly, the absorption cross section is

$$A = 2^n \rho_0 c_0 S_a R_T / |Z_R|^2 \quad \text{m}^2 \tag{8.46}$$

where n is 0 for resonators placed in free space, 1 if flush mounted in a wall, 2 if in a two-dimensional corner, and 3 if in a three-dimensional corner; $R_T = Z_m' + Z_{\text{rad}}$ is the total dissipative resistance.

At the resonance frequency f_0, where $Z_v'' + Z_m'' = 0$, the absorption cross section reaches its maximum value A_0:

$$A_0 = 2^n \frac{S_a \rho_0 c_0}{R_{\text{rad}}} \left[\frac{R_T / R_{\text{rad}}}{|1 + R_T / R_{\text{rad}}|^2} \right] \quad \text{m}^2 \tag{8.47a}$$

For a flush-mounted resonator with a round mouth area $S_a = \pi a^2$, for which $n = 1$, the radiation resistance is $R_{\text{rad}} = 2(\pi a)^2 \rho_0 c_0 / \lambda_0^2$, and Eq. (8.47a) yields

$$A_0 = \frac{1}{\pi} \lambda_0^2 \left[\frac{R_T / R_{\text{rad}}}{|1 + R_T / R_{\text{rad}}|^2} \right] \quad \text{m}^2 \tag{8.47b}$$

where λ_0 is the wavelength at the resonance frequency of the resonator. The maximum value of the absorption cross section is obtained by matching the loss resistance to the radiation resistance at the resonance frequency, yielding ($R_T = R_{\text{rad}}$)

$$A_0^{\max} = \frac{\lambda_0^2}{4\pi} \quad \text{m}^2 \tag{8.48}$$

According to Eq. (8.48), a matched resonator tuned to 100 Hz can achieve absorption cross section $A_0^{\max} = 0.92\ \text{m}^2$.

The frequency dependence of A according to Eq. (8.46) can be presented in generalized form as

$$\frac{A(f)}{A_0} = \frac{1}{1 + Q^2 \phi^2} \tag{8.49}$$

where the Q of the resonator and the normalized frequency parameter ϕ are

$$Q = \frac{Z''_m(\omega_0)}{Z'_m(\omega_0)} \qquad \phi = \frac{f}{f_0} - \frac{f_0}{f}$$

Figure 8.15 shows the normalized absorption cross section, A/A_0, as a function of the normalized frequency ϕ with Q as parameter, indicating that the normalized bandwidth of the absorption cross section f_0/Q corresponds to relative bandwidth $\Delta f/f_0 = 1/Q$.

Consequently, the choice of Q determines the relative shape of the absorption curve and the factor in the square brackets in Eq. (8.47b) determines the height of the curve at resonance. To obtain optimal absorption, characterized by high absorption at resonance A_0 and wide bandwidth, requires resistance matching $R_T = R_{\text{rad}}$ and a relatively low value of Q.

Nonlinearity and Grazing Flow

The oscillating air mass (that includes the end correction term) is a measure for the reversible kinetic energy of the acoustical resonator. Reversibility and spatial coherence diminish with increasing turbulence due to jet flow through the orifice caused by high-amplitude sound and by grazing flow of velocity U_∞. Both reduce the air mass that participates in the oscillating motion and consequently increase resonance frequency and increase the losses of the resonator.

Nonlinearity and grazing flow effect only the mouth impedance Z_m of the resonator orifice. Table 8.6 is a compilation of the nonlinear effects owing to high sound pressure level and grazing flow for a perforated plate. The round holes are regularly distributed. The porosity of the perforated plate is $\epsilon = \pi a^2/b^2$, where a is the hole radius and b the hole spacing. The amplitude nonlinearity is characterized by the particle-velocity-based Mach number $M_0 = v/c_0$, which is the ratio of the particle velocity in the resonator orifice and the speed of sound in air. The grazing flow nonlinearity is characterized by the flow-velocity-based Mach number

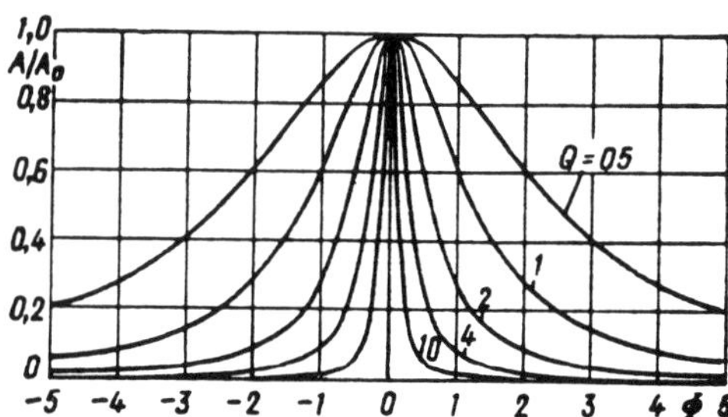

Fig. 8.15 Normalized absorption cross section A/A_0 as function of normalized frequency ϕ with Q as parameter.

TABLE 8.6 Nonlinearity of Orifice Impedance $Z_m = Z'_m + jZ''_m$ of Perforated-Plate Resonator Covers Caused by High Sound Pressure Level L_p and Grazing Flow U_∞[a]

Grazing Flow	Low SPL $L_p < L_{0l}$	Medium SPL $L_{0l} \le L_p \le L_{0h}$	High SPL $L_p > L_{0h}$
No Flow or Very Low Flow, $M_\infty \le 0.025$, $U_\infty \le 8$ m/s	$L_{0l} = 107 + 27 \log [4(1 - \epsilon^2)\omega\rho_0\nu(1 + l/2a)^2$ dB		$L_{0h} = L_{0l} + 30$ dB
	$Z'_m = R_0 \quad Z''_m = X_0(\delta)$	$Z'_m = \sqrt{R_h^2 - R_0^2} \quad Z''_m = X_0(\delta)$	$Z'_m = R_h \quad Z''_m = X_0(\delta)$
	$R_0 = \dfrac{\rho_0}{\epsilon}\sqrt{8\nu\omega}\left(1 + \dfrac{l}{2a}\right)$		$R_h = \dfrac{1}{\epsilon}\sqrt{2\rho_0(1 - \epsilon^2)} \times 10^{(-2.25 + 0.018L_p)}$
	$X_0 = \dfrac{\omega\rho_0}{\epsilon}\left[\sqrt{\dfrac{8\nu}{\omega}}\left(1 + \dfrac{l}{2a}\right) + l + \delta\right]$		
	$\delta = \delta_0 = 0.85(2a)\phi_0(\epsilon)$	$\delta = \delta_0\phi_1(M_0)$	$\delta = \delta_0\phi_1(M_0)$
	$\phi_0(\epsilon) = 1 - 1.47\sqrt{\epsilon} + 0.47\sqrt{\epsilon^3}$	$M_0 = \dfrac{10^{(-2.25 + 0.025L_p)}}{\sqrt{0.5\ \rho_0 c_0^2(1 + \epsilon^2)}}$	$\phi_1(M_0) = \dfrac{1 + 5 \times 10^3 M_0^2}{1 + 10^4 M_0^2}$
With Flow, $M_\infty \ge 0.025$, $U_\infty \ge 10$ m/s	$L_p \le L_{ml} \quad L_{ml} = 175 + 40 \log M_\infty$ dB	$L_{ml} \le L_p \le L_{mh}$	$L_p > L_{mh} \quad L_{mh} = L_{ml} + 18$ dB
	$Z'_m = R_m \quad Z''_m = X_0(\delta)$	$Z'_m = \sqrt{R_m^2 + R_h^2} \quad Z''_m = X_0(\delta)$	$Z'_m = R_h \quad Z''_m = X_0(\delta)$
	$R_m = 0.6\ \rho_0 c_0 \dfrac{1 - \epsilon^2}{\epsilon}(M_\infty - 0.025) - 40R_0(M_\infty - 0.05);\ M_\infty \le 0.05$		
	$R_m = 0.3\ \rho_0 c_0 \dfrac{1 - \epsilon^2}{\epsilon} M_\infty \quad M_\infty > 0.05$		
	$\delta = \delta_0\phi_2(M_\infty)$	$\delta = \delta_0\phi_2(M_\infty)$	$\delta = \delta_0\phi_1(M_\infty)$
	$\phi_2(M_\infty) = 1/(1 + 305M_\infty^3)$		

[a] L_p is octave-band SPL in dB *re* 20 μPa at the peak of the spectrum.

$M_\infty = U_\infty / c_0$, where U_∞ is the velocity of the grazing flow far from the wall. The formulas presented in Table 8.6 are analytical results adjusted to yield agreement with measured data [8]. The impedances in Table 8.6 are values averaged over the plate surface.

Loss Resistance of Resonators

The most difficult part of resonator design is the prediction of the loss resistance. In a previous section on acoustical impedance of resonators formulas are given to predict the friction loss resistance [first term in Eq. (8.42a)], which is valid only for rounded orifice perimeter. Additional resistance due to sharp edges or burrs cannot be predicted analytically. Table 8.6 contains formulas to predict resonator resistance due to nonlinear effects owing to high sound pressure level and grazing flow.

Should it be desirable to obtain higher loss resistance than provided by friction and nonlinear effects (e.g., for obtaining larger absorption bandwidth), porous materials such as a screen, felt, or a layer of fibrous sound-absorbing material must be placed in (or behind) the resonator orifice. Figure 8.16 shows some of the more frequently used ways to increase resonator resistance. In Fig. 8.16, R_1 is the flow resistivity of the porous material, d is its thickness, t_1 is its distance from the back side of the cover plate, t is the depth of the resonator cavity, $2a$ is the diameter of the holes, Δl is the end correction, and ϵ is the porosity of the cover plate.

Placing of flow-resistive materials in or behind the resonator opening reduces nonlinearity and sensitivity to grazing flow.

Spatial Average Impedance of Resonator Arrays

Locally reacting absorbers such as resonators are frequently used in the form of a surface array shown in Fig. 8.17, where the individual resonators are arranged at a raster. If the rasters (b or $\sqrt{S_b}$ in Fig. 8.17) are much smaller than the acoustical wavelength, it is not necessary to take into account the interaction of the individual elements (resonators) with the sound field. To characterize the absorber, it is sufficient in this case to compute the spatial average wall-impedance Z_1:

$$Z_1 = \frac{\langle p \rangle}{\langle v \rangle} = \frac{S_b\, p_a}{S_a\, v_a} = \frac{S_b}{S_a} Z_R = \frac{Z_R}{\epsilon} \qquad \mathrm{N \cdot s/m^3} \tag{8.50}$$

where Z_R is the impedance of the resonator as given in (Eq. 8.37) "measured" in the orifice of the resonator and $\epsilon = S_a/S_b$ is the surface porosity. The effective wall impedance Z_1 is then used in Eq. (8.18) or (8.25) to determine the absorption coefficient. The impedances Z_m in Table 8.6 are effective impedances averaged over the plate surface (resonator mass impedance divided by porosity).

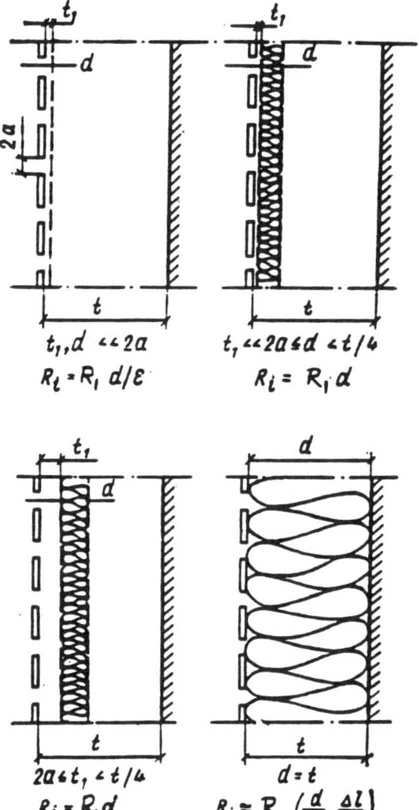

Fig. 8.16 Methods of increase resonator resistance and the achieved specific acoustical resistance R_i (R_1 = flow resistivity of porous material and d is its thickness).

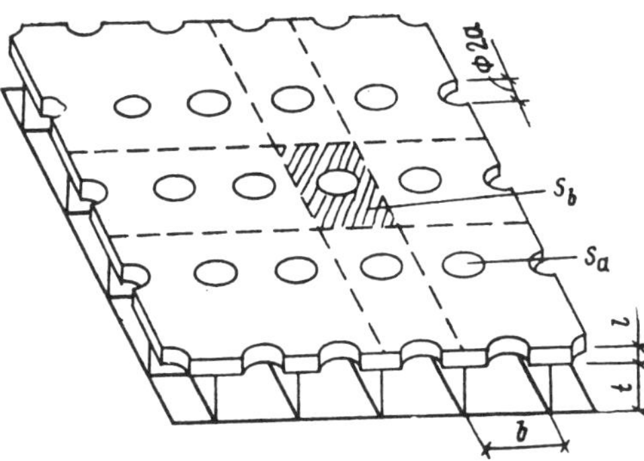

Fig. 8.17 Resonators arranged on a raster to form a surface array.

8.6 PLATE AND FOIL ABSORBERS

Plate absorbers are used in absorbing low-frequency tonal noise in situations where Helmholtz resonators are not feasible. The resonator consists of the mass of a thin plate and usually the stiffness of the airspace between the plate and the elastic mounting at the perimeter of the plate. Foils are a special case of a thin limp plate.

Limp Thin Plate or Foil Absorber

The resonance frequency of a foil absorber is

$$f_0 = 60/\sqrt{m''t} \qquad \text{Hz} \tag{8.51}$$

where t is the depth of the partitioned airspace in meters and m'' is the mass per unit area of the limp foil in kilograms per square meter. If the airspace depth is not small ($t > \frac{1}{8}\lambda$), there will be many resonance frequencies ω_n ($n = 0, 1, 2, \ldots$) given by

$$\frac{\omega_n t}{c_0} \tan \frac{\omega_n t}{c_0} = \frac{\rho_0 t}{m''} \tag{8.52}$$

For unpartitioned airspace behind the foil the resonance frequency depends on the angle of sound incidence θ_0. In this case t must be replaced in Eqs. (8.51) and (8.52) by $t \cos \theta_0$. If the airspace is filled (without obstructing the movement of the foil) with a porous sound-absorbing material, then use $(t/c_0)\Gamma_a''$ instead of t/c_0 in Eq. (8.52).

Foil-wrapped Porous Absorber

Sound absorbers consisting of a protective layer of thin foil, a porous layer, and airspace behind are frequently used where the porous material must be protected from dust, dirt, and water. The effect of the thin foil can be taken into account by a series impedance $Z_s = j\omega\, m''$ that is added to the wall impedance Z_1 [see Eqs. (8.20) and (8.22)]. It is essential that the foil is not stretched so that its inertia and not its membrane stresses control its response. Inserting a large mesh (≥ 1 cm) wire cloth between the porous sound-absorbing material and the protective foil is a practical way to assure this.

Elastically Supported Stiff Plate

A form of resonant plate absorber that can be easily treated analytically is the elastically supported stiff plate. The elastic support may be localized by discrete points to occur along the perimeter in the form of a resilient gasket strip. The effective stiffness of such a resonator is

$$s'' = s_e/S_p + \rho_0 c_0^2/t \qquad \text{N/m}^3$$

and the resonance frequency

$$f_0 = \frac{1}{2\pi}\sqrt{s''/m''} = \frac{1}{2\pi}\sqrt{\frac{s_e/S_p + \rho_0 c_0^2/t}{m''}} \qquad \text{Hz} \tag{8.53}$$

where $s_e = S_p \, \Delta p/\Delta x$ is the dynamic stiffness of resilient plate mounting and S_p is the surface area of the plate. Note that the elastic mounting increases the resonance frequency (compared with that obtainable with a limp foil of the same m''). This can be compensated for by an appropriate increase of m''. However, an increase of m'' increases the Q of the resonator ($Q = \sqrt{s''/m''}/R$) resulting in a narrower bandwidth.

Elastic Foil Absorber

Elastic foil absorbers consist of small air volumes enclosed in thin (200–400-μm) foil coffers. The coffers have typical dimensions of a few centimeter. A typical foil absorber is shown in Fig. 8.18. The wavelength of the incident sound is large compared with the typical dimensions of the coffers. Consequently, the incident sound periodically compresses the air in the individual coffers by exiting all of the volume-displacing modes of vibration of the various coffer walls. Because of the small thickness of the coffer walls, these resonances fall into the frequency range from 200 to 3150 Hz, which is of primary interest in building applications.[9] The foil material can be plastic (PVC) or metal. It is beneficial to emboss the foil because this leads to a more even distribution of the resonance frequencies. Figure 8.19 shows the measured random-incidence sound absorption coefficient obtained

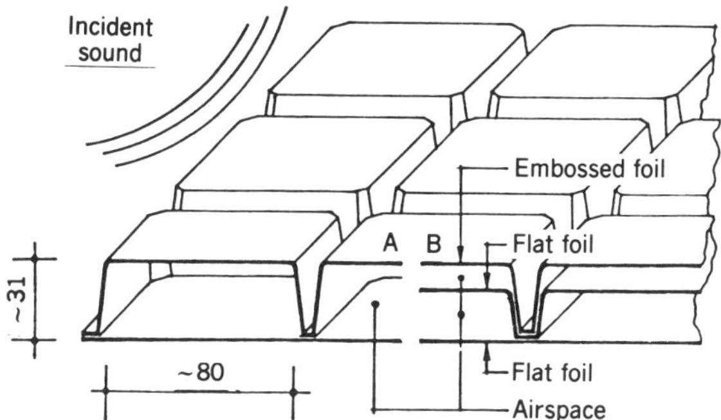

Fig. 8.18 Construction of a foil absorber made of cold-drawn PVC foil.

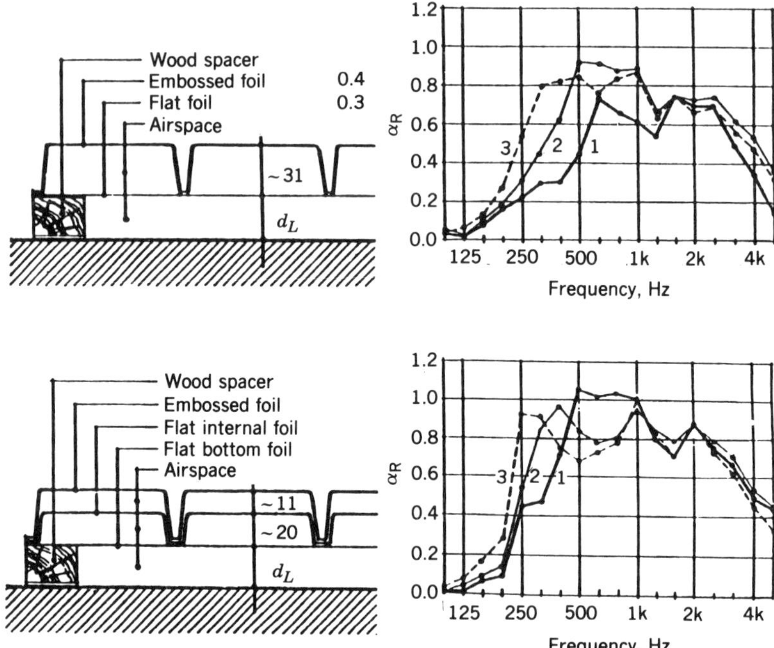

Fig. 8.19 Random-incidence sound absorption coefficient α_R of absorbers made of cold-drawn PVC foil: (1) $d_L = 0$; (2) $d_L = 25$ mm; (3) $d_L = 50$ mm.

with two different foil absorber configurations, indicating that significant broad-band sound absorption is obtainable.

The key advantage of these foil absorbers is that they are nonporous, do not support bacterial growth, are lightweight, and can be made light transparent. Their main application is breweries, packaging plants, hospitals, and computer chip manufacturing areas, where high emphasis is placed on hygiene and dust cannot be tolerated. They also lend themselves as sound absorbers in high-moisture environments such as swimming pools and as muffler baffles in cooling towers. Their light transparency is a distinct advantage in industrial halls. Considerable experience has been gained in such applications in Europe [9].

REFERENCES

1. R. R. Sullivan, "Specific Surface Measurements on Compact Bundles of Fibers," *J. Appl. Phys.* **13**, 725–730 (1942).
2. K. Attenborough, "The Influence of Microstructure on Propagation in Porous Fibrous Absorbants," *J. Sound Vib.* **16**, 419 (1971).
3. H. L. Kuntz, "High-Intensity Sound in Air-saturated, Fibrous Bulk Porous Materials," Ph.D. Thesis, University of Texas, Austin, 1982.

4. F. P. Mechel, ''Design Charts for Sound Absorber Layers,'' *J. Acoust. Soc. Am.* **83**(3), 1002–1013 (1988).
5. F. P. Mechel, ''Akustische Kennwerte von Faserabsorbern,'' [Acosutic Parameters of Fibrous Sound Absorbing Materials], Vol. I, Bericht BS 85/83; Vol. II, Bericht BS 75/82, Fraunhofer Inst. Bauphysik, Stuttgart.
6. J. Royar, ''Untersuchungen zum Akustishen Absorber-Kanten-Effekt an einem zweidimensionalen Modell'' [Investigation of the Acoustic Edge-Effect Utilizing a Two-Dimensional Model], Ph.D. Thesis, Faculty of Mathematics and Nature Sciences, University of Saarbrueken, 1974.
7. F. Mechel, ''Sound Absorbers and Absorber Functions,'' in G. L. Osipova and E. J. Judina, (eds.), *Reduction of Noise in Buildings and Inhabited Regions* (in Russian) Strojnizdat Publisher, Moskow, 1987.
8. J. L. B. Coelho, ''Acoustic Characteristics of Perforate Liners in Expansion Chambers,'' Ph.D. Thesis, Institute of Sound Vibration, Southampton, 1983.
9. F. Mechel and N. Kiesewetter, ''Schallabsorber aus Kunstoff-Folie,'' [Plastic-Foil Sound Absorber], *Acustica* **47**, 83–88 (1981).

BIBLIOGRAPHY

P. F. Mechel, *Sound Absorbers*, Vol. I, *Exterior Sound Fields, Interactions*; Vol. II, *Internal Sound Fields, Structures, Applications*; Vol. III, *Computer Programs* (in German), S. Hirzel Publisher, Stuttgart, 1989, 1991.

Chapter 9

Interaction of Sound Waves with Solid Structures

ISTVÁN L. VÉR
Bolt Beranek and Newman, Inc.
Cambridge, Massachusetts

The response of structures to dynamic forces or dynamic pressures is the subject of structural dynamics and acoustics. Structural dynamics is concerned predominantly with dynamic stresses severe enough to endanger structural integrity. Structural acoustics deals with low-level dynamic processes in structures resulting from excitation by forces, moments, and pressure fields. The primary interest in airborne and structureborne noise problems is the prediction of

1. power input into the structure,
2. the response of the excited structure,
3. propagation of structureborne sound to connecting structures, and
4. sound radiated by the vibrating structure.

The subject matter of this chapter is specific to noise control problems and thus is restricted to the audible frequency range and to air as the surrounding medium, though many of the concepts are directly applicable to liquid media or to higher or lower frequencies.

A typical noise control problem is illustrated in Fig. 9.1. A resiliently supported floor slab in the room to the left (source room) is excited by the periodic impacts

Noise and Vibration Control Engineering: Principles and Applications, Edited by Leo L. Beranek and István L. Vér.
ISBN 0-471-61751-2 © 1992 John Wiley & Sons, Inc.

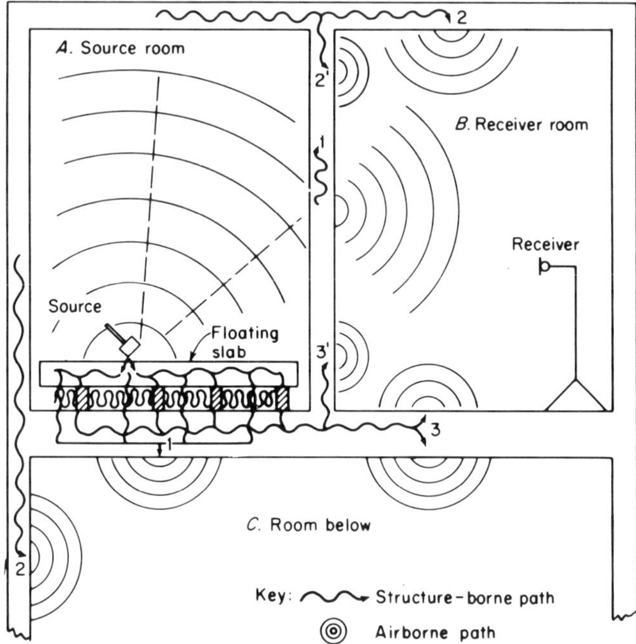

Fig. 9.1 Sound transmission paths between an impact source in room A and a receiver in room B. Also shown are the paths to room C below.

of a tapping machine, while the microphone in the receiving room to the right registers the resulting noise. A part of the vibrational energy is dissipated in the floating slab, a part is radiated directly as sound into the source room, and the remainder is transmitted through the resilient layer into the building structure.

The radiated sound energy builds up a reverberant sound field in the source room that in turn excites the walls. The vibrations in the wall separating the two rooms, identified as path 1 in Fig. 9.1, radiate sound directly into the receiver room. The vibrations in the other partitions of the source room travel in the form of structureborne sound to the six partitions of the receiving room and radiate sound into the receiver room. The structure-borne paths are identified by waves, 2, 2′ and 3, 3′.

To reduce the noise level in the receiving room by a desired amount, the acoustical engineer must estimate the sound power transmitted from the source room to the receiver room through each path and then design appropriate measures to (1) reduce power input to the floating slab, (2) increase power dissipation in the slab, (3) increase vibration isolation, (4) reduce the structureborne noise along its propagation path between source and receiver room, (5) reduce the sound radiated by the partitions of the source room, and (6) reduce the reverberation buildup of the sound in the receiver room by increasing sound absorption.

The aim of this chapter is to provide the information needed to solve such typical noise control engineering problems.

9.1 TYPES OF WAVE MOTION IN SOLIDS

Wave motion in solids can store energy in shear as well as in compression. The types of waves possible in solids include compressional waves, flexural waves, shear waves, torsional waves, and Rayleigh waves. Compressional waves are of practical importance in gases and liquids, which can store energy only in compression. The different types of waves in solids result from different ways of stressing. For a wave to propagate in solids, liquids, and gases, the medium must be capable of storing energy alternatively in kinetic and potential form. Kinetic energy is stored in any part of a medium that has mass and is in motion, while potential energy is stored in parts that have undergone elastic deformation.

Solid materials are characterized by the following parameters:

Density	ρ_M	kg/m^3
Young's modulus	E	N/m^2
Poisson's ratio	ν	
Loss factor	η	

The shear modulus G is related to Young's modulus as

$$G = \frac{E}{2(1 + \nu)} \qquad \text{N/m}^2 \tag{9.1}$$

The sketches on the left of Table 9.1 illustrate the deformation pattern typical for compressional, shear, torsional, and bending waves in bars and plates. Acoustically important parameters of solid materials are compiled in Table 9.2. Note that all tables in this chapter are placed at the end of text.

In dealing with structureborne sound, we must distinguish between velocity and propagation speed. The term *velocity* refers to vibration velocity of the structure (i.e., the time derivative of the local displacement), which is linearly proportional to the excitation and will be designated by the letter v. As illustrated on the left of Table 9.1, the displacement and velocity are in the direction of wave propagation for longitudinal waves and perpendicular to it for shear and bending waves.

The term *speed* refers to the propagation speed of structureborne sound, which is a characteristic property of the structure for each type of wave motion and is independent of the strength of excitation provided that the deformations are small enough to avoid nonlinearities. Propagation speeds will be designated by the letter c. We must distinguish between *phase speed* and *group speed* (or energy speed).

The phase speed c is defined in terms of (1) the wavelength λ, which is the distance in the propagation direction for which the phase of sinusoidal wave changes 360°, and (2) the frequency f of the sinusoid. Consequently, $c = \lambda f$. Formulas to calculate phase speed of the various wave types are given on the right side of Table 9.1. Note that the phase speed for longitudinal, shear, and torsional waves is independent of frequency. Consequently, they are referred to as nondispersive waves. This means that the time history of an impulse, such as caused by a hammer blow striking axially on one end of a semi-infinite bar, as illustrated in Fig. 9.2*a*, will have the same shape regardless of where the axial motion is sensed. Because the pulses sensed at different locations are time-delayed versions of each other, the propagation speed of the longitudinal wave can be determined experimentally as the ratio of the axial distance and the difference of arrival time. Since all frequency components travel at the same speed, the phase speed (which is defined only for a steady-state sinusoidal excitation) is identical with the speed of energy transport.

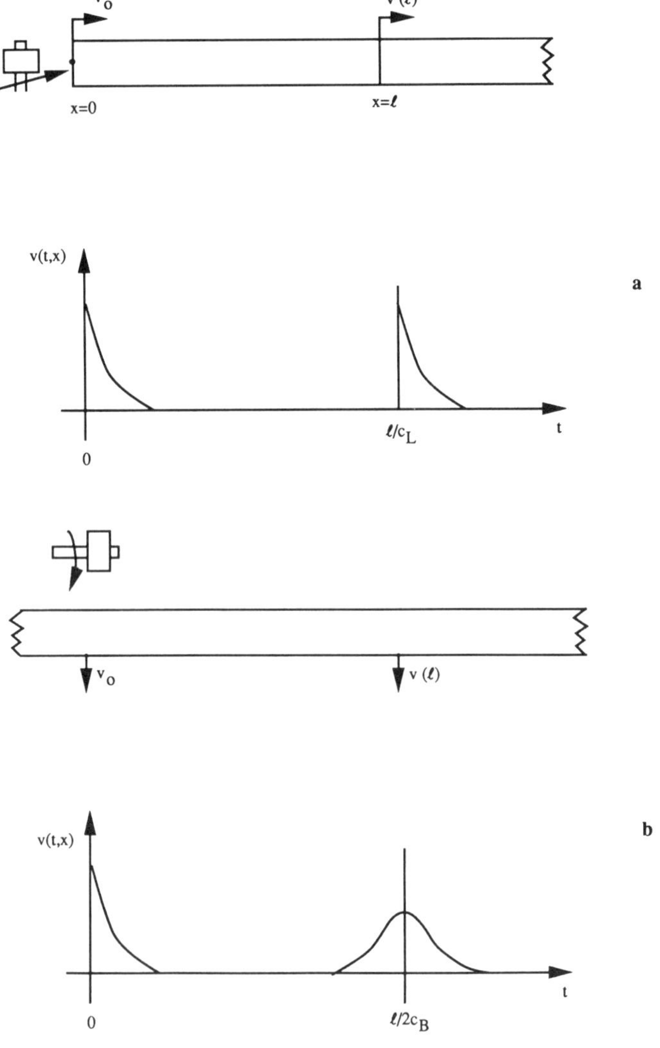

Fig. 9.2 Time history of beam motion for impulsive excitation: (*a*) axial impact generating nondispersive compressional waves; (*b*) normal impact generating dispersive bending waves.

The phase speed of the bending wave can be easily measured by exciting the infinite beam with a steady-state sinusoidal force acting normal to the beam axis and measuring the gradient of the phase $d\phi/dx$ along the beam. The phase velocity c_B is defined as

$$c_B = \frac{2\pi f}{|d\phi/dx|} \qquad \text{m/s} \tag{9.2}$$

However, the situation is much more complicated in the case of a complex waveform such as illustrated in Fig. 9.2*b*, where an infinite beam is impacted normally and a bending-wave pulse that contains a wide band of frequencies is created. Because of their higher phase speed, high-frequency components speed ahead of the low-frequency component. Consequently, the width of the pulse increases as it travels along the beam. The speed of energy transport for the bending-wave pulse is not obvious. In case of light fluid loading (when the energy carried in the surrounding fluid is negligible), it is meaningful to define the energy transport speed as the ratio of transmitted power W and the energy density E'' (per unit length of beam). For a pulse whose spectrum peaks at frequency f the energy speed c_{BG} is twice the phase speed[1]:

$$c_{BG}(f) = 2c_B(f) \qquad \text{m/s} \tag{9.3}$$

Accordingly, the power transmitted by a plane bending wave in the direction of propagation per unit area of a structure normal to that direction is

$$W_S = c_{BG}E'' = 2c_B\rho_M v^2 \qquad \text{W/m}^2 \tag{9.4}$$

where v is the rms value of the bending-wave velocity and ρ_M is the density of the material. According to Eq. (9.4), the bending-wave power in an infinite beam of cross section S is given by

$$W = 2c_B S\rho_M v^2 \qquad \text{W} \tag{9.5}$$

and power transmitted by a plate of thickness h across a line length l aligned parallel to the wave front is

$$W = 2c_B hl\rho_M v^2 \qquad \text{W} \tag{9.6}$$

9.2 MECHANICAL IMPEDANCE AND POWER INPUT

Except for the initial transient, the velocity at the excitation point is zero if a structure is subjected to a static force. Accordingly, a static force does not transmit power into a fixed structure. For dynamic excitation by a point force or moment [the moment $M = Fl_m$ is defined as a pair of forces (F) of equal magnitude but opposite direction acting simultaneously at a short distance ($\frac{1}{2}l_M$) on both sides of the excitation point], there is always a finite dynamic velocity or dynamic rotation at the excitation point. If the velocity at the excitation point has a component that is in phase with the force, or the angular velocity has a component that is in phase with the moment, then power is fed continuously into the structure.

Mechanical Impedance

The mechanical impedance is a measure of how a structure ''resists'' outside forces or moments. A thorough understanding of mechanical impedances is a necessary requirement for understanding and for solving most noise control problems. The concept of mechanical impedance is best introduced by considering a harmonic force $F = \hat{F}e^{j\omega t}$ acting on an ideal lumped-parameter system such as a rigid unrestrained mass, a massless spring, or a dashpot. The velocity response is also a harmonic function of the form $v = \hat{v}e^{j\omega t}$. The force impedance Z_F is defined as

$$Z_F = \frac{\hat{F}}{\hat{v}} \quad \text{N} \cdot \text{s/m} \tag{9.7}$$

and the moment impedance Z_M as

$$Z_M = \frac{\hat{M}}{\hat{\dot{\theta}}} \quad \text{N} \cdot \text{s/m} \tag{9.8}$$

where M is the exciting moment (torque) and $\dot{\theta}$ is the angular velocity at the excitation point.

The point force impedance for these basic lumped elements are listed below:

Element	Z_F	
mass	$j\omega m$	F m
spring	$-js/\omega$	F s
dashpot	r	F r

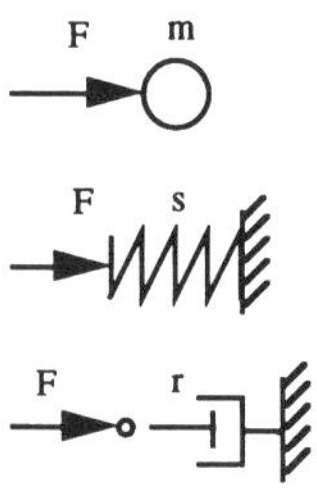

where m is the mass of a rigid body, s is the stiffness of the spring, and r is the resistance of the dashpot. Note that the direction of the force must go through the center of gravity of the mass so that the velocity response is free of rotation and is in the direction of the force. For the rigid mass and the massless spring the force impedance is imaginary, indicating that the force and velocity are in quadrature and the power input is zero.

Formulas for computing the point force and moment impedances of infinite structures are given in Tables 9.3 and 9.4, respectively. Both force and moment impedance are idealized concepts, defined for point forces and point moments. Forces or moments can be considered applied at a point, provided that the surfaces over which they act are smaller than one-sixth of the structural wavelength. Naturally, the excited surface area must be large enough so that the force does not cause plastic deformation of the structure. There is proportionality between force and moment impedance in the form of

$$Z_M = l_M^2 Z_F, \quad \text{N} \cdot \text{m/s} \tag{9.9}$$

where l_M is the effective length given in Table 9.5.

Power Input

For the dashpot the force impedance is real. The velocity is in phase with the force and the power that is fed continuously into the dashpot with resistance and dissipated in it is given by $\frac{1}{2}|\hat{F}|^2/r$. More generally, for a system with a complex force impedance Z_F, the input power is

$$W_{\text{in}} = \frac{\omega}{2\pi}\int_0^{2\pi/\omega} F(t)v(t)\,dt = \tfrac{1}{2}|\hat{F}|\,|\hat{v}|\cos\phi = \tfrac{1}{2}|\hat{F}|^2\text{Re}\left\{\frac{1}{Z_F}\right\} \quad \text{W} \tag{9.10}$$

where $\hat{F}$ and $\hat{v}$ are the peak amplitude of the exciting force and the velocity, ϕ is the relative phase between $F(t)$ and $v(t)$, and Re{ } refers to the real part of the bracketed quantity. Equation (9.10) is valid for any situation, not only for the dashpot.

Rigid masses, massless springs, and dashpots are useful abstractions that represent the dynamic properties of structural elements. At sufficiently low frequencies, the housing of a resiliently mounted pump behaves like a rigid mass, while the resilient rubber mount that supports it behaves like a spring. This lumped-parameter characterization of mechanical elements becomes invalid when, with increasing frequency, the structural wavelength becomes smaller than the largest dimensions of the structural element.

To deal with finite structures that are large compared with the structural wavelength we need another useful abstraction. Regarding their input impedance—which governs power input—it is useful to consider such structures as infinite. The strategy for predicting the vibration response and sound radiation for these structures is to estimate the power input as if they were infinite and then predict the average vibration that such input power will cause in the actual finite structure. Because most structures are made of plates and beams, our strategy requires us to characterize the impedance of infinite plates and beams. The abstraction of infinite size is invoked so that no structural waves reflected from the boundaries come back to the excitation point. In this respect, structures can be considered infinite if no reflected waves that are coherent with the input come back to the excitation point because the boundaries are completely absorbing or because there is sufficient damping so that the response attributable to the reflected waves is small compared to that attributable to the local excitation.

In respect to point force excitation, infinite structures are characterized by their point force impedance $\tilde{Z}_{F\infty}$ or point input admittance (also called mobility) $\tilde{Y}_{F\infty}$ defined as

$$\tilde{Z}_{F\infty} = \frac{\tilde{F}_0}{\tilde{v}_0} \quad \text{N} \cdot \text{s/m} \tag{9.11}$$

$$\tilde{Y}_{F\infty} = \frac{1}{\tilde{Z}_{F\infty}} = \frac{\tilde{v}_0}{\tilde{F}_0} \quad \text{m/N} \cdot \text{s}, \tag{9.12}$$

where $\tilde{F}_0$ is the applied point force and $\tilde{v}_0$ is the velocity response at the excitation point. The tilde above the symbols (which we will drop in further considerations) signifies that F_0 and v_0 are complex scalar quantities characterized by a magnitude and phase. Since $\tilde{v}_0$ usually has a component that is in phase with $\tilde{F}_0$ and one that is out of phase with $\tilde{F}_0$, the parameters $\tilde{Z}_F$ and $\tilde{Y}_F$ usually have both real and imaginary parts.

The power input to infinite structures excited by a point force or by a point velocity source (i.e., a vibration source of high internal impedance) are, respectively,

$$W_{\text{in}} = |\tfrac{1}{2}\hat{F}_0^2| \text{Re}\left\{\frac{1}{Z_{F\infty}}\right\} \quad \text{W} \tag{9.13a}$$

$$W'_{\text{in}} = G_{FF} \text{Re}\left\{\frac{1}{Z_{F\infty}}\right\} \quad \text{W/Hz} \tag{9.13b}$$

$$W_{\text{in}} = (\tfrac{1}{2}\hat{v}_0^2) \text{Re}\{Z_{F\infty}\} \quad \text{W} \tag{9.14a}$$

$$W'_{\text{in}} = G_{VV} \text{Re}\{Z_{F\infty}\} \quad \text{W/Hz} \tag{9.14b}$$

where $\hat{F}_0$ and $\hat{v}_0$ are peak amplitudes of the exciting sinusoidal force and velocity and G_{FF} and G_{VV} are the force and velocity spectral densities where the excitation is of broadband random nature. For excitation by a moment of peak amplitude $\hat{M}$ or by an enforced angular velocity of peak amplitude $\hat{\theta}$, the corresponding expressions are

$$W_{\text{in}} = |\tfrac{1}{2}\hat{M}_0^2| \text{Re}\left\{\frac{1}{Z_M}\right\} \quad \text{W} \tag{9.15a}$$

$$W'_{\text{in}}/\text{Hz} = G_{MM} \text{Re}\left\{\frac{1}{Z_M}\right\} \quad \text{W/Hz} \tag{9.15b}$$

$$W_{\text{in}} = |\tfrac{1}{2}\hat{\theta}| \operatorname{Re}\{Z_M\} \qquad \text{W} \tag{9.16a}$$

$$W'_{\text{in}} = G_{\theta\theta} \operatorname{Re}\{Z_M\} \qquad \text{W/Hz} \tag{9.16b}$$

Table 9.3 lists the point force impedance $Z_{F\infty}$ and Table 9.4 the moment impedance $Z_{M\infty}$ for semi-infinite and infinite structures. Table 9.5 lists the effective length l_M that connects the point force and moment impedances according to Eq. (9.9). Information regarding power input to infinite structures for force, velocity, moment, and angular velocity excitation is compiled in Table 9.6. The most complete collection of impedance formulas, which incudes beams, box beams, orthotropic and sandwich plates with elastic or honeycomb core, grills, homogeneous and rib-stiffened cylinders, spherical shells, plate edges, and plate intersections and transfer impedances of various kinds of vibration isolators is presented in reference 12.

For approximate calculations or for cases where no formulas are given in Tables 9.3–9.6, the power input to infinite structures can be estimated according to Heckl[10, 11] as

$$W_{\text{in}} = \frac{(\frac{1}{2}\hat{F}_0^2)}{Z_{\text{eq}}} \qquad \text{W} \tag{9.17}$$

$$W_{\text{in}} = (\tfrac{1}{2}\hat{v}_0^2) Z_{\text{eq}} \qquad \text{W} \tag{9.18}$$

where Eq. (9.17) is used for localized force excitation (i.e., low-impedance vibration source) and Eq. (9.18) for localized velocity excitation (i.e., high-impedance vibration source) and Z_{eq} is estimated as

$$Z_{\text{eq}} = \omega\rho_M S\epsilon[\alpha\lambda] \qquad \text{N}\cdot\text{s/m} \quad \text{(beam)} \tag{9.19a}$$

$$Z_{\text{eq}} = \omega\rho_M h\epsilon[\pi(\alpha\lambda)^2] \qquad \text{N}\cdot\text{s/m} \quad \text{(plate)} \tag{9.19b}$$

$$Z_{\text{eq}} = \omega\rho_M \epsilon[\tfrac{4}{3}\pi(\alpha\lambda)^3] \qquad \text{N}\cdot\text{s/m} \quad \text{(halfspace)} \tag{9.19c}$$

where ρ_M is the density of the material, S is the cross-sectional area of the beam, h is the plate thickness, λ is the wavelength of the motion excited most strongly (bending, shear, torsion, or compression), and ϵ is 1 when the structure is excited in the "middle" and 0.5 when it is excited at the end or at the edge (semi-infinite structure) and α is a number in the range of 0.16–0.6. If not known, it is customary to use $\alpha = 0.3$.

The equivalent impedance Z_{eq} in Eq. (9.19) has an instructive and easy-to-remember interpretation, namely that the magnitude of the point input impedance is roughly equal to the impedance of a lumped mass that lies within a sphere of radius $r \simeq \frac{1}{3}\lambda$ centered at the excitation point. This portion of the structure is shown in the third column of Table 9.3. This principle can be extended to composite structures[10] such as beam-stiffened plate where the combination of Eqs.

(9.19a) and (9.19b) yields

$$Z_{\text{eq}} \cong \omega \rho_B S_B (\tfrac{1}{3}\lambda_{BB}) + \omega \rho_p h (\tfrac{1}{3}\lambda_{BP}) \qquad \text{N} \cdot \text{s/m} \tag{9.20}$$

Another useful rule of thumb is that, regarding power input, excitation by a moment $\hat{M}$ and by an enforced angular velocity $\hat{\dot{\theta}}$ can be represented by an equivalent point force $\hat{F}_{\text{eq}}$ or equivalent velocity $\hat{v}_{\text{eq}}$,[11]

$$\hat{F}_{\text{eq}} \cong \frac{\hat{M}}{0.2\lambda_B} \qquad \text{N} \tag{9.21}$$

$$\hat{v}_{\text{eq}} \cong \hat{\dot{\theta}}(0.2\lambda_B) \qquad \text{m/s} \tag{9.22}$$

If the excitation force extends over an area that is large compared with the bending wavelength, the power transmitted into the structure becomes substantially less than it would be for a concentrated force. For extended velocity source the power transmitted becomes significantly larger than it would be for a concentrated velocity excitation.

According to reference 5, the real part of the point force impedance, Re $\{Z_{F\infty}\}$, can be predicted according to Bode's theorem as

$$\text{Re}\{Z_{F\infty}\} \cong |Z_{\text{eq}}| \cos(\tfrac{1}{2}\epsilon\pi) \qquad \text{N} \cdot \text{s/m} \tag{9.23}$$

where ϵ is defined here as an exponent of the Z_{eq}-versus-frequency curve (i.e., $Z_{\text{eq}} \sim \omega^{\epsilon}$) where $\epsilon = 0$ for homogeneous isotropic thick plates, $\epsilon = 0.5$ for beams in bending, and $\epsilon = \frac{2}{3}$ for cylinders below the ring frequency.

Parameters Influencing Power Input. When the excitation point is near a structural junction, part of the incident vibration wave is reflected. This reflected wave influences the velocity at the excitation point and thereby also the driving point impedance and consequently the power input. Fig. 9.3*a* shows the effect of the vicinity of a T-junction comprised of identical, anechoically terminated beams on the point force impedance $Z_2(x_0)$ as a function of the distance x_0 between the junction and the excitation point.[13] As expected, the impedance increases rapidly with decreasing distance when $x_0 < \frac{1}{3}\lambda_B$. The junction has a considerable effect even for large x_0, causing approximately a 2 : 1 fluctuation in the magnitude of the impedance. Figure 9.3*b* shows the measured distribution of the beam vibration velocity $|v(x)|$ along each of the three beam branches, indicating that the observed strong standing-wave pattern is limited to that part of beam 2 that lies between the junction and the excitation point.

The impedance forumulas presented in Tables 9.3–9.6 are strictly valid only when the surrounding fluid medium has no appreciable effect on the response of the structure. This is generally true for air as the surrounding medium but not for water, which is 800 times more dense than air. The effect of fluid loading at low

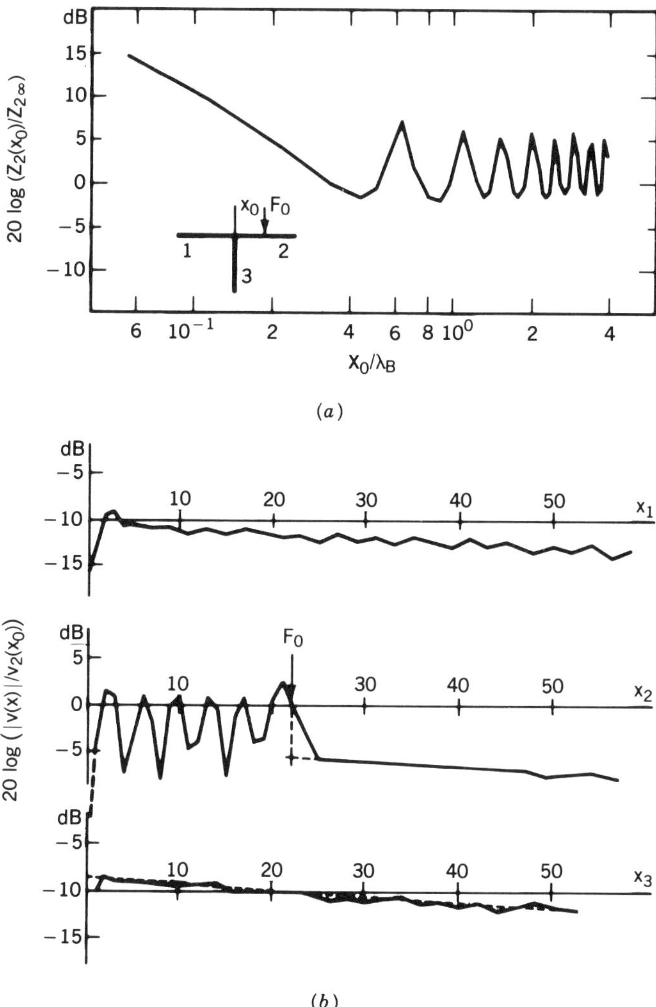

Fig. 9.3 Effect of the vicinity of T-junction on the measured point force impedance and velocity response[13]: (*a*) normalized force impedance $Z_1(x_0)/Z_{1\infty}$ as a function of the normalized distance x_0/λ_B and x_i is the distance from the junction along beam i; (*b*) distribution of normalized velocity, $v(x)/v_0$, along the three beam branches as numbered in (*b*).

frequencies is equivalent[14] to adding a virtual mass per unit area ρ_s',

$$\rho_s' = \frac{\rho_0 c_B(f)}{\omega} \cong \rho_0[\tfrac{1}{6}\lambda_B(f)] \qquad \text{kg/m}^2 \tag{9.24}$$

to the mass per unit area of the plate. In Eq. (9.24), ρ_0 is the density of the fluid and $c_B(f)$ and $\lambda_B(f)$ are the bending-wave speed and the wavelength of the free bending waves in the unloaded plate (see Table 9.1). Because ρ_s' is inversely proportional to $\omega^{1/2}$, fluid loading results in greater reduction of the response at the excitation point at low than at high frequencies. Fluid-loading effects are treated in references 14–17.

9.3 POWER BALANCE AND RESPONSE OF FINITE STRUCTURES

For finite structures, the response at the excitation point, and accordingly also the input admittance, $Y = 1/Z$, depends on the contribution of the waves reflected from the boundaries or discontinuities. Consequently, Y varies with the location of the excitation point and with frequency. However, both the space-averaged and freuqency-averaged input admittances of the finite structure $\langle Y(x, f)\rangle_x$ and $\langle Y(x, f)\rangle_f$ equal the point input admittance of the equivalent infinite structure [18], namely,

$$\langle Y(x, f)\rangle_x = \langle Y(x, f)\rangle_f = Y_\infty \qquad \text{m/N} \cdot \text{s} \tag{9.25}$$

Figures 9.4*a* and 9.4*b* show the typical variation of the real and imaginary parts of the point input admittance of a finite plate with location x and frequency f respectively. This particular behavior has considerable practical importance:

1. The power introduced into a finite structure by a point force of random-noise character can be well approximated by the power that the same force would introduce into an equivalent infinite structure.
2. The power introduced into a finite structure by a large number of randomly spaced point forces can be approximated by the power the same forces would introduce into an equivalent infinite structure.

Resonant Modes and Modal Density

The peaks in Fig. 9.4*a* correspond to resonance frequencies of the finite structure, where waves reflected from the boundaries travel in closed paths such that they arrive at their starting point in phase. The spatial deformation pattern that corresponds to such a particular closed path is referred to as the *mode shape* and the frequency where it occurs as the *eigenfrequency* or *natural frequency* of the finite structure. The importance of such resonances lies in the high transverse velocity caused by the in-phase superposition of the multiple reflections that may result in increased sound radiation or fatigue.

Exact calculation of the natural frequencies is possible only for a few highly idealized structures. Fortunately, the modal density $n(f)$, which is defined as the average number of natural frequencies in a 1-Hz bandwidth, depends not too strongly on the boundary conditions. Accordingly, one can make a reliable statistical prediction of the modal density using the formulas obtained for the equivalent idealized system.

For example, the modal density of a thin, flat, homogeneous, isotropic plate of

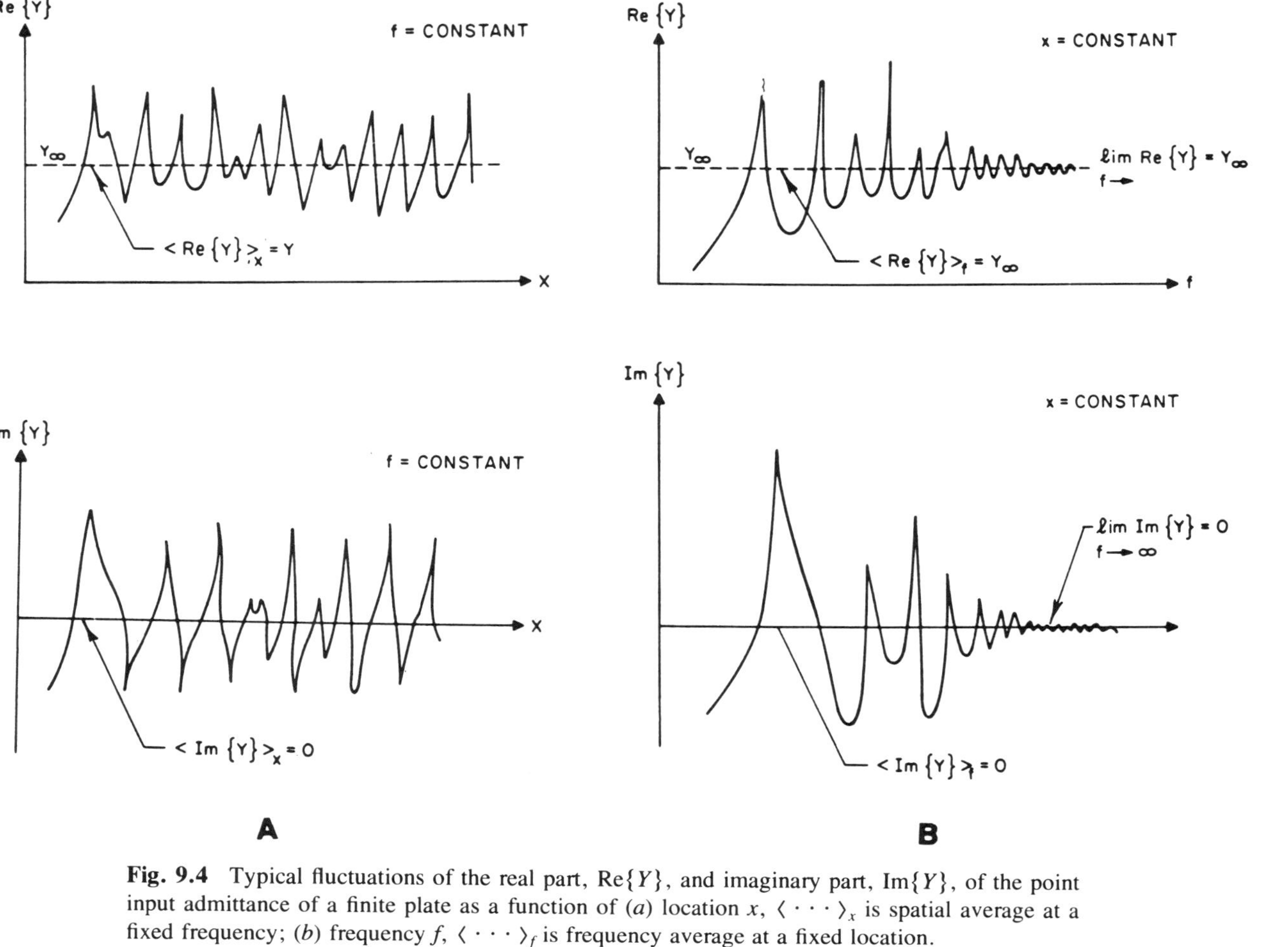

Fig. 9.4 Typical fluctuations of the real part, $\mathrm{Re}\{Y\}$, and imaginary part, $\mathrm{Im}\{Y\}$, of the point input admittance of a finite plate as a function of (*a*) location x, $\langle \cdots \rangle_x$ is spatial average at a fixed frequency; (*b*) frequency f, $\langle \cdots \rangle_f$ is frequency average at a fixed location.

not too high aspect ratio two octaves above the first plate resonance is already well approximated by the modal density of a rectangular plate of equal surface area, which is given by[1]

$$n(f) \approx \frac{\sqrt{12}\,S}{2c_L h} \qquad \text{s} \tag{9.26}$$

where S is the area (one side) and h is the thickness. Note that $n(f)$ is independent of frequency and that it is large for large and thin plates.

The modal density of rooms of volume V one octave above the first room resonance is usually well approximated by the first term of the modal density of a rectangular room with hard walls,

$$n(f) = \frac{4\pi V}{c_0^3} f^2 + \frac{\pi S}{c_0^2} + \frac{L}{8c_0} \qquad \text{s} \tag{9.27}$$

where S is the total wall surface area and L is the longest dimension of the rectangular room. The first resonance frequency, modal densities, and mode shapes of a number of finite structures and the point input impedance (the inverse of the admittance) of the equivalent infinite structure are compiled in Table 9.7. More detailed information on mode shapes, natural frequencies, and modal densities are given in reference 19. The following important relationship exists between the modal density and the real part of the point admittance of the equivalent infinite system[1]:

$$\text{Re}\{Y_\infty\} = \frac{n(f)}{4M} \qquad \text{m/N} \cdot \text{s} \tag{9.28}$$

where M is the total mass of the finite system.

Power Balance

The power balance given in Eq. (9.29),

$$W_{\text{in}} = W_d + W_{\text{tr}} + W_{\text{rad}} \qquad \text{W} \tag{9.29}$$

states that, in steady state, the power introduced, W_{in}, equals the power dissipated in the structure, W_d, the power transmitted to connected structures, W_{tr}, and the power radiated as sound into the surrounding fluid, W_{rad}. If the excitation is by a point force of peak amplitude $\hat{F}_0$ and the structure is a homogeneous isotropic thin plate of thickness h, area S (one side), density ρ_M, with longitudinal wave speed c_L, immersed in a fluid of density ρ_0, and speed of sound c_0, Eq. (9.29) takes the form

$$\tfrac{1}{2}\hat{F}^2 \text{Re}\{Y_F\} = \langle v^2 \rangle S[\rho_s \omega \eta_d + \rho_s \omega \eta_{\text{tr}} + 2\rho_0 c_0 \sigma] \qquad \text{W} \tag{9.30}$$

where $\rho_s = \rho_M h$ is mass per unit area of the plate, $\omega = 2\pi f$ is the radian frequency, η_d and η_{tr} are the dissipative and transmissive loss factors, and σ is the sound radiation efficiency of the plate (see Section 9.6). Combining dissipative and transmissive losses into a single composite loss factor $\eta_c = \eta_d + \eta_{tr}$ and assuming $\mathrm{Re}\{Y_F\} = Y_{F\infty} = 1/Z_{F\infty} = 1/(2.3\rho_s c_L t)$, as given in Table 9.3, and solving for the space–time averaged mean-square plate velocity $\langle v^2 \rangle$ yields

$$\langle v^2 \rangle = \frac{\hat{F}_0^2}{4.6\rho_s^2 c_L t\omega\eta_c S(1 + 2\rho_0 c_0 \sigma/\rho_s \omega\eta_c)} \quad \mathrm{m^2/s^2} \tag{9.31}$$

Example 9.1 Predict the space–time averaged velocity response $(\langle v^2 \rangle)^{1/2}$ of a 2-mm-thick, 1×2-m steel plate to a point force $\hat{F}_0 = 10$ N peak amplitude at a frequency of 10 kHz, when $\sigma = 1$ and $\eta_c = 0.01$. Compare $\langle v^2 \rangle$ with the mean-square velocity at the excitation point v_0^2. The input parameters to be used in connection with Eq. (9.31) and Table 9.2 are

$$\hat{F}_0 = 10\ \mathrm{N} \qquad \rho_s = \rho_M t = 7700\ \mathrm{kg/m^3} \times 2 \times 10^{-3}\ \mathrm{m} = 15.4\ \mathrm{kg/m^2}$$

$$c_L = 5050\ \mathrm{m/s} \qquad \omega = 2\pi f = 2\pi \times 10^4\ \mathrm{rad/s} \qquad \eta_c = 0.01$$

$$S = 2\ \mathrm{m^2} \qquad \rho_0 = 1.2\ \mathrm{kg/m^3} \qquad c_0 = 340\ \mathrm{m/s} \qquad \sigma = 1 \text{ yielding}$$

$$\langle v^2 \rangle = \frac{10^{-2}}{4.6 \times (15.4)^2 \times 5.05 \times 10^3 \times 2 \times 10^{-3} \times 2\pi \times 10^4 \times 10^{-2}} \times \frac{1}{2(1 + 2 \times 1.2 \times 340/15.4 \times 2\pi \times 10^4 \times 2 \times 10^{-2})}$$

$$\langle v^2 \rangle = 6.7 \times 10^{-6}\ \mathrm{m/s^2} \qquad \sqrt{\langle v^2 \rangle} = 2.6 \times 10^{-3}\ \mathrm{m/s}$$

$$v_0^2 = \frac{1}{2}\hat{F}_0^2 Y_\infty = \frac{\hat{F}^2}{4.6\rho_s c_L t} = \frac{10^2}{4.6 \times 15.4 \times 5.03 \times 10^3 \times 2 \times 10^{-3}}$$

$$= 0.14\ \mathrm{m/s}$$

$$\frac{v_0^2}{\langle v^2 \rangle} = \frac{(0.14)^2}{6.7} \times 10^{-6} = 2925 \qquad \sqrt{\frac{\langle v^2 \rangle}{v_0^2}} = 54$$

9.4 REFLECTION AND TRANSMISSION OF SOUND AT PLANE INTERFACES

When a plane sound wave traveling in a homogeneous medium encounters a plane interface with another medium, it may be (1) totally reflected, (2) partially reflected, or (3) totally transmitted depending upon the angle of incidence, the propagation speed of sound, and the density of the materials on both sides of the interface. Interfaces of practical importance are between air and water, between air and solid materials, between air and porous materials (such as ground or sound-absorbing materials), and between layers of fibrous sound-absorbing material such as treated in Chapter 8.

The simplest case is when the wave front of the incident plane wave is parallel to the plane of the interface (i.e., the wave propagates normal to the interface), as shown schematically in Fig. 9.5. In this simple case the transmitted wave p_{tran} retains the propagation direction of the incident wave p_{inc} and the following relationships apply:

$$\frac{p_{\text{tran}}}{p_{\text{inc}}} = \frac{4\text{R}_e\{Z_2\}/\text{R}_e\{Z_1\}}{\left|\dfrac{Z_2}{Z_1} + 1\right|^2} = \frac{2Z_2}{Z_2 + Z_1} \tag{9.32}$$

$$\frac{p_{\text{ref}}}{p_{\text{inc}}} = 1 - \frac{W_{\text{tran}}}{W_{\text{ref}}} = \frac{Z_2 - Z_1}{Z_2 + Z_1} \tag{9.33}$$

and for the power transmission and reflections

$$\frac{W_{\text{tran}}}{W_{\text{inc}}} = 1 - \left|\frac{Z_2 - Z_1}{Z_2 + Z_1}\right|^2 \tag{9.34}$$

$$\frac{W_{\text{ref}}}{W_{\text{inc}}} = \left|\frac{Z_2 - Z_1}{Z_2 + Z_1}\right|^2 \tag{9.35}$$

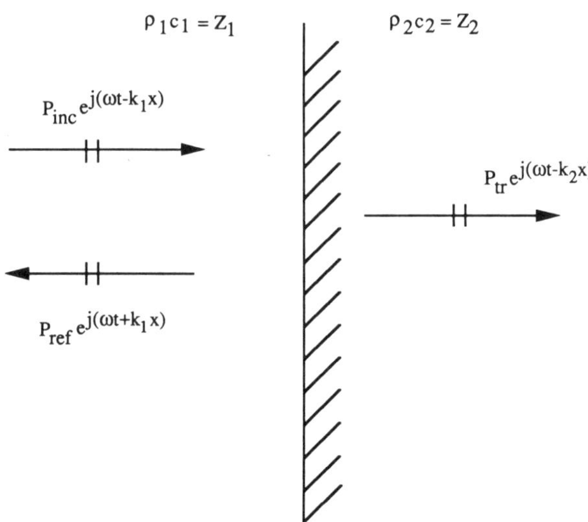

Fig. 9.5 Reflection and transmission of a plane sound wave normally incident on the plane interface of an infinitely thick medium: p_{inc}, p_{ref}, p_{tr}, peak amplitudes of incident, reflected, and transmitted pressure; ρ, c, density and speed of sound of respective media.

where $Z_i = \rho_i c_i$ is the characteristic impedance of media i for plane waves and ρ_i and c_i are the density and sound speed of the medium. For an air–steel interface and for a water–steel interface where $Z_2/Z_1 = 3.9 \times 10^7/4.1 \times 10^2$ and $Z_2/Z_1 = 3.9 \times 10^7/1.5 \times 10^6$, respectively, Eqs. (9.32)–(9.35) yield

Interface	p_{tran}/p_{inc}	p_{ref}/p_{inc}	W_{tran}/W_{inc}	W_{ref}/W_{inc}
Air/steel	1.99998	0.99998	0.00004	0.99996
Water/steel	1.927	0.927	0.141	0.859

indicating that plane waves normally incident from air onto bulk solid materials transmit only an extremely small portion of the incident energy while those incident from liquids are able to transmit an important fraction of the incident sound energy.

If the plane wave arrives at the plane interface at an oblique angle ϕ_1 ($\phi_1 = 0$ is normal incidence), the angle of the reflected wave $\phi_r = \phi_i$ but the angle of the transmitted wave ϕ_2 depends on the ratio of the propagation speeds in the two materials according to Snell's law:

$$\frac{c_1}{c_2} = \frac{\sin \phi_1}{\sin \phi_2} \tag{9.36}$$

If the plane interface is between two semi-infinite fluids or between a fluid and a porous sound-absorbing material, only compressional waves are generated. However, at interfaces between a fluid and a solid the energy transmitted into the semi-infinite solid contains both compressional and shear waves.[27] As illustrated in Fig. 9.6, the transmitted wave brakes toward the normal of the interface if $c_2 < c_1$ and away from the normal if $c_2 > c_1$.

If $c_2 < c_1$, there is always a transmitted wave even for grazing incidence ($\phi_1 = 90°$), and for lossless media the power transmission reaches 100% at an oblique-incidence angle ϕ_1 when

$$\left(\frac{Z_1}{Z_2}\right)^2 = \frac{1 - (c_2/c_1)^2 \sin^2\phi_1}{\cos^2\phi_1} \tag{9.37}$$

If $c_2 > c_1$ sound transmission occurs only in a limited incidence angle range $0 < \phi_1 < \phi_{1L} = \sin^{-1}(c_1/c_2)$. For angles $\phi_1 > \phi_{1L}$ there is a total reflection and the sound wave penetrates into the second medium only in the form of a near field that exponentially decays with distance from the interface. The pressure reflection coefficient for oblique-incidence sound is given by

$$R(\phi_1) = \frac{p_{ref}(\phi_1)}{p_{inc}(\phi_1)} = \begin{cases} \dfrac{Z_2/\sqrt{1 - [(c_2/c_1)\sin \phi_1]^2} - Z_1/\cos \phi_1}{Z_2/\sqrt{1 - [(c_2/c_1)\sin \phi_1]^2} + Z_1/\cos \phi_1} \\ \quad \text{for } c_2 < c_1 \text{ or } c_2 > c_1 \text{ and } \phi_1 < \phi_{1L} \\ 1 \\ \quad \text{for } c_2 > c_1 \text{ and } \phi_1 > \phi_{1L} \end{cases} \tag{9.38}$$

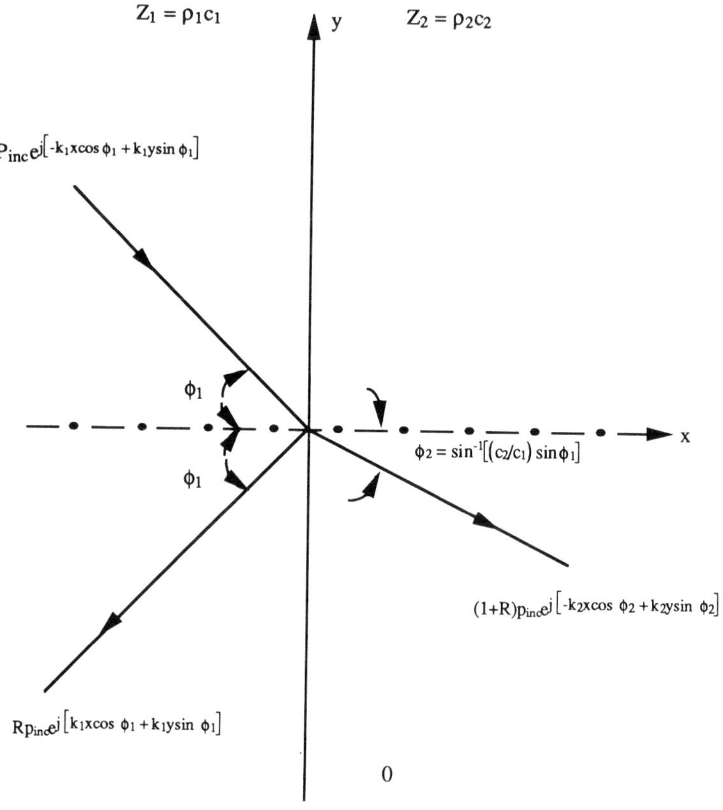

Fig. 9.6 Reflection and transmission of an oblique incidence plane sound wave at the plane interface of two semi-infinitely thick medium, $k_1 = \omega/c_1$, $k_2 = \omega/c_2$.

The limiting angle for plane-wave sound transmission from air to steel is $\phi_{LC} = \sin^{-1}(c_0/c_s) = 3.8°$ for compressional waves and $\phi_{LS} = \sin^{-1}(c_0/c_s) = 4.5°$ for shear waves. For sound transmission from water to steel, the corresponding limiting angles are $\phi_{LC} = 13°$ and $\phi_{LS} = 15°$, respectively. For angles larger than these, there is total reflection and only exponentially decaying near fields exist in the solid.

9.5 POWER TRANSMISSION BETWEEN STRUCTURAL ELEMENTS

In the preceding sections, the power lost to a connected structure was considered only as an additional mechanism that increases the loss factor of the excited structure. In many practical problems, however, the power transmitted to a neighboring structure is the prime reason for a noise reduction program.

The power balance equation states that the power introduced into the directly excited structure is either dissipated in it or is transmitted to neighboring structures.

Accordingly, if in a noise reduction problem the power is to be *confined* to the excited structure, the power *dissipated* in the structure must greatly exceed the power *transmitted* to the neighboring structures. This requires a *high loss factor* for the excited structure and a construction that *minimizes the power transmission* to neighboring structures. Methods to achieve high damping are the subject of Chapter 12.

Reduction of Power Transmission through a Change in Cross-sectional Area

The simplest construction that causes a partial reflection of an incident compression or bending wave is a sudden change in cross-sectional area as shown schematically in Fig. 9.7.

Reflection Loss of Compression Waves. The *reflection loss* ΔR_L, defined as the logarithmic ratio of the incident to the transmitted power (for both sections of the same material), is calcuated as[1]

$$\Delta R_L = 20 \log \left[\tfrac{1}{2}(2\sigma^{1/2} + \sigma^{-1/2})\right] \quad \text{dB} \tag{9.39}$$

where $\sigma = S_2/S_1$ is ratio of the cross-sectional areas (see Fig. 9.7).

The reflection loss as a function of cross-sectional area ratio is plotted as the dashed line in Fig. 9.7. Since Eq. (9.39) is symmetrical for S_1 and S_2, the reflection loss is independent of the direction of the incident wave. This equation is also valid for plates where $\sigma = h_2/h_1$ is the ratio of the thicknesses. Note that a 1 : 10

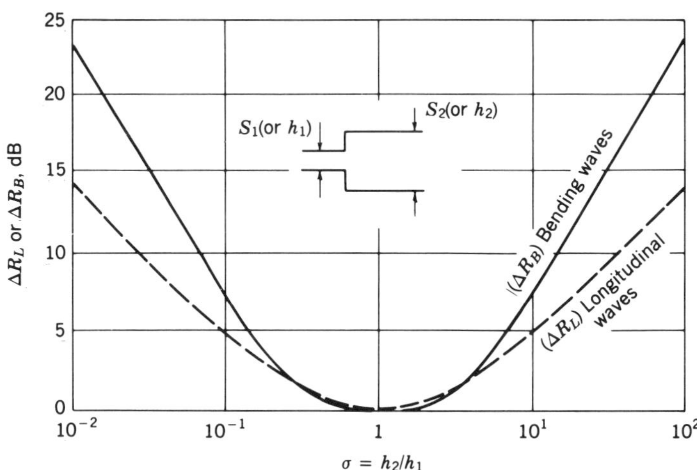

Fig. 9.7 Attenuation at discontinuity in cross section as a function of thickness ratio. (After ref. 1.)

change in cross-sectional area yields only a 4.8-dB reflection loss. To achieve a 10-dB reflection loss, a 1 : 40 change in cross-sectional area would be necessary!

Reflection Loss of Bending Waves. The reflection loss for bending waves of perpendicular incidence at low frequencies is independent of frequency and is given by[1]

$$\Delta R_B = 20 \log \frac{\frac{1}{2}(\sigma^2 + \sigma^{-2}) + (\sigma^{1/2} + \sigma^{-1/2}) + 1}{(\sigma^{5/4} + \sigma^{-5/4}) + (\sigma^{3/4} + \sigma^{-3/4})} \quad \text{dB} \qquad (9.40)$$

The equation is also plotted in Fig. 9.7 (solid line).

We conclude from Fig. 9.7 that a change in cross-sectional area is not a practical way to achieve high reflection loss in load-bearing structures.

Reflection Loss of Free Bending Waves at an *L*-Junction

Structural elements that necessitate a change in the direction of a bending wave play an important role in structures. We consider here normally incident bending waves at a junction between two plates (or beams) at right angles: for low frequencies both the transmitted and reflected energy is predominantly in the form of bending waves. In this frequency range the reflection loss (logarithmic ratio of incident to transmitted power) for plates and beams of the same materials is given by[1]

$$\Delta R_{BB} = 20 \log \left[\frac{\sigma^{5/4} + \sigma^{-5/4}}{\sqrt{2}} \right] \quad \text{dB} \qquad (9.41)$$

This equation is plotted in Fig. 9.8. Because ΔR_{BB} is symmetric in σ, the reflection factor does not depend upon whether the original bending wave is incident from the thicker or from the thinner beam or plate. Note that the lowest reflection loss of 3 dB occurs for equal thicknesses ($\sigma = 1$). If the two plates or beams constituting the junction are of different material, replace the ratio $\sigma = h_2/h_1$ by

$$\sigma = \left(\frac{B_2 c_{B_1}}{B_1 c_{B_2}} \right)^{2/5} \qquad (9.42)$$

where B and c_B are the bending stiffness and propagation speed of free bending waves, respectively (see Table 9.1). At higher frequencies the incident bending wave also excites longitudinal waves in the second structure.[1]

Reflection Loss of Bending Waves Through Cross Junctions and T-junctions

Other structures that may provide a substantial reflection of an incident bending wave are the *cross junction* of walls shown schematically in Fig. 9.9 and the

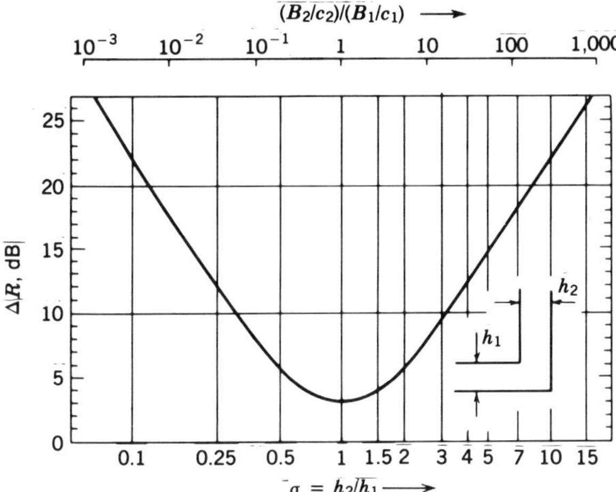

Fig. 9.8 Attenuation of bending waves at corners (in absence of longitudinal wave interactions) as a function of thickness ratio. (After ref. 1.)

T-junction in Fig. 9.10. If a bending wave of perpendicular incidence reaches the cross junction from plate 1, it is partially reflected and partially transmitted to the other plates. The transmitted power splits into a number of different wave types, namely, bending waves in plate 3 and longitudinal and bending waves in plates 2 and 4. Because of the symmetry in the geometry, plates 2 and 4 will have the same excitation.[1]

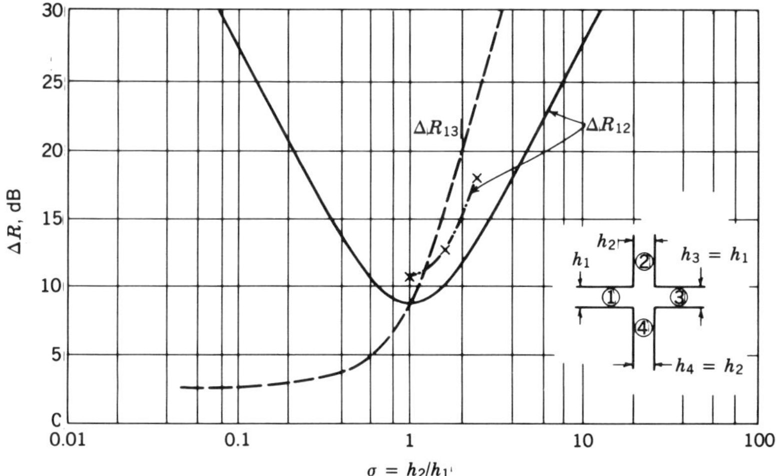

Fig. 9.9 Attenuation of bending waves at plate intersections (in absence of longitudinal wave interactions) as a function of thickness ratio. (After ref. 1.) (— · —) ΔR_{12} for random incidence computed by Kihlman.[20]

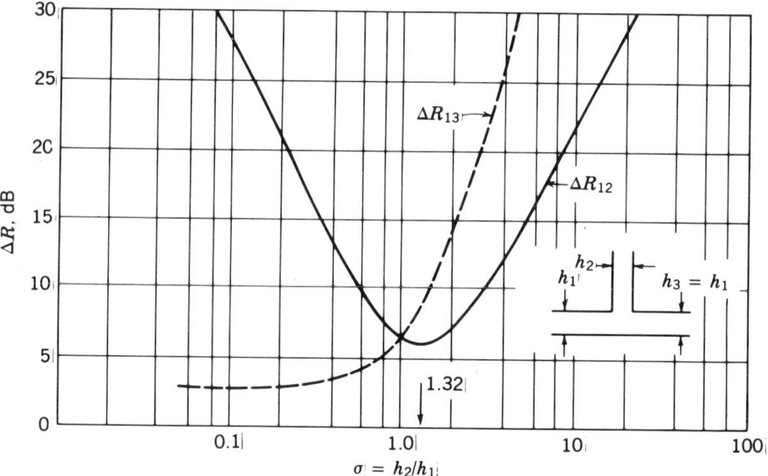

Fig. 9.10 Attenuation of bending waves at plate intersections (in absence of longitudinal wave interactions) as a function of thickness ratio. (After ref. 1.)

The reflection loss (defined as the logarithmic ratio of the power in the incident to that in the transmitted bending wave) is given as a function of the ratio of the plate thickness for plates or beams of the same material and is shown in Figs. 9.9 and 9.10. When the plates are made of different material, the ratio $\sigma = h_2/h_1$ is given by Eq. (9.42). The amplitudes of the bending waves transmitted without a change in direction are restrained by the perpendicular plate, and the reflection loss in this direction, ΔR_{13}, increase monotonically with increasing thickness of the restraining plate, h_2. Since this plate effectively stops the vertical motion of the horizontal plate at the junction, even for very thin vertical walls, ΔR_{13} remains level at 3 dB, indicating that only the power carried by the bending moment can pass the junction. For those bending waves that change direction at the junction, the reflection loss becomes a minimum ($\Delta R_{12} = 9$ dB) at a thickness ratio $\sigma = h_2/h_1 = 1$ for the cross junction; corresponding numbers for the T-junction are 6.5 dB at a thickness ratio $\sigma = h_2/h_1 = 1.32$. The reflection loss then increases symmetrically for increasing or decreasing thickness ratio (h_2/h_1).

The transmission of free bending waves at cross junctions for random incidence has been computed and the reflection loss for a number of combinations of dense and lightweight concrete plates determined.[20] The results for ΔR_{12} are plotted in Fig. 9.9 (as x's), which indicate that ΔR_{12} for random incidence is somewhat higher than that for normal incidence. It was also found that for random incidence ΔR_{12} is independent of frequency but ΔR_{13} decreases with increasing frequency.

Power Transmission from a Beam to a Plate

The structural parts of a modern building frequently include columns and structural floor slabs. Consequently, the power transmission from a beam to a plate, which models this situation, is of practical interest. Let us first consider the reflection loss for longitudinal and bending waves incident *from* the beam onto an infinite homogeneous plate.

Reflection Loss for Longitudinal Waves in a Beam. When a longitudinal wave in the beam reaches the plate, its energy is partly reflected back up the beam and is partly transmitted to the plate in the form of a bending wave. The reflection loss (equal to the logarithmic ratio of the incident to the transmitted power) is given by[21]

$$\Delta R_L = -10 \log \left(1 - \left|\frac{Y_b - Y_p}{Y_b + Y_p}\right|^2\right) \quad \text{dB} \tag{9.43}$$

where Y_b = admittance of semi-infinite beam for longitudinal waves, $= 1/Z_b$, $\text{m}/\text{N} \cdot \text{s}$
Y_p = point admittance of infinite plate, $1/Z_p$, $\text{m}/\text{N} \cdot \text{s}$

Both Y_p and Y_b are real and frequency independent and can be found for infinite beams and plates from Table 9.3 by taking the reciprocal of the impedances, that is, $Y_b = 1/\rho_b c_{L_b} S_b$ and $Y_p = 1/2.3\rho_p c_{L_p} h^2$.

Complete Power Transmission between Beam and Plate. Inspecting Eq. (9.43), we note that the reflection factor is zero (all the incident energy is transmitted to the plate) when $Y_b = Y_p$. Equating the values for Y_p and Y_p given above yields the requirements for complete power transmission from beam to the plate:

$$S_b = 2.3 \frac{\rho_p c_{L_p} h^2}{\rho_b c_{L_b}} \quad \text{m}^2 \tag{9.44}$$

If the column and the slab are of the same material so that $\rho_p c_{L_p} = \rho_b c_{L_b}$, Eq. (9.44) simplifies to

$$S_b = 2.3h^2 \quad \text{m}^2 \tag{9.45}$$

This equation says that for perfect power transfer from a beam of square cross section to a large plate, the cross dimension of the beam must be 1.52 times the thickness of the plate and for a beam of circular cross section the radius must be 0.86 times the plate thickness. Actually, this is well within the range of slab thicknesses–column cross section ratios commonly found in architectural structures. The reflection factor for different geometries of steel beam and plate connections has been measured.[21] The results for a substantial mismatch (*a*) and for a near matching (*b*) are plotted in Fig. 9.11.

Reflection Loss for Bending Waves in a Beam. When a beam carries a free bending wave, a part of the energy carried by the wave is transmitted to the plate

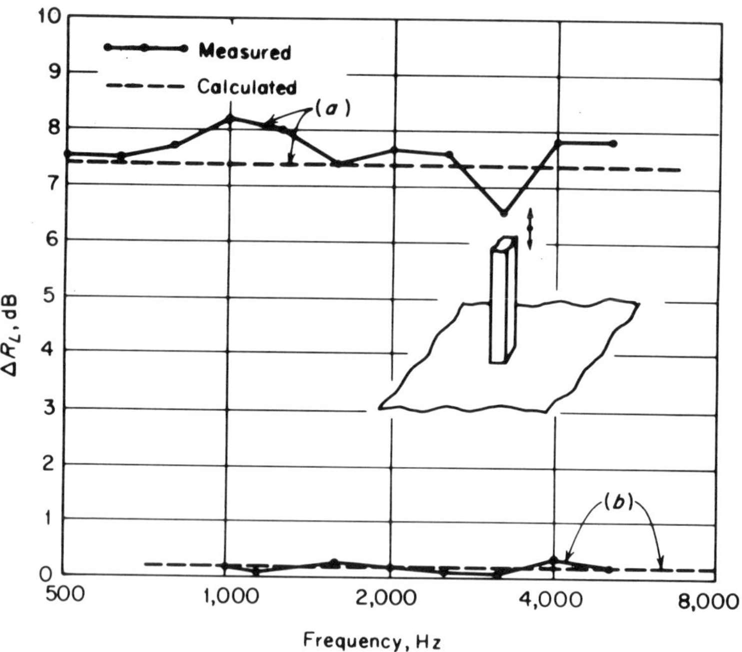

Fig. 9.11 Reflection loss for longitudinal waves, ΔR_L, for a steel beam plate system: (*a*) plate thickness 2 mm, beam cross section 10 × 20 mm; (*b*) plate thickness 4 mm, beam cross section 5 × 10 mm. (After ref. 21.)

by the effective bending moment and excites a radially spreading free bending wave in the plate. A part of the incident energy is reflected from the junction. Here the reflection loss is determined by the respective moment impedances[1,21] of the plate and the beam:

$$\Delta R_b = -10 \log \frac{Y_b^M - Y^M}{Y_b^M + Y_p^M} \quad \text{dB} \tag{9.46}$$

where the moment admittances Y_b^M and Y_p^M are given by[1,20]

$$Y_b^M = \frac{2}{1 + j} \frac{k_b^2}{\rho_b S_b c_{B_b}} \quad \text{m/N} \cdot \text{s} \tag{9.47}$$

and

$$Y_b^M = \frac{\omega}{16B_p} \left[1 + j \frac{4}{\pi} \ln \frac{1.1}{k_p a} \right] \quad \text{m/N} \cdot \text{s} \tag{9.48}$$

where k = wavenumber, m^{-1}

B_p = bending stiffness per unit width of the plate, N · m

a = effective distance of pair of point forces making up moment on plate, m; for rectangular and circular beam cross section $a_r = \frac{1}{3}d$ and $a_c = 0.59r$

d = side dimension of rectangular beam cross section (in direction of bending), m

r = radius of circular beam cross section, m

The reflection loss obtained[21] for the bending-wave excitation of a steel rod of 1 × 2 cm cross-sectional area attached to a 0.2-cm-thick semi-infinite steel plate for two perpendicular directions of bending of the rod is plotted in Fig. 9.12.

Complete Power Transmission for Bending Waves. Since the moment impedances of the plate and beam are both frequency dependent, complete power

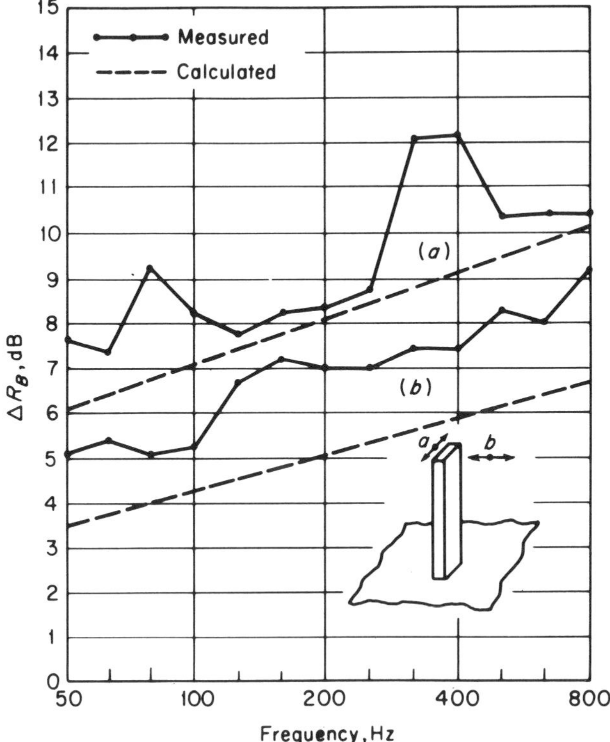

Fig. 9.12 Reflection loss for bending waves, ΔR_B, for a steel beam plate system for two different directions of bending of the beam (a and b). Plate thickness 2 mm, beam cross section 10 × 20 mm. (After ref. 21.)

transmission ($\Delta R_B = 0$) can occur only at a single frequency. The criteria for perfect power transmission is achieved when both the real and imaginary parts of the moment impedances of the beam and plate are equal, which requires that

$$\lambda_b = 0.39 \frac{B_b}{B_p} \qquad \text{m} \tag{9.49}$$

$$\lambda_p = 2.6a \qquad \text{m} \tag{9.50}$$

where B_p = bending stiffness of beam, N · m²
λ_b = bending wavelength in beam, m
λ_p = bending wavelength in plate, m

Reduction of Power Transmission between Plates Separated by Thin Resilient Layer

In architectural structures it is customary to provide a so-called vibration break by inserting a thin layer of resilient material between structural elements. The geometry of such a vibration break is shown in Fig. 9.13. Often this construction serves also as an expansion joint.

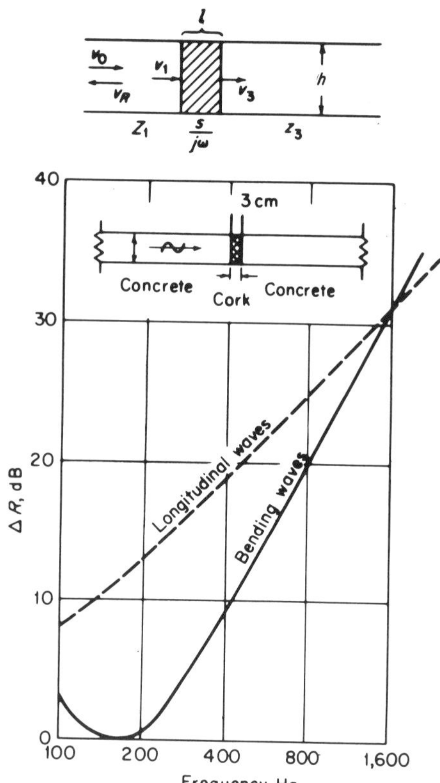

Fig. 9.13 Attenuation due to an elastic interlayer of 3-cm-thick cork between 10-cm-thick slabs of concrete as a function of frequency. (After ref. 1.)

Reflection Loss for Compression Waves. The reflection loss is given by[1]

$$\Delta R_L = 10 \log \left[1 + \left(\frac{\omega Z_1}{2s_i} \right) \right] \quad \text{dB} \tag{9.51}$$

where Z_1 = impedance of solid structure for compressional waves, N · s/m
s_i = stiffness of resilient layer in compression, N/m

Above a certain frequency ($\omega = 2s_i/Z_1$) the ΔR_L-versus-frequency curve increases with a 20-dB/decade slope with increasing frequency. Below this frequency the resilient layer transmits the incident wave almost entirely. As an example,[1] the ΔR_L-versus-frequency curve for a 3-cm-thick layer of cork inserted between two 10-cm-thick concrete slabs (or columns) is shown in Fig. 9.13.

In order to achieve a high reflection loss, the resilient layer must be as soft as is permitted by the load-bearing requirements ($s_i/\omega \ll Z_1$). However, the stiffness of the layer cannot be reduced indefinitely by increasing the thickness. For frequencies where the thickness of the layer is comparable with the wavelength of compressional waves in the resilient material, the layer can no longer be considered as a simple spring characterized by its stiffness alone. The reflection loss in this frequency region is given by[1]

$$\Delta R_L = 10 \log \left[\cos^2 k_i l + \frac{1}{4} \left(\frac{Z}{Z_i} + \frac{Z_i}{Z} \right)^2 \sin^2 k_i l \right] \quad \text{dB} \tag{9.52}$$

where k_i = wavenumber for compressional waves in the resilient material, m^{-1}
Z = impedance of equivalent infinite structure for compression waves, $= Z_1 = Z_3$ N · s/m
Z_i = impedance of equivalent infinite (length) resilient material for compression waves, N · s/m
l = length of resilient layer, m

The impedance Z, Z_i and the wavenumber k_i are assumed real; thus Eq. (9.52) does not account for wave damping in the resilient material. As expected, $\Delta R_L = 0$ for $Z_i = Z$. The maximum of ΔR_L is reached when $k_i l = (2n + 1)(\frac{1}{2}\pi)$; here the reflection loss becomes

$$\Delta R_{L,\max} = 20 \log \frac{Z_i^2 + Z^2}{2ZZ_i} \quad \text{for } l = \tfrac{1}{2}(2n + 1)\lambda_i \tag{9.53}$$

For $k_i l = n\pi$ ($l = \frac{1}{2}n\lambda_i$), the denominator of Eq. (9.51) becomes unity independent of the magnitude Z, yielding

$$\Delta R_{L,\min} = 0 \quad \text{for } l = \tfrac{1}{2}n\lambda_i$$

Finally for $k_i l \ll 1$ and $Z_i \ll Z$, Eq. (9.52) simplifies to Eq. (9.51).

Reflection Loss for Bending Waves. The geometry in Fig. 9.13 suggests that for bending waves of perpendicular incidence the moments and forces acting on both sides of the junction must be equal. However, the transverse velocity and angular velocity on both sides of the junction are different because of shear and compressional deformation of the resilient layer. The elastic layer behaves quite differently for bending waves than it does for compressional waves.[1] The most striking difference is the *complete transmission* of the incident bending wave at a certain frequency and a *complete reflection* of it at another, higher frequency. Unfortunately, it turns out that the frequency of complete transmission for architectural structures of interest usually occurs in the audio frequency range. As an example, Fig. 9.13 shows the reflection factor for bending waves as a function of frequency for a layer of 3-cm-thick cork between 10-cm-thick concrete slabs. The transmission is complete at a frequency of 170 Hz and then decreases with increasing frequency. The reflection loss for bending waves can be approximated by[1]

$$\Delta R_B \cong 10 \log \left[1 + \frac{1}{4} \left(1 - \frac{E}{E_i} \frac{2\pi^3 l h^2}{\lambda_B^3} \right)^2 \right] \quad \text{dB} \tag{9.54}$$

where E = Young's modulus of structural material
E_i = Young's modulus of elastic material
l = length of elastic layer, m
h = thickness of structure, m
λ_B = wavelength of bending waves in structure, m

The bending wavelength in the structure for which the elastic layer provides a complete transmission of bending waves is given by

$$\lambda_{B,\text{trans}} = \pi \left(\frac{E}{E_i} h^2 l \right)^{1/3} \quad \text{m} \tag{9.55}$$

If one wishes to reduce the frequency of complete transmission, the ratio E/E_i and the length of the resilient layer must be large. However, the length of the resilient material should always be small compared with the wavelength of bending waves in the elastic layer to avoid resonances.

A complete reflection of the incident bending wave occurs when the bending wavelength in the plate equals π times the plate thickness, $\lambda_{Bs} = c_b/f_s = \pi h$. Consequently, the frequency where *complete reflection* occurs is *independent* of the dynamic properties and the length of the elastic layer and is given by the thickness and by the dynamic properties of the plate or beam.

9.6 SOUND RADIATION

Vibration of rigid and elastic structures forces the surrounding fluid or gas particles at the interface to oscillate with the same velocity as the vibrating structure and thus causes sound. The sound waves propagate in the form of compressional waves that travel with the speed of sound in the surrounding medium.

Infinite Rigid Piston

Conceptually the simplest sound-radiating structure is an infinite plane rigid piston. The motion of the piston forces the fluid particles to move along parallel lines that are perpendicular to the plane of the piston. There is no divergence that could lead to inertial reaction forces, such as those that can occur along the edges of a finite piston where the fluid can move to the side. Consequently, the reaction force per unit area (i.e., the sound pressure) is fully attributable to compression effects. If the piston vibrates with velocity $\hat{v}\cos\omega t$, it generates a plane sound wave traveling perpendicular to the plane of the piston. The sound pressure as a function of distance is

$$p(x, t) = \hat{v}\rho_0 c_0 \cos(\omega t - k_0 x) \qquad \mathrm{N/m^2} \tag{9.56}$$

and the radiated sound power per unit area is

$$W'_{\mathrm{rad}} = 0.5\hat{v}^2\rho_0 c_0 = \langle v^2\rangle_t \rho_0 c_0 \qquad \mathrm{W/m^2} \tag{9.57}$$

where ρ_0 and c_0 are density and speed of sound of the medium, $\omega = 2\pi f$ is the radian frequency, $k_0 = \omega/c_0 = 2\pi/\lambda_0$ is the wavenumber, $\lambda_0 = c_0/f$ is the wavelength of the radiated sound, and $\langle v^2\rangle_t$ is the time-averaged mean-square velocity (i.e., $v = v_{\mathrm{rms}}$).

Infinite Plate in Bending

If a plane bending wave of velocity amplitude $\hat{v} = \sqrt{2}v$ and bending wave speed c_B travels on a thin plate in the positive x direction, the sound pressure as a function of x and perpendicular distance z is given by[1]

$$\hat{p}(x, y) = \frac{j\hat{v}\rho_0 c_0 e^{j\omega t}}{\sqrt{(k_B/k_0)^2 - 1}}\, e^{-jk_B z} \exp\left(-z\sqrt{k_B^2 - k_0^2}\right) \qquad \mathrm{N/m^2} \tag{9.58}$$

where $k_B = 2\pi f/c_B = 2\pi/\lambda_B$ and $k_0 = 2\pi f/c_0 = 2\pi/\lambda_0$ are the bending wavenumber in the plate and the wavenumber in the air, respectively. Inspecting Eq. (9.58), one finds that for $c_B < c_0$, $(k_B/k_0 > 1)$, the sound pressure is 90° out of phase with the velocity at the interface so that no sound power is radiated by the plate. The sound pressure constitutes a near field that decays exponentially with increasing z. For $c_B > c_0$ $k_B/k_0 < 1$, Eq. (9.58) has the form of

$$\hat{p}(x, y) = \frac{\hat{v}\rho_0 c_0 e^{-j\omega t}}{\sqrt{1 - (k_B/k_0)^2}}\, e^{-jk_B x} \exp\left(-jk_0 z\sqrt{1 - (k_B/k_0)^2}\right) \qquad \mathrm{N/m^2} \tag{9.59}$$

where the pressure and velocity are in phase at the interface ($z = 0$) and the sound power radiated by a unit area of the plate is

$$W'_{\text{rad}} = \begin{cases} \dfrac{0.5\hat{v}^2 \rho_0 c_0}{\sqrt{1 - (\lambda_0/\lambda_B)^2}} & \text{W/m}^2 \quad \text{for } \lambda_B > \lambda_0 \\ 0 & \qquad\quad \text{for } \lambda_B < \lambda_0 \end{cases} \tag{9.60}$$

which, with increasing frequency ($\lambda_0/\lambda_B << 1$), approaches that of an infinite rigid piston as given in Eq. (9.57).

Pulsating Sphere

A pulsating sphere represents an idealized model that describes the sound radiation properties of many small practical sound sources that vibrate in a manner that results in a net volume displacement. Assume that a sphere of radius a pulsates with a normal surface velocity $\hat{v}(a)$ at a frequency $f = \omega/2\pi$. The sound pressure produced $\hat{p}(r)$ decreases with increasing distance r as

$$\hat{p}(r) = \hat{p}(a)\left(\frac{a}{r}\right) e^{-jk_0 r} \qquad \text{N/m}^2 \tag{9.61}$$

The particle velocity $v(r)$, which points in the radial direction, is

$$\hat{v}(r) = \frac{1}{j\omega\rho_0}\frac{dp}{dr} = \hat{p}(a)\frac{1}{j\omega\rho_0}\left[\frac{1}{r^2} + jk_0 r\right] e^{-jk_0 r} \qquad \text{m/s} \tag{9.62}$$

where $k_0 = \omega/c_0$ is the wavenumber and ρ_0 and c_0 are the density and speed of sound, respectively. Evaluating Eq. (9.62) at the surface of the sphere ($r = a$) and solving for $p(a)$ yields

$$\hat{p}(a) = \frac{\hat{v}(a)}{1/\rho_0 c_0 + 1/j\omega\rho_0 a} = \hat{v}(a)\rho_0 c_0 \frac{(k_0 a)^2 + jka}{1 + (k_0 a)^2} = \hat{v}(a) Z_{\text{rad}} \qquad \text{N/m}^2 \tag{9.63}$$

where Z_{rad} is the radiation impedance of the pulsating sphere. Equation (9.63) indicates that at low frequencies where $ka << 1$ ($\omega\rho_0 a << \rho_0 c_0$), the vibration velocity $\hat{v}(a)$ produces a sound pressure $\hat{p}(a) << \hat{v}(a)\rho_0 c_0$ and that only a fraction of this already small sound pressure is in phase with the velocity. The physical reasons for this behavior are as follows:

1. At low frequencies the fluid is pushed out of the way slowly and diverges along radial lines so that the particle velocity decreases with increasing distance just as it would for a steady-state flow in a diverging duct; the reaction force is small and attributable mostly to the inertia of the fluid and compression remains low.

2. With increasing frequency the diversion process must take place more rapidly and the reaction force increases as does the fraction of it that is attributable to compression.
3. At high frequencies, it becomes easier to compress the fluid than to accelerate it to accomplish the divergence process and the reaction force becomes fully attributable to compression effects.

This is why pulsating bodies (of any shape) that are small compared with the wavelength radiate sound much less efficiently than bodies that are large compared with the wavelength. The sound power radiated by a sphere pulsating with a peak surface velocity $\hat{v}(a)$ is

$$W_{\text{rad}} = \tfrac{1}{2}(4\pi a^2)\hat{v}^2(a)Z_{\text{rad}} = \tfrac{1}{2}\hat{v}^2(a)\rho_0 c_0 S \frac{(k_0 a)^2}{1 + (k_0 a)^2} \quad \text{W} \tag{9.64}$$

where $S = 4\pi a^2$ is the radiating surface area.

Radiation Efficiency

It is customary to define the radiation efficiency of a vibrating body as

$$\sigma_{\text{rad}} = \frac{W_{\text{rad}}}{\langle v_n^2 \rangle \rho_0 c_0 S} \tag{9.65}$$

where $\langle v_n^2 \rangle$ is the normal component of the space–time average mean-square vibration velocity of the radiating surface of area S and W_{rad} is the radiated sound power. With this definition, Eqs. (9.57), (9.60), and (9.63) yield

$$\sigma_{\text{rad}} = \begin{cases} 1 & \text{for infinite rigid piston} \\ 0 & \text{for } \lambda_B < \lambda_0 \text{ for infinite plate in bending} \\ [1 - (\lambda_0/\lambda_B)]^{-1/2} & \text{for } \lambda_B > \lambda_0 \text{ for infinite plate in bending} \\ (k_0 a)^2/[1 + (k_0 a)^2] & \text{for pulsating sphere} \end{cases} \tag{9.66}$$

It is important to know that the radiation efficiency depends not only on the size of the radiating body as compared with the wavelength but also on the manner the body is vibrating. If a sphere vibrates back and forth instead of pulsating, then the net volume displacement is zero and the radiation efficiency is[22]

$$\sigma_{\text{rad}} = \frac{(k_0 a)^4}{4 + (k_0 a)^4} \tag{9.67}$$

Comparing Eqs. (9.66) and (9.67) reveals that at low freuencies a rigid body vibrating in translation radiates much less (approximately $\frac{1}{4}k_0 a$ times) sound power than a pulsating body of the same surface area. This is because the fluid displaced by the forward face of the sphere has sufficient time to move (with a speed less than the speed of sound) to the void left by the backward face and this inertial reaction takes place with little compression. This phenomenon is often referred to as an acoustical short circuit. Radiation efficiencies of some typical structural elements are given in Table 9.8 (see refs. 22 and 23 for a more extensive collection). Radiation efficiencies for oscillating three-dimensional bodies of near-unity aspect ratio (such as a sphere or cube) are plotted in Fig. 9.14 and those of pipes of circular rods oscillating as rigid bodies in Fig. 9.15. Note that with increasing frequency, when the distance between the neighboring, out-of-phase moving parts of the vibrating body becomes larger than the wavelength of the radiated sound (i.e., $2\pi a > \lambda_0$ for oscillating bodies and $\lambda_B > \lambda_0$ for plates in bending), the air must be compressed to yield rapidly enough, and the radiation efficiency approaches unity.

Point-excited Infinite Thin Plate

When a very large homogeneous isotropic thin plate is excited by a point force of amplitude $\hat{F}_0$ or by an enforced local velocity of amplitude $\hat{v}_0$, free bending waves with a propagation speed of $c_B(f) \sim \sqrt{f}$ propagate radially from the excitation

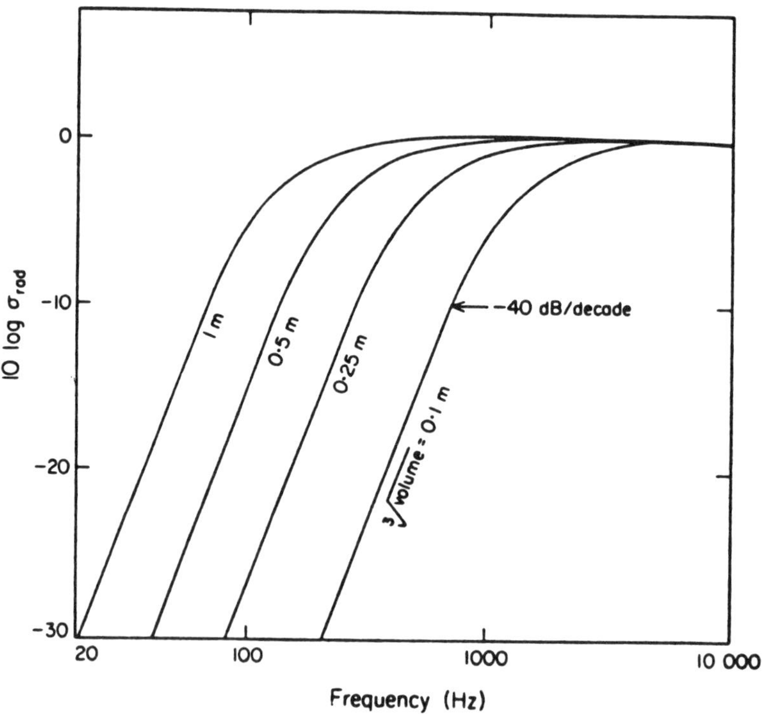

Fig. 9.14 Radiation efficiency of oscillating three-dimensional bodies. (After ref. 22.)

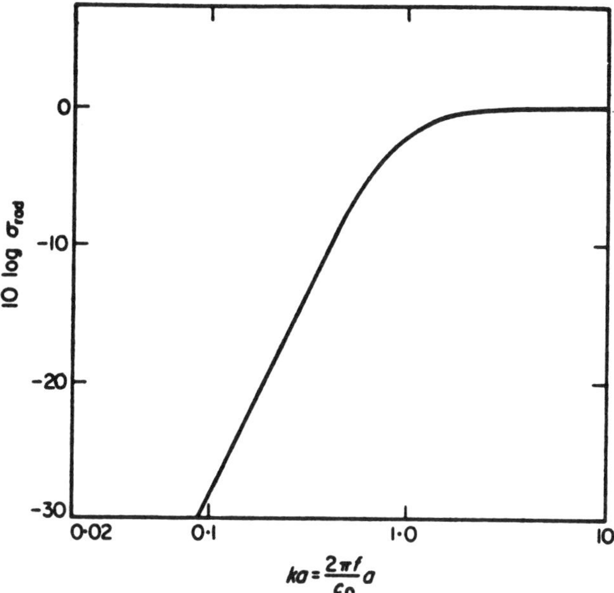

Fig. 9.15 Radiation efficiency of oscillating rigid pipes and rods of radius a. (After ref. 22.)

point. At low frequencies, where $c_B < c_0$, the free bending waves radiate no sound. When the wave front approaches a straight line [see Eq. (9.58)], the sound radiation that does occur is due to the in-phase vibration of the plate in the vicinity of the excitation point. The sound intensity radiation pattern has a $\cos^2\phi$ dependence, where ϕ is the angle to the normal of the plate. The mechanical power input W_{in}, the radiated sound power W_{rad}, and the acoustical-mechanical conversion efficiency $\eta_{\text{am}} = W_{\text{rad}}/W_{\text{in}}$, are listed below.

	Point force excitation	Point velocity excitation	
$W_{\text{in}} =$	$\frac{1}{2}\hat{F}_0^2 \dfrac{1}{2.3\rho_s c_L h}$	$(\frac{1}{2}\hat{v}_0^2)2.3\rho_s c_L h$	(9.68)
$W_{\text{rad}}(f < f_c) =$	$\hat{F}^2 \dfrac{\rho_0}{4\pi\rho_s^2 c_0}$	$\hat{v}^2 \dfrac{c_L^2 h^2 \rho_0}{2.38 c_0}$	(9.69)
$\eta_{\text{am}}(f < f_c) =$	$0.37 \dfrac{\rho_0}{\rho_M}\dfrac{c_L}{c_0}$	$0.37 \dfrac{\rho_0}{\rho_M}\dfrac{c_L}{c_0}$	(9.70)

where $\rho_s = \rho_M h$ is the mass per unit of the plate, ρ_M is the density and c_L the speed of longitudinal waves in the plate material, h is the plate thickness, and ρ_0 and c_0 are the density and speed of sound in the surrounding fluid.

Equations (9.68)–(9.70) contain the following, quite surprising, information:

1. W_{in}, W_{rad}, and η_{am} are independent of frequency and plate loss factor.
2. For point force excitation, the radiated sound power depends only on the mass per unit area of the plate ($W_{rad} \sim 1/\rho_s^2$) and not on stiffness.
3. For point velocity excitation, the radiated sound power depends only on stiffness ($W_{rad} \sim c_L^2 h^2$) and not on the plate material density.
4. The acoustical–mechanical conversion efficiency is the same for point force or point velocity excitation, and it is independent of plate thickness h and is a material constant [$\eta_{am} \sim (c_L/\rho_M)(\rho_0/c_0)$].

For noise control engineers, observations 2 and 3, embodied in Eq. (9.69), are very important. To minimize sound radiation from highly damped, point-excited, thin, platelike structures, the plate should have large mass per unit area for force excitation (e.g., by a low-impedance vibration source) and low stiffness (low E/ρ_M) for velocity excitation (by a high-impedance vibration source). Equation (9.70) supplies the rationale for violin makers, who want to convert as large a portion of the mechanical power of the vibrating string as possible into radiated sound, to use wood ($\eta_{am} = 0.024$) and not steel ($\eta_{am} = 0.0023$) or lead ($\eta_{am} = 0.0004$) for the body of the violin.

Note also that the radiated sound power given in Eq. (9.69) usually represents the minimum achievable for a finite-size plate since it accounts only for the sound radiated from the vicinity of the excitation point.

Point-excited Finite Plates

For finite-size plates, the sound power radiated has two components. The first component is radiated from the vicinity of the excitation point given in Eq. (9.69). The second component is radiated by the free bending waves as they interact with plate edges and discontinuities. The contribution of these two components to total radiated noise is represented by the first and second terms in Eqs. (9.71a) and (9.71b), where the first equation is valid for point force and the second for point velocity excitation:

$$W_{rad}^F \cong \hat{F}^2 \left[\frac{\rho_0}{4\pi\rho_s^2 c_0} + \frac{\rho_0 c_0 \sigma_{rad}}{4.6\rho_s^2 c_L h \omega \eta_c} \right] \quad \text{W} \tag{9.71a}$$

$$W_{rad}^v \cong \hat{v}^2 \left[\frac{\rho_0 c_L^2 h^2}{2.38 c_0} + 1.15 \left(\frac{c_L h}{\omega \eta_c} \right) \rho_0 c_0 \sigma_{rad} \right] \quad \text{W} \tag{9.71b}$$

where η_c is the composite loss factor and σ_{rad} is the radiation efficiency of the plate for free bending waves. The second term in Eqs. (9.71) has been derived by the power balance (see Section 9.3) of the finite plate, assuming that the mechanical power input to the finite plate can be well approximated by the power input to the equivalent infinite plate.

Equations (9.71a) and (9.71b) can be used to assess the useful upper limit of the composite loss factor ($\eta_c^{\max}$) which, if exceeded, results only in added expense but no meaningful reduction of the radiated noise. This is done by equating the first and second terms in the square brackets and solving for η_c.

9.7 SOUND EXCITATION AND SOUND TRANSMISSION

The most common problem that noise control engineers have to deal with is the transmission of sound through solid partitions such as windows, walls and floor slabs, etc. The problem may be either prediction or design. The prediction problem is typically this: Given a noise source, a propagation path up to the partition, and the size and construction of the partition and the room acoustics parameters of the receiver room, predict the noise level in the receiver room. The design problem is typically stated as: Given a source, a propagation path, the room acoustics parameters of the receiver room, and a noise criterion (in the form of octave-band sound pressure, levels), determine the construction of those partitions that would assure that the noise criterion are met with a sufficient margin of safety. Sound propagation from the source to the source-side face of the partition and from the receiver-side face of the partition to the receiver are treated in Chapters 5 and 7.

This section deals with the transmission coefficient τ and sound transmission loss R, which characterize sound transmission through partitions. These parameters are defined as

$$\tau(\phi, \omega) = \frac{W_{\text{trans}}(\phi, \omega)}{W_{\text{inc}}(\phi, \omega)} \tag{9.72}$$

$$R(\phi, \omega) = 10 \log \frac{1}{\tau(\phi, \omega)} = 10 \log \frac{W_{\text{inc}}(\phi, \omega)}{W_{\text{trans}}(\phi, \omega)} \quad \text{dB} \tag{9.73}$$

where $W_{\text{inc}}(\phi, \omega)$ is the sound power incident at angle ϕ at frequency $\omega = 2\pi f$ on the source side and $W_{\text{trans}}(\phi, \omega)$ is the power transmitted (radiated by the receiver side).

Though knowledge of the sound transmission loss of a window or curtain wall for a particular angle of incidence may be desirable sometimes, it is a more common problem to characterize the transmission of sound between two adjacent rooms where the sound is incident on the separating partition from all angles with approximately equal probability. In such ''random incident'' sound fields, the sound intensity (energy incident on a unit area) I_{random} is related to the space-averaged mean-square sound pressure in the source room $\langle p^2 \rangle$ as

$$I_{\text{random}} = \frac{\langle p^2 \rangle}{4\rho_0 c_0} \quad \text{W/m}^2 \tag{9.74}$$

Laboratory procedures ASTM E90-70[24] and ISO 140/3-1978,[25] adopted for measuring the sound transmission loss of partitions, are based on such a diffuse

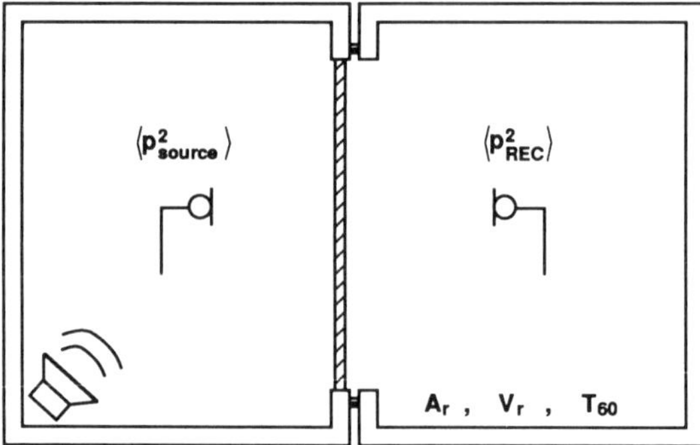

Fig. 9.16 Experimental setup for laboratory measurements of the sound transmission loss of partitions (see text for definition of symbols).

sound field obtained by utilizing large reverberation rooms (see Chapter 7) as source and receiver room. The measurement procedure depicted in Fig. 9.16 has three steps. The first step is to measure the sound power incident on the source-side face of the test partition of area S_w:

$$W_{\text{inc}} = I_{\text{inc}} S_w = \frac{\langle p^2_{\text{source}} \rangle S_w}{4\rho_0 c_0} \quad \text{W} \tag{9.75}$$

by measuring the space–time average mean-square sound pressure level $\langle p^2_{\text{source}} \rangle$ in the source room by spatially sampling the sound field. The second step is to measure the transmitted sound power W_{trans} from the power balance of the receiver room,

$$W_{\text{trans}} = \frac{\langle p^2_{\text{rec}} \rangle A_r}{4\rho_0 c_0} \quad \text{W} \tag{9.76}$$

yielding the laboratory sound transmission loss

$$R_{\text{lab}} = 10 \log \frac{W_{\text{inc}}}{W_{\text{trans}}} = \langle L_p \rangle_S - \langle L_p \rangle_R + 10 \log \frac{S_w}{A_r} \quad \text{dB} \tag{9.77}$$

where A_r is the total absorption in the receiving room. The third step is to determine A_r from the known volume and the measured reverberation time, T_{60}, of the receiver room, as described in Chapter 7.

Once R_{lab} has been measured, it can be used for predicting the mean-square sound pressure in a particular receiver room acoustically characterized by its total absorption A_r through a partition of surface area S_w for an incident sound field of intensity I_{inc} as

$$\langle p_r^2 \rangle = \frac{\bar{\tau} I_{\text{inc}} S_w (4\rho_0 c_0)}{A_r} = \frac{\bar{\tau} S_w \langle p^2_{\text{source}} \rangle}{A_r} \quad \text{N}^2/\text{m}^4 \tag{9.78a}$$

$$\langle \text{SPL}_R \rangle = \langle \text{SPL}_S \rangle - R_{\text{lab}} + 10 \log \frac{S_w}{A_r} \quad \text{dB} \tag{9.78b}$$

provided that the partition is not much smaller than tested and that edge conditions are not much different. Curves of measured random-incidence sound transmission loss versus frequency for standard windows, doors, and walls are available from manufacturers and should be used in design and prediction work. Measured sound transmission losses of some selected partitions are given by Bies and Hansen in their Table 8.1 (see the Bibliography).

The purpose of the discussion that follows is to identify the physical processes and the key parameters that control sound transmission through partitions and to provide analytical methods that further the development of an informed judgment needed for working with data obained in laboratory measurements. Most importantly, however, this will focus on predicting sound transmission loss for situations that are different from those employed in the standardized laboratory measurements (i.e., for near-grazing incidence) and for the task of designing nonstandard partitions for unique applications.

Excitation of structures by an incident sound wave is significantly different from excitation by localized forces, moments, enforced velocities, or angular velocities. As discussed in the previous sections, the response of thin, platelike structures to localized excitation results in radially spreading free bending waves. The propagation speed of these waves, $c_B(f)$, is as unique a characteristic of the plate as is the period of a pendulum. The thinner the plate and the lower the frequency, the lower is the propagation speed.

If the structure is excited by an incident sound wave, forcing occurs simultaneously over the entire exposed surface of the plate. The incident sound enforces its spatial pattern on the plate, causing it to instantaneously conform to its trace. The trace "runs" along the plate with a speed that approaches the speed of sound in the source-side media when the sound runs nearly parallel to the plate (grazing incidence) and approaches infinity as the sound incidence approaches normal. This "sound-forced" response of the thin plate to sound excitation is referred to as the "forced wave." In contrast to the free bending waves, the speed of the forced bending wave is independent of frequency, plate thickness, and mass per unit area (though the amplitude of the response depends on them). Because of their supersonic speed, forced waves radiate sound very efficiently at all frequencies (e.g., their radiation efficiency $\sigma_F \geq 1$), except for panels that are small compared with the acoustical wavelength.

Transmission of Normal-Incidence Plane Sound Waves through an Infinite Plate

It is advisable to introduce the complex process of sound transmission by considering first the least complicated case when a plane sound wave is normally incident on a uniform homogeneous, isotropic, flat plate of thickness h, as shown in Fig. 9.17. Because the pressure exerted on the plate is in phase over the entire surface

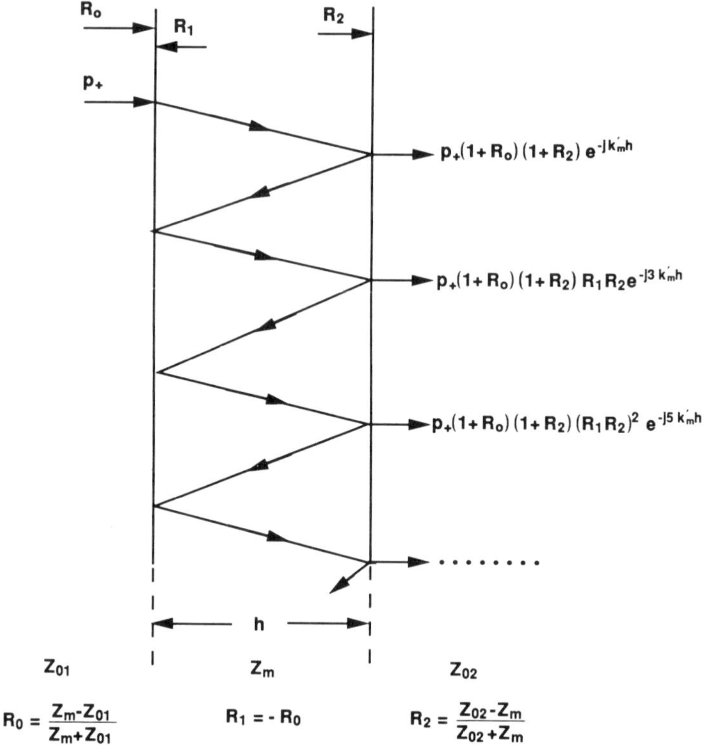

Fig. 9.17 Transmission of a normally incident sound wave through a flat, homogeneous, isotropic plate. The transmitted pressure p_T is the sum of all the infinite components on the right. (See text for definition of symbols.)

of the plate, only compressional waves are excited. The sound wave of pressure amplitude p_+ propagates in a gas of characteristic plane-wave impedance $Z_{01} = \rho_{01}c_{01}$. It encounters the solid plate of characteristic impedance $Z_M = \rho c_L$ and on the transmitting side (right) will radiate sound into another gas of characteristic impedance $Z_{02} = \rho_{02}c_{02}$. The multiple reflection and transmission phenomena at the interfaces are governed by Eqs. (9.32) and (9.33). The amplitude of the transmitted sound pressure p_T in the receiver-side media is obtained by the summation of the transmitted components as illustrated in Fig. 9.17, yielding[26]

$$p_T = p_+ \frac{(1 + R_0)(1 + R_2)e^{-jk_m h}}{1 - R_1 R_2 e^{-j2k_m h}} \qquad \text{N/m}^2 \tag{9.79}$$

where R_0, R_1, and R_2 are the reflection factors at the interfaces and directions indicated in Fig. 9.17 and h is the plate thickness. Propagation losses can be taken into account by a complex wavenumber $k'_m = k_m(1 + \frac{1}{2}j\eta)$, where $k_m = \omega/c_L$ is the wavenumber of compressional waves in the plate and η is the loss factor. The normal-incidence sound transmission loss is defined as

$$R_N = 10 \log \frac{W_{\text{inc}}}{W_{\text{rad}}} = 10 \log \frac{p_+^2}{p_T^2} \quad \text{dB} \tag{9.80a}$$

Assuming that the same gas is on both sides of the plate, Eq. (9.79) yields

$$R_N = 10 \log \left[\cos^2 k'_m h + 0.25 \left(\frac{Z_0}{Z_m} + \frac{Z_m}{Z_0} \right)^2 \sin^2 k'_m h \right] \quad \text{dB} \tag{9.80b}$$

which for $|k'_m h| \ll 1$ yields the simple expression

$$R_N \cong 10 \log \left[1 + \left(\frac{\rho_s \omega}{2\rho_0 c_0} \right)^2 \right] \quad \text{dB} \tag{9.80c}$$

known as the normal-incidence mass law. In Eq. (9.80c), $\rho_s = \rho h$ is the mass per unit area of the plate and $\rho_0 c_0 = Z_0$ is the characteristic impedance of the gas, assumed the same on both sides.

Figure 9.18 shows the computed normal-incidence sound transmission loss of a 0.6-m-(2-ft) thick dense concrete wall. At low frequencies where the plate thickness is less than one-sixth of the compressional wavelength ($f < c_L/6h$) the sound transmission loss follows the normal-incidence mass law of Eq. (9.80c), increasing by approximately 6 dB for each doubling of the frequency or the mass per unit area. At high frequencies compressional wave resonances in the plate occur and

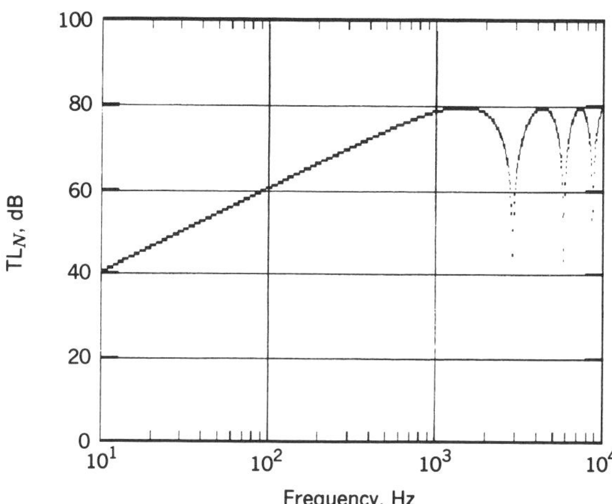

Fig. 9.18 Normal-incidence sound transmission loss R_0 of a 0.6-m-thick dense concrete wall computed according to Eq. (9.80b) assuming $\eta = 0$.

the R_N-versus-frequency curve exhibit strong minima at $f_n = nc_L/2h$, yielding

$$R_N^{\min} = 20 \log \left(1 + \frac{\pi}{4} \frac{Z_m}{Z_0} \eta n\right) \quad \text{dB} \tag{9.81a}$$

and the maximal achievable sound transmission loss can be approximated as

$$R_N^{\max} \cong 20 \log \frac{Z_m}{2Z_0} \quad \text{dB} \tag{9.81b}$$

Equation (9.81a) indicates complete transmission of the incident sound ($R_N^{\min}$ = 0 dB) at compressional wave resonances if $\eta = 0$. However, Eq. (9.81a) indicates that even a small loss factor of $\eta = 0.001$ ensures that the normal-incidence transmission loss of a concrete partition does not dip below 24 dB at compressional wave resonances. Note, however, that no such dips are observed for random incidence because different incident angles correspond to different frequencies where such dips occur. As indicated in Eq. (9.81b), the maximum achievable normal-incidence sound transmission loss of homogeneous, isotropic single plates of any thickness is limited by the ratio of the characteristic impedances and is 80 dB for dense concrete, 94 dB for steel, and 68 dB for wood.

Transmission of Oblique-Incidence Plane Sound Waves through an Infinite Plate

The transmission of oblique-incidence sound through infinite plane plates can be formulated either in terms of shear and compressional waves, where the bending of the plate is considered as a superposition of these two wave types, or by utilizing the bending-wave equation of the plate. Both fomulations are discussed below.

Combined Compressional and Shear Wave Formulation. When the plane sound wave is incident on the plate at an oblique angle ϕ_I (to the normal of the plate) that is smaller than the limiting angle $\phi_{IL} = \sin^{-1}(c_0/c_L)$ given in Section 9.4, the sound wave excites both longitudinal and shear waves in the plate material. These two wave types travel between the two faces of the plate in a similar manner as the normally incident wave depicted in Fig. 9.17, except that the compressional and shear waves run at oblique angles dictated by the respective ratios c_0/c_L and c_0/c_s.

At oblique angles $\phi_I > \phi_{IL}$ (i.e., $\phi_I > 3.8°$ for steel plate and air), the incident sound produces waves that travel parallel to the surface of the plate with wave speed equal to the trace of the incident sound wave, $c_{\text{TRACE}} = c_0/\sin\phi_I$. If the plate is infinitely thick, these waves will "penetrate" the plate only in the form of near fields, where the pressure wave in the plate decays exponentially with distance from the surface. For a thin plate, however, the near field consists of two components: one that decays exponentially and the other that grows exponentially with distance away from the excited surface of the plate. The normal velocity of the two plate surfaces is determined by the combined effect of these two near-field components.[27]

At high frequencies ($\frac{1}{2}k_s h \geq 1$), the two surfaces of the plate vibrate at different velocities, and a complex structureborne standing-wave pattern develops in the plate. At low frequencies ($\frac{1}{2}k_s h \ll 1$), the normal velocity of the receiver side of the plate surface is nearly equal, in magnitude and phase, to that of the source-side surface. Since these near fields appear along the receiver side of the plate at the same trace speed as the trace speed of the incident sound of the source side, the radiated sound has the same direction as the incident sound. If the normal velocity of the receiving-side surface of the plate, resulting from the exponentially growing and exponentially decaying near fields, is the same as the normal component of the particle velocity of the incident sound, complete transmission occurs.

The problem of the transmission of an oblique-incidence plane sound wave through a solid, homogeneous, isotropic, plane-parallel, infinite, solid plate was treated in 1938 by Reissner,[27] providing a complete solution for the plate-internal structureborne sound and for the transmitted fluid-borne sound. His solution, which is applicable to both thick and thin plates, can be presented in a form similar to that given by[28]

$$\begin{aligned} R(\phi) &= 10 \log \left(\frac{p_{\text{in}}^2}{p_{\text{trans}}^2} \right) \\ &= -20 \log \left| \left[\frac{\rho_s \omega}{2\rho_0 c_0 / \cos\phi} \left(\frac{A^2}{k_{yc} k_s^4} \tan \frac{k_{yc} h}{2} + \frac{4 k_{ys} k_0^2 \sin^2\phi}{k_s^4} \tan \frac{k_{ys} h}{2} \right) - j \right]^{-1} \right. \\ &\quad \left. + \left[\frac{\rho_s \omega}{2\rho_0 c_0 / \cos\phi} \left(\frac{A^2}{k_{yc} k_s^4} \cot \frac{k_{yc} h}{2} + \frac{4 k_{ys} k_0^2 \sin^2\phi}{k_s^4} \cot \frac{k_{yc} h}{2} \right) + j \right]^{-1} \right| \quad \text{dB} \end{aligned} \tag{9.82}$$

where; p_{in} and p_{trans} are the incident and transmitted sound pressures; ρ_M is the density of the plate material; c_s is the speed of shear waves in the plate material; h is the plate thickness; $\rho_s = \rho_M h$; $k_0 = \omega / c_0$ is the wavenumber, ρ_0 the density, and c_0 the speed of sound in the fluid on both sides of the plate; ϕ is the incidence angle ($\phi = 0$ is normal incidence); $k_s = \omega / c_s$ and $k_c = \omega / c_L$ are the wavenumbers for shear and longitudinal waves in the plate material; $k_{yc}^2 = k_c^2 - k_0^2 \sin^2\phi$; $k_{ys}^2 = k_s^2 - k_0^2 \sin^2\phi$; and $A = 2k_0^2 \sin^2\phi - k_s^2$.

Equation (9.82) predicts total transmission $[R(\phi) = 0]$ at bending-wave coincidence when $c_0 / \sin\phi = c_B(f) = \sqrt{1.8 c_L h f}$ and at longitudinal wave resonances $f = n c_L / 2h$, $n = 1, 2, 3, \ldots$ and $\phi = 0$. For plates that are much thinner than the shear wavelength ($k_s / h \ll 1$), Eq. (9.82) is well approximated by

$$R(\phi) = 10 \log \left(\frac{1}{\tau} \right) \cong 10 \log \left\{ 1 + \left| \frac{\rho_s \omega}{2\rho_0 c_0 / \cos\phi} \left(1 - \frac{\omega^2 h^2 c_L^2}{12 c_0^4} \sin^4\phi \right) \right|^2 \right\} \quad \text{dB} \tag{9.83}$$

Considering now that the bending stiffness of the plate is $B = Eh^3/[12(1 - \nu^2)]$, $\rho_M h = \rho_s$ and $c_L^2 = E/\rho$, and the frequency where the speed of free bending waves in the plate, $c_B(f_c)$, equals the speed of sound in the media at the critical frequency

$$f_c = \frac{\omega_c}{2\pi} = \left(\frac{c_0^2}{2\pi}\right)\left(\frac{\rho_M h}{B}\right)^{1/2} \tag{9.84}$$

Equation (9.83) can be rewritten in the following form that was derived in a different way by[26]

$$R(\phi) = 10 \log \frac{1}{\tau} \cong 10 \log \left\{ 1 + \left| \frac{\rho_s \omega}{2\rho_0 c_0/\cos\phi} \left[1 - \left(\frac{f}{f_c}\right)^2 \sin^4\phi \right] \right|^2 \right\} \quad \text{dB} \tag{9.85}$$

Equations (9.82) and (9.83) are valid only for homogeneous isotropic plane-parallel plates of any thickness. On the other hand, Eq. (9.85) is valid also for inhomogeneous, isotropic plates, but the plates must be thin compared with the shear wavelength. Observing Eq. (9.85), note that the infinite isotropic plate has zero sound transmission loss under certain conditions:

$$R(\phi, \omega) = 0 \begin{cases} \text{at all frequencies for } \phi = 90^\circ \\ \text{at } f = 0 \text{ for all angles of incidence} \\ \text{at } f = f_c' = f_c/\sin\phi \text{ the coincidence, where trace matching occurs,} \\ \quad \text{i.e., where } c_0/\sin^2\phi = c_B(f) \end{cases}$$

Separation Impedance Formulation. Sound transmission through plates can be conveniently characterized by the separation impedance Z_s defined as[1,26]

$$Z_s = \frac{p_s - p_{\text{rec}}}{v_n} \qquad \text{N} \cdot \text{s/m}^3 \tag{9.86}$$

where p_s is the complex amplitude of the sound pressure at the source-side face of the plate representing the sum of the incident and reflected pressures ($p_s = p_{\text{inc}} + p_{\text{refl}}$), p_{rec} is the complex amplitude of sound pressure on the receiver-side face, and v_n is the complex amplitude of normal velocity of the receiver-side face of the plate. In the case of single panels, it is generally assumed that both faces of the panel vibrate in phase with the same velocity. The sound transmission coefficient of the plate τ and its sound transmission loss R are given by[1,26]

$$\tau(\phi) = \frac{I_{\text{trans}}}{I_{\text{inc}}} = \left| 1 + \frac{Z_s \cos\phi}{2\rho_0 c_0} \right|^2 \tag{9.87}$$

$$R(\phi) = 10 \log \frac{1}{\tau(\phi)} = 10 \log \left\{ \left| 1 + \frac{Z_s}{2\rho_0 c_0/\cos\phi} \right|^2 \right\} \quad \text{dB} \tag{9.88}$$

where ϕ is the angle of sound incidence ($\phi = 0$ for normal incidence).

The formulation of the sound transmission in terms of the separation impedance of Eq. (9.88) lends itself exceptionally well to predicting the sound transmission loss of multilayer partitions where the constituent layers may be thin plates separated by airspaces with and without porous sound-absorbing material (such as double and triple walls), windows, and so on. The reason for this is that the separation impedance of such multilayer partitions can be expeditiously obtained by appropriately combining the separation impedances of the constituent layers.

The separation impedance for a thick isotropic plate is obtained by solving the bending-wave equation of the plate as formulated by Mindlin[29]

$$\left(\nabla^2 - \frac{\rho_M}{G} \frac{\partial^2}{\partial t^2} \right) \left(B\nabla^2 - \frac{\rho_M h}{12} \frac{\partial^2}{\partial t^2} \right) \xi(x, y) + \rho_M h \frac{\partial^2}{\partial t^2} \xi(x, y)$$
$$= \left(1 - \frac{B}{Gh} \nabla^2 + \frac{\rho_M h^2}{12G} \frac{\partial^2}{\partial t^2} \right) \Delta p\ (x, y, 0) \tag{9.89}$$

where $\Delta p(x, y, 0)$ is the pressure differential across the plate, $\zeta(x, y)$ is the plate displacement in the z direction normal to the plate surface, $\nabla^2 = \partial^2/\partial x^2 + \partial^2/\partial y^2$ is the Laplacian operator, ρ_M is the density, G is the shear modulus, B is the bending stiffness, and h is the thickness of the panel. The solution of Eq. (9.89) for a plane sound wave of incidence angle ϕ yields[30]

$$Z_s = \frac{j\left\{ \left[\rho_M h + \left(\frac{\rho_M h^3}{12} + \frac{\rho_M B}{G} \right) \frac{\omega^2}{c_0^2} \sin^2\phi \right] \omega - \left[\frac{B}{c_0^4} \sin^4\phi + \frac{\rho_M^2 h^3}{12G} \right] \omega^3 \right\}}{\left[1 + \frac{B\omega^2 \sin^2\phi}{G c_0^2 h} - \frac{\rho_M h^2 \omega^2}{12G} \right]} \quad \text{N} \cdot \text{s/m}^3 \tag{9.90}$$

Equation (9.90) can be approximated by

$$Z_s \cong Z_m + \left(\frac{1}{Z_B} + \frac{1}{Z_{\text{sh}}} \right)^{-1} \quad \text{N} \cdot \text{s/m}^3 \tag{9.91}$$

where $Z_m = j\omega\rho_M h$ is the mass impedance of the plate per unit area and $Z_B = -jB\omega^3 \sin^4\phi/c_0^4$ and $Z_{\text{sh}} = -jGh\omega \sin^2\phi/c_0^2$ are the bending-wave and shear wave impedances of the plate per unit area.

If $|Z_{\text{sh}}| \ll |Z_B|$, the plate responds predominantly in shear and $Z_s \cong Z_m + Z_{\text{sh}}$. In this case, Eq. (9.90) yields

$$R_{\text{sh}}(\phi) = 20 \log \left| 1 + \frac{j\omega\rho_M h \left[1 - \left(\frac{c_s}{c_0} \right)^2 \sin^2\phi \right]}{2\rho_0 c_0/\cos\phi} \right| \quad \text{dB} \tag{9.92}$$

Equation (9.92) indicates that trace coincidence between the incident sound wave and the free shear waves in the plate (which would lead to complete transmission) can be avoided at all incident angles provided that the panel is specifically designed to yield a low shear wave speed such that $c_s^2 = G/\rho < c_0^2 = P_0\kappa/\rho_0$. Unfortunately, this desirable condition cannot be met with homogeneous plates made of any construction-grade material. As we will discuss later, however, it is possible to satisfy the $c_s < c_0$ criterion with specially designed[31] inhomogeneous sandwich panels without compromising the static strength of the panel and thereby preserving the mass law behavior described by

$$R_{\text{sh}}(\phi) \cong R_{\text{mass}}(\phi) = 20 \log \left[1 + \frac{\rho_s \omega \cos \phi}{2\rho_0 c_0} \right] \quad \text{dB} \tag{9.93}$$

which is representative of a limp panel of the same mass per unit area $\rho_s = \rho_M h$ as the shear panel.

Thin homogeneous panels are easier to bend than to shear so that $Z_B << Z_{\text{sh}}$. It follows that $Z_s \cong Z_m + Z_B$ and Eqs. (9.88) and (9.90) yield

$$R(\phi) = 10 \log \frac{1}{\tau(\phi)} = 10 \log \left| 1 + \frac{j\rho_s \omega[1 - (f/f_c)^2 \sin^4\phi]}{2\rho_0 c_0/\cos \phi} \right|^2 \quad \text{dB} \tag{9.94}$$

Note that Eq. (9.94) is identical with Eq. (9.85) obtained by a different derivation. Equations (9.85) and (9.94) indicate that above the critical frequency f_c given in Eq. (9.84), trace coincidence between the incident sound wave and the free bending waves in the plate occurs when $f = f_c/\sin^2\phi$ (which is called the coincidence frequency) and would result in complete transmission if the plate had no internal damping.

It is customary to account for internal damping in Eqs. (9.82) and (9.89) by introducing a complex Young's modulus $E' = E(1 + j\eta)$ that results in complex wavenumbers $k_c' = k_c(1 + \frac{1}{2} j\eta)$ and $k_s' = k_s(1 + \frac{1}{2} j\eta)$ and yields the following modified form of the sound transmission loss

$$R(\phi) = 10 \log \frac{1}{\tau(\phi)} = 10 \log \left\{ \left| 1 + \frac{\rho_s \omega}{2\rho_0 c_0/\cos \phi} \cdot \left\{ \eta + j \left[1 - \left(\frac{f}{f_c} \right)^2 \sin^4\phi \right] \right\} \right|^2 \right\} \quad \text{dB} \tag{9.95}$$

Equation (9.95) indicates that in the vicinity of $f = f_c/\sin^2\phi$ the curves of sound transmission loss versus frequency exhibit a minimum that is controlled by the damping.

Figure 9.19 shows the curve of sound transmission loss versus frequency for a 4.7-mm- ($\frac{3}{16}$-in.-) thick glass plate for normal ($\phi = 0°$), $\phi = 45°$, and near-grazing ($\phi = 85°$) angles of incidence computed according to Eq. (9.95). Figure 9.19

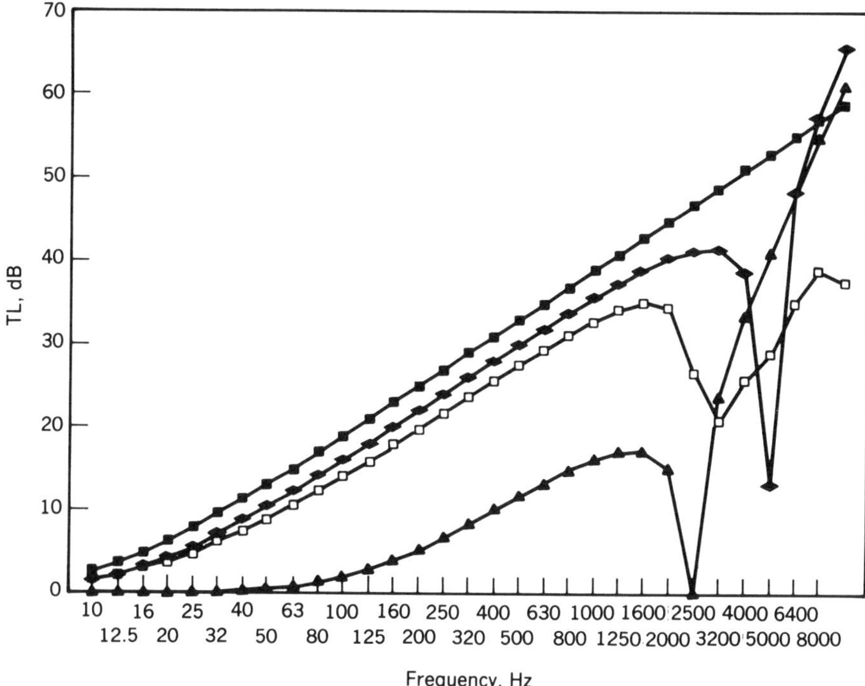

Fig. 9.19 Computed TL-versus-frequency curve of a 3.7-mm- ($\frac{3}{16}$-in.-) thick infinite glass plate for various angles of incidence: (■) normal ($\phi = 0°$); (♦) $\phi = 45°$; (▶) $\phi = 85°$; (□) random.

illustrates the decrease in sound transmission loss that occurs with increasing angle of incidence (owing to the cos ϕ term) and the trace-matching dip that occurs at $f = f_c/\sin^2\phi$.

Transmission of Random-Incidence (Diffuse) Sound through an Infinite Plate

A plane wave impinging on the plate at one particular angle is not a typical problem. The sound field in a room is better modeled as a diffuse sound field, which is an ensemble of plane sound waves of the same average intensity traveling with equal probability in all directions. A region of unit area on the plate will be exposed, at any instant, to plane sound waves incident from all areas on a hemisphere whose center is the area on the plate. These waves are uncorrelated and have equal intensity. The sound intensity incident on the unit area of the plate from any particular angle will be the intensity of the plane wave at angle I_{inc} multiplied by the cosine of the angle of incidence. The total transmitted intensity is then

$$I_{trans} = \int_{\Omega} \tau(\phi) I_{inc} \cos\theta \, d\Omega \qquad \text{W/m}^2 \tag{9.96}$$

The integration is over a hemisphere of solid angle Ω, where $d\Omega = \sin\phi \, d\phi \, d\theta$. Because I_{inc} is the same for all plane waves and τ is independent of the polar angle θ, an average transmission coefficient may be defined by

$$\bar{\tau} = \frac{\int_0^{\phi_{lim}} \tau(\phi)\cos\phi \sin\phi \, d\phi}{\int_0^{\phi_{lim}} \cos\phi \sin\phi \, d\phi} \tag{9.97}$$

where ϕ_{lim} is the limiting angle of incidence of the sound field. For random incidence, ϕ_{lim} is taken as $\frac{1}{2}\pi$, or 90°. The sound transmission coefficient $\tau(\phi)$ is that given in Eq. (9.95), and the random-incidence sound transmission loss is given by

$$R_{random} = 10 \log \frac{1}{\bar{\tau}} \quad \text{dB} \tag{9.98}$$

At low frequencies ($f << f_c$) the random-incidence sound transmission loss (for $TL_N > 15$ dB) is found by averaging the argument of Eq. (9.95) over a range of ϕ from 0° to 90° to yield[32, 33]

$$R_{random} \cong R_0 - 10 \log(0.23 R_0) \quad \text{dB} \tag{9.99}$$

which is commonly referred to as the random-incidence mass law.

It has become common practice to use the *field-incidence mass law*, which is defined (for $R_0 \geq 15$ dB) as[34]

$$R_{field} \cong R_0 - 5 \quad \text{dB} \tag{9.100}$$

This result, which yields better agreement with measured data than Eq. (9.99), approximates a diffuse incident sound field with a limiting angle ϕ_{lim} of about 78° in Eq. (9.97).[34]

The mass law transmission losses R_0, R_{field}, and R_{random}, valid for frequencies well below coincidence, are plotted versus $f\rho_s$ in Fig. 9.20.

Field-Incidence Sound Transmission for Thin Isotropic Plates, R_{field}. Equations (9.96) and (9.97) must be solved by numerical integration. The results of such an integration of the transmission coefficient between the angles 0° and 78° and application of Eq. (9.98) to give the *field-incidence transmission loss* are presented in Fig. 9.21 for all values of f/f_c. The ordinate is the difference between the field-incidence transmission loss R_{field} and the normal-incidence mass law transmission loss at the critical frequency $R_0(f_c)$. The latter is easily determined from Eq. (9.80c) or from Fig. 9.20 when the mass per unit area and the critical frequency of the panel are known. Note that predicted transmission losses of less than about 15 dB for $f << f_c$ or less than 25 dB for $f \simeq f_c$ from Fig. 9.21 are not accurate.

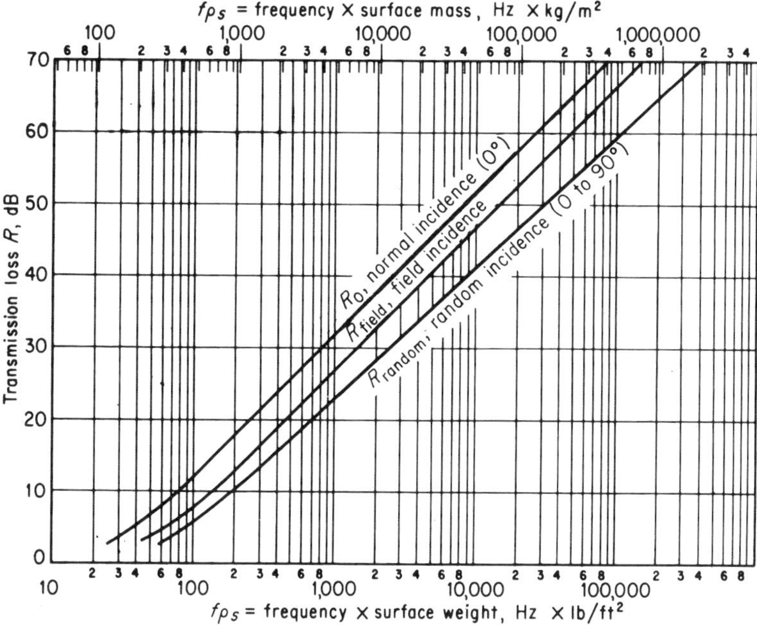

Fig. 9.20 Theoretical sound transmission loss of large panels for frequencies well below coincidence ($f \leq 0.5f_c$). Field incidence assumes a sound field that allows all angles of incidences up to 78° from normal.

Example 9.2 Calculate the normal-incidence mass law for an aluminum panel weighing 10 lb/ft² at a frequency of 500 Hz. Also determine the random-incidence and field-incidence mass laws. What is R_{field} at 2800 Hz when $\eta = 10^{-2}$?

SOLUTION The normal-incidence mass law is given by Eq. (9.80c) and the upper curve of Fig. 9.20. We have $f\rho_s = 500 \times 10 = 5000$ Hz · lb/ft². From Fig. 9.20, $R_0 = 45.5$ dB. The random-incidence mass law is given by Eq. (9.99) and the lower curve of Fig. 9.20; that is, $R_{\text{random}} = 35$ dB. The field-incidence mass law is given by Eq. (9.100) and the middle curve of Fig. 9.20, that is, $R_{\text{field}} = 45.5 - 5 = 40.5$ dB.

From Table 9.2, $\rho_s f_c = 7000$ Hz · lb/ft²; $f_c = 7000/10 = 700$ Hz and $R_0(f_c) = 48$ dB from Fig. 9.20. Evaluating Fig. 9.21 at $f/f_c = 2800/700 = 4$ and $\eta = 0.01$, we get $R_f(f) - R_0(f_c) = -6$ dB, yielding $R_r(f) = [R_r(f) - R_0(f_c)] + R_0(f) = -6 + 48 = 42$ dB.

Sound Transmission for Orthotropic Plates. Sound transmission for orthotropic plates differs from that of isotropic plates because orthotropic plates have markedly different bending stiffnesses in the different principal directions. The difference in bending stiffness for plane plates may result from the anisotropy of the plate material (such as for wood caused by grain orientation) or from the construction of the plate such as corrugations, ribs, cuts, and so on. Consequently,

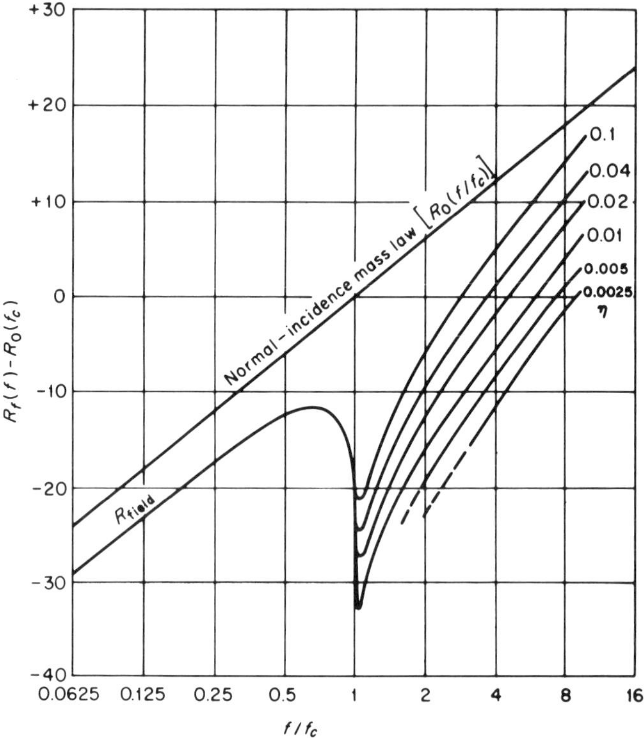

Fig. 9.21 Field-incidence forced-wave transmission loss. The ordinate is the difference between the field-incidence transmission loss at the frequency f and the normal-incidence transmission loss at the *critical frequency* ($f/f_c = 1$). Note that for a predicted transmission loss of less than 15 dB or for the dashed areas on the figure, the transmission loss depends on both the surface weight and the loss factor, and the curves provide only a lower bound estimate to the actual transmission loss. Use of curve: (1) determine $\rho_s f_c$ from Table 9.2, (2) determined f_c, (3) determine $R_0(f_c)$ from Fig. 9.20 or Table 9.2, (4) read $R_f(f) - R_0(f_c)$ from Fig. 9.21 at the required η, and (5) $R_f(f) = [R_f(f) - R_0(f_c)] + R_0(f_c)$. Top curve is the normal-incidence mass law defined in Eq. (9.80c).

the speed of free bending waves is different for these two directions and the orthotropic panel has two coincidence frequencies given by[35]

$$f_{c1} = \frac{c_0^2}{2\pi}\sqrt{\frac{\rho_s}{B_x}} \quad \text{Hz} \tag{9.101a}$$

$$f_{c2} = \frac{c_0^2}{2\pi}\sqrt{\frac{\rho_s}{B_y}} \quad \text{Hz} \tag{9.101b}$$

where B_x is the bending stiffness for the stiffest direction and B_y the direction perpendicular to this. The random-incidence sound transmission loss of an orthotropic plate is predicted by[38]

$$R_{\text{random}} \cong \begin{cases} 10 \log \left[\left(\dfrac{\rho_s \omega}{2\rho_0 c_0} \right)^2 \right] - 5 & \text{for } f \ll f_{c1} \\[2ex] 10 \log \left[\left(\dfrac{\rho_s \omega}{2\rho_0 c_0} \right)^2 \right] - 10 \log \left[\dfrac{1}{2\pi^3 \eta} \dfrac{f_{c1}}{f} \sqrt{\dfrac{f_{c1}}{f_{c2}}} \left(\ln \dfrac{4f}{f_{c1}} \right)^4 \right] & \\ & \text{for } f_{c1} < f < f_{c2} \\[2ex] 10 \log \left[\left(\dfrac{\rho_s \omega}{2\rho_0 c_0} \right)^2 \right] - 10 \log \dfrac{\pi f_{c2}}{2\eta f} & \text{for } f > f_{c2} \end{cases} \tag{9.102}$$

where η is the loss factor. For corrugated plates, as shown in Fig. 9.22, the bending stiffnesses can be approximated by

$$B_y = \frac{Eh^3}{12(1 - \nu^2)} \quad \text{Nm} \tag{9.103a}$$

$$B_x = B_y \left(\frac{s}{s'} \right) \quad \text{Nm} \tag{9.103b}$$

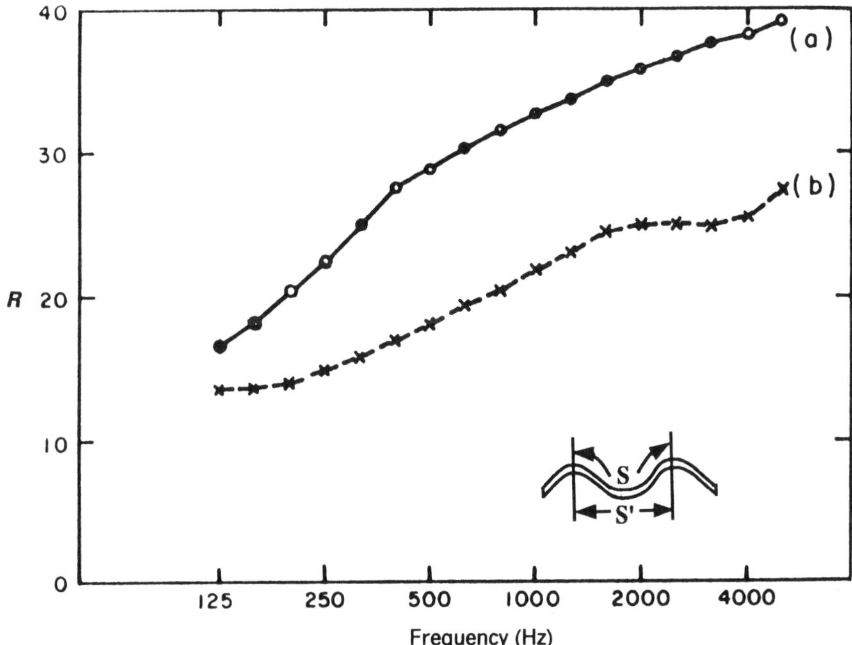

Fig. 9.22 Measured sound transmission loss of steel plates (after ref. 28): (*a*) plane plate, $\rho_s = 8 \text{ kg/m}^2$; (*b*) corrugated plate, $\rho_s = 11 \text{ kg/m}^2$; s, distance between corrugations along surface; s', distance along straight line.

where s and s' are defined in Fig. 9.22. Note that the increase in bending stiffness caused by corrugations, ribs, and stiffeners always results in a reduction of the sound transmission loss, while measures such as partial-depth saw cuts, which decrease bending stiffness, result in an increase of the sound transmission loss of plates.

Sound Transmission Loss for Inhomogeneous Plates. Sound transmission loss for inhomogeneous plates, such as appropriately designed sandwich panels, can be substantially higher than for homogeneous panels of the same mass per unit area, provided that such plates favor the propagation of the free shear waves (with frequency-independent propagation speed) rather than free bending waves for which the propagation speed increases with increasing frequency. However, they must be designed such that the shear wave speed remains below the speed of sound in air so that no trace coincidence occurs. Consequently, the sound transmission loss of such so-called shear wall panels closely approximates the field-incidence mass law. Information for designing such panels is given in reference 31. However, ordinary sandwich panels are very poor sound barriers because of their low mass and high bending stiffness that result in a coincidence frequency that usually falls in the middle of the audio frequency range. Dilation resonance, which occurs at the frequency where the combined stiffness impedance of the face plate and that of the enclosed air equals the mass impedance of the plate, also leads to further deterioration of the sound transmission loss of sandwich panels.

Sound Transmission through a Finite-Size Panel

For most architectural applications, where the first resonance frequency of typical platelike partitions is well below the frequency range of interest and the plate size is much larger than the acoustical wavelength, Eq. (9.95) or Fig. 9.21 (which are strictly valid only for infinitely large panels) can be used to predict the sound transmission loss of finite panels. In many industrial applications, the finite size of the panel must be taken into account.

In finite panels the sound-forced bending waves encounter the edges of the plate and generate free bending waves, such that the sum of the incident forced bending wave and the generated free bending wave satisfies the particular plate edge condition (e.g., zero displacement and angular displacement at a clamped edge). Consequently, the sound-forced bending waves continuously feed free bending wave energy into the finite panel and build up a reverberant, free-bending wave field. The mean-square vibration velocity of this free bending wave field, $\langle v_{\mathrm{FR}}^2 \rangle$ can be obtained using a power balance for the finite plate. The power introduced into the finite plate at the edges equals the power lost by the plate owing to viscous losses in the plate material, energy flow into connected structures, and sound radiation. The transmitted sound radiated by the finite panel is given by

$$W_{\mathrm{rad}} = \langle v_{\mathrm{FO}}^2 \rangle \rho_0 c_0 S \sigma_{\mathrm{FO}} + \langle v_{\mathrm{FR}}^2 \rangle \rho_0 c_0 S \sigma_{\mathrm{FR}} \quad \mathrm{W} \tag{9.104}$$

where $\langle v_{\mathrm{FO}}^2 \rangle$ is the mean-square velocity of the sound-forced supersonic bending waves, $\sigma_{\mathrm{FO}} \geq 1$ is the radiation efficiency of the forced waves, S is the surface area of the panel, and σ_{FR} is the radiation efficiency of the free bending waves. Since $\sigma_{FR} << 1$ below the critical frequency of the panel ($f << f_c$), it is frequently the case that the vibration response of the panel is controlled by the free bending waves (i.e., $\langle v_{FR}^2 \rangle >> \langle v_{FO}^2 \rangle$) but the sound radiation is controlled by the less intense but more efficiently radiating forced waves.

The classical definition of sound transmission loss is $R = 10 \log(W_{inc}/W_{trans})$ where W_{inc} is the sound power incident at the source side and W_{trans} is that radiated from the receiver side of the panel. If the incident sound is a plane wave arriving at an incident angle ϕ ($\phi = 90°$ for grazing incidence), then it is assumed that

$$W_{inc} = 0.5 \; |\hat{p}_{in}|^2 S \cos \phi / \rho_0 c_0 \qquad \text{W} \tag{9.105}$$

This assumption leads to the dilemma that at grazing incidence ($\phi = 90°$) no power is incident on the panel. It is common knowledge that grazing incidence sound excites the panel to forced vibrations, and the panel radiates sound into the receiver room when the forced bending waves in the panel and the sound wave at the receiver side, which run parallel to the panel, encounter the edges of the finite panel. This unresolved conceptual problem has been avoided[34] by limiting the incident angle range to 78°, in computing the field incidence sound transmission loss for the infinite panel according to Eq. (9.100), so as to yield reasonable agreement with laboratory measurements for panel sizes typically employed in such tests.

Obviously, it is not the incident sound power but the mean-square sound pressure on the source side that is forcing the panel. Since this quantity is proportional to the sound energy density in the source room $E_s = \langle p_s^2 \rangle / \rho_0 c_0^2$, it has been suggested[37,38] that the sound transmission loss of a finite partition be defined as

$$R_E \equiv 10 \log \left(\frac{E_s}{E_R} \frac{S}{A} \right) \qquad \text{dB} \tag{9.106}$$

where $E_R = 4W_{trans}/c_0 A$ is the energy density in the receiver room, S is the surface area of the panel (one side), and A is the total absorption in the receiver room. The transmitted sound power $W_{trans} = \frac{1}{4} c_0 A E_R$ is owing to the velocity of the plate. The forced response is dominated by the mass-controlled separation impedance $Z_s \simeq j\omega\rho_s$. The sound radiation of the sound-forced finite plate is controlled by its radiation impedance $Z_{rad} \cong \text{Re}\{Z_{rad}\} = \rho_0 c_0 \sigma_F$. Consequently, the low-frequency sound transmission loss of the finite partition is predicted as[38]

$$R_E \cong R_0 - 3 - 10 \log \sigma_F \qquad \text{dB} \tag{9.107}$$

where R_0 is the normal-incidence mass law sound transmission loss given in Eq. (9.80c) and σ_F is the forced-wave radiation efficiency of the finite panel given in Table 9.8. Note that σ_F depends on panel size as well as on incidence angle and can be smaller or larger than unity. This implies that the sound transmission loss of finite panels can be larger than the normal-incidence mass law even for grazing incidence if the size of the panel is small compared with the wavelength. When the panel size is much larger than the acoustical wavelength, σ_F approaches $1/\cos\phi$. For predicting the sound transmission loss of finite partitions over the entire low-frequency range ($f \ll f_c$), Eq. (9.107) should be used.

According to reference 38, the classical sound transmission formulas for infinite panels can be used to approximate the sound transmission loss of finite partitions by substituting $1/\sigma_F$ instead of $\cos\phi$ and $\sqrt{1 - 1/\sigma_F^2}$ instead of $\sin\phi$ in Eq. (9.95) and carrying out the integration in Eq. (9.97) from $\phi = 0$ to $\phi = 90°$ to obtain an estimate of the random-incidence sound transmission loss that agrees well with laboratory measurements. There is no need to resort to limiting the incident angle range to 78°. The radiation efficiency for random-incidence (diffuse-field) sound-forced excitation is given in Table 9.8, and this expression should be used in Eq. (9.107). As shown in Fig. 9.23, the random-incidence sound transmission loss of partitions of approximately 4 m^2 surface area, which are typically used in laboratory measurements, yield predicted values that are 5 dB below the normal-incidence mass law, giving theoretical justification for using the field-incidence mass law defined in Eq. (9.100) for partitions of this size. Note, however, that for small partitions Eq. (9.107) is more accurate. It should be pointed out that all of the sound transmission loss prediction formulas are valid only in the frequency region well above the first bending-wave resonance of the panel where the

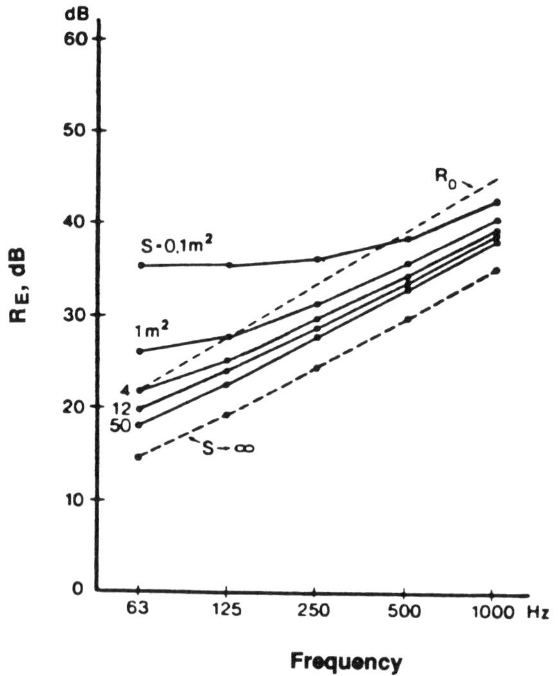

Fig. 9.23 Random-incidence sound transmission loss of 25-kg/m^2 panels as a function of frequency with panel surface area S as a parameter, mass-controlled low-frequency region; computed according to Eq. (9.109). (After refs. 37 and 38.)

forced response is mass controlled. For small, very stiff partitions the frequency range of interest may extend into the stiffness-controlled region below the frequency of the first bending-wave resonance. In this case, the volume compliance of the panel, as described in Chapter 13, should be used to predict sound transmission.

Empirical Method for Predicting Sound Transmission Loss of Single Partitions. An alternate technique useful in preliminary design is illustrated in Fig. 9.24. In essence, it considers the loss factor of the material to be determined completely by the material selection and substitutes a "plateau" or horizontal line for the peak and valley of the forced-wave analysis in the region of the critical frequency.[34,39] Its use will be demonstrated in Example 9.3.

Example 9.3 Calculate the transmission loss of a $\frac{1}{8}$-in.- (3-mm-) thick, 5 × 6.5-ft (1.52 × 2-m) aluminum panel by the alternate (plateau) method.

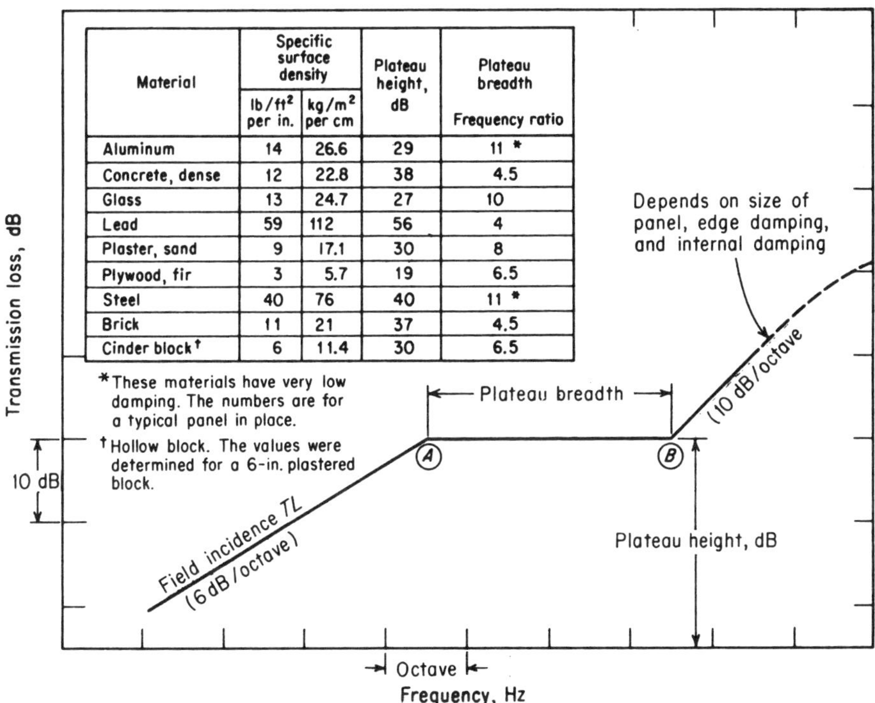

Material	Specific surface density lb/ft² per in.	Specific surface density kg/m² per cm	Plateau height, dB	Plateau breadth Frequency ratio
Aluminum	14	26.6	29	11 *
Concrete, dense	12	22.8	38	4.5
Glass	13	24.7	27	10
Lead	59	112	56	4
Plaster, sand	9	17.1	30	8
Plywood, fir	3	5.7	19	6.5
Steel	40	76	40	11 *
Brick	11	21	37	4.5
Cinder block†	6	11.4	30	6.5

*These materials have very low damping. The numbers are for a typical panel in place.

†Hollow block. The values were determined for a 6-in. plastered block.

Fig. 9.24 Approximate design chart for estimating the sound transmission loss of single panels. The chart assumes a reverberant sound field on the source side and approximates the behavior around the critical frequency with a horizontal line or plateau. The part of the curve to the left of A is determined from the field-incidence mass law curve (Fig. 9.20). The plateau height and length of the line from A to B are determined from the table. The part above B is an extrapolation. This chart is fairly accurate for large panels. Length and width of the panel should be at least 20 times the panel thickness.[34]

SOLUTION From Table 9.2, the product of surface mass and critical frequency is $\rho_s f_c = 34{,}700$ Hz · kg/m^2. The $\frac{1}{8}$-in.-thick aluminum plate has a surface density of 8.5 kg/m^2. From Fig. 9.20 we find that the normal-incidence transmission loss at the critical frequency is $R_0(f_c) = 48.5$ dB and that the field-incidence transmission loss at 1000 Hz is R_{field} (8500 Hz · kg/m^2) = 31.5 dB.

The procedure by the plateau method is as follows[34,39]:

1. Using semilog paper (with coordinates decibels versus log frequency), plot the field-incidence mass law transmission loss as a line with a 6-dB/octave slope through the point 31.5 dB at 1000 Hz.
2. From Fig. 9.24, the plateau height for aluminum is 29 dB. Plotting the plateau gives the intercept of the plateau with the field-incidence mass law curve at approximately 750 Hz.
3. From Fig. 9.24, the plateau width is a frequency ratio of 11. The upper frequency limit for the plateau is therefore 11 × 750 Hz = 8250 Hz.
4. From the point 29 dB, 8250 Hz, draw a line sloping upward at 10 dB/octave. This completes the plateau method estimate (see curve b in Fig. 9.33).

Sound Transmission through Double- and Multilayer Partitions

The highest sound transmission loss obtainable by a single partition is limited by the mass law. The way to "break this mass law barrier" is to use multilayer partitions such as double walls, where two solid panels are separated by an airspace that usually contains fibrous sound-absorbing material and double windows where light transparency requirements do not allow the use of sound-absorbing materials.

The transmission of sound through a multilayer partition can be computed in a similar manner as the sound absorption coefficient of multilayer sound absorbers treated in Chapter 8 (see the discussion of multilayer absorbers in Ref. 4 of Chapter 8). Figure 9.25 shows the situation where a plane sound wave of frequency $f = \omega/2\pi$ is incident at an angle ϕ on a panel that has N layers and $N + 1$ interfaces. The important boundary conditions are that the wavenumber component parallel to the panel surface $k_x = k \sin\phi$ must be the same in all of the layers and that the acoustical pressure and particle velocity at the interfaces of the layers must be continuous.[40–42]

The layers are characterized by their wave impedance formula, which relates the complex wave impedance at the input-side interface Z_I with that at the termination-side interface Z_T and by their pressure formula that relates the complex sound pressure at the input-side interface p_I to that at the terminal-side interface p_T.

The impedance and pressure formulas for an impervious orthotropic thin plate are given by[42]

$$Z_I = Z_T + Z_S$$

$$= Z_T + j\left[\omega\rho_s - \frac{1}{\omega}(B_x k_x^4 + 2B_{xy}k_x^2 k_y^2 + B_y k_y^4)\right] \quad \text{N} \cdot \text{s/m}^3 \qquad (9.108)$$

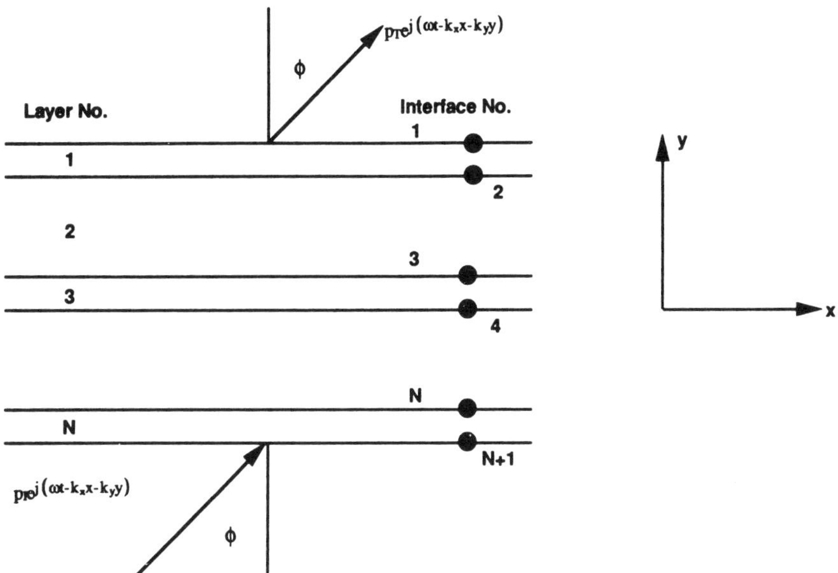

Fig. 9.25 Transmission of a plane oblique incident sound wave through an infinite lateral dimension multilayer panel.

$$p_I = p_T \frac{Z_I}{Z_T} \qquad \text{N/m}^2 \tag{9.109}$$

where ρ_s is the mass per unit area and B is the bending stiffness of the orthotropic plate.

For isotropic plates the second term in Eq. (9.108) reduces to $Z_s = Z_m + Z_B = j(\omega\rho_s - Bk_x^4/\omega)$. If the impervious layer is a composite of an isotropic homogeneous plate with a bonded damping material of thickness h_D, Young's modulus E_D, Poisson ratio ν_D, and damping loss factor η_D and the plate characteristics are thickness h_p, Young's modulus E_p, Poisson ratio ν_p, and loss factor η_p, then the complex bending stiffness $B = B_{\text{comp}}(1 + j\eta_{\text{comp}})$ is obtained[1,42] from

$$B_{\text{comp}} = \frac{E_p h_p^3}{12(1 - \nu_p^2)} + \frac{E_D h_D (h_p + h_D)^2}{4(1 - \nu_D^2)} \qquad \text{Nm} \tag{9.110}$$

$$\eta_{\text{comp}} = \tfrac{1}{4} B_{\text{comp}} \, (\eta_p E_p h_p + \eta_D E_D h_D) \, (h_p + h_D)^2 \tag{9.111}$$

For a porous sound-absorbing layer of thickness h the impedance and pressure formulas are given by[42]

$$Z_I = Z_a \frac{k_a}{k_{ay}} \frac{(1 + Z_a\Gamma_a/Z_T\Gamma_{ay})e^{j\Gamma_{ay}h} + (1 - Z_a\Gamma_a/Z_T\Gamma_{ay})e^{-j\Gamma_{ay}h}}{(1 + Z_a\Gamma_a/Z_T\Gamma_{ay})e^{j\Gamma_{ay}h} - (1 - Z_a\Gamma_a/Z_T\Gamma_{ay})e^{-j\Gamma_{ay}h}} \tag{9.112}$$

$\text{N} \cdot \text{s/m}$

$$p_T = \frac{p_I}{2}\left[\left(1 + \frac{Z_a\Gamma_a}{Z_I\Gamma_{ay}}\right)e^{-j\Gamma_{ay}h} + \left(1 - \frac{Z_a\Gamma_a}{Z_I\Gamma_{ay}}\right)e^{j\Gamma_{ay}h}\right] \quad \text{N/m}^2 \qquad (9.113)$$

where $Z_a = \rho_0 c_0 Z_{an}$ is the complex characteristic acoustical impedance of the porous bulk material for plane waves and Z_{an} is its normalized value while $\Gamma_a = \Gamma_{an} k_0$ is the complex wavenumber of plane sound waves in the bulk porous material; $\Gamma_{ay}^2 = \Gamma_a^2 - k_x^2$. Formulas for computing Γ_{an} and Z_{an} on the basis of the flow resistivity of the porous material R_1 are given in Chapter 8 [see Eqs. (8.12a) and (8.12b)]. The simpler approximate formulas for Γ_a and Z_a given in reference 42 are less accurate and their use is not recommended.

In the special case of an air layer without porous sound-absorbing material, substitute $Z_a = \rho_0 c_0$ and $k_a = k_0$ into Eqs. (9.112) and (9.113).

The computation of the sound transmission loss of a layered partition proceeds as follows:

1. Set the termination impedance at the receiver-side interface (interface 1 in Fig. 9.25) to $Z_T = \rho_0 c_0/\cos\phi$.
2. Apply the appropriate impedance formula for layer 1 and compute the input impedance at interface 2.
3. Using the impedance computed in step 2 as the termination impedance for layer 2, compute the input impedance at interface 3 and proceed down the chain of impedance calculations until the input impedance at the source-side interface (interface $N + 1$ in Fig. 9.25), Z_{N+1}, is obtained.
4. Compute the sound pressure at the source-side interface as the sum of the incident and reflected pressures $p_{N+1} = p_I[2\alpha/(\alpha + 1)]e^{j(\omega t - k_x x - k_y y)}$, where $\alpha = Z_{N+1}/(\rho_0 c_0/\cos\phi)$.
5. Apply the appropriate pressure formulas in succession until the sound pressure at the receiver-side interface, p, is obtained.
6. Determine the transmission coefficient $\tau(\omega, \phi) = p_1^2/p_I^2$ for all frequencies of interest.
7. Perform computational steps 1–6 for the incident angle range from $\phi = 0°$ to $\phi = 90°$ in one-third degree increments.
8. Compute the random-incidence sound transmission coefficient for isotropic layers (see ref. 42 for orthotropic impervious layers) as

$$\tau_R(\omega) = \int_0^{\pi/2} \tau(\omega, \phi)\sin 2\phi \, d\phi$$

9. Compute the random-incidence sound transmission loss of the infinite layered partition as

$$R_{\text{random}}(\omega) = 10 \log \frac{1}{\tau_R(\omega)}$$

Figure 9.26 shows the random-incidence sound transmission loss of an infinite three-layer partition as well as that of its constituent layers, computed according to Eqs. (9.108)–(9.113). The partition consists of two 1-mm-thick steel plates and a 100-mm-thick airspace that may or may not contain a fibrous sound-absorbing material. Figure 9.26 illustrates the benefit of using sound-absorbing layers. Note that below 500 Hz the double wall without sound-absorbing material in the airspace provides substantially lower random-incidence sound transmission loss than a single 1-mm steel plate or the single plate combined with the sound-absorbing layer. The highest sound transmission loss is obtained when the airspace is filled with the porous sound-absorbing material. In practical situations, leaks and structure-borne connections between the face plates at edges of the partition usually limit the maximally achievable sound transmission loss at high frequencies to a range of 40–70 dB. At low frequencies the finite size of the partition results in

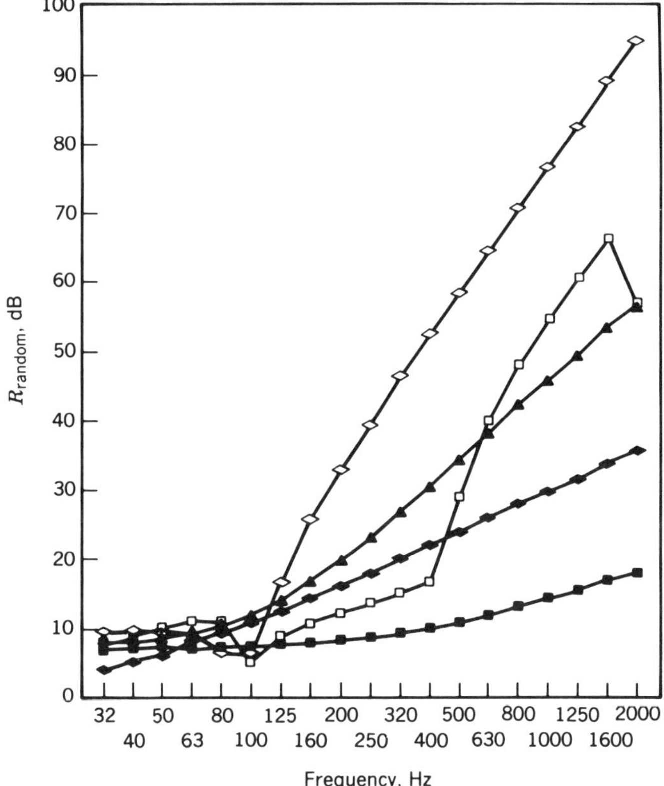

Fig. 9.26 Computed random-incidence sound transmission loss of a double wall and its constituent layers: (■) 100-mm-thick fibrous absorber, flow resisitivity $R_1 = 16{,}700$ N/s · m^4; (♦) 1-mm-steel plate; (▲) 1-mm steel plate and 100-mm fibrous absorber. Double wall consisting of two 1-mm-thick steel plates: (□) 100-mm airspace, no fibrous absorber; (◇) 100-mm airspace with fibrous absorber.

higher values than predicted for infinite partitions. The effect of the sound-absorbing material in the airspace results in refraction of the oblique-incidence sound toward the normal, thereby reducing the dynamic stiffness of the air between the plates. The sound-absorbing material also prevents high sound energy buildup in the cavity. These result in a substantial increase in sound transmission loss. The flow resistivity of the sound-absorbing material should be about $R_1 = 5000$ $N \cdot s/m^4$.[43] Higher values of R_1 yield only diminishing returns.

The filling of the airspace with a gas of 50% lower speed of sound than air (such as SF_6 or CO_2) has the same effect as the sound-absorbing material[44,45] as illustrated in Fig. 9.27. Using a light gas such as helium, which has three times higher speed of sound than air, also improves sound transmission loss to the same extent as a heavy gas fill. In this case, the improvement is due to the higher speed of sound in the gas fill, which makes it easier to push the gas tangentially than to compress it. Double windows, which can be hermetically sealed and must be light transparent, are partitions where this beneficial effect can be exploited.

Empricial Method for Predicting Sound Transmission Loss of Double Partitions. Goesele[46] has proposed a simplified method to predict the sound transmission loss R of a double partition when the measured sound transmission losses of the two constituent single partitions R_{I} and R_{II} are available, there are no structure-borne connections, and the gap is filled with porous sound-absorbing material. The prediction is given in Eq. (9.114) as

$$R \cong R_{\mathrm{I}} + R_{\mathrm{II}} + 20 \log \left(\frac{4\pi f \rho_0 c_0}{s'} \right) \quad \text{dB} \tag{9.114}$$

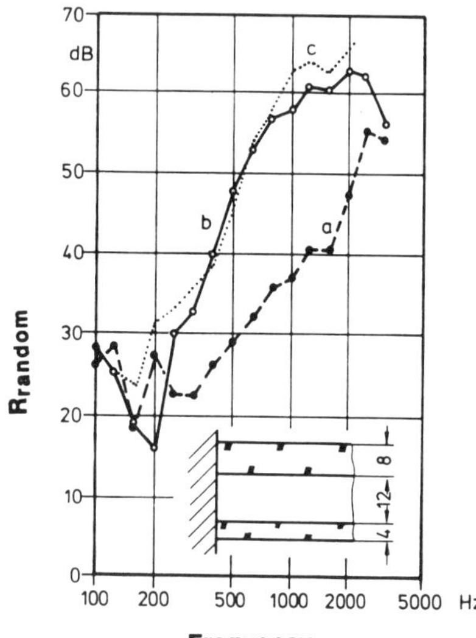

Fig. 9.27 Improvement of the sound transmission loss of a double glass partition (no contact at the edges) owing to heavy gas (SF_6) fill of the gap: *a*, measured with air-filled gap; *b*, measured with SF_6-filled gap; *c*, computed for mineral wool fill. (After ref. 44.)

where

$$s' = \begin{cases} \dfrac{\rho_0 c_0^2}{d} & \text{for } f < f_d = \dfrac{c_0}{2\pi d} \quad \text{N/m}^3 \quad (9.115a) \\ 2\pi f \rho_0 c_0 & \text{for } f > f_d \quad \text{N/m}^3 \quad (9.115b) \end{cases}$$

is the dynamic stiffness of the gap and d is the gap thickness.

If no measured sound transmission loss data for the constituent single partitions is available, the sound transmission loss of the double wall made of two identical panels can be predicted on the basis of material properties as

$$R(f_R < f < f_c) \cong 20 \log \frac{\pi f \rho_{s1}}{\sqrt{2}\,\rho_0 c_0} + 40 \log \left(\frac{\sqrt{2} f}{f_R} \right) \quad \text{dB} \qquad (9.116a)$$

$$R(f > f_R, f > f_c) \cong 40 \log \left[\frac{\pi f \rho_{s1}}{\rho_0 c_0} \sqrt{2\eta} \left(\frac{f}{f_c} \right)^{1/4} \right] + 20 \log \frac{4\pi f \rho_0 c_0}{s'} \quad \text{dB} \qquad (9.116b)$$

where f_c is the critical frequency of the panels and f_R is the double-wall resonance frequency

$$f_R = \frac{1}{2\pi} \sqrt{\frac{2\sqrt{2}\,s'}{\rho_{s1}}} \quad \text{Hz} \qquad (9.117)$$

Figure 9.28 shows that Eq. (9.114) yields good agreement with measured data in the entire frequency region, while Eqs. (9.116a) and (9.116b) give good agreement only well below and well above the critical frequency but fail in the frequency

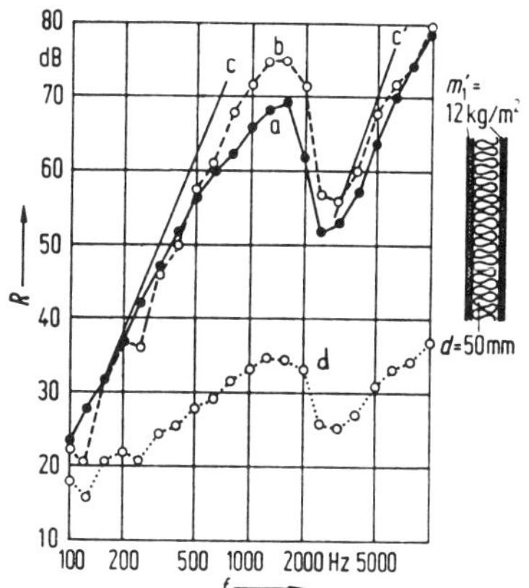

Fig. 9.28 Sound transmission loss of a double partition consisting of two identical, 12.5-mm-thick gypsum boards separated by a 50-mm-thick gap filled with fibrous sound-absorbing material: *a*, measured values; *b*, predicted by Eq. (9.116); *c*, predicted by Eqs. (9.117a) and (9.117b); *d*, measured sound transmission loss of a single gypsum board wall. (After ref. 46.)

region near the critical frequency. Prediction methods for the sound transmission loss of double walls with point and line bridges are given by reference 30 and by Bies and Hansen (see Bibliography).

Sound Transmission Loss of Ducts and Pipes

Pipes and ducts that carry high-intensity internal sound are excited into vibration and radiate sound to the outside. This sound transmission in the breakout direction (i.e., from inside to outside) is characterized by breakout sound transmission loss R_{i0}, which is a measure of the rate at which sound energy from the interior of the duct radiated to the outside. When pipes and ducts traverse areas of high-intensity sound such as found in mechanical equipment rooms, the exterior sound field excites ductwall vibrations and the vibrating walls generate an internal sound field that can travel to distant quiet areas. This sound transmission from the outside to the inside direction is characterized by the break-in sound transmission loss R_{0i}, which is a measure of the rate at which sound energy from the exterior sound field enters the duct.

The sound power level radiated by a duct or pipe of length l, L_w^{i0}, is predicted as[47]

$$L_w^{i0}(l) = L_w^i(0) - R_{i0} + 10 \log \left(\frac{Pl}{S}\right) + C \qquad \text{dB } re\ 10^{-12}\ \text{W} \qquad (9.118a)$$

where

$$C = \frac{1 - e^{-(\tau + \beta) l}}{(\tau + \beta) l} \qquad (9.118b)$$

$$\tau = \frac{P}{S} \times 10^{-R_{i0}/10} \qquad (9.118c)$$

$$\beta = \frac{\Delta L_1}{4.34} \qquad (9.118d)$$

where $L_w^i(0)$ is the sound power level in the duct at the source side, S is the area in square meters, P is the perimeter of the duct cross section in meters, and ΔL_1 is the sound attenuation in decibels per unit length inside the duct due to porous lining. Equation (9.118a) contains only measurable quantities and is used as a basis for the experimental evaluation of the breakout sound transmission loss R_{i0}, by measuring the sound power $W_{0i}(l)$ radiated by a test duct of length l into a reverberation room, the sound power in the duct at the source side, $W_i(0)$, and the sound attenuation inside the duct, ΔL_1, and solving Eq. (9.118a) for R_{i0} by iteration.

The sound power level of the sound propagating in one direction, $L_w^{i0}(l)$ when a duct of length l traverses through a noisy area (and sound breaks into the duct) is predicted by[47]

$$L_w^{i0}(l) = L_w^{\text{inc}} - R_{i0} - 3 + C \qquad \text{dB } re\ 10^{-12}\ \text{W} \qquad (9.119)$$

where L_w^{inc} is the sound power level of the sound incident on the duct of length l, R_{0i} is the break-in sound transmission loss, and C is as defined in Eq. (9.118b). On the basis of reciprocity,[48] the following relationship exists between breakout and break-in sound transmission loss:

$$R_{0i} \cong R_{i0} - 10 \log \left\{ 4\gamma \left[1 + 0.64 \frac{a}{b} \left(\frac{f_{\text{cut}}}{f} \right)^2 \right] \right\} \quad \text{dB} \tag{9.120}$$

where a and b are the larger and smaller sides of a rectangular duct cross section, f is the frequency, f_{cut} is the cutoff frequency of the duct, and γ is 1 below cutoff and 0.5 above cutoff. Empirical methods for predicting the breakout sound transmission loss of unlined, unlagged rectangular sheet metal ducts are given in references 48–50. Table 9.9 contains predicted values of octave-band breakout sound transmission loss versus frequency for rectangular sheet metal ducts of sizes most frequently used in low-velocity HVAC systems. Sound transmission loss predictions for round and flat-oval ducts are given elsewhere.[49,50] Figure 9.29 shows the break-in sound transmission loss of an unlined, unlagged sheet metal duct. The solid curve was obtained by directly measuring the break-in sound transmission loss while the open circles represent data points obtained by applying the reciprocity relationship embodied in Eq. (9.120) for the measured breakout sound transmission loss. The importance of Eq. (9.120) is that it makes it unnecessary to measure both R_{0i} and R_{i0} separately because if one has been measured, the other can be predicted.

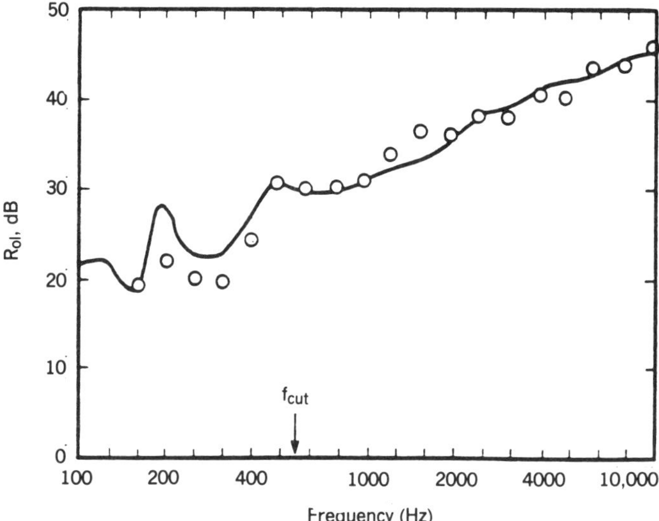

Fig. 9.29 Break-in sound transmission loss R_{0i} of a 0.3 m × 0.91 mm thick sheet metal duct (after ref. 48): (——) measured; (○○○) predicted from measured R_{i0} using Eq. (9.120).

Sound Transmission through a Composite Partition

Partitions that separate adjacent rooms frequently consist of areas that have different sound transmission losses, such as a wall that contains a door with an uncovered key hole. If all parts of the composite partition are exposed to the same average sound intensity, I_{inc}, on the source side, then the sound power transmitted is

$$W_{\text{trans}} = I_{\text{inc}} \sum_{i=1}^{n} S_i \times 10^{-R_i/10} = I_{\text{inc}} \sum_{i=1}^{n} S_i \tau_i \qquad \text{W} \tag{9.121a}$$

and the transmission loss of the composite partition is

$$R_{\text{comp}} = 10 \log \frac{W_{\text{inc}}}{W_{\text{trans}}} = -10 \log \sum_{i=1}^{n} \frac{S_i}{S_{\text{tot}}} \times 10^{-R_i/10} \qquad \text{dB} \tag{9.121b}$$

where S_i is the surface area of each component, S_{tot} is the total area of all components, and R_i is the sound transmission loss of the ith component. The sound transmission loss of a small hole radius $a \ll \lambda_0$ in a thin plate of thickness h is well approximated by[26]

$$R_{\text{hole}} \cong 20 \log \frac{h + 1.6a}{\sqrt{2}a} \qquad \text{dB} \tag{9.122}$$

indicating that small holes in thin partitions ($h < a$) yield a frequency-independent sound transmission loss of $R_{\text{hole}} \simeq 0$ dB. Note, however, that Eq. (9.122) is valid only for small round holes. Long narrow slits can have a "negative sound transmission loss."[51] The sound transmission loss of holes and slits can be increased substantially by sealing them with either porous sound-absorbing material or an elastomeric material or designing them as silencer joints. Prediction of the sound transmission loss of such acoustically sealed openings are given in references 52 and 53.

Flanking Sound Transmission

The sound transmission loss of partitions is measured in acoustical laboratories where sound transmission from the source room to the receiver room occurs only through the partition under test. However, if the same partition constitutes a part of a building, then sound can be transmitted through many paths, as shown schematically in Fig. 9.30. Path 1 represents the primary path, which is characterized by sound transmission loss of the separating partition R_1, usually available from laboratory measurements. The sound transmission loss for each of the $n = 4$ paths (1 the direct and 2, 3, and 4 the flanking paths) is defined as[54]

$$R_n \equiv 10 \log \frac{W_1^{\text{inc}}}{W_{\text{TRAN}}^{n}} \qquad \text{dB} \tag{9.123}$$

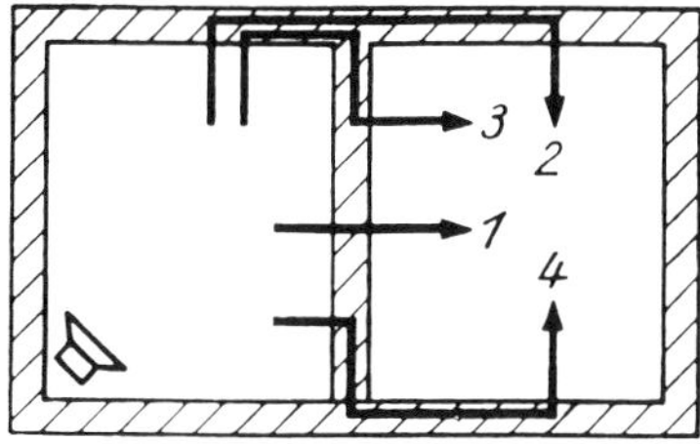

Fig. 9.30 Sound transmission paths between two adjacent rooms in a typical building (after ref. 54): 1, primary path; 2–4, flanking paths.

and the composite sound transmission loss, which combines sound transmission along each of the four paths, as

$$R_{\text{comp}} \equiv -10 \log \left(10^{-R_1/10} + \sum_{n=2}^{4} \sum_{m=1}^{4} 10^{-R_{mn}/10} \right) \quad \text{dB} \qquad (9.124)$$

where $m = 4$ represents the four sound-excited flanking partitions (i.e., the two side walls, floor, and ceiling of the receiver room) each of which transmit sound along each of the three flanking paths $n = 2, 3, 4$. Usually flanking path $n = 2$ contributes as much to the receiver room sound power as do the two other flanking paths $n = 3$ and $n = 4$ together. The contribution of the back wall of the source and receiver room is usually negligible. The process of flanking transmission along flanking paths $n = 2, 3, 4$ is as follows: (1) the sound field in the source room excites the flanking walls to vibration, (2) the vibration is transmitted through the wall junctions to the receiver room walls, and (3) the receiver room walls radiate sound power into the receiver room that adds to that transmitted by the separating partition through the direct path. If the source and receiver room have no common wall, the entire sound transmission takes place through flanking paths. For adjacent rooms with a common wall, the composite sound transmission loss given in Eq. (9.124) should be used to predict the sound pressure level in the receiver room. The component flanking transmission losses R_{mn}, for homogeneous isotropic single-wall construction can be approximated by[54]

$$R_{mn} = 10 \log \left(\frac{1}{4\pi\sqrt{12}} \left(\frac{\rho_M}{\rho_0} \right)^2 \frac{c_L h^3 \omega^2}{c_0^4} \eta_m \right) + \Delta L_{\text{JUNCT}}$$
$$+ 10 \log \frac{S_1}{S_{\text{rad}}} - 10 \log (\sigma_m \sigma_{\text{rad}}) \quad \text{dB} \qquad (9.125)$$

where ρ_M is density and c_L the longitudinal wave speed of the wall material, h is the thickness, η_m is the composite loss factor of the mth partition in the source room, and ΔL_{JUNCT} is the attenuation of the structure-borne sound amplitude at the wall junction along the transmission path n. The symbol S_1 is the surface area of the separating partition and S_{rad} is the surface area of the partition in the receiver room involved in the transmission along path n. The symbols σ_m and σ_{rad} are the radiation efficiencies of the mth partition in the source room and the radiation efficiency of the partition in the receiver room that radiates sound owing to the sound-induced vibration of the mth partition in the source room transmitted through the nth path. A more accurate prediction of the effect of flanking paths on sound transmission loss can be made utilizing the statistical energy analysis method discussed in the next section.

9.8 STATISTICAL ENERGY ANALYSES

Statistical energy analysis (SEA) is a point of view in dealing with the vibration of complex resonant structures. It permits calculation of the energy flow between connected resonant structures, such as plates, beams, and so on, and between plates and the reverberant sound field in an enclosure.[55-59]

System of Modal Groups

In respect to the energy E stored in a structure, it may be thought of as a *system of* resonant modes or resonators. First, let us consider the power flow between two groups of resonant modes of two coupled structures having their modal resonance frequencies within a narrow frequency band $\Delta\omega$ (see Fig. 9.31).

We assume that each resonant mode of the first system (box 1 in Fig. 9.31) has the same energy. Also, assume that the coupling of the individual resonant modes of the first system with each resonance mode of the second system is approximately the same.

If we further assume that the waves carrying the energy in one system are uncorrelated with the waves carrying the energy gained through coupling to the other system, we can separate the power flow (each equation is written for a narrow frequency band $\Delta\omega$) as

$$W'_{12} = E_1 \omega \eta_{12} \quad \text{W} \tag{9.126}$$

$$W'_{21} = E_2 \omega \eta_{21} \quad \text{W} \tag{9.127}$$

where W'_{12} = power system 1 transmits to system 2, W
W'_{21} = power system 2 transmits to system 1, W
E_1 = total energy in system 1, m · kg/s
E_2 = total energy in system 2, m · kg/s
ω = center frequency of band, rad/s

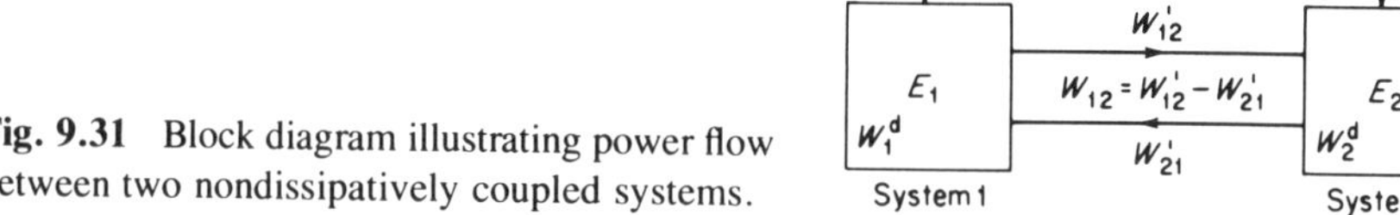

Fig. 9.31 Block diagram illustrating power flow between two nondissipatively coupled systems.

η_{12} = coupling loss factor from system 1 to system 2, as defined in Eq. (9.126)

η_{21} = coupling loss factor from system 2 to system 1, as defined in Eq. (9.127)

The net power flow between the two systems is, accordingly,

$$W_{12} = W'_{12} - W'_{21} = E_1 \omega \eta_{12} - E_2 \omega \eta_{21} \qquad \text{W} \tag{9.128}$$

Modal Energy E_m

Let us define modal energy as

$$E_m = \frac{E(\Delta\omega)}{n(\omega)\,\Delta\omega} \quad \text{W} \cdot \text{s/Hz} \tag{9.129}$$

where $E(\Delta\omega)$ = total energy in system in angular frequency band $\Delta\omega$

$n(\omega)$ = modal density, =number of modes in unit bandwidth ($\Delta\omega = 1$) centered on ω, the angular frequency

$\Delta\omega$ = bandwidth, rad/s

If the previously made assumptions about the equal distribution of energy in the modes and the same coupling loss factor are valid, then it may be shown that

$$\frac{\eta_{21}}{\eta_{12}} = \frac{n_1(\omega)}{n_2(\omega)} \tag{9.130}$$

where $n_1(\omega)$ = modal density of system 1 at frequency ω, s

$n_2(\omega)$ = modal density of system 2 at frequency ω, s

Equation (9.130) implies that for equal total energies in the two systems, $E_1 = E_2$, the system that has the *lower* modal density [lower $\eta(\omega)$] transfers more energy to the second system than is transferred from the second to the first system.

Combining Eqs. (9.128) and (9.130) yields*

$$W_{12} = \omega \eta_{12} n_1(\omega) [E_{m1} - E_{m2}]\, \Delta\omega \quad \text{W} \tag{9.131}$$

where W_{12} = net power flow between systems 1 and 2 in band $\Delta\omega$, centered at ω, W

E_{m1}, E_{m2} = modal energies for systems 1 and 2, respectively [see Eq. (9.129)], W · s/Hz

This equation is positive if the first term in the brackets is greater than the second.

*Note that Eq. (9.130) is a necessary requirement if Eq. (9.131) is to obey the consistency relationship $W_{12} = -W_{21}$.

The principle of the SEA method is given by Eq. (9.131), which is a simple algebraic equation with energy as the independent dynamic variable. It states that *the net power flow between two coupled systems in a narrow frequency band, centered at frequency* ω*, is proportional to the difference in the modal energies of the two systems at the same frequency. The flow is from the system with the higher modal energy to that with the lower modal energy.*

It may help to understand Eq. (9.131) if we use the thermodynamical analogy of heat transfer between two connected bodies of different temperature, where the heat flow is from the body of higher temperature to that of lower temperature and the net heat flow is proportional to the difference in temperature of the two bodies. Consequently, the modal energy E_m is analogous to temperature, and the net power flow W_{12} is analogous to heat flow. The case of equal modal energies of the two systems where the net power flow is zero is analogous to the equal temperature of the two bodies.

Equal Energy of Modes of Vibration

Equal energy of the modes within a group will usually exist if the wave field of the structure is diffuse. Also, since the frequency-adjacent resonance modes of a structure are coupled to each other by scattering and damping, there is always a tendency for the modal energy of resonant modes to equalize within a narrow frequency band even if the wave field is not diffuse.

Noncorrelation between Waves in the Two Systems

In sound transmission problems usually only one system is excited. The power W'_{12} transmitted to the nonexcited system builds up a semidiffuse vibration field in that system. Accordingly, the waves that carry the transmitted power W'_{21} back to the excited system are almost always sufficiently delayed and randomized in phase with respect to the waves carrying the incident power W'_{12} that there is little correlation between the two wave fields.

Realization of Equal Coupling Loss Factor

Equality of the coupling loss factors between individual modes within a group is a matter of grouping modes of similar nature. If the coupled system is a plate in a reverberant sound field, the *acoustically slow edge and corner modes* and *acoustically fast surface modes* are grouped separately.

Composite Structures

Composite structures generally consist of a number of elements such as plates, beams, stiffeners, and so on. We may divide a complex structure into its simpler member provided the wavelength of the structure-borne vibration is small compared with the characteristic dimensions of the elements. Where this is true, the modal density of a complex structure is approximately that of the sum of the modal densities of its elements. If the power input, the various coupling loss factors, and the power dissipated in each element are known, the power balance equations will yield the vibrational energy in the respective elements of the structure.

The dissipative loss factor for an element of a structure is obtained by separating that element from the rest of the structure and measuring its decay rate as discussed in Chapter 12. The coupling loss factor can be determined experimentally from Eq. (9.131); however, the procedure is difficult. Theoretical solutions are available for the coupling loss factors of a few simple structural connections.[59] When the coupling loss factor between a sound field and a simple structure is desired, such a loss factor can be calculated from Eq. (9.130) if the radiation ratio of the structure is known, as shown in the next section.

Power Balance in a Two-Structure System

The power balance of the simple two-structure system of Fig. 9.31 is given by the following two algebraic equations:

$$W_1^{\text{in}} = W_1^d + W_{12} \quad \text{W} \tag{9.132}$$

$$W_2^{\text{in}} = W_{12} + W_2^d \quad \text{W} \tag{9.133}$$

where W_1^{in} = input power to system 1, W
W_1^d = power dissipated in system 1, W
W_{12} = net power lost by system 1 through coupling* to system 2, $= W'_{12} - W'_{21}$, W
W_2^{in} = input power to system 2, W
W_2^d = power dissipated in system 2, W

*As in our previous analysis, we assume that the coupling is nondissipative.

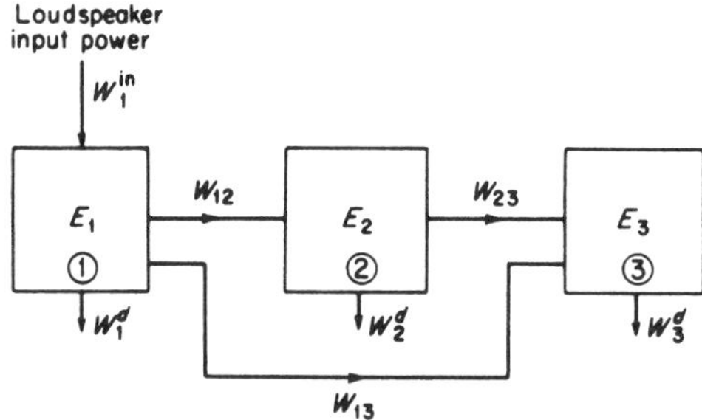

Fig. 9.32 Block diagram illustrating the power flow in three-way coupled systems: W_{13}, transmission of sound by those modes whose resonance frequency lies outside the source band. The "nonresonant" modes are important below the critical frequency.

The power dissipated in a system is related to the energy stored by that system, E_i, through the dissipative loss factor η_i, namely,

$$W_i^d = E_i \omega \eta_i \quad \text{Watts} \tag{9.134}$$

where E_i is energy stored in system i in newton-meters.

Assuming that the second system does not have direct power input ($W_2^{\text{in}} = 0$), the combination of Eqs. (9.129)–(9.131), (9.133), and (9.134) (with $i = 2$) yields the ratio of the energies stored in the two respective systems

$$\frac{E_2}{E_1} = \frac{n_2}{n_1} \frac{\eta_{21}}{\eta_{21} + \eta_2} \tag{9.135}$$

If the coupling loss factor is very large compared to the loss factor in system 2, that is, if $\eta_{21} >> \eta_2$, Eq. (9.135) yields the equality of the modal energies $(E_1/n_1 \, \Delta\omega = E_2/n_2 \, \Delta\omega)$.

Diffuse Sound Field Driving a Freely Hung Panel

Let us now examine the special case of the excitation of a homogeneous panel (system 2) that hangs freely, exposed to the diffuse sound field of a reverberant room (system 1).

For this case, the total energies for each system are given by

$$E_1 = DV = \frac{\langle p^2 \rangle}{\rho_0 c_0^2} V \quad \text{N} \cdot \text{m} \tag{9.136}$$

$$E_2 = \langle v^2 \rangle \rho_s S \quad \text{N} \cdot \text{m} \tag{9.137}$$

where D = average energy density in reverberant room, N/m^2
$\langle p^2 \rangle$ = mean-square sound pressure (space-time average), N^2/m^4
V = room volume, m^3
$\langle v^2 \rangle$ = mean-square plate vibration velocity (space–time average), m^2/s^2
S = plate surface area (one side), m^2
ρ_s = mass per unit area of panel, kg/m^2

To find the coupling loss factor η_{21}, we must first recognize that W'_{21} equals the power that the plate, having been excited into vibration, radiates back into the room. Thus

$$W_{\text{rad}} = W'_{21} = 2\langle v^2 \rangle \rho c \sigma_{\text{rad}} S \equiv E_2 \omega \eta_{21} = \langle v^2 \rangle \rho_s S \omega \eta_{21} \quad \text{W} \tag{9.138}$$

where σ_{rad} = radiation ratio for plate, dimensionless
W_{rad} = acoustical power radiated by both sides of plate, which accounts for factor 2, W

Solving for η_{21} yields

$$\eta_{21} = \frac{2\rho_0 c_0 \sigma_{\text{rad}}}{\rho_s \omega} \tag{9.139}$$

The modal density of the reverberant sound field in the room $n_1(\omega)$ and that of the thin homogeneous plate $n_2(\omega)$ are given in Table 9.7, namely,

$$n_1(\omega) = \frac{\omega^2 V}{2\pi^2 c_0^3} \quad \text{s} \tag{9.140}$$

$$n_2(\omega) = \frac{\sqrt{12}\, S}{4\pi c_L h} \quad \text{s} \tag{9.141}$$

where c_L = propagation speed of longitudinal waves in the plate material, m/s
h = plate thickness, m

Inserting Eqs. (9.136)–(9.141) into Eq. (9.135) yields the desired relation between the sound pressure and plate velocity:

$$\langle v^2 \rangle = \langle p^2 \rangle \frac{\sqrt{12}\,\pi c_0^2}{2\rho_0 c_0 h c_L \rho_s \omega^2} \frac{1}{1 + \rho_s \omega \eta_2 / 2\rho c \sigma_{\text{rad}}} \quad \text{m}^2/\text{s}^2 \tag{9.142}$$

The mean-square (space–time average) acceleration of the panel is simply

$$\langle a^2 \rangle = \omega^2 \langle v^2 \rangle = \langle p^2 \rangle \frac{\sqrt{12}\,\pi c_0^2}{2\rho_0 c_0 h c_L \rho_s} \frac{1}{1 + \rho_s \omega \eta_2 / 2\rho_0 c_0 \sigma_{\text{rad}}} \quad \text{m}^2/\text{s}^4 \tag{9.143}$$

It can be shown that as long as the power dissipated in the plate is small compared with the sound power radiated by that plate ($\rho_s \omega \eta_2 \ll 2\rho_0 c_0 \sigma_{\text{rad}}$), the equality of the modal energies of the sound field and the plate yields the proper plate velocity and acceleration. Also, under this condition, the ratio of the mean-square plate acceleration to the mean-square sound pressure is independent of frequency. In general, *the plate response is always smaller than that calculated by the equality of the modal energies by the last factor on the right of Eq. (9.142) or (9.143), which equals the ratio of the power loss by acoustical radiation to the total power loss*. In dealing with the excitation of structures by a sound field, the concept of equal modal enegy often enables one to give a simple estimate for the upper bound of the structure's response.

Example 9.4 Calculate the rms velocity and acceleration of a 0.005-m- ($\frac{1}{10}$-in.-) thick homogeneous aluminum panel resiliently suspended in a reverberant room. The space-averaged sound pressure level $\overline{L}_p$ = 100 dB ($\sqrt{\langle p^2 \rangle}$ = 2 N/m^2) as measured in a one-third-octave band centered at a frequency $f = \omega/2\pi$ = 1000

Hz. The appropriate constants of the panel and surrounding media are $\rho_s = 13.5$ kg/m^2; $c_L = 5.2 \times 10^3$ m/s; $h = 5 \times 10^{-3}$ m; $\rho_0 = 1.2$ kg/m^3; $c_0 = 344$ m/s; and $\eta_2 = 10^{-4}$.

SOLUTION First, calculate the factor

$$\frac{\rho_s \omega \eta_2}{2\rho_0 c_0 \sigma_{\text{rad}}} = \frac{13.5 \times 2\pi \times 10^3 \times 10^{-4}}{2 \times 1.2 \times 344 \times \sigma_{\text{rad}}} = \frac{8.5}{820\sigma_{\text{rad}}} \ll 1$$

According to the above inequality the mean-square acceleration of the panel given by Eq. (9.143) simplifies to

$$\langle a^2 \rangle \approx \langle p^2 \rangle \frac{\sqrt{12}\,\pi c_0^2}{2\rho_0 c_0 h c_L \rho_s} = 19 \text{ m}^2/\text{s}^4$$

or $a_{\text{rms}} = \sqrt{\langle a^2 \rangle} = 4.34$ m/s^2, an acceleration level of 113 dB *re* 10^{-5} m/s^2

The mean-square velocity is

$$\langle v^2 \rangle = \frac{\langle a^2 \rangle}{\omega^2} = \frac{19}{4\pi^2 \times 10^6} = 4.76 \times 10^{-7} \text{ m}^2/\text{s}$$

or $v_{\text{rms}} = \sqrt{\langle v^2 \rangle} = 6.9 \times 10^{-4}$ m/s, a velocity level of 97 dB *re* 10^{-8} m/s.

Sound Transmission Loss of a Simple Homogeneous Structure by the SEA Method

The SEA method may be used to analyze the transmission of sound between two rooms coupled to each other by a single common, thin homogeneous wall.[58] (i.e., there are no flanking paths). System 1 is the ensemble of modes of the diffuse, reverberant sound field in the source room, resonant within the frequency band $\Delta\omega$. System 2 is an appropriately chosen group of vibration modes of the wall. System 3 is the ensemble of modes of the diffuse reverberant sound field in the receiving room, resonant within the frequency band $\Delta\omega$. A loudspeaker in the source room is the only source of power, and the power dissipated in each system is assumed to be large compared with the power lost to the other two systems through the coupling (see Fig. 9.32).

The procedure is as follows:

1. Relate W_1^{in} to the power lost by sound absorption in the room. This yields $\langle p_1^2 \rangle$, the space-averaged mean-square sound pressure in the source room.
2. Calculate $E_1 = \langle p_1^2 \rangle V_1/\rho c^2$, where V_1 is the volume of the source room.
3. The reverberant sound power incident on the dividing wall of area S_2 is $W_{\text{inc}} = E_1 c S_2 / 4V_1$.

4. From the power balance of the wall for resonant modes within the bandwidth, $\Delta\omega$ is next determined, that is, $W_{12} = W_2^d + W_{23}$, so that

$$W_{12} = \text{Eq. (9.134)} + 0.5\ [\text{Eq. (9.138)}]$$

This sum equals Eq. (9.131). Because η_{12} of Eq. (9.131) is not well known, it is replaced by using Eq. (9.130) and the definition of loss factor η_{21}, which yields $\eta_{21} = \rho_0 c_0 \sigma_{\text{rad}} / \rho_s \omega$.
5. The vibrational energy of the wall is $E_2 = \langle v^2 \rangle \rho_s S_2$.
6. Combining steps 3, 4, and 5 yields the mean-square wall velocity $\langle v^2 \rangle$ as a function of mean-square source room pressure $\langle p_1^2 \rangle$.
7. The power radiated into the receiving room is

$$W_{23} = \rho_0 c_0 S_2 \sigma_{\text{rad}} \langle v^2 \rangle$$

8. Finally, the resonance transmission coefficient τ_r is found by dividing step 7 by step 3.
9. The *resonance transmission loss*, defined as $R_r = 10 \log(1/\tau_r)$, is computed from step 8 using Eq. (9.85) and assuming that $\rho_s \omega \eta_2 >> 2\rho_0 c_0 \sigma_{\text{rad}}$ to yield

$$R_r = 20 \log \left(\frac{\rho_s \omega}{2\rho_0 c_0} \right) + 10 \log \left(\frac{f}{f_c} \frac{2}{\pi} \frac{\eta_2}{\sigma_{\text{rad}}^2} \right) \quad \text{dB} \qquad (9.144)$$

The first term in Eq. (9.144) is approximately the normal-incidence mass law transmission loss R_0, so that Eq. (9.144) becomes

$$R_r = R_0 + 10 \log \left(\frac{f}{f_c} \frac{2}{\pi} \frac{\eta_2}{\sigma_{\text{rad}}^2} \right) \quad \text{dB} \qquad (9.145)$$

where f_c = critical frequency [see Eq. (9.84)], Hz
η_2 = total loss factor of wall, dimensionless
σ_{rad} = radiation ratio for wall, dimensionless

Thus we have obtained the transmission loss between two rooms separated by a common wall using the SEA method.

Below the critical frequency and when the dimensions of the wall are large compared with the acoustical wavelength, the radiation factor σ_{rad} can be taken from Table 9.8.

It is important to note that if the sound transmission loss of an equivalent *infinite* wall is compared with the data measured and predicted by the SEA method, it is found that *above the critical frequency*, the transmission loss R for the infinite wall yields the same results as Eq. (9.145), which takes into account only the resonance transmission of a finite wall.

Below the critical frequency the sound transmission loss of a finite panel is more controlled by the contribution of those modes that have their resonance frequencies outside of the frequency band of the excitation signal than by those with resonance frequencies within that band. Since only the contributions of the latter are included in the previous SEA calculation, Eq. (9.145) usually overestimates the sound transmission loss of a finite panel below the critical frequency. Figure 9.33 shows that below the critical frequency for a $\frac{1}{8}$-in.-thick aluminum panel the sound transmission loss of the resonant modes alone (curve *a*) is approximately 10 dB higher than that measured on the actual panel (curve *d*).

A *composite transmission factor* that approximately takes into account both the forced and resonance waves is closely approximated by

$$\frac{1}{\tau} = \frac{W_{\text{inc}}}{W_{\text{forced}} + W_{\text{res}}}$$

$$= \frac{(\langle p^2 \rangle / 4\rho c) S_2}{\langle p^2 \rangle (\pi \rho c S_2 / \rho_s^2 \omega^2) + \langle p^2 \rangle (\sqrt{12} \pi c^3 \rho \sigma_{\text{rad}}^2 S_2 / 2\omega^3 \rho_s^2 c_L h \eta_2)} \tag{9.146}$$

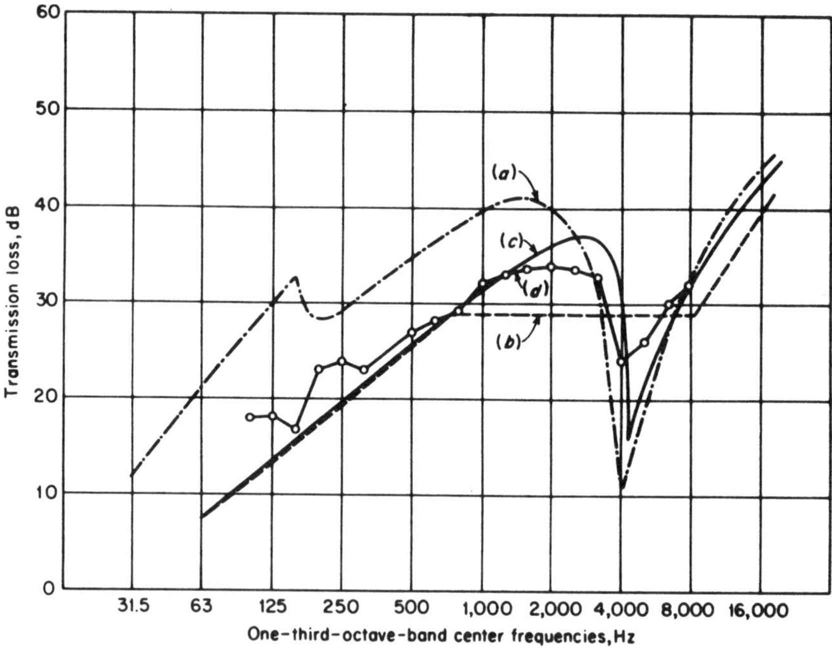

Fig. 9.33 Comparison of experimental and theoretical transmission loss of a 5 ft × 6.5 ft × 1/8 in. aluminum panel. The theoretical calculations are based on (*a*) resonance mode calculation, (*b*) plateau calculation, and (*c*) forced-wave calculation. Curve *d* shows the experimental results. (After ref. 58.)

At low frequencies where the first term in the denominator dominates, the transmission factor becomes

$$\frac{1}{\tau} \approx \frac{1}{\pi}\left(\frac{\rho_s \omega}{2\rho c}\right)^2 \qquad (9.147)$$

and the sound transmission loss is [see Eq. (9.100)]

$$R = 10 \log \frac{1}{\tau} \approx R_0 - 5 = R_{\text{field}} \quad \text{dB} \qquad (9.148)$$

At high frequencies where the second term in the denominator becomes dominant, the sound transmission loss is given by Eq. (9.145).

9.9 RECIPROCITY AND SUPERPOSITION

The principles of reciprocity and superposition apply to linear systems with time-invariant parameters. Not only solid structures but fluid volumes at rest fall into this category. Consequently, reciprocity and superposition applies to systems that consist of solid structures surrounded by acoustical spaces and can be used to great advantage not only in structureborne noise and airborne noise but also in structural acoustics, which deals with the interaction of sound waves with solid structures.

The principle of superposition, illustrated in Fig. 9.34, allows the use of the simplest excitation sources such as a point force source or a point-monopole sound source to explore the response to more complex excitation sources such as a moment acting on a structural element or an acoustical dipole radiating into an acoustical volume. The principle of reciprocity, which can be traced back to Lord Rayleigh,[60] is illustrated in the upper three sketches in Fig. 9.35. It states that

$$\mathbf{F}_2\mathbf{v}_2 = \mathbf{F}_{\text{I}}\mathbf{v}_{\text{I}} \quad \text{watts} \qquad (9.149\text{a})$$

The symbol $\mathbf{F}_{\text{I}}$ is (generalized) force when point 1 is source and point 2 is receiver and $\mathbf{v}_{\text{I}}$ is (generalized) velocity when point 1 is receiver and point 2 is source. The vector product $\mathbf{F}_{\text{I}}\mathbf{v}_{\text{I}}$ must yield the instantaneous power, or in complex notation Re $\{\frac{1}{2}\mathbf{F}_{\text{I}}\mathbf{v}_{\text{I}}^*\}$ = Re $\{\frac{1}{2}\mathbf{F}_2\mathbf{v}_2^*\}$ must yield the time-averaged power. Note that $\mathbf{F}$ and $\mathbf{v}$ are vector quantities as signified in the figure by the arrow above the symbols. If $\mathbf{v}_{\text{I}}$ and $\mathbf{F}_{\text{I}}$ and $\mathbf{v}_2$ and $\mathbf{F}_2$ are measured or applied in the same direction, as illustrated by the sketch in the lower left side of Fig. 9.35, the vector notation can be exchanged for the less complicated scalar notation, where force and velocity are characterized by a magnitude and phase (i.e., $\tilde{F} = Fe^{j\phi_f}$, $\tilde{v} = ve^{j\phi_v}$). In this case, the reciprocity takes the form of the equality of transfer functions

$$\frac{\tilde{v}_2}{\tilde{F}_{\text{I}}} = \frac{\tilde{v}_{\text{I}}}{\tilde{F}_2} \quad \text{m/N} \cdot \text{s} \qquad (9.149\text{b})$$

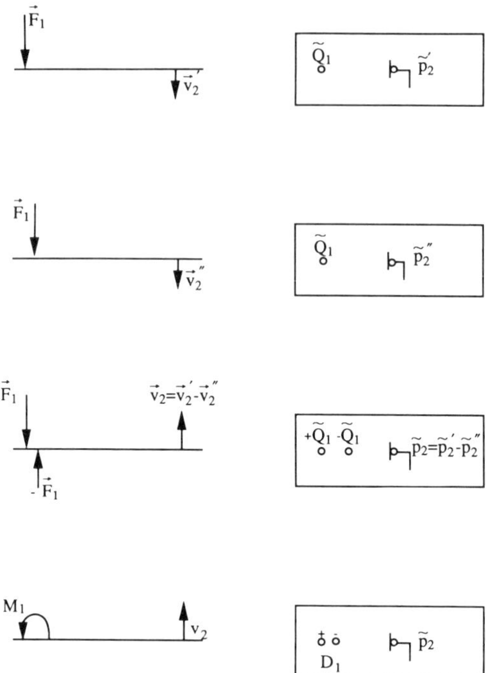

Fig. 9.34 Use of superposition to predict the response of structures and acoustical spaces (v_2 and p_2) to complex excitation sources such as moments (M_1) and dipoles (D_1) on the basis of point force (F_1) and acoustical monopole (Q_1). Note that the structure and the acoustical space may be of arbitrary shape and the acoustical space may be unbound or bound by elastic or sound-absorbing boundaries and may contain an arbitrary number and size of rigid or elastic scatterers.

This should be kept in mind in our latter deliberations, where the special vector notation is not carried through. Since monopole strength $\tilde{Q}$ and sound pressure $\tilde{p}$ are scalar quantities (defined by their magnitude and phase), no directional restraints exist in the acoustical case illustrated in Fig. 9.35*b*.[61,62] However, dipole and quadrupole sound sources (constructed from adjacent out-of-phase monopoles) have highly directional radiation characteristics and must be connected to directional quantities of the sound field such as pressure gradient $\mathbf{dp}/dx$ and $\mathbf{d}^2\mathbf{p}/d^2x$ measured in the same direction relative to the orientation of the dipole or quadrupole sound source for the reciprocity to apply.

Table 9.10 contains useful reciprocity relationships applicable to higher order excitation sources and responses. These relationships automatically follow from the joint application of superposition and reciprocity to the appropriate combination of simple sources such as point forces and acoustical monopoles.

The principle of reciprocity can be used to considerable advantage in both experimental and analytical work. In experimental work it is difficult and cumbersome to excite complex structures by point forces and moments and to measure

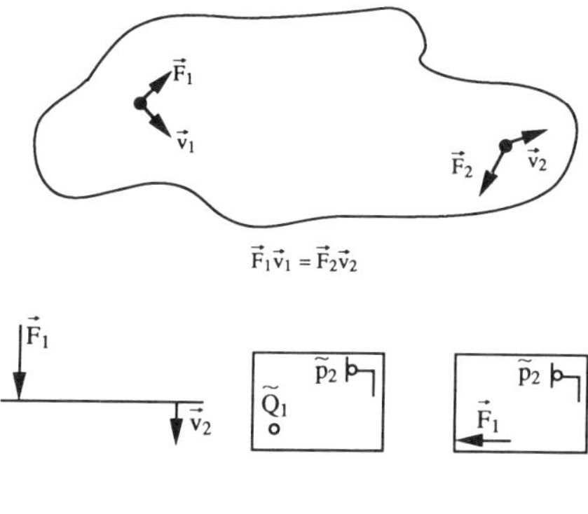

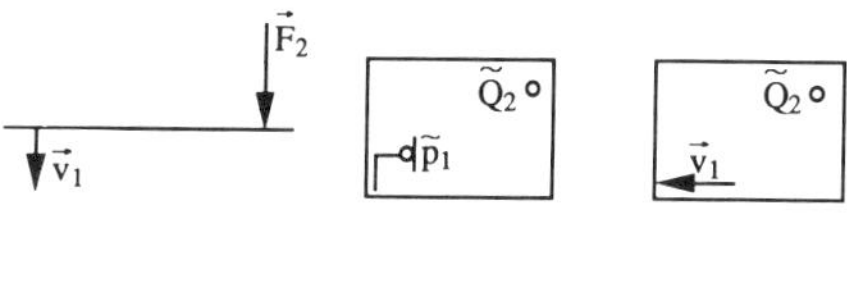

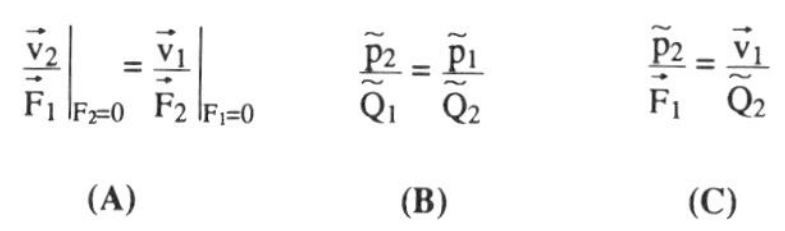

$$\left.\frac{\vec{v}_2}{\vec{F}_1}\right|_{F_2=0} = \left.\frac{\vec{v}_1}{\vec{F}_2}\right|_{F_1=0} \qquad \frac{\tilde{p}_2}{\tilde{Q}_1} = \frac{\tilde{p}_1}{\tilde{Q}_2} \qquad \frac{\tilde{p}_2}{\vec{F}_1} = \frac{\vec{v}_1}{\tilde{Q}_2}$$

(A) (B) (C)

Fig. 9.35 Principle of reciprocity as applied to a complex structure. Top sketch illustrates general principle. Lower sketches apply in special situations: (*a*) structure; (*b*) an acoustical space; (*c*) a structure coupled to an acoustical space. Symbols: **F**, point force vector; **v**, velocity response vector (measured by the same direction as **F**); $\tilde{Q}$, volume velocity of acoustical point source; $\tilde{p}$, sound pressure.

the sound pressure such excitation causes in the interior of a vehicle. It is almost always easier to obtain the sought transfer function between the acoustical pressure and the exciting force or moment by placing a small acoustical source of known volume velocity at the microphone location and measuring the vibration response at the position and in the direction of the applied force and applying the reciprocity relationship illustrated in Fig. 9.35*c*.

The form of reciprocity illustrated in Fig. 9.35*c* is the most useful in structural acoustics. Its practical application is described in reference 62. In the direct experiment the force $\tilde{F}_1$ is applied to the structure by a shaker (and its magnitude and phase is measured by a force gauge inserted between the shaker and the structure), and the magnitude and phase of the sound pressure p_2 is measured by a microphone. The phase of F_1 and p_2 are referenced to the voltage U applied to the shaker. In the reciprocal experiment—which is much easier to perform than the direct experiment—a small omnidirectional sound source (an enclosed loudspeaker whose diameter is smaller than one-quarter acoustical wavelength) with calibrated volume velocity response Q is placed at the former microphone location and the velocity response of the structure at the former excitation point v_2 is measured (in the same direction as the force was applied) by a small accelerometer. The phases of Q and v_2 are referenced to the voltage U applied to the loudspeaker sound source. The volume velocity calibration of the sound source (Q/U) is obtained by placing it in an anechoically terminated rigid tube, baffling it so it radiates only toward the anechoic termination, sweeping the loudspeaker voltage through the frequency range of interest, and measuring the transfer function (p/U), where U is the voltage applied to the loudspeaker and p is the sound pressure measured by a microphone located two tube diameters or further away from the source. The sought volume velocity calibration of the source is then computed as[64] $|Q/U| = |P/U|(S/\rho_0 c_0)$, where S is the cross-sectional area of the tube, ρ_0 is the density of air, and c_0 is the speed of sound in air. When phase information is important, the sound source can be calibrated in an anechoic chamber by measuring the sound pressure $p(r)$ at a large distance $r >> \lambda_0$ away from the source and computing the volume velocity calibration $Q/U = [p(r)/U](4\pi r^2/\rho_0 c_0)e^{-j2\pi fr/c_0}$.

Figure 9.36 illustrates the application of reciprocity on a complex structural acoustical problem, namely, the prediction of the interior noise of an aircraft cabin to point force excitation of the fuselage.[62] First, the tail end of the aircraft fuselage was excited by a point force, and sound pressure generated in the cabin at a specific point was measured to obtain the direct transfer function $\tilde{p}/\mathbf{F}$ identified by the dotted line. Next, the reciprocal experiment was carried out by placing a point sound source of known volume velocity $\tilde{Q}$ at the former location of the microphone and measuring the vibration velocity response of the fuselage at the former location of the applied force. The transfer function $\mathbf{v}/\tilde{Q}$ obtained this way is shown as the solid curve in Fig. 9.36. Referencing the phases of $\tilde{p}$ and $\mathbf{F}$ to the voltage applied to the shaker and that of $\mathbf{v}$ and $\tilde{Q}$ to the voltage applied to the loudspeaker source, not only the magnitude but also the phase of the reciprocal transfer functions

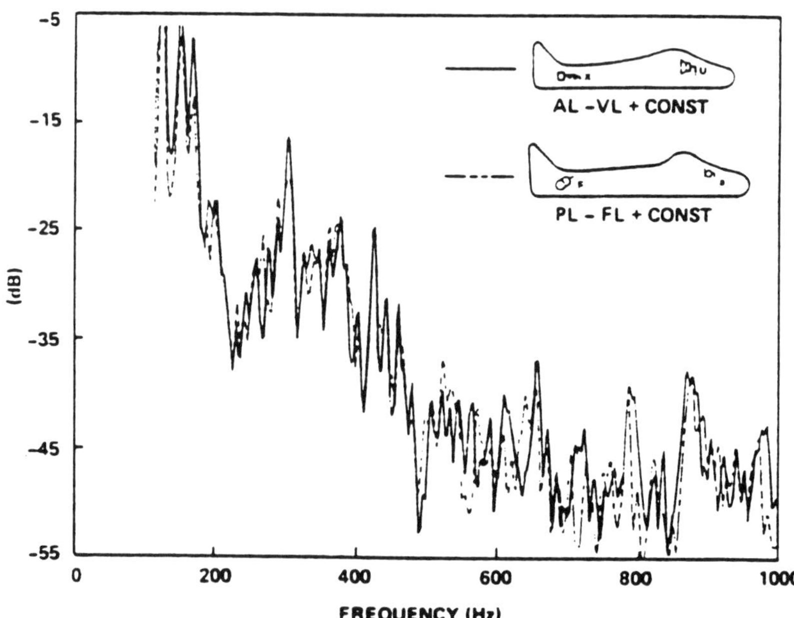

Fig. 9.36 Direct transfer function $20 \log|\tilde{p}/\tilde{F}| = PL - FL + \text{const}$ and reciprocal transfer function $20 \log|\tilde{v}/\tilde{Q}| = AL - VL + \text{const}$, measured on an aircraft fuselage. (After ref. 62.)

$\tilde{p}/\mathbf{F}$ and $\mathbf{v}/\tilde{Q}$, can be retained so that the interior noise caused by many simultaneously acting forces and moments can be predicted.[64]

Prediction of Noise Caused by Multiple Correlated Forces

The use of reciprocity and superposition in the case of multiple, correlated force input is illustrated in Fig. 9.37. The problem is to predict the sound pressure p_R in the receiver room below after installation of a vibrating machine in the source room above. The building is constructed, and the machine manufacturer provides the magnitude, the direction, and the mutual phase ϕ_{12} of forces $\mathbf{F}_1$ and $\mathbf{F}_2$ the machine—when installed on soft springs—will impart to the floor. The prediction of p_R proceeds as illustrated in the lower part of Fig. 9.37 by measuring the reciprocal transfer functions $-\mathbf{v}_1/\tilde{Q}_R = \tilde{p}_R/\mathbf{F}_1$ and $-\mathbf{v}_2/\tilde{Q}_R = \tilde{p}_{R2}/\mathbf{F}_2$ and utilizing the principle of superposition. The forces and velocities must be measured in the

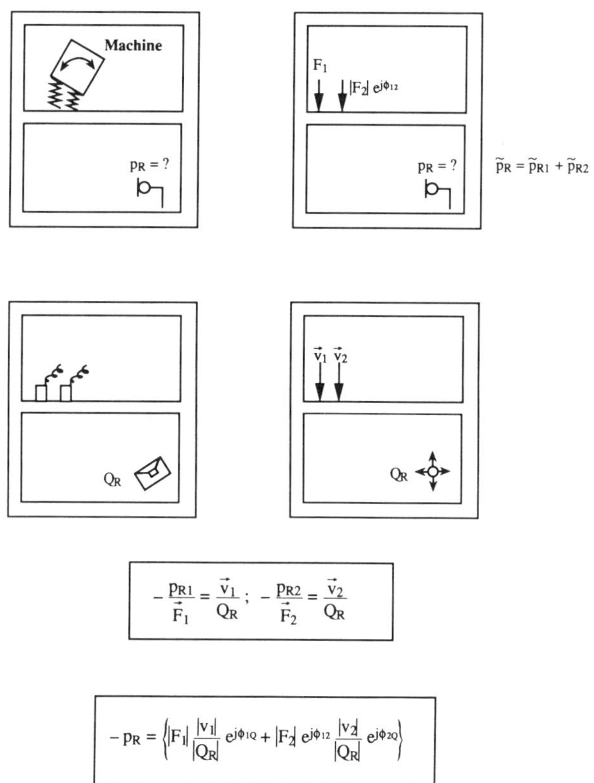

Fig. 9.37 Use of reciprocity and superposition to predict the sound pressure in a room caused by two correlated forces $\tilde{F}_1$ and $\tilde{F}_2$ acting on the building structure. Top sketch represents actual situation and lower sketch the reciprocity prediction: ϕ_{1Q} and ϕ_{2Q} represent the phase of the transfer functions $\tilde{v}_1/\tilde{Q}_R$ and $\tilde{v}_2/\tilde{Q}_R$, respectively; each is conveniently referenced to the excitation voltage of the loudspeaker sound source.

same direction. Performing the reciprocity prediction for a number of different loudspeaker positions in the receiver room, the spatial variation of p_R can also be predicted. The methodology can be easily extended to more than two simultaneously acting, correlated forces.[69]

The reciprocity is most useful in the early stages of design of aircraft and ground vehicles, well before a flightworthy version of the aircraft or a roadworthy version of the ground vehicle is available. Reciprocity helps the noise control engineer to find answers to some difficult questions:

1. What will be the contribution of the structureborne noise at the engine firing rate to the noise in the passenger compartment?
2. Which engine mount transmits most of the structure-borne noise?
3. Which direction of vibration force is most critical?
4. Which mutual phasing of forces acting on the individual engine mounting points is most critical?
5. Most importantly, what is the effect of changing the design of engine mounting brackets on cabin noise?

All these questions can be answered without applying known forces in three orthogonal directions to each of the engine-mounting brackets.

Source Strength Identification by Reciprocity

When it is not feasible to measure directly the strength of noise and vibration sources during the operation of vehicles, equipment, and machinery, reciprocity can be used to obtain them indirectly. This is accomplished by measuring the noise or vibration during the operation of the equipment at a distant, accessible receiver location and—when the equipment is not operating—exciting it at these distant receiver locations and measuring the acoustical or structural response at the source location, which is now accessible. The principle is illustrated in Fig. 9.38. Common in the three problems shown in Fig. 9.38 is the knowledge of the location and nature of the excitation sources. Unknown is their magnitude and mutual phase, which must be determined by observation of response to these sources at distant locations and by reciprocity experiments as described below.

The upper left-hand side of Fig. 9.38 represents the case where the sound field in an enclosure is excited by two monopole sound sources of unknown strength and mutual phase $\tilde{Q}_1(?)$ and $\tilde{Q}_2(?)$ (e.g., the openings of the inlet pipe leading to two cylinders of a reciprocating compressor). The first experiment, illustrated in the upper sketch, is the measurement of the magnitude and mutual phase of the sound pressure $\tilde{p}_3$ and $\tilde{p}_4$ at accessible distant locations 3 and 4 obtained when both sources were operating simultaneously. The reciprocal experiments, illustrated in the two lower sketches, are performed when the sources are not operational by placing a monopole sound source of known volume velocity Q at the former microphone locations 3 and 4 and measuring the magnitude and phase of the sound

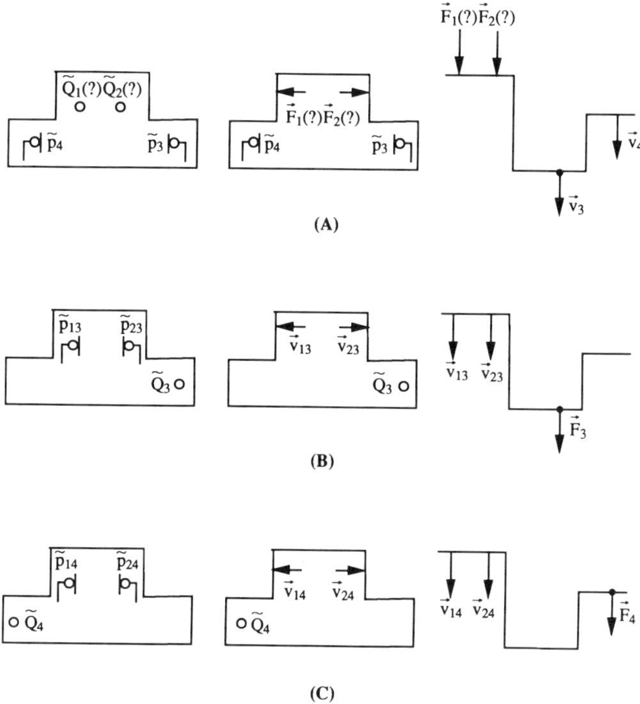

Fig. 9.38 Source strength identification utilizing reciprocity and superposition: (*a*) response measurements at distant locations with sources active and (*b*) and (*c*) reciprocity measurements, sources inactive. (See text for explanation of symbols.)

pressure produced at the two former source locations $\tilde{p}_{13}$, $\tilde{p}_{23}$, $\tilde{p}_{14}$, and $\tilde{p}_{24}$. The phase of the sound pressure is referenced to the loudspeaker voltage.

The sketches in the middle column in Fig. 9.38 illustrate a case where the sound field in an enclosure is produced by two forces $\tilde{F}_1$ and $\tilde{F}_2$ of known location and direction but unknown magnitude and mutual phase, both acting simultaneously at the enclosure wall (e.g., forces caused by a vibration isolation-mounted reciprocating or rotating machine). In this case, the reciprocal experiment yields the vibration velocity responses at the former force application points $\tilde{v}_{13}$, $\tilde{v}_{23}$, $\tilde{v}_{14}$, and $\tilde{v}_{24}$, measured in the same direction as the force.

The situation shown in the sketch on upper right-hand side in Fig. 9.38 illustrates the situation when the determination of the unknown magnitude and mutual phase of the two forces $\tilde{F}_1(?)$ and $\tilde{F}_2(?)$ must be diagnosed (e.g., forces transmitted to a structural floor by a vibration-isolated machine). In this case, the reciprocal experiment is carried out by exciting the building structure at the two distant observation points 3 and 4 by known forces $\tilde{F}_3$ and $\tilde{F}_4$ and measuring the velocity responses $\tilde{v}_{13}$, $\tilde{v}_{23}$, $\tilde{v}_{14}$, and $\tilde{v}_{24}$ at the former force excitation points 1 and 2. Utilizing the principle of reciprocity and superposition yields the following pairs of linear equations:

$$\tilde{p}_3 = \tilde{Q}_1(?) \left[\frac{\tilde{p}_{13}}{\tilde{Q}_3}\right] + \tilde{Q}_2(?) \left[\frac{\tilde{p}_{23}}{\tilde{Q}_3}\right] \quad \text{N/m}^2 \tag{9.150a}$$

$$\tilde{p}_4 = \tilde{Q}_1(?) \left[\frac{\tilde{p}_{14}}{\tilde{Q}_4}\right] + \tilde{Q}_2(?) \left[\frac{\tilde{p}_{24}}{\tilde{Q}_4}\right] \quad \text{N/m}^2 \tag{9.150b}$$

which can be solved for the two unknowns $\tilde{Q}_1(?)$ and $\tilde{Q}_2(?)$. Similarly, the set of equations for obtaining $\tilde{F}_1(?)$ and $\tilde{F}_2(?)$ in the situation illustrated in the center sketches in Fig. 9.38 are

$$\tilde{p}_3 = \tilde{F}_1(?) \left[\frac{\tilde{v}_{13}}{\tilde{Q}_3}\right] + \tilde{F}_2(?) \left[\frac{\tilde{v}_{23}}{\tilde{Q}_3}\right] \quad \text{N/m}^2 \tag{9.151a}$$

$$\tilde{p}_4 = \tilde{F}_1(?) \left[\frac{\tilde{v}_{14}}{\tilde{Q}_4}\right] + \tilde{F}_2(?) \left[\frac{\tilde{v}_{24}}{\tilde{Q}_4}\right] \quad \text{N/m}^2 \tag{9.151b}$$

and for that illustrated in the sketches on the right

$$\tilde{p}_3 = \tilde{F}_1(?) \left[\frac{\tilde{v}_{13}}{\tilde{F}_3}\right] + \tilde{F}_2(?) \left[\frac{\tilde{v}_{23}}{\tilde{F}_3}\right] \quad \text{N/m}^2 \tag{9.152a}$$

$$\tilde{p}_4 = \tilde{F}_1(?) \left[\frac{\tilde{v}_{14}}{\tilde{F}_4}\right] + \tilde{F}_2(?) \left[\frac{\tilde{v}_{24}}{\tilde{F}_4}\right] \quad \text{N/m}^2 \tag{9.152b}$$

In the case of n unknown excitation sources, the prediction equations represent an $n \times n$ matrix.

Extension of Reciprocity to Sound Excitation of Structures

As illustrated in Fig. 9.39, the reciprocity relationship can be extended for surface excitation of structures (e.g., by an incident sound wave). Consider first a small part of the surface of a cylindrical body (such as an aircraft fuselage) with surface area dA exposed to a local sound pressure of $\tilde{p}_1$ as illustrated in the upper sketch in Fig. 9.39 resulting in a local force of $\tilde{F}_1 = \tilde{p}_1\, dA$. For this force the reciprocity relationship shown in Fig. 9.35c yields $\Delta\tilde{p}_{R1}/\tilde{F}_1 = \Delta\tilde{v}_{1R}/\tilde{Q}_R$, which for $\tilde{F}_1 = \tilde{p}_1\, dA$ and $\Delta\tilde{Q}_{1R} = \Delta\tilde{v}_{1R}\, dA$ becomes

$$\frac{\Delta\tilde{p}_R}{\tilde{p}_1} = \frac{\Delta\tilde{Q}_{1R}}{\tilde{Q}_R} \tag{9.153a}$$

When the structure is exposed to a complex sound field distribution $\tilde{p}_1, \tilde{p}_2, \ldots,$ $\tilde{p}_n$, as illustrated in the lower sketch, the resulting interior sound pressure at the receiver location, $\tilde{p}_R$, is given by

$$\tilde{p}_R = \sum_{i=1}^{n} \tilde{p}_i \left[\frac{\Delta\tilde{Q}_{iR}}{\tilde{Q}_R}\right] \quad \text{N/m}^2 \tag{9.153b}$$

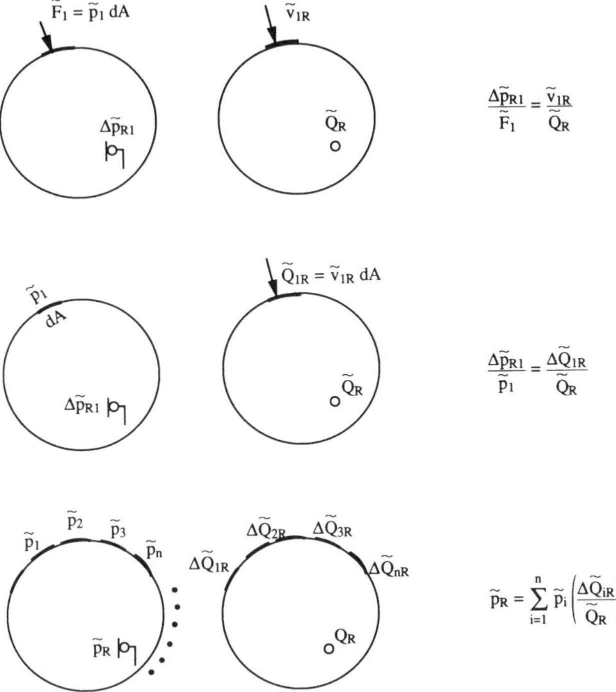

Fig. 9.39 Extension of reciprocity to sound excitation of structures. Top: Reciprocity relationship for external pressure, $\tilde{p}_1$, acting on a small area (dA) of the structure and the resulting internal sound pressure, $\Delta \tilde{p}_{R1}$, and structural response v_{1R} produced by an internal point sound source of volume velocity Q_R located at the former receiver position. Middle: Substitution of volume velocity response $\Delta \tilde{Q}_{1R} = d\tilde{v}_{1R}\, dA$ into the reciprocity relationship. Bottom: Reciprocity relationship for external incident sound excitation p_i and internal sound pressure $\tilde{p}_R$ and volume velocity responses of the structure $\Delta \tilde{Q}_{1R}$ produced by an internal point sound source of volume velocity $\tilde{Q}_R$.

where the transfer functions $\Delta \tilde{Q}_{1R}/\tilde{Q}_R$ represent the reciprocity calibration of the structure as a transducer. This extension of reciprocity by Fahy[63] has the advantage that the reciprocity calibration of the structure (in the form of discretized transfer functions $\Delta \tilde{Q}_{iR}/\tilde{Q}_R$) can be carried out with a capacitive transducer that directly measures the structure's local volume displacement $\Delta \tilde{Q}_{iR}/j\omega$. To obtain sufficient resolution, the side length of the square-shaped capacitive transducer used in measuring the volume displacement must not exceed one-eighth of the acoustical wavelength. Note that the bending wavelength in thin, platelike structures is usually much smaller than the transducer size so that the capacitive transducer acts as a wavenumber filter, accounting only for those components of the vibration field that results in a net volume displacement. The high-wavenumber components, which result only in local near fields, are "averaged out." The additional advantage of the capacitive transducer is that it does not influence the vibration response of the structure.

Reciprocity can also be used in predicting the sound pressure attributable to the complex vibration pattern of a vibrating body in cases where the radiated sound cannot be measured directly (e.g., other correlated vibration sources dominate the sound field). The reciprocity prediction proceeds in two steps. First the vibration pattern of the body is mapped during the operation of the equipment by measuring the vibration velocity $\tilde{v}_i$ at a large number of locations. The phase of the velocity responses is referenced to the velocity measured at a designated reference location. Next the machine is shut off and a point sound source of known volume velocity Q is placed at the receiver location where the sound pressure should be predicted and the sound pressure $\tilde{p}_i$ produced by the point sound source at the various locations along the stationary surface of the body is measured. The phase of the pressure responses is conveniently referenced to the voltage applied to the loudspeaker sound source. The sound pressure at the receiver location, $\tilde{p}_R$, attributable to the periodic vibration of body is predicted as

$$\tilde{p}_R = \sum_{i=1}^{n} \tilde{v}_i \, dA_i \frac{\tilde{p}_i}{Q} \cong \sum_{i=1}^{n} \Delta \tilde{Q}_i \frac{\tilde{p}_i}{Q} \qquad \text{N/m}^2 \tag{9.154}$$

where dA_i is the area and $\Delta \tilde{Q}_i = \tilde{v}_i \, dA_i$ is the volume velocity of the ith sample of the vibrating surface.

The solid line in Fig. 9.40 represents the directly measured sound pressure $\tilde{p}_R$ at a specific location in a room when a thin plate was excited by a shaker to a

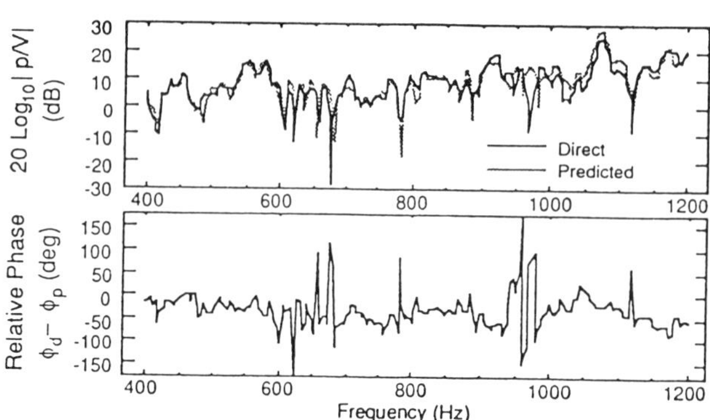

Fig. 9.40 Sound pressure response at a specific location in a room produced by a thin plate excited by a shaker to complex vibration pattern. Solid line: Directly measured transfer function between sound pressure $\tilde{p}_R$ and excitation force **F**, phase is referenced to shaker voltage $\tilde{U}$. Dotted line: Transfer function predicted by measuring the magnitude and relative phase of the plate response, $\Delta \tilde{Q}_i = \tilde{v}_i$ dA, at $i = 81$ position during shaker excitation and the magnitude and phase of the sound pressure $\tilde{p}_i$ at the surface of the stationary plate when a point sound source of volume velocity $\tilde{Q}_R$ is placed at the former receiver position and applying the reciprocity relationship $\tilde{p}_R = \Sigma_{i=1}^{81} \Delta \tilde{Q}_i(\tilde{p}_i/\tilde{Q}_R)$ (after ref. 63). Top curve, magnitude of pressure; lower curve, phase spectrum of directly measured minus the phase spectrum of the predicted pressure.

complex vibration pattern. The dotted curve represents the reciprocity prediction according to Eq. (9.154) utilizing $n = 81$ sampling points on the plate.[63] The experimental results indicate the feasibility of using reciprocity predictions in engineering applications.

Reciprocity in Moving Media

Reciprocity requires that exchanging the function of the source and receiver should not result in any change in the sound propagation path. This is true only if the acoustical medium is at rest. As illustrated in Fig. 9.41, the propagation path between source and receiver remains the same if the exchange of source and receiver positions is accompanied by a reversal of the direction of the uniform mean flow. In this case, or in the case of low mach number potential flow where the shear layer is small compared with the acoustic wavelength, the reciprocity also applies in moving media.[65] Reversal of potential flow is usually easy to accomplish in analytical calculations. However, in many experimental situations where the shear layer is not small compared with the acoustic wavelength, reciprocity does not apply.

Fundamental questions regarding reciprocity are dealt with in references 66–68. Analytical applications of reciprocity and superposition for predicting power

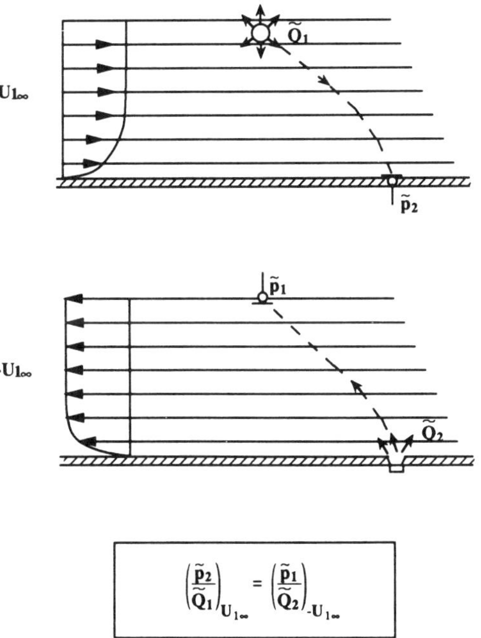

$$\left(\frac{\tilde{p}_2}{\tilde{Q}_1}\right)_{U_{1\infty}} = \left(\frac{\tilde{p}_1}{\tilde{Q}_2}\right)_{-U_{1\infty}}$$

Fig. 9.41 Reciprocity in moving media. Reversal of the function of source and receiver must be accompanied by reversing the flow direction. Streamlines must remain unchanged to assure that the propagation path between source and receiver remains the same.

input structures excited by complex airborne or structure-borne sources is presented in references 69–71. Ship acoustics application of reciprocity are treated in references 72 and 73.

9.10 IMPACT NOISE

There are many practical cases where the excitation of a structure can be represented reasonably well by the periodic impact of a mass on its surface. Footfall in dwellings, punch presses, and forge hammers fall into this category. This section deals only with footfall noise in buildings. For the prediction and control of impact noise of machines and equipment, the reader is referred to a series of 10 papers[74–84] that covers all aspects of impact noise of machinery.

Standard Tapping Machine

A standard tapping machine[84] is used to rate the impact noise isolation of floors in dwellings. This machine consists of five hammers equally spaced along a line, the distance between the two end hammers being about 40 cm. The hammers successively impact on the surface of the floor to be tested at a rate of 10 times per second. Each hammer has a mass of 0.5 kg and falls with a velocity equivalent to a free-drop height of 4 cm. The area of the striking surface of the hammer is approximately 7 cm^2; the striking surface is rounded as though it were part of a spherical surface of 50 cm radius. The impact noise isolation capability of a floor is rated by placing the standard tapping machine on the floor to be tested and measuring the one-third-octave-band sound pressure level L_p' averaged in space in the room below.

$$L_n \equiv L_p - 10 \log \frac{A_0}{S\overline{\alpha}_{S,\text{ab}}} \qquad \text{dB } re\ 2 \times 10^{-5}\ \text{N/m}^2 \tag{9.155}$$

where L_p = $\frac{1}{3}$-octave-band sound pressure level as measured, dB
$S\overline{\alpha}_{S,\text{ab}}$ = total absorption in receiving room (see Chapter 7), m^2
A_0 = reference value of absorption, = 10 m^2

The physical formulation of the problem of impact noise is that of the excitation of a plate by a periodic force impulses. Such periodic forces can be presented by a Fourier series consisting of an infinite number of discrete-frequency components, each with amplitude F_n, given by

$$F_n = \frac{2}{T_r} \int_0^{T_r} F(t) \cos \frac{2\pi n}{T_r} t\, dt \qquad \text{N} \tag{9.156}$$

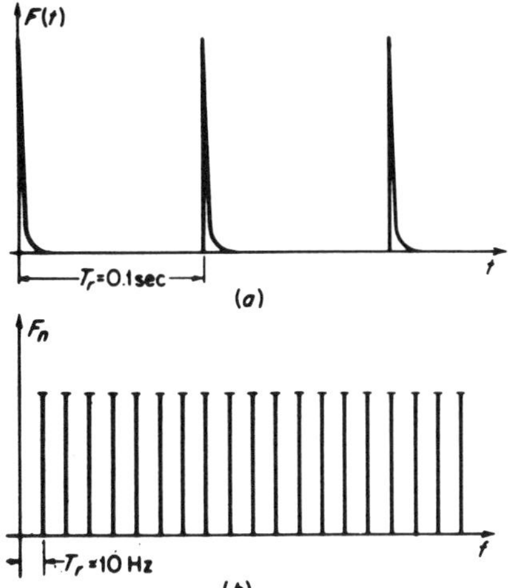

Fig. 9.42 (*a*) Time function and (*b*) Fourier components of the force that a standard tapping machine exerts on a massive rigid floor.

where $T_r = 1/f_r = 0.1$ is the time interval between hammer strikes and $n = 1, 2, 3, \ldots$. The curve in Fig. 9.42*a* shows the time function of the force $F(t)$, and that in Fig. 9.42*b* shows the amplitude of its Fourier components.

It is an experimental fact that when the hammer strikes a hard concrete slab, the duration of the force impulse is small compared even with the period of the highest frequency of interest in impact testing. For less stiff structures, like wooden floors, this assumption is not valid and the exact shape of $F(t)$ has to be determined and used in Eq. (9.156). For a thick concrete slab the effective length of the force impulse is short enough so that $\cos[(2\pi n/T_r)t] \approx 1$, and all components have the same amplitude. Because the integral in Eq. (9.156) is the momentum of a single hammer blow (assuming no rebound) equal to mv_0 (in kg · m/s), the amplitudes of the Fourier components of the force for a repetition frequency f_r are

$$F_n = 2f_r m v_0 \qquad \text{N} \tag{9.157}$$

The velocity of the hammer at the instant of impact is

$$v_0 = \sqrt{2gh} \qquad \text{m/s} \tag{9.158}$$

where h = falling height of hammer, m
g = acceleration of gravity (9.8 m/s^2)

Let us define a mean-square-force spectrum density S_{f0} that when multiplied by the bandwidth will yield the value of the mean-square force in the same bandwidth,

$$S_{f0} = \tfrac{1}{2} T_r F_n^2 = 4 f_r m^2 gh \qquad \text{N}^2/\text{Hz} \tag{9.159}$$

For the standard tapping machine the numerical value of S_{f0} is 4 N^2/Hz.

Accordingly, the mean-square force in an octave band $\Delta f_{\text{oct}} = f/\sqrt{2}$ is

$$F^2_{\text{rms}}(\text{oct}) = \frac{4}{\sqrt{2}} f \qquad \text{N}^2 \tag{9.160}$$

The octave-band sound power level radiated by the impacted slab (which is assumed to be isotropic and homogeneous) into the room below is calculated by inserting Eq. (9.160) into Eq. (9.71a), which yields

$$L_w(\text{oct}) \approx 10 \log_{10} \left(\frac{\rho c \sigma_{\text{rad}}}{5.1 \rho_p^2 c_L \eta_p t^3} \right) + 120 \qquad \text{dB } re\ 10^{-12}\ \text{W} \tag{9.161}$$

where ρ = density of air, kg/m^3
c = speed of sound in air, m/s
σ_{rad} = radiation factor of slab
ρ_p = density of slab material, kg/m^3
c_L = propagation speed of longitudinal waves in slab material, m/s
η_p = composite loss factor of slab
t = thickness of slab, m

Note that the *sound power level is independent of the center frequency of the octave*, that *doubling the slab thickness decreases the level of the noise radiated into the room below by 9 dB*, and that the *sound power level decreases with increasing loss factor*.

Improvement of Impact Noise Isolation by an Elastic Surface Layer

Experience has shown that the impact noise level of even an 8–10-in.-thick dense concrete slab is too high to be acceptable. A further increase of thickness to reduce impact noise is not economical.

Impact noise may be reduced effectively by an elastic surface layer, much softer than the surface of the slab, applied to the structural slab. The resilient layer changes the shape of the force pulse and the amount of mechanical power introduced into the slab by the impacting hammer, as shown in Fig. 9.43.

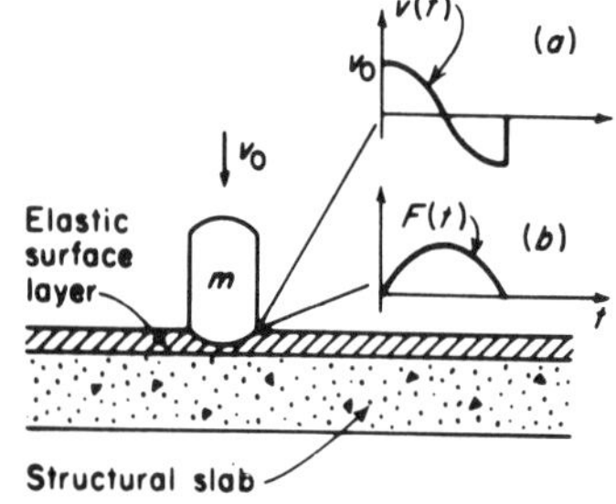

Fig. 9.43 Velocity and force pulse of a single hammer blow on an elastic surface layer over a rigid slab: (*a*) velocity pulse; (*b*) force pulse.

We would expect, if the elastic layer is linear and nondissipative, that the velocity will be at its maximum v_0 at the instant of impact $t = 0$. It will then decrease to zero and the mass will rebound to nearly the same velocity (it is assumed the hammer is not permitted to bounce a second time) according to the function shown by curve a of Fig. 9.43. The force function is shown by curve b.

The improvement in impact noise isolation achieved by the addition of the soft surface layer is defined in terms of the logarithmic ratio[89]

$$\Delta L_n = 20 \log \frac{F}{F'} = 20 \log \left(\left| \frac{1 - nf_r/f_0}{\cos\left(\dfrac{\pi}{2} n \dfrac{f_r}{f_0}\right)} \right| \right) \quad \text{dB} \tag{9.162}$$

where

$$n = 1, 2, 3, \ldots \tag{9.163}$$

$$f_0 = \frac{1}{2\pi} \sqrt{\frac{A_h}{m}} \sqrt{\frac{E}{h}} \quad \text{Hz} \tag{9.164}$$

and F', F = forces acting on slab with and without resilient surface layer, respectively, N
A_h = striking area of hammer, m^2
m = mass of hammer, kg
E = dynamic Young's modulus of elastic material, N/m
h = thickness of layer, m

The characteristic frequency f_0 of an elastic surface layer for the standard tapping machine is plotted in Fig. 9.44 as a function of E/h.

Equation (9.162), which assumes no damping, is plotted in Fig. 9.45 as a function of the normalized frequency f/f_0. Below $f/f_0 = 1$, the improvement is zero. Above $f/f_0 = 1$ the improvement increases with an asymptotic slope of 40 dB/decade.

Figures 9.44 and 9.45 (use the 40-dB/decade asymptote) permit one to select an elastic surface layer to achieve a specified ΔL_n.

Example 9.5 The required improvement in impact noise isolation should be 20 dB at 300 Hz. Design a resilient covering for the concrete slab.

SOLUTION From Fig. 9.45 we obtain $f/f_0 \approx 3$, which gives $f_0 = 100$ Hz. Entering Fig. 9.44 with this value of f_0 yields $E/h = 2.8 \times 10^8$ N/m^3 (or $E/h \approx 1000$ psi/in.) Any material having this ratio of Young's modulus to thickness will provide the required improvement. If we wish to select a 0.31-cm- ($\frac{1}{8}$-in.-) thick layer, the dynamic modulus of the material should be 8.7×10^5 N/m^2 (8000 psi). Since the dynamic modulus of most elastic materials is about twice the sta-

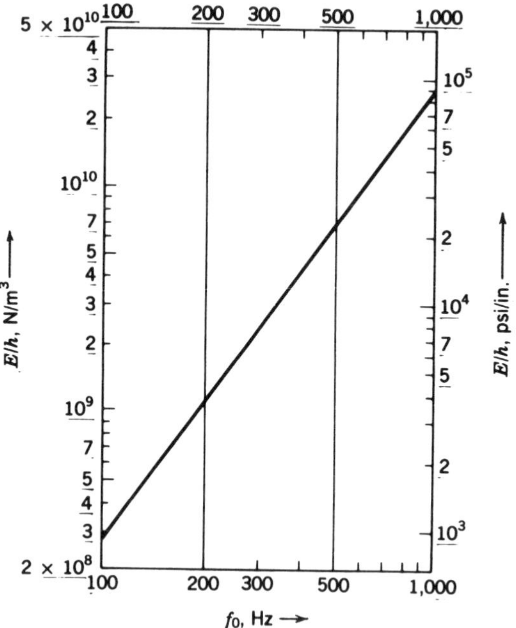

Fig. 9.44 Chart for the selection of an elastic surface layer, where f_0 = characteristic frequency, E = Young's modulus, and h = thickness of layer.

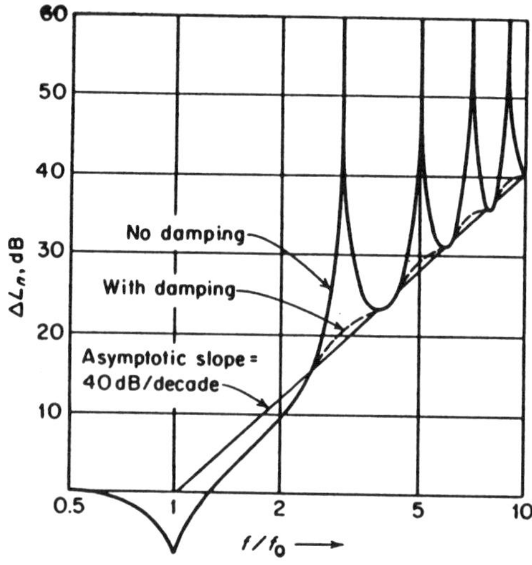

Fig. 9.45 Improvement in impact noise isolation ΔL_n versus normalized frequency for a resilient surface layer (select f_0 to yield desired improvement).

tiscally measured Young's modulus,[85] a material with $E \leq 4.35 \times 10^5$ N/m^2 (4000 psi) should be selected.

Frequently used materials for the elastic surface layer are rubberlike materials, vinyl-cork tile, or carpet. The impact isolation improvement curve (ΔL_n vs. frequency) has been measured and reported[91,92] for a large variety of elastic surface layer configurations.

The expected normalized impact sound level in the room below (see Fig. 9.1) for a composite floor (with a heavy structural slab) is that of the bare concrete structural floor minus the improvement caused by the elastic surface layer, i.e.,

$$L_{n,\text{comp}} = L_{n,\text{bare}} - \Delta L_n \tag{9.165}$$

where

$$L_{n,\text{bare}}(\text{oct}) = 116 + 10 \log \left(\frac{\rho c \sigma_{\text{rad}}}{5.1 \rho_p^2 c_L \eta_p t^3} \right) \quad \text{dB } re \; 2 \times 10^{-5} \text{ N/m}^2 \tag{9.166}$$

for homogeneous isotropic slabs.

Measured values of impact noise isolation of a large number of floor constructions and improvement of impact noise isolation by various surface layers are presented in reference 94.

Improvement through Floating Floors. It is often more practical to use a floating floor above a structural slab than a soft resilient surface layer. The advantages are that (1) both the impact noise isolation and the airborne sound transmission loss of the composite floor are improved and (2) the walking surface is hard. For analysis, floating floors can be categorized as either (1) locally reacting or (2) resonantly reacting, as defined below.

Locally Reacting Floating Floors. A locally reacting floor is one where the impact force of the hammer on the upper slab (slab 1) is transmitted to the structural slab (slab 2), primarily in the immediate vicinity of the excitation point, and where there is no spatially homogeneous reverberant vibration field on slab 1. In this case the bending waves in the floating slab are highly damped. If the Fourier amplitude of the force acting on plate 1 is given by Eq. (9.157), the reduction in transmitted sound level is[85]

$$\Delta L_n = 20 \log \left[1 + \left(\frac{f}{f_0} \right)^2 \right] \approx 40 \log \frac{f}{f_0} \tag{9.167}$$

where

$$f_0 = \frac{1}{2\pi} \sqrt{\frac{s'}{\rho_{s1}}} \tag{9.168}$$

and ρ_{s_1} = mass per unit area of floating slab, kg/m^2
s' = dynamic stiffness per unit area of the resilient layer between slab 1 and slab 2 including trapped air, N/m

Resonantly Reacting Floating Floors. If the floating slab is thick, rigid, and lightly damped, the impact force of the hammers excites a more-or-less spatially homogeneous reverberant bending-wave field.

The improvement in impact noise isolation at higher frequencies, where the power dissipated in slab 1 exceeds the power transmitted to slab 2, can be approximated by[87,89]

$$\Delta L_n \approx 10 \log \frac{2.3\rho_{s_1}^2 \omega^3 \eta_1 c_{L_1} h_1}{n' s^2} \tag{9.169}$$

where h_1 = thickness of floating slab, m
c_{L_1} = propagation speed of longitudinal waves in floating slab, m/s
ρ_{s_1} = mass per unit area of floating slab, kg/m^2
η_1 = loss factor of floating slab
n' = number of resilient mounts per unit area of slab, m^2
s = stiffness of mount, N/m

Equation (9.169) indicates that, in contrast to the locally reacting case where ΔL_n increases at a rate of 40 dB for each decade increase in frequency, the increase is only 30 dB/decade if the loss factor of the floating slab η_1 is frequency independent. Another difference is the marked dependence of ΔL_n on this loss factor. The loss factor is determined both by the energy dissipated in the slab material itself and by the energy dissipated in the resilient mounts.

Figure 9.46 shows the improvement of a floating-floor system under impacting by a standard tapping machine and high-heel shoes, respectively.[88] The negative improvement in the vicinity of the resonance frequency f_0 can be observed.

Impact Noise Isolation versus Sound Transmission Loss

Whether a floor is excited by the hammers of a tapping machine or by an airborne sound field in the source room, it will in both cases radiate sound into the receiving room. There is a close relation between the airborne sound transmission loss R and the normalized impact noise level L_n for a given floor.

In the case of acoustical excitation, the sound power transmitted to the receiver room comprises the contribution of forced waves and of resonance waves. The forced waves usually dominate below the critical frequency of the slab and the resonance waves above. The sound power transmitted by exciting the slab by a standard tapping machine is made up of the contributions of the near-field component and of the reverberant component.

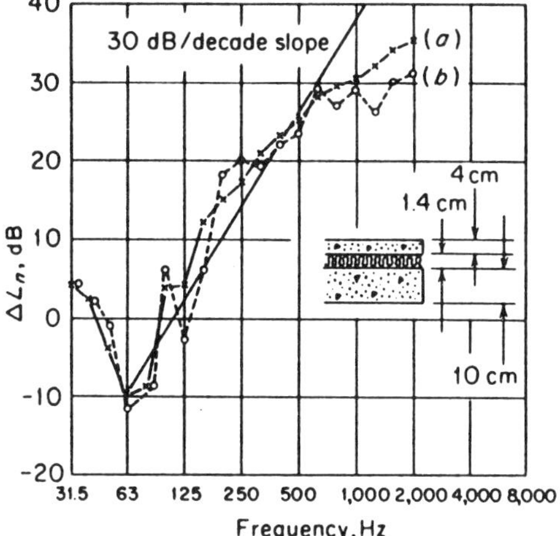

Fig. 9.46 Improvement in impact noise isolation ΔL_n for a resonantly reacting floating floor for excitation by (*a*) a standard tapping machine and (*b*) high-heeled shoes. Note the negative ΔL_n in the vicinity of the resonance frequency. (After ref. 88.)

The relation between sound transmission loss R and normalized impact noise level, assuming measurement in octave bands, is[93]

$$L_n + R = 84 + 10 \log \left[\frac{S_{f0} f}{\sqrt{2}} \left(\frac{\dfrac{\rho}{2\pi\rho_s^2 c} + \dfrac{\rho c \sigma_{\text{rad}}}{2.3\rho_s^2 c_L \omega \eta_p h}}{\dfrac{\pi\rho c}{\omega^2\rho_s^2} + \dfrac{\pi\sqrt{12}c^3\rho\sigma_{\text{rad}}^2}{2\rho_s^2 c_L \omega^3 \eta_p h}} \right) \right] \quad \text{dB} \qquad (9.170)$$

where ρ_s = mass per unit area of slab, kg/m^2
c_L = propagation speed of longitudinal waves in slab, m/s
σ_{rad} = radiation factor of slab
h = thickness of slab, m
η_p = composite loss factor of slab
S_{f0} = mean-square force spectrum density as given in Eq. (9.159), N^2/Hz

In the special case of a thick, lightly damped slab,

$$L_n + R = 43 + 30 \log f - 10 \log \sigma_{\text{rad}} - \Delta L_n \qquad (9.171)$$

where ΔL_n represents the effect of the surface layer only. For a bare structural slab, by definition, $\Delta L_n = 0$.

Equation (9.171) states that the *sum of the airborne sound transmission loss and the normalized impact noise level is independent of the physical character-*

istics of the structural slab above the critical frequency of the slab where $\sigma_{rad} \approx 1$.[90]

Below the coincidence frequency where the forced waves control the airborne sound transmission loss but the impact noise isolation is still controlled by the resonant vibration of the impacted slab, Eq. (9.170) yields

$$R + L_n = 39.5 + 20 \log f - \Delta L_n - 10 \log \frac{\eta_p}{f_c \sigma_{rad}} \quad \text{dB} \qquad (9.172)$$

where ΔL_n = effect of surface layer only (zero for structural slab), dB
f_c = critical frequency of structural slab, Hz

In this case, the sum $R + L_n$ decibels depends on the physical characteristics of the slab and frequency. Figure 9.47 shows the measured sound transmission loss

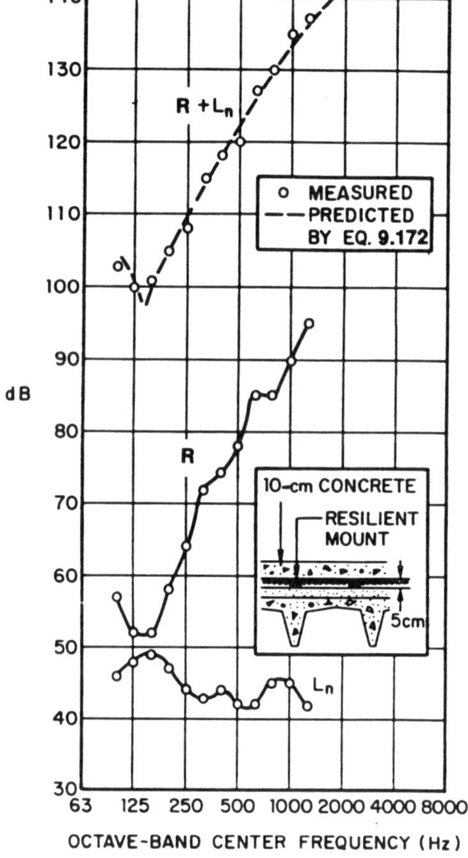

Fig. 9.47 Measured sound transmission loss R and normalized impact sound level L_n and their sum ($R + L_n$) of a resonantly reacting floating floor assembly. Dotted curve: $R + L_n$ predicted by Eq. (9.171). (After ref. 93.)

R and normalized impact sound level L_n as well as their sum for a typical floating floor. The measured and predicted values for the sum are in good agreement, indicating that the precautionary measures taken to eliminate flanking have been successful.

In checking out the performance of floating floors in the field, it is advisable to measure both R and L_n. The discrepancy between the measured R and that calculated from Eq. (9.172) is a direct indication of flanking. By measuring the acceleration level on the wall surfaces in the source and receiving rooms during acoustical and impact excitation, the flanking paths can be immediately identified.

APPENDIX: SOUND TRANSMISSION CLASS AND IMPACT ISOLATION CLASS

Determination of Sound Transmission Class

The purpose of the sound transmission class (STC) classification is to provide a single-number rating that can be used for comparing partitions for general building design purposes. The procedure described here has been adopted formally by the American Society for Testing and Materials.[95] The procedure adopted by the International Standardization Organization yields identical results.[96] A general discussion of sound transmission loss is given in Section 9.7 and in reference 97.

To determine the STC of a test specimen, the sound transmission losses R for the specimen are measured[98] in series of 16 one-third-octave test bands and these data are compared with the points on a reference STC contour, as described below.

Graphical Determination of STC. If the measured one-third-octave-band transmission losses for the test specimen are plotted on the specified graph paper (see Fig. 9.48 legend) as a function of frequency, the sound transmission class may be determined by comparison with a standard transparent overlay on which the STC contour of Fig. 9.48 is drawn. The STC contour is shifted vertically relative to the test curve until some of the measured transmission loss R values for the test specimen fall below those of the STC contour and the following two conditions are fulfilled: the maximum deficiency in any single one-third-octave band shall not exceed 8 dB and the sum of the deficiencies (i.e., the deviations below highest value (in integral decibels) that meets the above requirements, the STC for the specimen is the R value corresponding to the intersection of the contour and the 500-Hz ordinate (read on the vertical scale of the R-versus-frequency plot of the test specimen).

Presentation of Results. It is recommended that the test data (for one-third-octave bands) be plotted on the specified graph paper together with the corresponding STC contour obtained as described above. In this way attention is drawn to the frequency regions that limit the STC of the test specimen. Whenever possible, the ordinate scale should start at 0 dB.

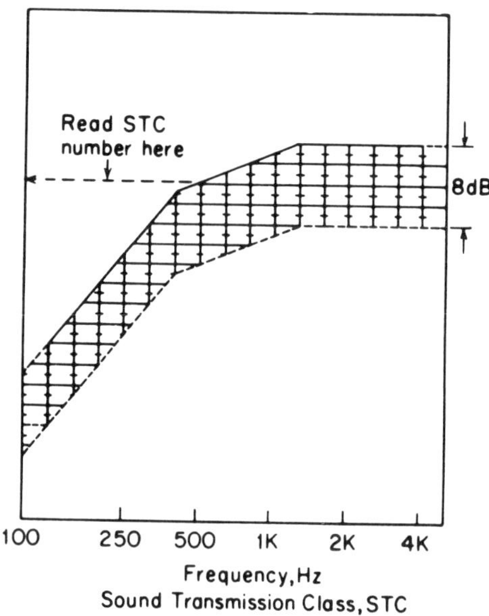

Fig. 9.48 Form of overlay for STC determination. See text for details of positioning of the solid curve above. The STC equals the *R* value on ordinate of the transmission loss curve that lies beneath the arrow above. Use this overlay with one-third-octave sound transmission loss data plotted on graph paper that meets these specifications: 5 cm = 25 dB; 5 cm = 1 decade of frequency.

Limitation. The STC rating[99] is useful when designing walls that provide insulation against the sounds of speech, music, radio, or television. It is not valid for noise sources with spectra that differ markedly from the above, for example, industrial processes, aircraft, motor vehicles, power transformers, and the like.

Determination of Impact Isolation Class

The impact isolation class (IIC) is a single-number rating that provides a means for comparing the acoustical performance of floor–ceiling assemblies when excited by impacts produced by a standard tapping machine.[100]

The *impact sound transmission levels* resulting from the operation of a standard tapping machine on a test floor are characterized by the band spectrum of the space averaged sound pressure levels produced in the receiving room beneath the test floor. The $\frac{1}{3}$-octave band sound pressure levels are measured with an indicating device that reads rms sound pressure in decibels *re* 2×10^{-5} N/m^2.

The impact sound transmission levels L_{IST} are generally normalized to remove the effects of varying degrees of sound absorption that might occur from one test receiving room to another. The relation between L_{IST} and the measured band level is

$$L_{\text{IST}} = L_2 + 10 \log_{10} \frac{A}{A_0}$$

where L_2 = space-averaged, one-third-octave-band sound pressure level measured in the receiving room, dB

A = actual total sound absorption in receiving room at center frequency of band, $= \Sigma S_i \alpha_i$, m^2 (see Chapter 7).

A_0 = reference sound absorption, $= 10$ m^2

Frequency Range of Measurements. The space-averaged sound pressure levels should be determined in 16 contiguous frequency bands, each one-third octave wide, and covering the range of center frequencies from 100 to 3150 Hz. If possible, the measurements should be made at higher and lower frequencies, although those values would not enter into the IIC rating.

Statement of Results. The measured one-third-octave-band sound pressure levels should be normalized to a standard reference absorption of $A_0 = 10$ m^2, and the resulting L_{IST} should be plotted, for all frequencies of measurement, as a curve on the specified graph paper (see Fig. 9.49 legend).

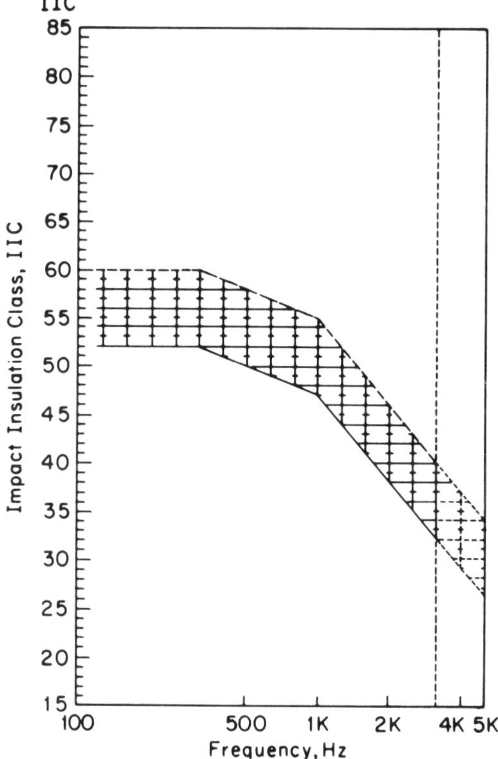

Fig. 9.49 Form of overlay for IIC determination. See text for details of positioning of the solid curve above. The IIC equals the value on the above ordinate that lies over the L_{ISP} = 60 dB value on the impact sound transmission curve beneath the overlay. Use this overlay with one-third-octave impact sound transmission data plotted on graph paper that meets these specifications: 5 cm = 25 dB; 5 cm = 1 decade of frequency.

Improvement of Impact Sound Insulation. For measuring the improvement of impact sound insulation, such as the improvement owing to floor coverings or floating floors,[101] it is often possible to measure the impact sound transmission level before and after treatment, particularly in laboratory tests. In such cases, the improvement (lowering of levels) owing to the treatment may be denoted by ΔL_{IST} and should be given in the form of a curve on the specified graph paper.

Impact Isolation Class (IIC). The purpose of this single-number rating is to permit easy comparison of the impact noise isolation performance of floor–ceiling assemblies for general guidance in building design. The procedure for assigning the single-number rating IIC is to plot the measured impact sound transmission levels L_{IST} on the specified graph paper. The transparent overlay of Fig. 9.49 is then placed on the graph, aligned with the frequency scale, and adjusted so that all data points lie on or below the broken-line contour. This procedure ensures initially that single deviations are less than or equal to a maximum of 8 dB. Then the deviations above the solid-line contour are summed. The total must not exceed 32 dB; if greater, the overlay is adjusted upward until the total equals 32 dB. The IIC value, *read from the overlay*, is that overlay value that lies over (corresponds to) the impact sound pressure level of 60 dB on the graph scale.[97]

Other single-number rating procedures in the literature include impact noise rating (INR)[102] and impact sound insulation class I_i.[103]

The significance of impact sound transmission levels produced by the standard tapping machine and the corresponding single-number rating IIC is the subject of current debates in several national and international standard organizations.[104] The reader is advised to keep abreast of documents of ASTM, ANSI, and ISO for information on current standards for tapping machines and single-number rating procedures.

REFERENCES

1. L. Cremer, M. Heckl, and E. E. Ungar, *Structureborne Sound*, 2nd ed., Springer Verlag, 1988.
2. M. Heckl, "Compendium of Impedance Formulas," Bolt Beranek and Newman Report No. 774, May 26, 1961.
3. I. Dyer, "Moment Impedance of Plate," *J. Acoust. Soc. Am.* **32**, 247–248 (1960).
4. E. Eichler, "Plate Edge Admittances," *J. Acoust. Soc. Am.* **36**(2), 344–348 (1964).
5. H. G. D. Goyder and R. G. White, "Vibrational Power Flow from Machines into Builtup Structures, Part I: Introduction and Approximate Analyses of Beam and Plate-Like Foundations," *J. Sound Vib.* **68**(1), 59–75 (1980); "Part II: Wave Propagation and Power Flow in Beam-stiffened Plates," *J. Sound Vib.* **68**(1), 77–96 (1980); "Part III: Power Flow through Isolation Systems," *J. Sound Vib.* **68**(1), 97–117 (1980).
6. U. J. Kurze, "Low Noise Design XII; Mechanical Impedance" (Laermarm Konstruieren XII; Mechanische Impedance), Bundessanstalt fuer Arbeitschutz, FB Nr. 398, Wirtschaftverlag NW, Bremerhaven, 1985 (in German).

7. K.-P. Schmidt, "Low Noise Design (III); Reducing Noise by Changing the Input Impedance (Laermarm Konstruieren (III); Aenderung der Eingangsimpedanz als Massnahme zur Laermminderung)," Bundesanstalt fuer Arbeitschutz, FB Nr. 169, Wirtschaftsverlag NW, Bremerhaven, 1979 (in German).
8. VDI 3720, Blatt 6 (Preliminary, July 1984), "Noise Abatement by Design; Mechanical Input Impedance of Structural Elements, Especially Standard-Section Steel," VDI-Verlag Duesseldorf, 1984 (in German).
9. M. Heckl, "Bending Wave Input Impedance of Beams and Plates," Proceedings 5th ICA, Liège, Paper L67, 1965, (in German).
10. M. Heckl, "A Simple Method for Estimating the Mechanical Impedance," Proceedings DAGA, pp. 828–830, VDI Verlag, 1980 (in German).
11. M. Heckl, "Excitation of Sound in Structures," *Proc. INTER-NOISE*, pp. 497–502, 1988.
12. R. J. Pinnington, "Approximate Mobilities of Builtup Stuctures," Report No. 162, Institute of Sound and Vibration Research, Southampton, England, 1988.
13. H. Ertel and I. L. Vér, "On the Effect of the Vicinity of Junctions on the Power Input into Beam and Plate Structures," *Proc. INTER-NOISE*, pp. 684–692, 1985.
14. P. W. Smith Jr., "The Imaginary Part of Input Admittance: A Physical Explanation of Fluid-loading Effects on Plates," *J. Sound Vib.* **60**(2), 213–216 (1978).
15. M. Heckl, "Input Impedance of Plates with Radiation Loading," *Acustica* **19**, 214–221 (1967/68) (in German).
16. D. G. Crighton, "Force and Moment Admittance of Plates under Arbitrary Fluid Loading," *J. Sound Vib.* **20**(2), 209–218 (1972).
17. D. G. Crighton, "Point Admittance of an Infinite Thin Elastic Plate under Fluid Loading," *J. Sound Vib.* **54**(3), 389–391 (1977).
18. R. H. Lyon, "Statistical Analysis of Power Injection and Response in Structures and Rooms," *J. Acoust. Soc. Am.* **45**(3), 545–565, (1969).
19. R. D. Blevins, "Formulas for Natural Frequency and Mode Shape," Van Nostrand Reinhold Co., New York, 1979.
20. T. Kihlman, "Transmission of Structureborne Sound in Buildings," Report No. 9, National Swedish Institute of Building Research, Stockholm, 1967.
21. M. Paul, "The Measurement of Sound Transmission from Bars to Plates," *Acustica* **20,** 36–40 (1968) (in German).
22. E. J. Richards et al, "On the Prediction of Impact Noise, II: Ringing Noise," *J. Sound Vib.* **65**(3), 419–451 (1979).
23. R. Timmel, "Investigation on the Effect of Edge Conditions on Flexurally Vibrating Rectangular Panels on Radiation Efficiency as Exemplified by Clamped and Simply Supported Panels," *Acustica* **73**, 12–20 (1991) (in German).
24. ASTM E90, "Laboratory Measurement of Airborne-Sound Transmission Loss of Building Partitions," American Society for Testing and Materials, Philadelphia, PA.
25. ISO 140/3-1978 "Laboratory Measurement of Airborne Sound Insulation of Building Elements," American National Standards Institute, New York (currently under revision).
26. L. Cremer, *Lectures on Technical Acoustics* (*Vorlesungen ueber techische Akustik*), Springer-Verlag, Berlin, 1971.
27. H. Reissner, "Transmission of a Normal and Oblique-Incidence Plane Compressional Wave Incident from a Fluid on a Solid, Plane, Parallel Plate (Der senkrechte und

schräge Durchtritt einer in einem fluessigen Medium erzeugten ebenen Dilations-(Longitudinal)—Welle durch eine in diesem Medium benfindliche planparallele feste Platte,'' *Helvatia Physica Acta*, **11**, 140–145 (1938) (in German).

28. M. Heckl, ''The Tenth Sir Richard Fairey Memorial Lecture: Sound Transmission in Buildings,'' *J. Sound Vib.* **77**(2), 165–189 (1981).
29. R. D. Mindlin, ''Influence of Rotary Inertia and Shear on Flexural Motion of Elastic Plates,'' *J. Appl. Mech.* **18**, 31–38 (1951).
30. B. H. Sharp, ''A Study of Techniques to Increase the Sound Insulation of Building Elements,'' Wyle Laboratories Report WR-73-5, HUD Contract No. H-1095, June 1973.
31. G. Kurtze and B. G. Watters, ''New Wall Design for High Transmission Loss or High Damping,'' *J. Acoust. Soc. Am.* **31**(6), 739–748 (1959).
32. H. Feshbach, ''Transmission Loss of Infinite Single Plates for Random Incidence,'' Report No. TIR 1, Bolt Beranek and Newman, October 1954.
33. L. L. Beranek, ''The Transmission and Radiation of Acoustic Waves by Structures,'' (the 45th Thomas Hacksley Lecture of the British Institution of Mechanical Engineers), *J. Inst. Mech. Eng.* **6**, 162–169 (1959).
34. L. L. Beranek, *Noise Reduction*, McGraw-Hill, New York, 1960.
35. M. Heckl, ''Investigations of Orthotropic Plates,'' (Untersuchungen an Orthotropen Platten), *Acustica* **10**, 109–115 (1960) (in German).
36. M. Heckl, ''Sound Transmission Loss of the Homogeneous Finite Size Single Wall (Die Schalldaemmung von homogene Einfachwaenden endlicher Flaeche),'' *Acustica* **10**, 98–108 (1960).
37. H. Sato, ''On the Mechanism of Outdoor Noise Transmission through Walls and Windows; A Modification of Infinite Wall Theory with Respect to Radiation of Transmitted Wave,'' *J. Acoust. Soc. Jpn* **29**, 509–516 (1973) (in Japanese).
38. J. H. Rindel, ''Transmission of Traffic Noise through Windows; Influence of Incident Angle on Sound Insulation in Theory and Experiment,'' Ph.D. Thesis, Techn. Univ. Denmark, Report No. 9, 1975 (also see *Proc. DAGA*, 509–512), (1975); also private communications.
39. B. G. Watters, ''Transmission Loss of Some Masonry Walls,'' *J. Acoust. Soc. Am.* **31**(7), 898–911 (1959).
40. Y. Hamada and H. Tachibana, ''Analysis of Sound Transmission Loss of Multiple Structures by Four-Terminal Network Theory,'' *Proc. INTER-NOISE*, pp. 693–696, 1985.
41. A. C. K. Au and K. P. Byrne ''On the Insertion Loss Produced by Plane Acoustic Lagging Structures,'' *J. Acoust. Soc. Am.* **82**(4), 1325–1333 (1987).
42. A. C. K. Au and K. P. Byrne, ''On the Insertion Loss Produced by Acoustic Lagging Structures which Incorporate Flexurally Orthotropic Impervious Barriers,'' *Acustica* **70,** 284–291 (1990).
43. K. Goesele and U. Goesele, ''Influence of Cavity-Volume Damping on the Stiffness of Air Layer in Double Walls,'' *Acustica* **38**(3), 159–166 (1977) (in German).
44. K. W. Goesele, W. Schuele, and B. Lakatos, ''Gas Filling of Isolated Glass Windows'' (Glasfuellung bei Isolierglasscheiben) Forschunggemeinschaft Bauen und Wohen, FBW Blaetter 4, 1982 (in German).

45. H. Ertel and M. Moeser, "Effects of Gas Filling on the Sound Transmission Loss of Isolated Glass Windows in the Double Wall Resonance Frequency Range," FBW Blatter, Stuttgart, 5/1984 (in German).

46. K. Goesele, "Prediction of the Sound Transmission Loss of Double Partitions (without Structureborne Connections)," *Acustica*, **45**, 218–227 (1980) (in German).

47. I. L. Vér, "Definition of and Relationship between Breakin and Breakin Sound Transmission Loss of Pipes and Ducts," *Proc. INTER-NOISE*, pp. 583–586, 1983.

48. L. L. Vér, "Prediction of Sound Transmission through Duct Walls: Breakout and Pickup," ASHRAE Paper 2851 (RP-319), *ASHRAE Trans.* **90**(Pt. 2), 391–413 (1984).

49. Anon, "Sound and Vibration Control," in *ASHRAE Handbook, Heating, Ventilating, and Air-Conditioning Systems and Applications*, American Society of Heating, Refrigerating and Air Conditioning Engineers, Atlanta, GA, 1987, Chapter 52.

50. A. Cummings, "Acoustic Noise Transmission through Duct Walls," *ASHRAE Trans.* **91**(Pt. 2A), 48–61, (1985).

51. M. C. Gomperts and T. Kihlman, "Transmission Loss of Apertures in Walls," *Acustica* **18**, 140 (1967).

52. F. P. Mechel, "The Acoustic Sealing of Holes and Slits in Walls," *J. Sound Vib.* **111**(2), 297–336 (1986).

53. F. P. Mechel, "Acoustical Insulation of Silencer Joints (Die Schalldaemmung von Schalldaempfer-Fugen," *Acustica* **62**, 177–193 (1987) (in German).

54. W. Fasold, W. Kraak, and W. Schirmer, "Pocketbook Acoustic," Part 2, Section 6.2, VEB Verlag Technik, Berlin, 1984 (in German).

55. R. H. Lyon and G. Maidanik, "Power Flow between Linearly Coupled Oscillators," *J. Acoust. Soc. Am.* **34**(5), 623–639 (1962).

56. E. Skudrzyk, "Vibrations of a System with a Finite or Infinite Number of Resonances," *J. Acoust. Soc. Am.* **30**(12), 1114–1152 (1958).

57. E. E. Ungar, "Statistical Energy Analysis of Vibrating Systems," *J. Eng. Ind., Trans. ASME, Ser. B*, **87**, 629–632 (1967).

58. M. J. Crocker and A. J. Price, "Sound Transmission Using Statistical Energy Analysis," *J. Sound Vib.* **9**(3), 469–486 (1969).

59. R. H. Lyon, *Statistical Energy Analysis of Dynamical Systems; Theory and Application*, MIT Press, Cambridge, MA, 1975.

60. Lord Rayleigh, *Theory of Sound*, Vol. I, Dover Publications, New York, 1945, pp. 104–110.

61. M. Heckl, "Some Applications of Reciprocity Principle in Acoustics," *Frequenz* **18**, 299–304 (1964) (in German).

62. I. L. Vér, "Uses of Reciprocity in Acoustic Measurements and Diagnosis," *Proc. INTER-NOISE*, pp. 1311–1314, 1985.

63. F. J. Fahy, "The Reciprocity Principle and its Applications in Vibro-Acoustics," *Proceedings of the Institute of Acoustics (UK)*, **12**(1), 1–20 (1990).

64. I. L. Vér, "Reciprocity as a Prediction and Diagnostic Tool in Reducing Transmission of Structureborne and Airborne Noise into an Aircraft Fuselage," Vol. 1, "Proof of Feasibility," Bolt Beranek and Newman (BBN), Report No. 4985 (April 1982); Vol. 2, "Feasibility of Predicting Interior Noise Due to Multiple, Correlated Force Input," BBN Report No. 6259 (May 1986), NASA Contract No. NAS1-16521.

65. L. M. Lyamshev, "On Certain Integral Relations in Acoustics of a Moving Medium," *Dokl. Akad. Nauk SSSR* **138**, 575–578 (1961); *Sov. Phys. Dokl.* **6**, 410 (1961).
66. Yu I. Belousov and A. V. Rimskii-Korsakov, "The Reciprocity Principle in Acoustics and Its Application to the Calculation of Sound Fields of Bodies," *Sov. Phys. Acoust.* **21**(2), 103–109 (1975).
67. L. Cremer, "The Law of Mutual Energies and Its Application to Wave-Theory of Room Acoustics," *Acustica* **52**(2), 51–67 (1982/83) (in German).
68. M. Heckl, "Application of the Theory of Mutual Energies," *Acustica* **58**, 111–117 (1985) (in German).
69. I. L. Vér, "Use of Reciprocity and Superposition in Predicting Power Input to Structures Excited by Complex Sources," *Proc. INTER-NOISE*, pp. 543–546, 1989.
70. J. M. Mason and F. J. Fahy, "Development of a Reciprocity Technique for the Prediction of Propeller Noise Transmission through Aircraft Fuselages," *Noise Control Eng. J.* **34**, 43–52 (1990).
71. P. W. Smith, "Response and Radiation of Structures Excited by Sound," *J. Acoust. Soc. Am.* **34**, 640–647 (1962).
72. H. F. Steenhock and T. TenWolde, "The Reciprocal Measurement of Mechanical-Acoustical Transfer Functions," *Acustica* **23**, 301–305 (1970).
73. T. TenWolde, "On the Validity and Application of Reciprocity in Acoustical, Mechano-Acoustical and other Dynamical Systems," *Acustica* **28**, 23–32 (1973).
74. E. J. Richards, M. E. Westcott, and R. K. Jeyapalan, "On the Prediction of Impact Noise, I: Acceleration Noise," *J. Sound Vib.* **62**, 547–575 (1979).
75. E. J. Richards, M. E. Westcott, and R. K. Jeyapalan, "On the Prediction of Impact Noise, II: Ringing Noise," *J. Sound Vib.* **65**, 419–451 (1979).
76. E. J. Richards, "On the Prediction of Impact Noise, III: Energy Accountancy in Industrial Machines," *J. Sound Vib.* **76**, 187–232 (1981).
77. J. Cuschieri and E. J. Richards, "On the Prediction of Impact Noise, IV: Estimation of Noise Energy Radiated by Impact Excitation of a Structure," *J. Sound Vib.* **86**, 319–342 (1982).
78. E. J. Richards, I. Carr, and M. E. Westcott, "On the Prediction of Impact Noise, Part V: The Noise from Drop Hammers," *J. Sound Vib.* **88**, 333–367 (1983).
79. E. J. Richards, A. Lenzi, and J. Cuschieri, "On the Prediction of Impact Noise, VI: Distribution of Acceleration Noise with Frequency with Applications to Bottle Impacts," *J. Sound Vib.* **90**, 59–80 (1983).
80. E. J. Richards and A. Lenzi, "On the Prediction of Impact Noise, VII: Structural Damping of Machinery," *J. Sound Vib.* **97**, 549–586 (1984).
81. J. M. Cuschieri and E. J. Richards, "On the Prediction of Impact Noise, VIII: Diesel Engine Noise," *J. Sound Vib.* **102**, 21–56 (1985).
82. E. J. Richards and G. J. Stimpson, "On the Prediction of Impact Noise, IX: The Noise from Punch Presses," *J. Sound Vib.* **103**, 43–81 (1985).
83. E. J. Richards and I. Carr, "On the Prediction of Impact Noise, X: The Design and Testing of a Quietened Drop Hammer," *J. Sound Vib.* **104**(1), 137–164 (1986).
84. ISO 140, Pt. VI (1978), "Laboratory Mesurements of Impact Sound Isolation of Floors," American National Standards Institute, New York.
85. K. Gösele, "Die Bestimmung der Dynamischen Steifigkeit von Trittschall-Dämmstoffen," Boden, Wand und Decke, Heft 4 and 5, Willy Schleunung, Markt Heidenfeld, 1960.

86. L. Cremer, ''Theorie des Kolpfschalles bei Decken mit Schwimmenden Estrich,'' *Acustica* **2**(4), 167–178 (1952).

87. I. L. Vér, ''Acoustical and Vibrational Performance of Floating Floors,'' Report No. 72, Bolt Beranek and Newman, October 1969.

88. R. Josse and C. Drouin, ''Étude des impacts lourds a l'interieur des bâtiments d'habitation,'' Rapport de fin d'étude, Centre Scientifique et Technique du Bâtiment, Paris, February 1, 1969, DS No. 1, 1.24.69.

89. I. L. Vér, ''Impact Noise Isolation of Composite Floors,'' *J. Acoust. Soc. Am.* **50**(4)(Pt. 1), 1043–1050 (1971).

90. M. Heckl and E. J. Rathe, ''Relationship between the Transmission Loss and the Impact-Noise Isolation of Floor Structures,'' *J. Acoust. Soc. Am.* **35**(11), 1825–1830 (1963).

91. K. Gösele, ''Trittschall-Entstehung und Dämmung,'' VDI-Berichte VDI Verlag, Duesseldorf, Band 8, pp. 23–28, 1956.

92. K. Gösele, ''A Simple Method to Calculate the Impact Noise Isolation of Floors,'' Boden Wand und Decke, Vol. 11, Willy Schleunung, Markt Heidenfeld, 1964 (in German).

93. I. L. Vér, ''Relationship between Normalized Impact Sound Level and Sound Transmission Loss,'' *J. Acoust. Soc. Am.* **50**(6)(Pt. 1), 1414–1417 (1971).

94. I. L. Vér and D. H. Sturz ''Structureborne Sound Isolation,'' in C. M. Harris (ed.), *Handbook of Acoustical Measurements and Noise Control*, 3rd ed., McGraw-Hill, New York, 1991, Chapter 32.

95. ASTM E 90-83 ''Laboratory Measurement of Airborne Sound Transmission Loss of Building Partitions,'' American Society for Testing and Materials, Philadelphia, PA.

96. ISO 717-1982, ''Rating of Sound Insulation in Buildings and of Building Elements—Part I: Airborne Sound Insulation in Buildings and of Interior Building Elements.'' American National Standards Institute, New York.

97. ASTM E 989-89 ''Standard Classification for Impact Insulation Class (IIC)'' American Society for Testing and Materials, Philadelphia, PA.

98. The test procedures are specified in either ASTM E 336-1984, ''Measurement of Airborne Sound Insulation in Buildings,'' or ISO 140, Part 1, 1978.

99. A large number of measured values of STC may be found in ''Solution to Noise Problems in Apartments and Hotels,'' Owens-Corning Fiberglas Corporation, 1969; in ''Sweet's Catalog,'' Sweet's Construction Division, McGraw-Hill Information Systems Company; in ''Sound Isolation of Wall, Floor and Door Constructions,'' *Natl. Bur. Std. Nomograph* **77**, 1964; and in ref. 97.

100. The standard tapping machine and the measurement procedure for laboratory and field evaluation of impact sound transmission loss are described in ISO 140, Parts 6–7, 1978. Practical measurement details are given in ref. 97.

101. For the measurement of the improvement of insulation owing to a floating floor or a floor covering, ISO 140, Part 8, 1978 specifies that a standard reinforced concrete floor of thickness 12 + 2 cm be used. ISO 140, Part 9, 1985, specifies a 14-cm slab, weighting approximately 350 kg/m^2, and approximately 10 m^2 in surface area.

102. ''Impact Noise Control in Multifamily Dwellings,'' FHA Rept. 750, Federal Housing Adminstration, Washington, DC, January 1963.

103. ISO 717-1982, Part 2: "Impact Sound Insulation," American National Standards Institute, New York.

104. See, for example, T. Mariner and H. W. W. Hehmann, "Impact-Noise Rating of Various Floors," *J. Acoust. Soc. Am.* **41**, 206–214 (1967).

BIBLIOGRAPHY

Beranek, L. L. *Acoustical Measurements*, rev. ed., Acoustical Society of America, Woodbury, NY, 1988.

Bies, D. A. and C. H. Hansen, *Engineering Noise Control*, Unwin Hyman, London/Boston/Sydney, 1988.

Cremer, L., M. Heckl, and E. E. Ungar, *Structureborne Sound*, 2nd ed., Springer-Verlag, Berlin, Heidelberg, New York, London, Paris, Tokyo, 1988.

Fahy, F., *Sound and Structural Vibration*, Academic Press, London, 1989, revised, corrected paperback edition.

Fasold, W., W. Kraak, and W. Shirmer, *Taschenbuch Akustik* [Pocket Book Acoustics] Vols. 1 and 2, VEB Verlag Technik, Berlin, 1984 (in German).

Harris, C. M. (ed.), *Handbook of Acoustical Measurements Noise and Control*, 3rd ed., McGraw-Hill, New York, 1991.

Junger, M. C. and D. Feit, *Sound Structures and Their Interaction*, 2nd ed., MIT Press, Cambridge, MA, 1986.

Reichardt, W. *Technik-Woerterbuch, Technische Akustik* [Dictionary of Technical Acoustics; English, German, French, Russian, Spanish, Polish, Hungarian, and Slowakian], VEB Verlag, Technik, Berlin, 1978.

Table 9.1 Speeds of Sound in Solids, Deformation of Solids for Different Wavetypes and Formulas for Propagation Speed

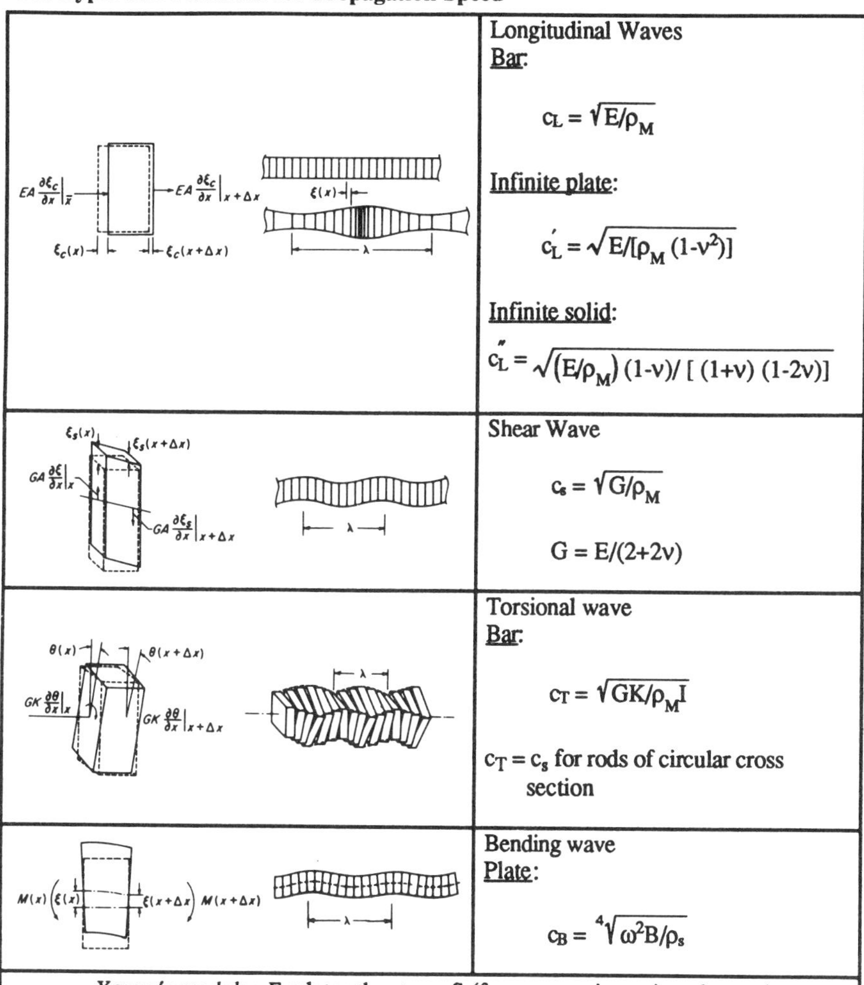

Deformation	Wave type and propagation speed
	Longitudinal Waves Bar: $c_L = \sqrt{E/\rho_M}$ Infinite plate: $c_L' = \sqrt{E/[\rho_M(1-\nu^2)]}$ Infinite solid: $c_L'' = \sqrt{(E/\rho_M)(1-\nu)/[(1+\nu)(1-2\nu)]}$
	Shear Wave $c_s = \sqrt{G/\rho_M}$ $G = E/(2+2\nu)$
	Torsional wave Bar: $c_T = \sqrt{GK/\rho_M I}$ $c_T = c_s$ for rods of circular cross section
	Bending wave Plate: $c_B = \sqrt[4]{\omega^2 B/\rho_s}$

Young's modulus E relates the stress S (force per unit area) to the strain (change in length per unit length), N/m^2. *Poisson's ratio* ν is the ratio of the transverse expansion per unit length of a circular bar to its shortening per unit length, under a compressive stress, dimensionless. It equals about 0.3 for structural materials and nearly 0.5 for rubber-like materials. The *density* of the material is ρ_M, kg/m^3. ρ_s is mass per unit area (kg/m^2) for plates and mass per unit length (kg/m) for bars, rods or beams. The *shear modulus G* is the ratio of shearing stress to shearing strain, N/m^2. *I* is the *polar moment* of inertia, m^4. The *torsional stiffness factor K* relates a twist to the shearing strain produced, m^4. The *bending stiffness* per unit width *B* equals $Eh^3/[12(1-\nu^2)]$ for a homogeneous plate, N-m, where h is the thickness of the bar (or plate) in the direction of bending, m. For rectangular rods $B = Eh^3w/12$ where h is the cross section dimension in the plane of bending and w that perpendicular to it (and width), m.

Table 9.2 Acoustically Important Parameters of Solid Material†

Material	Density ρ_M kg/m^3	Young's Modulus E N/m^2	Poisson Ratio ν	Speed of Sound c_L m/sec	Product of Surface Density and Critical Frequency $\rho_s f_c$		TL at Critical Frequency R (f_c) dB	Internal Damping Factor for Bending at 1000 Hz η*	Acoustical-Mechanical Conversion Efficiency η_{am} Eq. 9.70
					Hz-kg/m^2	Hz-lb/ft^2			
Aluminum	2700	7.16×10^{10}	0.34	5,150	34,700	7,000	48.5	10^{-4}-10^{-2}**	2.5×10^{-3}
Copper	8900	4.6×10^{10}	0.35	2,273					3.3×10^{-4}
Glass	2500	6.76×10^{10}		5,200	38,000	7,800	49.5	0.001-0.01**	2.7×10^{-3}
Lead (chemical or tellurium)	11000	1.58×10^{10}	0.43	1,200	605,000	124,000	73.5	0.015	1.4×10^{-4}
Plexiglas or Lucite	1150	3.73×10^{9}		1,800	35,400	7,250	49.0	0.002	2×10^{-3}
Steel	7700	1.96×10^{11}	0.31	5,050	97,500	20,000	57.5	10^{-4}-10^{-2}**	8.5×10^{-3}
Brick	1900 to 2300				217 to 373	7,000 to 12,000	48.5 to 53	0.01	
Concrete, dense poured	2300	2.61×10^{10}		3,400	43,000	9,000	50.5	0.005-0.02	1.9×10^{-3}
Concrete (Clinker) slab, plastered on both sides 5 cm thick	1500	-	-	-	48,800	10,000	51.5	0.005-0.02	
Masonry Block	750	-	-	-	23,200	4,750	45.0	0.005-0.02	
Hollow cinder with 1.6 cm sand plaster, nom. thickness 15 cm (6 in.)	900	-	-	-	25,500	5,220	46.0	0.005-0.02	
Hollow dense concrete, nom. 15 cm (6 in.) thick	1100	-	-	-	23,000	4,720	45.0	0.007-0.02	

Table 9.2 (*Continued*)

Material	Density ρ_M kg/m^3	Young's Modulus E N/m^2	Poisson Ratio ν	Speed of Sound c_L m/sec	Product of Surface Density and Critical Frequency $\rho_s f_c$		TL at Critical Frequency R (f_c) dB	Internal Damping Factor for Bending at 1000 Hz η*	Acoustical-Mechanical Conversion Efficiency η_{am} (Eq. 9.70)
					Hz-kg/m^2	Hz-lb/ft^2			
Hollow dense concrete, sand-filled voids, nom. 15 cm (6 in.) thick	1700				42,200	8,650	50.0	Varies with frequency	
Solid dense concrete nom. 10 cm (4 in.) thick	1700				54,100	11,100	52.5	0.012	
Gypsum board 1.25 to 5 cm (1/2 to 2 in.) thick	650			6,800	20,000	4,500	45.0	0.01-0.03	
Plaster, solid, on metal or gypsum lathe	1700				24,500	5,000	45.5	0.005-0.01	
Fir timber	550			3,800	4,880	1,000	31.5	0.04	9×10^{-3}
Plywood 0.6 cm to 3.12 cm (1/2 to 2 in.) thick	600				12,700	2,600	40	0.01-0.04	
Wood waste material bonded with plastic 23 kg/m^2 (5 lb/ft^2)	750				73,200	15,000	55.0	0.005-0.01	

*The range in values of η are based on limited data. The lower values are typical for material alone while the higher values are the maximum observed on panels in place.

**The loss factors for structures of these materials are sensitive to construction techniques and edge conditions.

†Where boxes are blank the parameter is either not meaningful or is not available.

Table 9.3 Driving Point Force Impedance of Infinite Structures (from Refs. 1–12)

Element	Picture	Equivalent Mass (see Eq. 9.19)	Driving Point Force Impedance	Range of Validity	Auxiliary Expressions and Notes
Beam, in compression					$c_L = \sqrt{\frac{E}{\rho_M}}$
Semi-Infinite	F, S	$\lambda_L/2\pi$	$Z_F = \rho_M c_L S$		S = cross section area
Infinite	S	S, λ_L/π	$Z_F = 2\rho_M c_L S$	$S < (\lambda_s/4)^2$	$\lambda_S = \frac{1}{f}\sqrt{\frac{E}{2(1+\nu)\rho_M}}$
Beam in Bending					r = radius of contact area
Semi Infinite	F, S, 2r	$\frac{\sqrt{2}}{2\pi}\lambda_B$, S	$Z_F = \frac{1}{2}(1+j)\rho_M S c_B$	$S < (\lambda_B/6)^2$	$c_B = \left(\frac{EI\omega^2}{\rho_M S}\right)^{1/4}$
					ν = Poisson Ratio
Infinite	F, S, 2r	$\frac{\sqrt{2}}{\pi}\lambda_B$	$Z_F = 2(1+j)\rho_M S c_B$	$2r > 9S/\lambda_S$	$\lambda_B = c_B/f$
Thin Infinite Plate			$Z_F = 8\sqrt{B'\rho_M h}$	$h < \lambda_B/6$	$B' = \left(\frac{Eh^3}{12(1-\nu^2)}\right)$
Vertical Force	h, F, 2r	$\lambda_B/2$, h	$= 2.3\rho_M c_L h^2$	$2r > 3h$	
				$\lambda_S >> 2\pi r > 10h$	$\lambda_S = \frac{1}{f}\sqrt{\frac{E}{2(1+\nu)\rho_M}}$
Horizontal Force (In Plane)	F, h, 2r		$\frac{1}{Z_F} = \frac{\pi f}{4Gh}\left[\frac{3-\mu}{2} + jH\right]$		$H = K + L$
					$K = (1-\mu)\,\ell n\left(\frac{\lambda_L}{\pi r}\right)$
					$L = 2\ell n\left(\frac{\lambda_s}{\pi r}\right)$

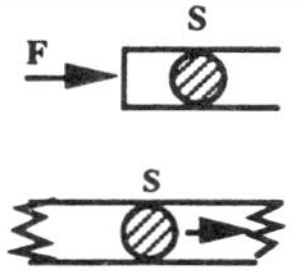

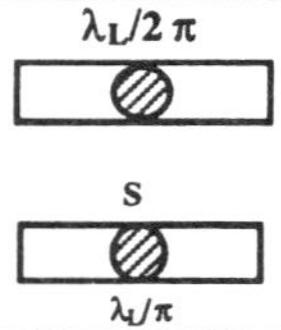

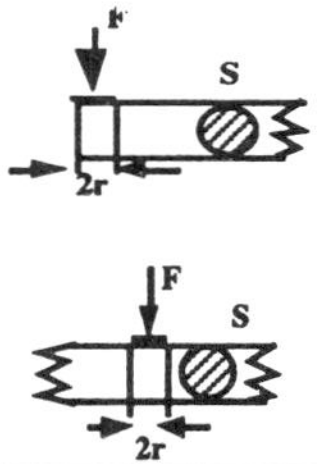

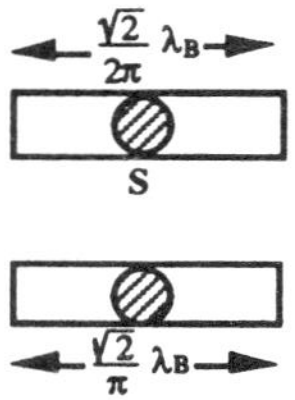

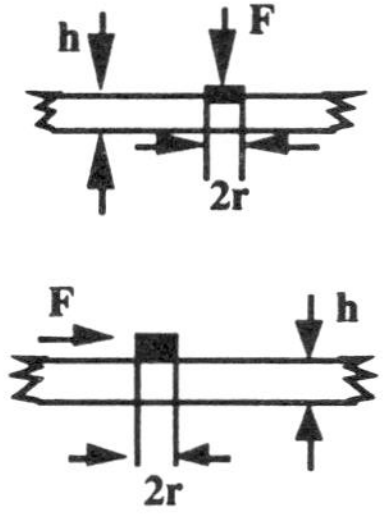

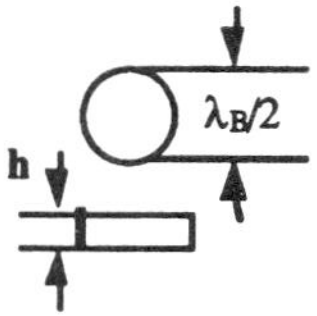

Table 9.3 (*Continued*)

Element	Picture	Equivalent Mass (see Eq. 9.19)	Driving Point Force Impedance	Range of Validity	Auxiliary Expressions and Notes
Thin Semi Infinite Plate	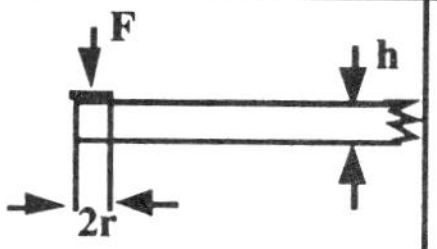 F, h, 2r	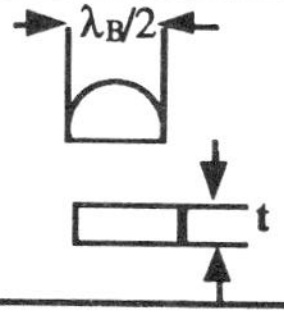 $\lambda_B/2$, t	$Z_F = 3.5\sqrt{B'\rho_M h}$ $\approx \rho_M c_L h^2$	$h < \lambda_B/6$ $2r > 3h$	
Beam-Stiffened Infinite Thin Plate	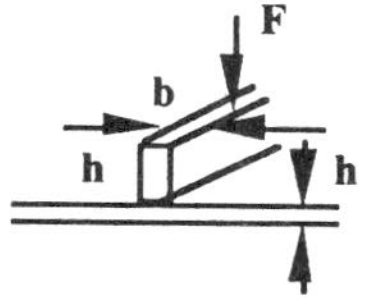 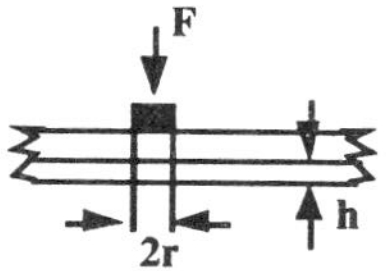 F, b, h, h; F, h, 2r	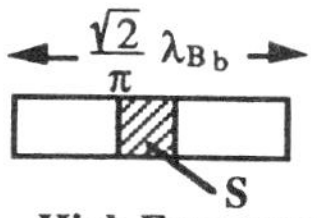 $\frac{\sqrt{2}}{\pi}\lambda_{B_b}$, S **High Frequency Approximation**	$Z_F = \frac{(1-j)\,k'}{4\rho_s\,\omega}$ $Z_F \cong Z_{FB} = 2(1+j)\rho_M S_b c_B$ **High Frequency Approximation**	$S_b < (\lambda_B/6)^2$ $2r > 9S_b/\lambda_B$	$\rho_s' = \rho_M S_B + 2\rho_M h/k_p$ $k' = \left(\frac{\rho_s'}{B}\right)^{1/4} \omega^{1/2} A$ $A = 1 - j\,\frac{\rho_M h}{2\rho_s' k_p}$ k_p = plate bending wave number $B = Eh^3b/3$ $s = k_b/k_p$
Infinte String	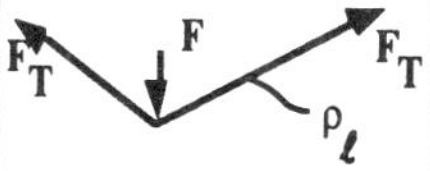 F_T, F, F_T, ρ_ℓ	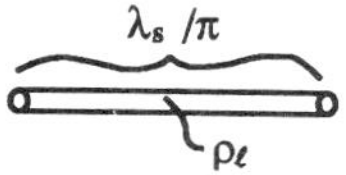 λ_s/π, ρ_ℓ	$Z_F = 2\sqrt{\rho_\ell F_T}$		$c_s = \sqrt{F_T/\rho_\ell}$ ρ_ℓ = mass per unit length F_T = Tension Force

Table 9.3 (*Continued*)

Element	Picture	Equivalent Mass (see Eq. 9.19)	Driving Point Force Impedance	Range of Validity	Auxiliary Expressions and Notes
Semi-Infinite Plate In-Plane Edge Force	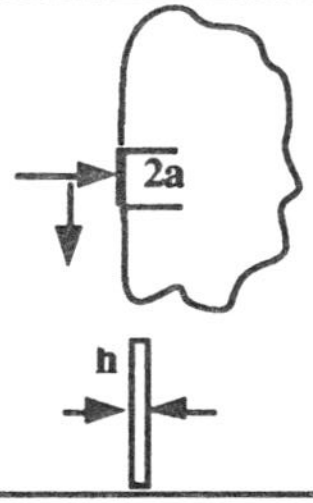 2a; h	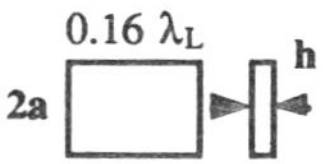 $0.16\,\lambda_L$; 2a; h	$Z_F \cong \rho_M h c_L (2a)$		
Infinite Corrugated-Plate	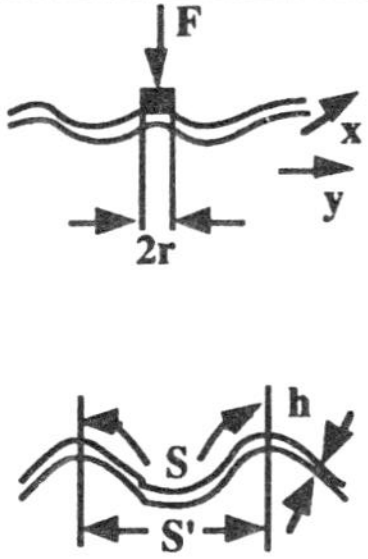 F; x; y; 2r; h; S; S'	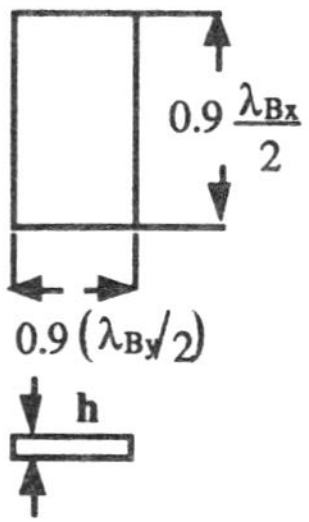 $0.9\,\frac{\lambda_{Bx}}{2}$; $0.9\left(\lambda_{By}/2\right)$; h	$Z_F = 8\left[(\rho_M h)^2 B_x B_y\right]^{1/4}$	$S << \lambda_B$	$B_y = \frac{Eh^3}{12(1-\nu^2)}$ $B_x \cong \left(\frac{1-\nu}{1+\nu}\right)^2 \frac{s}{c} B_y$ $f_{c1} = \frac{c_o^2}{2\pi}\sqrt{\frac{\rho_M h}{B_x}}$ $f_{c2} = f_{c1}\sqrt{\frac{B_x}{B_y}}$
Elastic Halfspace	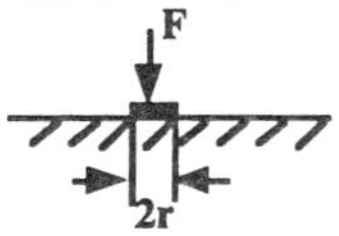 F; 2r		$Z = \frac{-j0.64\,Gr}{f\,(1-\nu/2)} + 1.79\,r^2\sqrt{\frac{4G\rho_M}{(1-\nu)}}$	$2r < \lambda_s/6$	ρ_M = density

Table 9.3 (*Continued*)

Element	Picture	Equivalent Mass (see Eq. 9.19)	Driving Point Force Impedance	Range of Validity	Auxiliary Expressions and Notes
Infinite Cylindrical Shell	F, h, D		$Z_\infty \cong 2.3\, \rho_M c_L h$	$f > 1.5\, f_R$	$f_R = c_L/\pi D$ ring frequency
			$\mathrm{Re}\{1/Z_\infty\} = \frac{\sqrt{12}}{8\rho_M c_L h^2}\left(\frac{f}{f_R}\right)^{2/3}$	$f < 0.7\, f_R$	

Table 9.4 Moment Impedance Z_M of Infinite and Semi-infinite Structures

Element	Picture	Driving Point Moment Impedance	Auxiliary Expressions and Notes
Semi-Infinite Beam			ρ_ℓ = mass per unit length (kg/m)
Free End	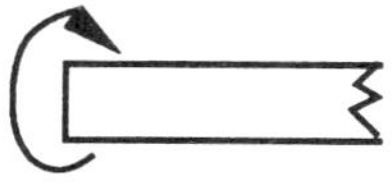	$\frac{(1-j)\rho_\ell c_B^3(f)}{8\pi^2 f^2}$	$c_B(f)$ = bending wave speed (m/sec) in bending $c_B(f) = \sqrt{2\pi f}\,(EI/\rho_\ell)^{1/4}$
Pinned End		$\frac{(1-j)\rho_\ell c_B^3(f)}{4\pi^2 f^2}$	E = Young's Modulus (N/m^2) I = Area moment of inertia in bending (m^4) f = frequency ρ_M = material density (kg/m^3)
Infinite Beam	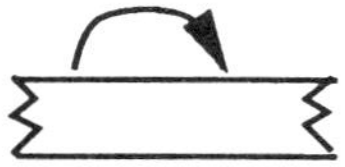	$\frac{(1-j)\rho_\ell c_B^3(f)}{2\pi^2 f^2}$	
Infinite Homogeneous Isotropic Plate		$\frac{16\rho_M h c_L^2 k_B^2(f)}{2\pi f\left[1-j\,1.27\,\ell n(k_B a/2.2)\right]}$	$k_B(f) = 2\pi f/c_B(f)$ bending wavenumber h = plate thickness

Table 9.4 (*Continued*)

Element	Picture	Driving Point Moment Impedance	Auxiliary Expressions and Notes
Semi-Infinite Homogeneous Isotropic Plate		$\dfrac{12\rho_M h c_\ell^2 k_B^2(f)}{2\pi f\left[1-j\,3.35\,\ell n(kr/3.5)\right]}$	
At Joint of Homogeneous, Isotropic Plates		$\dfrac{\rho_M c_\ell^2 h^3}{75.4f}\left\{16\left[\dfrac{1+j\,1.27\,\ell n(kb/2.2)}{1+(1.27\,\ell n\,kb/2.2)^2}\right]+12\left[\dfrac{1+j\,3.35\,\ell n(kr/3.5)}{1+(3.35\,\ell n\,kr/3.5)^2}\right]\right\}$	

Table 9.5 Effective Length l_M Connecting Force and Moment Impedance $l_M^2 Z_F$ for Infinite Beams and Plates (after Ref. 6)

Element	Picture	Equivalent Force Pair	Effective Length ℓ_M	Auxiliary Expressions
Beam in Bending	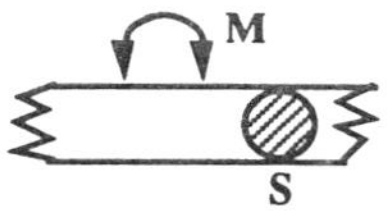 M, S	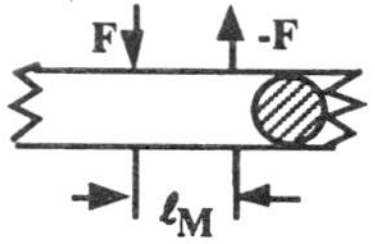 F, -F, ℓ_M	$\frac{\lambda_B/\sqrt{j}}{2\pi}$	$j = \sqrt{-1}$
Beam in Torsion	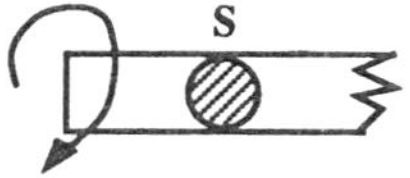 S	 F, -F, ℓ_M	0.79i	For Hollow Beams $i = \sqrt{\frac{d_o^2 - d_i^2}{2}}$ d_o = outside diameter d_i = inside diameter
Plate, Vertical Moment	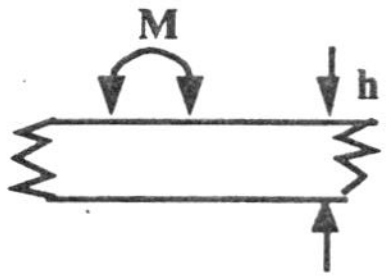 M, h	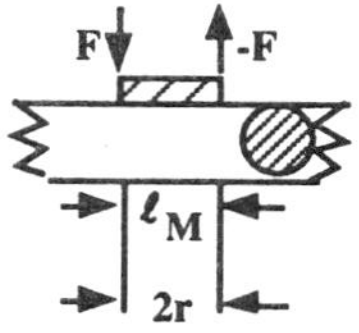 F, -F, ℓ_M, 2r	$\frac{\lambda_B/\sqrt{8\pi j}}{\sqrt{\ell n(\lambda_B/\pi r)}}$	Bending Waves $3h < 2r << \lambda_B/\pi$
Plate, In-Plane Torsional Moment	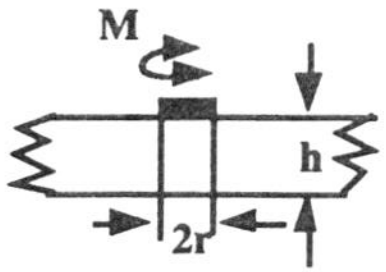 M, h, 2r	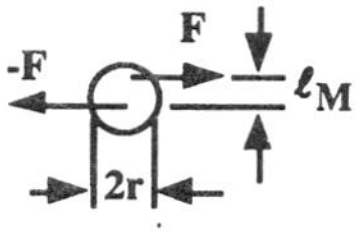 F, -F, ℓ_M, 2r	$\frac{r}{\sqrt{\ell n(\lambda_s/\pi r)}}$	Shear Waves $3h < 2r << \lambda_s/\pi$

Table 9.6 Power Input to Structures (after Refs. 5, 6, 10, 12, 69)

Element	Picture	Power Input to Infinite Element		Finite Elements		Auxiliary Expressions
		Force or Moment Excitation	Velocity or Angular Velocity Excitation	Onset of Infinite Behavior	W_{fin}/W_{inf}	
Beam in Longitudinal Wave Motion; Force or Velocity Excitation	$\hat{F},\hat{v}$; S	$\frac{\lvert\hat{F}\rvert^2}{4\,\rho_M S c_L}$	$4\lvert\hat{v}\rvert^2 S\rho_M c_L$	$\omega > \frac{\pi c_L}{\eta \ell}$	$\frac{4}{\pi\eta}$	$c_L = \sqrt{E/\rho_M}$ ℓ = length η = loss factor Q = torsion constant
Beam in Torsion Moment or Angular Velocity Excitation	$\hat{M},\hat{\dot{\theta}}$; S	$\frac{\lvert\hat{M}\rvert^2}{4\,GQJ}$	$4\lvert\hat{\dot{\theta}}\rvert^2\sqrt{GQJ}$	$\omega > \frac{\pi c_T}{\eta \ell}$	$\frac{4}{\pi\eta}$	G – shear modulus J = mass moment of inertia per unit length
Beam in Bending; Force or Velocity Excitation	$\hat{F},\hat{v}$; S	$\frac{\lvert\hat{F}\rvert^2}{8\,\rho_M S c_B(f)}$	$\lvert\hat{v}\rvert^2 S\rho_M c_B(f)$	$\omega > \frac{4\pi c_B(f)}{\eta \ell}$	$\frac{4\sqrt{2}}{\pi\eta}$	$c_T = \sqrt{\frac{E/\rho_M}{GQ/J}}$ = torsional wave speed ρ_M = density
Beam in Bending; Moment or Angular Velocity Excitation	$\hat{M},\hat{\dot{\theta}}$; S	$\frac{\lvert\hat{M}\rvert^2 c_B(f)}{8EI}$	$\frac{\lvert\hat{\dot{\theta}}\rvert^2 EI}{c_B(f)}$	$\omega > \frac{4\pi c_B(t)}{\eta \ell}$	$\frac{2\sqrt{2}}{\pi\eta}$	E = Young's Modulus I = second moment of area $\dot{\theta}$ = angular velocity
Plate in Bending; Force or Velocity Excitation	$\hat{F},\hat{v}$; h	$\frac{\lvert\hat{\tilde{F}}^2\rvert}{16\sqrt{B_p\rho_M h}} = \frac{\lvert\hat{\tilde{F}}^2\rvert}{4.6\rho t^2 c_L}$	$4\,\hat{\tilde{v}}^2\sqrt{B_p\,\rho_M h} = 1.15\,\hat{\tilde{v}}^2\rho_M h^2 c_L$	$\omega > \frac{8}{\eta \ell_1 \ell_2 \sqrt{\frac{B_p}{\rho_M h}}}$	$\frac{32\ell_1\ell_2}{\pi^2\eta(\ell_1^2+\ell_2^2)}\frac{\omega}{c_B}$	$c_B = \sqrt{\omega}\sqrt{\frac{E/I}{\rho_M S}}$ bending wave speed $B_p = \frac{h^3 E}{12(1-\nu^2)}$ S = area

Table 9.6 *(Continued)*

Element	Picture	Power Input to Infinite Element		Finite Elements		Auxiliary Expressions
		Force or Moment Excitation	Velocity or Angular Velocity Excitation	Onset of Infinite Behavior	W_{fin}/W_{inf}	
Plate in Bending; Moment or Angular Velocity Excitation	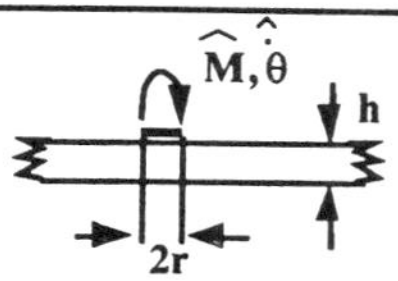 $\hat{M}, \hat{\dot{\theta}}$; h; 2r	$\sim \dfrac{\omega \lvert \hat{M} \rvert^2}{16\,B_p}$ for $r > h$	$\dfrac{4 \lvert \hat{\dot{\theta}} \rvert^2 B_p}{\omega \left\{ 1 + \left[\dfrac{4}{\pi} \ln \left(\dfrac{\omega r}{c_B} \right) - \dfrac{8}{\pi(1-\nu)} \left(\dfrac{h}{\pi r} \right)^2 \right]^2 \right\}}$	$\omega > \dfrac{8}{\eta \ell_1 \ell_2} \sqrt{\dfrac{B_p}{\rho_M h}}$		
Thin Walled Pipe in Bending; Force Excitation	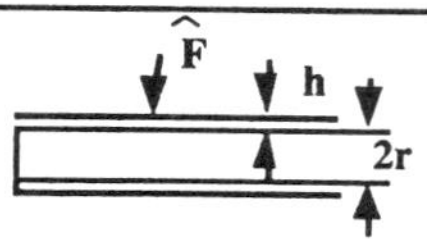 $\hat{F}$; h; 2r	$\tilde{F}^2 / \left(16\pi\rho_M r\, h \sqrt{c_L r \omega}\right)$, for $f < 0.123\, c_L h/r$ $\tilde{F}^2 \sqrt{V/(2+V)} / \left(\omega \rho_M 2\lambda^2/\pi^2\right)$, for $f > 0.123\, c_L h/r$				$V = \omega r/c_L$ λ_p = bending wave-length in equivalent thickness plate
Plate in Bending; Multiple Force Excitation, Equally Spaced	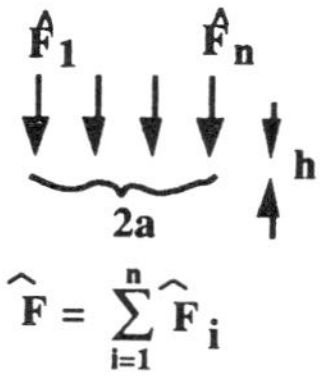 $\hat{F}_1$ … $\hat{F}_n$; 2a; h; $\hat{F} = \sum_{i=1}^{n} \hat{F}_i$	$\tilde{F}^2 [2J_1(z)/z]^2 / \left(16\sqrt{B_p \rho_M h}\right)$				$Z = 2\pi a/\lambda_B$ J_1 = Bessel Function of order one

Table 9.6 (*Continued*)

Element	Picture	Power Input to Infinite Element		Finite Elements		Auxiliary Expressions
		Force or Moment Excitation	Velocity or Angular Velocity Excitation	Onset of Infinite Behavior	W_{finite}/W_{inf}	
Plate in Bending; Large Area Velocity Excitation	$\hat{v}$, t, 2r	$\tilde{v}^2 \frac{\pi}{2} \rho_M c_B (r+0.8\lambda_B)\, h$				
Elastic Halfspace; Single Force	$\hat{F}$	$\frac{48\, \tilde{F}^2}{\omega \rho_M \pi\, \lambda_s^3}$				$\lambda_s = \sqrt{G/\rho_M}/f$ shear wavelength
Elastic Halfspace Equal Multiple Forces Along a Line, Equally Spaced	$\hat{F} = \Sigma \hat{F}_i$, ℓ	$\frac{16\, \tilde{F}^2}{\omega \rho_M \ell\, \lambda_s^2}$; $\ell > \lambda_s/2$				

Table 9.6 (*Continued*)

Element	Picture	Power Input to Infinite Element		Finite Elements		Auxiliary Expressions
		Force or Moment Excitation	Velocity or Angular Velocity Excitation	Onset of Infinite Behavior	W_{finite}/W_{inf}	
Infinite Plate Excited by Nearby: Monopole	y, Q, h	$W_{MON} \cong \frac{Q^2(\rho_0 c_0)^2 \operatorname{Re}\{Y_\infty\}}{(f_c/f) - 1} e^{-4\pi y \frac{\sqrt{ff_c}}{c_0}\sqrt{1-(f/f_c)}}$; $f < f_c$				$Q =$ rms volume velocity $Y_\infty =$ point force admittance $k_B =$ free plate bending wave number $f_c =$ critical coincidence frequency
Lateral Dipole	d, +, -, y	$W_{LD} \cong 0.5\,(k_B d)^2\, W_{MON}$		; $f < f_c$		
Perpendicular Dipole	+, -, d, y	$W_{PD} \cong (k_B d)^2[1-(f/f_c)]^2 W_{MON}$ (for $f < f_c$ and light fluid loading)				

Table 9.7 First Resonance Frequency, Mode Shape, and Modal Density of Finite Elements (after Ref. 19)

Element	Picture	Boundary Conditions*	First Resonance Frequency	Mode Shape ϕ (x,y,z)	Modal Density† $n(\omega)$	Auxiliary Formulas
Beam in Compression	S, ℓ	f - f c - c	$c_L/2\ell$	$\cos(n\pi x/\ell)$ $\sin(n\pi x/\ell)$	$\ell/\pi c_L$	$\kappa = \sqrt{I/S}$ radius of gyration
Beam in Bending	S, ℓ	p - p f - f c - c c - f	$(\pi/2)\,(\kappa c_L/\ell^2)$ $(1/2\pi)\,(4.73/\ell)^2\,\kappa c_L$ $(1/2\pi)\,(4.73/\ell)^2\,\kappa c_L$ $(1/2\pi)\,(1.875/\ell)^2\,\kappa c_L$	$\sqrt{2}\sin(k_n x)$ } See Ref. 19	$\ell/(2\pi\kappa c_L)$	$k_n = \sqrt{2\pi f_n/c_L\kappa}$
Rectangular Plate in Bending	S, a, b, h	ffff ssss cccc	a/b: 1, 1.5, 2.5 C_1: 3.33, 3.31, 2.13 (ffff); 4.88, 5.28, 7.1 (ssss); 8.89, 10.0, 14.6 (cccc) $f_1 = 10^3 C_1 (h/S)(c_L/c_{L_{st}})$	See Ref. 19	$\frac{\sqrt{12}\,S}{4\pi c_L h}$	a/b = aspect ratio h = plate thickness, m S = area, m^2 $c_{L_{st}}$ = 5050 m/sec c_L = longitudinal wave speed
Membrane	S, F′		See Ref. 18	See Ref. 19	$\frac{S}{2\pi c_m}$	F′ = tension per unit length ρ_s = mass per unit area $c_m \sqrt{F'/\rho_s}$
String	ℓ	c - c	$\pi c_s/\ell$	$\sin(n\pi x/\ell)$	$\ell/\pi c_s$	$c_s = \sqrt{F'/\rho_\ell}$
Rectangular Air Volume	ℓ_x, ℓ_y, ℓ_z	hard walls	$c_o/2\ell_{max}$	$\cos\left(\frac{n_x\pi x}{\ell_x}\right)\cos\left(\frac{n_y\pi y}{\ell_y}\right)\cos\left(\frac{n_z\pi z}{\ell_z}\right)$	$\frac{\omega^2 V}{2\pi c_o^3}$	F = tension force ρ_ℓ = mass per unit length $V = \ell_x\ell_y\ell_z$ c_o = speed of sound n = 1,2,3,...

*Boundary conditions: f = free; s = simply supported; c = clamped; p = pinned
†$n(\omega) = n(f)/2\pi$

Table 9.8 Radiation Efficiency of Vibrating Bodies (after Refs. 1, 22, 23, 38)

Body	Picture	σ_{rad}	Auxiliary Expressions
Small Pulsating Body	 V	$\frac{(ka)^2}{1 + (ka)^2}$	c_o = speed of sound $k_o = 2\pi f/c_o$ a = source radius
Small Oscillating Rigid Body	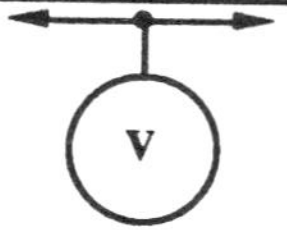 V	$\frac{(ka)^4}{4 + (ka)^4}$ See Fig. 9.14	
Pulsating Pipe	 V 2a	$2/\pi k_o a \lvert H_1(k_o a)\rvert^2$ for $(\pi/2)k_o a \le 2/\pi$	H_1 = Hankel Function second kind, order one $k_o = 2\pi f/c_o$
Oscillating Pipe or Rod	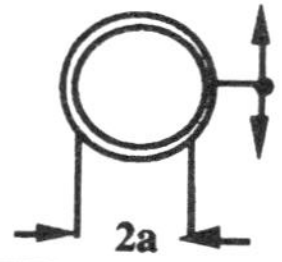 2a	$2/\pi k_o a \lvert H_1'(k_o a)\rvert^2$ See Fig. 9.15	H_1' = first derivative of H_1 in respect of its argument
Circular Pipes in Bending	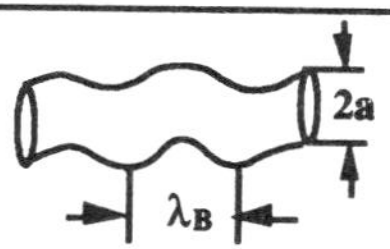 2a λ_B	zero ; $f < f_c$ $(k_o a)^3 [1-(f_c/f)]$; $f > f_c$; $k_d a << 1$ 1 ; $f > f_c$; $k_o a >> 1.5$	$k_d^2 = k_o^2 - k_B^2$ $k_B = 2\pi/\lambda_B$ f_c = critical frequency where $k_B = k_o$

Table 9.8 (*Continued*)

Body	Picture	σ_{rad}	Auxiliary Expressions
Rectangular and Elliptic Beams in Bending		See Ref. 22	
Infinite Thin Plate Supporting Free Bending Waves		0, for $f < f_c$ $1/[1-(f_c/f)]^{1/2}$ for $f > f_c$	
Finite Thin Plate Supporting Free Bending Waves; Plate Surrounded by Rigid Baffle	L_{max} L_{min} S, P	$\frac{2\lambda_c}{S}\left[g_1(\beta) + \frac{P}{2\lambda_c} g_2(\beta)\right] C_1$; $f < f_c$ $0.45\,(P/\lambda_c)^{1/2}(L_{min}/L_{max})^{1/4}$; $f = f_c$ $(1-f_c/f)^{-1/2}$; $f \geq 1.3\, f_c$	f_c = critical frequency. See Eq. 9.84 $\lambda_c = c_o/f_c$ $S = L_{max} L_{min}$ = area (one side) $P = 2\,(L_{max}+L_{min})$ = perimeter $\beta = (f/f_c)^{1/2}$
		$g_1(\beta) = \begin{cases} (4/\pi^4)\left[(1-2\beta^2)/\beta(1-\beta^2)^{1/2}\right] ; & f < 0.5\, f_c \\ 0 ; & f > 0.5\, f_c \end{cases}$ $g_2(\beta) = (1/4\pi^2)\frac{(1-\beta^2)\ln[(1+\beta)/(1-\beta)] + 2\beta}{(1-\beta^2)^{3/2}}$; $C_1 = \begin{cases} 1 \text{ for simple supported edges} \\ \beta^2 \exp(10\lambda_c/P) \text{ for clamped edges} \end{cases}$	

Table 9.8 (*Continued*)

Body	Picture	σ_{rad}	Auxiliary Expressions
Thick Finite Plate Supporting Free Bending Waves		$0.45\sqrt{P/\lambda_o}$ for $f \le f_b$ 1 for $f >> f_b$	$f_b = f_c + \frac{5c_o}{P}$
Infinite Plate Sound-Forced Waves	$\sigma_F = 1/\cos\phi$		ϕ = incidence angle, degree
Finite Square Plate Oblique Incidence Plane Sound Wave Excitation	$\sigma_F = \min\begin{Bmatrix} A\left[(k_o/2)\sqrt{S}\right] \\ 1/\cos\phi \end{Bmatrix}$ for $0.1\,\lambda_o^2 < S < 0.4\,\lambda_o^2$ $\sigma_F = \min\begin{Bmatrix} \left[(0.5)^{(\phi/90)}\sqrt{k_o/2\sqrt{S}}\right] \\ 1/\cos\phi \end{Bmatrix}$ for $S > 0.4\,\lambda_o^2$		$A = (0.5)\,(0.8)^{(\phi/90)}$ $\alpha = 1 - 0.34\phi/90$ $k_o = 2\pi/\lambda_o = 2\pi f/c_o$
Finite Square Plate Diffuse Sound Field Excitation	$\sigma_F = 0.5\left[0.2 + \ln(k_o\sqrt{S})\right]$ for $k_o\sqrt{S} > 1$		

TABLE 9.9 Octave-Band Breakout Sound Transmission Loss, R_{i0}, of Rectangular Sheet Metal Ducts (after Ref. 48)

Duct Size in. (m)	Gauge	Octave Band Center Frequency, Hz 63	125	250	500	1000	2000	3000	4000
10 × 4	26*	18	26	29	32	35	38	41	45
(0.25 × 0.1)	22	25	24	31	34	36	49	43	45
12 × 8	26*	19	22	25	28	31	34	38	43
(0.3 × 0.2)	24	17	24	27	30	33	36	41	45
14 × 8 (0.36 × 0.2)	26*	21	23.5	27	30	33	36	41	45
14 × 10	22*	19	26	29	32	35	38	43	45
(0.36 × 0.25)	20	20	28	31	34	37	40	44	45
16 × 10 (0.4 × 0.25)	24*	20.5	23.5	26.5	29.5	32.5	35	40.5	45
18 × 12 (0.46 × 0.3)	24*	20	23	26	29	33	34.5	40.5	45
20 × 12	24*	19.5	22.5	25.5	28.5	31.5	34.5	40.5	45
(0.51 × 0.3)	18	26	29	32	35	38	40	45	45
24 × 10	24*	19	22	25	28	31	34.5	40.5	45
(0.61 × 0.25)	16	27.5	30.5	33.5	36.5	38	42	45	45
30 × 12	24*	18.5	21.5	24.5	27.5	30.5	34.5	40.5	45
(0.76 × 0.3)	16	27	30	33	36	39	43	45	45
36 × 12	22*	19.5	22.5	25.5	28.5	31.5	36.5	42.5	45
(0.91 × 0.3)	16	26	29	32	35	38	43	45	45
40 × 18	22*	19	22	25	28	30.5	36.5	42.5	45
(1 × 0.46)	16	25.5	28.5	31.5	34.5	37.5	43	45	45
48 × 20	22*	18	21	24	27	30.5	36.5	42.5	45
(1.22 × 0.51)	16	25	28	31	34	36	43	45	45
50 × 30	22*	17.5	20.5	23.5	26.5	30.5	36.5	42.5	45
(1.27 × 0.76)	16	24	27	30	33	37	43	45	45
60 × 24	22*	19	22	25	28	32	38	44	45
(1.52 × 0.61)	16	24	27	30	33	37	43	45	45
72 × 24	20*	18.5	21.5	24.5	27.5	32	38	44	45
(1.83 × 0.61)	16	23	26	29	32	37	43	45	45

*Designate gauges most frequently used for low velocity ducts

Thickness	Gauge	26	24	22	20	18	16
	mm	0.45	0.6	0.76	0.91	1.2	1.5

TABLE 9.10 Reciprocity Relationships for Higher Order Excitation Sources and Responses (after Ref. 63)

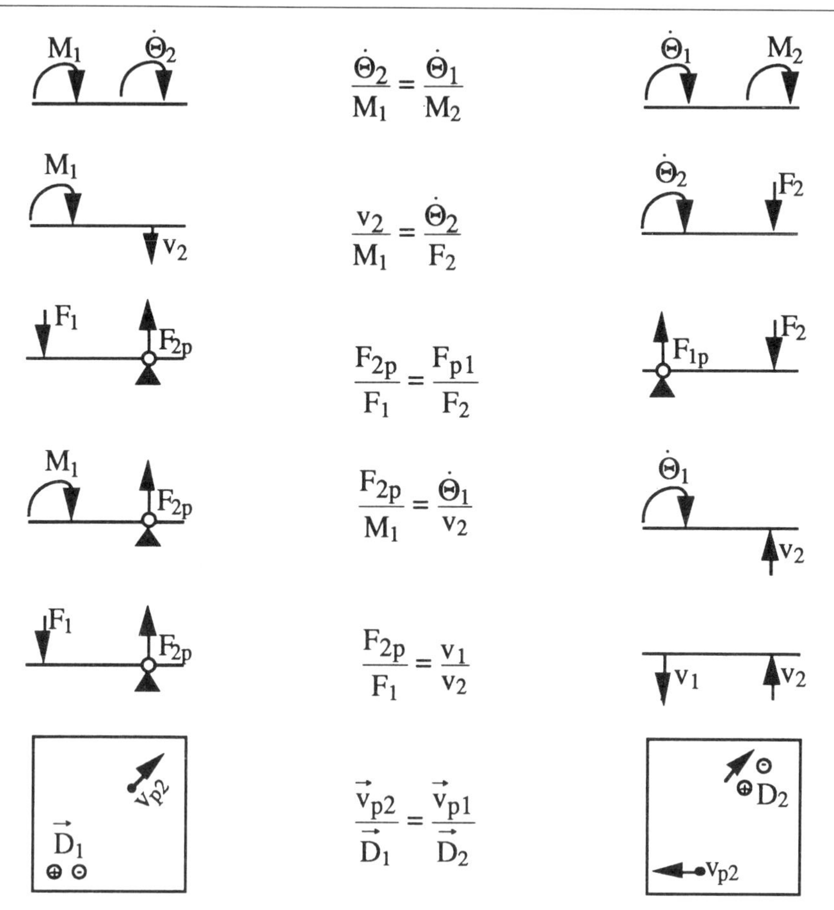

*M = Moment, v = velocity, v_p = particle velocity, $\dot{\Theta}$ = angular velocity, F = Force, F_p = Force acting at pinned restraints, D = dipole sound source.

CHAPTER 10

Passive Silencers and Lined Ducts

ANTHONY G. GALAITSIS and ISTVÁN L. VÉR
Bolt Beranek and Newman, Inc.
Cambridge, Massachusetts

10.1 INTRODUCTION

The noise generated by air/gas handling/consuming equipment, such as fans, blowers, and internal combustion engines, is controlled through the use of two types of devices: (1) passive silencers and lined ducts whose performance is a function of the geometric and sound-absorbing properties of their components and (2) active noise control silencers whose noise cancellation features are controlled by various electromechanical feed-forward and feedback techniques. This section focuses on passive noise control applications while active noise control is discussed in Chapter 15.

In the remainder of this section, the term "silencer" is often used generically to refer to any type of passive noise control device. Specific names are introduced primarily in cases where the use of *silencer* may lead to ambiguities.

Over the last three decades the silencers and lined ducts have been the focus of many research programs that improved our understanding of the basic phenomena and resulted in more accurate design methods. The groundwork for the behavior of lined ducts with no flow was laid even earlier by Morse[1] and Cremer.[2] Similarly, the first systematic evaluation of mufflers with no flow was conducted by Davis et al.[3] Those works were followed by numerous theoretical and experimental inves-

Noise and Vibration Control Engineering: Principles and Applications, Edited by Leo L. Beranek and István L. Vér.
ISBN 0-471-61751-2 © 1992 John Wiley & Sons, Inc.

tigations that addressed additional important issues such as uniform flow, temperature, gradient, and the behavior of new silencer components that are common in modern applications. Reviews of the major accomplishments in this area have been presented periodically in the form of individual articles in professional journals, chapters in engineering handbooks,[4–7] and more recently, a book by Munjal[8] solely devoted to mufflers.

The analytical work on this subject benefited substantially from the early adaptation of the transfer matrix approach to silencer modeling. This method, which was promoted to the description of mechanical systems early on,[9,10] was used along with electrical analogs to describe the behavior of basic silencer components[11,12] for plane-wave sound propagation in the absence of flow. Subsequent investigations began addressing the effects of flow on the response of elements consisting of area discontinuities[13,14] and generating transfer matrix models for silencer elements. The predicted performance of various reactive silencers using transfer matrices with convective and dissipative effects in the presence of flow agreed well with measured data.[15]

Concurrently, systematic studies were initiated on the behavior of perforated metals to take advantage of their dissipative properties[16–18] for general applications in the transportation industry. Such perforates are used in automotive applications as part of the two-pipe and three-pipe elements that contain one and two perforated pipes, respectively, within a larger diameter rigid-wall cylinder (cavity). Such configurations were investigated[19–24] by combining an orifice model with the transfer matrices of the axially segmented perforated tube or tubes and the unpartitioned cavity.

Other approaches used in silencer analysis and design include the finite-element,[25–29] boundary element,[30–31] and acoustical-wave finite-element[32] methods. Each has some inherent advantages over a limited range of applications, but on the average they are more complex to implement. Therefore, the transfer matrix method is presently the most widely used approach.

Experimental work conducted in parallel with the analysis validated the developed models and pointed out areas requiring further investigation. Methods for the direct measurement of transfer matrices were developed[33–36] to validate and refine the models of existing silencer elements. Substantial attention was also paid to the source impedance that was expected to influence the insertion loss of a silencer and the net radiated acoustical power of the system. The source impedance of a six-cylinder engine was measured with an impedance tube using pure tones,[37] and the impedance of electroacoustical drivers and multicylinder engines was subsequently measured[38–41] using the much quicker two-microphone methods.[42,43] More recent indirect methods characterize the source impedance through measurements using two, three, or four different acoustical loads.[44–47]

The research performed and the knowledge developed in the field of passive silencers and lined ducts is very extensive, and it cannot be adequately covered in a single chapter. Accordingly, this chapter reviews the basic background information, summarizes the methodologies for the design of silencers and lined ducts, demonstrates these approaches through selected examples, and discusses additional topics and references useful to general applications. A reader wishing to acquire more information on internal combustion engine silencers is urged to review references 48–50 for a summary of a general automotive silencer development approach and of typical predicted and measured results, 51 and 52 for condensed qualitative reviews of the accomplishments in this field, and 8 for an extensive bibliography on and a detailed discussion of most silencer-related topics.

10.2 GENERAL DISCUSSION

The majority of noise sources have an intake and an exhaust and, generally, require both intake and exhaust silencers. The two cases are characterized by different flow direction, average gas temperature, and average sound pressure levels, but the corresponding hardware is developed using the same principles and design methods. The following discussion is tailored to exhaust silencers but is equally applicable to intake silencers as well.

Silencer Selection Factors

The use of a silencer is prompted by the need to reduce the radiated noise of a source, but in most applications, the final selection is based on trade-offs between the predicted acoustical performance, mechanical performance, and volume/weight and cost of the resulting system.

The acoustical performance (insertion loss) of a candidate silencer is determined from the free-field sound pressure levels measured at the same relative locations with respect to the noise outlet of the unsilenced and silenced sources.

The impact of the silencer on the mechanical performance of the source is determined from the change in the silencer back pressure. For a continuous-flow source, such as a fan or a gas turbine, the impact is determined from the increase in the average back pressure; by contrast, for an intermittent-flow source, such as a reciprocating engine, the impact is a function of the increase in the exhaust manifold pressure when an exhaust valve is open.

Most silencers are subject to volume/weight constraints that also influence the silencer design process. Finally, the initial purchase/installation cost and the periodic maintenance cost are additional important factors that have a similar impact on the silencer selection process.

Factors Influencing Acoustical Performance

Reactive and Dissipative Silencers. The net change ΔW in the acoustical power radiated from a source can be expressed as

$$\Delta W = W_1 - W_2 = W_1 - (W_1' - W_d) = (W_1 - W_1') + W_d \qquad \text{N} \cdot \text{m/s} \tag{10.1}$$

where W_1 and W_2 correspond to the unsilenced and silenced source (Fig. 10.1), $W_1 - W_1'$ is the net change in the acoustical power output of the source resulting

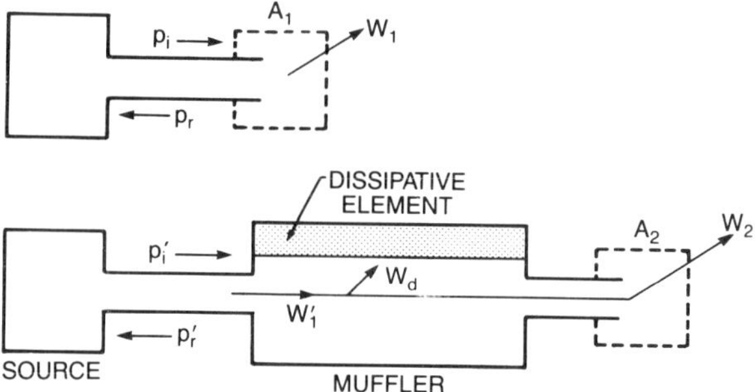

Fig. 10.1 Radiated noise reduction mechanisms of a silencer.

from changes in the silencer's reflection coefficient, and W_d is attributable to the dissipative properties of the silencer.

The reflection coefficient is manipulated primarily by introducing cross-sectional discontinuities in the piping system, and the dissipated part W_d depends largely on the frictional energy loss in the silencer elements. In practice, all silencer elements reflect and dissipate acoustical energy to some extent. However, traditionally, they have been categorized into reactive and dissipative silencers when their insertion loss is dominated by reflective and dissipative mechanisms, respectively. This convention has also been adopted in this chapter, but the reader should be aware that most common applications involve various degrees of overlap between the two extremes.

Flanking Paths and Secondary Source Mechanisms. The acoustical power radiated from a silenced exhaust includes airborne noise from the exhaust pipe outlet (W_{EX}); shell noise radiated from the vibrating walls of the silencer and exhaust line pipes (W_{SH}); and a contribution from additional sources (W_{AD}), such as source shell and source intake, located upstream of the silencer.

The shell noise results from the excitation of silencer components by the vibrating engine, by the intense internal acoustical pressure fields, and by aerodynamic forces. These contributions can be reduced substantially through various methods, but they cannot be totally eliminated, and eventually, they set the limit for the achievable insertion loss for high-performance silencers.

The acoustical power contributions W_{SH} and W_{AD} are excluded from further consideration in the remainder of the chapter; that is, the discussion is focused on the impact of silencers on the airborne noise contributions W_{EX}. However, one should be aware of their existence and significance, particularly in experimental applications where limited resources may lead to the measurement of W_{EX} in the presence of W_{SH} and W_{AD}.

Silencer System Modeling

Silencer Component Representation. The basic components of a typical silencer system comprise the noise source, the silencer, connecting pipes, and the

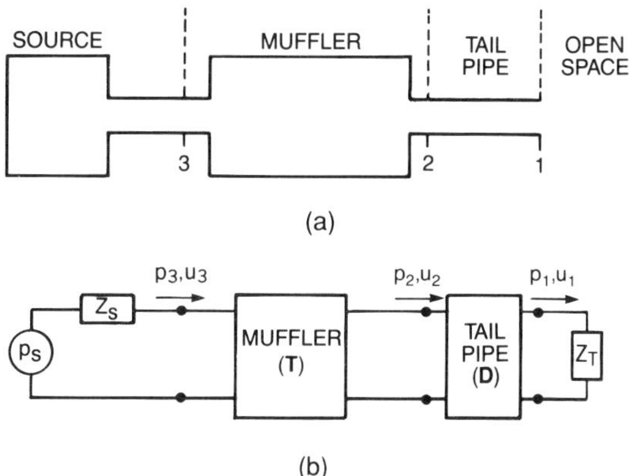

Fig. 10.2 (*a*) Acoustical and (*b*) electric anàlog components of a silencer.

surrounding medium (Fig. 10.2*a*). The pipe connecting the source to the silencer is treated as part of the source. Figure 10.2*b* shows the corresponding electric analog of the acoustical system that is widely used to facilitate the representation and handling of acoustical transmission lines. A comprehensive discussion of electric analogs for acoustical elements may be found in reference 53. This analog model, which uses acoustical pressure p and mass flow $\rho_0 Su$ instead of voltage and current, represents the noise source with a source of pressure p_s and an internal impedance Z_s, the silencer and pipe segment with four-pole elements (T_{ij} and D_{ij}), and the surrounding medium with a load impedance Z_T. Analytical expressions for the transfer matrices of pipes and selected silencer elements are presented in Section 10.3.

Let p_i, u_i, $i = 1, 2, 3$, designate the acoustical pressure and particle velocity at the interfaces of acoustical components of the source–silencer–load system illustrated in Fig. 10.2*b*. These quantities may be obtained by solving the system of equations describing the response of the components depicted in Fig. 10.2*b*, namely,

$$p_s = p_3 + \rho Z_s S_3 u_3 \tag{10.2}$$

$$\begin{bmatrix} p_3 \\ \rho S_3 u_3 \end{bmatrix} = \begin{bmatrix} T_{11} & T_{12} \\ T_{21} & T_{22} \end{bmatrix} \begin{bmatrix} p_2 \\ \rho S_2 u_2 \end{bmatrix} \tag{10.3}$$

$$\begin{bmatrix} p_2 \\ \rho S_2 u_2 \end{bmatrix} = \begin{bmatrix} D_{11} & D_{12} \\ D_{21} & D_{22} \end{bmatrix} \begin{bmatrix} p_1 \\ \rho S_1 u_1 \end{bmatrix} \tag{10.4}$$

$$p_1 = Z_T \rho S_1 u_1 \tag{10.5}$$

where ρ is the mean gas density and S_i is the duct cross-sectional area at the ith location. The calculated quantities p_i, $\rho S_i u_i$, $i = 1, 2, \ldots$, can then be combined with the selected performance criterion to determine the effectiveness of the silencer.

Acoustical Performance Criteria. The most frequently used performance criteria are the insertion loss (IL), noise reduction (NR), and transmission loss (TL). All three use a sound pressure level difference as a performance indicator. Therefore, they require no explicit knowledge of the source strength p_s. The number of required system parameters (source impedance Z_s, transfer matrix elements D_{ij}, and termination impedance Z_T), other than the silencer matrix T_{ij}, can be further reduced by the specific choice of the performance criterion.

The IL is defined as the change in the radiated sound pressure level resulting from the insertion of the muffler, that is, the replacement of a pipe segment of length l_1 (Fig. 10.3*a*) located downstream from the source by the silencer and a new tailpipe segment of length l_2 (Fig. 10.3*b*). Insertion loss is expressed as

$$\text{IL} = L_b - L_a \quad \text{dB}$$

where L_b and L_a are the measured sound pressure levels at the same relative location (distance and orientation) with respect to exhaust outlet (Fig. 10.3*a*) before and after the installation of the muffler. Mathematically,

$$\text{IL} = 20 \log \left| \frac{\tilde{T}_{11}Z_T + \tilde{T}_{12} + \tilde{T}_{21}Z_sZ_T + \tilde{T}_{22}Z_s}{D'_{11}Z_T + D'_{12} + D'_{21}Z_sZ_T + D'_{22}Z_s} \right| \tag{10.6}$$

where c is the speed of sound, S is the cross-sectional area of the reference pipe, $\tilde{T}_{ij}$ is the combined transfer matrix element of the silencer and its tailpipe ($\tilde{T}$ = TD), and D'_{ij} is the transfer matrix element of the replaced exhaust pipe segment. If the silencer is simply added to the end of the source ($l_1 = l_2 = 0$ in Figs. 10.3*a* and 10.3*b*), then both D' and D are identity matrices (see Section 10.3 for a discussion of transfer matrices for reactive elements) and Eq. (10.6) is reduced to

$$\text{IL}\ (l_1 = l_2 = 0) = 20 \log \left| \frac{T_{11}Z_T + T_{12} + T_{21}Z_TZ_s + T_{22}Z_s}{Z_T + Z_s} \right| \tag{10.7}$$

The noise reduction is given by

$$\text{NR} = L_2 - L_1 = 20 \log \left| \frac{p_2}{p_1} \right| \tag{10.8}$$

that is, it is the difference between the sound pressure levels measured at two points 1 and 2 (Fig. 10.3*c*) of the silenced system located upstream and downstream of the silencer, respectively. If D_{ij} and $\tilde{T}_{ij}$ represent the transfer matrix elements for the silencer portions downstream of points 1 and 2, respectively,

$$\text{NR} = 20 \log \left| \frac{\tilde{T}_{11}Z_T + \tilde{T}_{12}}{D_{11}Z_T + D_{12}} \right| \tag{10.9}$$

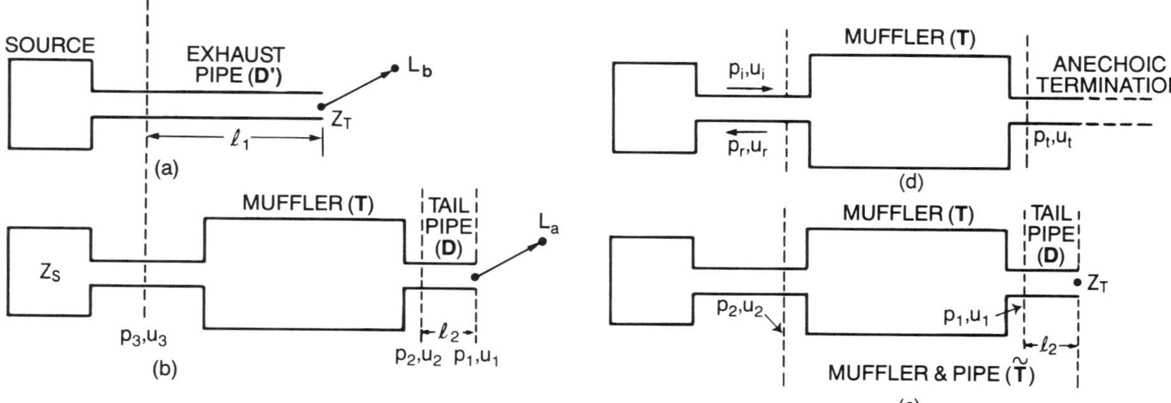

Fig. 10.3 Quantities used for the determination of (*a*, *b*) insertion loss, (*c*) noise reduction, and (*d*) transmission loss.

The TL is the acoustical power level difference between the incident and transmitted waves of an anechoically terminated silencer (see Fig. 10.3*d*). In terms of the corresponding transfer matrix,

$$\text{TL} = 20 \log \left| \frac{T_{11} + \dfrac{S}{c} T_{12} + \dfrac{c}{S} T_{21} + T_{22}}{2} \right| \qquad (10.10)$$

The IL is the most appropriate indicator of a silencer's performance because it is the level difference of the acoustical power radiated from the unsilenced and silenced system. It is easy to quantify from pre- and postsilencing data, but it is hard to predict because it depends on Z_s and Z_T, which vary from one application to another. By contrast, the TL is easy to predict but is only an approximation of the silencer's actual performance because it does not account for the source impedance and models all silencer outlets with anechoic terminations. Equations (10.6) and (10.10) show that the TL and IL of a silencer become identical when the source and silencer termination are anechoic.

The ultimate selection of an evaluation criterion is based on trade-offs between the desired accuracy in the predictions and the amount of available resources. For example, Z_s, required for IL predictions, can be determined experimentally through various methods,[37–47] but the procedure is too costly for the majority of applications. As a result, the silencer design is generally based on predicted TL, with a clear understanding of the associated approximations, while the final evaluation of the hardware during field tests is based on the measured IL.

10.3 REACTIVE SILENCERS

Reactive silencers consist typically of several pipe segments that interconnect with a number of larger-diameter chambers. These silencers reduce the radiated acoustical power primarily through the use of cross-sectional discontinuities that reflect the sound back toward the source. This section discusses the features and models for the most commonly encountered silencer components.

Basic Silencer Elements

Every silencer can be divided into a number of segments or elements each represented by a transfer matrix. The transfer matrices can then be combined to obtain the system matrix, which may then be substituted into Eq. (10.6), (10.9), or (10.10) to predict the corresponding acoustical performance for the silencer system.

The procedure is illustrated by considering the silencer of Fig. 10.4, which is divided into the basic elements, labeled 1–9, indicated by the dashed lines. Elements 1, 3, 5, 7, and 9 are simple pipes of constant cross section. Element 2 is a simple area expansion, element 4 is an area contraction with an extended outlet pipe, element 6 is an area expansion with an extended inlet pipe, and element 8 is

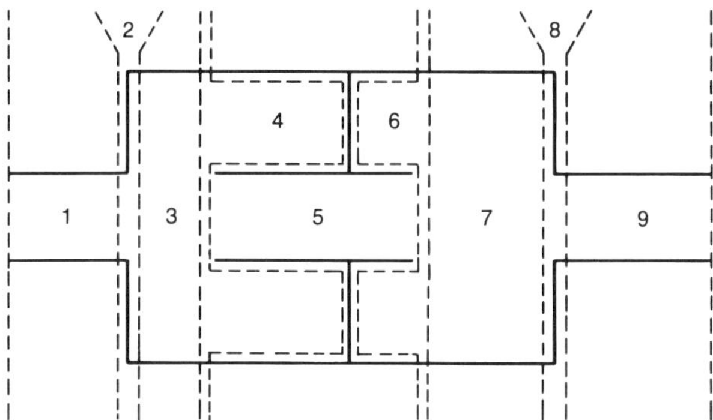

Fig. 10.4 Decomposition of a silencer into basic elements.

a simple area contraction. The nine elements are characterized by the transfer matrices $T^{(1)}, \cdots, T^{(9)}$. Therefore, the system matrix $T^{(S)}$ for the entire silencer is obtained from

$$T^{(S)} = T^{(1)}T^{(2)} \cdots T^{(9)} \tag{10.11}$$

through matrix multiplication. The matrices for each of the above elements may be derived from the formulas presented later in this section.

Common reactive muffler elements are shown in Fig. 10.5. In addition to the plane pipe, extended inlet expansion, and extended outlet contraction elements that were encountered above, these elements include a resonator, that is, a cavity-backed opening in the wall of the pipe (d), a reversal expansion element (e), and a reversal contraction element (f). The simple expansion and contraction elements of Fig. 10.4 are special cases of configurations (b) and (c) of Fig. 10.5.

Figure 10.6 depicts additional elements that are used extensively in passive silencers. All of these elements have perforated internal piping utilizing the resistive properties of the perforations to increase the noise reduction through acoustical energy dissipation. Typical elements include extended perforated inlet and outlet

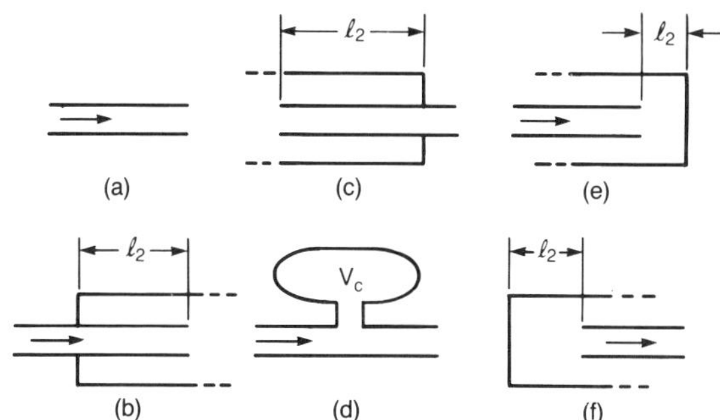

Fig. 10.5 Plain tubular silencer elements.

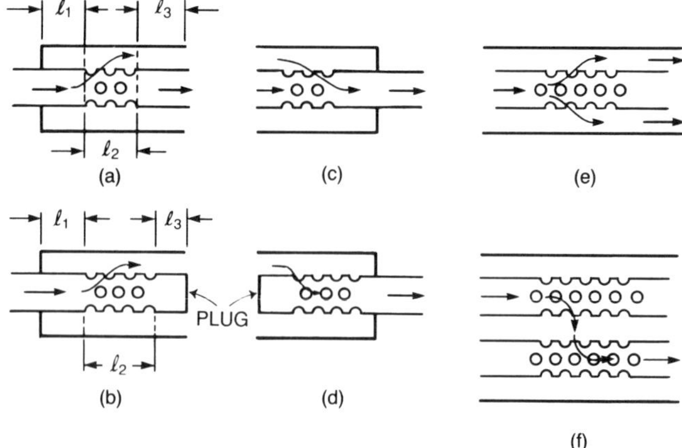

Fig. 10.6 Perforated tubular silencer elements.

pipes (a, c) and perforated-plug expansions and contractions (b, d), which can be thought as combinations of plain pipes and transition elements (Fig. 10.5) with the two-duct perforated element[19–23] (e). These basic elements can be used to generate a large number of special configurations[48] by adjusting the pipe lengths l_1, l_2, and l_3 as well as the porosity (size and density) of the perforated holes. More complicated configurations can be obtained using the three-duct element shown in Fig. 10.6f.

Transfer Matrices for Reactive Elements

The transfer matrix of a silencer element is a function of the element geometry, state variables of the medium, mean flow velocity, and properties of duct liners, if any. The results presented below correspond to linear sound propagation of plane waves in the presence of superimposed flow. In certain cases, the matrix may also be influenced by nonlinear effects, higher order modes, and temperture gradients; these latter effects, which can be included in special cases, are discussed qualitatively later in this section but are excluded from the analytical procedure described below. The following is a list of variables and parameters that appear in most transfer matrix relations of reactive elements:

p_i = acoustical pressure at ith location of element
u_i = particle velocity at ith location of element
ρ = mean density of gas, kg/m
c = sound speed, $= 331\sqrt{\theta/273}$ m/s
θ = absolute temperature K, K = °C + 273
S_i = cross section of element at ith location, m^2
$Y_i = c/S_i$
A, B = amplitudes of right- and left-bound fields

$k_c \cong k/(1 - M^2)$, assuming negligible frictional energy loss along straight pipe segments
$k = \omega/c = 2\pi f/c$
$\omega = 2\pi f$
f = frequency, Hz
$M = V/c$
V = mean flow velocity through S
$T_{i,j}$ = element of transmission matrix

Symbols without subscripts, such as V, c, and M, describe quantities associated with the reference duct.

Pipe with Uniform Cross Section. The acoustical pressure and mass velocity fields p, ρSu in a pipe (Fig. 10.7) with uniform cross section S and a mean flow V are given by

$$p = (Ae^{-jk_c x} + Be^{-jk_c x})e^{-(Mk_c x - \omega t)} \tag{10.12}$$

$$\rho Su = (Ae^{-jk_c x} - Be^{-jk_c x})\frac{e^{-(Mk_c x - \omega t)}}{Y} \tag{10.13}$$

where S is the cross section of the pipe. Equations (10.12) and (10.13) can be evaluated at $x = 0$ and $x = l$ to obtain the corresponding fields p_2, ρSu_2 and p_1, ρSu_1, respectively. Upon elimination of the constants A and B, one obtains

$$\begin{bmatrix} p_2 \\ \rho Su_2 \end{bmatrix} = \begin{bmatrix} T_{11} & T_{12} \\ T_{21} & T_{22} \end{bmatrix} \begin{bmatrix} p_1 \\ \rho Su_1 \end{bmatrix} \tag{10.14}$$

where the transmission matrix T_{pipe} is given by

$$T_{\text{pipe}} = \begin{bmatrix} T_{11} & T_{12} \\ T_{21} & T_{22} \end{bmatrix}_{\text{pipe}} = e^{-jMk_c l} \begin{bmatrix} \cos k_c l & jY_0 \sin k_c l \\ \dfrac{j}{Y} \sin k_c l & \cos k_c l \end{bmatrix} \tag{10.15}$$

In the transfer matrix of Eq. (10.15) the acoustical energy dissipation that may result from friction between the gas and the rigid wall as well as from turbulence

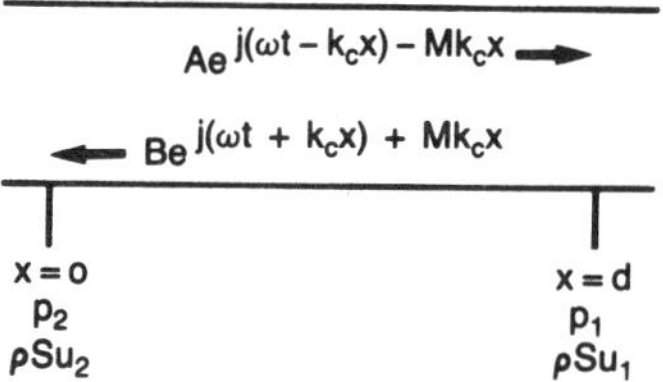

Fig. 10.7 Acoustical pressure fields in a plain tubular pipe.

is neglected. These effects, which would result in a slightly different matrix,[18] may be noticeable in very long exhaust systems but are negligible for most silencer applications.

Cross-Sectional Discontinuities. The transition elements used to model most cross-sectional discontinuities are shown in Fig. 10.5 and in the first column of Table 10.1. Using decreasing element-subscript values with distance from the noise source, the cross sectional areas upstream, at, and downstream of the transition (S_3, S_2, and S_1) are related through[8]

$$C_1 S_1 + C_2 S_2 + S_3 = 0 \tag{10.16}$$

where the constants C_1 and C_2 (Table 10.1) are selected so as to satisfy the mass conservation equation across the transition.

Table 10.1 also shows the pressure loss coefficient K for each configuration that accounts for the conversion of some mean-flow energy and acoustical field energy into heat at the discontinuities. As indicated, $K \leq 0.5$ for area contractions, while $K \to (S_1/S_3)^2$ for area expansions at large values of S_1/S_3.

Transfer matrices for cross-sectional discontinuities (csd) in the presence of mean flow that include terms proportional up to the fourth power (M^4) of the Mach number are presented in reference 8. However, in most silencer design applications $M << 1$. Therefore, terms of the form $1 \pm M^n$, $n \geq 2$, are set to unity. This simplification reduces the matrix T_{csd}, which relates the upstream and downstream acoustical fields, p_3, $\rho S u_3$ and p_1, $\rho S_1 u$, through

$$\begin{bmatrix} p_3 \\ \rho S_3 u_3 \end{bmatrix}_{\text{upstream}} = T_{\text{csd}} \begin{bmatrix} p_1 \\ \rho S_1 u_1 \end{bmatrix}_{\text{downstream}} \tag{10.17}$$

Table 10.1 Parameter Values of Transition Elements

ELEMENT TYPE	C_1	C_2	K
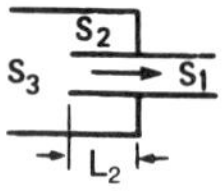 S_2, $S_3 \rightarrow S_1$, L_2	−1	−1	$\dfrac{\left(1-\dfrac{S_1}{S_3}\right)}{2}$
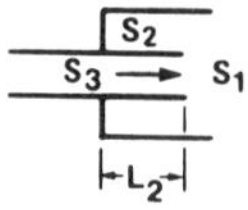 S_2, $S_3 \rightarrow S_1$, L_2	−1	1	$\left(\dfrac{S_1}{S_3}-1\right)^2$
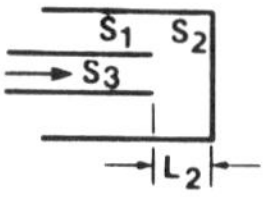 S_1 S_2, $\rightarrow S_3$, L_2	1	−1	$\left(\dfrac{S_1}{S_3}\right)^2$
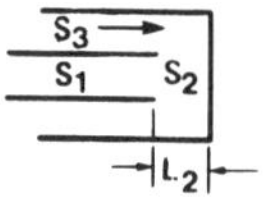 $S_3 \rightarrow$, S_1 S_2, L_2	1	−1	0.5

to

$$[T_{csd}] = \begin{bmatrix} 1 & KM_1Y_1 \\ \dfrac{C_2S_2}{C_1S_2Z_2 + S_2M_3Y_3} & \dfrac{C_2S_2Z_2 - M_1Y_1\{C_1S_1 + S_3K\}}{C_2S_2Z_2 + S_3M_3Y_3} \end{bmatrix} \tag{10.18}$$

where

$$Z_2 = -j\frac{c}{S_2}\cot kl_2 \tag{10.19}$$

l_2 being the length of extended inlet/outlet pipe in meters.

Resonators. A resonator is a cavity-backed opening in the sidewall of a pipe (Fig. 10.8). The opening may consist of a single hole on the pipe wall (*a*) or a closely distributed group of holes (*b*). The cavity behind this opening can be a straight pipe of cross-sectional area S_c and depth l_c (*c*), a concentric cylinder extending a length l_1 and l_2 upstream and downstream of the opening (*d*), or an odd-shaped chamber of total volume V_c (*e*).

The resonator opening in the wall of the main duct is assumed to have an axial dimension that is much smaller than the wavelength and typically smaller than a duct diameter. This requirement ensures that the duct–cavity interaction is in phase over the entire surface of the connecting opening. Multihole openings with substantial axial dimensions must be modeled using the elements of Fig. 10.6*a*.

The transfer matrix for a resonator

$$T_r = \begin{bmatrix} 1 & 0 \\ \dfrac{1}{Z_r} & 1 \end{bmatrix} \tag{10.20}$$

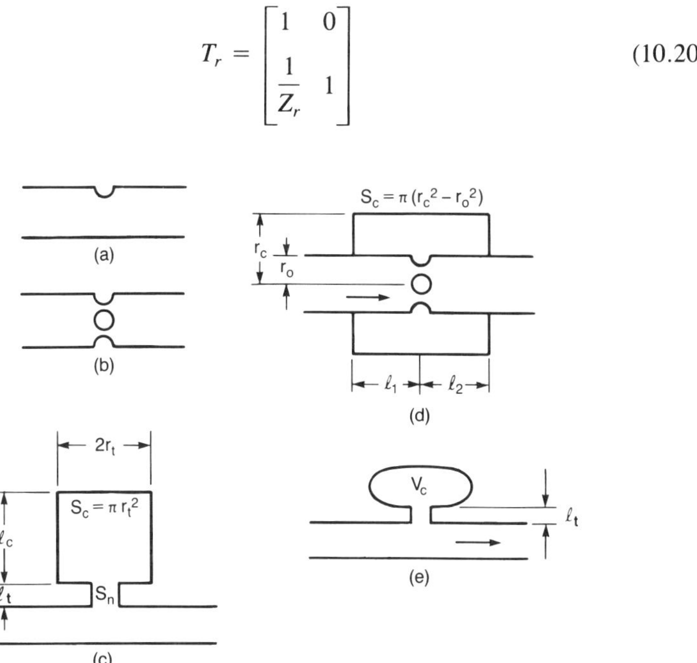

Fig. 10.8 Resonator components.

where $Z_r = Z_t + Z_c$, Z_t being the impedance of the throat connecting the pipe to the cavity and Z_c the impedance of the cavity. Here Z_c is independent of the flow in the main duct and is given by one of the following expressions:

$$Z_c = \begin{cases} Z_{\mathrm{tt}} = -j\dfrac{c}{S_c}\cot kl_c \\[2ex] Z_{\mathrm{cc}} = -j\dfrac{c}{S_c}\dfrac{1}{\tan kl_1 + \tan kl_2} \\[2ex] Z_{\mathrm{gv}} = -j\dfrac{c}{kV_c} \end{cases} \qquad \mathrm{m \cdot s^{-1}} \qquad (10.21)$$

where the subscripts tt, cc, and gv refer to the transverse tube, concentric cylinder, and general volume cavities, which are illustrated in Fig. 10.8 along with the lengths l_1, l_2, and l_c, the cross section S_c, and the volume V_c. It should be noted that at low frequencies ($kl_c \ll 1$ and kl_1, $kl_2 \ll 1$) both Z_{tt} and Z_{cc} are reduced to the expression of Z_{gv}.

The impedance Z_t of the throat connecting the duct to the cavity changes dramatically with grazing flow. Therefore, it is characterized by two sets of values. For $M = 0$, this quantity is given by

$$Z_t^{[M=0]} = \frac{1}{n_h}\left(\frac{ck^2}{\pi} + j\,\frac{ck(l_t + 1.7r_0)}{S_0}\right) \qquad \mathrm{m \cdot s^{-1}} \qquad (10.22)$$

where l_t is the length of the connecting throat, r_0 the orifice radius, S_0 the area of a single orifice, and n_h the total number of perforated holes. In the absence of mean flow this expression has yielded good agreement between predicted and measured results for single-hole and well-localized multihole resonators.

The presence of grazing flow ($M \neq 0$ in the duct) has a strong effect on the impedance Z_t of the resonator throat. Measurements conducted using single- and multihole throats[16,55] have led to the empirical expression

$$Z_t^{[M\neq 0]} = \frac{c}{n_h S_0}\,[7.3 \times 10^{-3}\,(1 + 72M) + j2.2 \times 10^{-5}[1 + 5l_t]\,[1 + 408\rho]] \qquad \mathrm{m \cdot s^{-1}} \qquad (10.23)$$

where the parameters l_t and r_0 must be furnished in meters. This equation is more appropriate for predictions in the presence of grazing flow.

Perforated Tubular Elements

The basic building blocks of perforated-tube silencers (Fig. 10.6) feature a cross flow between the interior and exterior spaces of the perforated tubes, which can vary significantly from one application to another (Fig. 10.9). Specifically, there

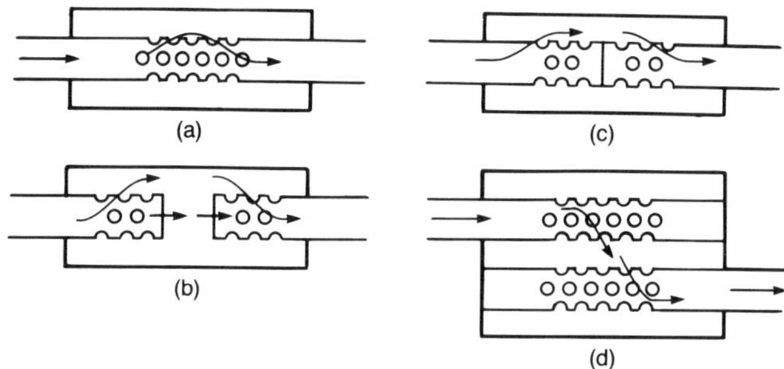

Fig. 10.9 Muffler designs utilizing perforated tubular elements.

is a position-dependent cross flow along the perforated segment of a concentric perforated-tube resonator (Fig. 10.9*a*) but its total net value is zero. By contrast, a perforated extended inlet or outlet (Fig. 10.9*b*) involves a partial but nonzero net cross flow through its wall, while the perforated plug inlet or outlet (Figs. 10.9*c*, *d*) forces the entire flow through its perforated segments.

The derivation of the transfer matrices for the perforated-tube elements has been presented in the literature,[8,24] but it is too lengthy and cannot be sufficiently condensed for inclusion in this chapter. Those interested in designing perforated-tube silencers are directed to Chapter 3 of reference 8, which discusses the subject extensively and includes a list of relevant references.

Termination and Source Impedances

The utilization of insertion loss as a criterion of silencer performance requires, in addition to the transfer matrix T of the silencer [Eq. (10.6)], the source impedance Z_s and the termination impedance Z_T.

The termination impedance $Z_R(M)$ can be expressed in terms of its zero-mean-flow counterpart. The derivation of $Z_T(0)$ of an unflanged pipe is given in reference 54. In the low-frequency limit $[(kr_d)^2 \ll 1]$, $Z_T(0)$ of a flanged pipe is approximated by[8]

$$Z_T(M) = Y_d \left\{ (\tfrac{1}{2}kr_d)^2 - M[1 + (kl_e)^2] + jkl_e \right\} \tag{10.24}$$

where $l_e = 0.61r_d$ is an end correction, $Y_d = c/S_d$, and r_d, S_d are the radius and cross-sectional area of the termination duct.

The characterization of the source impedance has been the subject of many investigations and has been measured through various methods.[37–47] The underlying objective of such measurements has been to use the insertion loss, which requires both termination and source impedance, for the proper characterization of the silencer performance. In most applications, however, the silencer design is based on the predicted transmission loss [Eq. (10.10)], which requires no knowledge of Z_s and Z_T. The insertion loss is used only in special cases where stringent noise requirements justify the extra cost of measuring and using the source impedance.

Discussion of Approximations

The matrix elements presented in Eqs. (10.14), (10.18), and (10.20) include the effects of element geometry, convection (wavenumber and discharge losses at discontinuities), and friction and radiation losses for the throat impedance of the resonators but exclude possible effects of high-order modes, nonlinear response, and temperature gradients. The temperature effects are either those associated with the elevated mean temperature or those resulting from the temperature gradients. An elevated mean temperature increases the sound speed and therefore the wavelengths corresponding to a given frequency. These effects are properly accounted for by substituting the mean temperature into the sound speed formula presented earlier in this section.

Large temperature drops (hundreds of degrees Kelvin) across a muffler are expected to have a significant effect because they result in substantially different sound speeds from one end to the muffler to the other. Proper accounting of these effects requires a substantial amount of analytical work as seen from the transfer matrix derivation for a plain pipe with a temperature gradient.[56,57] However, the effects of such gradients have been shown to be insignificant[58] in typical applications. Therefore, the features that influence silencer design include mainly the mean gas temperature.

The high sound pressure levels encountered in exhaust systems sometimes distort the linear response of silencer components. These effects tend to raise[59,60] the acoustical resistance of various elements (such as the throat resistance of a resonator), which broadens and flattens the spectral peaks of the silencer's acoustical performance. However, the overall character of the performance curve is not altered, and the initial silencer design can still be based on the linear theory.

The transmission matrices presented thus far apply to plane-wave sound propagation through a moving medium. Similar matrices may also be derived for higher order modes that are consistent with the boundary condition of each silencer element. Each higher mode has a cutoff frequency below which it is highly attenuated[61]. Therefore, the low-frequency performance of the silencer is dominated by the plane-wave mode, which is always present. Since the cutoff frequencies are inversely proportional to the duct diameter, it follows that higher modes appear in the large-diameter body of the muffler at much lower frequencies than in the smaller diameter inlet/outlet pipes.

Above its cutoff, a high-order mode may become as important as the plane-wave mode. The contributions of the first few modes can be reduced through an appropriate radial and angular shift of the inlet pipe with respect to the outlet pipe,[62,63] but these measures become less effective as the number of modes increases at higher frequencies. Under these circumstances, if the design is based on plane-wave theory, the designer should at least identify the cutoff frequencies and be prepared for a decreasing accuracy in the predictions above such cutoff frequencies.

Performance of Plain Tubular Element Mufflers

The major steps in the acoustical evaluation of a silencer design are demonstrated through a detailed analysis of a few basic designs composed of the plain tubular elements that are found in most commercially available mufflers. More complex designs, with additional plain and/or perforated-tube elements, which lead to a broader range of silencer configurations, can be analyzed through the same procedure.

The acoustical evaluation of the various designs is based on the predicted transmission loss (TL_p) rather than on the insertion loss (IL_p) to bypass the requirements for source impedance measurements. Usually, this assumption results in significant differences between TL_p and IL_m, the predicted and measured acoustical performance of prototype silencers, and this requires a design iteration after the experimental evaluation of the baseline silencer.

The silencer configurations selected for the examples (Fig. 10.10) are a transverse-tube resonator (*a*), a single-chamber muffler (*b*), and a double-chamber muffler with an internal connecting pipe (*c*). The frequency and the flow and geometric parameters are expressed in terms of normalized quantities to extend the validity of the results over a broader range of silencer sizes and operating conditions.

The normalized frequency is $kL = 2\pi fL/c$, where the length L corresponding to each configuration is shown in Fig. 10.10 and the flow parameter is the Mach number M. Each silencer involves additional geometric parameters that increase in number with the complexity of the design. Even for the relatively simple designs of Fig. 10.10 these parameters include $N = S_2/S_1$, R_2/L for the transverse tube; N, L_1/L, L_2/L, R_1/L for the single-chamber silencer; and N, L_1/L, L_2/L, R_1/L, L_3/L, L_4/L, and L_5/L for the double-chamber silencer. The parameter list of the latter two cases would increase further if the diameters of the inlet, outlet, and inner-connecting pipes were not identical.

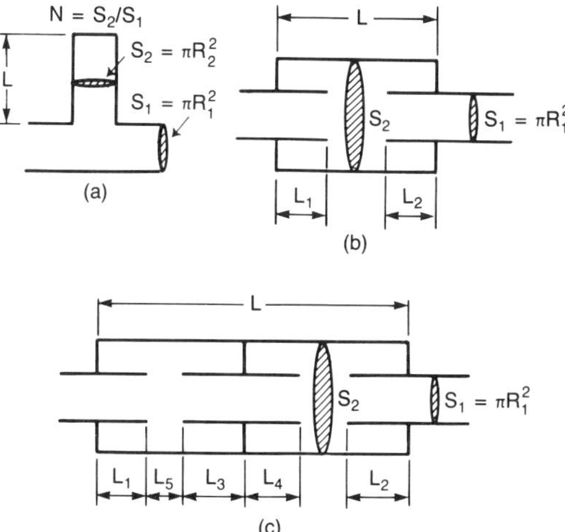

Fig. 10.10 Simple silencers composed of plain tubular elements.

***Transverse-tube Resonator:* $M = 0$.** The transverse-tube resonator offers a simple arrangement for the cancellation of noise over narrow frequency bands. Its transmission loss is obtained by combining Eq. (10.10) with Eqs. (10.20)–(10.23). In the absence of flow ($M = 0$) and as long as $kR_2 \ll 1$ and $R_2/L \ll 1$, the resonator response is dominated by the compliance of the air column in the tube [Eq. (10.21)] and the transmission loss displays sharp peaks at $kL = (n + \frac{1}{2})\pi$, $n = 0, 1, \ldots$. Figure 10.11 shows that the width of these peaks increases with the area ratio N.

The transverse-tube or quarter-wavelength resonator can be very effective in the cancellation of steady tones such as those produced by engines operated at constant speed. However, it has limited applications because most sources emit broadband and time-varying noise and because its performance can be affected significantly by grazing flow. The utilization of such undamped resonators in high-Mach-number environments is usually avoided to prevent possible conversion of steady-flow energy into acoustical energy—as in an organ pipe—by the aerodynamic interaction between the flow and the resonator cavity.

The results of Fig. 10.11 correspond to a resonator that is modeled exclusively by the impedance Z_{tt} of Eq. (10.21). If the cavity is connected to the main duct through a perforated facing, then the contribution of the perforations [Eq. (10.22) or (10.23)] must also be included in the transverse-tube impedance before the resonator matrix [Eq. (10.20)] is evaluated. Inclusion of the additional parameters is straightforward but increases the number of resulting combinations, as illustrated by the next two examples.

Single-chamber Silencer. The geometry of a single-chamber silencer (Fig. 10.10*b*) is defined by the parameters $N = S_2/S_1$, L_1/L, L_2/L, and R_1/L. Predic-

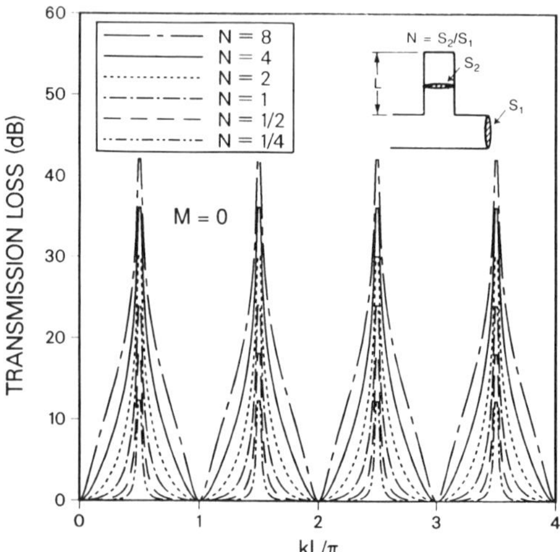

Fig. 10.11 Predicted transmission loss of a transverse tube resonator.

tion of the corresponding transmission loss utilizes three of these parameters along with the Mach number M and the normalized frequency kL; the fourth parameter, R_1/L, which affects the pipe end correction, has no effect because the corresponding correction is added to the assumed infinitely long inlet and outlet pipes. The selected configurations of the following examples include a set of designs with a simplified geometry and a few designs with particularly desirable performance features.

The first group, which corresponds to $L_1/L = L_2/L$, focuses the search on symmetrical designs, (i.e., with extended inlet and outlet pipes of equal length). The predicted transmission loss of this configuration for four values of L_2/L in the range of $0 \leq L_2/L \leq 0.3$ is presented in Fig. 10.12. Similar predictions performed for $M \leq 0.1$ showed that the transmission loss of these designs is not affected appreciably by flows at least as high as $M = 0.1$ (in the pipe); therefore, all Fig. 10.12 plots have been labeled with $M \leq 0.1$.

The four plots of Fig. 10.12 correspond to $L_2/L = 0, 0.2, 0.25$, and each curve corresponds to an area ratio in the range $2 \leq N \leq 64$. For $L_2/L = 0$ the predicted transmission loss displays minima and maxima at $(kL)_{min} = n\pi$ and $(kL)_{max} = (n + \frac{1}{2})\pi$, $n = 0, 1, \ldots$, which correspond to modes that are well and poorly supported, respectively, by the one-dimensional geometry of the chamber. For $L_2/L \neq 0$ the transmission loss also develops a dominant peak near $(kL)_{peak} = (n + \frac{1}{2})\pi L/L_2$, that is, when the extended pipe is equal to an odd number of quarter-wavelengths. Particularly desirable performance features are obtained by selecting L_2/L so as to align the $(kL)_{peak}$ with one of the $(kL)_{min}$, as indicated by the $L_2/L = 0.25$ case (Fig. 10.12*c*), which exhibits a significantly high and broadband peak. This effect is somewhat less pronounced in practical cases because of manufacturing tolerances that affect the hardware symmetry and the exact lengths of the extended inlet/outlet pipes.

The next group of single-chamber silencers feature extended inlet/outlet pipes of different lengths ($L_1 \neq L_2$). Therefore, their definition requires an additional parameter that increases further the number of realizable configurations. The present discussion is limited to a few specific designs with particularly favorable acoustical performance features.

As in the case of the symmetric silencers, the transmission loss of an asymmetric single-chamber silencer displays the same minima and maxima at $(kL)_{min} = n\pi$ and $(kL)_{max} = (n + \frac{1}{2})\pi$ that are associated with the expansion chamber length. Concurrently, it features pronounced peaks at $(kL)_{peak1} = (n + \frac{1}{2})\pi$ and $(kL)_{peak2} = (n + \frac{1}{2})\pi L/L_2$ for $n = 0, 1, 2, \ldots$. Consequently, one may obtain designs with favorable acoustical performance by using some of the peaks associated with $(kL)_{peak1}$ and/or $(kL)_{peak2}$ to fill up the valleys of $(kL)_{min}$.

The predicted transmission loss for a few asymmetric silencers is illustrated in Figs. 10.13 and 10.14. Figure 10.13 corresponds to an extended outlet muffler ($L_1 = 0$) with $M = 0$ and Fig. 10.14 corresponds to an extended outlet muffler with $4L_1 = 2L_2 = L$ and $M = 0$ or $M = 0.1$. Fig. 10.13*a* shows the typical form of the pronounced peak resulting when a $(kL)_{peak1}$ is aligned with a $(kL)_{max}$. Figures 10.13*b*–*c* show the predicted transmission loss for $L_1/L = 0.25, 0.5$, which offer

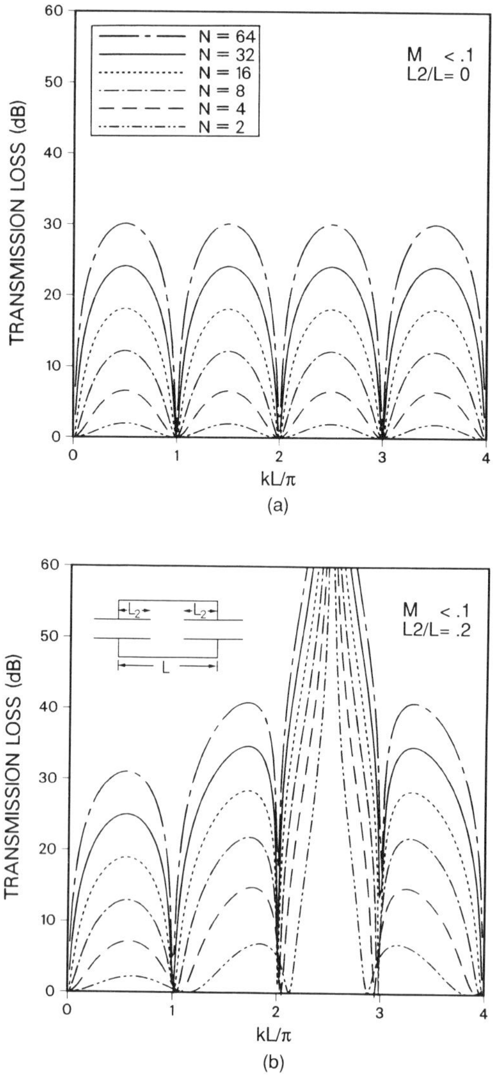

Fig. 10.12 Predicted transmission loss of symmetric single-chamber silencers.

good broadband performance by eliminating the valleys at $kL/\pi = 1, 2$, respectively.

Figure 10.14 shows the predicted transmission loss of a single-chamber silencer with $L_1/L = 0.5$ and $L_2/L = 0.25$ as well as its estimated performance variation with steady-flow changes and with geometric imperfections caused by manufacturing tolerances. Figure 10.14*a* shows the predicted transmission loss when $M = 0$; this design exhibits the best overall broadband performance features among all single-chamber silencers. When the Mach number is raised to $M = 0.1$, the

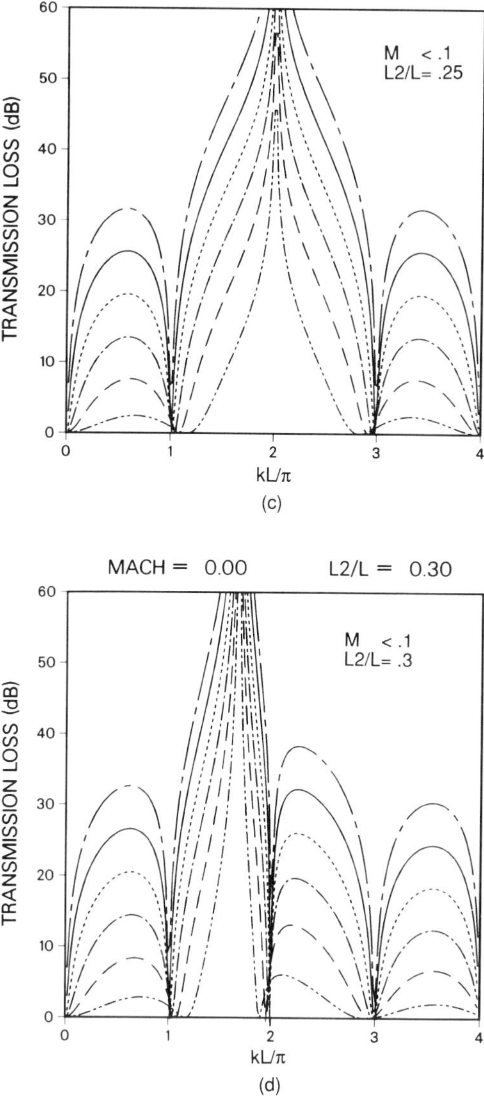

Fig. 10.12 *(Continued)*

transmission loss is distorted in the vicinity of $kL = n\pi$ and is decreased near $kL = 2\pi$ (Fig. 10.14*b*). This design exhibits a higher sensitivity to the Mach number than the previous single-chamber examples. Figure 10.14*c* is similar to Fig. 10.14*b* but has $L_2/L = 0.505$ instead of $L_2/L = 0.5$, which corresponds to a 1% manufacturing error in the length of the extended segment of the outlet pipe. This geometric imperfection introduces an additional distortion as indicated by the predicted transmission loss changes (Fig. 10.14*c*) in the vicinity of $kL/\pi = 3$.

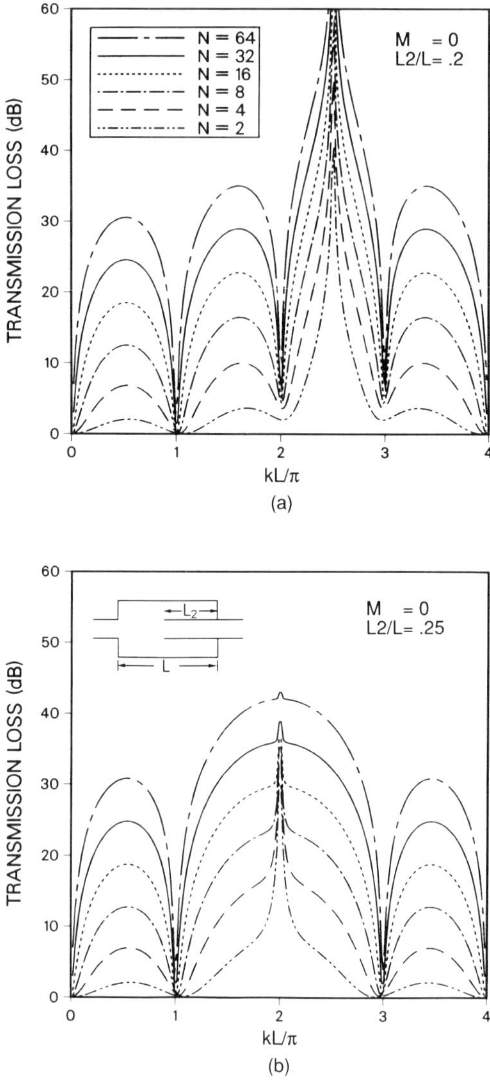

Fig. 10.13 Predicted transmission loss of asymmetric single-chamber silencers with extended outlets.

The variations shown in Fig. 10.14 are typical of the ones encountered in practical applications. Such variations coupled with modeling assumptions (e.g., omission of higher order modes) and procedural simplifications (e.g., use of transmission loss rather than insertion loss) tend to reduce some of the predicted benefits. Therefore, it is emphasized that predicted results such as those in Figs. 10.12–10.14 may be used to identify general trends, but in practical applications, the actual benefits must also be confirmed through direct measurements.

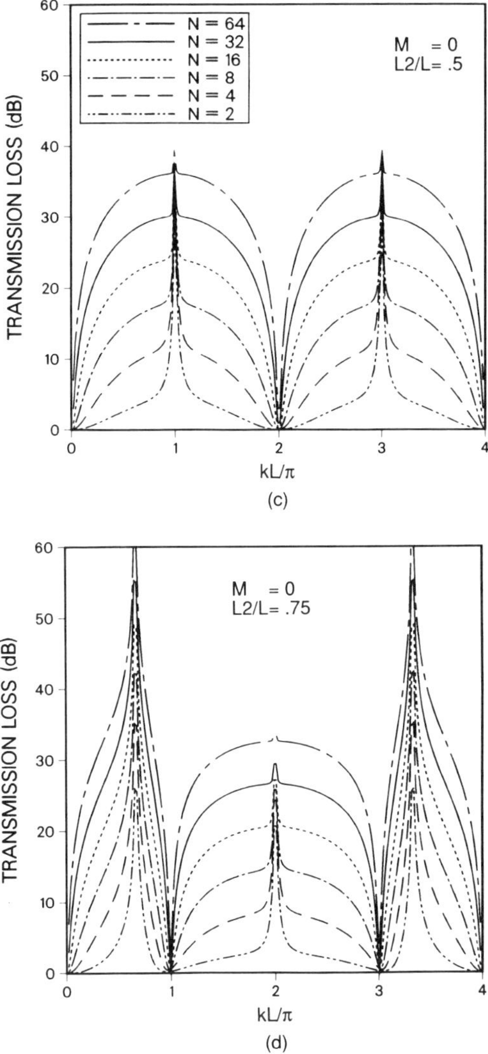

Fig. 10.13 *(Continued)*

Double-chamber Silencer with Inner Connecting Pipe. The parameters required to predict the transmission loss for a muffler of this design (Fig. 10.10*c*) include $N = S_2/S_1$, L_1/L, L_2/L, R_1/L, and M, which are the same as for the single-chamber muffler, and L_3/L, L_4/L, and L_5/L, which define the lengths of the inner-connecting pipe segments and the location of the inner partition. Some important performance features are illustrated by focusing on the examples shown in Fig. 10.15, which are characterized by $2(L_1 + L_5 + L_3) = L$, that is, corresponding to silencers with two chambers of equal lengths. The designs of Figs.

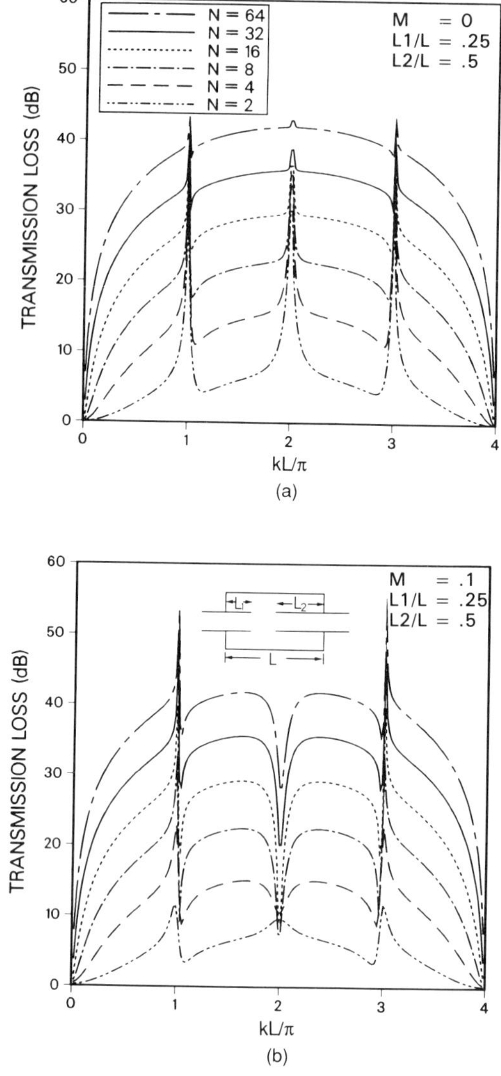

Fig. 10.14 Predicted transmission loss of a specially tuned, asymmetric, single-chamber silencer with $L_1/L = 0.25$

10.15*a*, *d* have $L_1 = L_2 = 0$ and $L_3 = L_4$ (i.e., are symmetric with respect to the inner partition) and those of Figs. 10.15*e–f* have $L_1 = L_4 = 0$ (i.e., have zero-length extended inlets in both chambers). Figures 10.15*a*, *d* offer some insight on the effect of M, L_3/L, and R_1/L on the predicted transmission loss, and Figs. 10.15*c–f* illustrate the favorable predicted performance features of certain configurations. Predictions for additional double-chamber designs are found in reference 4.

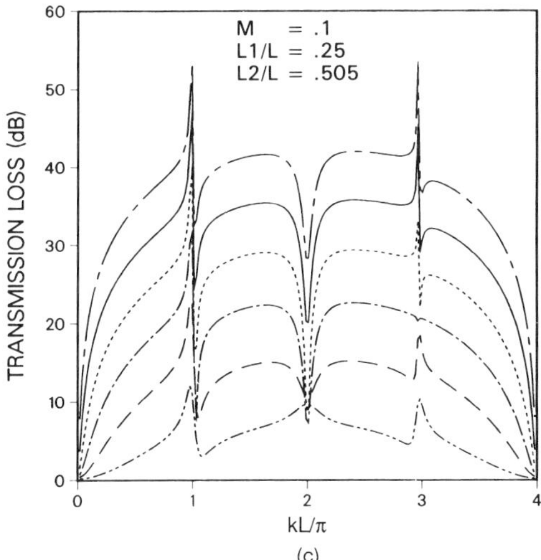

Fig. 10.14 (*Continued*)

The parameter R_1/L specifies the end correction δ,[3] which changes the effective "acoustical" length of the connecting pipe from $L_3 + L_4$ to $L_3 + L_4 + 2\delta$. The end correction may be omitted when $L_3 + L_4$ is greater than several pipe diameters ($L_3 + L_4 >> \delta$) and $k\delta << 1$ over the frequency range of interest but must be included when it becomes comparable to or greater than L_2 or at high frequencies. In fact, when the actual length of the connecting pipe vanishes ($L_3 + L_4 = 0$), its effective acoustical length becomes 2δ.

Figure 10.15*a* shows the transmission loss of a double-chamber silencer with $M = 0$, $L_1 = L_2 = L_3 = L_4 = 0$, and $\delta/L = 0.025$, which is used as a baseline design in the following discussion. The selected δ/L value is somewhat arbitrary but is within the range of magnitudes encountered in commercially available silencers. If the end correction is neglected, then all the predicted curves of Fig. 10.15*a* become identical to those of Fig. 10.12*a*. Comparison of these two figures over the indicated frequency range shows that as the frequency increases, the even-numbered lobes of the spectra grow in amplitude and spread over frequency at the expense of the odd-numbered lobes, which shrink and gradually disappear. These changes illustrate the importance of including the end correction when $L_3 + L_4 \rightarrow 0$.

Figure 10.15*b* is obtained by changing the Mach number of the baseline design from $M = 0$ to $M = 0.1$. This change tends to fill up the valleys between the first and second lobes and between the third and fourth lobes of the predicted transmission loss curves, which improves overall performance, particularly at high values of N over the indicated frequency range.

Figues 10.15*c*, *d* are for a symmetric double-chamber silencer with the inner-connecting pipe extending equally on both sides of the central partition while the

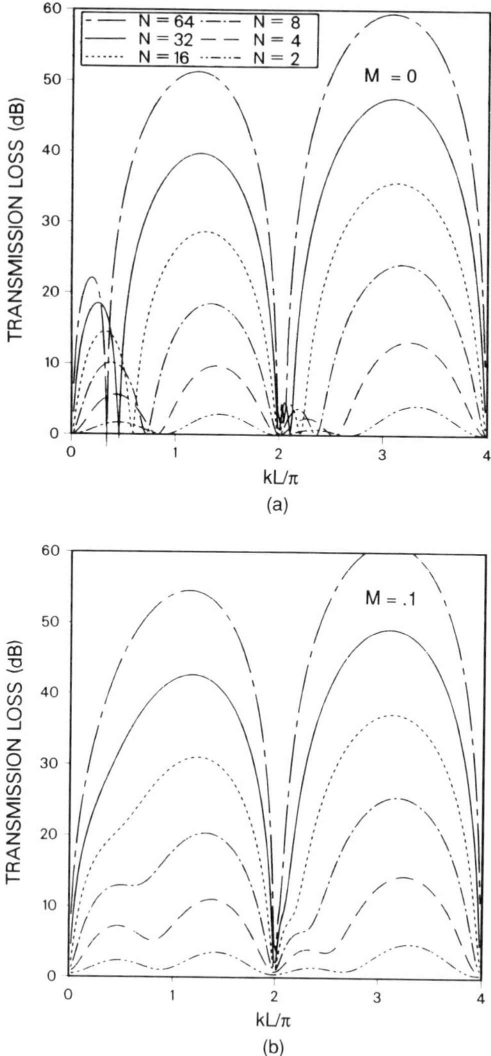

Fig. 10.15 Predicted performance of double-chamber silencers that (*a*, *b*) $L_1 = L_2 = L_3 = L_4 = 0$, $L_5/L = 0.5$, and $\delta/L = 0.025$; (*c*, *d*) $L_1 = L_2 = \delta = 0$ and $L_3 = L_4 = L_5 = \frac{1}{4}L$; (*e*, *f*) $L_1 = L_4 = \delta = 0$ and $L_3 = L_2 = L_5 = \frac{1}{4}L$ (see Fig. 10.10*c*).

second one (Figs. 10.15*e*, *f*) features extended outlets in both chambers. In all three cases the predicted results are shown for $M = 0$ and $M = 0.1$ and the end correction of the inner pipe is assumed to be zero.

The predicted transmission loss of the symmetric double-chamber silencer (Figs. 10.15*c*, *d*) displays a good broadband performance when the connecting pipe is approximately one-half the length of the silencer ($L_3 = L_4 = \frac{1}{4}L$). The word *approximately* expresses the fact that these ideal curves are distorted to a certain

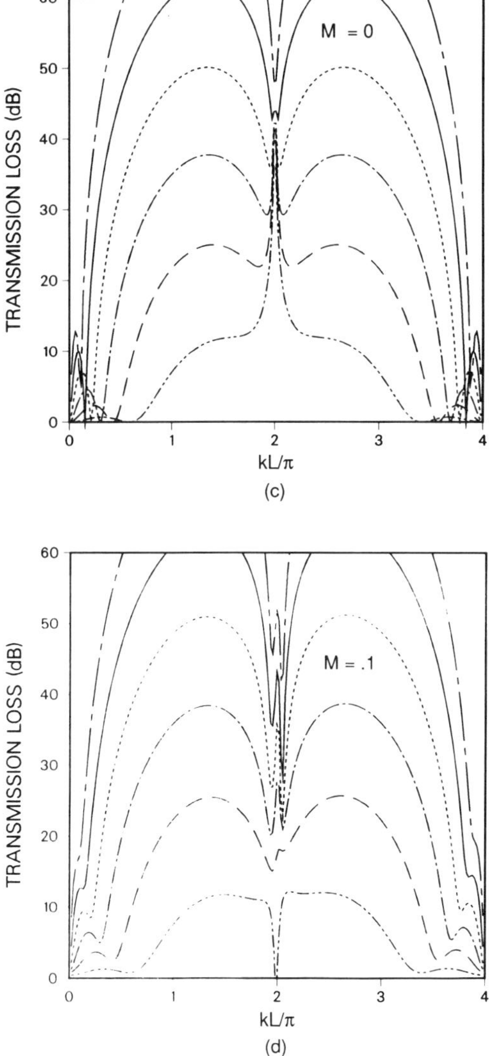

Fig. 10.15 (*Continued*)

extent when the exact end correction is included and, generally, when more effects associated with modeling assumptions and manufacturing tolerances are properly accounted for. A comparison of Figs. 10.15*c* and 10.15*d*, which correspond to $M = 0$ and $M = 0.1$, show that effects of flow, over the range $0 \leq M \leq 0.1$, are insignificant except near $2kL_3 = n\pi$, where they reduce the predicted performance of designs with low expansion ratios.

The design in Figs. 10.15*e*, *f* is obtained by removing the extended inlet pipe in the right-hand chamber in Fig. 10.15*c* and adding an extended outlet pipe of

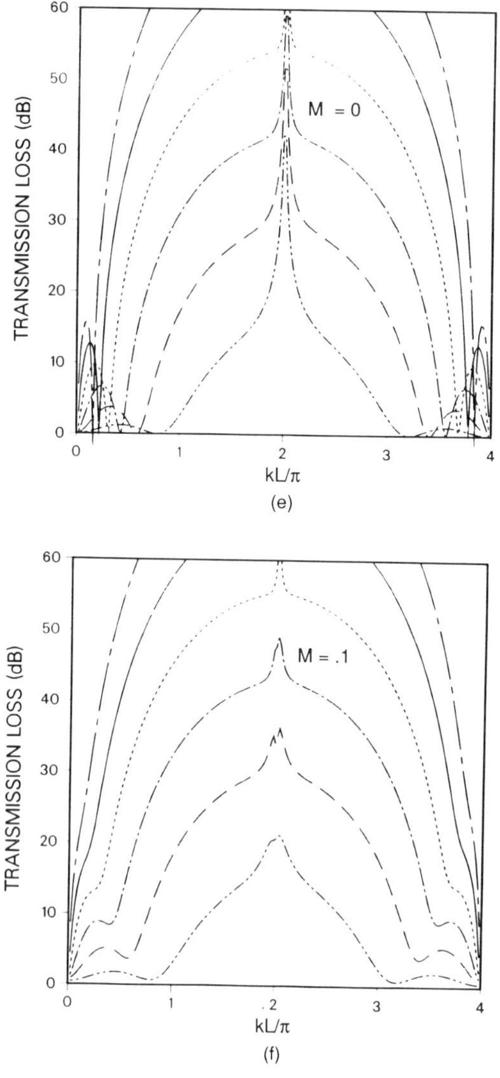

Fig. 10.15 *(Continued)*

equal length. This geometric permutation reduces the width of the main spectral lobes (compare Fig. 10.15*e* to 10.15*c*) but improves significantly the predicted performance in the vicinity of $kL = 2n\pi$, $n = 1, 2, \ldots$. As in the case of Fig. 10.15*c*, Mach numbers in the range $0 \leq M \leq 0.1$ do not affect the overall shape of the predicted transmission loss curves significantly.

Comments. It should be emphasized that the examples of Figs. 10.12–10.15 are presented as illustrations of (a) the prediction procedure, (b) typical performance trends with parameter changes, and (c) the large number of available silencer design options and not as the recommended solutions to any particular case(s). The final choice of a silencer design will depend on the source spectrum, the desired goal, and the various size/weight/cost trade-offs that govern most practical applications.

The expectations for the acoustical performance of a silencer formulated after reviewing Figs. 10.11–10.15 or other similar results should be tempered by the awareness of the assumptions made in developing these predictions. When these assumptions are not met, the corresponding results may be affected adversely. For example, Figs. 10.11–10.15 correspond to plane-wave modes, that is, to wavelengths much larger than the silencer pipe radius ($\sqrt{N}kR_1 = kR_2 << 1$). This assumption limits the maximum allowed area ratio ($N_{max} \approx 1/(kR_1)^2$) once the highest frequency of interest is selected. In other words, when the area ratio N becomes comparable or exceeds the maximum allowed value ($N \geq N_{max}$), one may no longer expect to achieve the N-dependence trends displayed by the curves in Fig. 10.11–10.15. Accordingly, the preliminary silencer design for a given application may be based on the analytical procedure presented above while the final silencer design should also be supported, whenever possible, by experimental data.

10.4 DISSIPATIVE SILENCERS

Dissipative silencers are the most widely used devices to attenuate the noise in ducts through which gas flows and in which the broadband sound attenuation must be achieved with a minimum of pressure drop across the silencer. They are frequently used in the intake and exhaust ducts of gas turbines, air conditioning and ventilation ducts connected to small and large industrial fans, cooling-tower installations, and the ventilation and access openings of acoustical enclosures, and they have an allowed pressure drop that typically ranges from 125 to 1500 Pa (0.5–6 in. of water). Unlike reactive silencers, which mostly reflect the incident sound wave toward the source and dissipate little sound energy, dissipative silencers attenuate sound by converting the acoustical energy propagating in the passages into heat caused by friction in the voids between the oscillating gas particles and the fibrous or porous sound-abosrbing materials.

The theories of dissipative silencers were developed long ago,[1,65–69] and they are highly complex. This chapter provides design information in a form that can be readily used by engineers not thoroughly trained in acoustics. Though there is a large variety of geometries used, the most common configurations include parallel-baffle silencers, round silencers, and lined ducts, as shown schematically in Fig. 10.16.

Figure 10.17 shows some of the baffle constructions that have been used in typical applications. Only the acoustically significant features are shown. Protective treatments, such as perforated facing, fiberglass cloth, and porous screens are omitted. If the same concepts depicted in Figs. 10.17*a–i* are applied as a duct lining, one-half of the baffles depicted in Fig. 10.17 constitutes the lining. The

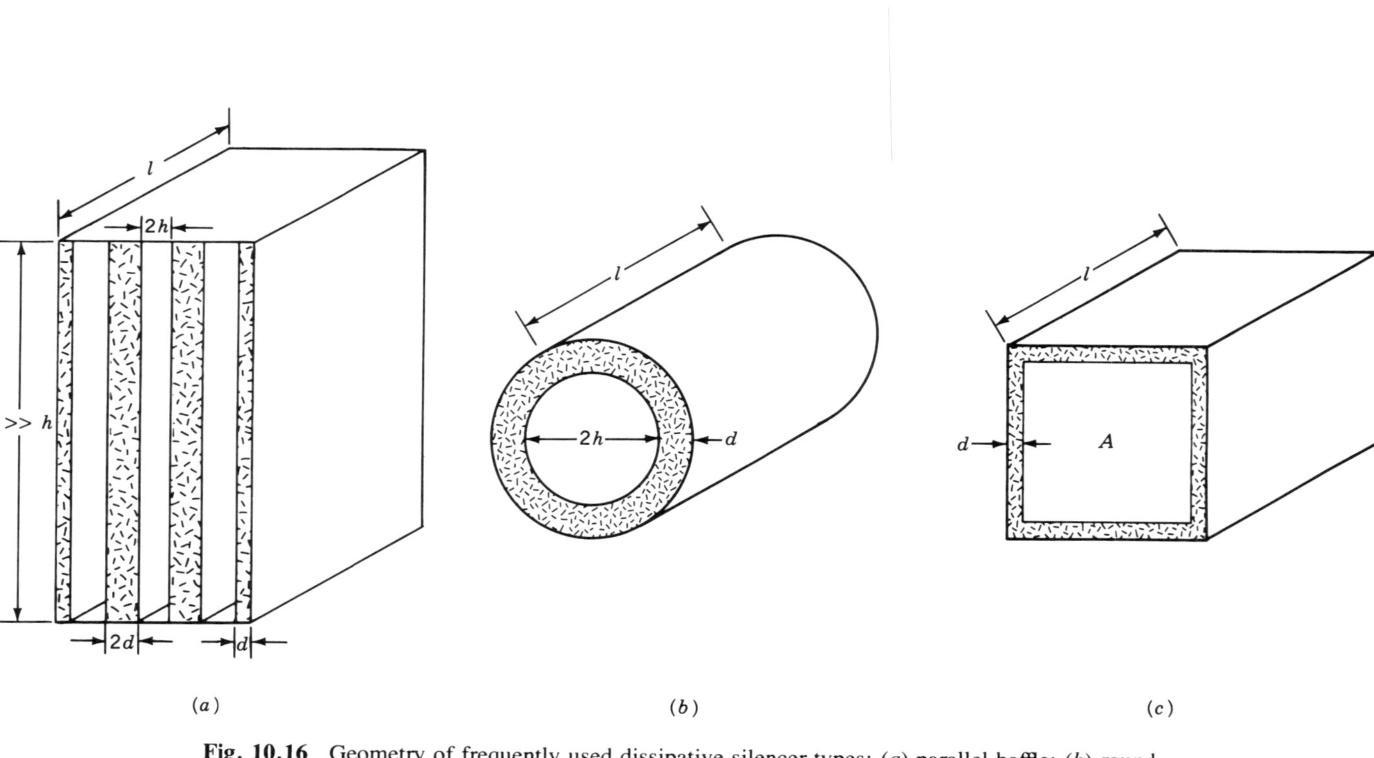

Fig. 10.16 Geometry of frequently used dissipative silencer types: (*a*) parallel baffle; (*b*) round (*c*) lined ducts.

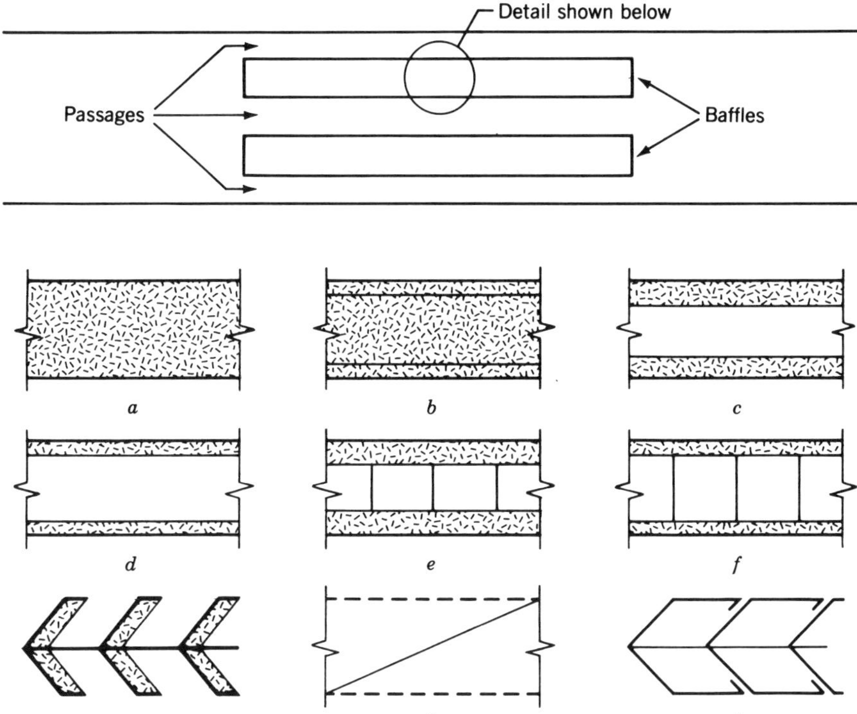

Fig. 10.17 Cross section of frequently used silencer baffle configurations: (*a*) full-depth porous; (*b*) porous center layers with thin resistive facing on both sides; (*c*) thick porous surface layers with unpartitioned center air space; (*d*) thin porous surface layers with unpartitioned air space; (*e*) thick porous layer with partitioned center air space; (*f*) thin porous surface layers with partitioned center air space; (*g*) tuned cavities (Christmas tree), porous material in cavities protected from flow; (*h*) small percentage open-area perforated-surface plates exposed to grazing flow, partitioned center airspace; (*i*) Helmholtz resonators.

full-depth porous baffle depicted in Fig. 10.17*a* is the most frequently used in parallel-baffle silencers. The other baffle configurations are used in special-purpose silencers custom designed to yield high attenuation in specific frequency ranges or to work in hostile environments of contaminated flow or high temperatures.

Key Performance Parameters

The key design parameters of silencers are acoustical insertion loss (IL), pressure drop (Δp), flow-generated noise, size, cost, and life expectancy. The challenge of silencer design is to obtain the needed insertion loss without exceeding the allowable pressure drop and size for a minimum of cost. These are frequently opposing requirements, and the optimal design represents a balanced compromise between them.

Insertion Loss. The insertion loss of a silencer is defined as

$$\mathrm{IL} = 10 \log \frac{W_0}{W_M} \tag{10.25}$$

where W_0 and W_M represent the sound power radiated from the duct without and with the silencer, respectively. Provided that structureborne flanking along the muffler casing and sound radiation from the casing is kept low, the sound power radiated with the silencer in place is given by

$$W_M = W_0 \times 10^{-(\Delta L_l + \Delta L_{\mathrm{ENT}} + \Delta L_{\mathrm{EX}})/10)} + W_{\mathrm{SG}} \tag{10.26}$$

where W is in watts (N · m/s); ΔL_l represents the attenuation of the silencer of length l; ΔL_{ENT} and ΔL_{EX} are the entrance and exit losses, respectively; and W_{SG} is the sound power generated by the flow exiting the silencer.

Combining Eqs. (10.25) and (10.26) yields

$$\mathrm{IL} = -10 \log \left(\frac{W_{\mathrm{SG}}}{W_0} + 10^{-(\Delta L_l + \Delta L_{\mathrm{ENT}} + \Delta L_{\mathrm{EX}})/10} \right) \tag{10.27}$$

In the extreme case of very high silencer attenuation, the second term on the right-hand side becomes comparable to the first, and the achievable insertion loss is affected by the self-generated noise of the silencer and in this case Eq. (10.27) is nonlinear). When the flow velocity in the silencer passages is sufficiently low, flow noise is negligible. In this case, Eq. (10.27) is linear and simplifies to

$$\mathrm{IL} \cong \Delta L_l + \Delta L_{\mathrm{ENT}} + \Delta L_{\mathrm{EX}} \quad \mathrm{dB} \tag{10.28}$$

Entrance Loss ΔL_{ENT}. In most dissipative silencers, the entrance losses ΔL_{ENT} are small if the incident sound energy is in the form of a plane wave normally incident on the silencer entrance. This is always the case for straight ducts at low frequencies. The small entrance loss should be considered as a safety factor. However, if the cross dimensions of the duct are much larger than the wavelength, the incident sound field is usually composed of a very large number of higher order modes. The conversion of the semidiffuse sound field in the entrance duct into a plane-wave field in the narrow silencer passages typically results in an entrance loss of 3–6 dB. The engineer can also assign any entrance loss between 0 dB at low frequencies and up to 8 dB at high frequencies on the basis of prior experience with similar situations or on the basis of scale model measurements. If no such information is available, Fig. 10.18 can be used to estimate the entrance loss.

Exit Loss ΔL_{EX}. Most exit losses ΔL_{EX} are generated when the silencer is located at the open end of a duct and the typical cross dimensions of the opening are small compared with the wavelength. In this case, the exit loss is predominantly determined by the end reflection. Exit losses for silencers inserted in ducts are usually small and can either be neglected or considered as part of the safety mar-

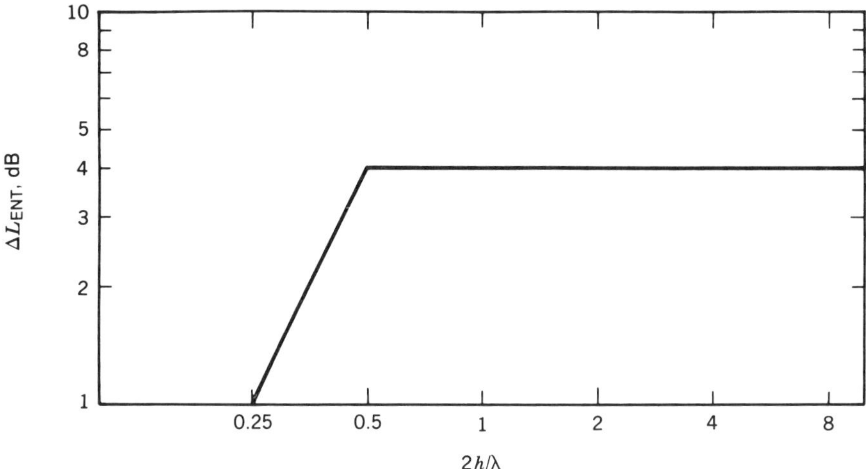

Fig. 10.18 Acoustical entrance loss coefficient, ΔL_{LENT}, of silencer in a large duct with a semi reverberant sound field in the entrance duct; $2h$ = silencer passage cross dimension.

gin. It should be noted that the relative importance of the entrance and exit losses diminishes as the silencer length increases because both quantities are independent of this length. Figure 10.19 shows qualitatively a typical sound pressure level versus distance recorded by a microphone traveling through the silencer and indicates the three components of insertion loss.

Silencer Attenuation ΔL_l**.** The silencer attenuation ΔL_l is proportional to its length l and to the lined perimeter of the passage,P, and inversely proportional to the cross-sectional area of passage, A. It can be expressed as

$$\Delta L_l = \frac{P}{A} lL_h \qquad \text{dB} \tag{10.29}$$

where L_n is the parameter that depends in a complex manner on the geometry of the passage and the baffle, on the acoustical characteristics of the porous sound-absorbing material filling the baffles, on frequency, and on temperature. The quantity L_h, which also depends on the velocity of the flow in the passage, is usually referred to as the attenuation per channel height. The major part of this chapter is devoted to the determination of this important normalized sound attenuation parameter.

Pressure Drop Δp**.** The total pressure drop Δp_T across a muffler is made up of entrance, exit, and friction losses:

$$\Delta p_T = \tfrac{1}{2}\rho v_P^2 \left(K_{\text{ENT}} + K_{\text{EX}} + \frac{P}{A} lK_F \right) = \Delta p_{\text{ENT}} + \Delta p_{\text{EX}} + \Delta p_F \qquad \text{N/m}^2 \tag{10.30}$$

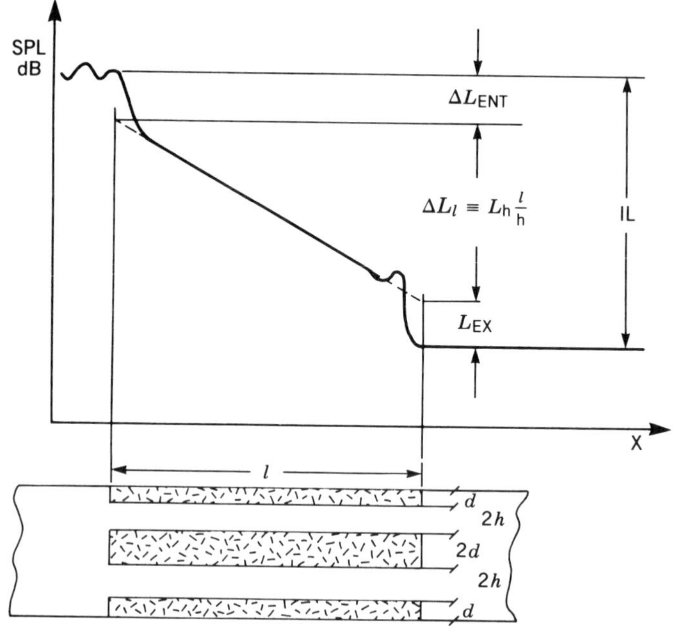

Fig. 10.19 Typical SPL-versus-distance curve obtained when microphone traverses through a silencer.

where ρ is the density of the gas and v_P its face velocity in the passage of the silencer. The constants K_{ENT} and K_{EX} are the entrance and exit head loss coefficients, which depend only on the geometry of the baffle/passage configuration. The third term on the right-hand side of Eq. (10.30) represents the friction losses, given by

$$\Delta p_F = \frac{P}{A} l(K_F \tfrac{1}{2}\rho v_P^2) \qquad \text{N/m}^2 \tag{10.31}$$

Comparing Eq. (10.29) with Eq. (10.31), one notes that baffle configurations that tend to yield high acoustical attenuation, ΔL_l, also tend to yield high frictional pressure losses, Δp_F. Both ΔL_l and Δp_F are proportional to $(P/A)l$, indicating that high silencer attenuation and low frictional pressure drop are contradictory requirements. This finding emphasizes the need to optimize L_h by beneficial choice of the acoustical parameters of the baffles before one resorts to obtaining increased sound attenuation by increasing the factor $(P/A)l$.

Parallel-Baffle Silencers

The parallel-baffle-type silencer shown in Fig. 10.16*a* is the most frequently used because of its good acoustical performance and low cost. The attenuation of such a silencer is proportional to the perimeter–area ratio P/A, the length l, and L_h. Therefore, it is maximized by maximizing $(P/A)L_h$. The largest perimeter–area ratio obtained for narrow passages is $1/h$. Allowing for entrance losses Eq. (10.29) yields the following simple formula for silencer attenuation:

$$\Delta L_l = L_h \frac{l}{h} + \Delta L_{\text{ENT}} \qquad \text{dB} \tag{10.32}$$

where ΔL_{ENT} can be approximated from Fig. 10.18. The following discussion shows how to obtain L_h from the geometric and acoustical parameters of the silencer.

Attenuation Performance and Prediction. The sound energy traveling in the passages of a parallel-baffle silencer, depicted in the top sketch in Fig. 10.17, is attenuated effectively in a wide bandwidth if (1) the sound enters the porous sound-absorbing material in the baffles and (2) a substantial part of the energy of the sound wave entering the baffle is dissipated before it can reenter the passage. Formulas for wall impedance that yield maximum attenuation in a narrow-frequency band are given in reference 2.

Requirement 1 is fulfilled if the passage height is small compared with wavelength (i.e., $2h < \lambda$) and the porous sound-absorbing material is open enough so that the sound wave enters the baffle rather than being reflected at the interface. This requires a "fluffy" material of low flow resistivity. Requirement 2 is fulfilled by a porous material of moderate flow resistivity. The requirements of easy sound penetration and high dissipation are contradictory unless the baffle is very thick and is packed with porous sound-absorbing material of low flow resistivity. Consequently, the choice of baffle thickness and flow resistivity of the porous sound-absorbing material is always a compromise.

Generally, the silencer geometry is controlled by the shape of the attenuation-versus-frequency curve we aim to achieve. To provide reasonable attenuation at the low end of the frequency spectrum, the baffle thickness $2d$ must be on the order of one-eighth of the wavelength. To provide reasonable attenuation at the high end of the frequency spectrum, the passage height $2h$ must be smaller than the wavelength. To allow reasonable penetration of the sound and yield the needed dissipation, the total flow resistance $R_1 d$ of a baffle of thickness $2d$ must be two to six times the characteristic impedance of the gas in the silencer passages at design temperature.

Quantitative Considerations. The normalized attenuation constant L_h is obtained by solving the coupled wave equation in the passage and in the porous material of the baffle and requiring that (1) the coupled wave, which propagates axially in both the passage and baffle, has a common propagation constant Γ_c and (2) both particle velocity and the sound pressure are continuous at the passage–baffle interface.

The coupled wave equation[1,65–68] can be solved by numerical iteration methods to yield the common propagation constant Γ_c. The normalized attenuation L_h is then obtained from

$$L_h = 8.68\text{h Re}\{\Gamma_c\} \quad \text{dB/m} \tag{10.33}$$

where Γ_c depends on the characteristic impedance ρc of the gas in the passage, on the characteristic impedance Z_a, and the propagation constant Γ_a of the porous material in the baffle and on the geometry.

The characteristic impedance Z_a and the propagation constant Γ_a of the porous sound-absorbing materials (which are complex quantities and vary with frequency) are generally not available. As discussed in Chapter 8, one can approximate these important parameters with reasonable accuracy if the flow resistivity of the porous sound-absorbing material R_1 is known. For fibrous sound-absorbing materials, the characteristic impedance Z_a and propagation constant Γ_a in the bulk porous material can be estimated using the empirical formulas presented in Chapter 8.

The most accurate characterization of the porous materials is achieved by measuring the characteristic impedance and propagation constant on a sufficiently large number of samples as a function of frequency at room temperature and by scaling these data to design temperature. Note that this attenuation, which is computed on the basis of the acoustical parameters of the porous material predicted from flow resistivity by employing the formulas given in Chapter 8, compares very well with experimental data if the porous sound-absorbing material is homogeneous.

Normalized Graphs to Predict Acoustical Performance of Parallel-baffle Silencers. The normalized attenuation L_h has been computed for various percentages of open area of the silencer cross section (i.e., for various h/d) and for various values of the normalized flow resistance $R = R_1 d/\rho c$ of isotropic porous sound-absorbing material in the baffles. It is presented in Figs. 10.20*a–c* for open-area ratios of 66, 50, and 33 percent, respectively. The effect of nonisotropic material is covered in reference 66. The vertical axis in Fig. 10.20 represents the normalized attenuation L_h in decibels. The lower horizontal scale represents the normalized frequency parameter $\eta = 2hf/c$. It is valid for all temperatures and gases provided that the speed of sound c is taken at the actual temperature. *The upper horizontal scale, which is valid only for air at room temperature*, represents the product of the half-passage height h in centimeters and the frequency f in kilohertz.

Figures 10.20*a–c* show that the attenuation starts to decrease rapidly above the frequency where the passage height $2h$ becomes large compared to the wavelength (i.e., $\eta > 1$). The attenuation in this frequency region can be increased by up to 10 dB by utilizing a two-stage silencer with staggered-baffle arrangement. Note that the bandwidth of appreciable attenuation increases with decreasing percentage of open area of the silencer cross section.

Figure 10.20 shows that in the range of $R = R_1 d/\rho c$ from 1 to 5, the attenuation does not depend strongly on the specific choice of the flow resistance; this coincidence is welcome because the present lack of adequate knowledge of and control

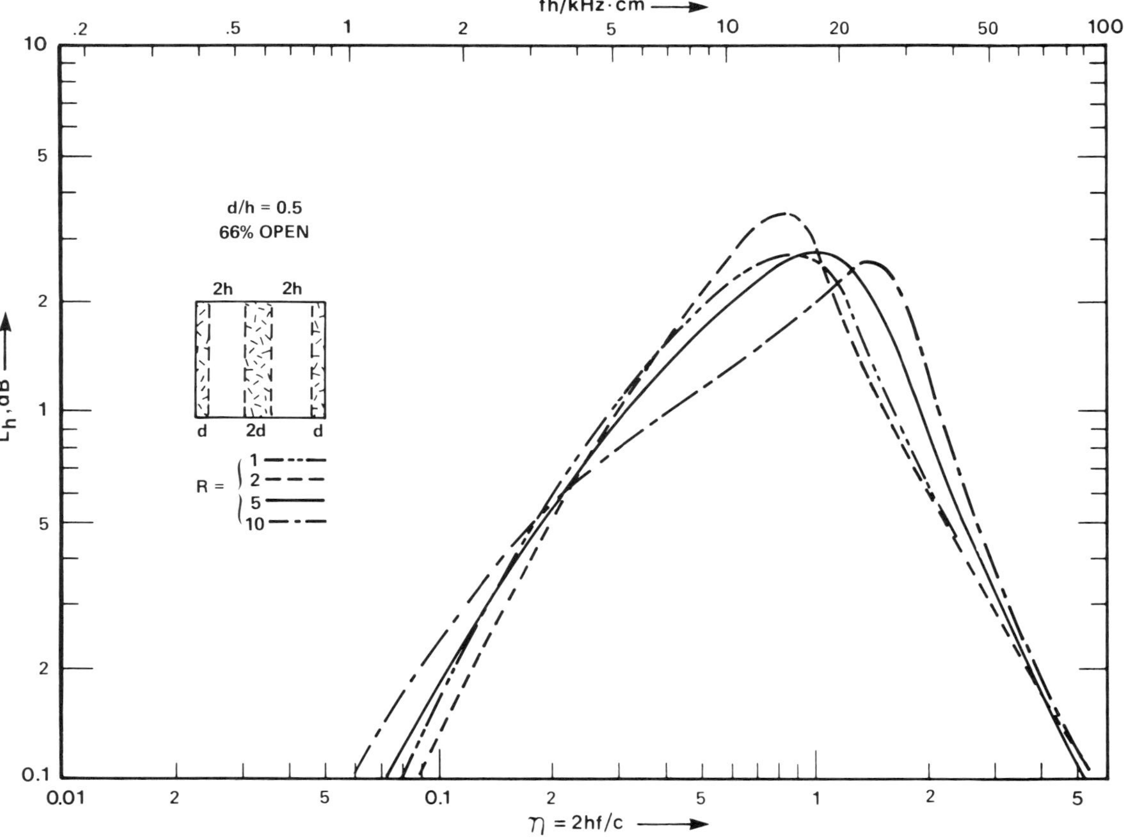

(*a*)

Fig. 10.20 Normalized attenuation-versus-frequency curves for parallel-baffle silencers with normalized baffle flow resistance $R = R_1 d/\rho c$ as parameter; (a) 66% open, (b) 50% open, (c) 33% open.

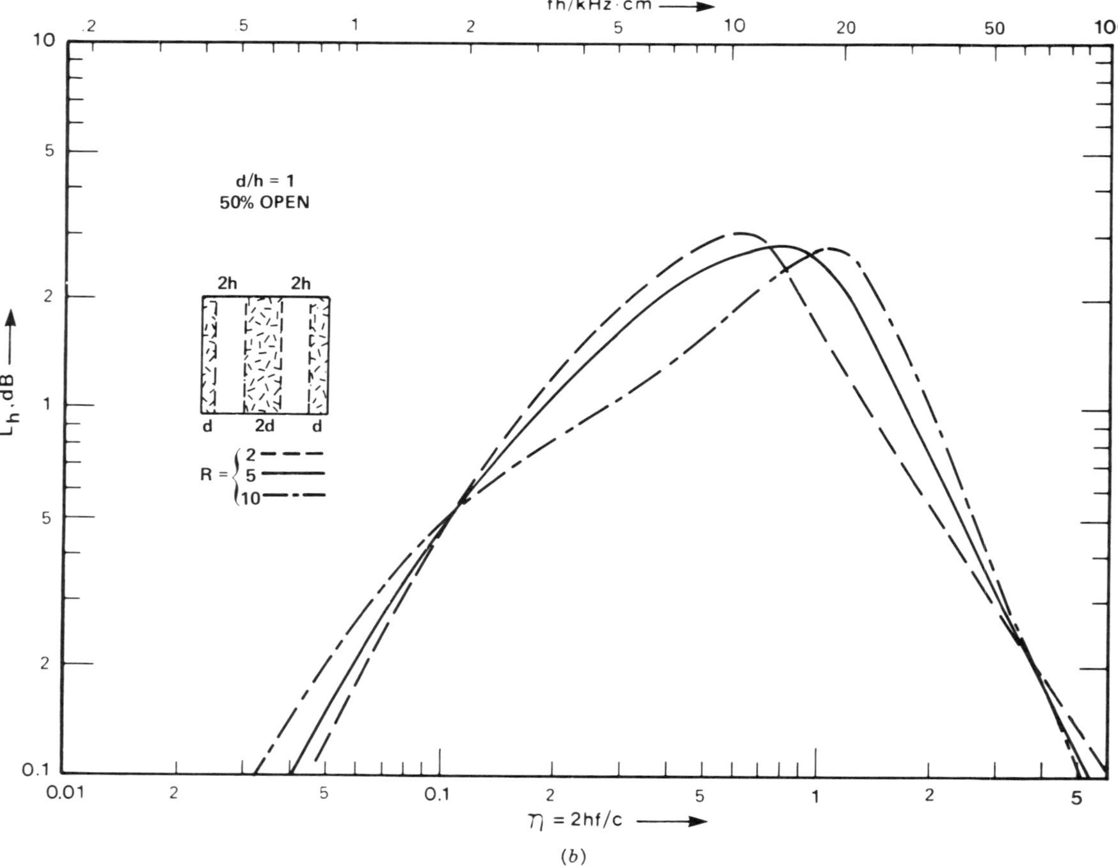

(b)

Fig. 10.20 *(Continued)*

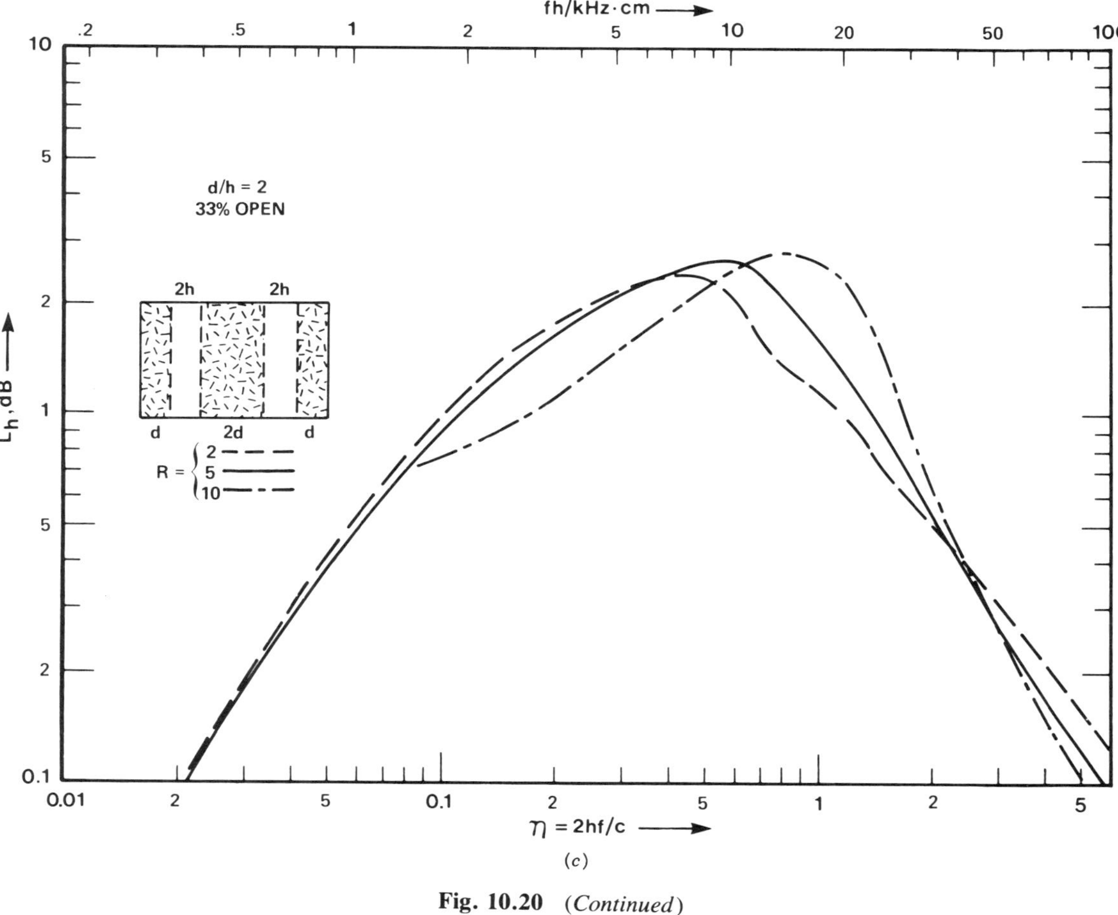

(c)

Fig. 10.20 *(Continued)*

over the material characteristics represents the weakest link in the prediction process. Note that if the normalized flow resistance becomes too large ($R \geq 10$), a substantial decrease of attenuation occurs in the frequency region from $\eta = 0.2$ to $\eta = 1$ accompanied by a modest increase of attenuation at very low and at high frequencies.

The normalized attenuation-versus-frequency curves presented in Fig. 10.20 correspond to zero flow; corrections to account for flow are presented in a later section.

Examples. The use of the design information presented in Fig. 10.20 is illustrated by a few examples.

Example 10.1: Performance Prediction Predict the attenuation-versus-frequency curve of a parallel-baffle silencer consisting of 200-mm- (8-in.-) thick parallel baffles 1 m (40 in.) long spaced 400 mm (16 in.) off center when the flow resistance of the baffle $R_1 d = 5\rho c$ and the duct carries a very low velocity air at 20°C; $h = 0.1$ m; $d = 0.1$ m; $l = 1$ m; $c = 340$ m/s; $\rho = 1.2$ kg/m^3; $R = R_1 d/\rho c = 5$.

SOLUTION

1. Determine frequency f^*, which corresponds to $\eta = 1$:

$$f^* = \frac{c}{2h} = \frac{340 \text{ m/s}}{0.2 \text{ m}} = 1700 \text{ Hz}$$

2. Determine $l/h = 1 \text{ m}/0.1 \text{ m} = 10$.
3. Determine applicable normalized attenuation-versus-frequency curve. For $d/h = 1$ and $R = 5$, the solid curve in Fig. 10.20*b* is applicable.
4. Mark the frequency $f^* = 1700$ Hz on the horizontal scale of a sheet of transparent graph paper that has the same horizontal and vertical scales as Fig. 10.20 and align it with $\eta = 1$ in Fig. 10.20*b*.
5. Shift the transparent graph paper vertically until the mark $L_h = 1$ in Fig. 10.20*b* corresponds to $10(l/h = 10)$ on the transparent overlay.
6. Copy the solid curve in Fig. 10.20*b* that corresponds to $R = 5$ on the overlay.
7. The copied curve then corresponds to the attenuation-versus-frequency curve of the silencer according to $\Delta L_l = L_h(l/h)$; the above procedure is sketched in Fig. 10.21.

Example 10.2: Design to Yield Specified Attenuation Design a parallel-baffle silencer that yields the attenuation listed below:

f, Hz	100	200	500	1000	2000	4000
ΔL_l, dB	4	9	19	26	10	5

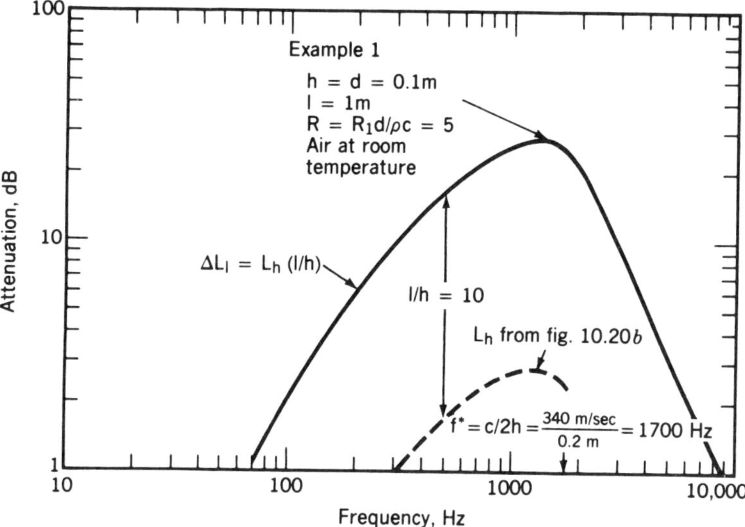

Fig. 10.21 Attenuation prediction for parallel-baffle silencer; Example 10.1.

SOLUTION

1. Plot the required attenuation-versus-frequency values listed in the table above on a transparent log-log paper of the same scale factor as the design graphs presented in Fig. 10.20. These are the circles in Fig. 10.22.

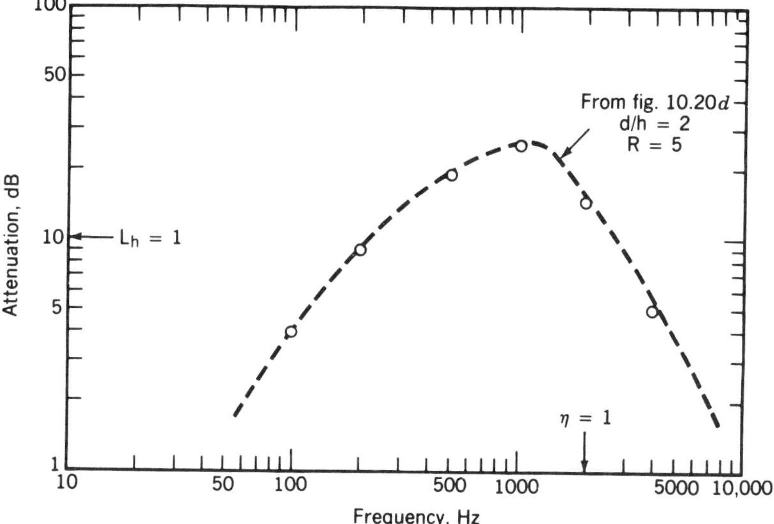

Fig. 10.22 Acoustical design of a parallel-baffle silencer; Example 10.2: (○) design requirements (-------) curve of matching attenuation versus frequency from Fig. 10.20*c*.

2. Find the graph in Fig. 10.20 that best matches the shape of the desired attenuation-versus-frequency curve plotted in Fig. 10.21. Overlay the transparent paper of Fig. 10.21 on the curve that yields the best match and shift the transparent overlay horizontally and vertically until all of the desired attenuation-versus-frequency points fall below the chosen normalized attenuation-versus-frequency curve. In this case, the solid curve in Fig. 10.20*c* gives the best match.
3. On the overlay, mark the frequency f^* that corresponds to $\eta = 1$ on the horizontal scale of the appropriate design curve under the overlay, and on the vertical scale, mark the attenuation ΔL^*, which corresponds to $L_h = 1$ on the design curve below, as shown in Fig. 10.22. In this case, these will be $f^* = 2000$ Hz and $\Delta L^* = 10$ dB.
4. From the design curve below the overlay, note the values of the parameters d/h and R that correspond to the curve that provides the best match. In this case, these are $d/h = 2$ and $R = R_1 d/\rho c = 5$.
5. On the basis of the information obtained in steps 3 and 4, one obtains the geometric and acoustical parameters of the silencer that will yield the specified attenuation as follows:

Passage height 2h:

$$\eta = 1 = \frac{2hf^*}{c} \quad \text{yields} \quad 2h = \frac{340 \text{ m/sec}}{2000 \text{ 1/sec}} = 0.17\text{m}$$

Baffle thickness 2d:

$$2d = 2(2h) = 2 \times 0.17 \text{ m} = 0.34 \text{ m}$$

Silencer length:

$$\Delta L^* = 10 = \frac{l}{h} \quad \text{yields} \quad l = \Delta L^* h = 10 \times \frac{0.17}{2} \text{ m} = 0.85 \text{ m}$$

Flow resistance per unit thickness of porous material:

$$\frac{R_1}{\rho c} = 5 \text{ yields } R_1 = \frac{5\rho c}{d} = \frac{5}{0.17 \text{ m}} \rho c = 29.4\rho c/\text{m}$$

or

$$R_1 = 0.7\rho c/\text{in.} \quad \text{or} \quad R_1 = 1.2 \times 10^4 \text{ N} \cdot \text{s/m}^4$$

Cross-sectional area: The cross-sectional area of the muffler is determined by the maximum allowable pressure drop and self-generated noise as discussed later.

Effect of Temperature. The design temperature affects the acoustical and aerodynamic performance of the silencers because the following key parameters depend on the temperature; (1) speed of sound, (2) density of the gas, and (3) viscosity. The effects of temperature are taken into account as follows:

$$c(T) = c_0 \sqrt{\frac{273 + T}{530}} \quad \text{m/s} \tag{10.34}$$

$$\rho(T) = \rho_0 \frac{293}{273 + T} \quad \text{kg/m}^3 \tag{10.35}$$

$$R(T) = R_0 \left(\frac{273 + T}{293} \right)^{1.2} \tag{10.36}$$

where T is the design temperature in degrees Celsius, c_0 is the speed of sound, ρ_0 is the density of the gas (usually air) at 20°C, and

$$R_0 = \frac{R_1(20°\text{C})d}{\rho_0 c_0} \tag{10.37}$$

where $R_1(20°\text{C})$ is the flow resistivity of the porous bulk material at 20°C. This material parameter is usually provided by the manufacturer or is measured.

Example 10.3: Effect of Temperature To illustrate how to account for the effect of temperature, let us predict the attenuation provided by the silencer of Example 10.1 at $T = 260°\text{C}$ (500°F).

SOLUTION The effect of temperature must be accounted for in the flow resistivity and in the speed of sound according to Eqs. (10.36) and (10.34), yielding: $c(T) = 457$ m/s and $R(T) = 10$. From now on, the solution proceeds according to the same steps followed in Example 10.1,

1. $f^* = c(T)/2h = 457 \text{ m/s}/0.2 \text{ m} = 2285$ Hz.
2. $l/h = 1 \text{ m}/0.1 \text{ m} = 10$.
3. The applicable normalized attenuation curve that corresponds to $d/h = 1$ and $R = 10$ is the dash–dot curve in Fig. 10.20*b*.

This results in the attenuation-versus-frequency curve shown as the solid line in Fig. 10.23. The dotted curve in Fig. 10.23 is the attenuation of the same silencer at room temperature (20°C), as determined in Example 10.1.

Comparing the solid curve obtained for 260°C with the dotted curve obtained for 20°C, one notes a shift in the attenuation-versus-frequency curve toward higher frequencies with increasing temperature. This shift is mainly due to the increase

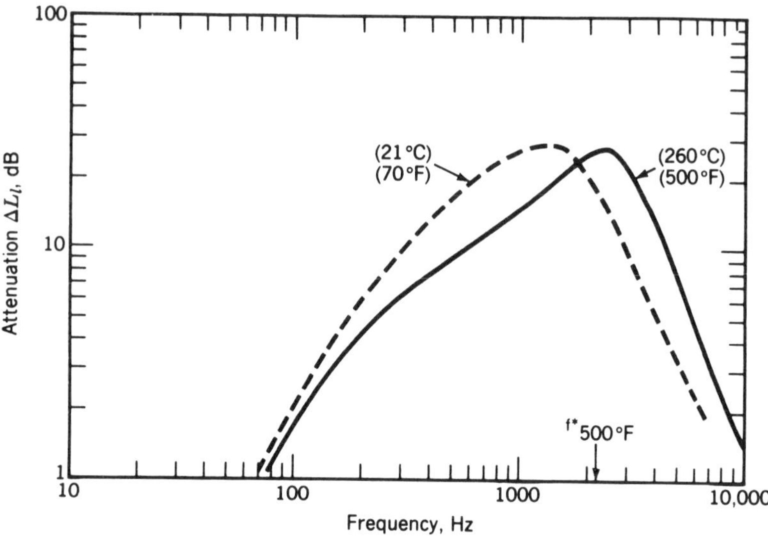

Fig. 10.23 Attenuation of silencer of Example 10.1 at 260°C (500°F) temperature; Example 10.3.

of propagation speed of sound with increasing temperature. In addition, one also observes distortion in the shape of the attenuation-versus-frequency curve. This is caused by the increase in the flow resistivity of the porous material, which in turn is due to the increasing viscosity with increasing temperature.

Baffle Thickness Considerations. A particular percentage of open area of a silencer can be accomplished either by a small number of thick baffles or by a large number of thin baffles. Figure 10.24 shows the attenuation-versus-frequency curves computed for a 2-m- (6.5-ft-) long silencer of 50% open area with baffle thickness $2d = 2h$ of 152 mm (5 in.), 203 mm (8 in.), 254 mm (10 in.), and 305 mm (12 in.). The baffles are filled with a fibrous sound-absorbing material that has a flow resistivity $R_1 = 51{,}500\ \text{N} \cdot \text{s/m}^4$ at 500°C ($1\rho c$/in. at 20°C). The outstanding feature of the data presented in Fig. 10.24 is that in the mid frequency region, the sound attenuation decreases with increasing baffle thickness. At 500 Hz, the 152-mm- (6-in.-) thick baffles yield 25 dB attenuation while the 305-mm- (12-in.-) thick baffles yield only 11 dB. This is because the sound does not fully penetrate into the fibrous sound-absorbing material in the thick baffles. Consequently, the material and space in the center of the thick baffles is wasted.

Figure 10.25 shows the predicted sound attenuation of a silencer of the same geometry as that in Fig. 10.24, but the baffles in this case are filled with fibrous sound-absorbing material of low flow resistivity such that the total normalized flow resistance of each baffle was $R = R_1 d/\rho c = 2$, which allows full penetration of the sound into even the thickest baffle. Consequently, the sound attenuation at low and mid frequencies depends only slightly on the baffle thickness. Comparing Figs. 10.24 and 10.25 reveals that using a few thick baffles (which is more economical

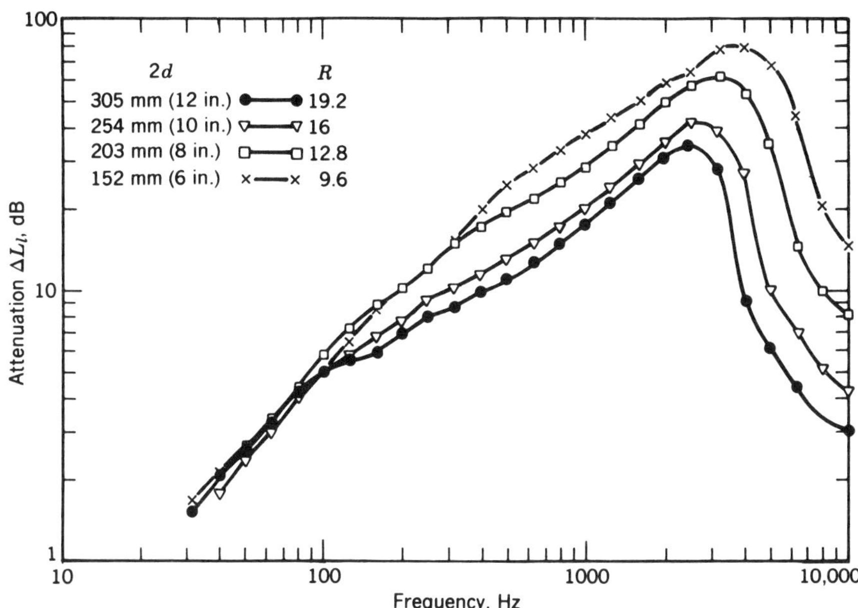

Fig. 10.24 Computed attenuation-versus-frequency curves of a 2-m- (6.6-ft-) long parallel-baffle silencer of 50% open area with baffle thickness $2d$ as parameter; design temperature $T = 500°C$. Fibrous fill has a flow resistivity $R_1 = 51{,}500\ N \cdot s/m^4$ ($1\rho c$/in. at 20°C).

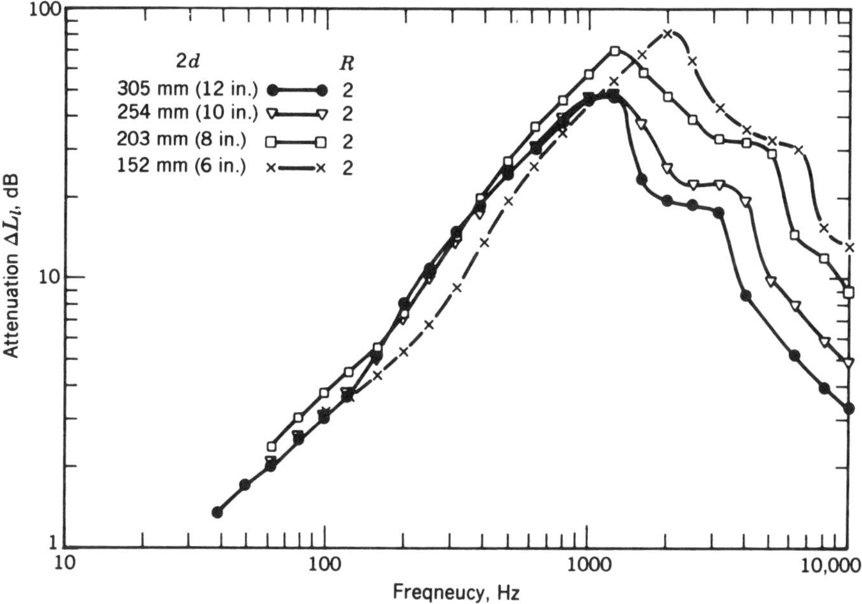

Fig. 10.25 Computed attenuation-versus-frequency curves of the same silencer as Fig. 10.24 but fibrous fill is chosen to yield $R = R_1 d/\rho c = 2$ for all baffle thicknesses.

than using many thinner baffles) results in decreased sound attenuation at mid frequencies unless the baffles are filled with a sound-absorbing material of low enough resistivity so that the normalized baffle flow resistance at design temperature, $R = R_1 d/\rho c$, is much less than 10. Fibrous sound-absorbing materials that fulfill this requirement for thick baffles used in elevated temperatures may not be readily available. Consequently, silencer baffles of traditional design in high-temperature applications should be kept at a thickness that seldom exceeds $2d = 200$ mm (8 in.).

Note that the attenuation-versus-frequency curve of parallel-baffle silencers increases monotonically up to a frequency $f = c/2d$, where the passage width corresponds to a wavelength, and decreases sharply above it.

Round Silencers

Round silencers, as depicted in Fig. 10.16*b*, are used in connection with round ducts. The curvature of the duct casing results in a high form stiffness that yields high sound transmission loss of the silencer wall at low frequencies. The acoustical performance of round silencers with respect to normalized attenuation L_h is very similar to that of parallel-baffle silencers. The diameter of the round passage, $2h$, and the thickness of the homogeneous isotropic lining, d, resemble the passage width $2h$ and the half-baffle thickness d of a parallel-baffle silencer. The silencer attenuation $\Delta L^r(l)$ of a round muffler of length l is obtained from

$$\Delta L^r(l) = L_h^r \frac{l}{h} \tag{10.38}$$

The sound attenuation in round ducts has been studied by Scott[65] and Mechel,[68]. Their work forms the theoretical foundations for computing their performance.

Fig. 10.26 shows curves of normalized attenuation L_h^r versus normalized frequency $\eta = 2hf/c$ for round silencers with homogeneous isotropic lining for thickness–passage radius ratios $d/h = 0.5$, 1, 2 with the normalized lining flow resistance $R = R_1 d/\rho c$ as parameter. The lower horizontal scale is valid for all temperatures, while the upper scale is valid only for air at room temperature. Figure 10.26 shows that the normalized attenuation increases monotonically with frequency until the wavelength corresponds to the diameter of the passage (i.e., $\eta = 2hf/c = 2h/\lambda = 1$). Above this frequency the attenuation decreases rapidly with increasing frequency, as was the case for parallel-baffle silencers. Note that the maximum normalized attenuation is about $L_{h,\max}^r \simeq 6$ dB for round silencers compared to $L_{h,\max}^p \cong 3$ dB for parallel-baffle silencers. This is because a round passage has twice as high perimeter–area ration $(2/h)$ than a narrow passage of a parallel-baffle silencer $(1/h)$. As expected, the attenuation bandwidth increases toward low frequencies with increasing lining thickness d. The curves presented in Fig. 10.26 are used in the same way as the corresponding curves for parallel-

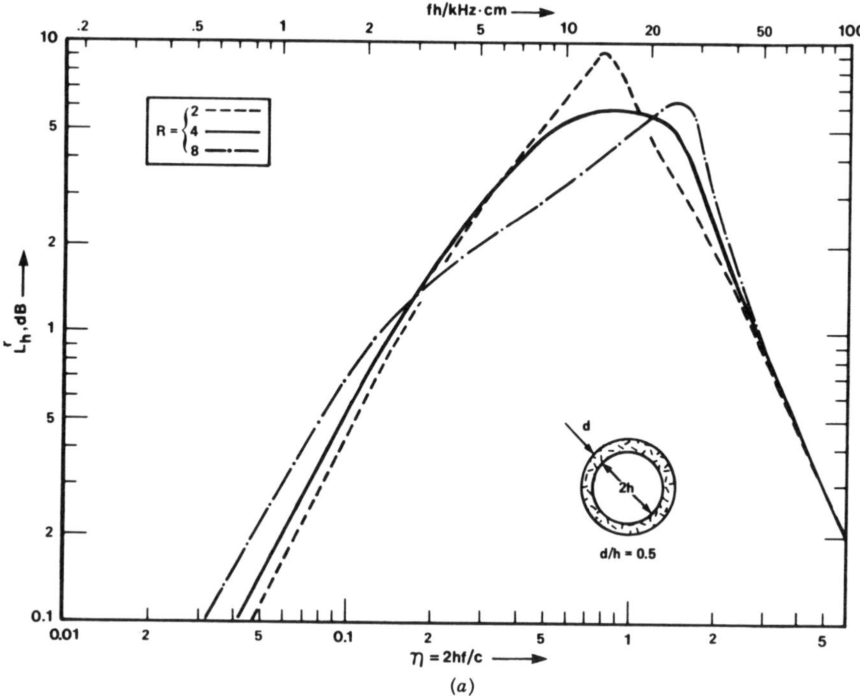

(*a*)

Fig. 10.26 Curves of normalized attenuation L_h^r versus normalized frequency for round silencers with normalized lining flow resistance $R = R_1 d/\rho c$ as parameter: (*a*) $d/h = 0.5$; (*b*) $d/h = 1$; (*c*) $d/h = 2$.

baffle silencers by applying the design steps listed earlier in the Examples 10.1–10.3.

Round silencers have a generic disadvantage of providing poor high-frequency performance when the passage diameter is large compared with the wavelength. This disadvantage can be overcome by inserting a center body into the passage. With the center body in place, the round silencer has a narrow annular passage, just like a parallel-baffle silencer, and the attenuation will continue to increase monotonically until the wavelength of the sound becomes equal to the width of the narrow annular passage, as illustrated in the example of Fig. 10.27. Both the rigid and absorbing center bodies result in a modest increase in attenuation at low and midfrequencies, which is mostly due to reduction of the passage cross-sectional area. The attenuation of this silencer without center body decreases sharply above 560 Hz, the frequency where the wavelength equals the diameter of the passage. With the rigid center body the attenuation continues to increase up to 1130 Hz and with the porous center body up to 2260 Hz.

If center bodies are impractical, the high-frequency attenuation of round silencers can also be increased by inserting parallel baffles into the round passage.

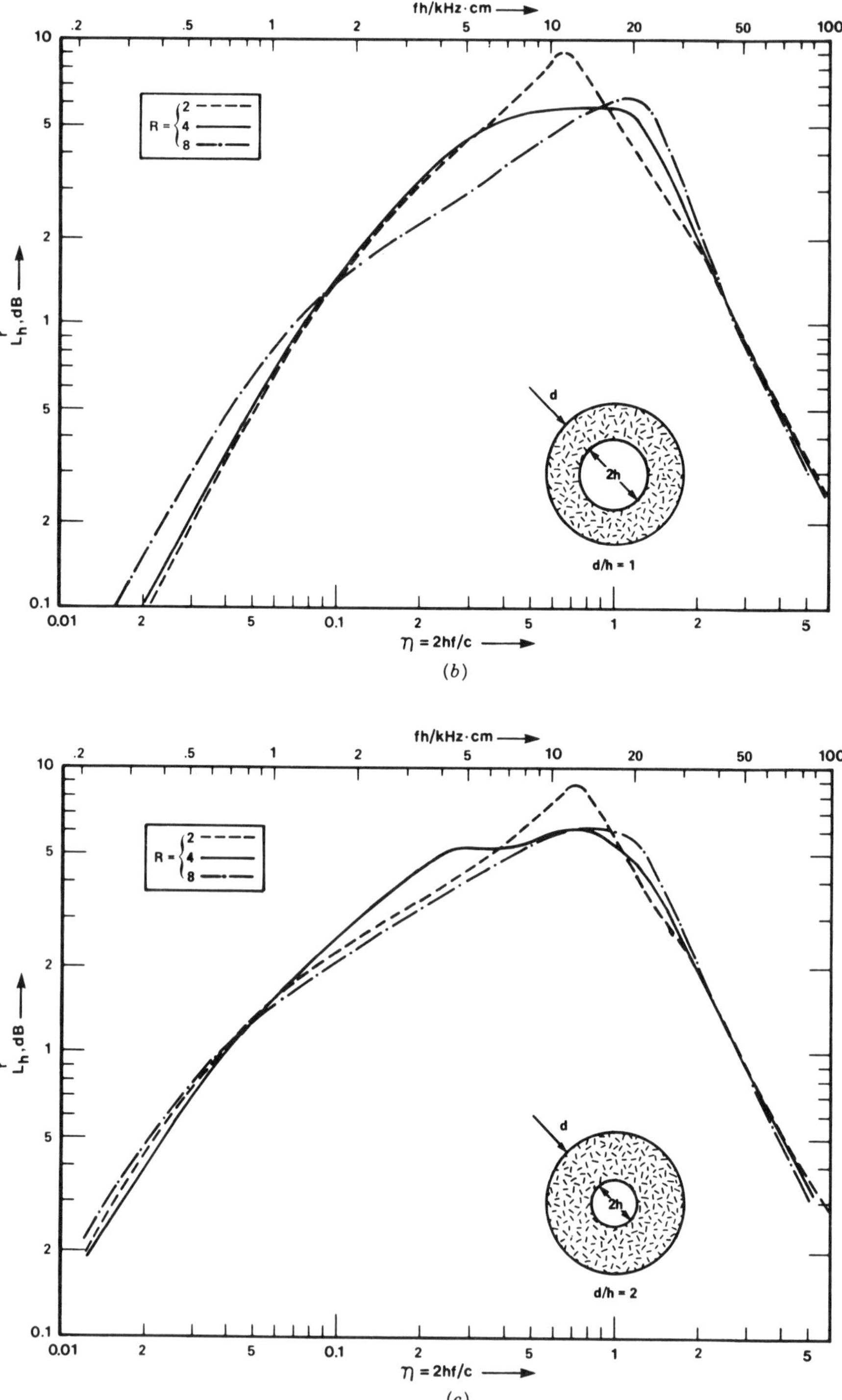

Fig. 10.26 (*Continued*)

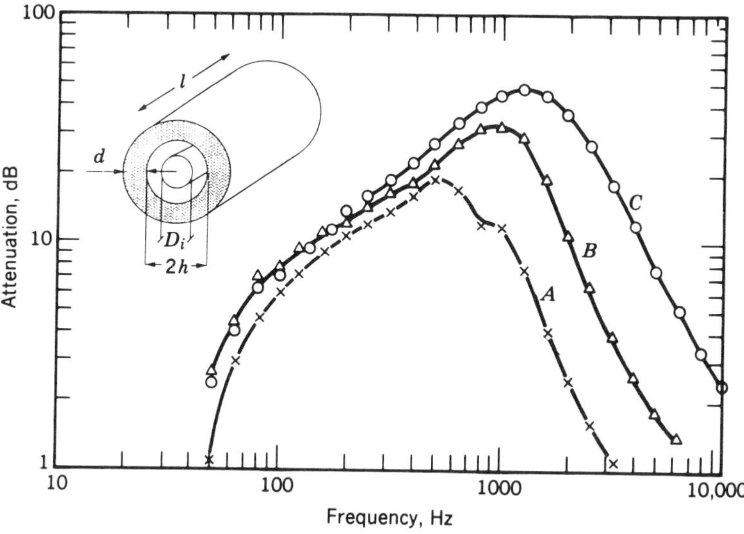

Fig. 10.27 Effect of center body on attenuation-versus-frequency curve of a round silencer, passage diameter $2h = 0.6$ m (24 in.), center-body diameter-0.3 m (12 in.), lining thickness $d = 0.2$ m (8 in.), length $l = 1.2$ m (48 in.), $T = 20°\mathrm{C}$ (68°F), $R_l = 16{,}000$ N · s/m^4 (1 ρc/in.): A, no center body; B, rigid center body; C, absorbing center body.

Lined Ducts

Lined ducts, as depicted in Fig. 10.16*c*, have a 2.5-cm- (1-in.-) or 5-cm-(2-in.-) thick lining on the interior surfaces. The liner consists of a fibrous bulk material that is glued to the duct walls and a very thin facing that protects the bulk material against flow erosion and provides the smooth surface needed to keep the friction losses low. The protective surface layer may consist of cloth, sprayed-on coating of plastic or neoprene or both.

The acoustically important feature of this coating is that it represents a thin-flow resistive layer spaced away from the rigid duct wall by the fibrous bulk material behind the resistive layer, which also resists the propagation of sound in the axial direction. As measured by Kuntz and Hoover,[69] the flow resistance of this protective surface layer is typically between 25 and 100 N · s/m^3 (0.06ρc to 0.25ρc), while the flow resistivity of the fibrous material ranges from 4800 to 12,000 N · s/m^4 (0.3ρc/in. to 0.74ρc/in.). Because the resistive surface layer is furthest away from the rigid duct wall, it is exposed to the highest particle velocity at low frequencies and practically determines the attenuation of sound at low frequencies. The fibrous lining material behind it functions as a spacer.

Estimation of Lined Duct Sound Attenuation. From extensive experimental data, Kuntz and Hoover[69] determined that the attenuation-versus-frequency curve for lined ducts can be predicted by empirical asymptotic formulas. The low-frequency asymptote (125–800 Hz) is

$$\Delta L_{\text{low}} = \frac{(P/A)l[\alpha_N(500)^{0.75}(H^{0.36})(f^{1.17+K_2\rho_L})]}{K_1\rho_L^{2.3}} \quad \text{dB} \qquad (10.39\text{a})$$

or

$$\Delta L_{\text{low}} = \frac{(P/A)l(H^{0.36})(d^{1.08})[f^{(1.17+K_2\rho_L)}]}{K_3\rho_L^{2.3}} \quad \text{dB} \qquad (10.39\text{b})$$

where Eq. (10.39a) is used when $\alpha_N(500)$ is available and Eq. (10.39b) when it is not. The high-frequency asymptote (800 Hz to 10 kHz) is

$$\Delta L_{\text{high}} = \frac{(P/A)lf^{[K_5 - 1.61\log_{10}(P/A)]}}{W^{2.5}H^{2.7}} K_4 \quad \text{for } l \leq 10 \text{ ft} \qquad (10.40)$$

and values for $l > 10$ ft are obtained by setting $l = 10$ in Eq. (10.40), where P = lined perimeter of the lined duct in millimeters (in.), A = cross-sectional area of the lined duct in square millimeters (in.^{2}), l = duct length in meters (ft), $\alpha_N(500)$ = normal-incidence sound absorption coefficient of the lining at 500 Hz, d = lining thickness in millimeters (in.), H = smaller inside cross dimension of the duct in millimeters (in.), f = frequency in hertz, ρ_L = density of the lining material in kg/m^3 (lb/ft^3), and W = larger inside dimension of the duct in millimeters (in.). The constants are as tabulated below.

	SI Units	English Units
K_1	0.0214	329
K_2	0.012	0.19
K_3	5.46×10^{-3}	1190
K_4	3.32×10^{18}	2.11×10^9
K_5	−3.79	−1.53

Typical duct liner normal-incidence sound absorption coefficients at 500 Hz, measured in reference 69, are listed below.

Thickness		Density		
mm	in.	kg/m^3	lb/ft^3	$\alpha_N(500)$
25	(1)	24	(1.5)	0.17
		48	(3)	0.19
50	(2)	24	(1.5)	0.47
		48	(3)	0.53

Prediction Procedure. The simplified prediction method includes the following steps:

1. Use Eq. (10.39) to calculate ΔL_{low} at 100 and 800 Hz.
2. Use Eq. (10.40) to calculate ΔL_{high} at 800 and 10,000 Hz.

3. On a double logarithmic paper plot these four points and join the $\Delta L_{\text{low}}(100)$ and $\Delta L_{\text{low}}(800)$ and subsequently the $\Delta L_{\text{high}}(800)$ and $\Delta L_{\text{high}}(10{,}000)$ points with straight lines
4. Draw a horizontal line at $\Delta L = 40$ dB.

The lowest value envelope defined by the three straight lines represents the predicted attenuation-versus-frequency curve for the lined duct.

Example 10.4 Estimate the attenuation-versus-frequency curve for a 3-m (10-ft.-) long duct with 50-mm- (2-in.-) thick lining when the free inside duct dimensions are 514 × 304 mm (24 × 12 in.). Input parameters in inch-pound units are $l = 10$ ft, $H = 12$ in., $W = 24$ in., $\rho_L = 3\ \text{lb/ft}^3$, $d = 2$ in., $\alpha_N(500) = 0.53$, $K_1 = 329$, $K_2 = 0.19$, $K_3 = 2.11 \times 10^9$, $K_4 = 2.11 \times 10^9$, and $K_5 = -1.53$. With these values, Eqs. (10.39a) and (10.40) yield

$$\Delta L_{\text{low}}(100) = 2.8 \qquad \Delta L_{\text{low}}(800) = 103$$

$$\Delta L_{\text{high}}(800) = 54 \qquad \Delta L_{\text{high}}(10{,}000) = 13$$

The solid lines in Fig. 10.28 represent the predicted attenuation-versus-frequency curves. Measured values are shown for comparison.

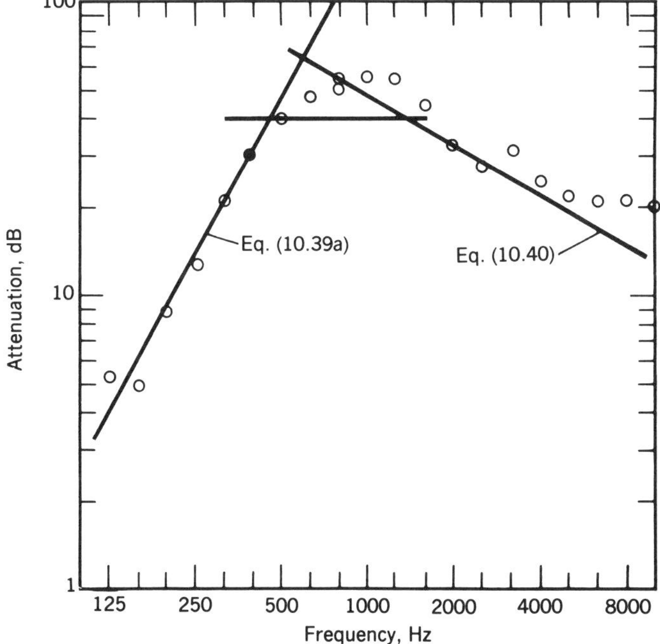

Fig. 10.28 Comparison of predicted and measured attenuation-versus-frequency curve of lined duct treated in example 10.4. (After ref. 69.)

Effect of Flow on Silencer Attenuation

The flow affects the attenuation of sound in silencers in three ways:

1. It changes slightly the effective propagation speed of the sound.
2. By creating a velocity gradient near the passage boundaries, it refracts the sound propagating in the passage toward the lining if the propagation is in the flow direction (exhaust silencers) and "focuses" the sound toward the middle of the passage if the propagation is opposite to the flow direction (intake silencers).
3. It increases the effective flow resistance of the baffle.

Figure 10.29*a* shows the effect of flow on the attenuation performance of an exhaust silencer when the flow direction coincides with the direction of sound propagation. Note that the attenuation is decreased at low frequencies and is increased slightly at high frequencies. In most cases, a Mach number $M > 0.1$ is not permissible because of material deterioration or self-noise.

Figure 10.29*b* shows what happens when the sound propagates against the flow. In this case the attenuation at low frequencies increases because it takes the sound a longer time to traverse through the muffler passage. The high-frequency attenuation decreases because the velocity gradients in the passage "channel" the sound toward the center of the passage. The effect of flow on attenuation can be approximated by appropriate shift of the attenuation-versus-frequency curve obtained without flow ($M = 0$).

The attenuation in the presence of flow is obtained by shifting the no-flow ($M = 0$) attenuation curve as illustrated in Fig. 10.30*a*, when the direction of sound propagation and the gas flow are the same and Fig. 10.30*b* when they are opposite.

Figure 10.31 shows the attenuation-versus-frequency curve of a silencer with $M = 0.15$ flow in the direction of sound propagation. The values represented by the crosses are those directly computed for $M = +0.15$ (dotted line in Fig. 10.29*a*) and those represented by the open circles were obtained by shifting the $M = 0$ curve in Fig. 10.29*a* according to the guidelines discussed in conjunction with Fig. 10.30*a*. The agreement between the curve obtained by shifting and by computations (that takes into account flow effects in the wave equation) is good.

The empirical flow correction procedure illustrated in Fig. 10.30 is based on experience with parallel-baffle silencers. We have no experience at present to gauge its applicability and accuracy for other silencer geometries.

Flow-generated Noise. At present there is no universally accepted method for predicting the flow-generated noise of silencers. Information provided by silencer manufacturers shall be used whereever available. The empirical predictive scheme presented here is based on a broad range of experimental data on flow-generated noise of duct silencers.[70]

The octave-band sound power level of the flow-generated noise, which is ap-

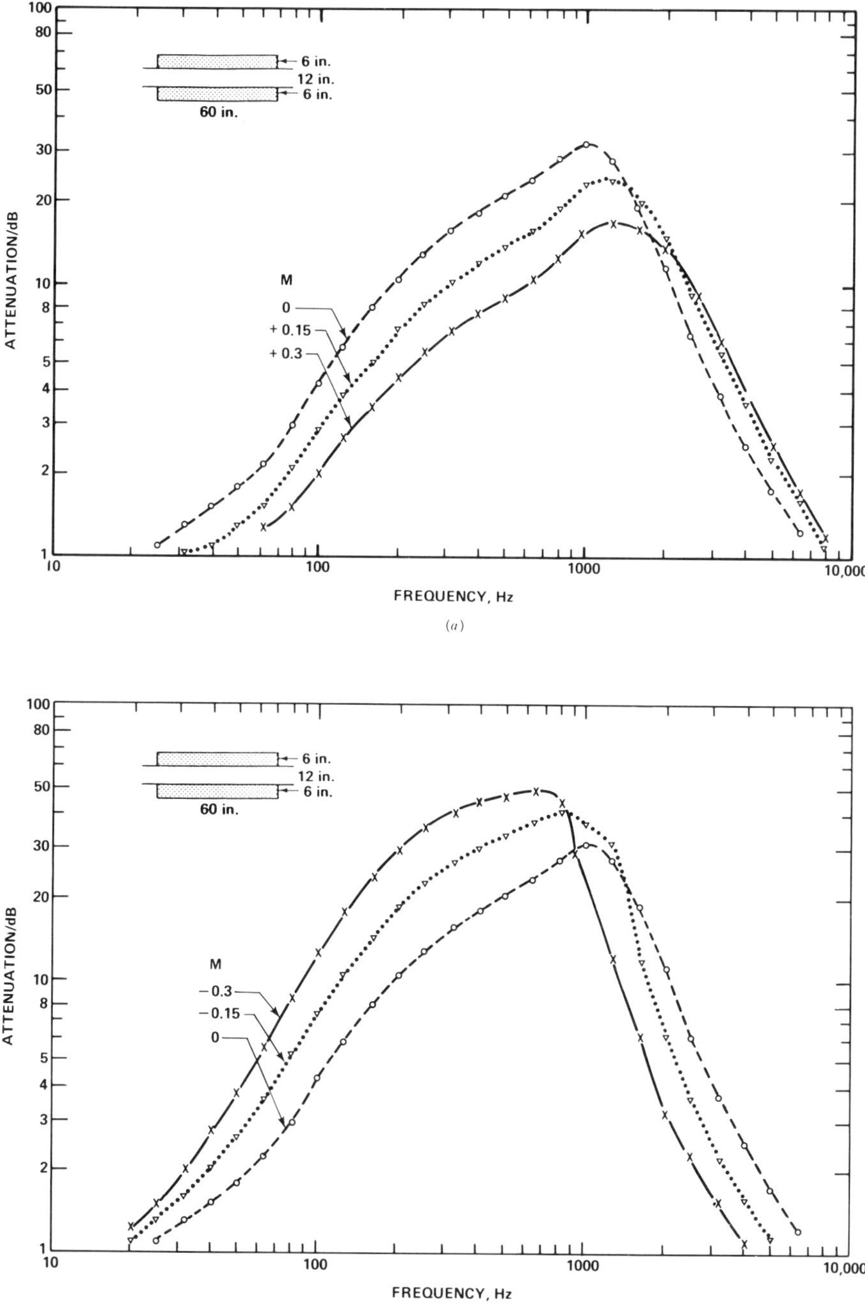

Fig. 10.29 Effect of flow on sound attenuation for Mach numbers $M = 0, 0.15, 0.3$: (*a*) sound propagation in flow direction; (*b*) sound propagation against flow direction.

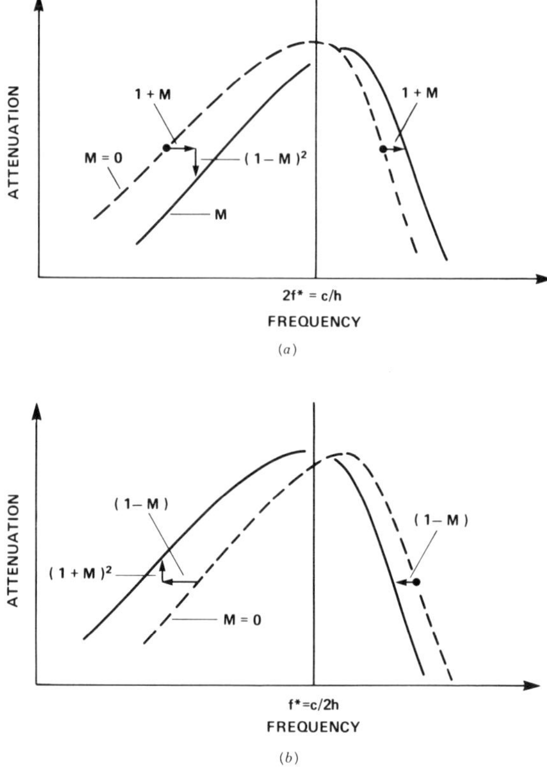

Fig. 10.30 Rules for shifting the no-flow ($M = 0$) attenuation-versus-frequency curve to account for flow effects: (*a*) sound propagation in flow direction; (*b*) sound propagation against flow direction.

proximately flat over the entire frequency region of interest, may be estimated by

$$L_{W_{\text{oct}}} = \begin{cases} -145 + 55 \log[V_F(\text{ft/min})] + 10 \log[A_F(\text{ft}^2)] \\ \quad - 45 \log(\text{P}_{\text{OA}}/100) - 25 \log \dfrac{460 + T(^\circ\text{F})}{530} \\ 8.4 + 55 \log[V_F(\text{m/s})] + 10 \log A_F(\text{m}^2) \\ \quad - 45 \log(\text{P}_{\text{OA}}/100) - 25 \log \dfrac{273 + T(^\circ\text{C})}{294} \\ \text{dB } re\ 10^{-12}\ \text{W} \end{cases} \qquad (10.41)$$

where V_F is the face velocity of the silencer (i.e., volume flow/face area), A_F is the face area of the silencer, P_{OA} is the percentage open area of the silencer (i.e., $P = 50$ for 50% open area), and T is the temperature of the air. Flow-generated

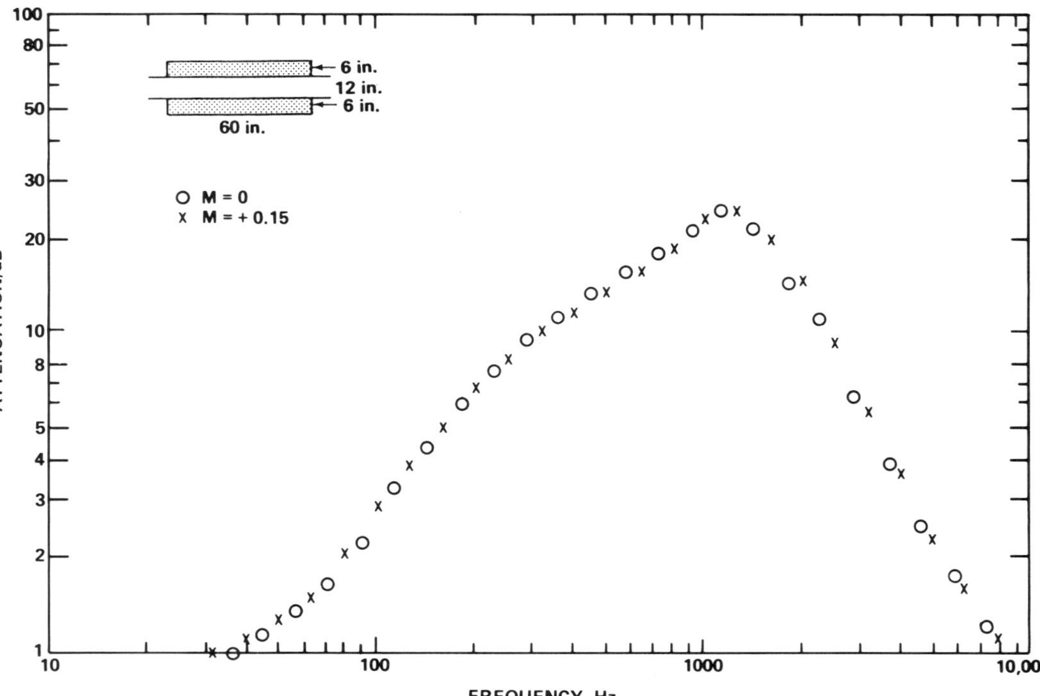

Fig. 10.31 Attenuation-versus-frequency curve of silencer for sound propagation in flow direction, $M = +0.15$: ($\times$) computed; ($\bigcirc$) Obtained by shifting the $M = 0$ attenuation-versus-frequency curve according to Fig. 10.30a.

noise should be predicted, and if found to be too high, it must be reduced to tolerable levels by choosing a larger open area for the silencer.

Prediction of Silencer Pressure Drop

The maximum permissible pressure drop at design flow rates and at design temperatures, together with the flow-generated noise, determines the cross-sectional dimensions of a silencer. It is important that the silencer designer make good use of all the available pressure drop, though with an adequate factor of safety. The maximum permissible pressure drop $\Delta p_{\max}$ must be carefully allocated among the pressure drop of transition ducts (Δp_{TRANS}) and the total pressure drop of the silencer, Δp_T:

$$\Delta p_{\max} \geq F_s[\Delta p_{\text{TRANS}} + \Delta p_T] \qquad \text{N/m}^2 \tag{10.42}$$

The specific choice of the safety factor ($F_s > 1$) is influenced by the degree of inhomogeneity of the inflow and by guaranty obligations regarding pressure drop performance of the silencer system. The silencer system usually includes the transition ducts both upstream and downstream of the silencer.

Detailed information on how to predict the pressure drop of inlet and exit transitions has been compiled by Idel'chik.[71] The silencer pressure losses are expressed as a product of the dynamic head in the muffler passage, $0.5\rho v_P^2$, and a loss coefficient K, as indicated in Eq. (10.30). The terms K_{ENT}, K_F, and K_{EX} represent the entrance, friction, and exit loss coefficients. They are predicted according to the formulas given in Table 10.2.

The required silencer face area is determined by an iteration process. First, Eq. (10.30) is solved for the passage velocity v_P. The initial face area A'_F is obtained as

$$A'_F = \frac{Q}{v_p}\frac{P_{\text{OA}}}{100} \qquad \text{m}^2 \tag{10.43}$$

where Q is the volume flow rate through the silencer (in m^3/s), v_P is the passage velocity (in m/s), and P_{OA} is the percentage of open area of the silencer cross section (e.g., $P_{\text{OA}} = 50$ for $h/d = 1$ for parallel baffles). Based on this initial value of $A_F = A'_F$, the pressure loss of the inlet and exit transitions is calculated and added to the muffler pressure drop. The total $F_S(\Delta p_{\text{TRANS}} + \Delta p_T)$ is compared with the maximum permissible pressure drop $\Delta p_{\max}$ according to Eq. (10.42). If $F_S(\Delta p_{\text{TRANS}} + \Delta p_T) > \Delta p_{\max}$, the face area of the silencer must be increased. This results in a decrease of Δp_T and a relatively smaller increase in Δp_{TRANS}. The iteration is continued until the inequality, expressed in Eq. (10.42), is satisfied.

Economic Considerations. For large silencers such as used in electric power plants, the product of the pressure drop and volume flow rate, $Q\,\Delta p$, represents a substantial power that is lost (converted to heat). The cost to produce this power

Table 10.2 Pressure Loss Coefficients for Parallel Baffle Silencers

Geometry	Loss Coefficient
Square Edge Nose 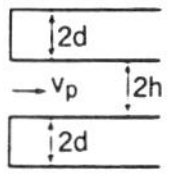	$K_{ENT} = \frac{0.5}{1 + h/d}$
Rounded Nose 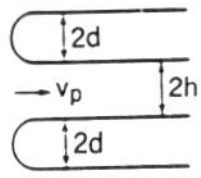	$K_{ENT} \simeq \frac{0.05}{1 + h/d}$
Typical Perforated Metal Facing 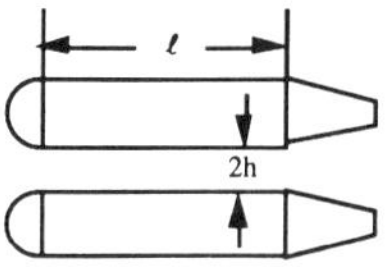	$K_F \simeq 0.0125 \frac{\ell}{h}$ ℓ = baffle length, tail and nose not included
Square Tail 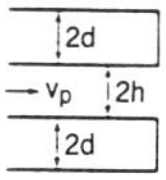	$K_{EX} = \left(\frac{1}{1 + h/d}\right)^2$
Rounded Tail 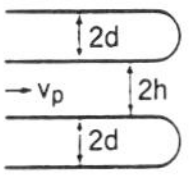	$K_{EX} = 0.7\left(\frac{1}{1 + h/d}\right)^2$
Faired Tail 7.5° 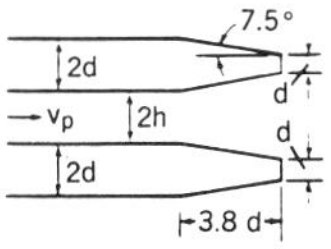	$K_{EX} = 0.6\left(\frac{1}{1 + h/d}\right)^2$

during the entire design life of the installation usually far exceeds the purchase cost of the silencer. Therefore, it is important to specify a silencer pressure drop that yields the lowest total cost. The total cost includes the present worth of revenue requirements to purchase and install the silencer (this decreases with increasing pressure drop allowance) and the present worth of revenue requirements of the energy cost caused by operating the silencer (which increases with increasing pressure drop allowance). The optimal pressure drop is that which yields the lowest total cost. Information on how to predict the optimal silencer pressure drop is given in reference 72.

REFERENCES

1. P. Morse, "Transmission of Sound in Pipes," *J. Acoust. Soc. Am.* **11**, 205–210 (1939). (See corrections in L. L. Beranek, "Sound Absorption in Rectangular Ducts," *J. Acoust. Soc. Am.* **12**, 288-231, 1940.)

2. L. Cremer, "Theory of Attenuation of Airborne Sound in a Rectangular Duct with Sound-absorbing Walls and the Maximum Achievable Attenuation," (in German), *Acustica* **3,** 249–263 (1953).
3. D. Davis, G. Stokes, D. Moore, and G. Stevens, "Theoretical and Experimental Investigation of Mufflers with Comments on Engine Exhaust Design," NACA Report 1192, 1954.
4. D. Davis, "Acoustical Filters and Mufflers," in C. Harris (ed.), *Handbook of Noise Control*, McGraw-Hill, New York, 1957, Chapter 21.
5. N. Doelling, "Dissipative Mufflers," in L. Beranek (ed.), *Noise Reduction*, McGraw-Hill, New York, 1960, Chapter 17.
6. T. Embleton, "Mufflers," in L. Beranek (ed.), *Noise and Vibration Control*, McGraw-Hill, New York, 1971, Chapter 12.
7. L. Eriksson, "Silencers," in D. Baxa (ed.), *Noise Control in Internal Combustion Engines*, Wiley, New York, 1982, Chapter 5.
8. M. Munjal, *Acoustics of Ducts and Mufflers*, Wiley, New York, 1987.
9. L. Pipes, "The Matrix Theory of Four Terminal Networks," *Philos. Mag.* **30**, 370 (1940).
10. C. Molloy, "Use of Four Pole Parameters in Vibration Calculations," *J. Acoust. Soc. Am.* **29**, 842–853 (1957).
11. J. Igarashi and M. Toyama, "Fundamentals of Acoustical Silencers(I)," Report No. 399, Aeronautical Research Institute, University of Tokyo, December 1958, pp. 223–241.
12. M. Fukuda and J. Okuda, "A study on Characteristics of Cavity-Type Mufflers," *Bull. Japanese Soc. of Mechanical Engineers* **13**(55), 96–104 (1970).
13. J. Alfredson and P. O. A. L. Davies, "Performance of Exhaust Silencer Components," *J. Sound Vib.* **15**(2), 175–196 (1971).
14. M. Munjal, "Velocity Ratio Sum Transfer Matrix Method for the Evaluation of a Muffler with Mean Flow," *J. Sound Vib.* **39**(1), 105–119 (1975).
15. P. Thawani and A. Doige, "Effect of Mean Flow and Damping on the Performance of Reactive Mufflers," *Can. Acoust.* **11**(1), 29–47 (1983).
16. D. Ronneberger, "The Acoustical Impedance of Holes in the Wall of Flow Ducts," *J. Sound Vib.* **24**, 133–150 (1972).
17. T. Melling, "The Acoustic Impedance of Perforates at Medium and High Sound Pressure Levels," *J. Sound Vib.* **29**(1), 1–65 (1973).
18. P. Dean, "An In Situ Method of Wall Acoustic Impedance Measurement in Flow Ducts," *J. Sound Vib.* **34**(1), 97–130 (1974).
19. J. Sullivan and M. Crocker, "Analysis of Concentric-Tube Resonators Having Unpartitioned Cavities," *J. Acoust. Soc. Am.* **64**(1), 207–215 (1978).
20. J. Sullivan, "A Method for Modeling Perforated Tube Muffler Components. I. Theory," *J. Acoust. Soc. Am.* **66**(3), 772–778 (1979).
21. J. Sullivan, "A Method for Modeling Perforated Tube Muffler Components. II. Applications," *J. Acoust. Soc. Am.* **66**(3), 779–788 (1979).
22. K. Jayaraman and K. Yam, "Decoupling Approach to Modeling Perforated Tube Muffler Components," *J. Acoust. Soc. Am.* **69**(2), 390–396 (1981).
23. P. Thawani and K. Jayaraman, "Modeling and Applications of Straight-through Resonators," *J. Acoust. Soc. Am.* **73**(4), 1387–1389 (1983).

24. M. Munjal, K. Rao, and A. Sahasrabudhe, ''Aeroacoustic Analysis of Perforated Muffler Components,'' *J. Sound Vib.* **114**(2), 173–188 (1987).

25. C. Young and M. Crocker, ''Prediction of Transmission Loss in Mufflers by the Finite Element Method,'' *J. Acoust. Soc. Am.* **57**(1), 144–148 (1975).

26. A. Craggs, ''A Finite Element Method for Damped Acoustic Systems: An Application to Evaluate the Performance of Reactive Mufflers,'' *J. Sound Vib.* **48**(3), 377–392 (1976).

27. Y. Kagawa and T. Omote, ''Finite Element Simulation of Acoustic Filters of Arbitrary Profile with Circular Cross Section,'' *J. Acoust. Soc. Am.* **60**(5), 1003–1013 (1976).

28. D. F. Ross, ''A Finite Element Analysis of Perforated Component Acoustic Systems,'' *J. Sound Vib.* **79**(1), 133–143 (1981).

29. R. Bernhard, ''Shape Optimization of Reactive Mufflers,'' *Noise Control Eng. J.* **27**(1), 10–17 (1986).

30. T. Tanaka, T. Fujikawa, T. Abe, and H. Utsumo, ''A Method for the Analytical Prediction of Insertion Loss of a Two-Dimensional Muffler Model Based on the Transfer Matrix Derived from the Boundary Element Method,'' *Trans.* ASME **107,** 86–91 (1985).

31. A. Seybert and C. Cheng, ''Application of Boundary Element Method to Acoustic Cavity Response and Muffler Analysis,'' *J. Vib. Acoust. Stress Reliabil. Des.* **109**, 15–21 (1987).

32. W. Eversman, ''A Systematic Procedure for the Analysis of Multiply Branched Acoustic Transmission Lines,'' *Trans. ASME* **109**, 168–177 (1987).

33. C. To and A. Doige, ''A Transient Testing Technique for the Determination of Matrix Parameters of Acoustic Systems, I: Theory and Principles,'' *J. Sound Vib.* **62**(2), 207–222 (1979).

34. C. To and A. Doige, ''A Transient Testing Technique for the Determination of Matrix Parameters of Acoustic Systems, II: Experimental Procedures and Results,'' *J. Sound Vib.* **62**(2), 223–233 (1979).

35. C. To and A. Doige, ''The Application of a Transient Testing Method to the Determination of Acoustic Properties of Unknown Systems,'' *J. Sound Vib.* **71**(4), 545–554 (1980).

36. T. Lung and A. Doige, ''A Time-averaging Transient Testing Method for Acoustic Properties of Piping Systems and Mufflers with Flow,'' *J. Acoust. Soc. Am.* **73**(3), 867–876 (1983).

37. A. Galaitsis and E. Bender, ''Measurement of Acoustic Impedance of an Internal Combustion Engine,'' *J. Acoust. Soc. Am.* **58**(Suppl. 1), 1975.

38. M. Prasad and M. Crocker, ''Insertion Loss Studies on Models of Automotive Exhaust Systems,'' *J. Acoust. Soc. Am.* **70**(5), 1339–1344 (1981).

39. M. Prasad and M. Crocker, ''Studies of Acoustical Performance of a Multi-cylinder Engine Exhaust Muffler System,'' *J. Sound Vib.* **90**(4), 491-508 (1983).

40. M. Prasad and M. Crocker, ''Acoustical Source Characterization Studies on a Multi-cylinder Engine Exhaust System,'' *J. Sound Vib.* **90**(4), 479–490 (1983).

41. D. Ross and M. Crocker, ''Measurement of the Acoustic Source Impedance of an Internal Combustion Engine,'' *J. Acoust. Soc. Am.* **74**(1), 18–27 (1983).

42. A. F. Seybert and D. F. Ross, ''Experimental Determination of Acoustic Properties Using a Two-Microphone Random Excitation Technique,'' *J. Acoust. Soc. Am.* **61**, 1362–1370 (1977).

43. J. Chung and D. Blaser, "Transfer Function Method of Measuring In-Duct Acoustic Properties II. Experiment," *J. Acoust. Soc. Am.* **68**, 914–921, (1980).

44. H. Alves and A. Doige, "A Three-Load Method for Noise Source Characterization in Ducts," *Proc. NOISE-CON*, pp. 329–334, 1987.

45. M. Prasad, "A Four Load Method for Evaluation of Acoustical Source Impedance in a Duct," *J. Sound Vib.* **114**(2), 347–356 (1987).

46. H. Boden, "Measurement of the Source Impedance of Time Invariant Sources by the Two-Microphone Method and the Two-Load Method," Department of Technical Acoustics Report, TRITA-TAK-8501, Royal Institute of Technology, Stockholm, 1985.

47. H. Boden, "Error Analysis for the Two-Load Method Used to Measure the Source Characteristics of Fluid Machines," *J. Sound Vib.* **126**(1), 173–177 (1988).

48. P. Thawani and R. Noreen, "Computer-aided Analysis of Exhaust Mufflers," ASME paper 82-WA/NCA-10, American Society of Mechanical Engineers.

49. L. Eriksson, P. Thawani, and R. Hoops, "Acoustical Design and Evaluation of Silencers," *J. Sound Vib.* **17**(7), 20–27 (1983).

50. L. Eriksson and P. Thawani, "Theory and Practice in Exhaust System Design," SAE Technical paper 850989, Society of Automotive Engineers, 1985.

51. A. Jones, "Modeling the Exhaust Noise Radiated from Reciprocating Internal Combustion Engines—A Literature Review," *Noise control Eng. J.* **23**(1), 12–31 (1984).

52. M. Munjal, "Recent Advances in the Analysis of Exhaust Mufflers," *Shock Vib. Dig.* **17**(3), 3–16 (1985).

53. L. L. Beranek, *Acoustics*, Acoustical Society of America, Woodburg, NY, 1986.

54. H. Levine and J. Schwinger, "On the Radiation of Sound from an Unflanged Circular Pipe," *Phys. Rev.* **73**(4), 383–406 (1948).

55. K. Rao and M. Munjal, "Experimental Evaluation of Perforates with Grazing flow," *J. Sound Vib.* **108**(2), 283–295 (1986).

56. M. Prasad and M. Crocker, "Evaluation of Four-Pole Parameters for a Straight Pipe with a Mean Flow and a Linear Temperature Gradient," *J. Acoust. Soc. Am.* **69**(4), 916–921 (1981).

57. M. Munjal and M. Prasad, "On Plane-Wave Propagation in a Uniform Pipe in the Presence of a Mean Flow and a Temperature Gradient," *J. Acoust. Soc. Am.* **80**(5), 1501–1506 (1986).

58. K. Peat "The transfer matrix of a uniform duct with a linear temperature gradient," *J. Sound Vib.* **123**(1), 43–53 (1988).

59. K. Ingard and H. Ising, "Acoustic Nonlinearity of an Orifice," *J. Acoust. Soc. Am.* **42**, 6–17 (1967).

60. K. U. Ingard and S. Labate, "Acoustic Circulation Effects and the Nonlinear Impedance of Orifices," *J. Acoust. Soc. Am.* **22**, 211–218 (1950).

61. L. Eriksson, "Higher Order Mode Effects in Circular Ducts and Expansion Chambers," *J. Acoust. Soc. Am.* **68**(2), 545–550 (1980).

62. L. Eriksson, "Effect of Inlet/Outlet Locations on Higher Order Modes in Silencers," *J. Acoust. Soc. Am.* **72**(4), 1208–1211 (1982).

63. J. Ih and B. Lee, "Theoretical Prediction of the Transmission Loss of Circular Reversing Chamber Mufflers," *J. Sound Vib.* **112**(2), 261–272 (1987).

64. J. Lamansuca, "The Transmission Loss of Double Expansion Chamber Mufflers with Unequal Size Chambers," *Appl. Acoust.* **24**, 15–32 (1988).

65. R. A. Scott, "The Propagation of Sound between Walls of Porous Material," *Proc. Phys. Soc. Lond.* **58**, 358–368 (1946).

66. U. J. Kurze and I. L. Vér, "Sound Attenuation in Ducts Lined with Non-Isotropic Material," *J. Sound Vib.* **24**, 177–187 (1972).

67. F. P. Mechel, "Explicit Formulas of Sound Attenuation in Lined Rectangular Ducts," *Acustica* **34**, 289–305 (1976). (in German).

68. F. P. Mechel, "Evaluation of Circular Silencers," *Acustica* **35**, 179–189 (1996) (in German).

69. H. L. Kuntz and R. M. Hoover, "The Interrelationships between the Physical Properties of Fibrous Duct Lining Materials and Lined Duct Sound Attenuation," ASHRAE Paper 3082 (RP-478), 1987.

70. I. L. Vér, "Prediction Scheme for Self-generated Noise of Silencers," *Proc. INTER-NOISE*, pp. 294–298, 1972.

71. I. D. Idel'chik, "Handbook of Hydraulic Resistance, Coefficients of Local Resistance and of Friction," translation from Russian by the Israel Program for Scientific Translations, Jerusalem 1966, U.S. Department of Commerce, Nat. Tech. Inf. Service AEC-TR-6630.

72. I. L. Vér and E. J. Wood, "Induced Draft Fan Noise Control;" Vol. 1 "Design Guide," Vol. 2 "Technical Report," Empire State Electric Energy Research Corporation, Research Report EP 82-15, January 1984.

CHAPTER 11

Vibration Isolation

ERIC E. UNGAR

Bolt Beranek and Newman, Inc.
Cambridge, Massachusetts

11.1 USES OF ISOLATION

Vibration isolation refers to the use of comparatively resilient elements for the purpose of reducing the vibratory forces or motions that are transmitted from one structure or mechanical component to another. The resilient elements, which may be visualized as springs, are called vibration isolators. Vibration isolation is generally employed (1) to protect a sensitive item of equipment from vibrations of the structure on which it is supported or (2) to reduce the vibrations that are induced in a structure by a machine it supports. Vibration isolation may also be used to reduce the transmission of vibrations to structural components whose attendant sound radiation one wishes to control.

11.2 CLASSICAL MODEL

Mass–Spring–Dashpot System

Many aspects of vibration isolation can be understood from analysis of an ideal, linear, one-dimensional, purely translational mass–spring–dashpot system like that

Noise and Vibration Control Engineering: Principles and Applications, Edited by Leo L. Beranek and István L. Vér.
ISBN 0-471-61751-2 © 1992 John Wiley & Sons, Inc.

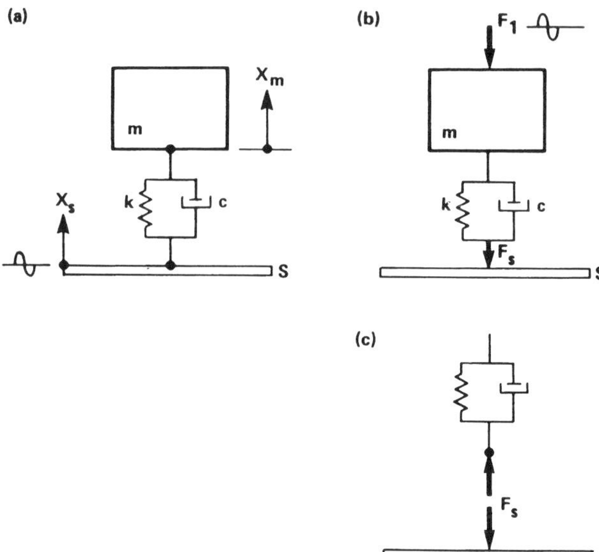

Fig. 11.1 One-dimensional translational mass–spring–dashpot system: (*a*) excited by support motion; (*b*) excited by force acting on mass; (*c*) force F_s between isolator and support.

sketched in Fig. 11.1. The isolator is represented by the parallel combination of a massless spring of stiffness k (which produces a restoring force proportional to the displacement) and a massless damper with viscous damping coefficient c (which produces a retarding force proportional to the velocity and opposing it). The rigid mass m, which here is taken to move only vertically and without rotation, corresponds either to an item to be protected (Fig. 11.1*a*) or to a machine frame on which a vibratory force acts (Fig. 11.1*b*).

Transmissibility

Although the mass–spring–dashpot diagrams of Figs. 11.1*a* and *b* are similar, they describe physically different situations. In Fig. 11.1*a*, the support S is assumed to vibrate vertically with a prescribed amplitude X_s at a given frequency; the purpose of the isolator is to keep the displacement amplitude X_m of mass m acceptably small. In Fig. 11.1*b*, a force of a prescribed amplitude F_1 at a given frequency is assumed to act on mass m; the purpose of the isolator is to keep the amplitude F_s of the force that acts on the support within acceptable limits, thereby also keeping the support's motion adequately small.

For the situation represented by Fig. 11.1*a*, the ratio $T = X_m/X_s$ of the amplitude of the mass displacement to that of the disturbing motion of the support S is called the (motion) transmissibility. For the situation represented by Fig. 11.1*b*, the ratio $T_F = F_s/F_1$ of the amplitude of the force transmitted to the support S to that of the disturbing force is called the force transmissibility. In many practical instances corresponding to force excitation as represented by Fig. 11.1*b*, the support is so stiff or massive that its displacement may be taken to be zero. It turns out that the force transmissibility T_{F0} obtained with an immobile support (for the case of Fig. 11.1*b*) is given by the same expression as the motion transmissibility (for the case of Fig. 11.1*a*),* namely,[1,2]

$$T = T_{F0} = \sqrt{\frac{1 + (2\zeta r)^2}{(1 - r^2)^2 + (2\zeta r)^2}} \tag{11.1}$$

where $r = f/f_n$ is the ratio of the excitation frequency to the natural frequency of the mass–spring system and $\zeta = c/c_c$, which is called the *damping ratio*, is the ratio of the system's viscous damping coefficient c to its critical damping coefficient c_c.

The critical damping coefficient is given by $c_c = 2\sqrt{km} = 4\pi f_n m$. The natural frequency f_n of a spring–mass system obeys[1,2]

$$2\pi f_n = \sqrt{\frac{k}{m}} = \sqrt{\frac{kg}{W}} = \sqrt{\frac{g}{X_{\text{st}}}} \tag{11.2}$$

where g denotes the acceleration of gravity, W the weight associated with the mass m, and X_{st} the static deflection of the spring due to this weight. In customary units,

$$f_n\ (\text{Hz}) \approx \frac{15.76}{\sqrt{X_{\text{st}}\ (\text{mm})}} = \frac{3.13}{\sqrt{X_{\text{st}}\ (\text{in.})}} \tag{11.3}$$

Figure 11.2 is a plot of Eq. (11.1). For small frequency ratios, $r \ll 1$, the transmissibility T is approximately equal to unity; the motion or force is transmitted essentially without attenuation or amplification. For values of r near 1.0, T becomes large (at $r = 1$ or $f = f_n$, $T = 1/2\zeta$); the system responds at resonance, resulting in amplification of the motion or force. All curves pass through unity at $r = \sqrt{2}$.

For $r > \sqrt{2}$, T is less than unity and decreases continually with increasing r. In this high-frequency range, which may be called the *isolation range*, the inertia of the mass plays the dominant role in limiting the mass excursion and thus in limiting the mass response to support displacement or to forces acting on the mass. Thus, if one desires to achieve good isolation, which corresponds to small transmissibility, one needs to choose an isolator with the smallest possible f_n (i.e., with the largest practical static deflection X_{st}) in order to obtain the greatest value of $r = f/f_n$ for a given excitation frequency.

*This equality is a consequence of the principle of reciprocity.[3] It holds not only for the simple system of Fig. 11.1 but also for any array of masses, linear springs, and dampers in which motion can occur only parallel to one axis. Because of this equality, the literature generally does not differentiate between the two types of transmissibility.

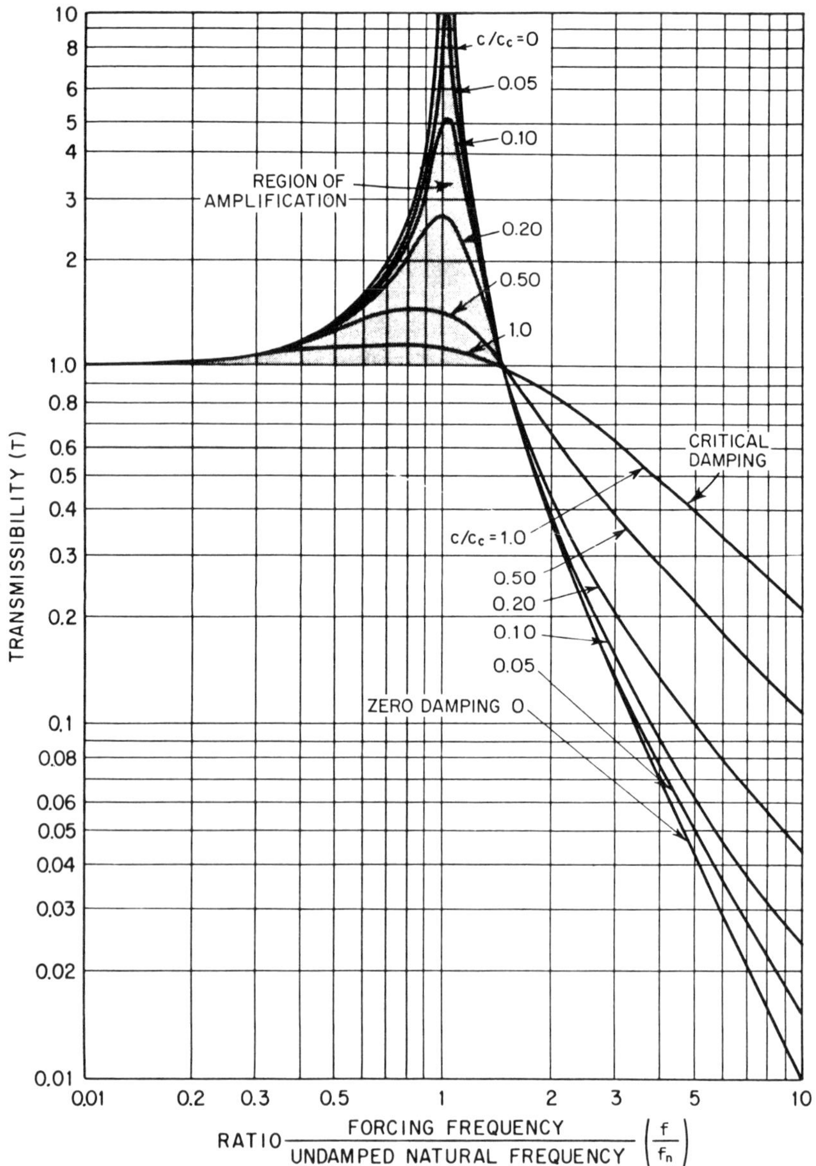

Fig. 11.2 Transmissibility of mass–spring–dashpot system of Fig. 11.1 from Eq. (11.1). (After ref. 4.)

Isolation Efficiency

The performance of an isolation system is sometimes characterized by the isolation efficiency I, which is given by $I = 1 - T$. Whereas the transmissibility indicates the fraction of the disturbing motion or force that is transmitted, the isolation efficiency indicates the fraction by which the transmitted disturbance is less than the excitation. Isolation efficiency is often expressed in percent. For example, if the transmissibility is 0.0085, the isolation efficiency is 0.9915, or 99.15%, indicating that 99.15% of the disturbance does not "get through" the isolator.

Effect of Damping

In the isolation range, that is, for frequency ratios $r > \sqrt{2}$, increased damping results in increased transmissibility, as evident from Fig. 11.2 or Eq. (11.1). Although this fact is mentioned in many texts, it is of little practical consequence for two reasons. First, in practice one rarely encounters systems with damping ratios that are greater than 0.1 unless high damping is designed into a system on purpose—and small values of the damping ratio, and certainly small changes in that damping ratio, have little effect on transmissibility. Second, Eq. (11.1) and Fig. 11.2 pertain only to viscously damped systems, in which a damper produces a retarding force that is proportional to the velocity. Although such systems have been studied most extensively (largely because they are relatively easy to analyze mathematically), the retarding forces in practical systems generally have other parameter dependences.

The effect of damping in practical isolators is generally better represented by *structural damping* than by viscous damping. In structural damping, the retarding force acts to oppose the motion, as in viscous damping, but is proportional to the displacement. For structurally damped systems,[2]

$$T = F_{F0} = \sqrt{\frac{1 + \eta^2}{(1 - r^2)^2 + \eta^2}} \tag{11.4}$$

where η denotes the *loss factor* of the system.

If a structurally damped system has the same amplification at the natural frequency as a similar viscously damped system, then $\eta = 2\zeta$. The transmissibility of a structurally damped system in the isolation range increases much less rapidly with increasing damping than is indicated in Fig. 11.2, which pertains to systems with viscous damping.

In practical isolation arrangements the damping typically is very small, that is, $\zeta < 0.1$. For such small amounts of damping, the transmissibility in the isolation range differs little from that for zero damping, so that one may approximate the transmissibility in that range by

$$T = T_{F0} = \frac{1}{|r^2 - 1|} \approx \left(\frac{f_n}{f}\right)^2 \tag{11.5}$$

where the rightmost expression applies for $r^2 = (f/f_n)^2 >> 1$.

Effects of Inertia Bases

An isolated machine is often mounted on a massive support, generally called an *inertia base*, to increase the isolated mass. If the isolators are not changed as the mass is increased, then the natural frequency of the system is reduced, the frequency ratio r corresponding to a given excitation frequency is increased, and the transmissibility is reduced; that is, isolation is improved.

However, practical considerations, such as the load-carrying capacities of isolators, usually dictate that the isolator stiffness be increased* as the mass is increased, with the static deflection essentially remaining unchanged. (In fact, conventional commercial isolators typically are specified in terms of the loads they can carry and the corresponding static deflection.) As evident from Eq. (11.3), the natural frequency remains unchanged if the static deflection remains unchanged, and addition of an inertia base then does not change the transmissibility.

Thus, inertia bases in practice typically do not improve isolation. They are of some benefit, however: With a greater mass supported on stiffer springs, forces that act on the isolated mass produce smaller static and vibratory displacements of that mass.

Effect of Machine Speed

In a rotating or reciprocating machine, the dominant excitation forces generally are due to dynamic unbalance and occur at frequencies that correspond to the machine's rotational speed and/or multiples of that speed. Greater speed thus corresponds to increased excitation frequencies and larger values of r—and therefore to reduced transmissibility (or improved isolation). As evident from Eq. (11.1) or (11.5), the transmissibility—the ratio of the transmitted to the excitation force—becomes smaller as the speed (and the excitation frequency) is increased; in the isolation range, the transmissibility is very nearly inversely proportional to the square of the excitation frequency. However, the excitation forces associated with unbalance vary as the square of speed (and frequency) and thus increase with speed about as much as the transmissibility decreases. The net result is that the magnitudes of the forces that are transmitted to the supporting structure are virtually unaffected by speed changes, although speed changes do affect the frequencies at which these forces occur.

Limitations of Classical Model

Although the linear single-degree-of-freedom model provides some useful insights into the behavior of isolation systems, it obviously does not account for many aspects of realistic installations. Clearly, real springs are not massless and may be nonlinear, and real machine frames and supporting structures are not rigid. Resiliently supported masses generally move not only vertically but also horizontally—and they also tend to rock.

In simple classical analyses, furthermore, the magnitude of the exciting force or motion is taken as constant and independent of the resulting response, whereas

*Note that the single spring and dashpot in the schematic diagrams of Fig. 11.1 represent the entire isolation system, which in reality may consist of many isolators. Addition of isolators amounts to an increase in the stiffness of the isolation system.

in actual situations the excitation often depends significantly on the response, as discussed in Section 11.4 under "Loading of Sources."

11.3 ISOLATION OF THREE-DIMENSIONAL MASSES

General

Unlike the simple model shown in Fig. 11.1, where the mass can only move vertically without rotation and the system has only one natural frequency, an actual three-dimensional mass has six degrees of freedom; it can translate in three coordinate directions and rotate about three axes. An elastically supported rigid mass thus has six natural frequencies; a nonrigid mass has many more. Obtaining effective isolation here requires that all of the natural frequencies fall considerably below the excitation frequencies of concern. Descriptions of the natural frequencies and responses of general isolated rigid masses are available[5] but are so complex that they provide little practical insight and tend to be used only rarely for design purposes.

Coupling of Vertical Motion and Rocking

Figure 11.3 is a schematic diagram of a mass m supported on two different isolators (or on two rows of isolators extending in the direction perpendicular to the plane of the paper) with stiffnesses k_1 and k_2 located at distances a_1 and a_2 from the mass center of gravity. One may visualize easily that a downward force applied at the center of gravity of the mass in general would produce not only a downward displacement of the center of gravity but also a rotation of the mass, the latter due to the moment resulting from the isolator forces. Similarly, purely vertical up-and-down motion of the support S would in general result in rocking of the mass, in addition to its vertical translation. Vertical and rocking motions here are said to be "coupled."

The natural frequency f_v at which pure vertical vibration would occur if the mass would not rock (i.e., the "uncoupled" vertical natural frequency) is given by

$$2\pi f_v = \sqrt{\frac{k_1 + k_2}{m}} \tag{11.6}$$

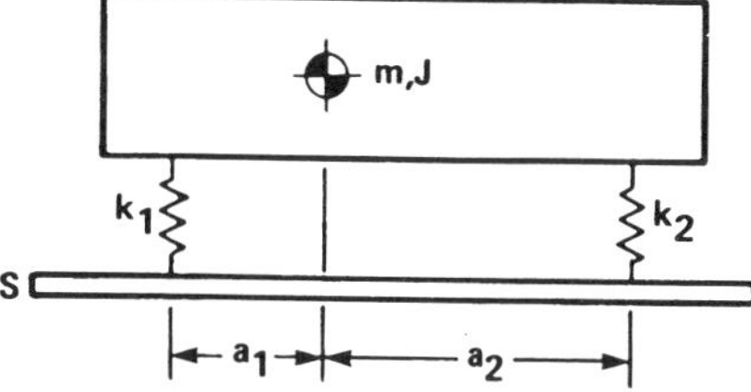

Fig. 11.3 Mass m with moment of inertia J supported on two isolators.

The natural frequency f_r at which the mass would rock if its center of gravity would not move vertically (i.e., the uncoupled rocking frequency) is given by

$$2\pi f_r = \sqrt{\frac{k_1 a_1^2 + k_2 a_2^2}{J}} \tag{11.7}$$

where J denotes the mass polar moment of inertia about an axis through the mass center of the gravity and perpendicular to the plane of the paper.

The rocking frequency f_r may be either smaller or larger than f_v. In the general case, the two natural frequencies of the system differ from both of these frequencies and correspond to motions that combine rotation and vertical translation (but without rotation out of the plane of the paper). These "coupled" motion frequencies always lie above and below the foregoing uncoupled motion frequencies; that is, one of the coupled motion frequencies lies below both f_v and f_r and the other lies above both f_v and f_r; coupling in effect increases the spread between the natural frequencies.[6,7]

The presence of coupled motions complicates the isolation problem because one needs to ensure that both of the coupled natural frequencies fall considerably below the excitation frequencies of concern. In order to eliminate this complication, one may select the locations and stiffnesses of the isolators so that the forces produced by them when the mass moves downward without rotation result in zero net moment about the center of gravity. This situation occurs if the isolators are designed so that they have the same static deflection under the static loads to which they are subject. (Selection of isolators that have the same unloaded height as well as the same static deflection results also in keeping the isolated equipment level.) Then f_v and f_r of Eqs. (11.6) and (11.7) are the system's actual natural frequencies, and vertical forces at the center of gravity or vertical support motions produce no rocking.

Effect of Horizontal Stiffness of Isolators

In the foregoing discussion, horizontal translational motions and the effects of horizontal stiffnesses of the isolators were neglected. However, any real isolator that supports a vertical load also has a finite horizontal stiffness, and in some systems, separate horizontally acting isolators may be present. Figure 11.4 is a schematic diagram of a mass on two vertical isolators (or on two rows of such isolators extending perpendicular to the plane of the paper) with stiffnesses k_1 and k_2. The effects of the horizontal stiffnesses of these isolators or of separate horizontally acting isolators are represented by the horizontally acting spring elements with stiffnesses h_1 and h_2, here assumed to act in the same plane at a distance b below the center of gravity.

The system of Fig. 11.4 has three degrees of freedom and therefore three natural frequencies. If the vertically acting isolators are selected and positioned so that they have the same static deflection, then (as discussed in the foregoing section)

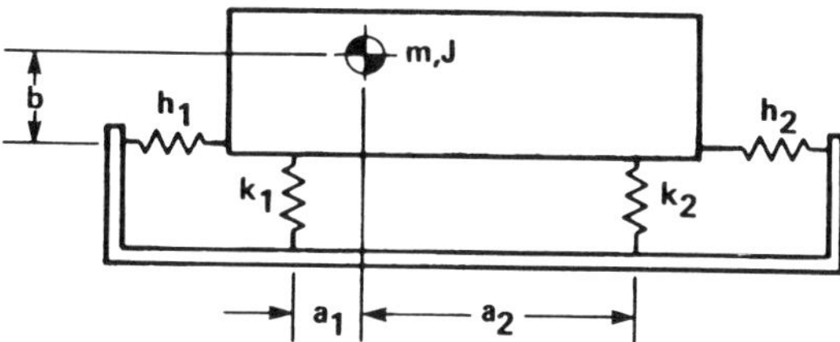

Fig. 11.4 Mass m with moment of inertia J supported by two vertically and two collinear horizontally acting isolators.

there is very little coupling between the vertical translational and the rocking motions, and the natural frequency corresponding to vertical translation is given by Eq. (11.6). The other two natural frequencies, which are associated with combined rotation and horizontal translation, are given by the two values of f_H one may obtain from[6,7]

$$\frac{f_H}{f_v} = (N \pm \sqrt{N^2 - SB})^{1/2} \tag{11.8}$$

where

$$N = \frac{1}{2}\left[S\left(1 + \frac{b^2}{r^2}\right) + B\right] \qquad B = \frac{a_1^2 k_1 + a_2^2 k_2}{r^2(k_1 + k_2)} \qquad S = \frac{h_1 + h_2}{k_1 + k_2} \tag{11.8a}$$

Here, $r^2 = J/m$ represents the square of the radius of gyration of the mass about an axis through its center of gravity.

For a rectangular mass of uniform density, the center of gravity is in the geometric center and the radius of gyration r obeys $r^2 = \frac{1}{12}(H^2 + L^2)$, where H and L represent the lengths of the vertical and horizontal edges of the mass.

Nonrigidity of Mass or Support

If the mass of Fig. 11.1*a* is flexible instead of rigid and has a resonance at a certain frequency, then it will deflect considerably in response to excitation at that frequency, resulting in large transmissibility and thus in poor isolation. An analogous situation occurs if the system of Fig. 11.1*b* is excited at a resonance frequency of the support.

Figure 11.5 is a schematic diagram representing the vertical translational motion of a machine isolated from a nonrigid support, where the support is represented by a spring–mass system, corresponding, for example, to the static stiffness of a building's floor and the effective mass that participates in the floor's vibration at its fundamental resonance. Isolator and floor damping have little effect on the off-resonance vibrations and may be neglected for the sake of simplicity. If f_M denotes the natural frequency of the isolated machine on a rigid support and f_S

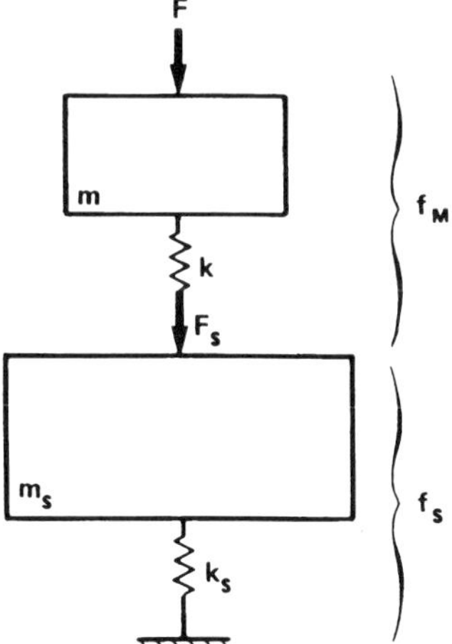

Fig. 11.5 Schematic representation of isolated machine on nonrigid support structure. Machine is represented by mass m subject to oscillatory force F; isolator is represented by spring k. Support structure is represented by effective mass m_s on spring k_s.

represents the natural frequency of the support without the machine in place, so that

$$2\pi f_M = \sqrt{\frac{k}{m}} \qquad 2\pi f_S = \sqrt{\frac{k_s}{m_s}} \tag{11.9}$$

then the force transmissibility, that is, the ratio of the magnitude of the force F_S transmitted to the support to that of the excitation force F, may be written as

$$T_F = \frac{F_S}{F} = \left| \frac{1 - R^2}{(1 - R^2)(1 - R^2G^2) - R^2/M} \right| \tag{11.10}$$

where $R = f/f_S$, $G = f_S/f_M$, and $M = m/m_s$ with f representing the excitation frequency.

Figure 11.6 shows a plot of the transmissibility calculated from Eq. (11.10) for the illustrative case where $M = \frac{1}{2}$, $G = 2$, together with a corresponding plot of the transmissibility that would be obtained with an immobile (infinitely rigid) support. For excitation frequencies that fall between the two resonance frequencies of the system with the nonrigid support, the transmissibility obtained with the nonrigid support is less than that with the rigid support; the reverse is true for excitation frequencies above the upper of the two resonance frequencies. For sufficiently high excitation frequencies, the difference between the two transmissibilities becomes negligible.

The two resonance frequencies f_c of the system with the nonrigid support may be found from

$$\left(\frac{f_c}{f_M}\right)^2 = P \pm \sqrt{P^2 - G^2} \qquad P = \tfrac{1}{2}(1 + G^2 + M) \tag{11.11}$$

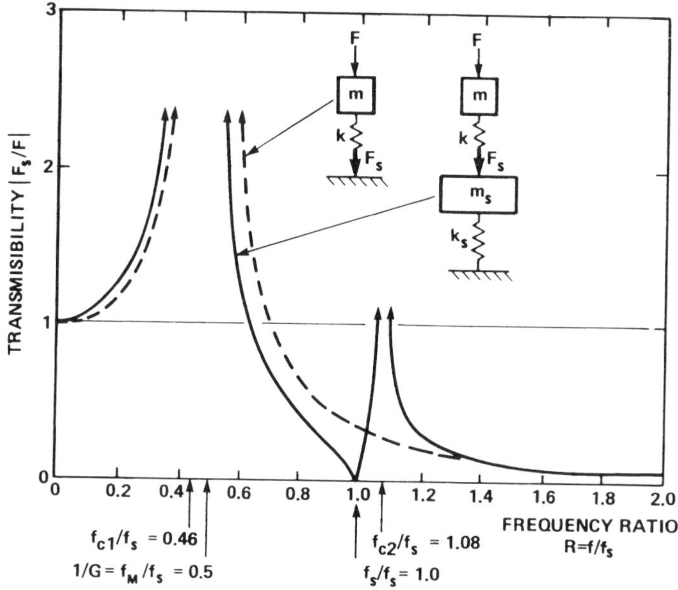

Fig. 11.6 Comparison of transmissibilities obtained with rigid and nonrigid supports. Curves were obtained from Eqs. (11.1) and (11.10) for zero damping and for mass ratio $M = m/m_s = \frac{1}{2}$ and system frequency ratio $G = f_S/f_M = 2$. Resonance frequencies f_{c1} and f_{c2} of coupled system and f_n and f_S of uncoupled machine and support are indicated on frequency scale.

with G and M defined as indicated after Eq. (11.10). The higher resonance frequency, obtained with the plus sign in Eq. (11.11), always lies above both f_M and f_S; the lower resonance frequency, obtained with the minus sign, always lies below both f_M and f_S.

11.4 HIGH-FREQUENCY CONSIDERATIONS

Although the relations discussed in the foregoing sections usually suffice for analysis and design in cases where only relatively low frequencies are of concern (i.e., for frequencies that are generally below the resonance frequencies of the machine and support structures themselves), at higher frequencies there occur complications that may need to be taken into account. At high frequencies, the isolated item and the support structure cease to behave as rigid masses, isolators may exhibit internal resonances, and vibration sources tend to be affected by "loading."

Loading of Sources

Loading of a source refers to the reduction of the source's vibratory motion that results from a force that opposes this motion. For example, the vibrations of a sheet metal enclosure tend to be reduced by the opposing forces produced when a hand is placed against it. Similarly, a factory floor that vibrates with a certain amplitude tends to vibrate less when a (nonvibrating) machine is bolted to it as the result of reaction forces produced by the machine.

In the latter example, if the machine is isolated from the floor, it is likely to produce smaller reaction forces on the floor than it would produce if it is bolted rigidly to the floor, and the floor may be expected to vibrate more if the machine is isolated than if it is rigidly fastened. Since use of isolation here increases the machine's excitation by the floor, one needs to evaluate how much protection of the machine the isolator provides in this case.

In order to be able to carry out such an evaluation, one needs a quantitative description of the source's response to loading. One may obtain such a description by measuring the motion produced by the source as it acts or several different masses or structures with different dynamic characteristics (impedances), which produce different (known or measurable) reaction forces. For prediction purposes, it often suffices to assume that the source's response to loading at any given frequency is "linear"; that is, the motion amplitude produced by the source decreases in proportion to the amplitude F_0 of the reaction force (which is equal to the amplitude of the force that the source generates). One may describe source motions at a given frequency equally well in terms of acceleration, displacement, or velocity, but use of velocity has become customary. For a linear source one may express the dependence of the source's velocity amplitude V_0 on the force as[8]

$$V_0 = V_{\text{free}} - M_s F_0 \tag{11.12}$$

Here V_{free} represents the velocity amplitude that the source generates if it is free of reaction forces, that is, if it produces zero force. (In general, the constants V_{free} and M_s for a given source may be different for different frequencies.) The quantity M_s, which indicates how rapidly V_0 decreases with increasing F_0, is known as the *source mobility* and may be found from $M_s = V_{\text{free}}/F_{\text{blocked}}$, where F_{blocked} denotes the force amplitude obtained if the source is "blocked" so that it has zero output velocity V_0.

One may readily verify that $M_s = 0$ corresponds to a *velocity source*, that is, to a source whose output velocity amplitude is constant, regardless of the magnitude of its output force F_0. Similarly, infinite M_s corresponds to a *force source*, whose output force F_0 is constant and independent of its output velocity.[8] (A rotating unbalanced mass generates forces that are virtually independent of its support motions and thus acts essentially like a force source. On the other hand, a piston driven by a shaft with a large flywheel moves at the same amplitude regardless of the force acting on it and thus behaves like a velocity source.)

Isolation Effectiveness[8]

In the presence of significant source loading, one cannot evaluate the performance of an isolation system on the basis of transmissibility because in the definition of transmissibility the magnitudes of the disturbances are prescribed and thus, in effect, are taken as constant. A measure of isolation performance that is useful in the presence of loading is the so-called *isolation effectiveness E*. Isolation effectiveness is defined as the ratio of the magnitude of the vibrational velocity of the item to be protected (called the "receiver") that results if the item is rigidly connected to the source to the magnitude of the receiver's velocity that is obtained if the isolator is inserted between the source and the receiver in place of the rigid connection. The definition of isolation effectiveness is analogous to that of insertion loss in airborne acoustics.

If receiver velocity V_R is proportional to the force F_R that acts on the receiver, so that $V_R = M_R F_R$, where M_R is called receiver mobility,* then the isolation effectiveness may be expressed in terms of a ratio of forces acting on the receiver as well as of a ratio of receiver velocities, namely,

$$E = \frac{V_{Rr}}{V_{Ri}} = \frac{F_{Rr}}{F_{Ri}} \tag{11.13}$$

where the added subscript r refers to the case in which a rigid connection replaces the isolator and the subscript i refers to the situation where the isolator is present.

Whereas small transmissibility T corresponds to good isolation, it is large values of E that imply effective isolation. For this reason, the reciprocal of the effectiveness is sometimes used to characterize the performance of an isolation system. Although this reciprocal differs from the transmissibility T in the general situation where the source is affected by loading, in the special case of sources that are unaffected by loading (i.e., for sources that generate load-independent velocity or force amplitudes), $E = 1/T$.

Effectiveness of Massless Linear Isolator

An isolator may be considered as "massless" if it transmits whatever force is applied to it. (Equal and opposite forces must act on the two sides of a massless isolator if it is not to accelerate infinitely.) A "linear" isolator is one whose extension or shortening is proportional to the applied force. The velocity difference across such an isolator at any frequency then is also proportional to the applied force, and the ratio of the magnitude of this velocity difference to that of the applied force is called the isolator's mobility M_I. Like the reciprocal of isolator stiffness, M_I is large for a soft isolator and zero for a rigid isolator.

For a massless linear isolator, the effectiveness obeys[8,11]

$$E = \left| 1 + \frac{M_I}{M_S + M_R} \right| \tag{11.14}$$

*Mobilities of receivers at their attachment points may be estimated analytically for simple configurations (e.g., refs. 9 and 10) or may be measured (as functions of frequency). The velocities and forces usually are expressed in terms of complex numbers, or *phasors*, which indicate both the magnitudes and the relative phases of sinusoidally varying quantities. Mobilities then also are complex quantities in general.

At a resonance of the receiver, the receiver's vibratory velocity V_R resulting from a given force F_R is large, so that the receiver's mobility is large. In view of Eq. (11.14), the effectiveness of an isolator is small in the presence of such a resonance.

Equation (11.14) also indicates that the effectiveness is small if the source mobility M_S is large. For a force source, for which M_S is infinite, the effectiveness is equal to unity, implying that the receiver vibrates just as much with the isolator in place as it does if the isolator is replaced by a rigid connection. This initially somewhat surprising result is correct: After all, the force source generates the same force, regardless of the velocity or displacement it produces, and the isolator transmits all of this force, with a softer isolator merely leading to greater displacement of the source at its output point.

Consequences of Isolator Mass Effects

Isolator mass effects may be neglected—that is, an isolator may be considered as massless—as long as the frequencies under consideration are appreciably lower than the first internal or *standing-wave* resonance frequency of the isolator.* Such standing-wave resonances tend to reduce the isolator's effectiveness severely, as illustrated by Fig. 11.7. This figure shows the calculated transmissibility of a leaf spring modeled as a uniform cantilever beam. The upper left-hand corner of the plot may be recognized as the usual transmissibility curve (similar to Fig. 11.2) in the vicinity of the resonance frequency $f_n = (1/2\pi)\sqrt{k/m}$ obtained for a massless isolator. With increasing excitation frequency the transmissibility does not decrease monotonically, as it would for a massless spring (see curve for $m_{\mathrm{sp}} = 0$); instead, there occur secondary peaks associated with standing-wave resonances. The frequency at which these peaks begin to occur increases as the ratio of the isolated mass m to the mass m_{sp} of the spring increases. Although in the figure only two peaks are shown for each mass ratio, there actually occurs a succession of peaks that become more closely spaced with increasing frequency. The magnitude of these peaks decreases with increasing damping.

In order to reduce the effects of standing-wave resonances, one thus needs to select an isolator with relatively high damping and a configuration for which the onset of standing-wave resonances occurs at comparatively high frequencies. This implies use of a material with high stiffness-to-weight ratio or, equivalently, with a high longitudinal wave velocity $\sqrt{E/\rho}$, where E denotes the material's modulus of elasticity and ρ its density, and also use of a configuration with small overall dimensions.

The isolation effectiveness at any specified frequency of a system in which iso-

*At lower frequencies, the only effect of the mass of the isolator is to reduce the fundamental resonance frequency of the system slightly. The modified resonance frequency may be calculated from the isolator stiffness and a mass consisting of the isolated mass plus a fraction of the mass of the isolator. If the isolator consists of a uniform spring or pad in compression or shear, the fraction is $\frac{1}{3}$; if the isolator consists of a uniform cantilever beam, the fraction is approximately 0.24.

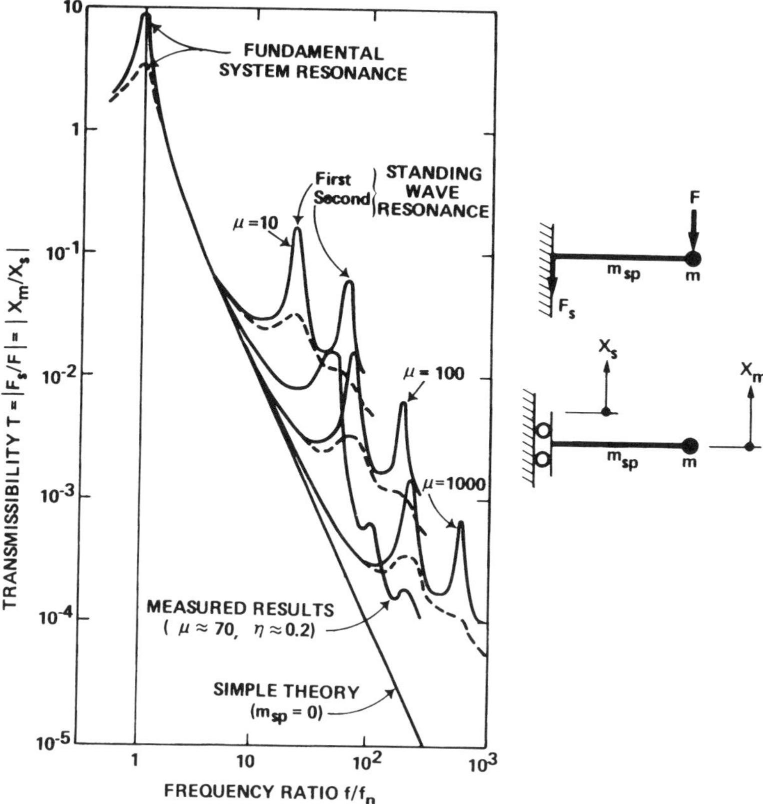

Fig. 11.7 Effect of isolator mass and damping on transmissibility of uniform cantilever. (After refs. 11 and 12.) For three values of ratio $\mu = m/m_{sp}$ of isolated mass to mass of cantilever spring. Solid calculated lines are for loss factor $\eta = 0.1$; dashed lines for $\eta = 0.6$. (Measured result shown is from ref. 13.) Frequency is normalized to f_n, the fundamental resonance frequency obtained with a massless spring.

lator mass effects are not negligible is given by[8]

$$
\begin{aligned}
E &= \left|\frac{\alpha}{M_S + M_R}\right| \left|1 + \frac{M_S}{M_{Isb}} + \frac{M_R}{M_{Irb}}\left(1 + \frac{M_S}{M_{Isf}}\right)\right| \\
\frac{1}{\alpha^2} &= \frac{1}{M_{Irb}}\left(\frac{1}{M_{Isb}} - \frac{1}{M_{Isf}}\right)
\end{aligned}
\qquad (11.15)
$$

where M_{Isb} denotes the isolator mobility (i.e., the velocity-to-force ratio) measured on the source side of the isolator if the receiver side of the isolator is "blocked" (i.e., prevented from moving), M_{Irb} denotes the mobility measured on the receiver side of the isolator if the source side is blocked, and M_{Isf} denotes the mobility measured on the source side if the receiver side is "free" or unconstrained. One may readily verify that Eq. (11.15) reduces to Eq. (11.14) for massless isolators, for which $M_{Isb} = M_{Irb} = M_I$ and M_{Isf} is infinite.

11.5 TWO-STAGE ISOLATION

A force applied to one side of a massless isolator, as has been mentioned, must be balanced by an equal and opposite force at the other side of the isolator. This is not the case for an isolator that incorporates some mass because the force applied to one side then is balanced by the sum of the inertia force and the force acting on the other side of the isolator. Thus, unlike for a massless isolator, the force transmitted by an isolator with mass can be less than the applied force.

One may realize such a force reduction benefit even at low frequencies, at which mass effects in isolators themselves are negligible, by adding a lumped mass "inside" an isolator. One may visualize this concept by considering a spring that is cut into two lengths with a rigid block of mass welded in place between the two parts, resulting in an isolator consisting of two lengths of spring with a mass between them. If a mass m is mounted atop this isolator, one obtains a system that may be represented by a diagram like that of Fig. 11.8*a*. Because this system consists of a cascade of two spring–mass systems, it is said to have two stages of isolation.

Transmissibility

The system of Fig. 11.8*a* has two natural frequencies f_b that may be found from

$$\left(\frac{f_b}{f_0}\right)^2 = Q \pm \sqrt{Q^2 - B^2} \qquad Q = \frac{1}{2}\left(B^2 + 1 + \frac{k_2}{k_1}\right) \tag{11.16}$$

where

$$B = \frac{f_I}{f_0} \qquad 2\pi f_I = \sqrt{\frac{k_1 + k_2}{m_I}} \qquad 2\pi f_0 = \frac{1}{\sqrt{m(1/k_1 + 1/k_2)}} \tag{11.16a}$$

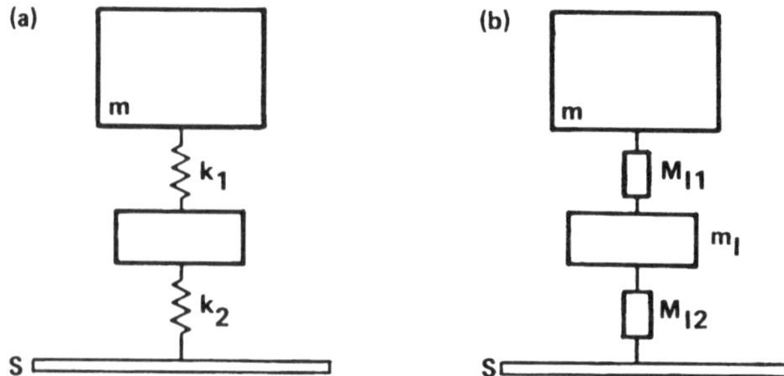

Fig. 11.8 Two-stage isolation with mass m_I in isolator system: (*a*) springs; (*b*) general isolation elements.

The frequency f_0 is the natural frequency of the system in the absence of any included mass m_I; that is, it is the natural frequency of a conventional simple single-stage system. The frequency f_I is the natural frequency of mass m_I moving between the two springs, with mass m held completely immobile. The upper frequency f_b, which one obtains if one uses the plus sign before the square root, always is greater than both f_0 and f_I; the lower frequency f_b, corresponding to the minus sign, always falls below both f_0 and f_I.

The transmissibility of a two-stage system like that of Fig. 11.8*a* obeys

$$\frac{1}{T} = \frac{1}{T_{F0}} = \frac{1}{B^2}\left(\frac{f}{f_0}\right)^4 - \left[1 + \frac{1 + k_2/k_1}{B^2}\right]\left(\frac{f}{f_0}\right)^2 + 1 \approx \left(\frac{f^2}{f_0 f_I}\right)^2 \quad (11.17)$$

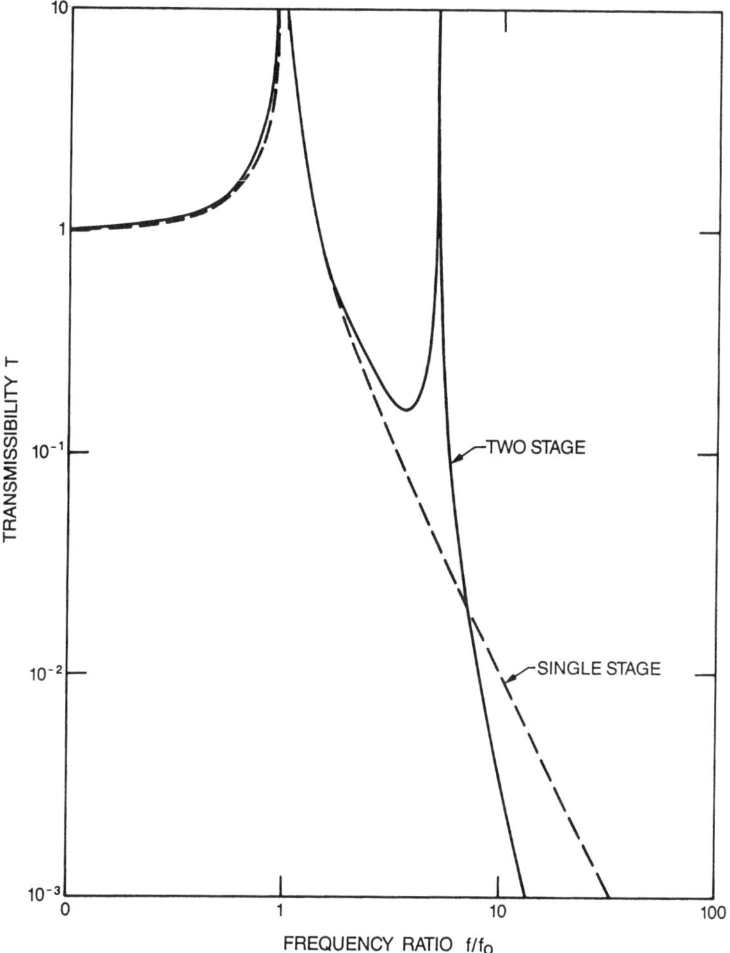

Fig. 11.9 Transmissibility of two-stage system. Calculated from Eq. (11.17) for $k_2/k_1 = 1$ and $B = f_I/f_0 = 5$.

where the last approximate expression applies for high frequencies, namely, for excitation frequencies that are much greater than both f_0 and f_I.

Figure 11.9 shows an illustrative plot of the transmissibility of a two-stage system with $B = f_I/f_0 = 5$ and $k_2/k_1 = 1$ together with a plot of the transmissibility of an (undamped) single-stage system. The second natural frequency of the two-stage system (at $f/f_0 \approx 5.1$) is clearly evident, as is the subsequent rapid decrease in that system's transmissibility with increasing frequency. At high frequencies, that is, a little above the aforementioned second natural frequency, the transmissibility of the two-stage system may be seen to be smaller than that of a single-stage system with the same fundamental natural frequency. As evident from Eq. (11.17), the high-frequency transmissibility of a two-stage system varies inversely as the fourth power of the excitation frequency, whereas Eq. (11.5) indicates that the transmissibility of a single-stage system varies inversely as only the second power of the excitation frequency.

Isolation Effectiveness

Although the foregoing results apply strictly only to isolators without damping, they also provide a reasonable approximation to the behavior of lightly damped systems, except near the natural frequencies. In order to account for high damping or more complicated linear isolator configurations (e.g., where each isolator is modeled by various series and parallel combinations of springs and dampers), it is convenient to represent each isolator by its mobility. A corresponding diagram appears in Fig. 11.8*b*.

As has been discussed in Section 11.4, the source mobility M_S is a measure of a vibration source's susceptibility to loading effects, and the isolation effectiveness E is a measure of the isolation performance, which unlike transmissibility takes loading effects into account. If one takes S in Fig. 11.8*b* to represent a general linear source and replaces the mass m by a general linear receiver with mobility M_R, one obtains a general linear two-stage system whose effectiveness one may write as[8]

$$E = |E_1 + \Delta E| \qquad E_1 = 1 + \frac{M_I}{M_S + M_R} \tag{11.18}$$

where

$$M_I = M_{I1} + M_{I2} \qquad \Delta E = \frac{(M_{I1} + M_S)(M_{I2} + M_R)}{M_m(M_S + M_R)} \tag{11.18a}$$

One may recognize E_1 as corresponding to the effectiveness of a single-stage system [see Eq. (11.14)], that is, to a two-stage system with zero included mass, with

M_I denoting the mobility of the two partial isolators in series. Thus, ΔE represents the effectiveness increase obtained by addition of the included mass m_I, whose mobility is M_m. Note that ΔE is inversely proportional to M_m, indicating that greater included masses generally result in greater effectiveness increases.

Optimization of Isolator Stiffness Distribution

Once one has selected the mobility M_I of the total isolator (or, equivalently, its compliance or stiffness), one needs to consider how to allocate this mobility among the components M_{I1} and M_{I2}. If one lets r_1 denote the fraction of the total mobility on the source side of the included mass, so that $M_{I1} = r_1 M_I$ and $M_{I2} = (1 - r_1)M_I$, it turns out that one may obtain the largest value of ΔE, namely,

$$\Delta E_{\max} = \frac{(M_I + M_s + M_R)^2}{4M_m(M_s + M_R)} \tag{11.19}$$

by making r_1 equal to its optimum value,*

$$r_{\text{opt}} = \frac{1}{2}\left(1 + \frac{M_R - M_S}{M_I}\right) \tag{11.20}$$

In view of Eq. (11.18a), in an efficient isolation system M_{I1} must be considerably grater than M_S, and also M_{I2} must be considerably greater than M_R. For such a system, one finds that $r_{\text{opt}} \approx \frac{1}{2}$ and that $\Delta E_{\max}$ may be approximated by replacing the expression in the parentheses of the numerator of Eq. (11.19) by M_I. Thus, if the total mobility of the isolator is sufficiently great, that is, if the total stiffness of the isolator is sufficiently small, one may generally obtain the greatest improvement $\Delta E_{\max}$ by allocating the same mobility or stiffness to the two isolator components.

It may be shown[8] that placing a given mass ''within'' the isolator as described above, so as to obtain a two-stage system with two like mobility components, results in greater effectiveness than placement of the mass directly at the receiver as long as $M_I >> M_R$. This inequality usually is likely to be satisfied in practice, except at resonances of the receiver at which M_R is very small. Similarly, as long as $M_I >> M_S$, placing the mass within the isolator results in greater effectiveness than placement of the mass directly at the source. The foregoing inequality generally is likely to be satisfied in practice, except for sources that behave essentially like force sources and thus have very high mobility.

*This result applies strictly only if the various mobilities or mobility ratios are real quantities. It suffices for development of an intuitive understanding, although more complicated expressions apply in the general case where the mobilities are represented by complex quantities.

11.6 PRACTICAL ISOLATORS

A great many different isolators are available commercially in numerous sizes and load capacities, with various attachment means, and with many types of specialized features. Details concerning such isolators typically may be found in suppliers' catalogs.

Most commercial isolators incorporate metallic or elastomeric resilient elements. Metallic elements most often are in the form of coil springs but also occur in the form of flexural configurations such as leaf springs or conical Belleville washers. Coil springs are predominantly used in compression, usually because such tensile springs tend to involve configurations that give rise to stress concentrations and thus have lesser fatigue life. Coil spring isolator assemblies often involve parallel and/or series arrangements of springs in suitable housings, are designed to have the same stiffness in the lateral directions as in the axial direction, and may also incorporate friction devices (such as wire mesh inserts) and in-series elastomeric pads for enhanced damping and isolation at high frequencies. Spring systems in housings need to be installed with care to avoid binding between the housing elements and between these elements and the springs.

Some commercial metallic isolators employ pads or woven assemblies of wire mesh to provide both resilience and damping. Others use arrangements of coils or loops of wire rope not only to provide damping but also to serve as springs.

A great many commercially available isolators that employ elastomeric elements have these elements bonded or otherwise attached to support plates or sleeves that incorporate convenient means for fastening to other components. The isolators may be designed so that the elastomeric element is used in shear, torsion, compression, or a combination of these modes. There are also available a variety of elastomeric gaskets, grommets, sleeves, and washers, intended to be used with bolts or similar fasteners to provide both connection and isolation.

Elastomeric pads often are used as isolators by themselves, as are pads of other resilient materials, such as cork, felt, fiberglass, and metal mesh. Such pads often are convenient and relatively inexpensive; their areas can be selected to support the required loads, and their thicknesses can be chosen to provide the desired stiffness.

In the design and selection of pads of solid (in contrast to foamed) elastomeric materials, one needs to take into account that a pad's stiffness depends not only on its thickness and area but also on its shape and constraints. This behavior is due to the incompressibility of elastomeric materials, which essentially prevents a pad from changing its volume as it is compressed and thus in essence does not permit a pad to be compressed if it is confined so that its edges cannot bulge outward. A pad's freedom to compress may be characterized by its shape factor, defined as the ratio of its loaded area to the total area of the edges that are free to bulge; the greater the shape factor, the greater is the pad's effective stiffness. However, a pad's freedom to deform while maintaining constant volume also is affected by how easily the loaded surfaces can slip relative to the adjacent surfaces; the more restricted this slippage, the greater the pad's effective stiffness. Some commercial isolation pads are furnished with top and bottom load-carrying surfaces bonded to metal or other stiff plates in order to eliminate the stiffness uncertainties due to unpredictable slippage.

In order to avoid the need for considering shape factor in pad selection, many commercial isolation pad configurations have a multitude of cutouts (e.g., closely spaced arrays of holes) or ribs, which provide roughly constant amounts of bulging area per unit surface area. If ribbed or corrugated pads or pads with cutouts are used in stacks, plates of a stiff material generally are used between pads to distribute the load on the load-bearing surfaces and to avoid having protrusions on one pad extending into openings on the adjacent pad.

So-called pneumatic, or air spring, isolators, which have found considerable use, obtain their resilience primarily from the compressibility of confined volumes of air. They may take the form of air-filled pillows of rubber or plastic, often with cylindrical or annular shapes, or they may consist essentially of piston-in-cylinder arrangements. Air springs can be designed to have small effective stiffnesses while supporting large loads and to have smaller heights than metal springs of equal stiffness. Practical air springs typically can provide resonance frequencies that may be as low as about 1 Hz. Some air spring configurations are laterally unstable under some load conditions and require the use of lateral restraints; some are available with considerable lateral stability.

Air springs of the piston-and-cylinder type can be provided with leveling controls, which automatically keep the isolated item's static position at a predetermined distance from a reference surface and (by use of several air springs and a suitable control system) at a predetermined inclination. The stiffness of a piston-type air spring is proportional to pA^2/V, where p denotes the air pressure, A the piston face area, and V the cylinder volume. The product pA is equal to the static load carried by the spring. Lower stiffnesses may be obtained with a given area at a given air pressure by use of larger effective volumes; for this reason, some commercial air spring isolators are available with auxiliary tanks that communicate with the cylinder volume via piping. In some instances, a flow constriction in this piping is used to provide low-frequency damping.

Pendulum arrangements often are convenient means for obtaining horizontally acting isolation systems with low natural frequencies. One may calculate the horizontal natural frequency of a pendulum system from Eq. (11.3) if one replaces X_{st} in that equation by the pendulum length. Some commercial isolation systems combine pendulum action for horizontal isolation with spring action for vertical isolation.

Various exotic isolation systems have also been investigated or employed for special applications. These include systems in which the spring action is provided by magnetic or electrostatic levitation or by streams or thin films of gases or liquids.

Active isolation systems have recently received increased attention. Such systems essentially are dynamic control systems in which the vibration of the item to be protected is sensed by an appropriate transducer whose suitably processed output is used to drive an actuator that acts on the item so as to reduce its vibration.

Active systems obviously are relatively complex, but they can provide better isolation than passive systems under some conditions, notably, in the presence of low-frequency disturbances, attenuation of which by passive means generally tends to be most difficult.

REFERENCES

1. W. T. Thomson, *Theory of Vibration with Applications*, 2nd ed., Prentice-Hall, Englewood Cliffs, NJ, 1981.
2. J. C. Snowdon, *Vibration and Shock in Damped Mechanical Systems*, Wiley, New York, 1968.
3. M. Harrison, A. O. Sykes, and M. Martin, "Wave Effects in Isolation Mounts," *J. Acoust. Soc. Am.* **24**, 62–72 (1952).
4. C. M. Harris, "Vibration Control Principles," in C. M. Harris (ed.), *Handbook of Noise Control*, 2nd ed., McGraw-Hill, New York, 1979, Chapter 19.
5. H. Himelblau, Jr. and S. Rubin, "Vibration of a Resiliently Supported Rigid Body," in C. M. Harris and C. E. Crede (eds.), *Shock and Vibration Handbook*, 2nd ed. McGraw-Hill, New York, 1976.
6. J. N. Macduff and J. R. Curreri, *Vibration Control*, McGraw-Hill, New York, 1958.
7. C. E. Crede and J. E. Ruzicka, "Theory of Vibration Isolation," in C. M. Harris and C. E. Crede, (eds.), *Shock and Vibration Handbook*, 2nd ed., McGraw-Hill, New York, 1976.
8. E. E. Ungar and C. W. Dietrich, "High-Frequency Vibration Isolation," *J. Sound Vib.* **4**, 224–241 (1966).
9. L. Cremer, M. A. Heckl, and E. E. Ungar, *Structure-borne Sound*, 2nd ed., Springer-Verlag, Berlin, 1988.
10. E. E. Ungar, "Mechanical Vibrations," in H. A. Rothbart (ed.), *Mechanical Design and Systems Handbook*, 2nd ed., McGraw-Hill, New York, 1985, Chapter 5.
11. D. Muster and R. Plunkett, "Isolation of Vibrations," in L. L. Beranek, (ed.), *Noise and Vibration Control*, McGraw-Hill, New York, 1971, Chapter 13.
12. E. E. Ungar, "Wave Effects in Viscoelastic Leaf and Compression Spring Mounts," *Trans. ASME Ser. B* **85**(3), 243–246 (1963).
13. D. Muster, "Resilient Mountings for Reciprocating and Rotating Machinery," Eng. Rept. No. 3, ONR Contract N70NR-32904, July 1951.

CHAPTER 12

Structural Damping

ERIC E. UNGAR

Bolt Beranek and Newman, Inc.
Cambridge, Massachusetts

12.1 THE EFFECTS OF DAMPING

The dynamic responses and sound transmission characteristics of structures are determined by essentially three parameters: mass, stiffness, and damping. Mass and stiffness are associated with storage of kinetic and strain energy, respectively, whereas damping relates to the dissipation of energy, or more precisely, to the conversion of the mechanical energy associated with a vibration to a form (usually heat) that is unavailable to the vibration.

Damping in essence affects only those vibrational motions that are controlled by a balance of energy in a vibrating structure; vibrational motions that depend on a balance of forces are virtually unaffected by damping. For example, consider the response of a classical mass–spring–dashpot system to a steady sinusoidal force. If this force acts at a frequency that is considerably lower than the system's natural frequency, the response is controlled by a quasi-static balance between the applied force and the spring force. If the applied force acts at a frequency that is considerably above the system's natural frequency, the response is controlled by a balance between the applied force and the mass's inertia. In both of these cases, damping has practically no effect on the responses. However, at resonance, where the excitation frequency matches the natural frequency, the spring and inertia ef-

Noise and Vibration Control Engineering: Principles and Applications, Edited by Leo L. Beranek and István L. Vér.
ISBN 0-471-61751-2 © 1992 John Wiley & Sons, Inc.

fects cancel each other and the applied force supplies some energy to the system during each cycle; as a result, the system's energy (and amplitude) increases until steady state is reached, at which time the energy input per cycle is equal to the energy lost per cycle *due to damping*.

In the light of energy considerations like the foregoing, one finds that increased damping results in (1) more rapid decay of unforced vibrations, (2) faster decay of freely propagating structure-borne waves, (3) reduced amplitudes at resonances of structures subject to steady periodic or random excitation with attendant reductions in stresses and increases in fatigue life, (4) reduced response to sound and increased sound transmission loss (reduced sound transmission) above the coincidence frequency (where the disturbing pressure wave moves along the structure in concert with the structural displacement), (5) reduced rate of buildup of vibrations at resonances, and (6) reduced amplitudes of "self-excited" vibrations, in which the vibrating structure accepts energy from an external source (e.g., wind) as the result of its vibratory motion.

12.2 MEASURES AND MEASUREMENT OF DAMPING

Most measures of damping are based on the dynamic responses of simple systems with idealized damping behaviors. Damping measurements typically involve observation of some characteristics of these responses.

Decay of Unforced Vibrations

Many aspects of the behaviors of vibrating systems can be understood in terms of the simple ideal linear mass–spring–dashpot system shown in Fig. 12.1. If this system is displaced by an amount x from its equilibrium position, the massless spring produces a force of magnitude kx tending to restore the mass m toward its equilibrium position, and the massless dashpot produces a retarding force of magnitude $c\dot{x}$. Here k and c are constants of proportionality; k is known as the spring constant and c as the *viscous damping coefficient*.

If this system is displaced from its equilibrium position by an amount X_0 and then released, the resulting displacement varies with time t as[1]

$$x = X_0 e^{-\zeta\omega_n t} \cos(\omega_d t + \phi) \tag{12.1}$$

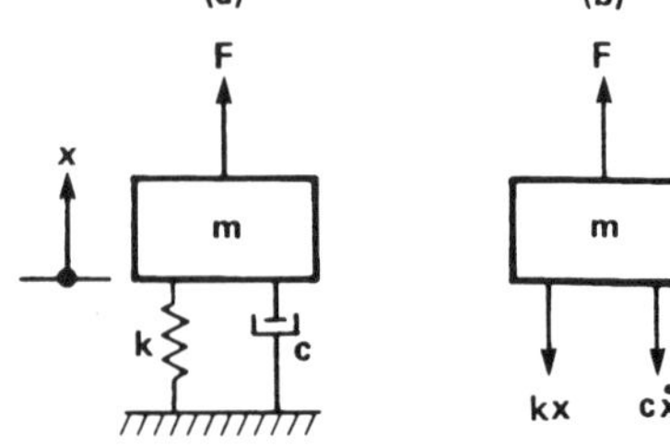

Fig. 12.1 System with single degree of freedom: (*a*) schematic representation of mass–spring–dashpot system or of vibrational mode of structure; (*b*) Free-body diagram of mass. Spring produces restoring force kx; dashpot produces retarding force cx.

provided that $\zeta < 1$. Here ϕ represents a phase angle, which depends on the velocity with which the mass is released, and ω_n and ω_d represent the undamped and damped radian natural frequencies of the system. These obey

$$\omega_n = \sqrt{\frac{k}{m}} = 2\pi f_n \qquad \omega_d = \omega_n\sqrt{1 - \zeta^2} \tag{12.2}$$

with f_n representing the cyclic (undamped) natural frequency. The constant ζ is called the *damping ratio* or *fraction of critical damping*; it is defined as

$$\zeta = \frac{c}{c_c} \qquad c_c = 2\sqrt{km} = 2m\omega_n \tag{12.3}$$

where c_c is known as the *critical damping coefficient*. For the small values of ζ one usually encounters in practice, ω_d is sufficiently close to ω_n, so that one rarely needs to distinguish between the damped and the undamped natural frequencies. Furthermore, the foregoing expression for ω_d applies only for viscous damping; other relations hold for other damping models.

The right-hand side of Eq. (12.1) represents a cosine function with an amplitude $X_0e^{-\zeta\omega_n t}$ that decreases as time t increases (see Fig. 12.2); its rate of decrease is $\zeta\omega_n$ and thus is proportional to ζ. However, Eq. (12.1) no longer applies for values of ζ that equal or exceed unity (or for values of c that equal or exceed c_c). For such large values of ζ or c one obtains a nonoscillatory decay represented by pure exponential expressions instead of the decaying oscillation represented by Eq. (12.1). The critical damping coefficient c_c constitutes the boundary between oscillatory and nonoscillatory decays.

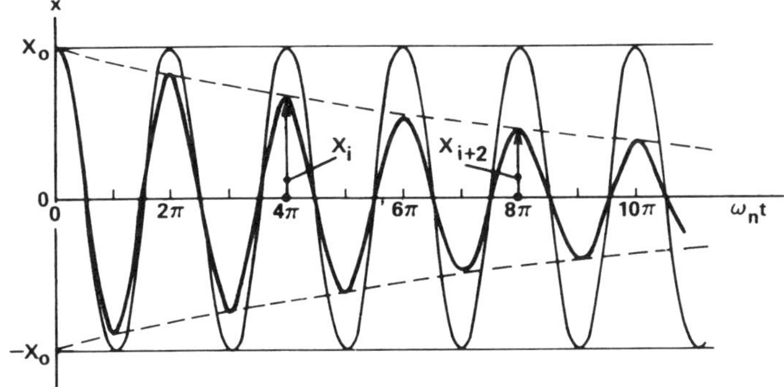

Fig. 12.2 Time variation of displacement of mass–spring–dashpot system released with zero velocity from initial displacement X_0. *Light curve*: Undamped system ($c = \zeta = 0$); amplitude remains constant at X_0. *Heavy curve*: Damped system ($0 < c < c_c$; $0 < \zeta < 1$); amplitude decreases according to $x = X_0e^{-\zeta\omega_n t}$, which is represented by upper dashed curve. Lower dashed curve corresponds to $x = -X_0e^{-\zeta\omega_n t}$. Amplitudes X_i and X_{i+2} illustrate values that may be used to calculate logarithmic decrement from Eq. (12.4) for $N = 2$.

The *logarithmic decrement* δ is a convenient, time-honored representation of how rapidly a free oscillation decays. It is defined by[1]

$$\delta = \frac{1}{N} \ln \frac{X_i}{X_{i+N}} \tag{12.4}$$

where X_i represents the value of x at any selected peak and X_{i+N} represents the value at the peak at N cycles from the aforementioned one. For a viscously damped system, it follows from Eq. (12.1) that $\delta = 2\pi\zeta$.

The utility of logarithmic measures of oscillatory quantities has long been recognized in acoustics, and definitions analogous to acoustical levels have come into use in the field of vibrations, particularly in regard to measurement. For example, one may define the displacement level L_x, in decibels, corresponding to an oscillatory displacement $x(t)$ in analogy to sound pressure level, as

$$L_x = 10 \log_{10} \frac{x^2(t)}{x_{\text{ref}}^2} \tag{12.5}$$

where x_{ref} denotes a (constant) reference value of displacement. One may then obtain a decay rate Δ, in decibels per second, and relate it to the damping ratio for a viscously damped system[2]:

$$\Delta = -\frac{dL_x}{dt} = 8.69\zeta\omega_n = 54.6\zeta f_n \tag{12.6}$$

Also in analogy to acoustics, one may define the reverberation time T_{60} as the time it takes for the displacement level to decrease by 60 decibels; thus,

$$T_{60} = \frac{60}{\Delta} = \frac{1.10}{\zeta f_n} \tag{12.7}$$

Because velocity and acceleration levels may be defined in full analogy to the definition of displacement level in Eq. (12.5), the decay rate and reverberation time expressions of Eqs. (12.6) and (12.7) also apply to these other vibration levels.

All of the foregoing measures of damping deal with quantities that characterize the decay of free vibrations, and not with energy as such. However, these quantities are related to energy. The total energy W of a mass–spring–dashpot system like that of Fig. 12.1 consists of the kinetic energy W_{kin} of the mass and the potential energy W_{pot} stored in the spring. For a system vibrating with amplitude X at its natural frequency ω_n, one finds[3]

$$W = W_{\text{kin}} + W_{\text{pot}} = \tfrac{1}{2}kX^2 = \tfrac{1}{2}m\omega_n^2 X^2 \tag{12.8}$$

The energy dissipated by the system of Fig. 12.1 corresponds to the work that is done on the dashpot; the energy D dissipated per cycle of a vibration at frequency ω_n and amplitude X may readily be found to obey $D = \pi\omega_n c X^2$.

The ratio of the energy dissipated per *cycle* to the energy present in the system is called the *damping capacity* ψ. For a viscously damped system it obeys $\psi = D/W = 4\pi\zeta$. The ratio of the average energy dissipated per *radian* to the energy in the system is called the *loss factor*. The loss factor η is equal to $1/2\pi$ times the damping capacity and, for a viscously damped system, is related to ζ as

$$\eta = \frac{D}{2\pi W} = 2\zeta \tag{12.9}$$

The expressions relating ψ and η to the damping ratio ζ are based on Eq. (12.8) and on the expression for D as well as on the assumption that the amplitude changes little during a cycle, that is, that the decay is slow and the damping is small. However, the definitions of damping capacity and loss factor are not limited to viscous damping or to small damping values.

If any extended structure that is not too highly damped vibrates in absence of external forces at one of its natural frequencies, all points on that structure move either in phase or in opposite phase with each other, and the structure is said to vibrate in one of its modes. In addition to the modal natural frequency, there corresponds to each mode a modal mass, a modal stiffness, and a modal damping value. With the aid of these parameters the behavior of a modal vibration may be described in terms of that of an equivalent simple mass–spring–dashpot system[1,4,5]. Thus, all of the foregoing discussion concerning this simple system also applies to structural modes.

Of course, extended structures also can exhibit unforced motions at a given frequency in which all points are not in or out of phase with each other. Such motions, which can be described in terms of propagating waves, also decrease due to damping. For flexural waves on a beam or for nonspreading (straight-crested) flexural waves on a plate, the *spatial decay rate* Δ_λ, defined as the reduction in vibration level per wavelength, obeys[3] $\Delta_\lambda = 13.6\eta$ in decibels per wavelength.

Steady Forced Vibrations

If the system of Fig. 12.1 is subject to a sinusoidal force $F = F_0 \cos \omega t$, then the displacement of the mass in the steady state is also sinusoidal and obeys $x = X\cos(\omega t - \phi) = \text{Re}\{\overline{X}e^{j\omega t}\}$, where $X = |\overline{X}|$ denotes the amplitude of the motion and $\overline{X} = Xe^{-j\phi}$ is the corresponding *phasor* or complex amplitude, with $j = \sqrt{-1}$. From the equation of motion for the system, it follows[1] that

$$\frac{X}{F_0/k} = \frac{|\overline{X}|}{X_{st}} = \left[\left(1 - \frac{\omega^2}{\omega_n^2}\right)^2 + \left(2\zeta\frac{\omega}{\omega_n}\right)^2\right]^{-1/2} \qquad \tan\phi = \frac{2\zeta\omega/\omega_n}{1 - \omega^2/\omega_n^2} \tag{12.10}$$

The deflection $X_{st} = F_0/k$, which was introduced in order to obtain a nondimensional expression, is the quasi-static or zero-frequency deflection that the system would experience due to a statically applied force of magnitude F_0. The ratio X/X_{st}, which indicates by what factor the amplitude under dynamic excitation exceeds the quasi-static deflection, is often called the *amplification*.

The foregoing response expressions (and related ones that involve velocity V or acceleration A instead of displacement) depend on damping; thus, damping data may be extracted from corresponding measurements. Two widely used measures of damping may be derived readily from the amplification expressions of Eq. (12.10), a plot of which is shown in Fig. 12.3. One is the *amplification at resonance*, conventionally represented by the letter Q and often simply called "the Q" of the system.[2] This corresponds to the value of X/X_{st} that results if the excitation frequency ω is equal to the natural frequency ω_n and is related to the viscous damping ratio by $Q = 1/2\zeta$. The second commonly used measure is the *relative bandwidth* $b = \Delta\omega/\omega_n \approx 1/Q$, where $\Delta\omega$ represents the difference between the two frequencies* (one below and one above ω_n; see Fig. 12.3) at which the amplification is equal to $Q/\sqrt{2}$.

In view of Eq. (12.10), the phase lag ϕ also provides a measure of damping. It is particularly convenient to use phase information in "Nyquist plots," that is, in plots of the real and imaginary parts of responses at a number of frequencies, as illustrated in Fig. 12.4. These plots are circles or nearly circles, with a diameter that is equal to Q if the plots are appropriately nondimensionalized.[6,7]

*These frequencies often are called the *half-power points* because at these the energy stored in the system (and that dissipated by it), which is proportional to the square of the amplitude [see Eq. (12.6)], is half of the maximum value.

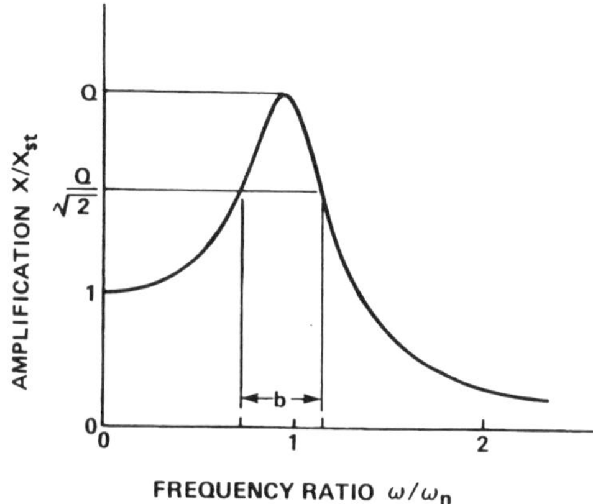

Fig. 12.3 Steady-state response of mass–spring–dashpot system to sinusoidal force. See Eq. (12.10): Q = value of amplification X/X_{st} at resonance. Relative bandwidth $b = \Delta\omega/\omega_n$ is determined from "half-power points," i.e., from frequencies at which response amplitude is $1/\sqrt{2}$ times the maximum.

(a)

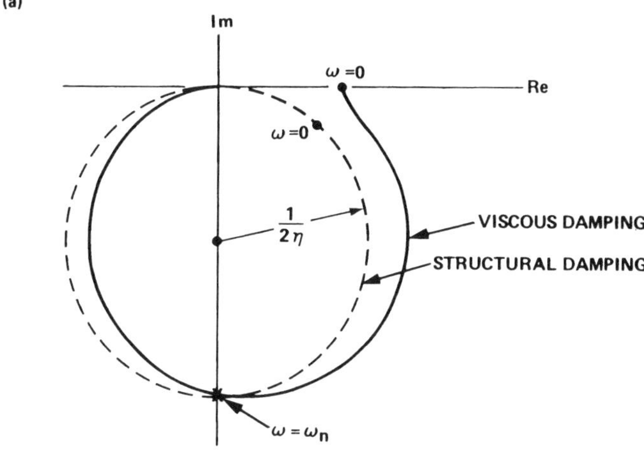

(b)

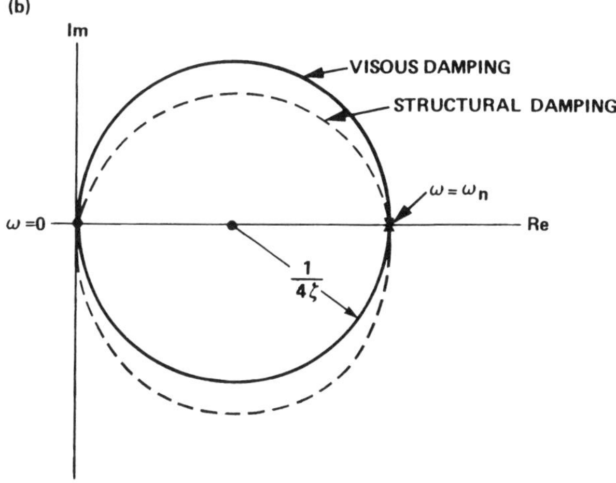

Fig. 12.4 Nyquist plots of nondimensional responses of viscously and structurally damped mass–spring–damper systems. (*a*) Real and imaginary parts of amplification X/X_{st}. (*b*) Real and imaginary parts of mobility $Vk/F_0\omega_n = Vc_c/2F_0$. Amplification plot for structural damping and mobility plot for viscous damping are exact circles; others are approximate circles that become more nearly circular with decreased damping. Diameter is exactly or approximately equal to Q. Figure plots correspond to $\zeta = 0.2$, $\eta = 0.4$.

If the system of Fig. 12.1 (or a structural mode modeled by it) is subject to a broadband force (or broadband modal force) rather than a single-frequency sinusoidal force, then the mean-square displacement $\bar{x}^2$ of the mass is given by[8]

$$\frac{\bar{x}^2}{\pi S_F(\omega)\omega_n/k^2} = \frac{1}{2\zeta} = \frac{1}{\eta} \tag{12.11}$$

where $S_F(\omega)$ denotes the spectral density of the force in terms of radian frequency (i.e., the value of excitation force squared per unit radian frequency interval). Note that the spectral density in cyclic frequency obeys $S_F(f) = 2\pi S_F(\omega)$. The foregoing equation is exact for excitations with spectral densities that are constant for all frequencies. It is a good approximation for excitations with spectral densities that vary only slowly in the vicinity of the system's natural frequency ω_n; the spectral density value to be used in the equation then is that corresponding to ω_n.

Interrelation among Measures of Damping

The previously defined measures are related to each other as follows[2,9]:

$$\eta = \frac{\psi}{2\pi} = 2\zeta = \frac{2.20}{f_n T_{60}} = \frac{\Delta}{27.3 f_n} = \frac{\delta}{\pi} = b = \frac{1}{Q} \tag{12.12}$$

The foregoing interrelations have been derived for viscous damping, that is, for damping that may be characterized by a retarding force proportional to velocity. However, they apply approximately also for any type of damping as long as the damping is not too great, say $\zeta < 0.1$ or $\eta < 0.2$, which encompasses most practical structures.

Measurement of Damping

Most approaches to measurement of the damping of structures are based on the previously discussed responses of simple systems, which, as has been mentioned, also correspond to those of structural modes. However, unlike mass–spring–dashpot systems, structures have a multiplicity of modes and corresponding natural frequencies. Therefore, many of the approaches applicable to simple systems can be applied only to structural modes whose responses can be separated adequately from those of all others because of differences in their natural frequencies or mode shapes.

Measurement of logarithmic decrement δ typically is applicable only to the fundamental modes of structures for which a clear record of the amplitude-versus-time trace can be obtained. If more than one mode is present, their decaying responses are superposed and the record becomes difficult to interpret.

The counting of peaks that is required for determination of the logarithmic decrement from Eq. (12.4) is not needed if one focuses on the decaying signal's envelope. For purposes of evaluating this envelope it is particularly convenient to use a display of the logarithm of the rectified amplitude versus time. Rectification is needed because the logarithm of negative numbers is undefined. In such a logarithmic display the envelope becomes a straight line whose slope is proportional to $\zeta\omega_n$ and thus to the decay rate. Not only does measurement of the slope enable one to evaluate the damping, but also observation of deviations of the envelope from a straight line permit one to judge whether the structure's damping is indeed amplitude independent and whether a superposition of responses with different decay rates is present.

Determination of decay rates is useful also in frequency bands in which a multitude of modes are excited. A typical measurement here involves excitation of a structure by a broadband force in a given frequency band, cutting off the excitation, then observing the envelope of the logarithm of the rectified signal obtained by passing the output of a transducer (usually an accelerometer) through a bandpass filter* tuned to the excitation band. The center frequency of this passband may be taken to represent ω_n for all modes in the band. Some judgment in interpreting the resulting envelopes and averaging of results from repeated measurements is generally required because different modes in the band may exhibit somewhat different decays.

A conceptually straightforward approach to measuring the damping of a structure involves the application of Eq. (12.9) to observed values of the energy dissipation and total vibrational energy W [Eq. (12.8)] present in a structure in the steady state. The structure is excited via an impedance head or a similar transducer arrangement that measures the force and motion at the excitation point. The instantaneous force and velocity values are multiplied and the product is time averaged to yield the average energy input per unit time, which is equal to the energy dissipated per unit time under steady-state conditions. For a given excitation frequency f, the energy D dissipated per cycle is equal to $1/f$ times the energy dissipated per unit time.

The energy W stored in the structure may be determined from its kinetic energy, which may be calculated from information on its mass distribution and from velocity values measured by a suitable array of accelerometers or other motion transducers. This measurement approach requires particular care in instrumentation selection and calibration but has a significant advantage: because it involves direct measurement of the dissipated energy, it does not rely on any particular model of dissipation. It also permits one, for example, to investigate how the loss factor varies with amplitude.

Force-and-motion transducer combinations may also be employed for the direct measurement of complex impedance or mobility (or of other force–motion ratios), from which damping information may be extracted, typically on the basis of Nyquist plots. Much corresponding specialized *modal testing* or *modal parameter extraction* instrumentation and software has recently become available.

Simple steady sinusoidal response measurements may also be made to measure Q and the half-power point bandwidth b directly on the basis of their definitions. These measurements require particular care to ensure that the near-resonance response of the mode of interest is not affected significantly by the responses of other modes with resonance frequencies near that of the mode of interest.

*The filter's response must be fast enough so that it can follow the decaying signal; otherwise one observes the decay of the filter response instead of that of the structural vibration. Filters with wider passbands need to be used to observe the more rapid decays associated with greater damping.

12.3 DAMPING MODELS

Analytical Models

All of the foregoing discussion has dealt with "viscous" damping, where energy dissipation results from a force that is proportional to the velocity of a vibrating system and that acts opposite to the velocity. This viscous model of damping action has been used most widely because it results in relatively simple linear differential equations of system motion and because it yields a reasonable approximation to the action of some real systems, particularly at small amplitudes.

Among the many other models that have received considerable attention, most also involve a motion-opposing force that is a function of velocity. In *dry friction* or *Coulomb* damping, the force is constant in magnitude (but changes its algebraic sign when the velocity does), and in *square law* and *power law* damping the force magnitude is proportional to the square or to some other power of velocity. Of course, modern numerical methods also permit one readily to analyze models involving other velocity dependences, such as one might obtain from corresponding experiments.

Viscoelastic Damping

In contrast to the previously mentioned models, so-called structural or viscoelastic damping involves a force that opposes the velocity but that is proportional to displacement. This model is widely held to represent the behavior of structures better than does viscous damping, and it has the advantage of analytical simplicity, particularly in complex (phasor) notation.

In a system undergoing sinusoidal motion described by $Xe^{j\omega t}$, the phasor corresponding to a *viscous* damping force is given by $cV = j\omega cX$, whereas the phasor corresponding to a *viscoelastic* damping force may be written as $j\eta kX$ in terms of the loss factor η and spring stiffness k. Thus, the sinusoidal response equation for a simple system may be written in phasor form as

$$\frac{X}{F} = \begin{cases} [-m\omega^2 + j\omega c + k]^{-1} & \text{for viscous damping} \\ [-m\omega^2 + j\eta k + k]^{-1} = [-m\omega^2 + \bar{k}]^{-1} & \text{for viscoelastic damping} \end{cases} \tag{12.13}$$

In the equation corresponding to viscoelastic damping the *complex stiffness* $\bar{k} = k(1 + j\eta)$ has been introduced. This is convenient because the real part of the stiffness indicates the system's strain energy storage capability and the imaginary part indicates its energy dissipation capability. The above expression pertaining to viscous damping leads directly to the dimensionless response relations of Eq. (12.10). One may obtain the corresponding dimensionless response equations for a viscoelastically damped system simply by replacing $2\zeta\omega/\omega_n$ by η in this equation.

It should be noted that the different frequency dependences of the two damping forces imply different system behaviors, except near the natural frequency. Furthermore, the viscoelastic model is often extended to loss factors that are functions of frequency*—particularly in order to represent measured frequency dependences of damping—and a loss factor with appropriate frequency variation may also represent viscous damping. With frequency-dependent loss factors, the viscoelastic damping model can represent the sinusoidal response of any system whose parameters are independent of amplitude.

Selection of Models

If one desires to determine the precise response of a system to a prescribed excitation, one generally needs to have a complete description of all forces, including the damping forces; that is, one needs a damping model that corresponds to the actual system. This is true, for example, if one wants to study the "wave shape" of the motion of a screeching brake or a chattering tool or if one needs to determine the response of a system to a transient, such as a shock.

In many practical instances, however, the details of the system's motion are of no interest and only the amplitudes are of concern. As has been mentioned, under steady resonant or free decay conditions (and in a few other situations), the amplitudes are established essentially by the energy in the system. For such conditions, the details of the damping model are unimportant as long as the model gives the correct energy dissipation per cycle. It is for this reason that measures of damping that involve only energy considerations have found wide acceptance.† One can use them for many practical purposes without investigating the time variation of the damping forces.

12.4 DAMPING MECHANISMS AND MAGNITUDES

Since damping involves the conversion of energy associated with a vibration to other forms that are unavailable to the vibration, there are as many damping mechanisms as there are ways to remove energy from a vibrating system. These include mechanisms that convert mechanical energy into heat as well as others that transport energy away from the vibrating system of concern.

Energy Dissipation and Transport

Material damping, *mechanical hysteresis*, or *internal friction* refer to the conversion of mechanical energy into heat that occurs within materials due to deformations that are imposed on them. This conversion may result from a variety of effects on the molecular, crystal lattice, or metal grain level, including magnetic,

*Caution is indicated here if one wants to calculate certain responses by transformation from the frequency to the time domain. Some frequency variations of the loss factor may lead to physically unrealizable results, e.g., to system motions that begin before a force is applied.[10]

†Equivalent viscous damping, defined as viscous damping that results in the same energy dissipation as the damping actually present in the system, is often used in analyses. This damping model obviously should not be used where details of the system motion are of concern.

thermal, metallurgical, and atomic phenomena.[11] Figure 12.5 indicates the ranges of the loss factors reported for some common materials.

Damping of a vibrating structure may also result from friction associated with relative motion between the structure and solids or fluids that are in contact with it. Also, an electrically conductive structure moving in a magnetic field is subject to damping due to eddy currents that result from the motion and that are converted into heat.

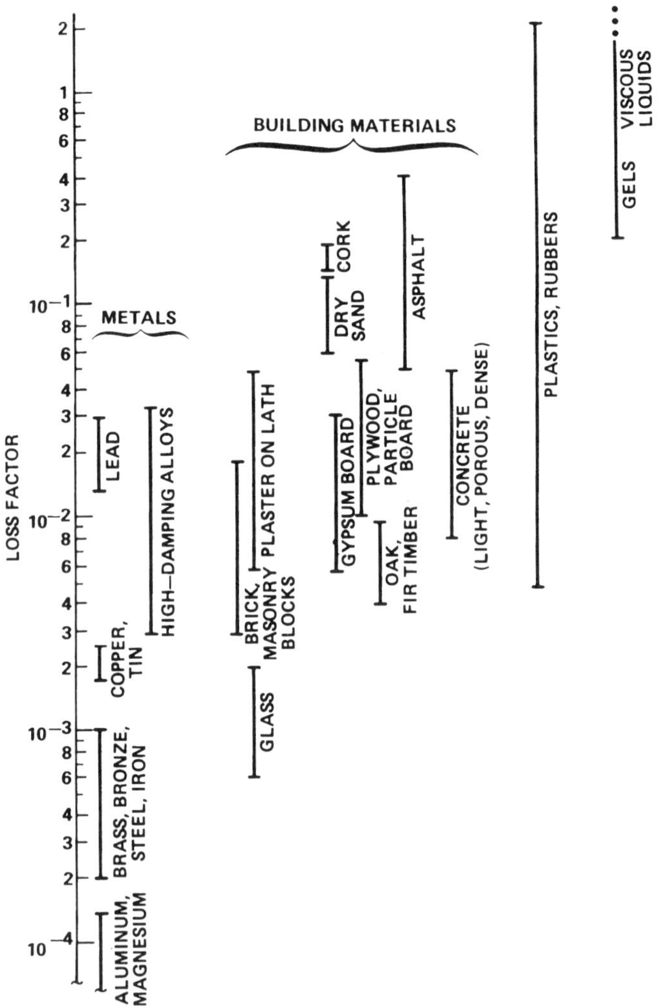

Fig. 12.5 Typical ranges of material loss factors at small strains, near room temperature, at audio frequencies. The loss factors of metals tend to increase with strain amplitude, particularly near the yield point, but the loss factors of plastics and rubbers tend to be relatively independent of strain amplitude up to strains of the order of unity. The loss factors of some materials, particularly those that can flow or creep, tend to vary markedly with temperature and frequency.

A granular material, such as sand, placed in contact with a vibrating structure tends to produce damping by two different mechanisms. At small amplitudes, damping results predominantly from interaction of asperities on adjacent grains and the attendant energy loss due to mechanical hysteresis. At large amplitudes, damping results predominantly from impacts between the structure and the grains or between grains; these impacts produce high-frequency vibrations of the structure and of the granular material, and the energy that goes into these vibrations (which eventually is converted to heat) is no longer available to the structural vibrations of concern.[12] Impact dampers, in which a small element is made to rattle against a vibrating structure, similarly rely on conversion of the energy of the vibrating structure to higher frequency vibrations.

A vibrating structure also experiences damping due to energy that is transported away from it to adjacent structures or to fluids in contact with it. For example, the damping of a panel that is part of a multipanel array (e.g., on an aircraft fuselage) typically is primarily due to energy that flows to adjacent panels; this energy transport generally makes it very difficult (or even practically impossible) to measure the dissipative damping of a structural component that is connected to others. Energy transport, together with other dissipation mechanisms that act at supports, usually make it difficult to measure the damping of lightly damped systems.

The *waveguide absorber* concept also involves conduction of energy away from a vibrating structure so as to increase its damping. Such an absorber essentially consists of a structural element (e.g., a long slender beam) along which waves can travel, which is provided with a termination or encapsulation that dissipates the energy transported by these traveling waves.[13] If such a waveguide absorber is attached to a vibrating structure, some of the structure's vibratory energy is used to generate waves on the absorber and thus is dissipated. To be effective, a waveguide absorber not only must be able to support waves in the frequency range of concern, but it also must be attached to the vibrating structure where this structure moves with considerable amplitude; it also must not reduce the structural motion at the attachment point excessively; that is, ideally the absorber should be impedance matched to the structure.

Damping Due to Boundaries and Reinforcements

For panels or other structural components that may be considered as uniform plates, one may estimate the loss factor η_b associated with energy loss at the panel boundaries, both due to energy transport to adjacent panels and due to dissipation at its boundaries, from information on the boundary absorption coefficients.[14] For a panel of area A vibrating at a frequency at which the flexural wavelength λ on the panel is considerably shorter than a panel edge, this loss factor is given by

$$\eta_b = \frac{\lambda}{\pi^2 A} \sum \gamma_i L_i \tag{12.14}$$

where γ_i denotes the absorption coefficient of the ith boundary increment whose length is L_i and where the summation extends over all boundary increments.

At frequency f, the flexural wavelength on a homogeneous plate of thickness h of a material with longitudinal wave speed c_L and Poisson's ratio ν is given by $\lambda = \sqrt{(\pi/\sqrt{3})hc_L/f(1 - \nu^2)}$.

The absorption coefficient γ of a boundary element is defined as the fraction of the panel bending-wave energy impinging on the boundary element that is not returned to the panel. Although the absorption coefficient values associated with a given boundary element rarely can be predicted well analytically, they can often be determined experimentally. For example, one might add a boundary element of length L_0 to a panel, measure the resulting loss factor increase $\Delta\eta$ at various frequencies, and calculate the absorption coefficient values γ_0 of this boundary element from $\gamma_0 = (\Delta\eta)\pi^2 A/\lambda L_0$, which follows from Eq. (12.14). Both Eq. (12.14) and the expression for γ_0 are based on the assumption that the absorption coefficient of a boundary element is independent of its length, an assumption that generally holds true if the wavelength λ is considerably smaller than the element length.

Equation (12.14) also permits one to account for the damping effects of linear discontinuities, such as seams or attached reinforcing beams, on panels, provided one knows the corresponding absorption coefficients. Since plate waves can impinge on both sides of a discontinuity located within the panel area, for such a discontinuity location one needs to use in Eq. (12.14) twice the actual discontinuity length.

The energy that panel beams or attached reinforcements can dissipate, and thus the damping they can produce, depends markedly on the fastening method used. Metal beams attached to metal panels or seams in such panels generally produce little damping if they are continuously welded or joined by means of a rigid adhesive. However, they can contribute significant damping if they are fastened by a flexible, dissipative adhesive or if they are fastened at only a number of points, for example, by rivets, bolts, or spot welds. At high frequencies, at which the flexural wavelength on the panel is smaller than the distance between fastening points, damping results predominantly not from interface friction but from an "air-pumping" effect produced as adjacent surfaces (at locations between the connection points) move away from and toward each other. Energy loss here is due to the viscosity of the air or other fluid present between the surfaces.[15-17]

The high-frequency absorption coefficient corresponding to a beam that is fastened to a panel at a multitude of points may be estimated from the experimental results summarized in Fig. 12.6, which shows how the beam's reduced absorption coefficient γ_r varies with reduced frequency f_r or with the ratio d/λ of fastener spacing to plate flexural wavelength. These reduced quantities, definitions of which are given in the figure, account for the absorption coefficients' dependence on beam width w, plate thickness h, fastener spacing d, and the longitudinal wave-speed c_L in the plate material by relating these to reference values of these parameters (indicated by the subscript 0). It should be noted that Fig. 12.6 pertains to panels immersed in air at atmospheric pressure; lesser absorption coefficient values apply for panels at reduced pressure and higher values for panels at greater pressures. A theory is available[15] to account for the effects of atmospheric pressure changes and for other gases or liquids present between the contacting surfaces.

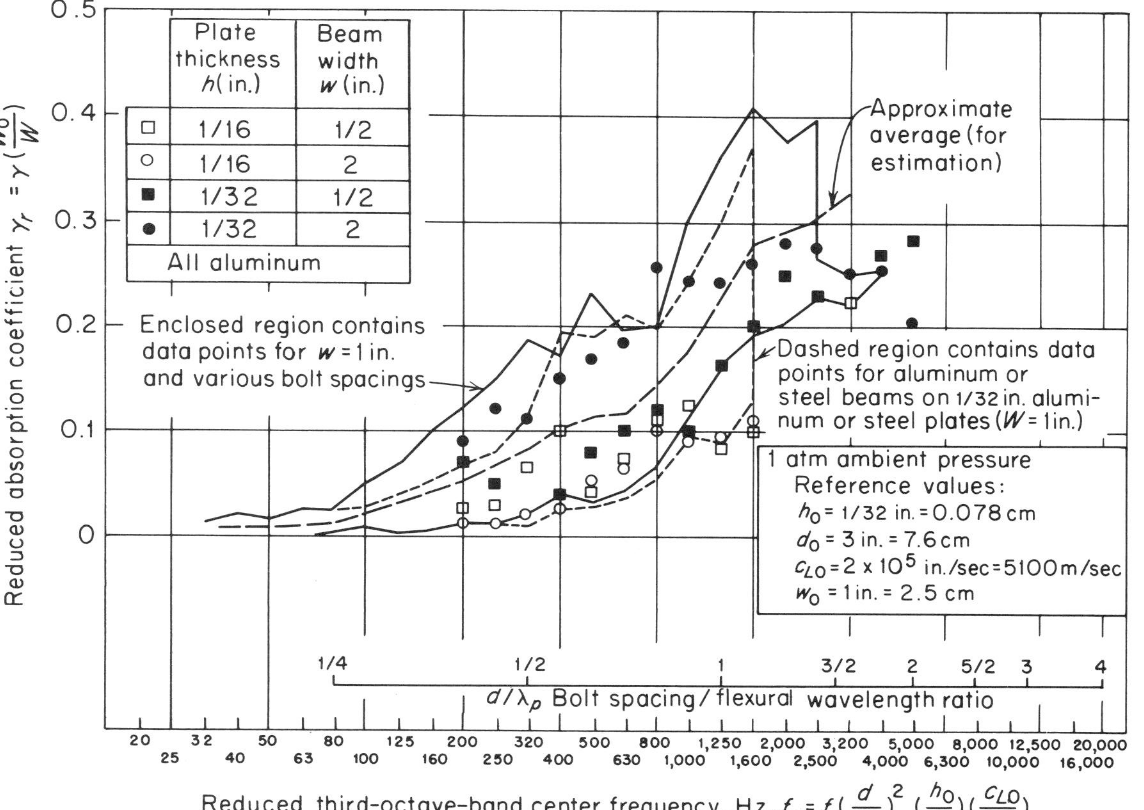

Fig. 12.6 Summary of reduced absorption coefficient data for beams fastened to plates by rows of rivets, bolts, or spot welds.[17]

Damping Due to Acoustical Radiation

Sound radiated by a vibrating structure transports energy from the structure and thus contributes damping. For a homogeneous panel of thickness h and material density ρ_p, one may calculate the panel's loss factor η_R at frequency f due to sound radiation from one side of the panel from its radiation efficiency σ by use of the relation[16]

$$\eta_R = \frac{\rho}{\rho_p} \frac{c}{2\pi f h} \sigma \qquad (12.15)$$

where ρ and c denote the density of the ambient medium and the speed of sound in it, respectively. If the panel can radiate from both of its sides, η_R is twice as great as indicated by Eq. (12.15). The magnitude of the radiation efficiency σ depends on the vibratory velocity distribution on the panel as well as on frequency and thus generally is different for different excitation distributions.

For a plate that has little inherent damping and that is excited at a single point, the radiation efficiency σ obeys[9]

$$\sigma = \begin{cases} (U/\pi^2 A f_c)\sqrt{f/f_c} & \text{for } f \ll f_c \\ 0.45\sqrt{U f_c / c} & \text{for } f = f_c \\ 1.0 & \text{for } f \gg f_c \end{cases} \qquad (12.16)$$

where A denotes the panel's surface area (one side), U its circumference, and f_c the coincidence frequency. This frequency, which is defined as that at which the plate flexural wavelength is equal to the acoustical wavelength in the ambient medium, is given by $f_c \approx c^2/1.8hc_L$, where c_L represents the longitudinal wave speed in the plate material. More detailed information on radiation efficiency is provided in Chapter 9. Equation (12.16) may be used for the general estimation of radiation efficiency values for plates that are not too highly damped. This equation also provides a reasonable estimate of the radiation efficiency of rib-stiffened plates if twice the total rib length is included in the circumference U.

12.5 VISCOELASTIC DAMPING TREATMENTS

Viscoelastic Materials and Material Combinations

Materials that have both damping (energy dissipation) and structural (strain energy storage) capability are called "viscoelastic." Although virtually all materials fall into this category, the term is generally applied only to materials, such as plastics and elastomers, that have relatively high ratios of energy dissipation to energy storage capability.

Structural materials with high strength-to-weight ratios typically have little inherent damping, as is evident from Fig. 12.5, whereas plastics and rubbers that are highly damped tend to have relatively low strength. This circumstance has led to the consideration of combinations of high-strength materials and high-damping viscoelastic materials for applications where both strength and damping are required. Additions of viscoelastic materials to structural elements have come to be known as viscoelastic damping treatments.

If a composite structure is deflected, it stores energy via a variety of deformations (such as shear, tension and compression, flexure) in each structural element. If η_i denotes the loss factor corresponding to the ith element deformation and W_i represents the energy stored in that deformation, then the loss factor η of the entire structure obeys[3]

$$\eta = \sum \eta_i \frac{W_i}{W_T} \tag{12.17}$$

where $W_T = \Sigma\, W_i$ denotes the total energy stored in the structure. The foregoing expression indicates that the loss factor η of the composite structure is equal to a weighted average of the loss factors corresponding to all of the element deformations, with the energy storages serving as the weighting factors. This expression also leads to an important conclusion: An element deformation can make a significant contribution to the total loss factor only if (1) the loss factor associated with it is significant *and* (2) the energy storage associated with it is a significant fraction of the total energy storage.

Mechanical Properties of Viscoelastic Materials

Because viscoelastic materials have both energy storage and energy dissipation capability, it is convenient to describe their behavior in terms of elastic and shear moduli that are complex quantities, in analogy to the definition of the complex stiffness introduced in Eq. (12.13). The complex Young's modulus $\overline{E}$ of a material, defined as the ratio of the stress phasor to the strain phasor, may be written as[11] $\overline{E} = E_R + jE_I = E_R(1 + j\eta_E)$, where the real part E_R is called the storage modulus, the imaginary part E_I is called the loss modulus, and the loss factor η_E associated with Young's modulus is equal to E_I/E_R. A completely analogous definition applies to the complex shear modulus.

For plastics and elastomers, the viscoelastic materials of greatest practical interest, the real and imaginary moduli as well as the loss factors vary considerably with frequency and temperature. However, these parameters usually vary relatively little with strain amplitude, preload, and aging.[11] The loss factor associated with the shear modulus typically is equal to that associated with the Young's modulus for all practical purposes, so that one generally need not distinguish between the two. Also, since most of the viscoelastic materials of practical interest are virtually incompressible, the shear modulus value is essentially equal to one-third of the corresponding Young's modulus value. One may also note that often $\eta_E^2 << 1$, so that $|\overline{E}| \approx E_R$.

Figure 12.7 shows how the (real) shear modulus and loss factor of a typical viscoelastic material varies with frequency and temperature. At low frequencies and/or high temperatures, the material is soft and mobile enough for the strain to follow an applied stress without appreciable phase shift so that the damping is small; the material is said to be in its "rubbery" state. At high frequencies and/

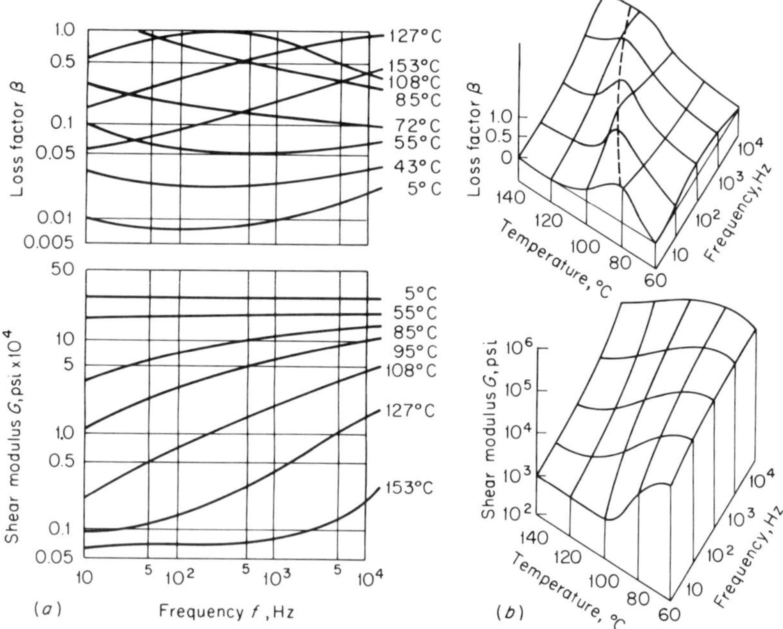

Fig. 12.7 Dependence of shear modulus and loss factor of a polyester plastic on frequency and temperature,[18]: (*a*) functions of frequency at constant temperature; (*b*) isometric plots on temperature–frequency plane.

or low temperatures, the material is stiff, immobile, may tend to be brittle, is relatively undamped, and behaves somewhat like glass; it is said to be in its "glassy" state. At intermediate frequencies and temperatures, the modulus takes on intermediate values and the loss factor is highest; the material is said to be in its "transition" state.

This material behavior may be explained on the basis of the interactions of the long-chain molecules that constitute polymeric materials. At low temperatures, the molecules are relatively inactive; they remain "locked together," resulting in high stiffness, and because they move little relative to each other, there is little intermolecular "friction" to produce damping. At high temperatures, the molecules become active; they move easily relative to each other, resulting in low stiffness, and because they interact little, there is again little energy dissipation due to intermolecular friction. At intermediate temperatures, where the molecules have intermediate relative motion and interaction, the stiffness also takes on an intermediate value and the loss factor is greatest. A similar discussion applies to the effect of frequency on the material properties, with the inertia of the molecules leading to their decreasing mobility and interaction with increasing frequency.

The observation that there exists a *temperature–frequency equivalence*, namely that an appropriate temperature decrease produces the same effect as a given frequency increase, has led to the development of convenient plots in which data for the frequency and temperature variations of each material modulus collapse onto single curves.[11,19] This collapse is achieved by plotting the data against a reduced frequency $f_R = f\alpha(T)$, where $\alpha(T)$ is an appropriately selected function of temperature T. In presentations of data in this form the function $\alpha(T)$ may be given analytically, in a separate plot or, as has recently come into vogue, in the form of a nomogram that is superposed on the data plot. Figure 12.8 is an illustration of such a plot and nomogram; its use is explained in the figure's legend.

Data on the properties of damping materials are available from knowledgeable suppliers of these materials. Compilations of data appear in references 11 and 21. Key information on some of these materials appears in Table 12.1 in a form that is useful for preliminary material comparison and selection for specific applications in keeping with the concepts discussed in the later portions of this chapter. For each listed material the table shows the greatest loss factor value η_{max} exhibited by the material and the temperatures at which this value is obtained at three frequencies. The table also lists three values of the modulus of elasticity: E_{max}, the greatest value of Young's modulus, applies at low temperatures (i.e., temperatures considerably below those corresponding to η_{max}); E_{min}, the smallest value of E, applies at high temperatures; the transition value E_{trans} applies in the η_{max} range; and $E_{I,max} \approx \eta_{max} E_{trans}$, the maximum value of the loss modulus, applies in the transition range.

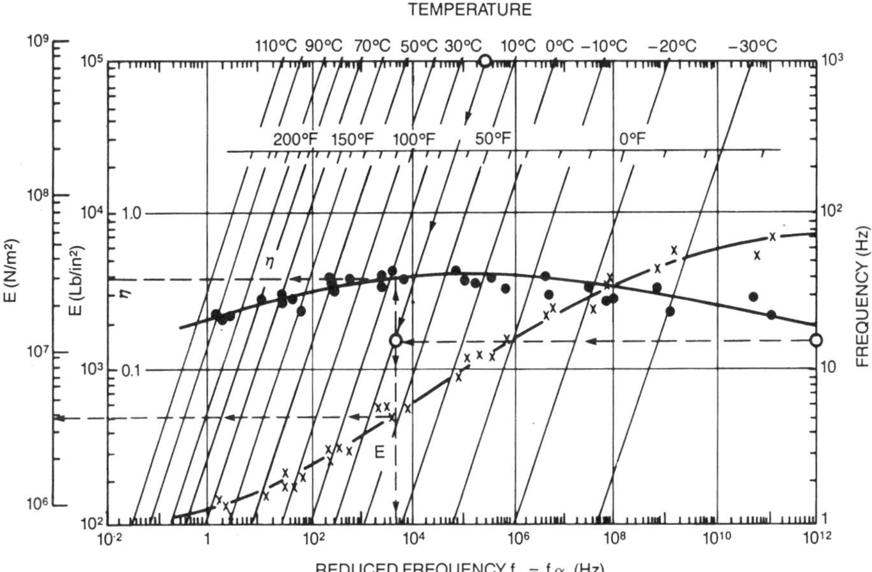

Fig. 12.8 Reduced frequency plot of elastic modulus E and loss factor η of "Sylgard 188" silicone potting compound. (After ref. 20.) Points indicate measured data to which curves were fitted. Nomograph superposed on data plot facilitates determination of reduced frequency f_R corresponding to frequency f and temperature T. Use of nomogram is illustrated by dashed lines: For $f = 15$ Hz and $T = 20°C$, one finds $f_R = 5 \times 10^3$ Hz and $E = 3.8 \times 10^6$ N/m², $\eta = 0.36$.

TABLE 12.1 Properties of Some Commercial Damping Materials*

	Maximum Loss Factor	Temperature (°F)** for η_{max} at			Elastic Moduli (PSI)***			
	η_{max}	10 Hz	100 Hz	1000 Hz	E_{max}	E_{min}	E_{trans}	E_{Imax}
Antiphon-13	1.8	25	75	120	3E5	1.2E3	1.9E4	3.E3E
Blachford Aquaplas	0.5	50	80	125	1.6E6	3E4	2.2E5	1.1E5
Barry Controls H-326	0.8	-40	-25	-10	6E5	3E3	4.2E4	3.4E4
Dow Corning Sylgard 188	0.6	60	80	110	2.2E4	3E2	2.6E3	1.5E3
EAR C-1002	1.9	23	55	90	3E5	2E2	7.7E3	1.5E4
EAR C-2003	1.0	45	70	100	8E5	6E2	2.2E4	2.2E4
Lord LD-400	0.7	50	80	125	3E6	3.3E3	1E5	7E4
Soundcoat DYAD 601	1.0	15	50	75	3E5	1.5E2	6.7E3	6.7E3
Soundcoat DYAD 606	1.0	70	100	130	3G5	1.2E2	6E3	6E3
Soundcoat DYAD 609	1.0	125	150	185	2E5	6E2	1.1E4	1.1E4
Soundcoat N	1.5	15	30	70	3E5	7E1	4.6E3	6.9E3
3M ISD-110	1.7	80	115	150	3E4	3E1	1E3	1.7E3
3M ISD-112	1.2	10	40	80	1.3E5	8E1	3.2E3	3.9E3
3M ISD-113	1.1	-45	-20	15	1.5E5	3E2	2.1E2	2.3E2
3M 468	0.8	15	50	85	1.4E5	3E1	2E3	1.6E3
3M ISD-830	1.0	-75	-50	-20	2E5	1.5E2	5.5E3	5.5E3
GE SMRD	0.9	50	80	125	E35	5E3	3.9E4	3.5E4

*Approximate values taken from curves in Ref. 11

**To convert to °C, use the formula °C = (5/9) (°F-32) or the approximate table below:

°F	-80	-60	-40	-20	0	20	40	60	80	100	120	140	160	180	200
°C	-62	-51	-40	-29	-18	-7	4	16	27	38	49	60	71	82	93

***Numbers shown correspond to storage (real) values of Young's modulus, except that E_{Imax} represents the maximum values of the loss (imaginary) modulus. E_{max} applies for low temperatures and/or high frequencies. E_{min} applies for high temperatures and/or low frequencies. E_{trans} and E_{Imax} applies in the range of η_{max}. Divide by 3 to obtain the corresponding shear modulus values.

To convert to N/m², multiply tabulated values by 7E3. The number following E represents the power of 10 by which the number preceding E is to be multiplied; e.g., 1.2E3 represents 1.2×10^3.

It is important to keep in mind that the mechanical properties of polymeric materials, including plastics and elastomers, tend to be more variable than those of metals and other classical structural materials. Some of this variability results from a polymer's molecular structure and molecular weight distribution, which depend not only on the material's chemical composition but also on its processing. Additional variability results from the various types and amounts of plasticizers and fillers that are added to most commercial materials for a number of practical purposes. Thus, it is quite common for nominally identical polymeric materials to exhibit considerably different mechanical behaviors. It may also occur that even material samples from the same production run have considerably different loss factors and moduli at frequencies and temperatures at which they are intended to be used, pointing toward the need for careful quality control and performance verification for critical applications.

Structures with Viscoelastic Layers

One may calculate the loss factor of a structure vibrating in a given mode by use of Eq. (12.17) if one knows the energies W_i stored in the various deformations of all of the component elements. Indeed, modern finite-element analysis methods[22-24] proceed by calculating the modal deflections, applying these to evaluate all the energy storage components and then using Eq. (12.17) to find the loss factor.

Analytical results have been developed for flexure of uniform beam and plate structures under conditions that are often approximated in practice. These results, which are extremely useful for design guidance and for development of an understanding of the important parameters, apply to structures whose deflection distributions are sinusoidal* and to structural configurations in which an insert or layer of viscoelastic material is subjected predominantly either to tension and compression or to shear.

Two-Component Beams

In flexure of a uniform beam with an insert or added layer or viscoelastic material, as illustrated by Fig. 12.9, the energy storage (and dissipation) associated with shear and torsional deformations may generally be neglected. If contact between the components is maintained without slippage at all surfaces and if the loss factor of the basic structural (nonviscoelastic) component is negligible, then the loss fac-

*The deflection distribution of any beam (or plate) vibrating in one of its natural modes is at least approximately sinusoidal (in one dimension for beams and in two dimensions for plates) at locations that are one wavelength or more from the boundaries, regardless of the boundary conditions. Therefore, the assumption of a sinusoidal deflection distribution is correct for a larger fraction of the structure as the frequency increases.

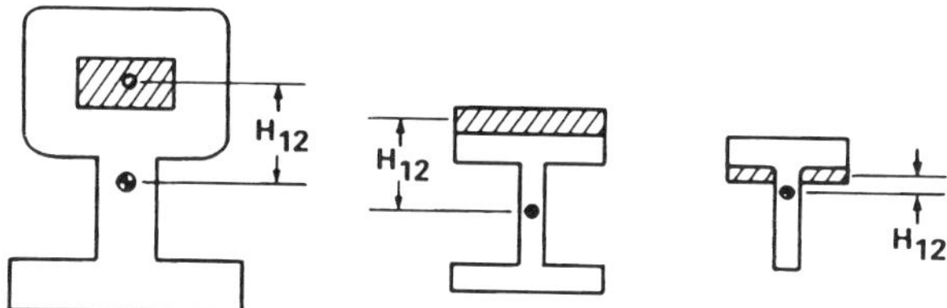

Fig. 12.9 End views of beams with viscoelastic inserts or added layers. Structural material is unshaded, viscoelastic material is shown shaded; H_{12} represents distance between neutral axis of structural component and that of viscoelastic component. Beam deflection is vertical, with wave propagation along the beam length, perpendicular to the plane of the paper.

tor η of the composite beam is related to the loss factor β of the material of the viscoelastic component by[25]

$$\frac{\eta}{\beta} = \left[1 + \frac{k^2(1 + \beta^2) + (r_1/H_{12})^2\alpha}{k[1 + (r_2/H_{12})^2\alpha]}\right]^{-1} \qquad (12.18)$$

where $\alpha = (1 + k)^2 + (\beta k)^2$ and H_{12} denotes the distance between the neutral axes of the two components. With subscript 1 referring to the structural (undamped) component and subscript 2 referring to the viscoelastic component, $k = K_2/K_1$, where $K_i = E_i A_i$ denotes the extensional stiffness of component i, expressed in terms of its Young's modulus (real part) E_i and cross-sectional area A_i. Furthermore, $r_i = \sqrt{I_i/A_i}$ represents the radius of gyration of A_i, where I_i denotes the centroidal moment of inertia of A_i.

For the often-encountered case where the structural component's extensional stiffness is much greater than that of the viscoelastic component, $k \ll 1$ and Eq. (12.18) reduces to

$$\eta \approx \frac{\beta E_2 I_T}{E_1 I_1 + E_2 I_T} \approx \frac{\beta E_2}{E_1}\frac{I_T}{I_1} \qquad (12.19)$$

where $I_T = I_2 + H_{12}^2 A_2 = A_2(r_2^2 + H_{12}^2)$ denotes the moment of inertia of A_2 about the neutral axis of A_1.

The last expression in Eq. (12.19) applies for $E_2 I_T \ll E_1 I_1$, which is generally true in practical structures where the area and elastic modulus of the viscoelastic component are small compared to those of the structural component. In this case the composite structure's neutral axis coincides very nearly with that of the structural component and the dominant energy storage is associated with flexure of the structural component (whose flexural stiffness is $E_1 I_1$). The dominant energy dissipation is associated with extension and compression of the viscoelastic component, with the average extension (equal to the extension at the viscoelastic component's neutral axis) given by the flexural curvature and the distance H_{12} between the neutral axes of the viscoelastic and the structural components.* The flexural curvature is greatest at the antinodes of the vibrating structure; most of the damping action thus occurs at these locations, with little damping resulting from the material near the nodes.

The second form of Eq. (12.19) contains two ratios; the first involves only material properties, and the second only geometric parameters. It indicates that the most important dynamic mechanical property of the viscoelastic material is its extensional loss modulus $E_l = \beta E_2$. In keeping with the conclusions based on the general energy expression [Eq. (12.17)], good damping of the composite structure

*A "spacer," a layer that is stiff in shear and soft in extension (e.g., like honeycomb), inserted between the structural and the viscoelastic component can increase H_{12} and thus the damping obtained with a given amount of viscoelastic material.

can be obtained only from a viscoelastic material that has not only a high loss factor but also a considerable energy storage capability.

Plates with Viscoelastic Coatings

A strip of a plate (see insert of Fig. 12.10) may be considered as a special case of a two-component beam, where the two components have rectangular cross sections. Thus Eqs. (12.18) and (12.19) apply, with $r_i = H_i/\sqrt{12}$ and $H_{12} = \frac{1}{2}(H_1 + H_2)$, where H_i denotes the component thickness. The energy storage and dissipation considerations that were discussed in the foregoing section as well as the foregoing remarks concerning the dominant damping material properties apply here also.

Figure 12.10 is a plot based on Eq. (12.18) for $\beta^2 << 1$. It shows that for small relative thicknesses $h_2 = H_2/H_1$, the loss factor ratio η/β is proportional to the viscoelastic layer thickness, whereas for very large relative thicknesses the loss factor ratio approaches unity; that is, the loss factor of the coated plate approaches

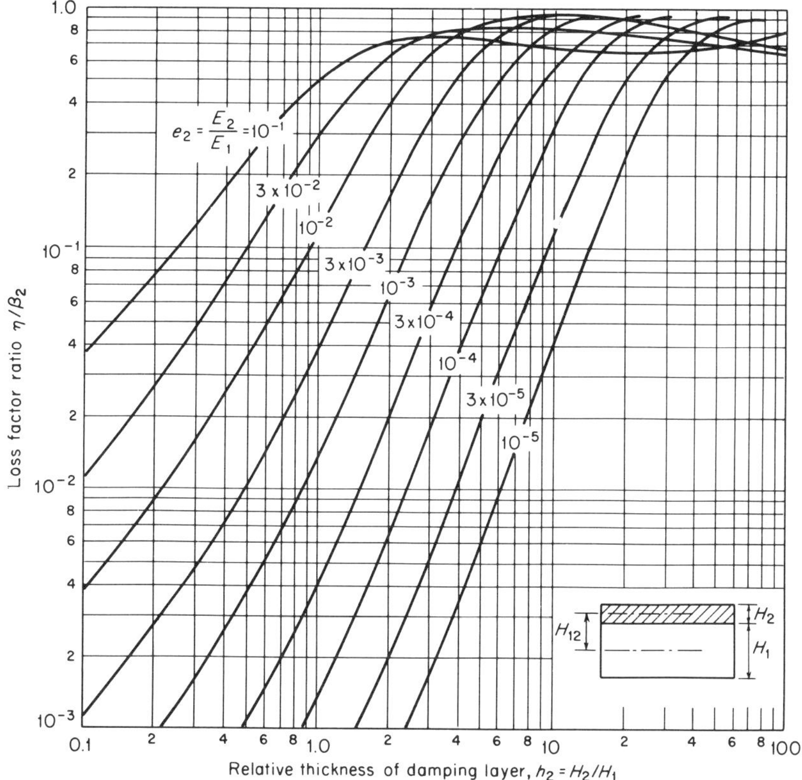

Fig. 12.10 Dependence of loss factor η of plate strip with added viscoelastic layer on relative thickness and relative modulus of layer.[26] Curves apply strictly only if loss factor β of viscoelastic material is small compared to unity.

that of the viscoelastic coating, as one would expect.* As also is evident from the figure, at small relative thicknesses the loss factor ratio is proportional to the modulus ratio $e_2 = E_2/E_1$. For small relative thicknesses, that is, in the regions where the curves of Fig. 12.10 are nearly straight, the loss factor of a coated plate may be estimated from[27]

$$\eta \approx \frac{\beta E_2}{E_1} h_2 (3 + 6h_2 + 4h_2^2) \tag{12.20}$$

where $h_2 = H_2/H_1$.

If two viscoelastic layers are applied to a plate, one layer on each side, then the loss factor of the coated plate may be taken as the sum of the loss factors contributed by the individual layers, with each contribution calculated as if the other layer were absent, provided that each viscoelastic layer has low relative extensional rigidity, that is, that $E_2H_2 \ll E_1H_1$. If this inequality is not satisfied, a more complex analysis is required.

Three-Component Beams with Viscoelastic Interlayers

Figure 12.11 illustrates uniform beams consisting of two structural (nonviscoelastic) components interconnected via a relatively thin viscoelastic component. Such three-component beams may be preferable to two-component beams for practical reasons because the viscoelastic material is exposed only at its edges; however, such beams can also be designed to have higher damping than two-component beams of similar weight.†

*For very thick viscoelastic coatings, deformation in the thickness direction (which are not considered in this simplified analysis) also may play a significant role, particularly at frequencies at which standing-wave resonances may occur in the viscoelastic material.[28]

†It should be noted that design changes to obtain increased damping generally also result in mass and stiffness changes, which tend to affect a structure's vibratory response and should be considered in the design process.[29,30]

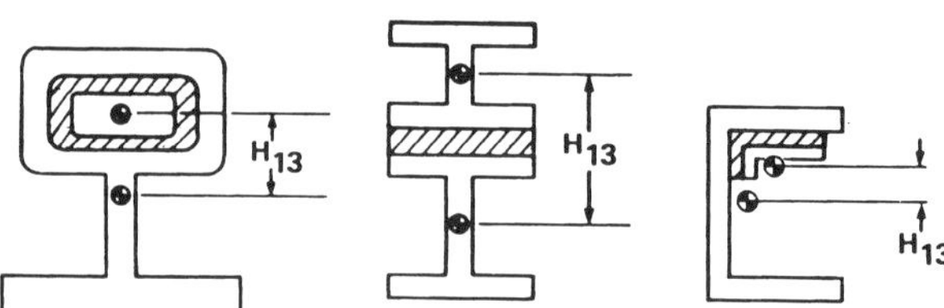

Fig. 12.11 End views of composite beams made up of two structural components (unshaded) joined via a viscoelastic component (shaded); H_{13} is distance between neutral axes of structural components. Beam deflection is vertical, with wave propagation along the beam length, perpendicular to the plane of the paper.

In flexure of a three-component beam with a viscoelastic layer whose extensional and flexural stiffnesses are small compared to those of the structural components, the dominant energy dissipation is associated with shear in the viscoelastic component and the most significant energy storage occurs in connection with extension/compression and flexure of the two structural components. The shear in the viscoelastic component is greatest at the vibrating structure's nodes. Thus, most of the energy dissipation occurs in the viscoelastic material near the nodes, with relatively little resulting in that near the antinodes. For efficient damping, it is important that the shearing action in the viscoelastic material not be restrained (particularly at and near nodes) by structural interconnections, such as bolts.

The loss factor η corresponding to a spatially sinusoidal deflection shape of such a three-component beam is related to the loss factor β of the viscoelastic material by[25]

$$\eta = \frac{\beta Y X}{1 + (2 + Y)X + (1 + Y)(1 + \beta^2)X^2} \tag{12.21}$$

where

$$X = \frac{G_2 b}{p^2 H_2} S \qquad \frac{1}{Y} = \frac{E_1 I_1 + E_3 I_3}{H_{13}^2} S \qquad S = \frac{1}{E_1 A_1} + \frac{1}{E_3 A_3} \tag{12.22}$$

Here subscripts 1 and 3 refer to the structural components and 2 to the viscoelastic component; E_i, A_i, and I_i represent, respectively, the Young's modulus, cross-sectional area, and moment of inertia of component i; H_{13} denotes the distance between the neutral axes of the two structural components; and G_2 represents the shear modulus (real part) of the viscoelastic material, H_2 the average thickness of the viscoelastic layer, and b its length as measured on a cross section through the beam. The wavenumber p of the spatially sinusoidal beam deflection obeys

$$\frac{1}{p^2} = \left(\frac{\lambda}{2\pi}\right)^2 = \frac{1}{\omega}\sqrt{\frac{B}{\mu}} \tag{12.23}$$

where λ represents the bending wavelength and B denotes the flexural rigidity and μ the mass per unit length of the composite beam.

The *structural parameter* Y of three-component structures depends only on the geometry and Young's moduli of the two structural components, whereas the *shear parameter* X depends also on the properties of the viscoelastic layer and on the wavelength of the beam deflection. The shear parameter X is proportional to the square of the ratio of the beam flexural wavelength to the *decay distance*,[31] that is, the distance within which a local shear disturbance decays by a factor of e, where $e \approx 2.72$ denotes the base of natural logarithms; thus, X also is a measure of how well the viscoelastic layer couples the flexural motions of the two structural components.

The (complex) flexural rigidity of a three-component beam is given by

$$\bar{B} = (E_1I_1 + E_3I_3)\left(1 + \frac{X^*Y}{1 + X^*}\right) \qquad X^* = X(1 - j\beta) \tag{12.24}$$

Its magnitude is $B = |\bar{B}|$. Thus, for small X, the flexural rigidity B of the composite beam is equal to the sum of the flexural rigidities of the structural components, that is, to the total flexural rigidity that the two components exhibit if they are not interconnected. For $X >> 1$, however, B approaches $1 + Y$ times the foregoing value, which is equal to the flexural rigidity of a beam with rigidly interconnected structural components 1 and 3.

For a given value of β and Y, the loss factor η of the composite beam takes on its greatest value,

$$\eta_{max} = \frac{\beta Y}{2 + Y + 2/X_{opt}} \tag{12.25}$$

at the optimum value of X, which is given by

$$X_{opt} = [(1 + Y)(1 + \beta^2)]^{-1/2} \tag{12.26}$$

With the aid of these definitions, one may rewrite Eq. (12.21) in terms of the ratio $R = X/X_{opt}$ in the following form, which provides a convenient view of the damping behavior of three-component beams:

$$\frac{\eta}{\eta_{max}} = \frac{2(1 + N)R}{1 + 2NR + R^2} \qquad N = (1 + \tfrac{1}{2}Y)X_{opt} \tag{12.27}$$

Figure 12.12, which is based on Eqs. (12.25) and (12.26), shows how η_{max}/β increases monotonically with Y, indicating the importance of selecting a configuration with a large value of Y in the design of highly damped composite structures.* Figure 12.13 gives approximate values of Y for some often encountered configurations. Figure 12.14 shows how η/η_{max} varies with X/X_{opt}, indicating the importance of making the operating value of X of a given design match X_{opt} as closely as possible in order to obtain a loss factor that approaches η_{max}.

If one knows the wavenumber $p = 2\pi/\lambda$ and the frequency associated with a given beam vibration, one may calculate the loss factor η of a composite beam simply by substituting the beam parameters and the material properties at any frequency (and temperature) of interest into Eqs. (12.21) and (12.22) or (12.25)–(12.27). The latter set of equations is particularly useful for judging how far from the optimum a given configuration may be operating.

*A shear-stiff, extensionally soft "spacer" (e.g., honeycomb) inserted between the viscoelastic and one or both structural components can serve to increase H_{13} and, therefore, the value of Y. See Eq. (12.22). For a given deflection of the composite structure, a spacer increases the damping by increasing the shear strain—and thus the energy storage and dissipation—in the viscoelastic component.[26]

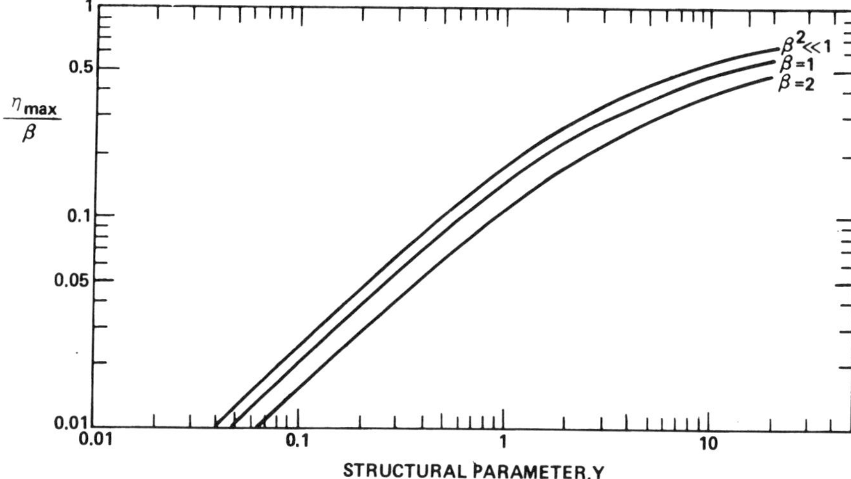

Fig. 12.12 Dependence of maximum loss factor η_{max} of three-component beams or plates on structural parameter Y and loss factor β of viscoelastic material.

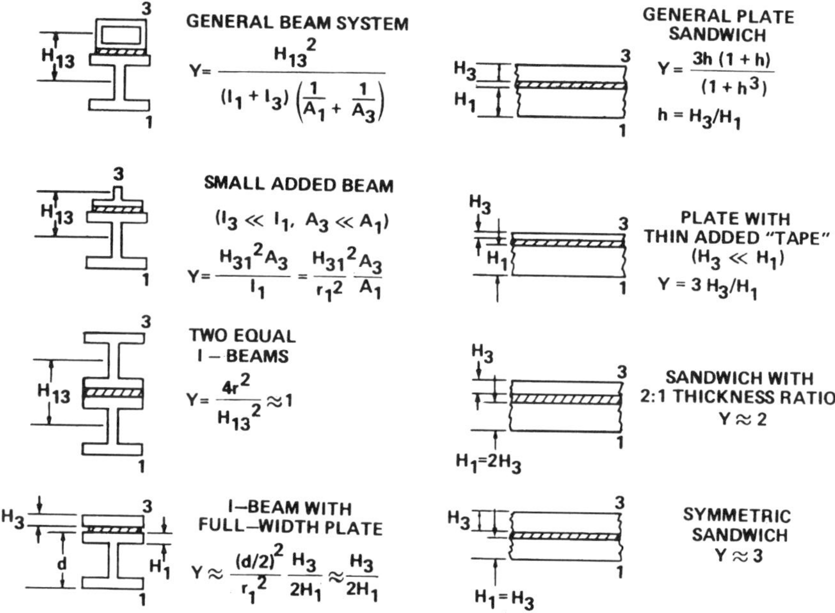

Fig. 12.13 Values of structural parameter Y for three-component beam and plate configurations with thin viscoelastic components and with structural components of the same material; viscoelastic component is shown cross-hatched. I, moment of inertia; A, cross-sectional area; r, radius of gyration; H_{13}, distance between neutral axes.

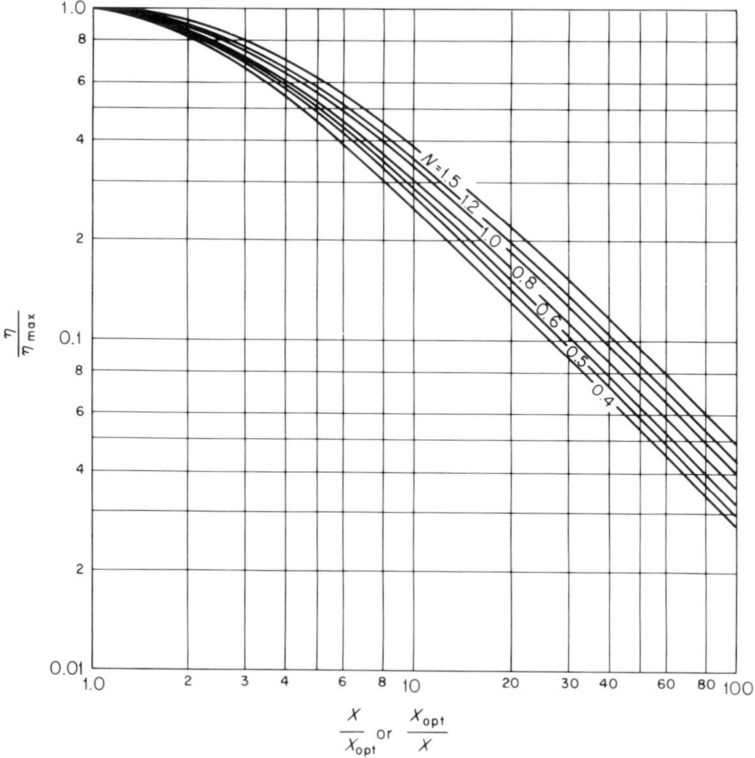

Fig. 12.14 Dependence of loss factor η of three-component beam or plate on shear parameter X [from Eq. (12.27)].

If one knows only the frequency and not the wavenumber p corresponding to a beam vibration, one needs to use Eq. (12.23) to determine p. Substitution of B as calculated from Eq. (12.24) into Eq. (12.23), followed by substitution of the result into the first of Eqs. (12.22), leads to a cubic equation in X. Although one may solve this numerically, it is often more convenient to determine X by use of an iteration procedure like that indicated in Fig. 12.15.

In contrast to the previously discussed two-component beam, the loss factor η of a three-component beam does not depend primarily on the loss modulus (i.e., on the product of the loss factor β and storage modulus E_2 or G_2) of the viscoelastic material. The loss factor of a three-component beam depends on β and G_2 separately, and the separate dependences must be taken into account in the design of such a beam. Design of a highly damped three-component beam structure requires (1) choice of a configuration with a large structural parameter Y, (2) selection of a damping material with a large loss factor β in the frequency and temperature range of interest, and (3) adjustment of the damping material thickness H_2 and length b so as to make X [as calculated from Eq. (12.22) for the value of G_2 applicable to the frequency and temperature of concern] approximately equal to

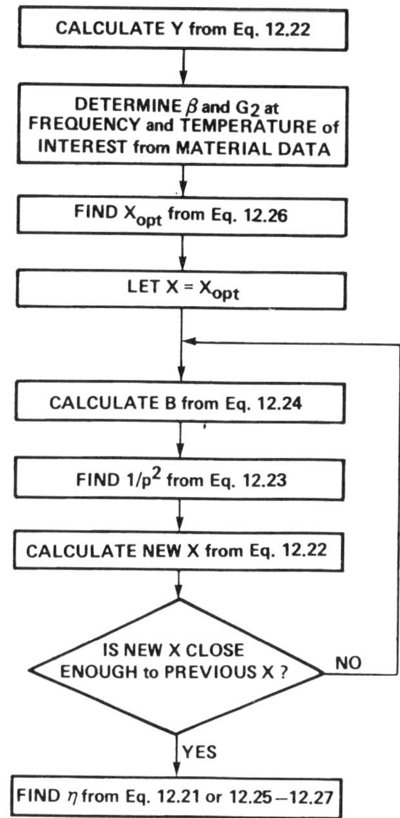

Fig. 12.15 Iteration procedure for determination of loss factor of three-component beams or plates for which wavelength is not known initially.

X_{opt} [given by Eq. (12.26)]. The resulting design then will have a loss factor approximately equal to η_{max} as given by Eq. (12.25).

The expected performance of any design should be checked for the frequency and temperature ranges of interest by means of the procedure described in the previous paragraph. Note that a given design may be expected to perform optimally—that is, to have X under operating conditions approximately equal to X_{opt}—only in a limited range of frequencies and temperatures, with reduced performance outside this range.

Plates with Viscoelastic Interlayers

A strip of a plate consisting of a viscoelastic layer between two structural layers may be considered as a special case of a three-component beam. In a plate strip, all components have rectangular cross sections of the same width; Eqs. (12.22) accordingly become

$$X = \frac{G_2}{p^2 H_2} S \qquad \frac{1}{Y} = \frac{E_1 H_1^3 + E_3 H_3^3}{12 H_{31}^2} S \qquad S = \frac{1}{E_1 H_1} + \frac{1}{E_3 H_3} \tag{12.28}$$

If, furthermore, $E_1I_1 + E_3I_3$ in Eq. (12.24) is replaced by $\frac{1}{12}(E_1H_1^3 + E_3H_3^3)$ and if μ in Eq. (12.23) is interpreted as the mass per unit surface area of the plate, then all of the foregoing discussion pertaining to beams also applies to plates.

REFERENCES

1. W. T. Thomson, *Theory of Vibration with Applications*, 2nd ed., Prentice-Hall, Englewood Cliffs, NJ, 1981.
2. R. Plunkett, "Measurement of Damping," in J. F. Ruzicka (ed.), *Structural Damping*, American Society of Mechanical Engineers, New York, 1959.
3. E. E. Ungar and E. M. Kerwin, Jr., "Loss Factors of Viscoelastic Systems in Terms of Energy Concepts." *J. Acoust. Soc. Am.* **34**(7), 954–957 (1962).
4. K. N. Tong, *Theory of Mechanical Vibration*, Wiley, New York, 1960.
5. E. E. Ungar, "Mechanical Vibrations," in H. A. Rothbart (ed.), *Mechanical Design and Systems Handbook*, 2nd ed., McGraw-Hill, New York, 1985, Chapter 5.
6. D. J. Ewins, *Modal Testing: Theory and Practice*, Research Studies Press Ltd., Letchworth, Hertfordshire, England, 1986.
7. V. H. Neubert, *Mechanical Impedance: Modelling/Analysis of Structures*, Josens Printing and Publishing Co., State College, PA, distributed by Naval Sea Systems Command, Code NSEA-SSN, 1987.
8. B. L. Clarkson and J. K. Hammond, "Random Vibration," in R. G. White and J. G. Walker (eds.), *Noise and Vibration*, by Wiley, New York, 1982, Chapter 5.
9. L. Cremer, M. Heckl, and E. E. Ungar, *Structureborne Sound*, 2nd ed., Springer-Verlag, New York, 1988.
10. S. H. Crandall, "The Role of Damping in Vibration Theory," *J. Sound Vib.* **11**(1), 3–18 (1970).
11. A. D. Nashif, D. I. G. Jones, and J. P. Henderson, *Vibration Damping*, Wiley, New York, 1985.
12. G. Kurtze, "Körperschalldämpfung durch Körnige Medien" [Damping of Structure-borne Sound by Granular Media], *Acustica* **6**(Beiheft 1), 154–159 (1956).
13. E. E. Ungar and L. G. Kurzweil, "Structural Damping Potential of Waveguide Absorbers," *Trans. Internoise 84*, pp. 571–574, December 1985.
14. M. A. Heckl, "Measurements of Absorption Coefficients on Plates," *J. Acoust. Soc. Am.* **34**, 308–808 (1962).
15. G. Maidanik, "Energy Dissipation Associated with Gas-Pumping at Structural Joints," *J. Acoust. Soc. Am.* **34**, 1064–1072 (1966).
16. E. E. Ungar, "Damping of Panels Due to Ambient Air," in P. J. Torvik, (ed.), *Damping Applications in Vibration Control*, American Society of Mechanical Engineers, AMD-Vol 38, New York, 1980, pp. 73–81.
17. E. E. Ungar and J. Carbonell, "On Panel Vibration Damping Due to Structural Joints," *AIAA J.* **4**, 1385–1390 (1966).
18. E. E. Ungar, "Damping of Panels," in L. L. Beranek, (ed.), *Noise and Vibration Control*, McGraw-Hill, New York, 1971, Chapter 14.
19. D. I. G. Jones, "A Reduced Temperature Nomogram for Characterization of Damping Material Behavior," *Shock Vib. Bull.* **48**(Pt. 2), 13–22 (1978).

20. D. I. G. Jones and J. P. Henderson, "Fundamentals of Damping Materials," Section 2 of Vibration Damping Short Course Notes, University of Dayton, M. L. Drake (ed.), 1988.

21. J. Soovere and M. L. Drake, "Aerospace Structures Technology Damping Design Guide," AFWAL-TR-84-3089, Flight Dynamics Laboratory, Wright-Patterson Air Force Base, OH, December 1985.

22. M. L. Soni and F. K. Bogner, "Finite Element Vibration Analysis of Damped Structures." *AIAA J.* **20**(5), 700–707 (1982).

23. C. D. Johnson, D. A. Kienholz, and L. C. Rogers, "Finite Element Prediction of Damping in Beams with Constrained Viscoelastic Layers," *Shock Vib. Bull.* **51**(Pt. 1), 71–82 (1981).

24. M. F. Kluesner and M. L. Drake, "Damped Structure Design Using Finite Element Analysis," *Shock Vib. Bull.* **52**(Pt. 5), 1–12 (1982).

25. E. E. Ungar, "Loss Factors of Viscoelastically Damped Beam Structures," *J. Acoust. Soc. Am.* **34**(8), 1082–1089 (1962).

26. E. M. Kerwin, Jr., "Damping of Flexural Waves in Plates by Spaced Damping Treatments Having Spacers of Finite Stiffness," *Proc. 3rd Int. Congr. Acoust. 1959*, Stuttgart, Elsevier, 1961, pp. 412–415.

27. D. Ross, E. E. Ungar, and E. M. Kerwin, Jr., "Damping of Plate Flexural Vibrations by Means of Viscoelastic Laminae," in J. Ruzicka (ed.), *Structural Damping*, American Society of Mechanical Enginers, New York, 1959, Section 3.

28. E. E. Ungar and E. M. Kerwin, Jr., "Plate Damping Due to Thickness Deformations in Attached Viscoelastic Layers," *J. Acoust. Soc. Am.* **36**, 386–392 (1964).

29. D. J. Mead, "Criteria for Comparing the Effectiveness of Damping Materials," *Noise Control* **1,** 27–38 (1961).

30. D. J. Mead, "Vibration Control (I)," in R. G. White and J. G. Walker (eds.), *Noise and Vibration*, Ellis Horwood Limited, Chichester, West Sussex, England, 1982.

31. E. M. Kerwin, Jr., "Damping of Flexural Waves by a Constrained Viscoelastic Layer," *J. Acoust. Soc. Am.* **31**, 952–962 (1959).

CHAPTER 13

Enclosures and Wrappings

ISTVÁN L. VÉR
Bolt Beranek and Newman, Inc.
Cambridge, Massachusetts, U.S.A.

Acoustical enclosures and wrappings constitute the most frequently used measures to reduce the noise radiated by machines, equipment, ducts, and pipes. Enclosures are also used to protect people or noise-sensitive equipment from exposure to high-intensity noise. Typical enclosure panels and wrappings are multilayer composite treatments, consisting of an impervious, exterior layer and a layer of porous sound-absorbing material, facing toward the interior. The massive, impervious exterior layer retains the sound energy radiated by the enclosed sound source. The porous sound-absorbing lining dissipates the retained sound energy.

The key difference between enclosures and wrapping is that in case of enclosures, the sound-absorbing layer is not in contact with the surface of the vibrating equipment, while in the case of wrappings, the porous sound-absorbing layer is in full surface contact with the vibrating body it surrounds. Because the porous sound-absorbing material of wrappings provides a full-surface, structureborne connection between the vibrating equipment and the exterior layer, it must not only be a good sound absorber but also be highly resilient so as to prevent the transmission of vibration to the outer impervious layer where it can be radiated as sound. Sound-absorbing materials used in enclosures, where there is no contact between the porous material and the vibrating equipment, can have a fairly rigid skeleton. Wrappings are most frequently used to decrease the sound radiation of vibrating surfaces

Noise and Vibration Control Engineering: Principles and Applications, Edited by Leo L. Beranek and István L. Vér.
ISBN 0-471-61751-2 © 1992 John Wiley & Sons, Inc.

such as ducts and pipes and sometimes also to gain extra sound attenuation of acoustical enclosures. Since fibrous sound-absorbing materials, such as glass fiber and mineral wool, are good heat insulators, properly designed wrappings can provide both substantial acoustical and heat insulation. On the other hand, the heat-insulating properties of the enclosure may be detrimental and require that provision be made for auxiliary cooling of the enclosure to prevent the buildup of excessively high temperatures. Only the acoustical design aspects of enclosures and wrappings are treated in this chapter.

13.1 ACOUSTICAL ENCLOSURES

Depending on their size (compared with the acoustical and bending wavelength) acoustical enclosures can be termed either small or large. The enclosure is considered small if both the bending wavelength is large compared with the largest wall panel dimensions and the acoustical wavelength is large compared with the largest interior dimension of the enclosure volume. In small acoustical enclosures, the interior volume has no acoustical resonances. If the largest dimension of the acoustical volume is $L_{\max} \leq \frac{1}{10}\lambda$, the sound pressure is evenly distributed within the volume. The enclosure is considered large if all of its interior dimensions are large compared with the acoustical wavelength and there are a large number of acoustical resonances in the interior volume in the frequency range of interest. Accordingly, even enclosures with large physical dimensions are acoustically small at every low frequencies while enclosures of small physical size are acoustically large at very high frequencies. In almost all acoustical enclosures the enclosure walls already exhibit numerous structure resonances in the frequency range where the first acoustical resonance occurs.

If the enclosure has no mechanical connections to the enclosed equipment, it is termed *free standing*. If there are mechanical connections, then the enclosure is *equipment mounted*. Enclosures that very closely surround the enclosed equipment and the volume of the machine is comparable to the volume of the enclosure are called *close fitting*. Enclosures without acoustically significant openings are referred to as *sealed enclosures*, and those with significant acoustical leaks (intentional or unintentional) as *leaky* acoustical enclosures. Figure 13.1 shows various configurations of sealed acoustical enclosures. This chapter deals with the acoustical design of enclosures. Nonacoustical aspects—such as ventilation, safety, and economy—are treated in a handbook by Miller and Montone (see Bibliography). Construction details and advice for writing purchase specifications are given in a VDI guideline (also see Bioblography).

Insertion Loss as Acoustical Performance Measure

The insertion loss is the most appropriate descriptor for the acoustical performance of enclosures of all types. The operational definition of the insertion loss (IL) of an acoustical enclosure is illustrated in Fig. 13.2. For noise sources that will be

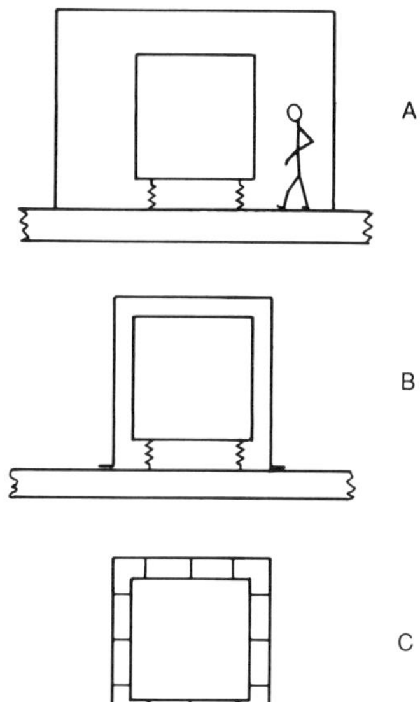

Fig. 13.1 Enclosure types: (*a*) free standing, large; (*b*) free standing, close fitting; (*c*) equipment mounted, close fitting.

positioned indoors, such as machinery in factory spaces, the sound-power-based insertion loss of the enclosure as indicated in Fig. 13.2*a* is the most meaningful. It is defined as

$$\text{IL}_w = 10 \log \left(\frac{W_0}{W_E} \right) = L_{w0} - L_{WE} \quad \text{dB} \tag{13.1}$$

where W_0 is the sound power radiated by the unenclosed source and L_{W0} the corresponding sound power level and W_E is the sound power radiated by the enclosed source and L_{WE} the corresponding sound power level. Both W_0 and W_E are measured in a reverberation room (see Chapter 4) or with the aid of a sound intensity meter.

For enclosures used with equipment deployed outdoors, a less precise but more

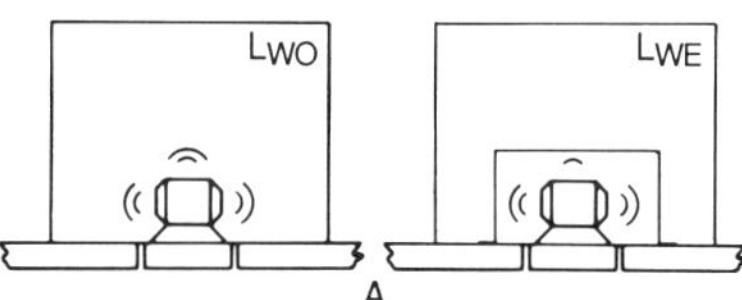

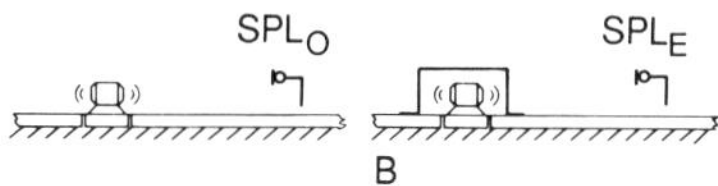

Fig. 13.2 Operational definition of enclosure insertion loss IL: (*a*) power based; $\text{IL}_w \equiv L_{w0} - L_{WE}$, (*b*) sound pressure based; $\text{IL}_P \equiv \text{SPL}_0 - \text{SPL}_E$.

easily implemented definition of insertion loss, the so-called pressure-based insertion loss, or IL_p, illustrated in Fig. 13.2*b* is most appropriate. It is defined as

$$IL_p = SPL_0 - SPL_E \quad dB \tag{13.2}$$

where SPL_0 is the average sound pressure level measured at a number of locations around the source without the enclosure and SPL_E is that measured with the source surrounded by the enclosure. The measurement positions may be chosen on a circle that is centered at the source location. The measurement distance should be at least three times the longest dimension of the enclosure. Equations (13.1) and (13.2) represent definitions readily implemented in the field or laboratory. Using these definitions one can readily determine whether a specific enclosure will or will not meet specific performance requirements stated in the form of a sound-power-level reduction, a sound-pressure-level reduction for a specified distance, or a sound-pressure-level reduction at a specified distance and direction. If the laboratory facilities are available, the sound-pressure-based insertion loss may also be measured in a semianechoic chamber.

Note that if the radiation patterns of both the unenclosed and enclosed source is omnidirectional, the two measures yield the same result ($IL_w = IL_p$).

Qualitative Description of Acoustical Performance

Figure 13.3 shows the typical shape of a curve of the acoustical insertion loss versus frequency of a free-standing, sealed acoustical enclosure. Region I is the small-enclosure region where neither the interior air volume nor the enclosure wall panels exhibit any resonances. In this frequency region the insertion loss is frequency independent and is controlled by the ratio of the volume compliance (inverse of volume stiffness) of the enclosure walls and that of the enclosed air volume.

Region II in Fig. 13.3 is the intermediate region where the insertion loss is controlled by the resonant interaction of the enclosure structure with the enclosed acoustical volume. The region is characterized by a number of alternating maxima and minima in the insertion loss. Typically, the first and most important minima

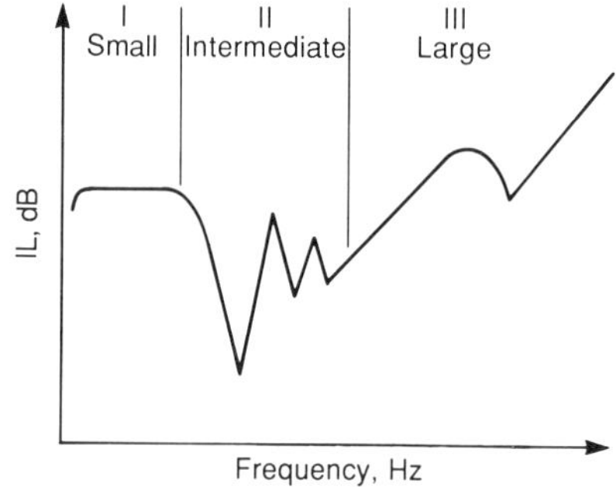

Fig. 13.3 Typical curve of insertion loss versus frequency of a sealed, free-standing acoustical enclosure. Region I: panel stiffness controlled; damping and interior absorption are ineffective. Region II: resonance controlled; damping and interior absorption are effective. Region III: controlled by sound transmission loss; sound transmission loss and interior absorption are effective; usually limited by leaks.

in the insertion loss occurs when the combined volume compliance of the interior air volume and the volume compliance of the wall panels matches the mass compliance of the wall panels. The insertion loss at this resonance frequency usually controls the low-frequency insertion loss of the enclosure and in some instances can become negative, signifying that the equipment with the enclosure may radiate more noise than without the enclosure. Additional minima of the insertion loss occur at acoustical resonances of the enclosure volume. Further minima in insertion loss occur when the frequencies of structural resonances of a wall panel and the frequencies of acoustical resonances of the enclosure volume coincide. It is imperative that enclosures designed for sound sources that radiate noise of predominantly tonal character (such as transformers, gears, reciprocating compressors and engines, etc.) should have no structural or acoustical resonances that correspond to the frequencies of the predominant components of the source noise.

Region III in Fig. 13.3 is the large-enclosures region where both the enclosure wall panels and the interior air volume exhibit a very large number of acoustical resonances. Here, statistical methods of room acoustics can be used to predict the sound field inside the enclosure (see Chapter 7) and the sound transmission through the enclosure walls (see Chapter 9). In this frequency region, the insertion loss is controlled by the interior sound absorption and by the sound transmission loss (R) of the enclosure wall panels. The dip in the curve of insertion loss versus frequency in region III corresponds to the coincidence frequency of wall panel (see Chapter 9), which for the most frequently used wall panels (1–2 mm steel or aluminum) falls above the frequency region of interest. The coincidence frequency may fall into the region of interest if the ratio of stiffness of the panel to its mass per unit area is high (e.g., a honeycomb panel).

The prediction of the insertion loss of small and large acoustical enclosures is treated in the following sections. Performance prediction in the intermediate frequency region requires a detailed finite-element analyses of the coupled mechanical acoustical system such as outlined in Chapter 6.

Small Sealed Acoustical Enclosures

For a small, sealed acoustical enclosure, where the sound pressure inside the cavity is evenly distributed, the insertion loss is given by[1-3]

$$\mathrm{IL}_{\mathrm{SM}} = 20 \log \left(1 + \frac{C_v}{\sum_{i=1}^{n} C_{wi}} \right) \quad \mathrm{dB} \tag{13.3}$$

where

$$C_v = \frac{V_0}{\rho c^2} \quad \mathrm{m^5/N} \tag{13.4}$$

is the compliance of the gas volume inside the enclosure, V_0 is the volume of the gas in the enclosure volume, ρ is the density of the gas, c is the speed of sound of the gas, and C_{wi} is the *volume compliance* of the ith enclosure wall plate defined as

$$C_{wi} = \frac{\Delta V_{pi}}{p} \quad \text{m}^5/\text{N} \tag{13.5}$$

where ΔV_{pi} is the volume displacement of the ith enclosure wall plate in response to the uniform pressure p. It is assumed here that the enclosure is rectangular and is made of n separate, homogeneous, isotropic plates, each with its own volume compliance.

At frequencies below the first mechanical resonance of the isotropic enclosure wall panel, the volume compliance of a homogeneous, isotropic panel C_{pi} is given by[2]

$$C_{wi} = \frac{10^{-3} A_{wi}^3 F(\alpha)}{B_i} \quad \text{m}^5/\text{N} \tag{13.6}$$

where A_{wi} is the surface area of the ith wall panel and $F(\alpha)$ is given in Fig. 13.4 as a function of the aspect ratio $\alpha = a/b$ of the panel, where a is the longest and

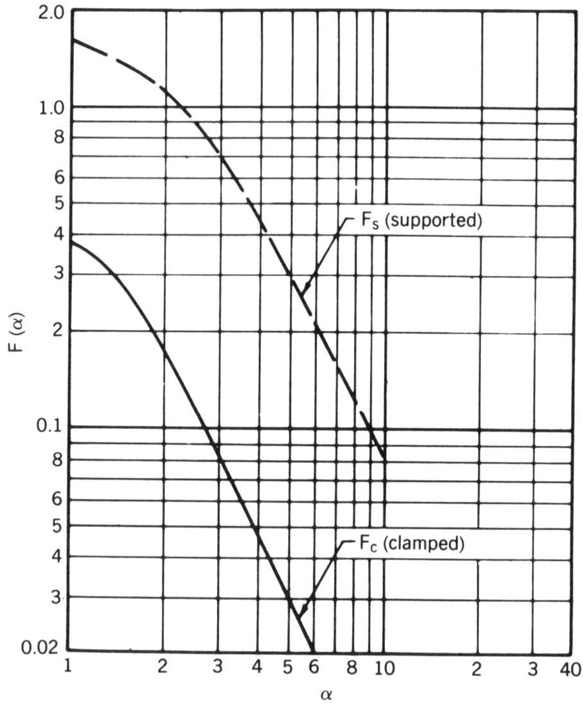

Fig. 13.4 Plate volume compliance function $F(\alpha)$ plotted versus the aspect ratio $\alpha = a/b$ for homogeneous isotropic panels either with clamped or simply supported edges. (After ref. 1.)

b is the smallest edge dimension of the wall panel. For homogeneous, isotropic wall panels, the bending stiffness per unit length of the panel is

$$B = \frac{Eh^3}{12(1 - \nu^2)} \quad \text{N} \cdot \text{m} \tag{13.7}$$

where E is Young's modulus and h is the thickness and ν the Poisson ratio of the wall panel. For a rectangular enclosure combining Eqs. (13.3)–(13.7) yields

$$\text{IL}_{\text{SM}} = 20 \log \left[1 + \frac{V_0 E h^3}{12 \times 10^{-3}(1 - \nu^2)\rho c^2} \sum_{i=1}^{6} \frac{1}{A_{wi}^3 F(\alpha_i)} \right] \quad \text{dB} \tag{13.8}$$

For the special case of a cubical enclosure with clamped walls of edge length a Eq. (13.8) yields

$$\text{IL}_s = 20 \log \left[1 + 41 \left(\frac{h}{a} \right)^3 \frac{E}{\rho c^2} \right] \quad \text{dB} \tag{13.9}$$

Equation (13.8) indicates that high insertion loss is achieved if the enclosure has small edge length, high aspect ratio, and large wall thickness, edge conditions are clamped, and the panels are made of a material of high Young's modulus. In short, for an enclosure with high insertion loss at low frequencies, the wall must be made as stiff as possible. According to Eq. (13.9), the low-frequency insertion loss of a cubical steel enclosure of edge length $a = 300$ mm, $E = 2 \times 10^{11} \cdot \text{N/m}^2$, $\rho = 1.2\ \text{kg/m}^3$, and $c = 340$ m/s is for two different wall thicknesses h:

Wall	Clamped		Simply Supported	
Thickness h	IL_s	f_0	IL_s	f_0
3 mm	35.5 dB	296 Hz	24 dB	162 Hz
1.5 mm	18.5 dB	148 Hz	9 dB	81 Hz

Here f_0 is the resonance frequency of each of the identical homogeneous isotropic clamped panels. The insertion loss values listed in the table above are valid only at frequencies well below f_0. Note in the table that much higher insertion loss is achieved with a clamped edge condition than with simply supported edge conditions. In practice, clamped edges are almost impossible to achieve. Consequently, one should use the simply supported edge condition in the initial design to ensure that the designed performance will be achieved.

It is of considerable practical interest to know which material should be used for a small sealed cubical enclosure to yield the highest insertion loss at low frequencies for the same enclosure volume and same total weight. Considering that the total mass M of a cubical enclosure of edge length a is $M = 6\rho_M a^2 h$, Eq.

(13.9) can be expressed as

$$\mathrm{IL}_s = 20 \log \left[1 + 0.19 \frac{M^3}{a^9 \rho c^2} \left(\frac{c_L}{\rho_M} \right)^2 \right] \quad \text{dB} \qquad (13.10)$$

where $c_L = \sqrt{(E/\rho_M)}$ is the speed of longitudinal waves in the bulk enclosure material and ρ_M is the density of the enclosure material. Equation (13.10) indicates that for all materials giving the same enclosure mass M, the material with the highest c_L/ρ_M ratio yields the highest insertion loss. Table 13.1 lists the c_L/ρ_M ratio and the normalized low-frequency insertion loss ΔIL for frequency used materials, where ΔIL is defined as the low-frequency insertion loss of an enclosure made out of a specific material minus the insertion loss of an enclosure of the same volume and weight built of steel. Table 13.1 indicates that, in regard to low-frequency insertion loss, aluminum and glass are superior to steel and that lead is the worst possible choice! Note, however, that this conclusion can be exactly opposite for a large enclosure if the coincidence frequency falls within the frequency range of interest.

Formstiff Small Enclosures. Because the insertion loss of small, sealed enclosures at very low frequencies ($L_{max} < \frac{1}{10}\lambda$) is controlled by the volume compliance of the enclosure walls, it is desirable to select constructions that provide the highest wall stiffness for the allowable enclosure weight. An enclosure consisting of a round, cylindrical body and two half-spherical end caps yields a very stiff construction and much higher insertion loss than a rectangular enclosure of the same volume and weight. Figure 13.5 shows the acoustical characteristics of such a small, cylindrical enclosure with no internal sound-absorbing treatment. Curve *A* represents insertion loss versus frequency measured with an external source, indicating that insertion loss at most frequencies exceeds 55 dB. Curve *B* represents the sound pressure levels in the enclosed air volume when excited with an external source, indicating the presence of strong acoustical resonances at 550 Hz and above. Curve *C* represents the sound-induced vibration acceleration of the cylindrical shell caused by an external sound source. Strong structural resonances are seen at 1.5, 3, and 3.75 kHz. Observing the three curves in Fig. 13.5 it is noted that the frequencies where the insertion loss is minimum coincide either with acoustical resonances of the interior volume or with structural resonances of the enclosure shell. The strong minima of the curve of insertion loss versus frequency in the vicinity of 1.5, 3,

TABLE 13.1 Normalized Low-Frequency Insertion Loss ΔIL and c_L/ρ_M Ratio for Different Construction Materials[a]

	Lead	Steel	Concrete	Plexiglass	Aluminum	Glass
c_L/ρ_M, (m^4/kg · s)	0.11	0.65	1.5	1.6	1.9	2.1
ΔIL, dB	−31	0	+14	+15	+19	+20

[a] ΔIL is for a cubical enclosure with identical sides.

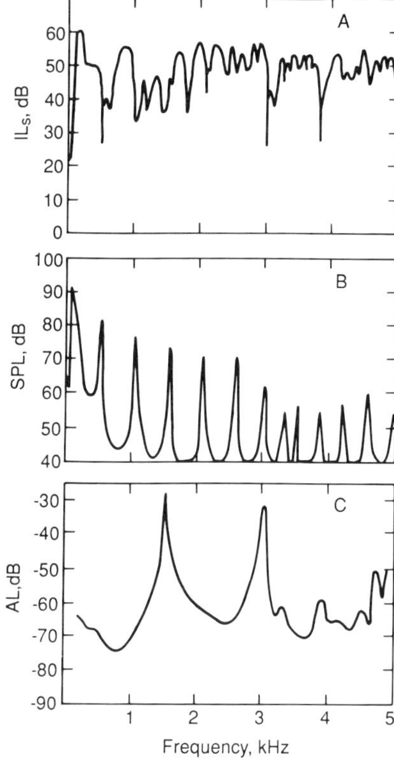

Fig. 13.5 Acoustical characteristics of a small, sealed, round cylindrical enclosure without interior absorption: (*a*) insertion loss IL; (*b*) resonant acoustical response of interior volume, SPL; (*c*) sound-induced vibration acceleration response of enclosure wall, AL.

and 3.75 kHz are caused by coincidence of the structural resonance of the shell with acoustical resonances of the interior air volume.

Another way to achieve a low wall compliance for a given total weight is to make the enclosure walls out of a composite material consisting of a honeycomb core sandwiched between two lightweight plates. A special form of such high-stiffness, low-weight panel, described by Fuchs, Ackermann, and Frommhold,[8] also provides high sound absorption at low frequencies in addition to the low volume compliance. In this case, the panel on its interior side has two double membranes. The first, thicker membrane, which is rigidly bonded to the honeycomb core, has round openings to make an inward-facing Helmholtz resonator out of each honeycomb cavity. A second, thin, membrane that is covering the first one is free to move over the resonator openings. The presence of this second, thin membrane lowers the resonance frequency of the Helmholtz resonators owing to the mass of the membrane covering the opening and increases the dissipation owing to air pumping. Such Helmholtz plates can be constructed from stainless steel, aluminum, or light-transparent plastic. They are fully sealed and can be hosed down for cleaning. Figure 13.6 shows the curve of insertion loss versus frequency of a cubical enclosure of 1 m edge length. The solid curve was obtained with enclosure walls made of the above-described Helmholtz plates of 10 cm thickness

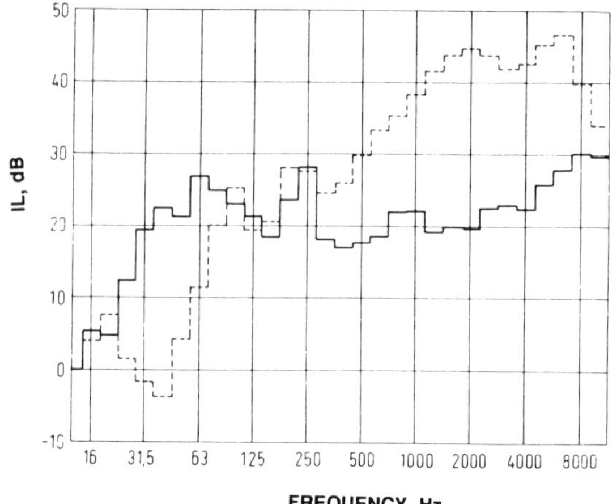

Fig. 13.6 Measured insertion loss of two 1 × 1 × 1-m enclosures. (After ref. 7.) Solid curve: Helmholtz plate walls, 10 cm thick, 8.5 kg/m^2, no sound-absorbing treatment. Dotted curve: Woodchip board walls, 1.3 cm thick, 5 cm thick sounding-absorbing layer, 10 kg/m^2.

with a mass per unit area of 8.5 kg/m^2 without any additional sound-absorbing lining. The dotted curve was obtained with an enclosure of the same size, made of 1.3-cm-thick wood chip board with 5-cm sound-absorbing lining, yielding a mass per unit area of 10 kg/m^2. The Helmholtz plate enclosure yields superior low-frequency performance and does not require extra sound-absorbing lining. However, the insertion loss of this enclosure at midfrequencies (above 250 Hz) is low because of the low coincidence frequency of the light and stiff panels.

Small Leaky Enclosures. Except for special constructions, such as hermetically sealed compressors, all enclosures are likely to be leaky and provide zero insertion loss as the frequency approaches zero. If an enclosure has a leak in the form of a round opening of radius a_L in a wall of thickness h, the leak represents a compliance $C_L = 1/j\omega Z_L$, where Z_L is the acoustical impedance of the leak given in Chapter 8, as

$$C_L = \frac{(\pi a_L^2)^2}{-\omega^2 \rho (h + \Delta h) \pi a_L^2 + j\omega R} \qquad \text{m}^5/\text{N} \tag{13.11}$$

where h is the plate thickness, $\Delta h \simeq 1.2 a_L$ represents the end correction, and R is the real part of the impedance of the leak. The sound pressure inside the small, leaky enclosure owing to the operation of a source of volume velocity q_0 is

$$p_{\text{ins}}^{\text{leaky}} = \frac{q_0}{j\omega} \frac{1}{C_L + C_v + \sum_i C_{wi}} \qquad \text{N/m}^2 \tag{13.12}$$

and the insertion loss of the small leaky enclosure is

$$\mathrm{IL}_{\text{leaky}} = \mathrm{IL}_L = 20 \log \frac{C_L + C_v + \sum_i C_{wi}}{C_L + \sum_i C_{wi}} \quad \text{dB} \qquad (13.13)$$

where C_v and C_{wi} are given in Eqs. (13.4) and (13.5), respectively. Since C_L approaches infinity as the frequency approaches zero, the insertion loss of leaky enclosures approaches zero for any form of leak. The insertion loss becomes negative in the vicinity of the Helmholtz resonance frequency of the compliant leaky enclosure given by

$$f_{0L} = \frac{ac}{2} \sqrt{\frac{1}{\pi(h + \Delta h)\left(V_0 + \rho c \sum_i C_{wi}\right)}} \quad \text{Hz} \qquad (13.14)$$

As $C_L / \Sigma\, C_{wi}$ becomes small with increasing frequency, the insertion loss of the small, leaky enclosure, IL_L, approaches that of the sealed enclosure, IL_s. This behavior is illustrated in Fig. 13.7.

Close-Fitting, Sealed, Acoustical Enclosures

A close-fitting acoustical enclosure is one in which a considerable portion of the enclosed volume is occupied by the equipment being quieted. Such enclosures are

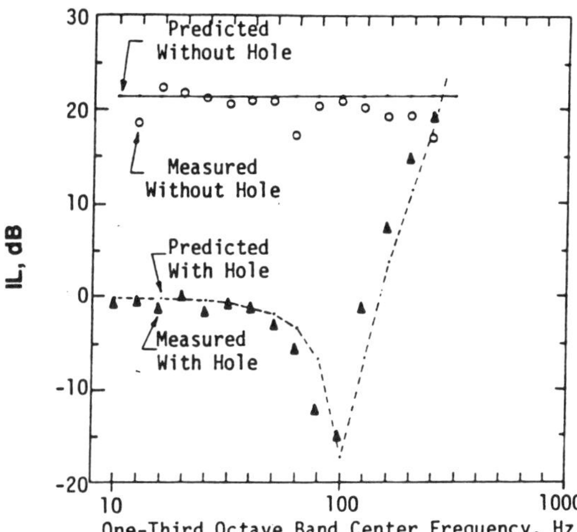

Fig. 13.7 Measured and predicted sound-pressure-based insertion loss (outside–inside direction) of a 50 × 150 × 300-mm aluminum enclosure of 1.5-mm-thick wall with a single 9.4-mm-diameter hole (after ref. 3): (——) predicted, sealed; (---) predicted, leaky; (○) measured, sealed; (▲) measured, leaky.

used in cases where the radiated noise must be reduced using a minimum of added volume. Vehicle engine enclosures, portable compressors, and transportable transformers fall into this category.

Free-standing close-fitting enclosures, such as shown in Fig. 13.1*b*, have no mechanical connections to the vibrating source and the enclosure walls are excited to vibration only by the airborne path. Conversely, the walls of machine-mounted close-fitting enclosures, such as depicted in Fig. 13.1*c*, are excited by both airborne and structure-borne paths.

The characteristic property of a close-fitting enclosure is that the air gap (the perpendicular distance between the vibrating surface of the machine and the enclosure wall) is small compared with the acoustical wavelength in most of the frequency range. The walls of the close-fitting enclosures are usually made of thin flat sheet metal with a layer of sound-absorbing material such as glass fiber or acoustical foam on the interior face. The purpose of the sound-absorbing lining is to damp out the acoustical resonances in the airspace.

Free-Standing, Close-Fitting, Acoustical Enclosures. Free-standing, close-fitting enclosures, which have no structure-borne connection to the vibrating equipment, always achieve higher insertion loss than a similar machine-mounted enclosure can provide. If the enclosure is perfectly sealed, then its insertion loss depends on (1) thickness, size, and material of the enclosure wall; (2) the edge-joining conditions of the wall plates and their loss factors; (3) the vibration pattern of the machine surface; (4) average thickness of the air gap between the wall and the machine; and (5) the type of sound-absorbing material in the air gap.

Figure 13.8 shows the measured dependence of the power-based insertion loss of a model close-fitting enclosure consisting of 0.6×0.4-m enclosure walls of thickness h and a variable spacing gap of thickness T. To obtain these data, an array of loudspeakers, which represented the vibrating surface of the enclosed machine, was phased to simulate various vibration patterns.[4] Figure 13.8*a* shows the dependence of the measured insertion loss on frequency obtained for 1-mm-thick aluminum and steel wall plates, respectively. Because the steel and aluminum plates have practically the same dynamic behavior, the higher insertion loss of the steel wall panel is attributable to its higher mass. In the frequency range between 300 Hz and 1 kHz, the 9-dB average difference corresponds to the difference in density of the two materials. Figure 13.8*b* shows the effect of wall thickness. Here again the insertion loss increases with increasing wall thickness. Below 250 Hz, where the insertion loss is controlled by the stiffness of the air and the stiffness and damping of the wall panels, the IL_w changes little with increasing frequency. In the frequency range between 200 Hz and 1 kHz, where the IL_w is controlled by the volume stiffness of the air and by the mass per unit area of the wall, the IL_w increases with a slope of 40 dB/decade. Above 1 kHz, the insertion loss is limited by the acoustical resonances in the airspace. Figure 13.8*c* shows that the insertion loss increases with increasing air gap thickness. Figure 13.8*d* shows the effect sound-absorbing material in the airspace and damping treatment of the enclosure wall on the achieved insertion loss. The sound-absorbing treatment in the airspace prevents the acoustical resonances in the airspace and results

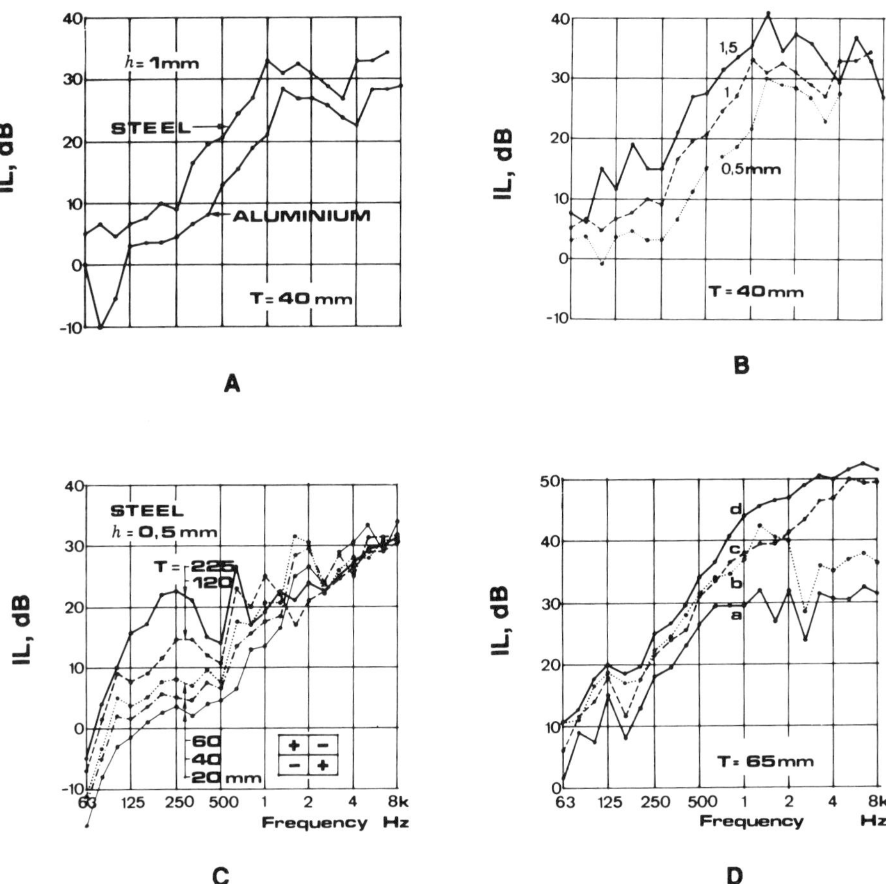

Fig. 13.8 Effect of key parameters on the power-based insertion loss of a model close-fitting enclosure. Wall panel 0.6 × 0.4. (After ref. 4.) (*a*) Effect of the wall material; 4-cm air gap, in-phase excitation. (*b*) Effect of wall panel thickness, 4-cm air gap, in-phase excitation. (*c*) Effect of average air gap thickness; 0.5-mm steel plate, quadrupole excitation. (*d*) Effect of sound-absorbing material and structural damping; 1-mm steel plate, 6.5-cm air gap, in-phase excitation: *a*, no damping, no absorption in airspace; *b*, with damping treatment; *c*, no damping, but absorption in airspace; *d*, with damping and absorption.

in a substantial increase of the insertion loss above 1 kHz. Structural damping helps to reduce the deleterious effect of the efficiently radiating plate resonance at 160 Hz. A combination of sound-absorbing treatment and structural damping results in a smooth, steeply-increasing insertion loss with increasing frequency and provides a very high degree of acoustical performance.

One-Dimensional Model. According to Bryne, Fischer, and Fuchs,[5] the insertion loss of free-standing, sealed, close-fitting acoustical enclosures can be pre-

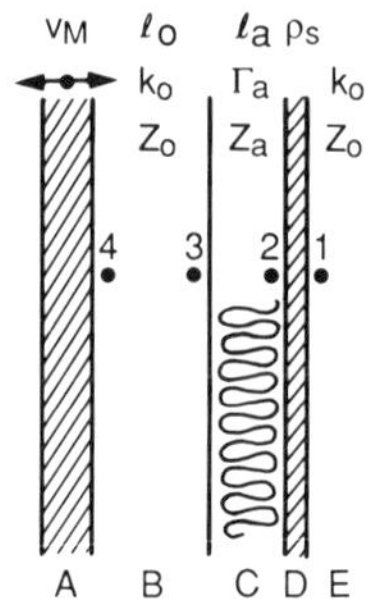

Fig. 13.9 One-dimensional model for predicting the power-based insertion loss IL of free-standing, sealed close-fitting acoustical enclosures; *A*, machine wall vibrating with velocity v_M; *B*, air gap thickness l_0; *C*, sound-absorbing layer thickness l_a; *D*, impervious enclosure wall of mass per unit area ρ_s; *E*, free space into which sound is radiated. (After ref. 5.)

dicted with reasonable accuracy with a simple one-dimensional model shown in Fig. 13.9. It is assumed that the machine wall (*A*) vibrates in phase (like a rigid piston) with velocity v_M and that the motion of the machine is not effected by the presence of the enclosure. The vibrating machine is surrouned by a flat, limp, impervious enclosure wall (*D*), which has an internal sound-absorbing lining (*C*) of thickness l_a and there is an air gap (*B*) of thickness l_0 between the machine and the sound-absorbing layer. The sound pressure on the outside surface of the enclosure (point 1 in Fig. 13.9) is given by

$$p_1 = p_2 \left(1 - \frac{j\omega\rho_s}{Z_0}\right) \quad \text{N/m}^2 \tag{13.15}$$

where $Z_0 = \rho c$ is the characteristic impedance of air. The following equations are used to obtain p_1 as a function of the machine wall vibration velocity v_M:

$$p_3 = 0.5p_2 \left[\left(1 + \frac{Z_a}{Z_3}\right) e^{-j\Gamma_a l_a} + \left(1 - \frac{Z_a}{Z_3}\right) e^{j\Gamma_a l_a}\right] \quad \text{N/m}^2 \tag{13.16}$$

$$p_4 = v_M Z_4 = 0.5p_3 \left[\left(1 + \frac{Z_0}{Z_4}\right) e^{-jk_0 l_0} + \left(1 - \frac{Z_0}{Z_4}\right) e^{jk_0 l_0}\right] \quad \text{N/m}^2 \tag{13.17}$$

$$Z_2 = Z_0 + j\omega\rho_s \quad \text{N} \cdot \text{s/m}^3 \tag{13.18}$$

$$Z_3 = Z_a \frac{(1 + Z_a/Z_2)e^{j\Gamma_a l_a} + (1 - Z_a/Z_2)e^{-j\Gamma_a l_a}}{(1 + Z_a/Z_2)e^{j\Gamma_a l_a} - (1 - Z_a/Z_2)e^{-j\Gamma_a l_a}} \quad \text{N} \cdot \text{s/m}^3 \tag{13.19}$$

$$Z_4 = Z_0 \frac{(1 + Z_0/Z_3)e^{jk_0 l_0} + (1 - Z_0/Z_3)e^{-jk_0 l_0}}{(1 + Z_0/Z_3)e^{jk_0 l_0} - (1 - Z_0/Z_3)e^{-jk_0 l_0}} \quad \text{N} \cdot \text{s/m}^3 \tag{13.20}$$

where $k_0 = 2\pi f/c$ is the wavenumber, ρ the density, and c the speed of sound air; Z_a is the complex characteristic impedance and Γ_a the complex propagation constant of the porous sound-absorbing material as given in Chapter 8.

The sound power radiated by the unenclosed machine wall is given by

$$W_0 = v_M^2 \rho c A \quad \text{W} \tag{13.21a}$$

and when it is enclosed with a free-standing, close-fitting enclosure

$$W_A = \frac{p_1^2}{\rho c} A \quad \text{W} \tag{13.21b}$$

yielding for the insertion loss

$$\text{IL}_F = 10 \log \frac{W_0}{W_A} = 20 \log \frac{v_M \rho c}{p_1^2} \quad \text{dB} \tag{13.22}$$

In deriving this equation, no account is taken of possible change in surface area or radiation efficiency between the enclosure surface and machine surface.

If the angle of sound incidence on the enclosure wall is known (i.e., a specific oblique angle or random), the power-based insertion loss of free-standing, close-fitting enclosures can be predicted on the basis of the two-dimensional model of sound transmission through layered media presented in Chapter 9.

Machine-mounted, Close-fitting Acoustical Enclosure. If the close-fitting enclosure is machine mounted, the rigid or resilient connections between the machine and the enclosure wall give rise to additional sound radiation of the enclosure wall. Assuming that the vibration velocity of the machine v_M is not affected by the connected enclosure, the additional sound power W_M radiated owing to the structural connections is

$$W_M = \sum_{i=1}^{n} F_i^2 \left(\frac{\rho}{\rho_s^2 c} + \frac{\rho c \sigma}{2.3 \rho_s^2 c_L h \omega \eta} \right) \quad \text{W} \tag{13.23}$$

where

$$F_i^2 = \frac{v_{Mi}^2}{|1/2.3\rho_s^2 c_L h + j\omega/s|^2} \quad \text{N}^2 \tag{13.24}$$

In Eq. (13.24), F_i is the force transmitted by the ith attachment point, v_{Mi} is the vibration velocity of the machine at the ith attachment point, n is the number of point attachments between the machine and the homogeneous, isotropic enclosure wall, ρ_s is the mass per unit area of the enclosure wall, h is the wall thickness, c_L is the speed of longitudinal waves in the wall material, η is the loss factor, σ is the radiation efficiency, and s is the dynamic stiffness of the resilient mount connecting the enclosure wall to the machine ($s = \infty$ for rigid point connections). To minimize structure-borne transmission, it is advantageous to select attachment points at those locations on the machine that exhibit the lowest vibration. The insertion loss of the machine-mounted, close fitting enclosure is

$$\text{IL}_{\text{MM}} = 10 \log_{10} \frac{W_0}{W_A + W_M} \quad \text{dB} \tag{13.25}$$

where W_0 and W_A are given in Eqs. (13.21a) and (13.21b).

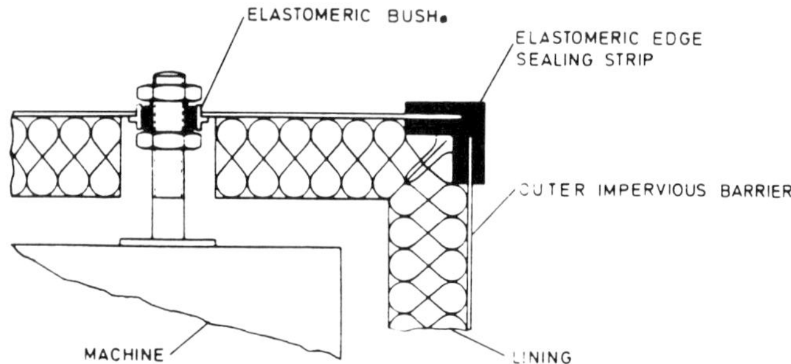

Fig. 13.10 Machine-mounted, close-fitting enclosure; resilient mounting, resilient panel edge connections. (After ref. 5.)

Figure 13.10 shows a close-fitting enclosure investigated experimentally and analytically by Byrne, Fischer, and Fuchs[5] in three configurations: (1) free standing, (2) rigidly machine mounted, and (3) resiliently machine mounted. Figure 13.11 shows the measured insertion loss obtained for each of the three configurations. The machine-mounted enclosure, which was supported from the machine at four points at each side, yield substantially lower-insertion loss at high frequencies than the free-standing enclosure. Elastic mounting yields higher insertion loss than rigid mounting.

As shown by reference 5, the simple one-dimensional analytical model based

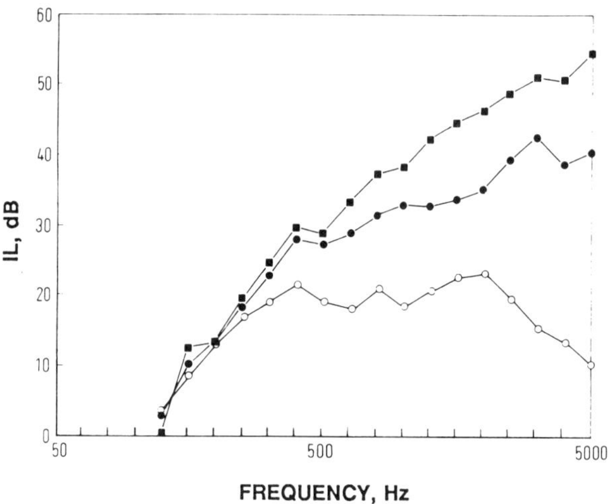

Fig. 13.11 Insertion Loss (IL_s) of a sealed, close-fitting enclosure made of 1-mm steel plates. (After refs. 6 and 7.) Dimensions: $1 \times 0.6 \times 0.8$ m; machine, enclosure wall distance 50 mm; interior lining 50 mm thick; flow resistivity $R_1 = 2 \times 10^4$ N · s/m^4. (■) Free standing; (●) machine mounted, resilient mounting (see Fig. 13.10); (○) machine mounted, rigid mounting.

on Eqs. (13.15)–(13.25) yields predictions that are in good agreement with measured data. The reason for this surprisingly good agreement is that the sound-absorbing treatment effectively prevents sound propagation (and the occurrence of acoustical resonances) in the plane parallel to the enclosure wall and also provides structural damping to the thin enclosure wall. Replacing the effective glass fiber sound-absorbing treatment with a thinner, less effective acoustical foam has resulted in substantial decrease in the insertion loss, indicating the crucial importance of an effective sound-absorbing treatment. Figure 13.12 shows how the specific choice of edge connections of the enclosure plates effected the insertion loss of the machine-mounted, close-fitting enclosure. Elastically sealed edges, such as shown in Fig. 13.10, yield higher insertion loss than rigidly connected (welded) edges. The better performance obtained with elastically sealed edges is due to the reduced coupling between the in-plane motion of one plate (which radiates no sound) with the normal motion of the connected plate (that radiates sound efficiently). The elastic edge seal also increases the loss factor of the enclosure plates, thereby reducing their resonant vibration response.

Intermediate-Size Enclosures

The intermediate frequency region, designated as region II in Figs. 13.3 and 13.13, is defined as the frequency region where either the enclosure walls, the enclosure air volume, or both exhibit resonances; but the resonances do not overlap so that statistical methods are not yet applicable.

Comparison of the solid and dotted curves in Fig. 13.13 show how the insertion loss, measured in the outside–inside direction (such as in a quiet control room in

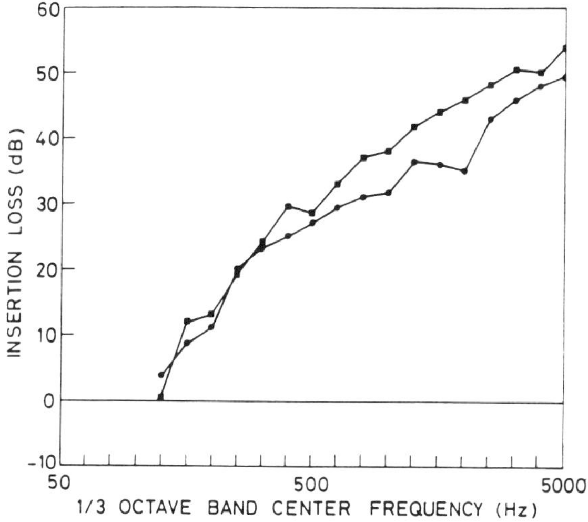

Fig. 13.12 Effect of plate edge connection on the insertion loss of sealed, resiliently machine-mounted, close-fitting enclosure: (■) elastically sealed edge; (●) welded edge.

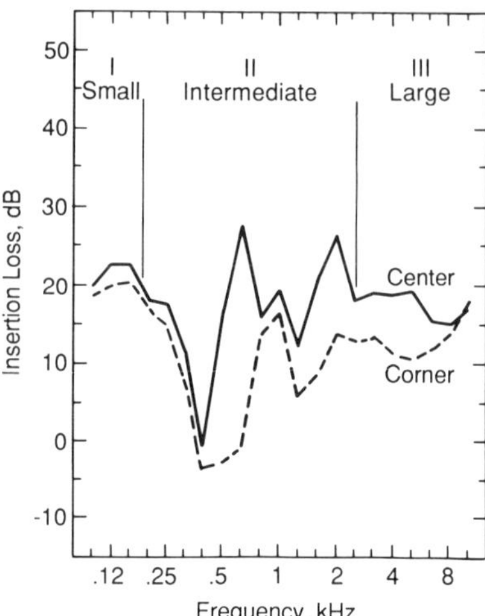

Fig. 13.13 Effect of microphone position on the sound pressure-based insertion loss measured in the outside–inside direction of a 50 × 150 × 300-mm (1.6-mm-thick) empty, plane, sealed, unlined aluminium acoustical enclosure. (After ref. 9.) Solid line: center microphone location. Dashed line: corner microphone location.

a noisy test facility), depends on the location of the microphone. Below 400 Hz, the sound pressure is uniformly distributed in the enclosure and the microphone in the corner measures the same sound pressure as that placed at the center. The first acoustical resonance occurs at 565 Hz, which results in a very low insertion loss at the enclosure corner in the 500- and 800-Hz center frequency one-third-octave bands, as illustrated by the dotted curve, but in a high insertion loss in the same bands for the enclosure center location, as depicted by the solid curve. Fluctuations owing to structural resonances can be reduced by application of damping treatment to the panels (see Chapter 12) and those owing to acoustical resonances by internal sound-absorbing lining (see Chapter 8).

Becuase the insertion loss in this intermediate frequency range fluctuates widely with frequency and position, it is very difficult to make accurate analytical predictions of the insertion loss. Frequently, involved finite-element analysis (see Chapter 6), model-scale or full-scale experiments, or crude approximations must be employed. For crude approximations one connects the low- and high-frequency predictions to cover the intermediate range.

Large Acoustical Enclosures

Acoustical enclosures are termed large at frequencies where both the enclosure wall panels and the enclosure volume exhibit a large number of resonant modes in a given frequency band and statistical methods for predicting the level of the interior sound field and the vibration response and sound radiation of the enclosure wall panels can be applied. Large acoustical enclosures used in industrial noise control have many paths for transmitting acoustical energy, as illustrated schematically in Fig. 13.14. These paths can be grouped into three basic categories: (1) through the enclosure walls, (2) through openings, and (3) through structure-borne paths. The first group is characterized by the sound excitation of the enclo-

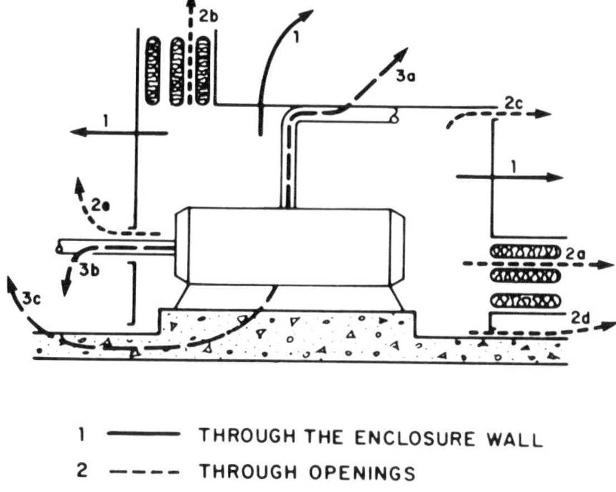

Fig. 13.14 Paths of noise transmission from a typical acoustical enclosure.

sure wall by the interior sound field, resulting in sound radiation from the exterior wall surfaces. Sound transmission by this path is relatively well understood, and in most cases the sound power transmitted can be predicted with good engineering accuracy (see Chapter 9). The second group is characterized by sound energy escaping through openings in the enclosure wall such as air intake and exhaust ducts, gaps between panels, and gaps around the gasketing at the floor and doors. These are also fairly well understood[10-12] but not so easily controlled. The third group is characterized by radiation of solid surfaces excited to vibration by dynamic forces, such as the enclosure walls when rigidly connected to the enclosed vibrating equipment, shafts and pipes that penetrate the enclosure wall, and the vibration of the uncovered portion of the floor. This path is difficult to predict without detailed information about the motion of the enclosed machine and its dynamic characteristics. Consequently, all possible efforts should be made to prevent any solid connections to the source. To obtain a balanced design, one must control each of these transmission paths and avoid overdesigning any one of them.

Analytical Model for Predicting Insertion Loss at High Frequencies. A sound source enclosed in a large, acoustically lined enclosure will radiate approximately the same sound power as it would in the absence of the enclosure. To achieve a high power-based insertion loss, a high percentage of this radiated sound power must be dissipated (i.e., converted into heat) within the enclosure proper. This is accomplished by providing walls of high sound transmission loss to contain the sound waves and by covering the interior with sound-absorbing treatment to convert the trapped acoustical energy into heat.

In analyzing the acoustical behavior of an enclosure at high frequencies, the first step is to calculate the space–time average mean-square pressure of the diffuse sound field, $\langle p^2 \rangle$, within the enclosure proper. Once the interior sound field is calculated, one can determine the sound power escaping from the enclosure through the various paths. The mean-square value of the diffuse sound pressure in the interior of the enclosure is obtained by balancing (a) the sound power injected into the enclosure from all the noise sources with (b) the power loss through dissipation (in the sound-absorbing lining, in the air, and in the wall structure) and the sound transmission through the walls of the enclosure and through diverse openings. According to reference 2, a sound source of power output W_0 generates in the enclosure a reverberant field with a space–time average mean-square sound pressure $\langle p^2 \rangle$ given by

$$\langle p^2 \rangle = W_0 \frac{4\rho c}{S_w \quad \alpha_w + \sum (S_i/S_w) 10^{-R_{wi}/10} + D} \times \frac{1}{S_i\alpha_i + \sum_k S_{sk} + mV + \sum_j S_{Gj} 10^{-R_{Lj}/10}} \quad \text{N}^2/\text{m}^4 \tag{13.26}$$

where

$$D = \left(\frac{4\pi\sqrt{12}\rho c^3 \sigma}{c_L h \rho_s \omega^2} \right) \left(\frac{\rho_s \omega \eta}{\rho_s \omega \eta + 2\rho c \sigma} \right) \tag{13.27}$$

and S_w is the total interior wall surface, S_{wi} is the surface area of the ith wall, α_w is the average energy absorption coefficient of the walls, R_{wi} is the sound transmission loss of the ith wall, ρ_s is the mass per unit area of a typical wall panel, η is the loss factor of a typical wall panel in place, σ is the radiation efficiency of a typical panel, c_L is the propagation speed of longitudinal waves in the plate material, h is the wall panel thickness, $S_i\alpha_i$ is the total absorption in the interior in excess of the wall absorption (i.e., the machine body itself), S_{Gj} is the area of the jth leak or opening, R_{Gj} is the sound transmission loss of the jth leak or opening, S_{sk} is face area of the kth silencer opening, assumed completely absorbing, m is the attenuation constant for air absorption, and V is volume of the free interior space. The sound power incident on the unit surface area is given by

$$W_{\text{inc}} = \frac{\langle p^2 \rangle}{4\rho c} = \frac{W_0}{\{\cdots\}} \quad \text{W} \tag{13.28}$$

where $\{\cdots\}$ represents the expression in the denominator in Eq. (13.26) and W_0 is the sound power output of the enclosed machine.

The terms in the denominator of Eq. (13.26) from left to right stand for (1) power dissipation by the wall absorption ($S_w\alpha_w$), (2) power loss through sound radiation of the enclosure walls ($\Sigma_i\ S_{wi} \times 10^{-R_{wi}/10}$), (3) power dissipation in the walls through viscous damping effects ($S_w D$), (4) power dissipation by sound-absorbing surfaces in addition to the walls ($S_i\alpha_i$), (5) sound power loss to silencer terminals ($\Sigma_k\ S_{sk}$), (6) sound absorption in air (mV), and (7) sound transmission to the exterior through openings and gaps ($\Sigma_j\ S_{Gj} \times 10^{-R_{Gj}/10}$).

The relative importance of the various terms may differ widely for different enclosures and even for the same enclosure in the different frequency ranges. For example, for a small airtight enclosure without any sound-absorbing treatment, the power dissipation in the wall panels may be of primary importance, while it is usually negligible for enclosures with proper interior sound-absorbing treatment. Air absorption is important in large enclosures at frequencies above 1000 Hz. The sound power transmitted through the enclosure walls, W_{TW}, that through gaps and openings, W_{TG}, and that through silencers, W_{TS}, is given in Eqs. (13.29)–(13.31), respectively

$$W_{\mathrm{TW}} = \frac{W_0 \left(\sum_i S_{wi} \times 10^{-R_{wi}/10} \right)}{\{\cdots\}} \quad \mathrm{W} \tag{13.29}$$

$$W_{\mathrm{TG}} = \frac{W_0 \left(\sum_j S_{Gj} \times 10^{-R_{Gj}/10} \right)}{\{\cdots\}} \quad \mathrm{W} \tag{13.30}$$

$$W_{\mathrm{TS}} = \frac{W_0 \left(\sum_k S_{sk} \times 10^{-\Delta L_k/10} \right)}{\{\cdots\}} \quad \mathrm{W} \tag{13.31}$$

where ΔL_k is the sound attenuation through the silencer over opening k. The power-based insertion loss of the enclosure in the inside–outside direction is defined as

$$\mathrm{IL} \equiv 10 \log_{10} \frac{W_0}{W_{\mathrm{TW}} + W_{\mathrm{TG}} + W_{\mathrm{TS}} + W_{\mathrm{SB}}} \quad \mathrm{dB} \tag{13.32}$$

is given by

$$\mathrm{IL} = 10 \log_{10} \frac{\{\cdots\}}{\sum_i S_{wi} \times 10^{-R_{wi}/10} + \sum_k S_{sk} \times 10^{-\Delta L_k/10}} \times \frac{1}{\sum_j S_{Gj} \times 10^{-R_{Gj}/10} + W_{\mathrm{SB}} \dfrac{\{\cdots\}}{W_0}} \quad \mathrm{dB} \tag{13.33}$$

where W_{SB} is the sound power transmitted through structureborne paths considered separately according to Eqs. (13.23) and (13.24).

It is well known that large sealed enclosures provide some insertion loss even if there is no internal sound-absorbing treatment because there is energy dissipation in the enclosure wall panels and there is air absorption. In addition, there exists a very thin acoustical boundary layer at the inner surfaces of the walls where sound power is dissipated by viscous losses for waves near grazing incidence by the change from adiabatic to isothermal compression as one approaches the walls. Reference 13 gives a formula to assess the average random-incidence sound absorption coefficient attributable to these effects as

$$\alpha_{\min} = 1.8 \times 10^{-4} \sqrt{f} \tag{13.34}$$

where f is the frequency in hertz. The insertion loss of unlined, sealed, free-standing enclosure, $\text{IL}_{\min}$, is obtained by substituting $\alpha_w = \alpha_{\min}$, $S_j = S_{Gi} = S_{sk} = W_{\text{SB}} = 0$, $R_{wi} = R_w$ into Eqs. (13.26) and (13.33), yielding

$$\text{IL}_{\min} = 10 \log_{10} \{1 + 10^{R_w/10}[\alpha_{\min} + D + mV/S_w]\} \quad \text{dB} \tag{13.35}$$

Figure 13.15 compares the insertion loss predicted for an unlined, sealed enclosure [using Eq. (13.8) at low frequencies and Eq. (13.35) at high frequencies] with experimental data obtained by reference 9. There is a reasonably good agreement between the predicted and measured values at both low and high frequencies not only for this specific enclosure but also for a large variety of unlined sealed enclosures of different size and wall panel thickness, indicating that the prediction formulas embodied in Eqs. (13.8) and (13.35) yield reasonable estimates for engineering design. Figure 13.15 also indicates that without internal sound-absorbing treatment the insertion loss remains very modest even at high frequencies, highlighting the importance of sound absorption in enclosure design.

The denominator of Eq. (13.33) reveals that if full advantage is to be taken of the high transmission loss of the enclosure walls, the airpaths through gaps, openings, and air intake and exhaust silencers and structureborne paths must be controlled to a degree that their contribution to the sound radiation is small compared with the sound radiation of the walls. If these paths are well controlled and the dissipation is achieved by the sound-absorbing treatment of the interior wall sur-

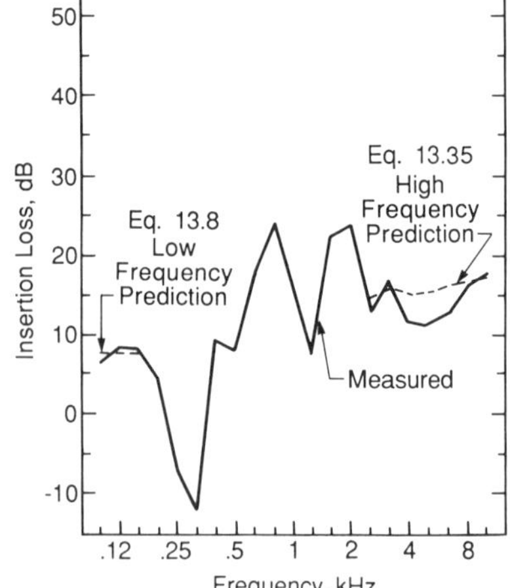

Fig. 13.15 Comparison of measured and predicted pressure-based IL of a 50 × 150 × 300-mm (0.8-mm-thick) sealed, unlined aluminum enclosure, with source outside. (After ref. 9.)

faces and by the absorption in the interior, then for $R_{wi} = R_w$ the insertion loss of the enclosure is well approximated by

$$\text{IL} \cong 10 \log \left(1 + \frac{S_w \alpha_w + S_i \alpha_i}{S_w} \times 10^{R_w/10} \right) \quad \text{dB} \tag{13.36}$$

If we further assume that the second term in Eq. (13.36) is much larger than unity and call the total absorption in the interior of the enclosure $S_w \alpha_w + S_i \alpha_i = A$, Eq. (13.36) simplifies to

$$\text{IL} \cong R_w + 10 \log \frac{A}{S_w} \quad \text{dB} \tag{13.37}$$

This approximate formula would yield negative insertion loss for very small values of interior absorption. It has been widely promulgated in the acoustical literature without mention of its limited validity. Note that the insertion loss of a large enclosure can approach the sound transmission loss of the enclosure wall panels only if there are no leaks and α_w approaches unity.

Key Parameters Influencing the Insertion Loss of Acoustical Enclosures. The primary paths for the transmission and dissipation of sound in large acoustical enclosures are shown in the block diagram in Fig. 13.16. The key pa-

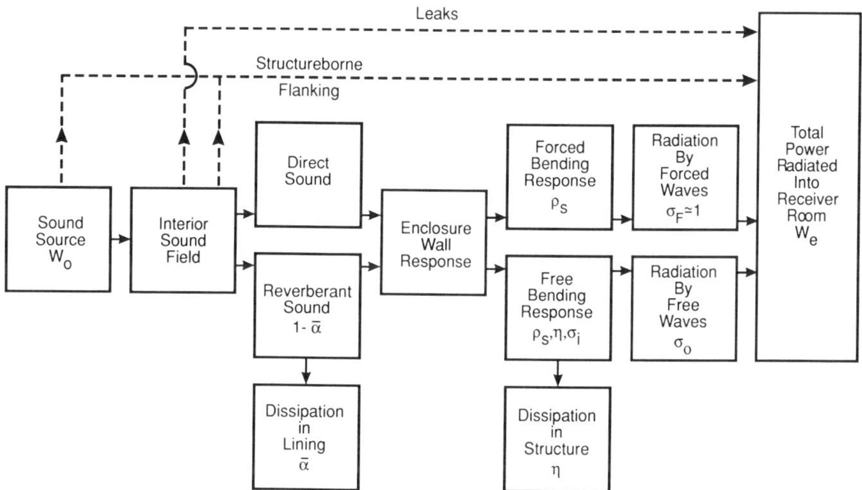

Fig. 13.16 Block diagram of key components controlling the sound attenuation process in acoustical enclosures: W_0, source sound power output; $\overline{\alpha}$, average sound absorption coefficient; ρ_s, mass per unit area; η, loss factor; σ_i, radiation efficiency of wall panels for radiation for inside direction; σ_0, radiation efficiency of wall panels for radiation for outside direction; σ_F, 1 radiation efficiency of forced waves; W_e, total sound power radiated by the enclosure.

TABLE 13.2 Effect of Key Parameters on Sound Insertion Loss of Large Enclosures

Parameter	Symbol	Effect on Insertion Loss
Absorption coefficient of lining	$\overline{\alpha}$	Increases IL by reducing reverberant buildup
Distance between machine and enclosure wall	d	If d decreases beyond a certain limit, IL decreases at low frequencies (close-fitting enclosure behavior)
Thickness of wall panel	h	Increases IL
Density of wall panel material	ρ_M	Increases IL
Speed of longitudinal waves in panel material	$c_L = \sqrt{E/\rho_M}$	Decreases IL below critical frequency
Loss factor of wall panel	η	Increases IL, especially near and above critical frequency and at first panel resonance
Critical frequency of wall panel	$f_c = c^2/1.8c_L h$	The higher is f_c for a given mass per unit area, the higher is IL
Radiation efficiencies of wall panels	σ_i, σ_0	The higher is the radiation efficiency for inside direction, σ_i, and for outside direction, σ_0, the smaller is IL
Stiffeners	—	Decrease IL by increasing σ
Leaks	—	Limit achievable IL; watch out for door gaskets and penetrations
Structureborne flanking	—	Limit achievable IL; watch out for solid connections between vibrating machine and enclosure and for floor vibration

rameter affecting enclosure insertion loss are summarized in Table 13.2. Fisher and Veres[6] and Kurtze and Mueller[14] have carried out systematic experimental investigations on how the (1) choice of wall panel parameters, (2) internal sound-absorbing lining, (3) leaks, and (4) vicinity of the enclosed machine to the enclosure walls and its vibration pattern influence the acoustical performance of enclosures. The model–machine sound source employed in reference 6 consisted of 1 × 1.5 × 2-m steel boxes of 1-mm wall thickness and were combined to yield a small (3 × 2 × 1.5-m) and a large (4 × 2 × 1.5-m) model machine. The walls of the model machines could be excited either by an internal loudspeaker or an internal tapping machine to simulate both sound and structure-borne excitation. The investigated rectangular walk-in enclosure had 4.5 × 2.5 × 2 m inside dimensions and was equipped with an operational personal-access door. The thickness and material of the wall panels as well as the thickness of the interior lining could be varied.

Wall Panel Parameters. Figure 13.17 shows the measured insertion loss versus frequency for three different values of the wall thickness h, indicating that above 1.5 mm, a further increase of the wall thickness brings only slight improvement at low frequencies and a slight deterioration at high frequencies near coincidence. Figure 13.17 also includes the field-incidence sound transmission loss for a 1.5-mm-thick steel plate for comparison, indicating that up to 500 Hz, the insertion loss matches well the field-incidence sound transmission loss of the panel.

Figure 13.18 compares the measured sound transmission loss of a typical 2.1 × 1.2-m wall panel and predicted sound transmission loss of an identical infinite-enclosure panel with the measured insertion loss obtained with an enclosure built with the same panels. The data indicate that the prediction formulas presented in Chapter 9 for infinite panels can be used to predict accurately the sound transmission loss of the finite panels of this specific size and that above 500 Hz the insertion loss is dominated by unintentional, small leaks. According to reference 6, the sound-absorbing interior liner provides sufficient structural damping of the wall panels so that additional damping treatment did not result in improved performance.

Stiffening the wall panels with exterior L-channels, which increase the radiation efficiency, resulted in slight deterioration of the acoustical performance. Using

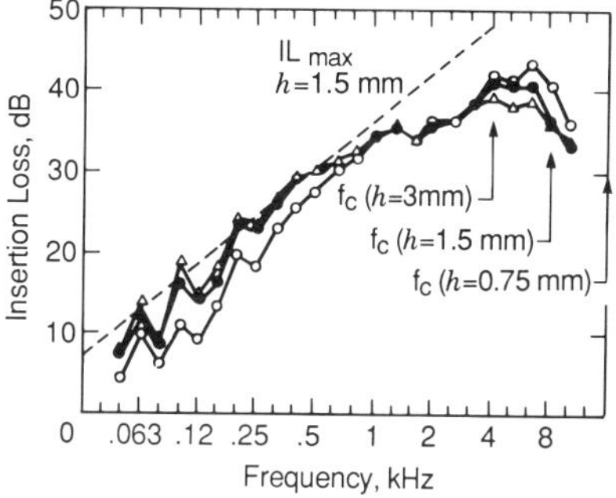

Fig. 13.17 Curves of power-based insertion loss versus frequency measured (inside–outside direction) for a large walk-in enclosure for various thicknesses of steel plate, h. (After ref. 6.) Sound-absorbing treatment, 70 mm thick, door sealed; f_c, coincidence frequency; $\mathrm{IL}_{\max} = R_f$ represents the maximum achievable insertion loss. (△) h = 3 mm; (●) h = 1.5 mm; (○) h = 0.75 mm.

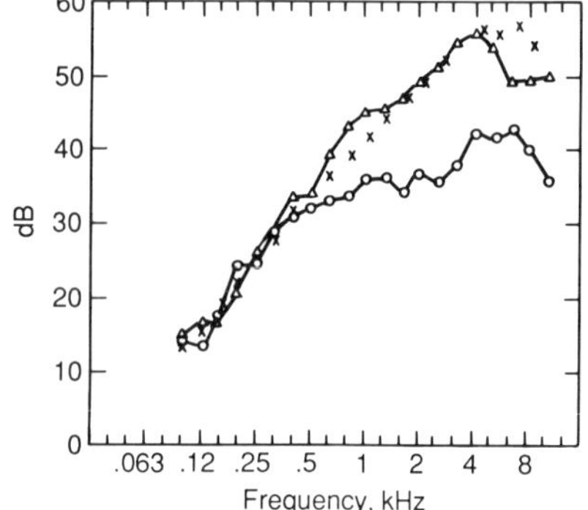

Fig. 13.18 Comparison of measured enclosure insertion loss IL and panel sound transmission loss R (after ref. 6): 1.5-mm-thick steel plate, 70-mm-thick sound absorbing treatment; ($\triangle$) R, measured; ($\times$) R, predicted according to Chapter 9; ($\bigcirc$) IL measured.

steel and wood chipboard of equal mass per unit area yielded different transmission loss but resulted in practically the same insertion loss because in the frequency range where the chipboard had its coincidence frequency, the insertion loss was already controlled by leaks.

As a general rule, wall panels should be selected to have *high enough coincidence frequency* to be above the frequency region of interest and be *large enough that their first structural resonance occurs below the frequency region of interest* but heavy enough to yield a field-incidence mass law sound transmission loss that matches the insertion loss requirements. Steel plates 1.5 mm thick usually fulfill most of these requirements.

Stiff, lightweight panels (such as honeycombs) usually yield a critical frequency as low as 500 Hz and provide a very low sound transmission loss in this frequency region. Consequently, such panels should not be used in enclosure design unless the enclosure is required to provide high insertion loss at only very low frequencies and mid- and high-frequency performance is not required.

Effect of Sound-absorbing Treatment. The proper choice of sound-absorbing treatment plays a more crucial role than the specific choice of wall material or wall thickness, as shown in Fig. 13.19. The sound-absorbing treatment helps to increase insertion loss by (1) reducing reverberant buildup in the enclosure at mid- and high frequencies, (2) increasing the transmission loss of the enclosure walls at high frequencies, and (3) covering up some of the unintentional leaks between adjacent panels and between panels and frames. Whenever feasible, the thickness of the interior sound-absorbing treatment should be chosen to yield a normal-incidence absorption coefficient $\alpha \geq 0.8$ in the frequency region of interest. As shown in Chapter 8, this can be achieved by a layer thickness $d \geq \frac{1}{10}\lambda$, where λ_L is the acoustical wavelength at the lower end of the frequency region, and by choosing a porous material of normalized flow resistance of $1.5 \leq R_1 d/\rho c < 3$.

Leaks. Leaks reduce the insertion loss of large enclosures just as they reduce it for small enclosures (see Eq. 13.13). The reduction in insertion loss due to leaks, $\Delta \mathrm{IL}_L$, is defined as

$$\Delta \mathrm{IL}_L \equiv \mathrm{IL}_s - \mathrm{IL}_L \cong 10 \log(1 + \beta \times 10^{R_w/10}) \quad \mathrm{dB} \qquad (13.38)$$

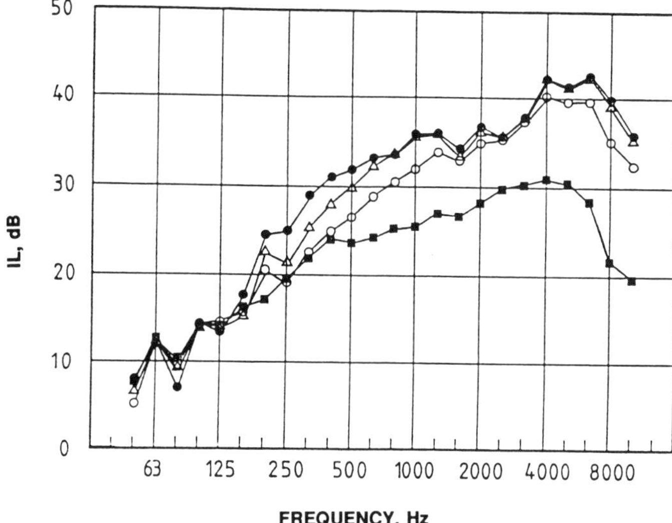

Fig. 13.19 Measured insertion loss of the enclosure with 1.5-mm-thick steel plates for various thicknesses of the sound-absorbing material (after ref. 6): (■) 0 mm (no absorption); (○) 20 mm; (△) 40 mm; (●) 70 mm.

where IL_s and IL_L is the insertion loss of the sealed and leaky enclosures, respectively, R_w is the sound transmission loss of the enclosure wall, and $\beta = (1/S_w) \Sigma_j S_{Gj} \times 10^{-R_{Gj}/10}$ is the leak ratio factor, S_{Gj} is the face area, and R_{Gj} is the sound transmission loss of the jth leak. The sound transmission loss of leaks can be positive or, in the case of longitudinal resonances in wide, rigid-walled gaps, also negative. For preliminary calculations it is customary to assume $R_{Gj} = 0$. In this case, $\beta = \Sigma_j S_{Gj}/S_w$ is the ratio of the total face area of the leaks and gaps and the surface area (one side) of the enclosure walls.

Figure 13.20 is a graphical representation of Eq. (13.38) indicating that the higher is the sound transmission loss of the enclosure walls, the higher is the reduction in insertion loss caused by leaks. For example, for an enclosure assembled from wall panels that provide a R_w of 50 dB, the total area of all leaks must be

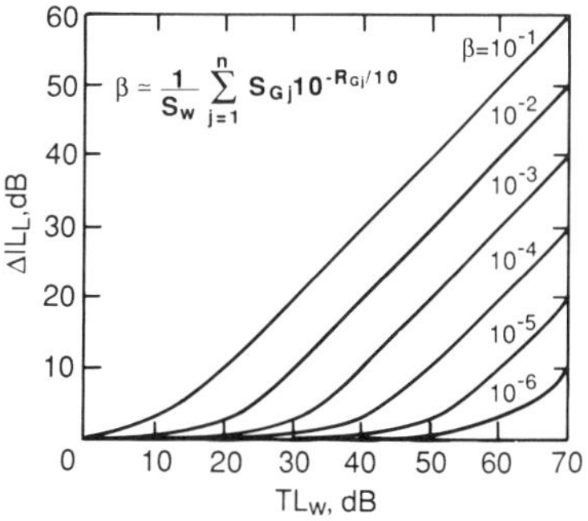

Fig. 13.20 Decrease of enclosure insertion loss, ΔIL, as a function of the wall sound transmission loss R_w with the leak ratio factor β as parameter; S_w is the wall surface area and R_{Gj} and S_{Gj} are the sound transmission loss and face area of the jth leak, respectively.

less than 1/1000 percent ($\beta = 10^{-5}$) of the total enclosure surface area if the insertion loss is not to be decreased by more than 3 dB!

Machine Position. Experimental investigations by Fischer and Veres[6] have shown that the insertion loss of an enclosure at low frequencies also depends on the specific position of the enclosed machine within the enclosure. At low frequencies, where the distance between the flat machine wall and the flat enclosure wall panel, d, becomes smaller than one-eighth of the acoustical wavelength ($d < \frac{1}{8}\lambda$), the observed decrease of the insertion loss (from its value obtained when the machine was positioned centrally so that all wall distances were large compared with $\frac{1}{8}\lambda$) was as high as 5 dB. Note that the machine–wall distance d includes the thickness of the internal lining. As a general rule, machines with omnidirectional sound radiation should be positioned centrally. It is especially important to avoid positioning the noisy side of machines very close to enclosure walls, doors, windows, and ventilation openings.

Machine Vibration Pattern. The insertion loss of an enclosure also depends on the specific vibration pattern of the machine although that dependence appears to not be very strong. Experimental investigations[6] showed that variation of the insertion loss was about ± 2.5 dB when measured with a loudspeaker inside the model machine and with excitation of the model machine by an ISO tapping machine.

Flanking Transmission through the Floor. The potential insertion loss of an enclosure can be limited severely by flanking if the floor is directly exposed to vibration forces, the internal sound field of the enclosure, or both. Figure 13.21 shows typical enclosure installations ranging from the best to the worst in regard to flanking transmission via the floor. For most equipment it is imperative to provide vibration isolation, a structural break in the floor, or both to reduce the transmission of structureborne excitation of the floor. Applicable prediction tools and isolation methods are described in Chapters 9 and 11. For the cases depicted in Figs. 13.21*c*, *d*, where the interior sound field directly impinges on the floor, the sound-induced vibration of the floor sets a limit to the achievable insertion loss of the enclosure. This limit can be estimated roughly from

$$\mathrm{IL}_L \cong R_F + 10 \log \sigma_F \quad \mathrm{dB} \tag{13.39}$$

where R_F is the sound transmission loss and σ_F is the radiation efficiency of the floor.

Relationship between Inside–Outside and Outside–Inside Transmission

Acoustical enclosures are most frequently used to surround a noisy equipment for reducing the noise exposure of a receiver located outside of the enclosure. Less frequently, the enclosure surrounds the receiver to reduce its noise exposure to a

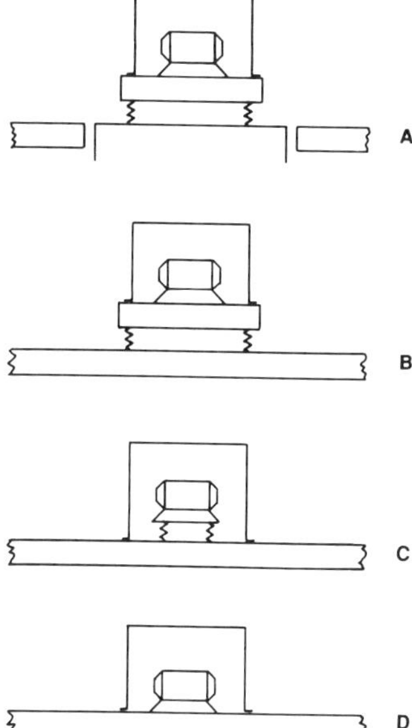

Fig. 13.21 Rating of enclosure installations in regard to flanking transmission of the floor: (*a*) best; (*b*) good; (*c*) bad; (*d*) worst.

sound source located outside of the enclosure. The acoustical performance of the enclosure in the former case is given by its insertion loss in the inside–outside direction, IL_{i0}, while the latter is given by its insertion loss in the outside–inside direction, IL_{0i}. All of our considerations in this chapter have dealt with sound transmission from the inside to the outside. According to the principle of reciprocity (see Chapter 9), exchanging the position of the source and receiver does not effect the transfer function. Though it is strictly true only for point sound sources and receivers, it also can be applied to extended sound sources and averaged observer positions in an enclosure, yielding $IL_{i0} = IL_{0i}$.

Example To demonstrate the use of the previously provided design information, predict the insertion loss of the large, acoustically lined, walk-in enclosure investigated experimentally in reference 6. The enclosure is $4.5 \times 2.5 \times 2.0$ m high and is constructed of 1.5-mm-thick steel plates and has a 70-mm-thick interior sound-absorbing lining. The sound transmission loss of the wall panels, R_w, given in Fig. 13.18, by the triangular data points and the sound absorption coefficient of the interior panel surfaces, $\overline{\alpha}$, taken from reference 6, as a curve that monotonically increases from 0.16 at 100 Hz and reaches a plateau of 0.9 at 600 Hz. The insertion loss is predicted by Eq. (13.33) using the following values: $S_{wi} = S_w = 39$ m^2, $\Sigma_j S_{Gj} \times 10^{-R_{Gj}/10} = 10^4$, $S_w = 3.9 \times 10^{-3}$, assuming that leaks account for 1/100 percent of the wall surface area, that there are no silencer openings ($S_S = 0$) and no structure-borne connections ($W_{SB} = 0$), and that air absorption (mV), the sound absorbed by the machine ($S_i\alpha_i$), and power lost due to dissipation in the panels (D) and due to sound radiation ($10^{-R_w/10}$) are small compared with the power dissipated in the sound-absorbing treatment on the interior wall surfaces.

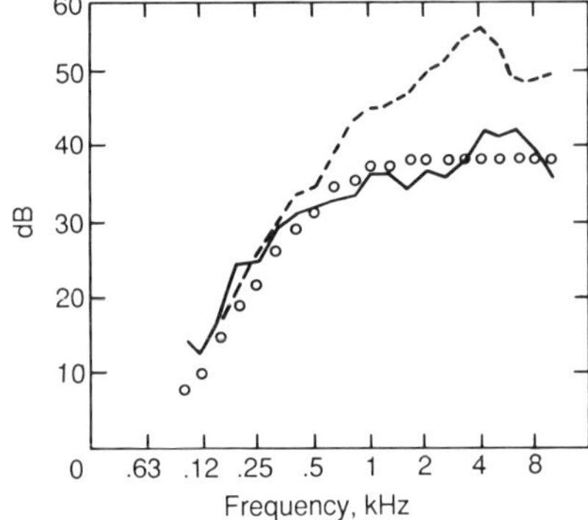

Fig. 13.22 Typical acoustical performance of a leaky enclosure: (——) IL, measured by ref. 6; (--------) R_w, measured by ref. 6; (○) IL, predicted by Eq. (13.40) for

$$\beta = \left(\sum_j \left(\frac{S_{Gj}}{S_w} \right) \times 10^{-R_{Gj}/10} \right) = 10^{-4}$$

Then Eq. (13.33) simplifies to

$$\mathrm{IL}_L \cong 10 \log \frac{S_w \overline{\alpha}}{\sum_i S_{wi} \times 10^{-R_{wi}/10} + \sum_j S_{Gj} \times 10^{-R_{Gj}/10}} \quad \mathrm{dB} \qquad (13.40)$$

The curve of insertion loss versus frequency predicted by Eq. (13.40) is represented by the open circles in Fig. 13.22, while the solid curve represents the measured data. The measured sound transmission loss of the wall panels, R_w, is the dotted line and represents the limiting insertion loss that could be achieved only if there would be no leaks and the random-incidence sound absorption coefficient $\overline{\alpha}_w$ would approach unity. Observing Fig. 13.22 note that the assumption of $1/100$ percent leaks ($\beta = 10^{-4}$) yielded a prediction that matches reasonably well the measured data, which indicates that even such a very small percentage of leaks, which can be realized only if extreme care is exercised during erecting the enclosure and careful caulking of all leaks detected during the initial performance checkout and careful adjustment of door gaskets, can substantially decrease the potential insertion loss of an enclosure.

Partial Enclosures

When the work process of the machine or safety and maintenance requirements do not allow a full enclosure, a partial enclosure (defined as those with more than 10% open area) is used to reduce radiated noise. When a partial enclosure is far enough from the sound source not to cause an increase in source sound power by hydrodynamic interactions of the near field of the source with edges of the opening and the walls of the partial enclosure are not rigidly connected to the vibrating machine to serve as a sounding board, a partial enclosure can provide a modest insertion loss of 5 dB or less. According to reference 15, the power-based insertion loss of partial enclosure can be estimated by

$$\mathrm{IL} = 10 \log \left[1 + \alpha \left(\frac{\Omega_{\mathrm{tot}}}{\Omega_{\mathrm{open}}} - 1 \right) \right] \quad \mathrm{dB} \qquad (13.41)$$

where Ω_{tot} is the solid angle of sound radiation of the unenclosed source and $\Omega_{open} = S_{open}/r$ is the solid angle the enclosed source (located at a distance r) "sees" the opening of area S_{open}. Reference 15 contains construction details, curves of measured insertion loss versus frequency, and cost information of 54 different partial enclosures.

13.2 WRAPPINGS

Many machines and pipes need thermal insulation to provide protection of operating personnel or to prevent excessive heat loss. Typically, they need a minimum of 25-mm- (1-in.-) thick glass or ceramic fiber blanket to achieve proper protection. Since many hot equipment, such as turbines, boiler feed pumps, compressor valves, and pipelines, are also sources of intense noise, it is frequently feasible to achieve both heat and acoustical insulation by providing a (preferably) limp, impervious surface layer on top of the porous blanket. The impervious surface layer also provides protection for the porous blanket. Acoustically, the combination of the resilient blanket and the heavy, limp, impervious surface layer provide a spring–mass isolation system. At frequencies where the mass impedance of the surface layer exceeds the combined stiffness impedance of the skeleton of these flexible blankets and the entrapped air, the vibration amplitude of the surface layer becomes smaller than that of the machine surface and wrapping results in an insertion loss that monotonically increases with increasing frequency. The insertion loss of the wrapping is defined the same way as the insertion loss of a close-fitting enclosure and for flat wrappings can be predicted in a manner similar to that of the close-fitting enclosure.

The major difference between wrappings and close-fitting enclosures is that for wrapping the skeleton of the porous layer is in full surface contact with the vibrating surface of the machine, but for the close-fitting enclosure there is no such contact. Consequently, for wrappings the vibrating machine surface transmits a pressure to the impervious surface layer not only through sound propagation in the voids between the fibers but also through the skeleton of the porous material. At low frequencies where the thickness of the porous layer, L, is much smaller than the acoustical wavelength, the porous layer can be represented by a stiffness per unit area, S_{tot}, that according to Mechel[16] can be estimated as

$$S_{tot} = S_M + S_L\left(1 - \frac{P}{A}\sqrt{\frac{\rho c^2}{L\pi f \gamma R_1 h'}}\right) \quad \text{N/m}^3 \tag{13.42}$$

where S_L is the stiffness of the trapped air,

$$S_{air} \cong \frac{\rho c^2}{\gamma L h'} \quad \text{N/m}^3 \tag{13.43}$$

γ the adiabatic exponent, h' the porosity, f the frequency, and R_1 the flow resistivity of the material, and S_M is given by

$$S_M = (2\pi f_0)^2 \rho_s, \qquad \text{N/m}^3 \tag{13.44}$$

In Eq. (13.44) the symbol f_0 represents the measured resonance frequency of a mass–spring system consisting of a small rectangular or square-shaped sample of the porous material covered with a metal plate of mass per unit area ρ_s. The resonance frequency is measured by putting the mass–spring system on top of a shaker table and performing a frequency sweep. The second term in Eq. (13.42) accounts for the air that escapes along the perimeter P of a test sample of surface area A, used in the experiment to determine the resonance frequency f_0 according to Eq. (13.44). Young's modulus of the porous material is then determined as $E = S_{\text{tot}}/L$, the speed of longitudinal waves in the porous layer as $c_L = \sqrt{E/\rho_M}$, where ρ_M is the density of the porous material, and the wavenumber as $k = 2\pi f/c_L$. Characteristics of some frequently used thermal insulation materials reported by Wood and Ungar[17] are given below:

Material	Dynamic Stiffness per Unit Area, S_M (N/m³)	Density, ρ_M (kg/m³)
Erco-Mat[a]	1.3×10^7	138
Erco Mat F[a]	2×10^6	104
Glass Fiber[b]	4×10^4	12

[a]Needled glass fiber insulation.
[b]Low-density Owens Corning Fiberglas Blanket.

Considering the porous heat insulating layer as a wave-bearing medium of density ρ and a complex Young's modulus $E' = E(1 + j\eta)$ and characterizing the impervious, limp surface layer by its mass per unit area ρ_s, Wood and Ungar[17] derived the analytical formula

$$\text{IL} = 20 \log \left| \cos(kL) - \frac{\rho_s}{\rho_M L} (kL) \sin(kL) \right| \qquad \text{dB} \tag{13.45}$$

where $k = \omega/\sqrt{E(1 + j\eta)/\rho_M}$ is the complex propagation constant of the porous layer, $\eta = 1/Q$ the loss factor, and $\rho_s\omega$ the mass impedance per unit area of the covering layer.

At very low frequencies, where $kL << 1$, Eq. (13.45) can be approximated by

$$\text{IL}_L = 20 \log \left[1 - \left(\frac{\omega}{\omega_n} \right)^2 \right] \qquad \text{dB} \tag{13.46}$$

where $\omega_n = E/\rho_s L$ is the resonance frequency of a system consisting of a massless spring of stiffness E/L and a mass ρ_s (per unit area). According to Eq. (13.46), the cladding provides no insertion loss for $\omega << \omega_n$ and yields a negative insertion loss (amplifies the radiated sound) if $\omega \cong \omega_n$.

For frequencies above the resonance frequency ($\omega >> \omega_n$) the blanket provides an attenuation that increases monotonically with increasing frequency. Noting that in Eq. (13.45) neither the cosine nor the sine term can exceed unity, an upper bound for the curve of insertion loss versus frequency at high frequencies is obtained by[17]

$$\mathrm{IL}_L \leq 20 \log \left[1 + \frac{\rho_s}{\rho_L} (kL)^2 \right] \quad \mathrm{dB} \tag{13.47}$$

Pipe Wrappings

Pipe wrappings consist of a resilient porous layer and an impervious jacket. The impervious jacket is usually sheet metal or loaded plastic. They achieve their acoustical performance in a similar manner as flat wrappings. However, there is an important difference. While flat wrappings do not increase the sound-radiating surface, when a wrapping is applied to a small-diameter pipe, the diameter of the impervious jacket can be substantially larger than that of the bare pipe. This increases both the radiating surface and the radiation efficiency. Accordingly, at low frequencies, below or slightly above the resonance frequency of the pipe wrapping, the insertion loss is negative. Positive insertion loss is usually achieved only above 200 Hz.

According to Michelson Fritz and Sazenhofen,[18] the maximal achievable insertion loss of wrappings can be estimated by the empirical formula

$$\mathrm{IL}_{\mathrm{max}} = \frac{40}{1 + \dfrac{0.12}{D}} \log \frac{f}{2.2 f_0} \quad \mathrm{dB} \tag{13.48}$$

where

$$f_0 = 60/\sqrt{\rho_s L} \quad \mathrm{Hz} \tag{13.49}$$

where L is the thickness of the porous resilient layer and D is the pipe diameter, both in meters, and ρ_s is the mass per unit area of the impervious jacket in kg/m^2.

Equation (13.49) is valid only if there are no structure-borne connections between the pipe and jacket and for frequencies $f \geq 2f_0$. Figure 13.23 shows curves of measured insertion loss versus frequency obtained in reference 18 for typical pipe wrapping consisting of a galvanized steel jacket, thickness 0.75–1 mm and a porous resilient layer with the following parameters:

Thickness L	30, 60, 80, and 100 mm
Density	85–120 kg/m^3
Flow resistivity	3×10^4 N · s/m^4
Dynamic Young's modulus	2×10^5 N/m^2

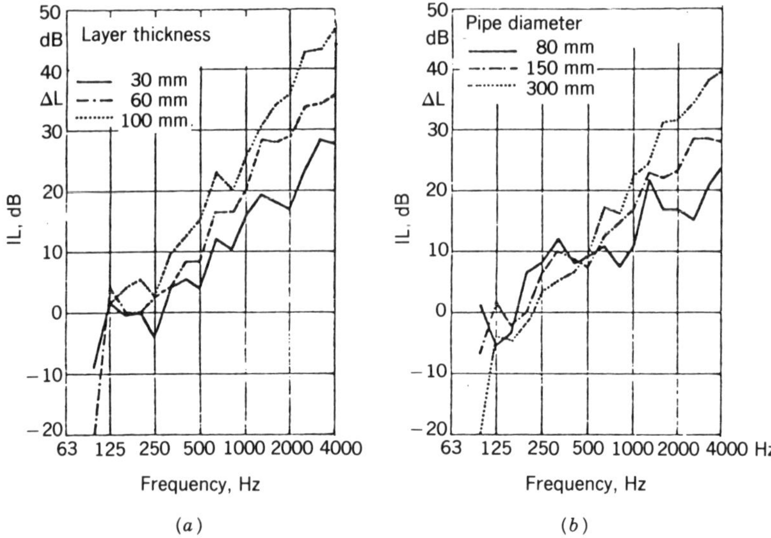

Fig. 13.23 Curves of measured insertion loss versus frequency of pipe wrappings. (After ref. 18.) (*a*) Effect of layer thickness. (*b*) Effect of pipe diameter.

Figure 13.23*a* shows the effect of layer thickness L on the insertion loss for a pipe of diameter $D = 300$ mm while Fig. 13.23*b* illustrates the effect of pipe diameter D for a constant layer thickness $L = 60$ mm. The insertion loss above 250 Hz increases with increasing thickness of the porous layer and with increasing diameter of the bare pipe. The standard deviation of the measured insertion loss around that predicted by Eq. (13.47) was 4 dB.

Note that spacers between the pipe and jacket result in insertion loss values that may be substantially lower than those predicted by Eq. (13.47) unless the spacers are less stiff dynamically than the porous layer.

REFERENCES

1. R. H. Lyon, "Noise Reduction of Rectangular Enclosures with One Flexible Wall," *J. Acoust. Soc. Am.* **35**(11), (1963), 1791–1797.
2. I. L. Vér, "Reduction of Noise by Acoustic Enclosures," *Isolation of Mechanical Vibration, Impact and Noise*, American Society of Mechanical Engineers, New York, Vol. 1, 1973, pp. 192–220.
3. J. B. Moore, "Low Frequency Noise Reduction of Acoustic Enclosures," *Proc. NOISE-CON 81*, pp. 59–64.
4. I. L. Vér and E. Veres, "An Apparatus to Study the Sound Excitation and Sound Radiation of Platelike Structures," *Proc. INTER-NOISE 80*, pp. 535–540.
5. K. P. Byrne, H. M. Fischer, and H. V. Fuchs "Sealed, Close-fitting, Machine-mounted Acoustic Enclosures with Predictable Performance," *Noise Control Eng. J.* **31**, 7–15 (1988).

6. M. H. Fischer and E. Veres, ''Noise Control by Enclosures'' (in German), Research Report 508, Bundesanstalt fuer Arbeitschutz, Dortmund, 1987; also Fraunhofer Institute of Building Physics Report, IBP B5 141/86, 1986.
7. M. H. Fischer, H. V. Fuchs, and U. Ackerman, ''Light Enclosures for Low Frequencies'' (in German), *Bauphysik* **11**(H.1), 50–60 (1989).
8. H. M. Fuchs, U. Ackermann, and W. Frommhold, ''Development of Membrane Absorbers for Industrial Noise Abatement'' (in German), *Bauphysik* **11**(H.1), 28–36 (1989).
9. J. B. Moreland, Westinghouse Electric Corporation, private communication.
10. K. Goesele, ''Sound Transmission of Doors'' (in German), *Berichte aus der Bauforschung*, H.63 Wilhelm Ernst & Sohn Publishers, Berlin, 1969, pp. 3–21.
11. F. P. Mechel, ''Transmission of Sound through Holes and Slits Filled with Absorber and Sealed,'' *Acustica* **61**(2), 87–103 (1986).
12. F. P. Mechel, ''The Acoustic Sealing of Holes and Slits in Walls,'' *J. Sound Vib.* **III**(2), 297–336 (1986).
13. L. Cremer, M. Heckl, and E. E. Ungar, *Structureborne Sound*, Berlin, Springer, 1988.
14. G. Kurtze and K. Mueller ''Noise Reducing Enclosures'' (in German), Research Report BMFT-FB-HA 85-005 Bundesministerium fuer Forschung und Technologie, 1985.
15. U. J. Kurze et al. ''Noise Control by Partial Enclosures; Shielding in the Nearfield'' (in German), Research Report No. 212, Bundesanstalt fuer Arbeitschutz und Unfallforshung, 1979.
16. F. P. Mechel, ''Sound Absorbers and Absorber Functions'' (in Russian), in *Reduction of Noise in Buildings and Inhabited Regions*, Strojnizdat, Moscow, 1987.
17. E. W. Wood and E. E. Ungar, BBN Systems and Technologies Corporation, private communications.
18. R. Michelsen, K. R. Fritz, and C. V. Sazenhofan, ''Effectiveness of Acoustic Pipe Wrappings'' (in German), *Proc. DAGA '80*, VDE-Verlag, Berlin 1980, pp. 301–304.

BIBLIOGRAPHY

Miller, R. K., and W. V. Montone, *Handbook of Acoustical Enclosures and Barriers*, Fairmont Press, Atlanta, GA, 1978.

''Noise Reduction by Enclosures'', VDI Guideline 2711 (in German), VDI-Verlag GmbH, Duesseldorf, Germany.

CHAPTER 14

Noise of Gas Flows

M. S. HOWE
Bolt Beranek and Newman, Inc.
H. D. BAUMANN
H. D. Baumann Associates

14.1 INTRODUCTION

The sound produced by unsteady gas flows and by interactions of gas flows with solid objects is called *aerodynamic sound*. The same flows often excite structural modes of vibration in surfaces bounding the flow and are then said to generate *structure-borne sound*. Unwanted flow-generated sound, noise, is a common by-product of most industrial processes. It also accompanies the operation of ships, automobiles, aircraft, rockets, and so on, and can adversely affect structural stability and be an important source of fatigue.

A practical understanding of the sources of aerodynamic noise is necessary over the whole range of mean-flow Mach numbers, from the very lowest (0.01 or less) associated with flows in air conditioning systems and underwater applications to the high supersonic range occurring in jet engines and high-pressure valves. In subsonic flows, the sound may be attributed to three basic aerodynamic source types: monopole, dipole, and quadrupole.[1]

These fundamental source types are discussed in this chapter, together with a survey of noise mechanisms associated with turbulent jets, spoilers and airfoils, boundary layers and separated flow over wall cavities, combustion, and valves.

Noise and Vibration Control Engineering: Principles and Applications, Edited by Leo L. Beranek and István L. Vér.
ISBN 0-471-61751-2 © 1992 John Wiley & Sons, Inc.

14.2 AEROACOUSTIC SOURCE TYPES

Aerodynamic Monopole

Monopole radiation is produced by the unsteady introduction of mass or heat into a fluid. Typical examples are pulse jets (where high-speed air is periodically ejected through a nozzle), turbulent flow over a small aperture in a large wall (where the flow induces pulsatile motion in the aperture), unsteady combustion processes, and heat release from boundaries and from a pulsed laser beam.

The radiation from a monopole source in an otherwise stationary fluid is equivalent to that produced by a pulsating sphere (Fig. 14.1*a*). Both the amplitude and phase of the acoustical pressure are spherically symmetric. When the monopole sound is generated by unsteady flow velocities, the dimensional relation between the radiated sound power and the flow parameters is

$$W_{\text{monopole}} \propto \frac{\rho L^2 U^4}{c} = \rho L^2 U^3 M \tag{14.1}$$

where W_{monopole} = radiated sound power, W
ρ = mean density of gas, kg/m^3
c = speed of sound in gas, m/s
U = flow velocity in source region, m/s
L = length scale of flow in source region, m
M = Mach number equal to U/c, dimensionless

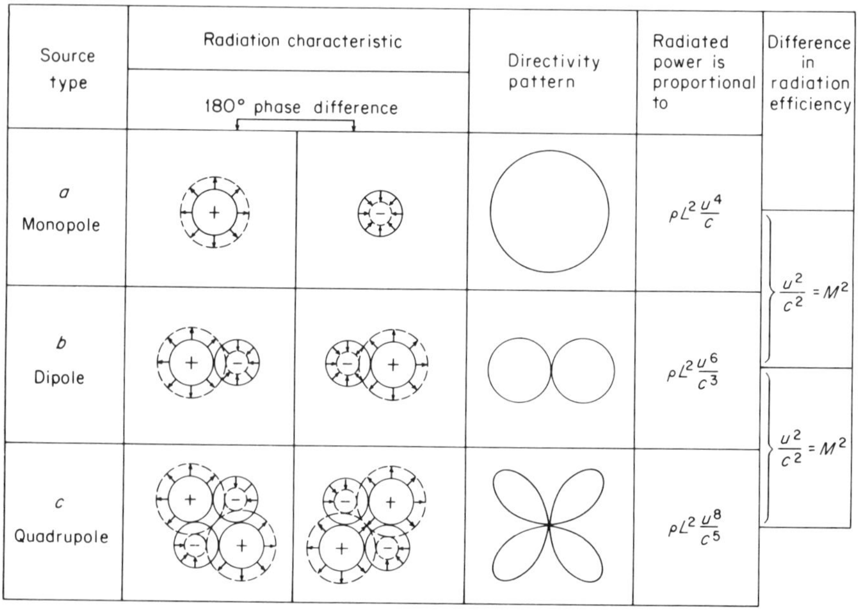

Source type	Radiation characteristic: 180° phase difference		Directivity pattern	Radiated power is proportional to	Difference in radiation efficiency
a Monopole	+	−		$\rho L^2 \frac{U^4}{c}$	
b Dipole	+ −	− +		$\rho L^2 \frac{U^6}{c^3}$	$\frac{U^2}{c^2} = M^2$
c Quadrupole	+ − / − +	− + / + −		$\rho L^2 \frac{U^8}{c^5}$	$\frac{U^2}{c^2} = M^2$

Fig. 14.1 Aeroacoustic source types and their dimensional properties in fluid of uniform mean density. See also Fig. 1.2.

Aerodynamic Dipole

Dipole sources arise when unsteady flow interacts with surfaces or bodies, when the dipole strength is equal to the force on the body, or when there are significant variations of mean fluid density in the flow. This source type is found in compressors where turbulence impinges on stators, rotor blades, and other control surfaces. Similarly, the unsteady shedding of vorticity from solid objects, such as telegraph wires, struts, and airfoils, generates ''singing'' tones that are also attributable to dipole sources. Other examples include the noise generated by hot jets exhausting into a cooler ambient medium and by the acceleration of temperature (or ''entropy'') inhomogeneities in a mean pressure gradient in a duct contraction.

The dipole is equivalent to a pair of equal monopole sources of opposite phase separated by a distance that is much smaller than the wavelength of the sound. Destructive interference between the radiations from the monopoles reduces the efficiency with which sound is generated by the dipole relative to a monopole and produces a double-lobed, figure-eight radiation field shape proportional to the cosine of the angle measured from the dipole axis (Fig. 14.1*b*). In fluid of uniform mean density, the dimensional dependence of the aerodynamic dipole sound power is

$$W_{\text{dipole}} \propto \frac{\rho L^2 U^6}{c^3} = \rho L^2 U^3 M^3 \tag{14.2}$$

This differs by a factor M^2 from the power output of the monopole. In subsonic flow ($M < 1$) the dipole is a less efficient source of sound.

If the specific entropy or temperature of the flow in the source region is not uniform (e.g., in the shear layer of a hot jet exhausting into a cooler ambient atmosphere), the density must also be variable. The relatively strong pressure fluctuations in the turbulent flow are then scattered by the density variations and produce sound of dipole type. The dipole strength is proportional to the difference between the actual acceleration of the density inhomogeneity in the turbulent pressure field and that which it would have experienced had the density been uniform.[2] The dimensional dependence of the corresponding sound power is

$$W_{\text{entropy}} \propto \frac{\rho L^2 (\delta T/T)^2 U^6}{c^3} = \rho L^2 \left(\frac{\delta T}{T}\right)^2 U^3 M^3 \tag{14.3}$$

where $(\delta T/T)^2$ is the mean-square fractional temperature fluctuation.

Aerodynamic Quadrupole

Quadrupole radiation is produced by the Reynolds stresses in a turbulent gas in the absence of obstacles. These arise from the convection of fluid momentum by the unsteady flow. The Reynolds stress forces must occur in opposing pairs, since the net momentum of the fluid is constant. Force pairs of this type are called quadrupoles and are equivalent to equal and opposing dipole sources (see Fig. 14.1*c*).

Aerodynamic quadrupoles and entropy dipoles are the dominant source types in high speed, subsonic, turbulent air jets. The quadrupole strength is larger where both the turbulence and mean velocity gradients are high, for example, in the turbulent mixing layer of a jet.

The dimensional dependence of the radiated quadrupole sound power is

$$W_{\text{quadrupole}} \propto \frac{\rho L^2 U^8}{c^5} = \rho L^2 U^3 M^5 \tag{14.4}$$

This differs from the dipole power by a factor M^2. At subsonic speeds ($M < 1$), the quadrupole radiation efficiency is lower than that of the dipole because of the double cancellation illustrated in Fig. 14.1*c*.

The monopole, dipole, and quadrupole sources decrease in their respective radiation efficiencies for subsonic flows, but the dependencies of their radiated sound powers on flow velocity show the opposite trend, that is, the total radiated sound power varies as the fourth, sixth, and eighth powers of flow speed U for monopole, dipole, and quadrupole sources, respectively. Thus, if the flow velocity is high enough, the radiation from the quadrupole sources may be the dominant source of sound even though the efficiency with which the sound is produced is small. This is usually the case for a jet engine at high subsonic exhaust velocities, although other internal sources caused by unsteady combustion or rough burning (predominantly monopole) or compressor noise (predominantly dipole) can also make a significant contribution to the total noise.

The value of the constant of proportionality for each type of source depends both on the sound-generating mechanism and the flow configuration. Thus, the constant for a singing wire differs from that for an edge tone, although both arise from unsteady surface forces (dipole sources). However, the proportionality relations (14.1)–(14.4) can be used to estimate the influence on the radiated sound power of changes in one or more of the source parameters. A twofold increase in the exhaust velocity U of a jet (quadrupole-type source) causes the sound power level to increase by 24 dB (eighth power of flow velocity), whereas a doubling of the exhaust nozzle area A (proportional to L^2) increases the sound power level by only 3 dB. Because the thrust of a jet engine is proportional to AU^2, the increase in the radiated sound will be smaller if a doubling of thrust is achieved by increasing the nozzle area by a factor of 2 rather than the exhaust velocity by a factor of 1.4.

Aerodynamic Sources of Fractional Order

The efficiency with which sound is produced by aerodynamic dipole sources on surfaces whose dimensions greatly exceed the acoustical wavelength frequently differs from those implied by Eqs. (14.2) and (14.3). For example, the net dipole strength associated with turbulent flow over a smooth, plane wall is zero: the radiation is the same as that generated by the quadrupole sources in the flow when the wall is regarded as a plane reflector of the sound. Similarly, the net strength of the dipoles induced by turbulence near the edge of a large wedge-shaped body varies with angle ϑ—being equivalent to that of a quadrupole when $\vartheta = 180°$ (i.e., for a plane wall), and to a *fractional* multipole order $\frac{3}{2}$ when $\vartheta = 0$ (knife edge). The latter case is important in estimating the leading and trailing edge noise of an airfoil, where at high enough frequency the sound power is proportional to $\rho L U^3 M^2$.

Influence of Source Motion

The directional characteristics of the radiation from acoustical sources are changed if the source moves relative to the fluid. Both the frequency and intensity of the sound are increased ahead of the source and decreased to its rear. It is usual to refer the observer to a coordinate system based on the source location at the time of emission of the received sound.

For a monopole source of strength $q(t)$ kilograms per second (equal to the product of the volume velocity and fluid density), moving at constant speed U in the direction illustrated in Fig. 14.2, the acoustical pressure p in the far field is

$$p_{\text{mon}} = \pm \left[\frac{dq/dt}{4\pi r(1 - M_{fr})^2} \right] \tag{14.5}$$

where the square brackets denote evaluation at the time of emission of the sound, (r, Φ) are coordinates defining the observer position relative to the source at the time of emission of the sound, and the plus and minus sign is taken according as $M_{fr} \lessgtr 1$, where

$$M_{fr} = \left(\frac{U}{c} \right) \cos \Phi \quad \text{(dimensionless)}$$

The frequency of the received sound is $f/(1 - M_{fr})$, where f, per second, is the frequency in a frame fixed relative to the source. The term $1/(1 - M_{fr})$ is called

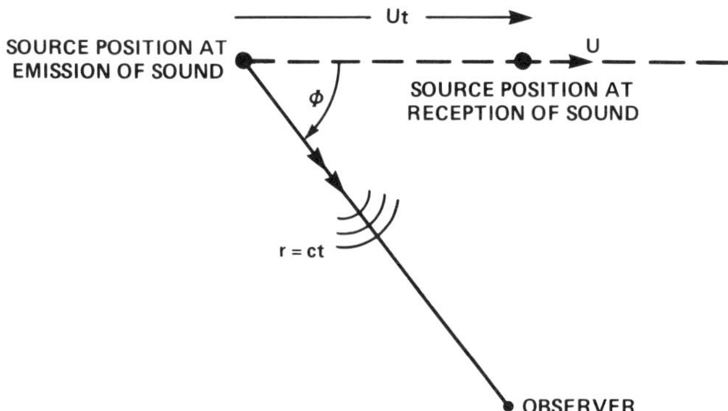

Fig. 14.2 Coordinates defining the observer position at the time of emission of the received sound from a moving source.

the *Doppler factor*. Its effect is to modify both the frequency and the amplitude of the acoustic field.

For a uniformly translating dipole of strength $f_i(t)$ (newtons), which is equivalent to an applied force in the i direction, and for a quadrupole $T_{ij}(t)$ (newton-meters), specified by directions i, j (and equivalent to a *couple* applied to the fluid), the respective acoustic pressure fields become

$$p_{\text{dipole}} = \pm\left[\frac{df_r/dt}{4\pi cr(1 - M_{fr})^2}\right] \qquad p_{\text{quad}} = \left[\frac{d^2T_{rr}/dt^2}{4\pi c^2 r|1 - M_{fr}|^3}\right] \tag{14.6}$$

where f_r, T_{rr} denote the components of the dipole and quadrupole source strengths in the observer direction at the time of emission of the sound and the plus and minus sign is taken according as $M_{fr} \lesssim 1$.

These expressions are valid for ideal, point sources moving at constant velocity. The influence of source motion on real, aerodynamic sources is often much more complicated. For example, a pulsating sphere is a monopole source when at rest. In uniform translational motion, however, the radiation is amplified by $\frac{7}{2}$ Doppler factors instead of the 2 of Eq. (14.5). This is because the monopole is augmented due to motion by a dipole whose strength is proportional to the convection Mach number M_f. The interaction of the volume pulsations with the mean flow over the sphere produces a net fluctuating force on the fluid.

14.3 NOISE OF GAS JETS

The sound generated by high-speed jets is usually associated with several different sources active simultaneously. *Jet mixing noise*, caused by the turbulent mixing of the jet with the ambient medium, and for imperfectly expanded supersonic jets *shock-associated noise*, produced by the convection of turbulence through shock cells in the jet, are the principal components of the radiation. The properties of sound sources in real jets differ considerably from those of the idealized models described in Section 14.2. Sources located within the "jet pipe" can also make a significant contribution to the noise. In the case of a gas turbine engine, the additional radiation includes combustion noise as well as tonal and broadband sound produced by interactions involving fans, compressors, and turbine systems. The following discussion is based on experimental data obtained and validated by several independent investigators and collated for prediction purposes by the Society of Automotive Engineers.[3] Formulas are given for predicting the *free-field* radiation from a jet in an ideal acoustic medium. In many applications it will be necessary to modify these predictions to account for atmospheric attenuation and interference caused by reflections from surfaces.

Jet Mixing Noise

Mixing noise is the most fundamental source of sound produced by a jet. The simplest free jet is an airstream issuing from a large reservoir through a circular

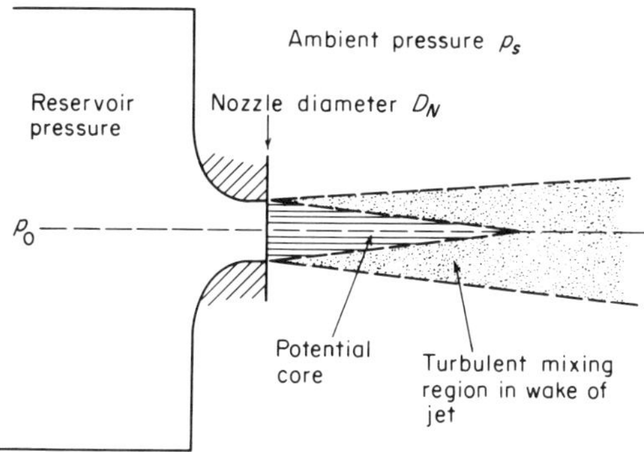

Fig. 14.3 Subsonic turbulence-free jet.

convergent nozzle (Fig. 14.3). The gas accelerates from near-zero velocity in the reservoir to a peak velocity in the narrowest cross section of the nozzle. Sonic flow occurs at the nozzle exit when the pressure ratio p_0/p_s exceeds 1.89, where p_0 is the steady reservoir pressure and p_s is the ambient pressure downstream of the nozzle. An increase above this critical pressure ratio leads to the appearance of a shock cell structure downstream of the nozzle and "choking" of the flow unless the convergent part of the nozzle is followed by a divergent section in which the pressure decreases smoothly to p_s.

For the idealized, shock-free jet, no interaction is assumed between the gas flow and solid boundaries. The noise is produced entirely by turbulent mixing in the shear layer. The sources extend over a considerable distance downstream of the nozzle. The high-frequency components of the noise are generated predominantly close to the nozzle, where eddy sizes are small. Lower frequencies are radiated from sources further downstream where eddy sizes are much larger. The sources may be regarded as quadrupoles, whose strength and directivity are modified by the influences of nonuniform fluid density (temperature) and convection by the flow.

The total radiated sound power W in watts of the mixing noise may be expressed in terms of the mechanical stream power of the jet, equal to

$$W_{\text{mech strm}} = \tfrac{1}{2}mU^2 \tag{14.7}$$

where m = mass flow of gas, kg/s
U = fully expanded mean jet velocity, m/s

The dependence of $W/\frac{1}{2}mU^2$ on jet Mach number $M = U/c$ and density ratio ρ_j/ρ_s, where c is the ambient sound speed and ρ_j, ρ_s are respectively the densities (kg/m^3) of the fully expanded jet and the ambient atmosphere, is illustrated in Fig. 14.4. For M less than about 1.05 the sound power increases as ρ_j/ρ_s decreases (i.e., as the jet temperature increases) and decreases at higher Mach numbers.

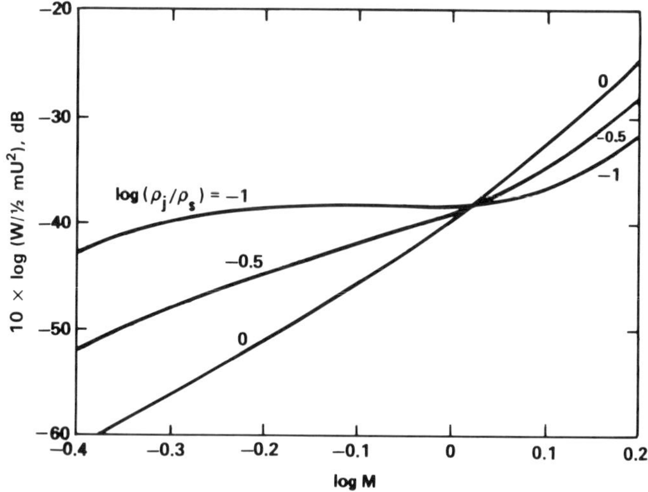

Fig. 14.4 Ten times the logarithm of the ratio of acoustic power of jet mixing noise to the mechanical stream power as a function of the fully expanded jet Mach number $M = U/c$ for different values of ρ_j/ρ_s.

When $M < 1$, W may be estimated from

$$\frac{W}{W_{\text{mech strm}}} \simeq \frac{4 \times 10^{-5}(\rho_j/\rho_s)^{(w-1)}M^{4.5}}{(1 - M_c^2)^2} \qquad M < 1 \tag{14.8}$$

where $M_c = 0.62U/c$ and w is the jet density exponent given by[4]

$$w = \frac{3M^{3.5}}{0.6 + M^{3.5}} - 1 \tag{14.9}$$

Equation (14.9) is applicable for M greater than about 0.35, including the supersonic region and is also used in formulas given below.

The overall sound pressure level (OASPL) at any point in space is defined as

$$\text{OASPL} = 20 \times \log_{10} \frac{p_{\text{rms}}}{20\ \mu\text{Pa}} \qquad \text{dB} \tag{14.10}$$

where p_{rms} is the rms sound pressure in pascals (N/m^2) (see Chapter 2). Source convection and refraction in the shear layers causes the sound field to be directive, the maximum noise being radiated in directions inclined at angles θ to the jet axis between about 30° and 45°.

Typical field directivities at a fixed observer distance for an air jet at three different jet Mach numbers $M = U/c$ are shown in Fig. 14.5. The curves give the variation of OASPL(θ) − OASPL(90°), the sound pressure level relative to its

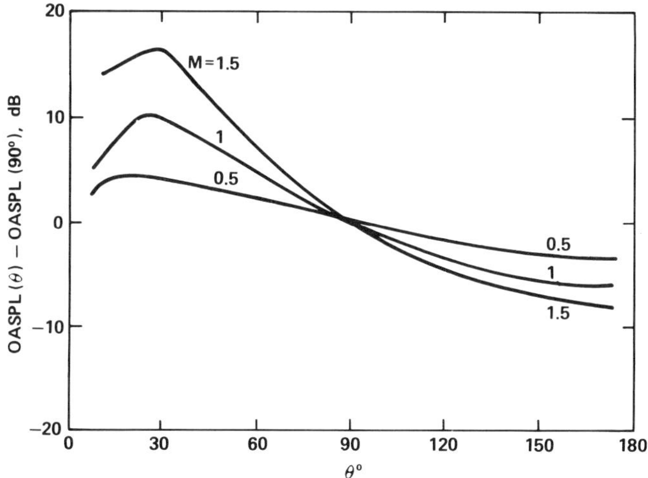

Fig. 14.5 Directivity of jet mixing noise.

value at $\theta = 90°$, and are approximated by the formula

$$\text{OASPL}(\theta) = \text{OASPL}(90°) - 30 \times \log\left[1 - \frac{M_c \cos\theta}{(1 + M_c^5)^{1/5}}\right]$$

$$- 1.67 \times \log\left[1 + \frac{1}{(10^{(40.56 - \theta')} + 4 \times 10^{-6})}\right] \quad (14.11)$$

where $M_c = 0.62U/c$
$\theta' = 0.26(180 - \theta)M^{0.1}$, degrees

The final term on the right of (14.11) is negligible unless $\theta < 180° - 150°/M^{0.1}$, that is, except in directions close to the jet axis.

At $\theta = 90°$ the overall sound pressure level of jet mixing noise can be calculated from the formula

$$\text{OASPL} = 139.5 + 10 \times \log\frac{A}{R^2} + 10 \times \log\left[\left(\frac{p_s}{p_{\text{ISA}}}\right)^2 \left(\frac{\rho_j}{\rho_s}\right)^w\right]$$

$$+ 10 \times \log\left(\frac{M^{7.5}}{1 - 0.1M^{2.5} + 0.015M^{4.5}}\right) \quad (14.12)$$

where A = fully expanded jet area, $= \frac{1}{4}\pi D_N^2$ for a subsonic jet, m^2
D_N = nozzle exit diameter, m
R = distance from center of nozzle exit, m
p_{ISA} = international standard atmospheric pressure at sea level, = 10.13 μPa

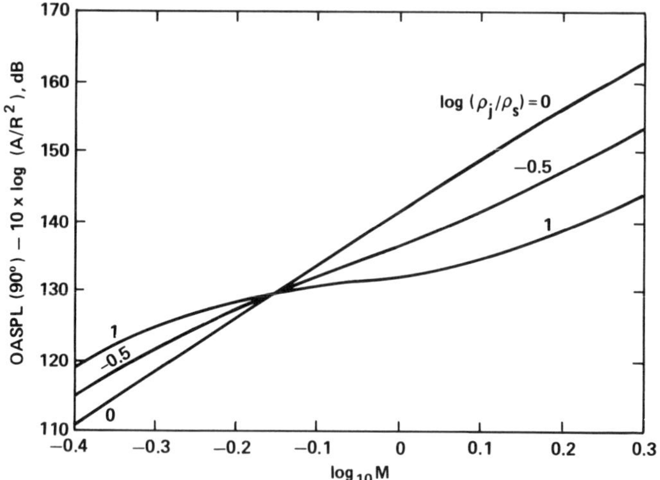

Fig. 14.6 Overall sound pressure level at $\theta = 90°$.

The dependence of OASPL(90°) $- \; 10 \times \log (A/R^2)$ on jet Mach number M and density ratio ρ_j/ρ_s is illustrated in Fig. 14.6.

The sound-pressure-level (SPL) frequency spectrum of jet mixing noise is broadband and peaks at a frequency $f = f_p$ that is a function of the radiation direction θ, jet Mach number M, and temperature ratio T_j/T_s, where T_j (K) is the fully expanded jet temperature and T_s (K) is the ambient gas temperature. The spectrum has a broad peak that occurs in the range $0.3 < S < 1$, where

$$S = \frac{f_p D_N}{U} \tag{14.13}$$

is the *Strouhal number* (dimensionless), tabulated in Table 14.1 for $\theta \geq 50°$ and for different values of T_j/T_s. For practical purposes the difference Δ = OASPL − SPL$_{\text{peak}}$ between the peak of the one-third-octave-band spectrum and the overall SPL at the same value of θ may be taken as 11 dB except as indicated in the table.

Figure 14.7 shows a typical relative one-third-octave-band SPL spectrum

TABLE 14.1 Values of Strouhal Number and Δ( ) for Different Values of T_j/T_s[a]

T_j/T_s	$\theta = 50°$	$\theta = 60°$	$\theta = 70°$	$\theta = 80°$	$\theta \geq 90°$
1	0.7 (11 dB)	0.8 (11 dB)	0.8 (11 dB)	1.0	0.9
2	0.5 (10 dB)	0.4 (10 dB)	0.6 (11 dB)	0.5	0.6
3	0.3 (9 dB)	0.4 (10 dB)	0.4 (10 dB)	0.4	0.5

[a] Strouhal number $S = fD_N/U$ at the peak of the one-third-octave-band spectrum of jet mixing noise for $U/c \leq 2.5$; Δ = OASPL − SPL$_{\text{peak}}$ ≃ 11 dB for all $\theta \geq 80°$ except as indicated in parentheses.

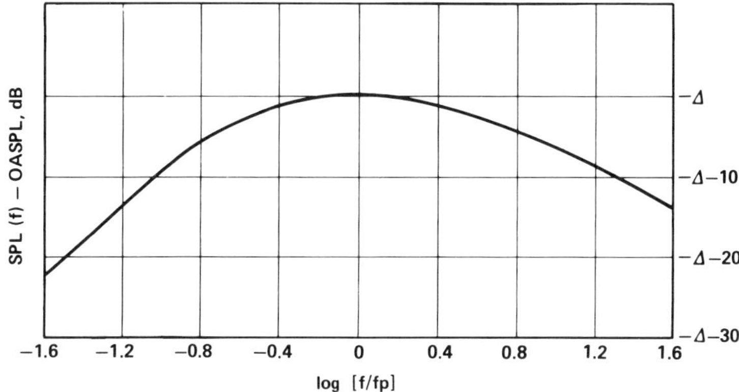

Fig. 14.7 One-third-octave-band SPL − OASPL (dB) for radiation directions $\theta \geq 90°$ and $\rho_j = \rho_s$.

SPL(f) − OASPL for $\rho_j = \rho_s$ and angles $\theta \geq 90°$. The characteristic shape is the same for all temperatures and angles, although there is a significant dependence on temperature and Mach number when θ is smaller than about 50° and $f > f_p$ when refraction of sound by the jet shear layer becomes important. For subsonic jets, the spectrum varies roughly as f^3 at low frequencies and decays like $1/f$ for $f > f_p$.

Figures 14.4–14.7 and Table 14.1, supplemented when necessary by Eqs. (14.8) and (14.9), comprise a general procedure for estimating the SPL, directivity, and spectrum of jet mixing noise.

Example 14.1 Find the total jet mixing noise sound-power- and sound-pressure-level spectrum at 60° from the jet axis and 3 m from the nozzle of a 0.03-m-diameter air jet. The jet exhausts into ambient air (15°C, p_s = 10.13 μPa, ρ_s = 1.225 kg/m^3) at sonic ($U/c = 1$) exit velocity, that is, 340 m/s.

SOLUTION The mechanical stream power $\frac{1}{2}mU^2$ is 1.70×10^4 W. The ratios ρ_j/ρ_s and U/c are unity. Therefore, from Fig. 14.3 [or Eqs. (14.8) and (14.9)] the ratio of sound power to mechanical stream power is 1.0×10^{-4}. The resulting overall sound power is 1.7 W. From Fig. 14.6 [or Eq. (14.12)] the OASPL at 90° to the jet axis and R = 3 m is 98.8 dB *re* 20 μPa. From Eq. (14.11) the OASPL at 60° to the jet axis and R = 3 m is 103.5 dB, and the peak Strouhal number is 0.8. The one-third-octave-band SPL spectrum is found from Fig. 14.6 by adding 103.5 dB to the ordinate with Δ = 11 dB (from Table 14.1).

Realistic jets exhausting from pipes and engine nozzles do not provide smooth, low-turbulence entrainment of flow but rather flow that has been disturbed or spoiled before leaving the nozzle. In this case, the above procedure is not valid unless the jet velocity U exceeds about 100 m/s. Below this speed major portions of the aerodynamically generated noise emanate from internal sources.

Noise of Imperfectly Expanded Jets

Supersonic, underexpanded, or "choked" jets contain shock cells through which the flow repeatedly expands and contracts (see inset to Fig. 14.8). Seven or more distinct cells are often visible extending up to 10 jet diameters downstream of the nozzle. They are responsible for two additional components of jet noise: *screech tones* and broadband *shock-associated noise*. Screech is produced by a feedback mechanism in which a disturbance convected in the shear layer generates sound as it traverses the standing system of shock waves. The sound propagates upstream through the ambient atmosphere and causes the release of a new flow disturbance at the nozzle exit. This is amplified as it convects downstream, and the feedback loop is completed when it encounters the shocks. A notable feature of the screech tones is that the frequencies are independent of the radiation direction, the fundamental occurring at approximately $f = U_c/L(1 + M_c)$, where $M_c = U_c/c$, U_c is the convection velocity of the disturbance in the shear layer, L is the axial length of the first shock cell, and c is the ambient speed of sound. Although screech is usually present in model-scale experiments and can be important for choked jet engines, it is usually easy to eliminate by minor modifications to nozzle design, for example, by notching.

In practice, it is the broadband, shock-associated noise that is important for supersonic jet engines. It can be suppressed by proper nozzle design but is always present for a convergent nozzle. The dominant frequencies are usually higher than the screech tones and can range over several octave bands. For predictive purposes it is permissible to estimate separately the respective contributions of mixing noise

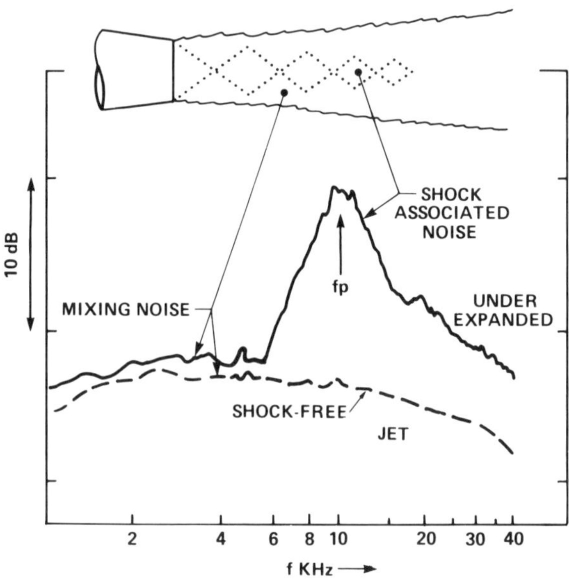

Fig. 14.8 Jet noise spectra for shock-free and underexpanded jets at $\theta = 90°$, $\beta = 1$.

and shock-associated noise to the overall sound power generated by the jet. The mixing noise predictions are the same as for a shock-free jet.

The overall SPL of shock associated-noise is approximately independent of the radiation direction and may be estimated from

$$\text{OASPL} = C_0 + 10 \times \log \frac{\beta^n A}{R^2}$$

$$C_0 = 156.5 \quad \text{(dB)} \qquad n = \begin{cases} 4 & (\beta < 1) \\ 1 & (\beta > 1) \end{cases} \qquad \frac{T_j}{T_s} < 1.1 \tag{14.14}$$

$$C_0 = 158.5 \quad \text{(dB)} \qquad n = \begin{cases} 4 & (\beta < 1) \\ 2 & (\beta > 1) \end{cases} \qquad \frac{T_j}{T_s} > 1.1$$

where T_j = total (reservoir) jet temperature, K
T_s = total ambient temperature, K
$\beta = \sqrt{M_j^2 - 1}$
M_j = fully (ideally) expanded jet Mach number, based on sound speed in jet

The shock-associated noise spectrum exhibits a well-defined peak in the vicinity of

$$f_p = \frac{0.9U_c}{D_N \beta (1 - M_c \cos\theta)} \tag{14.15}$$

where the convection velocity U_c meters per second is equal to $0.7U$ and θ is measured from the jet axis. When $T_j/T_s > 1.1$, the one-third-octave-band SPL (*re* 20 μPa) is given by

$$\begin{aligned} \text{SPL} = {} & 143.5 + 10 \times \log \frac{\beta^n A}{R^2} - 16.13 \times \log \left(\frac{5.163}{\sigma^{2.55}} + 0.096\sigma^{0.74} \right) \\ & + 10 \times \log \left\{ 1 + \frac{17.27}{N_s} \sum_{i=0}^{N_s - 1} \right. \\ & \left. \cdot \left[C(\sigma)^{(i^2)} \sum_{j=1}^{N_s - i - 1} \frac{\cos(\sigma q_{ij}) \sin(0.1158\sigma q_{ij})}{\sigma q_{ij}} \right] \right\} \end{aligned} \tag{14.16}$$

where $C(\sigma) = 0.8 - 0.2 \times \log_{10}(2.239/\sigma^{0.2146} + 0.0987\sigma^{2.75})$
$\sigma = 6.91\beta D_N f/c$, dimensionless
c = ambient speed of sound, m/s
N_s = 8 (number of shocks)
$q_{ij} = (1.7ic/U)\{1 + 0.06[j + \frac{1}{2}(i + 1)]\}[1 - 0.7(U/c)\cos\theta]$

and n is defined as in (14.14). In the case of a "cold jet," for which $T_j/T_s < 1.1$, the prediction (14.16) should be reduced by 2 dB.

The prediction procedure for shock-associated noise is applicable over all angles θ where shock cell noise is important (say, $\theta > 50°$). When used in conjunction with the prediction procedure for jet mixing noise, the one-third-octave-band sound pressure spectrum for the overall jet noise may be estimated by adding the respective contributions to each frequency band of the mean-square acoustical pressures of the mixing noise and the shock-associated noise. The influence of shock-associated noise is illustrated in Fig. 14.8, which compares the spectrum of the jet mixing noise of a fully expanded, shock-free supersonic jet from a convergent–divergent nozzle with that of the noise produced by an underexpanded jet exhausting from a convergent nozzle at the same pressure ratio. In the latter case the spectral peak occurs at the frequency f_p given by Eq. (14.15).

Flight Effects

The noise generated by a jet engine in flight is modified because of the joint effects of Doppler amplification (discussed in Section 14.2) and the reduction in mean shear between the jet flow and its environment. From an extensive series of experimental studies[3] empirical formulas have been developed to predict the SPL in flight in terms of the corresponding static levels.

For jet mixing noise the $\text{OASPL}(\theta)_{\text{flight}}$ is related to $\text{OASPL}(\theta)_{\text{static}}$ by

$$\text{OASPL}(\theta)_{\text{flight}} = \text{OASPL}(\theta)_{\text{static}} - 10 \times \log\left[\left(\frac{U}{U - V_f}\right)^{m(\theta)} (1 - M_f \cos\Phi)\right]$$

$$m(\theta) = \left[\left(\frac{6959}{|\theta - 125|^{2.5}}\right)^7 + \frac{1}{[31 - 18.5M - (0.41 - 0.37M)\theta]^7}\right]^{-1/7}$$

$$1.1 < M < 1.95 \qquad (14.17)$$

where θ = angle between jet axis and observer at retarded time of emission of sound, degrees
Φ = angle between flight direction and observer direction at retarded time of emission of sound, defined in Fig. 14.2
U = fully expanded jet velocity (relative to nozzle), m/s
V_f = aircraft flight speed, m/s
M_f = flight Mach number relative to sound speed in air, $= V_f/c$

The formula is applicable for $20° < \theta < 160°$. For $M < 1.1$ and $M > 1.95$, the relative velocity exponent $m(\theta)$ should be taken to be given by the above formula at $M = 1.1, 1.95$, respectively.

Estimates of the one-third-octave-band SPL spectrum in flight can be made by using Fig. 14.6 with the OASPL determined by (14.17) and the Strouhal number fD_N/U replaced by that based on the relative jet velocity: $fD_N/(U - V_f)$.

For shock-associated noise the influence of flight on both the spectrum and the

OASPL is approximately given by

$$\text{SPL}_{\text{flight}} = \text{SPL}_{\text{static}} - 40 \times \log(1 - M_f \cos \Phi) \tag{14.18}$$

14.4 MEASUREMENT CORRECTIONS FOR OPEN-JET WIND TUNNELS

Open-jet wind tunnels are frequently used to study sound generation by fluid–structure interactions or the effect of flight on noise sources. Measurements of the radiated sound are made with microphones in the stationary ambient medium, and it is necessary to introduce corrections to the measured results to account for the refraction that occurs when the sound traverses the open-jet free shear layer. These corrections involve both the amplitude and propagation direction of the received sound.

The simplest case of the problem is shown in Fig. 14.9. The observer in the ambient fluid and the source at a normal distance h within the jet flow are at rest and lie in the x–y plane, where the x direction is parallel to the mean flow at Mach number M_f and the y direction is normal to the shear layer. The mean fluid density is assumed to be constant across the shear layer. It is required to determine from the measured data at the observer position O, the pressure and propagation direction of the sound in terms of the emission–time coordinates (r, Θ_r) (c.f. Fig. 14.2, where $\Phi = 180° - \Theta_r$) of a hypothetical observer at C imagined to be within the

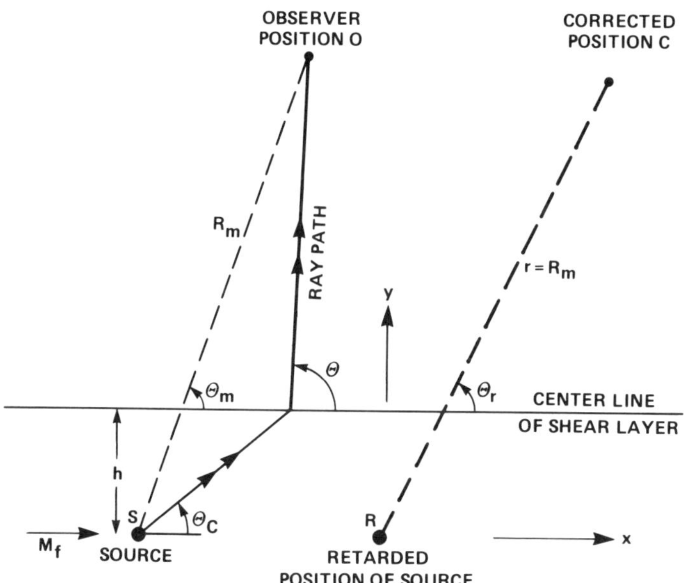

Fig. 14.9 Source and observer separated by a mean shear layer.

moving fluid when the flow extends to infinity in all directions. In other words, if the shear layer were absent and the flow extended to infinity, the center of a spherical wave front emitted by a source at S in Fig. 14.9 would be at R when the wave front reached point C.

Sound impinging on the shear layer is both refracted and changed in amplitude in propagating into the stationary fluid. The magnitude of the amplitude correction depends on the choice of the corrected observer position C. The corrections given below will assume that the retarded distance r of C from the source is equal to R_m, the rectilinear distance of the observer from actual source. The amplitude correction is given by[5]

$$\left|\frac{p_C}{p_m}\right| = \tfrac{1}{2}H(1 - M_f \cos\Theta)\left[\frac{\zeta}{\sin\Theta} + (1 - M_f\cos\Theta)^2\right]$$

$$H = \frac{\sin\Theta_m}{\sin\Theta}\left(\left\{1 + \frac{h}{R_m\sin\Theta_m}\left[\left(\frac{\sin\Theta}{\zeta}\right)^3 - 1\right]\right\}\right.$$

$$\left.\cdot\left[1 + \frac{h}{R_m\sin\Theta}\left(\frac{\sin\Theta}{\zeta} - 1\right)\right]\right)^{1/2} \tag{14.19}$$

and

$$\zeta^2 = (1 - M\cos\Theta)^2 - \cos^2\Theta \qquad \beta^2 = 1 - M^2$$

$$\cos\Theta_r = \frac{\cos\Theta}{1 - M\cos\Theta} \tag{14.20}$$

where p_m = measured acoustical pressure at observer position O, $\mathrm{N/m^2}$
p_C = corrected acoustical pressure at C, $\mathrm{N/m^2}$
Θ = propagation angle after refraction
Θ_M = measured angle of propagation
Θ_r = retarded propagation angle
R_m = actual and retarded distance of observer from source, $\equiv r$, m
h = distance of source from shear layer, m

When h is small compared to R_m, Θ and the measured angle Θ_m may be assumed to be the same. If this is not the case, Θ must be determined by solving the equations

$$R_m\cos\Theta_m = h\cot\Theta_C + (R_m\sin\Theta_m - h)\cot\Theta$$

$$\tan\Theta_C = \frac{\zeta}{\beta^2\cos\Theta + M} \tag{14.21}$$

where Θ_C is the propagation angle before refraction.

The correction formula ignores the influence of the finite thickness of the shear layer, but detailed studies indicate that the error involved is negligible at the wind tunnel Mach numbers (<0.5) relevant in practice.[5] The effect of finite values of h/R_m is small except when Θ_C is small (say, <20° when $M = 0.5$). Also neglected are the spreading of the shear layer with distance downstream and the possible influence of curvature of the interface (i.e., for an open jet of circular cross section). These can introduce errors at very low frequencies and if the change in the shear layer is finite over a distance of a wavelength or less.[6] Scattering of sound by shear layer turbulence can usually be neglected.

14.5 COMBUSTION NOISE OF GAS TURBINE ENGINES

Jet mixing noise and shock-associated noise of high-speed jet engines are the result of sources in the flow downstream of the nozzle. Their importance has progressively diminished in recent years with the introduction of large-diameter, high-bypass-ratio turbofan engines with much reduced mass efflux velocities. In consequence, greater attention has been given to noise generated within the engine (termed *core noise*), which tends to be predominant at frequencies less than 1 kHz. Combustion processes are a significant component of core noise, both directly in the form of thermal monopole sources and indirectly through the creation of temperature and density inhomogeneities ("entropy spots"), which behave as dipole sources when accelerated in nonuniform flow.

The noise prediction scheme outlined below is based on an analysis of combustion noise of turbojet, turboshaft, and turbofan engines as well as model-scale data[3]. Annular, can-type, and "hybrid" combustors were all included in the validation studies.

The overall sound power level (OAPWL, dB) is a function of the operating conditions of the combustor and the turbine temperature extraction and may be estimated from the equation

$$\text{OAPWL} = -60.5 + 10 \times \log \frac{mc^2}{\Pi_{\text{ref}}} + 20 \times \log \frac{(\Delta T/T_I)(P_I/p_{\text{ISA}})}{[(\Delta T)_{\text{ref}}/T_s]^2} \tag{14.22}$$

where
m = combustor mass flow rate, kg/s
P_I = combustor inlet total pressure, Pa
ΔT = combustor total temperature rise, K
$(\Delta T)_{\text{ref}}$ = reference total temperature extraction by engine turbines at maximum takeoff conditions, K
T_I = inlet total temperature, K
T_s = sea-level atmospheric temperature, 288.15 K
Π_{ref} = reference power, 10^{-12} W
p_{ISA} = sea-level atmospheric pressure, 10.13 μPa
c = sea-level speed of sound, 340.3 m/s

This formula is expected to yield predictions that are accurate to within ±5 dB.

The one-third-octave-band power level spectrum PWL(f) is given in terms of the OAPWL by

$$\mathrm{PWL}(f) = \mathrm{OAPWL} - 16 \times \log[(0.003037f)^{1.8509} + (0.002051f)^{-1.8168}] \quad \mathrm{dB} \tag{14.23}$$

where f is frequency (Hz) and the formula is applicable for 100 Hz $\leq f \leq$ 2000 Hz. The spectrum is essentially symmetric about a peak at f = 400 Hz (Fig. 14.10). In applications where the observed peak frequency f_p differs slightly from this value, the spectrum in the figure should be shifted, retaining its shape, so that the peak coincides with observation. In (14.23), f would be replaced by $400f/f_p$.

The one-third-octave-band sound pressure spectrum may now be determined from the formula

$$\mathrm{SPL} = -10.8 + \mathrm{PWL}(f) - 20 \times \log R - 2.5 \times \log\left[\frac{1}{(10^{(1.633-0.0567\theta)} + 10^{(19.43-0.233\theta)})^{0.4}} + 10^{(4.333-0.115\theta)}\right] \tag{14.24}$$

where θ = angle from jet exhaust axis, degrees
R = observer distance, m

The peak radiation is predicted by (14.24) to occur at $\theta \simeq 60°$, although at very low frequencies ($\leq$200 Hz) the peak may be shifted slightly toward the jet axis.

For an aircraft in flight at speed V_f, the modification of the predicted SPL may be estimated from Eq. (14.18).

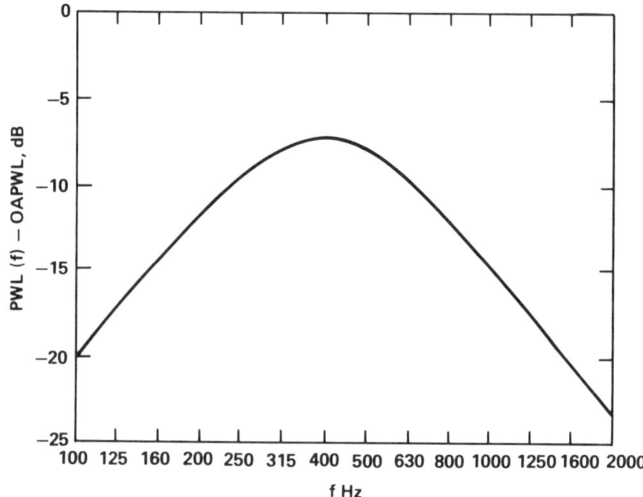

Fig. 14.10 One-third-octave frequency band power level spectrum of combustor noise.

14.6 TURBULENT BOUNDARY LAYER NOISE

The intense pressure fluctuations that can occur beneath a turbulent boundary layer are a source of sound and structural vibrations. The sound is produced directly by aerodynamic sources within the flow and indirectly by the *diffraction* at discontinuities in the wall (e.g., corners, ribs, support struts, etc.) of structural modes excited by the boundary layer pressures.

The pressure developed beneath a boundary layer on a *hard* wall is called the *blocked pressure* and is twice the pressure that a nominally identical flow would produce if the wall were absent. If the free-stream Mach number M is not small, both fluid compressibility and temperature variations associated with frictional heating are important, and when $M \geq 6$ (in *hypersonic* flow), chemical dissociation and ionization can occur. The rms wall pressure p_{rms} (Pa) can be estimated from the formula[7]

$$\frac{p_{\mathrm{rms}}}{q} \simeq \frac{\sigma}{\frac{1}{2}(1 + T_w/T) + 0.1(\gamma - 1)M^2} \tag{14.25a}$$

where $\sigma = 0.006$, dimensionless
$q = \frac{1}{2}\rho U^2$, Pa
$\rho =$ fluid density at outer edge of boundary layer, kg/m^3
$U =$ free-stream velocity at outer edge of boundary layer, m/s
$M =$ free-stream Mach number, $=U/c$ dimensionless
$c =$ speed of sound at outer edge of boundary layer, m/s
$T =$ temperature at outer edge of boundary layer, K
$T_w =$ temperature of wall, K
$\gamma =$ ratio of specific heats of gas, dimensionless

The quantity $\sigma \equiv (p_{\mathrm{rms}}/q)_{\mathrm{incompressible}}$, and (14.25a) represents a mapping from an incompressible flow to the compressible state by means of the *compressibility* factor

$$\epsilon_* = \frac{1}{\frac{1}{2}(1 + T_w/T) + 0.1(\gamma - 1)M^2} \tag{14.25b}$$

Recent measurements[8] suggest that the currently accepted value $\sigma = 0.006$ may be too low and that a better approximation is $\sigma = 0.01$.

In the particular case of an *adiabatic* wall (through which there is no heat flux), $T_w/T \simeq 1 + 0.45(\gamma - 1)M^2$, and (14.25a) becomes

$$\frac{p_{\mathrm{rms}}}{q} \simeq \frac{\sigma}{1 + 0.325(\gamma - 1)M^2} \tag{14.25c}$$

In many aeronautical applications wall pressure fluctuations are substantially higher if the flow *separates*. For example, when separation occurs at the compression corner at a ramp or at an expansion corner, the rms wall pressure can typically exceed about 2% of the local dynamic pressure $q = \frac{1}{2}\rho U^2$ (Pa).

The structural response of a flexible wall to forcing by the boundary layer depends on both the temporal and spatial characteristics of the pressure fluctuations. When the wall is locally plane, these can be expressed in terms of the *wall pressure wavenumber-frequency spectrum* $P(\mathbf{k}, \omega)$. This is the two-sided Fourier transform $(1/2\pi)^3 \int_{-\infty}^{\infty} R_{pp}(x_1, x_3, t) \exp\{-i(\mathbf{k} \cdot \mathbf{x} - \omega t)\}\, dx_1\, dx_3\, dt$ of the space–time correlation function of the wall pressure R_{pp}. By convention, coordinate axes (x_1, x_2, x_3) are taken with x_1, x_3 parallel and transverse to the mean flow respectively, x_2 measured outward from the wall, and the wavenumber $\mathbf{k} = (k_1, k_3)$ (in reciprocal meters) has components parallel to the x_1, x_3 axes only. The principal properties of the blocked-pressure spectrum $P_0(\mathbf{k}, \omega)$, say, are understood only for low-Mach-number flows ($M \ll 1$).

Wall Pressure Spectrum at Low Mach Number

$P_0(\mathbf{k}, \omega)$ is an even function of ω, and its general features are shown in Fig. 14.11[10] for $\omega > 0$. The strongest pressure fluctuations are produced by eddies in the *convective ridge* of the wavenumber plane that convect along the wall at about 70% of the free-stream velocity. At the center of the convective region

$$P_0(\mathbf{k}, \omega) \simeq \frac{10^3 \rho^2 v_*^6}{\omega^3} \qquad k_1 \simeq \frac{\omega}{U_c} \qquad k_3 = 0 \tag{14.26}$$

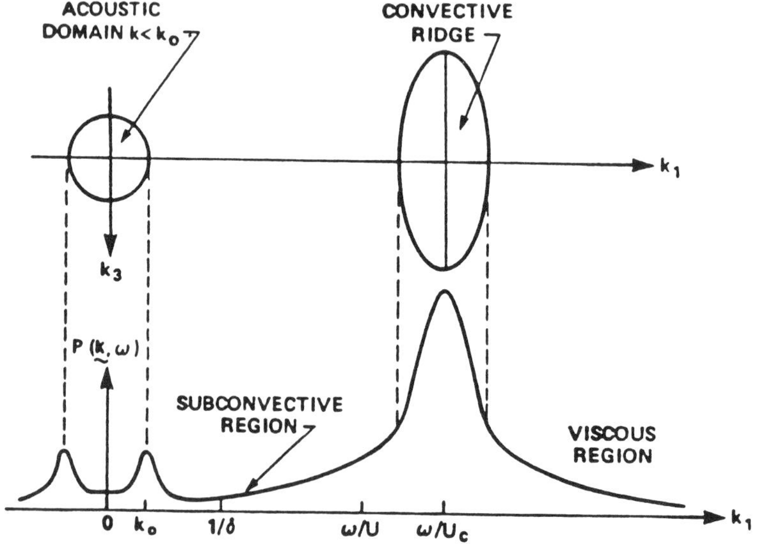

Fig. 14.11 Turbulent boundary layer wall pressure spectrum at low Mach number and for $\omega\delta/U \gg 1$, where δ is the boundary layer thickness m.

where ρ = mean fluid density, kg/m^3
v_* = friction velocity, $\simeq 0.035U$, m/s
$U_c \simeq 0.7U$, m/s
U = mainstream velocity at outer edge of boundary layer, m/s
ω = radian frequency, $= 2\pi f$, rad/s.

The region $k \equiv |\mathbf{k}| < \omega/c$ (c = speed of sound, m/s) is called the *acoustic domain*, and $k_0 = \omega/c$ is the acoustic wavenumber. The adjacent *subconvective* domain ($k_0 < k << \omega/U_c$) is important in determining the structural response of the wall to forcing by the turbulence pressures. In these regions $10 \times \log\{P_0(\mathbf{k}, \omega)\}$ is typically 30–60 dB below the levels in the convective domain.

Pressure fluctuations in the acoustic domain correspond to sound waves in the fluid. When the wall is smooth and flat, the generation of sound is dominated by quadrupole sources in the flow,[10] and the acoustic pressure frequency spectrum $\Phi(\omega)$ of sound produced by a fixed area of the wall (Fig. 14.12) is

$$\Phi(\omega) = 2 \frac{A}{R^2} k_0^2 \cos^2\theta \, P_0(k_0 \sin\theta \cos\phi, k_0 \sin\theta \sin\phi, \omega) \qquad (14.27)$$

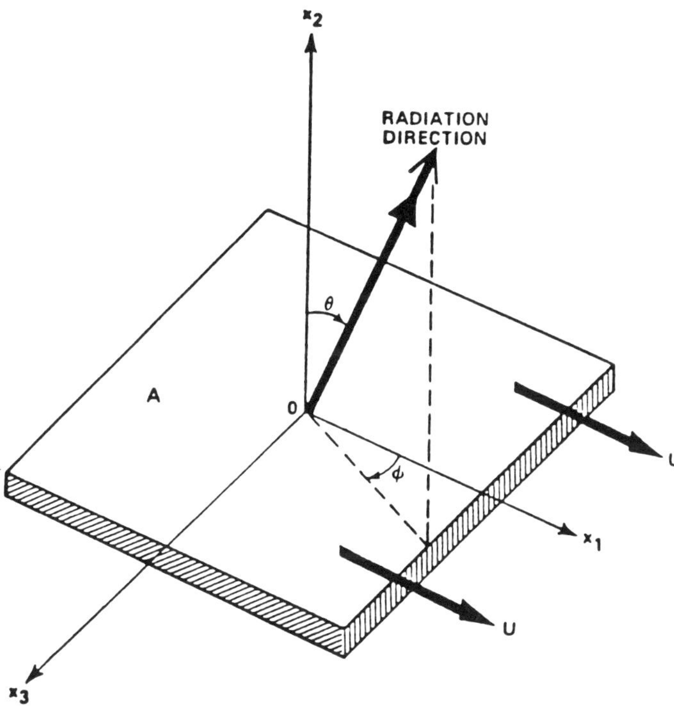

Fig. 14.12 Coordinates defining the radiation of sound from a region of the wall of area A.

where OASPL $= 10 \times \log [\int_0^\infty \Phi(\omega)\, d\omega/(20\ \mu\text{Pa})^2]$, dB
A = area of wall region, m^2
R = observer distance from center of A, m
(θ, ϕ) = polar angles of observer defined in Fig. 14.12

The behavior of $P_0(\mathbf{k}, \omega)$ in and near the acoustic domain is a matter of some contention. An upper bound is provided by Sevik's *wavenumber-white* approximations[11]

$$P_0(\mathbf{k}, \omega) = \frac{127\rho^2 v_*^4 U\delta^3}{c^2(\omega\delta/U)^{4.5}} \qquad k \le k_0 \qquad \frac{\omega\delta}{U} \gg 24$$

$$\frac{\Phi(\omega)}{\rho^2 v_*^3 \delta} = 254 \cos^2\theta \,\frac{A}{R^2}\frac{v^*}{U} M^4 \left(\frac{\omega\delta}{U}\right)^{-2.5} \qquad \frac{\omega\delta}{U} \gg 24 \tag{14.28}$$

where δ is the boundary layer thickness (distance from wall at which the mean flow velocity is equal to $0.99U$) in meters.

Chase[12] has proposed the following representation of the blocked-pressure spectrum that is applicable over the whole range of wavenumbers and is based on an empirical fit to experimental data in the convective and subconvective domains:

$$\frac{P_0(\mathbf{k}, \omega)}{\rho^2 v_*^3 \delta^3} = \frac{1}{[(k_+\delta)^2 + 1.78]^{5/2}} \left\{ \frac{0.1553(k_1\delta)^2 k^2}{|k_0^2 - k^2| + \beta^2 k_0^2} + 0.00078 \right.$$

$$\left. \times \frac{(k\delta)^2[(k_+\delta)^2 + 1.78]}{(k\delta)^2 + 1.78} \left(4 + \frac{|k_0^2 - k^2|}{k^2} + \frac{k^2}{|k_0^2 - k^2| + \beta^2 k_0^2}\right) \right\}$$

$$\frac{\omega\delta}{U} > 1 \tag{14.29}$$

where $k_+ = \sqrt{(\omega - U_c k_1)^2/9v_*^2 + k^2}$, m^{-1}
$\beta \simeq 0.1$, dimensionless

The first term in the curly brackets determines the behavior near the convective ridge and the second in the low-wavenumber and acoustic domains. Although the formula is formally applicable in the acoustic domain, it has not been validated there, and predictions are typically 20 dB or more lower than those given by Sevik's formula (14.28). The numerical coefficient β controls the height of the spectral peak (Fig. 14.11) at the boundary $k = k_0$ of the acoustic domain and determines the intensity of sound waves propagating parallel to the plane of the wall. This peak is predicted by theory, but its presence has not been unambiguously confirmed by experiment.

Example 14.2 Estimate the frequency spectrum $\Psi(\omega)$ of the power dissipated by flexural motion of a plane wall when the wall is excited by a low-Mach-number turbulent flow over a region of area A whose normal impedance $Z(k, \omega)$ (kg/m^2 s) is independent of the orientation of $\mathbf{k}$.

SOLUTION The impedance satisfies $Z(\mathbf{k}, \omega) = -p(\mathbf{k}, \omega)/v_2(\mathbf{k}, \omega)$, where p and v_2, respectively, denote the Fourier components of the pressure and normal velocity on the wall. Let $Z = R - iX$, where R and X are the resistive and reactive components of the impedance, respectively. The total power delivered to the flexural motions is equal to $\int_0^\infty \Psi(\omega)\, d\omega$, where

$$\Psi(\omega) = 2A \int_{k>|k_0|}^{\infty} \frac{R(\mathbf{k}, \omega) P_0(\mathbf{k}, \omega)\, d^2\mathbf{k}}{R(\mathbf{k}, \omega)^2 + [X(\mathbf{k}, \omega) + \rho\omega/\sqrt{k^2 - k_0^2}]^2}$$

The flexural mode wavenumbers are the roots $\mathbf{k} = \mathbf{k}_n$, $n = 1, 2, 3, \ldots$, of $X(\mathbf{k}, \omega) + \rho\omega/\sqrt{k^2 - k_0^2} = 0$. In practice, these usually lie in the low-wavenumber region where $P_0(\mathbf{k}, \omega) \equiv P_0(k, \omega)$, so that, when $R \ll X$ and Z is a function of $k \equiv \mathbf{k}$ and ω only,

$$\Psi(\omega) \simeq \sum_n \frac{4\pi^2 A k_n P_0(k_n, \omega)}{|(\partial/\partial k)[X(k, \omega) + \rho\omega/\sqrt{k^2 - k_0^2}]|_{k_n}}$$

When the wall consists of a vacuum-backed, thin elastic plate of bending stiffness B (kg $\cdot$ m^2/s^2), mass density m (kg/m^2) per unit area, and negligible damping $R = 0$, we have $X = -(Bk^4 - m\omega^2)/\omega$, and there is only one flexural mode that occurs at $k = k_* > |k_0|$ such that

$$\Psi(\omega) = \frac{4\pi^2 A\omega(k_*^2 - k_0^2) P_0(k_*, \omega)}{5Bk_*^4 - 4B(k_0 k_*)^2 - m\omega^2}$$

where k_* is the positive root of $Bk^4 - m\omega^2 - \rho\omega^2/\sqrt{k^2 - k_0^2} = 0$. Numerical estimates may be made by substituting for $P_0(k_*, \omega)$ from either (14.28) or (14.29). Except at very high frequencies $\omega\delta/U > 10^2$, Sevik's formula (14.29) will yield the greater estimate for the power fed into the flexural motions.

Wall Pressure Spectrum at High Mach Number

At high Mach numbers there is no clear separation of the wavenumber–frequency plane into acoustic and convective (incompressible) domains, and detailed properties of $P(\mathbf{k}, \omega)$ are largely unknown. The compressibility factor ϵ_* of (14.25b) can be used to define the following empirical representation of the wall pressure power spectral density $\Psi_{\text{wall}}(\omega) = \int_{-\infty}^{\infty} P(\mathbf{k}, \omega)\, d^2\mathbf{k}$:

$$\frac{\Psi_{\text{wall}}(\omega) U}{q^2 \delta*} = \frac{2\epsilon_*^{2(1+\mu)} \sigma^2}{\pi[1 + \epsilon_*^{4\mu}(\omega\delta*/U)^2]} \tag{14.30}$$

where p_{rms} = rms wall pressure, $\sqrt{\int_0^\infty \Psi_{\text{wall}}(\omega)\, d\omega}$, Pa

$\delta*$ = displacement thickness of boundary layer, $\simeq \frac{1}{8}\delta$, m

μ = -0.717, dimensionless

14.7 SPOILER NOISE

The term *spoiler noise* is used to characterize the noise produced by an obstacle or other obstruction that spans a duct carrying a mean flow. We shall discuss the case of air flowing in a duct or pipe that may be regarded as "semi-infinite" when it is desired to estimate the acoustical radiation from the open end. A spoiler-noise-generating system is shown schematically in Fig. 14.13. In practice the spoiler may be a strut, stringer, guide vane, or other flow control device. The nominally steady mean flow exerts unsteady lift and drag forces on the spoiler, which accordingly behaves as an acoustical source of *dipole* type.

The following idealizations will be made:

1. The spoiler can be of arbitrary shape, but its cross-flow dimensions are small relative to the pipe cross section. This ensures that flow speeds near the spoiler are not significantly greater than the pipe mean flow speed and therefore that turbulence mixing noise (quadrupole) can be neglected. Thus, the case of a valve, which produces a severe throttling of the flow, is excluded (see Section 14.10).
2. The peak frequency of the noise spectrum is below the "cutoff" frequency f_{co} of the pipe, which is given by

$$\begin{aligned} f_{\text{co}} &= 0.293c/r \quad \text{Hz} \quad \text{(circular pipe)} \\ &= 0.5c/w \quad \text{Hz} \quad \text{(rectangular pipe)} \end{aligned} \tag{14.31}$$

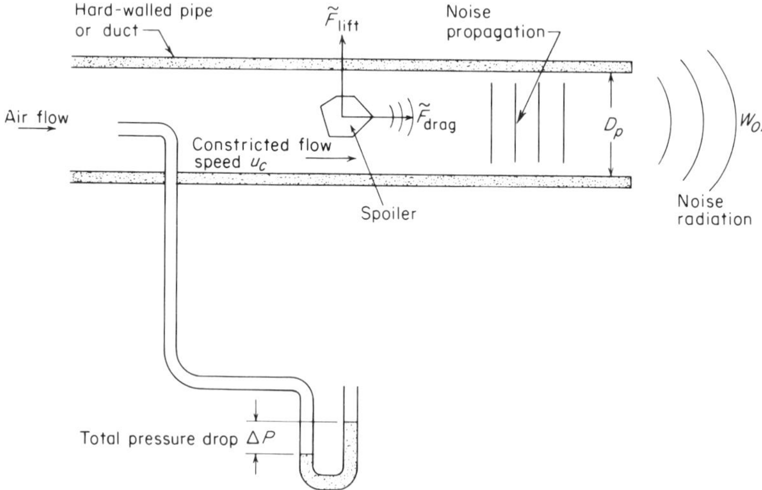

Fig. 14.13 Flow spoiler system. The sound power is generated by the unsteady drag force on the spoiler and radiated from the open end of the pipe.

where c = sound speed, m/s
r = radius of circular pipe, m
w = largest transverse dimension of rectangular pipe, m

If this condition is not fulfilled, the propagated sound power near and above the peak frequency (which determines most of the overall sound power) will be influenced by transverse modes of the pipe.

3. The pipe wall is acoustically rigid. The fluctuating lift forces on the spoiler are then canceled by images in the wall, and sound is produced solely by the unsteady drag, which corresponds to an acoustic dipole whose axis is parallel to the mean flow. A low-noise spoiler system should therefore use bodies of low drag, such as airfoils, rather than grids or other oddly shaped bodies.

For several elementary flow–spoiler configurations, such as the one shown in Fig. 14.13, the *fluctuating drag is directly proportional to the steady-state drag experienced by the body*. This depends on the pressure drop ΔP across the spoiler (as measured by an upstream and a downstream total-pressure probe), and the broadband noise radiated from the open end of the pipe is given by[13]

$$W_{\mathrm{OA}} = \frac{k(\Delta P)^3 D_p^3}{\rho^2 c^3} \tag{14.32}$$

where W_{OA} = overall radiated sound power, W
k = constant of proportionality, dimensionless
ΔP = total pressure drop across spoiler, Pa
D_p = pipe diameter, m
ρ = atmospheric density, $\mathrm{kg/m^3}$
c = atmospheric sound speed, m/s

Specific information about spoiler geometry is absent from (14.32) but is implicit in the pressure drop ΔP. From a variety of experimental spoiler configurations the constant k is found to be about 2.5×10^{-4} for air.

Equation (14.32) determines the *broadband* sound power. It should strictly be regarded as a lower bound that is applicable only if there are no *discrete-frequency* components of the noise. These are present in certain conditions (e.g., excitation of edge tones) and may well stand out against the broadband levels.[14] When such a mechanism has been identified, the discrete tones can usually be eliminated or controlled by detuning or rounding off sharp corners or edges or cutting feedback paths by treating reflecting surfaces with sound-absorbent materials.

The frequency spectrum measured outside the pipe for the noise generated by the spoiler exhibits a haystack structure (Fig. 14.14) with a peak at a frequency given by

$$f_p \approx bu_c/d \quad \mathrm{Hz} \tag{14.33}$$

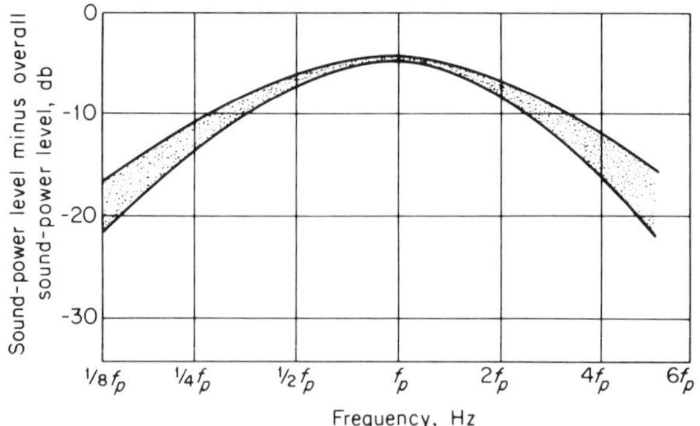

Fig. 14.14 Generalized octave-band spectrum for in-pipe-generated spoiler noise, referenced to the total power radiated from the open end and the peak frequency.

where u_c = constricted flow speed, m/s
d = projected width of spoiler, m
b = constant that equals 0.2 for pressure differences ΔP of the order 4000 Pa and 0.5 for ΔP of the order 40,000 Pa

The constricted flow speed u_c for cold air is given in Table 14.2 as a function of ΔP. Values in between those in the table can be interpolated.

Example 14.3 Find the sound power level spectrum of noise produced by a spoiler consisting of a flat plate of width 2 cm stretching across a circular pipe of internal diameter D_p = 5 cm. The pressure drop across the spoiler is ΔP = 10,000 Pa.

SOLUTION From Eq. (14.32) the total sound power $L_w \approx$ 101 dB *re* 10^{-12} W. When ΔP = 10,000 Pa, the peak frequency of the spectrum can be expected to be at $f_p \approx 0.35 u_c/d$.

From Table 14.2, we find u_c = 124 m/s; hence $f_p \approx$ 2170 Hz. The sound power spectrum in octave bands is obtained from Fig. 14.14 with 101 dB added to the ordinate.

14.8 GRID OR GRILLE NOISE

The characteristics of noise produced by flow through grids, grilles, diffusers, guide vanes, or porous plates, which often terminate air conditioning ducts, are similar

TABLE 14.2 Constricted Flow Speed u_c for Cold Air as Function of Pressure Drop

ΔP (Pa)	2500	5000	10,000	20,000	30,000	40,000
u_c, m/s	63	90	124	173	209	238

to those of spoiler-generated sound. The principal differences are that (1) the grid is located at the open end of the duct, (2) the duct cross-sectional area is typically quite large (say, 0.04–1 m^2), and (3) the velocity of the air in the duct is usually small, rarely exceeding 30 m/s. The speed of the "air jets" exhausting from the individual air passages (or orifices) in a diffuser is generally low enough (< 100 m/s) that jet mixing noise can be neglected. The dominant sound source is of dipole type and is associated with the interaction of the flow with the diffuser elements (e.g., guide vanes).

When a duct is terminated by a grid of circular rods (Fig. 14.15*a*), periodic shedding of vortices can occur producing a fluctuating lift force on each rod and an associated *tonal* component of the radiated sound. This source is a dipole (whose axis is parallel to the lift direction) and is in addition to the drag dipole. The frequency of the tone is given approximately by

$$f = \frac{0.2u}{D_R} \tag{14.34}$$

where u = mean flow speed, m/s
D_R = diameter of rods, m

Tonal oscillations produced by a feedback mechanism can also arise when the duct termination consists of a plate perforated with sharp-edged circular cylindrical apertures (Fig. 14.15*b*).

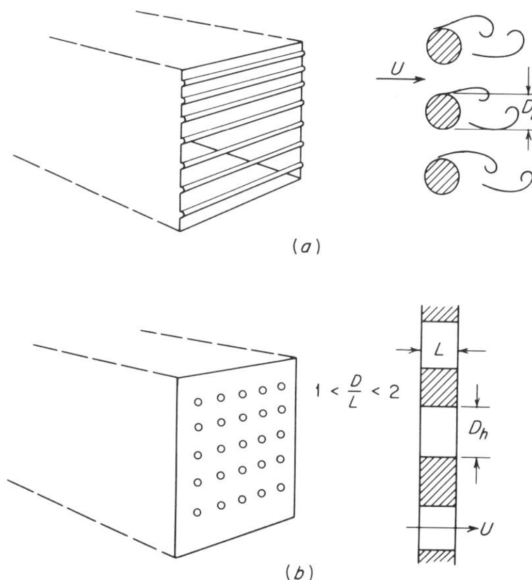

Fig. 14.15 Special cases of duct terminations: (*a*) circular rods; (*b*) sharp-edge circular cylindrical holes.

In general, however, the noise produced by typical air conditioner grids is almost always broadband and of conventional dipole character, with the sound power varying as the sixth power of the velocity. As in the case of spoiler noise, the overall sound power can be related to the pressure drop ΔP across the grid, independently of the specific grid geometry. To do this, we introduce the pressure drop coefficient

$$\xi = \frac{\Delta P}{\frac{1}{2}\rho u^2} \tag{14.35}$$

where ρ = density of air, $\mathrm{kg/m^3}$
u = mean flow speed in duct prior to grid, m/s

Three typical diffuser configurations together with their pressure drop coefficients ξ are illustrated in Fig. 14.16.[15] The values of ξ for similar diffuser configurations may be estimated from this figure.

The overall sound power level L_w from air conditioning diffusers can be estimated from the empirical formula[15]

$$L_w = 10 + 10 \times \log (S\xi^3 u^6) \qquad \text{dB } re\ 10^{-12}\ \text{W} \tag{14.36}$$

where S = area of duct cross section prior to diffuser, $\mathrm{m^2}$
ΔP = pressure drop through diffuser, Pa

The noise spectra of different diffusers do not exhibit identical shapes even when normalized to similar flow speeds and exhaust areas. Construction differences tend to emphasize different frequency regimes, and poorly designed diffusers will radiate discrete-frequency sound. In practical noise control problems, however, a general spectrum shape $L_w - 10 \times \log_{10}(S\xi^3)$ (dB re 10^{-12} W) can be used for each duct velocity u (m/s) that fits most diffuser noise spectra to within about ± 5 dB, as shown in Fig. 14.17.

To estimate the sound power spectrum of the radiation from a given diffuser, first determine the relevant curve from Fig. 14.17 for the particular flow speed. The ordinate in the figure is then increased by $10 \times \log(S\xi^3)$ decibels to yield the desired one-third-octave spectrum with a margin of error of about ± 5 dB.

14.9 SOUND GENERATION BY AIRFOILS AND STRUTS

Noise of a Strut in a Turbulent Stream

Long struts, airfoils, guide vanes, and so on, that offer negligible drag to a mean flow, frequently experience significant fluctuations in lift when exposed to a turbulent stream and behave as *dipole-type* sources of broadband sound. The strength of the radiation depends on the dimensions of the airfoil and on the turbulence intensity and correlation length of the velocity fluctuations.

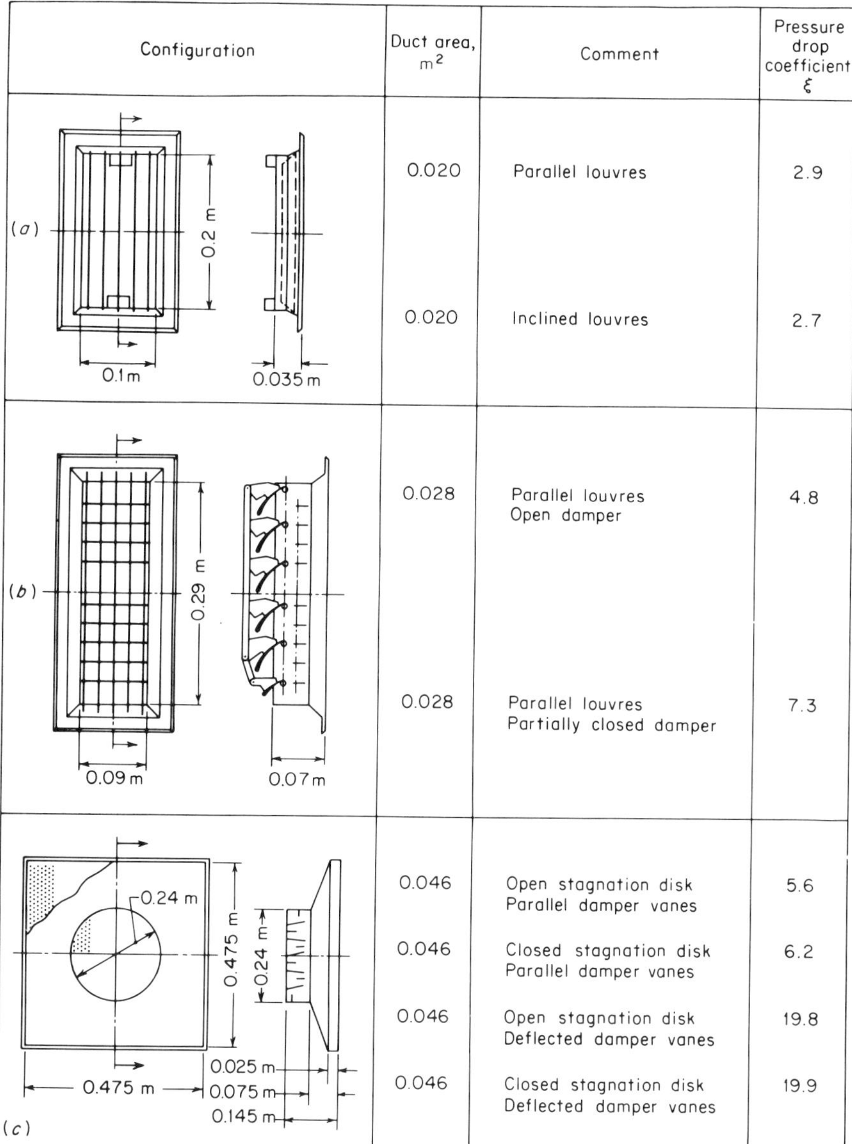

Configuration	Duct area, m^2	Comment	Pressure drop coefficient ξ
(a)	0.020	Parallel louvres	2.9
	0.020	Inclined louvres	2.7
(b)	0.028	Parallel louvres Open damper	4.8
	0.028	Parallel louvres Partially closed damper	7.3
(c)	0.046	Open stagnation disk Parallel damper vanes	5.6
	0.046	Closed stagnation disk Parallel damper vanes	6.2
	0.046	Open stagnation disk Deflected damper vanes	19.8
	0.046	Closed stagnation disk Deflected damper vanes	19.9

Fig. 14.16 Various duct terminations and their pressure drop coefficients $\xi = \Delta P/\frac{1}{2}\rho u^2$. The duct area S is in square meters. (After ref. 15.)

A simple description of the radiation can be given when the mean flow is of low Mach number (say, <100 m/s in air) and when the distance to the nearest boundary (e.g., duct wall) is at least of the order of the acoustic wavelength. In these circumstances the *frequency spectral density* of the total radiated sound power

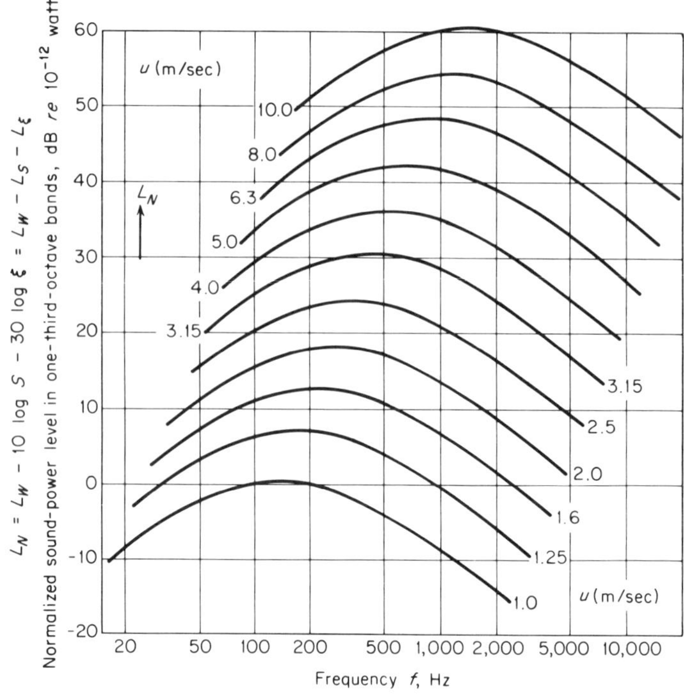

Fig. 14.17 Normalized one-third-octave sound power level spectra for noise radiated from diffusers of various flow velocities. (After ref. 15.)

$\Pi(\omega)$ (W · s) is given approximately by

$$\Pi(\omega) = \frac{\pi l a^2 \rho u^2 M^3 (\omega\Lambda/U)^4}{4(1 + \pi\omega a/U)\,[1 + (\omega\Lambda/U)^2]^{5/2}[1 + (\omega a/3c)^3]^{1/3}} \tag{14.37}$$

where ω = radian frequency, $=2\pi f$, f in Hz
a = chord of airfoil, m
l = span of airfoil, m
ρ = density of mean stream, kg/m^3
u = rms turbulence velocity in lift direction, m/s
U = velocity of mean stream, m/s
M = mean flow Mach number, $=U/c$, dimensionless
c = sound speed, m/s
Λ = integral scale of turbulence velocity, m

The total radiated sound power $W = \int_0^\infty \Pi(\omega)\, d\omega$, in watts, can be estimated from

$$W = \frac{1.78(la^2/\Lambda)\rho u^2 U M^3}{(1 + 10.71a/\Lambda)(1 + 1.79Ma/\Lambda)} \qquad 0 \le M \le 0.3 \tag{14.38}$$

At low frequencies the acoustical intensity exhibits the characteristic dipole *field shape*, proportional to $\cos^2 \theta$, where θ is measured from the direction of the mean lift. At higher frequencies (when the acoustic wavelength is much smaller than the chord of the airfoil but still exceeds the airfoil thickness) the field shape assumes a cardioid form, with a null in the forward direction and peaks close to the direction of the mean flow.

Airfoil Self-noise

In addition to gust/inflow turbulence-induced noise, an airfoil can also generate *self-noise* when turbulence that arises from the natural instability of the boundary layers on the surface of the airfoil is swept past the trailing edge. This is usually important when the hydrodynamic wavelength of the turbulence eddies is larger than the airfoil thickness at the trailing edge. The mechanism of noise production tends to be weaker than at the leading edge, because the violence of the unsteady motion at the trailing edge is alleviated by vortex shedding into the wake, and to be more prominent at higher frequencies[16].

The noise may be ascribed to a distribution of lift dipoles near the trailing edge. At low frequencies the acoustical intensity exhibits the characteristic dipole *field shape*, proportional to $\cos^2 \theta$, with a null in the plane of mean motion of the airfoil. At higher frequencies (when the acoustic wavelength is smaller than the chord of the airfoil) the field shape assumes a cardioid form, with a null in the downstream direction and the peak in the forward direction. In either case, when the trailing edge is at right angles to the mean flow, the frequency spectral density of the overall sound power $\Pi(\omega)$ (W · s) is given approximately by

$$\Pi(\omega) = \frac{\pi l a \omega^2}{24 \rho c^3 [1 + (\omega a/3c)^3]^{1/3}} \int_{-\infty}^{\infty} \frac{P_0(k_1, 0, \omega)}{|k_1|} dk_1 \qquad (14.39)$$

where $P_0(\mathbf{k}, \omega)$ = blocked-pressure wavenumber–frequency spectrum on airfoil just upstream of trailing edge (see Section 14.6)
k_1 = wavenumber component parallel to mean flow, m^{-1}

If the flow approaching the trailing edge is turbulent on both sides of the airfoil, $P_0(k_1, 0, \omega)$ in (14.39) should be replaced by the sum of the wall pressure spectra on the two sides.

When the Chase model (14.29) is used to estimate the integral in (14.39), the frequency spectral density of the sound power radiated by a section of the airfoil of spanwise length l (in meters) is

$$\Pi(\omega) = \frac{0.08 a l \delta \rho v_*^4 U_c (\omega\delta/U_c)^3}{c^3 [1 + (\omega a/3c)^3]^{1/3} [(\omega\delta/U_c)^2 + 1.78]^2} \qquad (14.40)$$

where δ = boundary layer thickness, m
v_* = friction velocity, m/s
$U_c \approx 0.7 \times$ velocity of mean stream, m/s

and other quantities are defined after Eq. (14.37). However, measured levels of $\Pi(\omega)$ frequently exceed predictions of this formula.[17] More detailed empirical formulas are given in the references.[17,18]

When the frequency is high enough [beyond the range of applicability of (14.40)] that the *Strouhal number* $\omega h/U \sim 1$, where h is the thickness of the trailing edge of the airfoil, the shedding of discrete vortices from the trailing edge can occur, provided the Reynolds number Ua/ν based on the chord a of the airfoil (dimensionless, ν is kinematic viscosity in m^2/s) does not exceed 10^6–10^7. This produces a distinct contribution to trailing-edge noise, often called "airfoil singing," whose amplitude is not easily predicted.[17]

14.10 AERODYNAMIC NOISE OF THROTTLING VALVES

The noise of gas control or throttling valves may generally be associated with two sources: (1) mechanical vibration of the trim and (2) aerodynamic throttling. Noise generation by these mechanisms rarely occurs simultaneously, but when it does, the cure of one is usually the cure of the other.

Aerodynamic Noise

Investigations of noise-induced pipe failures[19] have enabled maximum safe sound power levels to be established for given pipe sizes and wall thicknesses, as indicated in Fig. 14.18. The power levels shown in the figure correspond approximately to a sound level of 130 dBA at 1 m from the pipe wall downstream of the valve. Exceeding this level will most probably lead to piping failure, and a limit of 110 dBA is recommended for safety and to maintain the structural integrity of valve-mounted accessories.

The aerodynamic noise is determined by the mechanical stream power $W_{\text{mech strm}} = \frac{1}{2}mU^2$, (m being the mass flow and U the velocity; see Section 14.3) that is converted from potential energy (inlet pressure) to kinetic energy (velocity head) within the valve and subsequently to thermal energy (corresponding to reduced downstream pressure and an increase in entropy).[20] The pressure reduction (i.e., the conversion of kinetic into thermal energy) occurs via the generation of turbulence or, when the flow is supersonic, through shock waves. The sound power is equal to $\eta W_{\text{mech strm}}$, where η is an efficiency factor. Unlike the case of a jet discharging into the atmosphere, the jet from a valve cannot expand freely, and only a fraction of the kinetic energy is converted by turbulence into thermal energy. This is illustrated in Fig. 14.19, which is a schematic view of a throttling valve and the associated pressure profile for various downstream pressures P_2^*. On curve A the flow is subsonic, and turbulence-generated sound is predominantly of dipole type. At the critical pressure ratio ($P_1/P_2 = 1.89$ for air), curve B shows the presence of incipient shock waves followed by subsonic recompression, with both

*In the rest of this chapter all pressures are absolute static pressures.

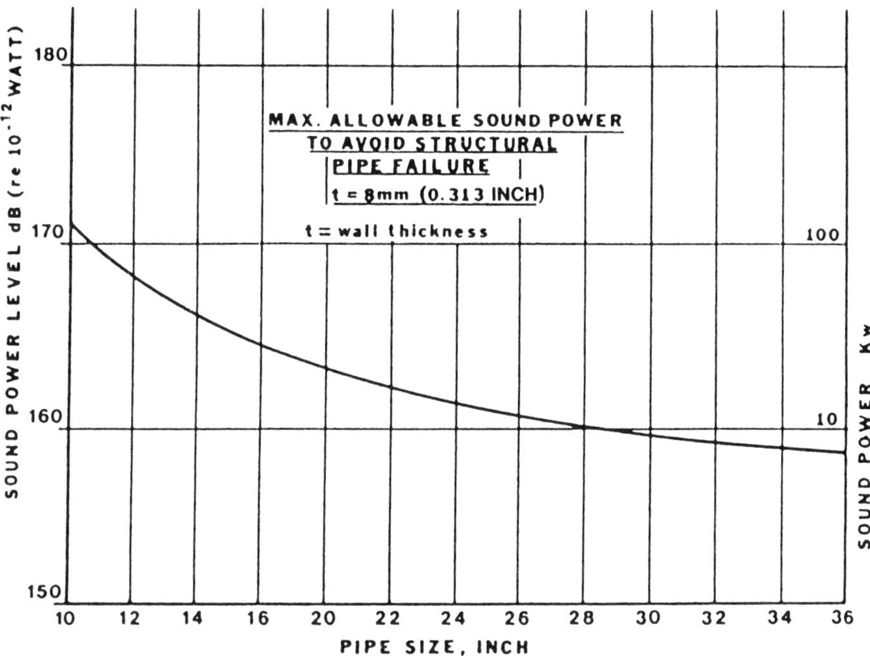

Fig. 14.18 Suggested sound power level and sound power (Kw) limits inside pipe to avoid structural pipe failure based on actual occurrence.[19]

turbulence and shocks as noise sources. At higher pressure ratios recompression tends to be nonisentropic, and sound is produced predominantly by shock waves (curve D).

Downstream of the vena contracta (D_j in Fig. 14.19) a portion of the velocity head is always recovered. The fraction lost is equal to F_L^2, where F_L is an experimentally determined pressure recovery coefficient that depends on valve type and size. The pressure drop $P_1 - P_2$ across the valve required to reach sonic velocity in the vena contracta is $F_L^2(P_1 - P_0)$, where P_0 is the static pressure in the vena contracta.[21] In subsonic flow the remaining portion of the jet power is recovered through isentropic recompression and is not converted into sound. If there were no pressure recovery at the valve orifice and assuming sonic flow at speed in c_0 m/s (curve D in Fig. 14.19),

$$W_{\text{mech strm}} = \tfrac{1}{2} m c_0^2 \quad \text{W} \tag{14.41}$$

where m (kg/s) is the mass flow, which is a function of the valve flow coefficient C_v (see Table 14.3) at the given valve opening, the specific gravity G_f of the gas or vapor (relative to air), and the valve inlet pressure P_1 (Pa) such that

$$W_{\text{mech strm}} = 7.7 \times 10^{-11} C_v F_L c_0^3 P_1 G_f \quad \text{W} \tag{14.42}$$

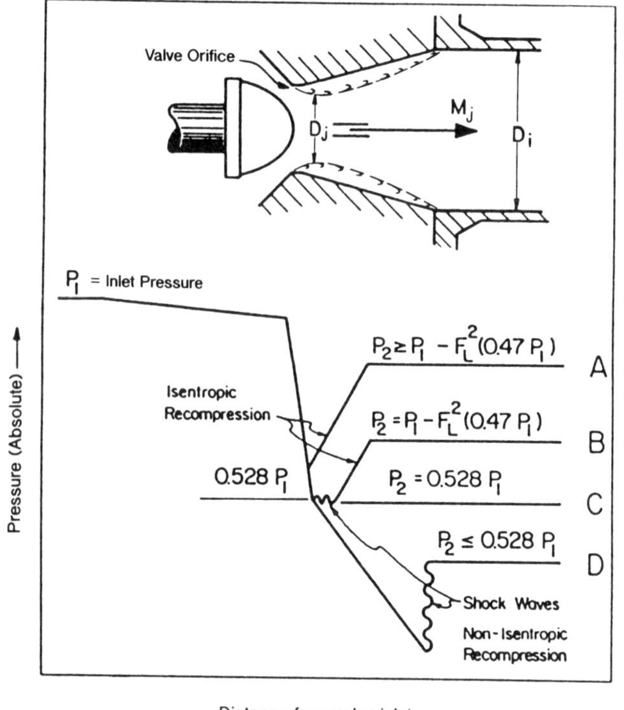

Fig. 14.19 Schematic flow profile and static pressure diagram of throttling valve for various downstream pressures (P_2).

The total sound power is then

$$W = \eta W_{\text{mech strm}} \qquad \text{W} \tag{14.43}$$

$$L_W = 10 \times \log \frac{\eta W}{10^{-12}} \qquad \text{dB} \tag{14.44}$$

When the pressure ratio is not sonic, η should be replaced by the acoustic power coefficient η_m determined below and (to simplify the calculations) should be used in conjunction with the sonic mechanical stream power given by (14.41). The acoustic power coefficient is further used in Eq. (14.50).

It may be remarked that the pressure recovery coefficient F_L can be used to predict the pressure P_0 in the vena contracta and also, to a satisfactory approximation, the area A_v of the vena contracta by

$$P_0 = P_1 - \frac{P_1 - P_2}{F_L^2} \tag{14.45a}$$

$$A_v = \frac{C_v F_L}{5.91 \times 10^4} \qquad \text{m}^2 \tag{14.45b}$$

TABLE 14.3 Factors for Valve Noise Prediction (Typical Values)

Valve Type	Flow to	Percentage of Capacity or Angle of Travel	C_v/D^2*	F_L	F_d
Globe, single-port parabolic plug	Open	100%	0.020	0.90	0.46
Globe, single-port parabolic plug	Open	75%	0.015	0.90	0.36
Globe, single-port parabolic plug	Open	50%	0.010	0.90	0.28
Globe, single-port parabolic plug	Open	25%	0.005	0.90	0.16
Globe, single-port parabolic plug	Open	10%	0.002	0.90	0.10
Globe, single-port parabolic plug	Close	100%	0.025	0.80	1.00
Globe, V-port plug	Open	100%	0.016	0.92	0.50
Globe, V-port plug	Open	50%	0.008	0.95	0.42
Globe, V-port plug	Open	30%	0.005	0.95	0.41
Globe, four-port cage	Open	100%	0.025	0.90	0.43
Globe, four-port cage	Open	50%	0.013	0.90	0.36
Globe, six-port cage	Open	100%	0.025	0.90	0.32
Globe, six-port cage	Open	50%	0.013	0.90	0.25
Butterfly valve, swing-through vane	—	75° open	0.050	0.56	0.57
Butterfly valve, swing-through vane	—	60° open	0.030	0.67	0.50
Butterfly valve, swing-through vane	—	50° open	0.016	0.74	0.42
Butterfly valve, swing-through vane	—	40° open	0.010	0.78	0.34
Butterfly valve, swing-through vane	—	30° open	0.005	0.80	0.26
Butterfly valve, fluted vane	—	75° open	0.040	0.70	0.30
Butterfly valve, fluted vane	—	50° open	0.013	0.76	0.19
Butterfly valve, fluted vane	—	30° open	0.007	0.82	0.08
Eccentric rotary plug valve	Open	50° open	0.020	0.85	0.42
Eccentric rotary plug valve	Open	30° open	0.013	0.91	0.30
Eccentric rotary plug valve	Close	50° open	0.021	0.68	0.45
Eccentric rotary plug valve	Close	30° open	0.013	0.88	0.30
Ball valve, segmented	Open	60° open	0.018	0.66	0.75
Ball valve, segmented	Open	30° open	0.005	0.82	0.63

*D is the internal pipe diameter in mm.

Acoustic Efficiency

At Mach 1 a free jet has an acoustic efficiency $\eta \approx 10^{-4}$. If it is assumed that the sources of valve noise are dipoles and quadrupoles of equal magnitude (5×10^{-5}) at the Mach 1 reference point, then their respective efficiencies are represented by curves A and B in Fig. 14.20, and both curves can be combined to yield a subsonic slope C that originates at 1×10^{-4} and varies as $U^{3.6}$. This curve can be modified further to account for the decrease in $W_{\text{mech strm}}$ ($\propto U^3$) with subsonic velocities, leading to the final curve D for the effective efficiency (i.e., acoustic power coefficient) η_m that varies as $U^{6.6}$ or $(P_1/P_0 - 1)^{2.57}$ when $U \propto (P_1/P_0 - 1)^{0.39}$, P_0 being the static pressure in the vena contracta. This approximation gives satisfactory results for $0.38 \lesssim M \leq 1$ (i.e., down to $P_1/P_0 = 1.1$).

For practical use it is convenient to express η_m in terms of P_1/P_2 = inlet pres-

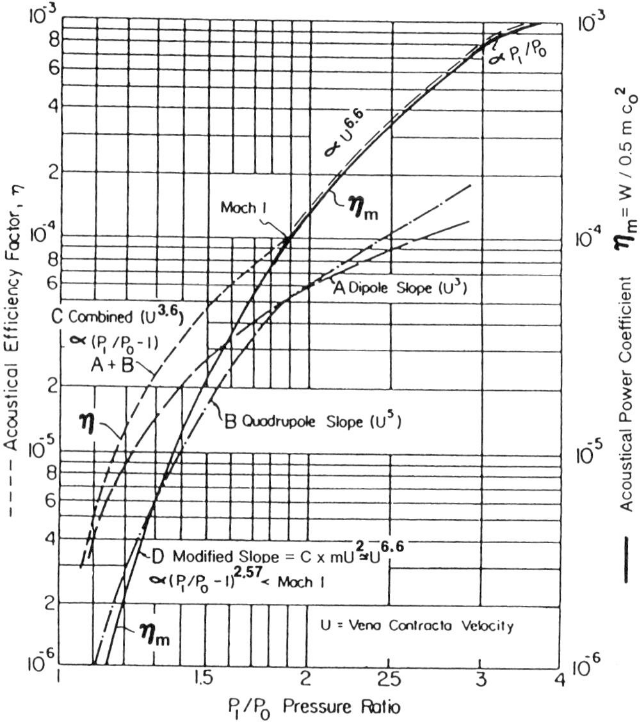

Fig. 14.20 Slopes of dipole and quadrupole acoustical efficiency curves, assuming each has equal strength at Mach 1; their combined slope and the η_m curve after subtracting the effects of change in mass flow m and orificial velocity U for regions below Mach 1 from the combined curve.

sure/outlet pressure. The following formulas[23,24] may be used for this purpose:

Regime I. $P_1/P_2 < P_1/P_{2\,\text{critical}}$ (subsonic):

$$\eta_{m\text{I}} = 10^{-4} F_L^2 \left(\frac{P_1 - P_2}{P_1 F_L^2 - P_1 + P_2} \right)^{2.6} \tag{14.46}$$

Regime II and III. $P_1/P_{2\,\text{critical}} < P_1/P_2 < 3.2\alpha$:

$$\eta_{m\text{II}} = 10^{-4} F_L^2 \left(\frac{P_1/P_2}{P_1/P_{2\,\text{critical}}} \right)^{3.7} \tag{14.47}$$

Regime IV. $3.2\alpha < P_1/P_2 < 22\alpha$ ($M_j > 1.4$):

$$\eta_{m\text{IV}} = 1.32 \times 10^{-3} F_L^2 \left(\frac{P_1/P_2}{P_1/P_{2\,\text{break}}} \right) \tag{14.48}$$

Regime V. $P_1/P_2 > 22\alpha$ (constant efficiency):

$$\eta_{m\mathrm{V}} = \text{maximum value from regime IV} \tag{14.49}$$

where
$$\alpha = (P_1/P_{2\,\mathrm{critical}})/1.89$$
$$P_1/P_{2\,\mathrm{break}} = \alpha\gamma^{\gamma/(\gamma-1)} \quad \text{at } M = \sqrt{2}$$
$$\gamma = \text{ratio of specific heats of gas}$$
$$P_1/P_{2\,\mathrm{critical}} = P_1/(P_1 - 0.5P_1F_L^2)$$

The values of these parameters may be taken from Table 14.4.

When the downstream piping is straight and there are no sudden changes in cross-sectional area, the internal sound pressure level L_{pi} (*re* 2×10^{-5} Pa) downstream of the valve is given by

$$L_{pi} = -61 + 10 \times \log \frac{\eta_m C_v F_L P_1 P_2 c_0^4 G_f^2}{D_i^2} \quad \text{dB} \tag{14.50}$$

where D_i = internal diameter of downstream pipe, m
G_f = specific gravity of gas relative to air at 20°C
$c_0 = \sqrt{\gamma P_0/\rho_0}$, m/s
γ = ratio of specific heats
ρ_0 = gas density at vena contracta, $\mathrm{kg/m^3}$

Pipe Transmission Loss Coefficient

Knowledge of the peak internal sound frequency f_p is crucial for a proper prediction of the pipe transmission loss coefficient T_L. The coefficient T_L does not vary significantly between the first cutoff frequency f_0 and the ring frequency f_r of the pipe (see lower curve in Fig. 14.21), but variations can be large at other frequencies. The slope of T_L is about -6 dB per octave below f_0 and $+6$ dB per octave above f_r.[23]

For Mach numbers in the pipe less than about 0.3 and for relatively heavy pipes (as found in typical process plants) it may be assumed that the *minimum* transmis-

TABLE 14.4 Important Pressure Ratios for Air*

F_L	0.5	0.6	0.7	0.8	0.9	1.0
P_1/P_2 critical	1.13	1.20	1.30	1.43	1.61	1.89
P_1/P_2 break	1.95	2.07	2.24	2.40	2.76	3.25
α Ratio	0.60	0.64	0.68	0.76	0.85	1.0
22α	13	14	15	16	19	22

P_1/P_0 critical = 1.89.
*May be used for other gases with reasonable accuracy.

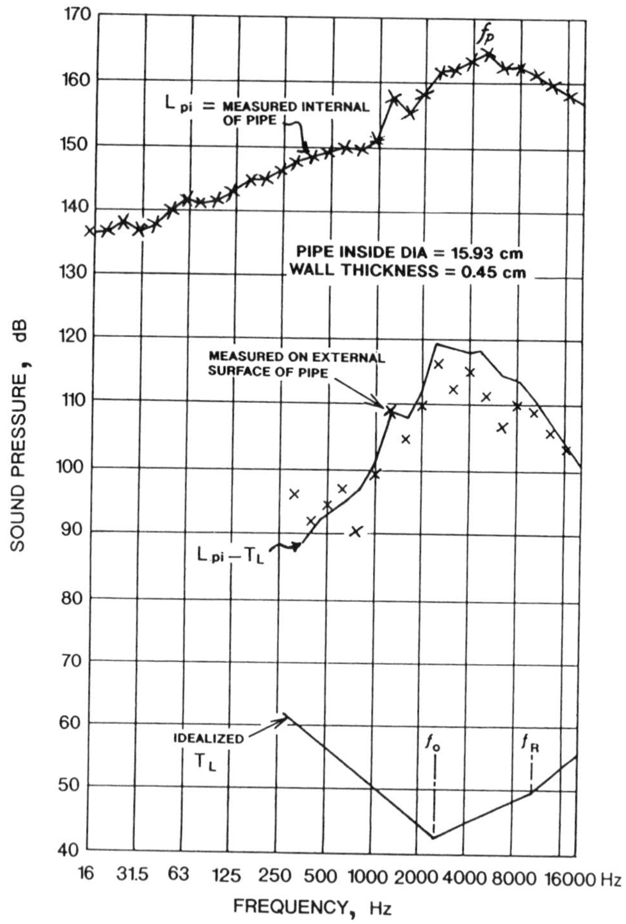

Fig. 14.21 Test data taken on a 150-mm globe valve showing good agreement using the stated transmission loss equation. Notice the resultant shift in the peak sound frequency (f_p) from ~4.8 kHz inside the pipe to the first cutoff frequency of the pipe (f_0) at 2.5 kHz outside the pipe.

sion loss occurs at f_0 and is given by[23]

$$T_{Lf_0} = 10 \times \log \left[9 \times 10^6 \frac{rt_p^2}{D_i^3} \left(\frac{P_2}{P_a} + 1 \right) \right] \quad \text{dB} \tag{14.51}$$

where r = distance from pipe center to observer, m
t_p = thickness of pipe wall, m
D_i = internal pipe diameter, m
P_2 = internal static pressure downstream of valve, Pa
P_a = external static pressure at same downstream distance, Pa

TABLE 14.5 Minimum Pipe Transmission Loss Coefficient T_{Lf_0}[a]

	Pipe Schedule	
Nominal Pipe Size (m)	40	80
0.025	72	76
0.050	65	69
0.080	64	68
0.100	60	64
0.150	57	61
0.200	54	58
0.250	52	57
0.300	51	56
0.400	50	55
0.500	48	53

[a]At 1 m distance from steel pipe (dB) for 200 kPa internal and 100 kPa external air pressure. Pipe schedules per ANSI B36.10.

Typical values are given in Table 14.5. For fluids with higher sonic velocities and for thinner pipe walls, the minimum value of T_{Lf_0} shifts toward the ring frequency f_r [see 14.59].

The total transmission loss coefficient for the valve is now expressed in the form

$$T_L = T_{Lf_0} + \Delta T_{Lf_p} \qquad \text{dB} \tag{14.52}$$

where the correction ΔT_{Lf_p} (dB) is determined by the peak noise frequency according to

$$\Delta T_{Lf_p} = 20 \times \log \frac{f_0}{f_p} \qquad f_p \leq f_0 \tag{14.53}$$

$$= 13 \times \log \frac{f_p}{f_0} \qquad f_p \leq 4f_0 \tag{14.54}$$

$$= 20 \times \log \frac{f_p}{4f_0} + 7.8 \qquad f_p > 4f_0 \tag{14.55}$$

The peak frequency is estimated from[24]

$$f_p = \frac{0.2 M_j c_0}{D_j} \qquad M_j < 1.4 \tag{14.56}$$

$$= \frac{0.28 c_0}{D_j \sqrt{M_j^2 - 1}} \qquad M_j > 1.4 \tag{14.57}$$

where c_0 = speed of sound at *vena contracta*, m/s
D_j = jet diameter at valve orifice

$$M_j = \left[\frac{2}{\gamma - 1}\left\{\left(\frac{P_1}{\alpha P_2}\right)^{(\gamma-1)/\gamma} - 1\right\}\right]^{1/2} \tag{14.58}$$

$$f_0 \approx \frac{5000}{4\pi D_i} \quad \text{for steel pipes and air} \tag{14.59}$$

For a gas other than air f_0 must be multiplied by $c_{0,\text{gas}}/c_{0,\text{air}}$.

The diameter D_j is difficult to determine because of the complex flow geometry in many valves. A reasonable approximation is to use the hydraulic diameter D_H as the jet diameter and make use of a valve style modifier F_d (see Table 14.3) to obtain

$$D_j \approx 4.6 \times 10^{-3} F_d \sqrt{C_v F_L} \quad \text{m} \tag{14.60}$$

This formula is *not* applicable if jet portions combine downstream (usually at higher pressure ratios). This is common with multiple-hole valve cages,[25] short-stroke parabolic valve plugs in the "flow to close" direction, and butterfly valves at larger openings.

The external sound level (measured at 1 meter from the wall) is now given by

$$L_a = 5 + L_{pi} - T_L + L_g \quad \text{dBa} \tag{14.61}$$

where L_g is a correction for fluid velocity in the pipe:

$$L_g = -16 \times \log\left(1 - \frac{1.3 \times 10^{-5} P_1 C_v F_L}{D_i^2 P_2}\right) \tag{14.62}$$

The laboratory test data shown in Fig. 14.21, taken with air on a DN150 globe valve, exhibits good agreement between predicted and measured external valve sound levels using T_L calculated from Eq. (14.52). In the absence of unusual phenomena such as jet "screech," Eq. (14.61) is found to be accurate to within ± 3 dB assuming that all the valve coefficients are well established.

Methods of Valve Noise Reduction

The two basic variables controlling the valve noise are the jet Mach number M_j and the internal peak frequency f_p. In subsonic flows lower jet velocities can be achieved by use of a valve trim that is less streamlined ($F_L \geq 0.9$, say).

At higher pressure ratios the only remedy is to use orifices or resistance paths in series, so that each operates subsonically. Such a "zig-zag" path valve is shown in Fig. 14.22. For example, for natural gas with an inlet pressure of 10000 kPa that is to be throttled down to 5000 kPa and assuming a valve pressure recovery

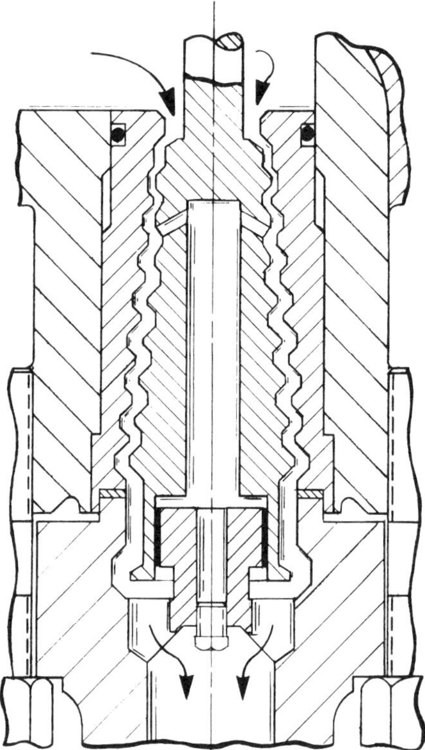

Single flow area,multi-step valve plug

Fig. 14.22 High-pressure reducing valve using multiple step trim with single flow path. Ideally each step will have subsonic orifice velocity.

factor $F_L = 0.9$, the internal pressure ratio P_1/P_0 for a single-stage reduction would be $P_1/P_2\alpha = 2.35 : 1$. However, using the trim from Fig. 14.22 with 10 reduction steps, the orificial pressure ratio for each step now is only $1.603 : 1$ with an acoustical power coefficient of only 1.1×10^{-7}. Ignoring the partial addition of 10 separate sound power sources, this represents a noise reduction of about 30 dB.

A less costly way to reduce "audible" valve noise is to increase the peak frequency by using multiorifice trim, as shown in Figure 14.23. The throttling is still single stage, but the jets are divided into a number of parallel ports. This reduces the value of the valve style modifier $F_d \approx 1/\sqrt{n_0}$, where n_0 is the number of equally sized parallel ports. Since F_d is proportional to jet diameter D_j, f_p can be shifted to higher levels. Thus, if $n_0 = 16$, as in Fig. 14.23, $F_d = 0.25$ and f_p is four times higher than for a single orifice ($F_d = 1$) with the same total flow area. If the peak frequency occurs in the mass-controlled region, an overall reduction in the external (audible) sound level of $\Delta T_{Lf_p} = 20 \times \log(f_{p_1}/f_{p_0}) \equiv 20 \times \log(1/0.25) \approx 12$ dB is achieved.

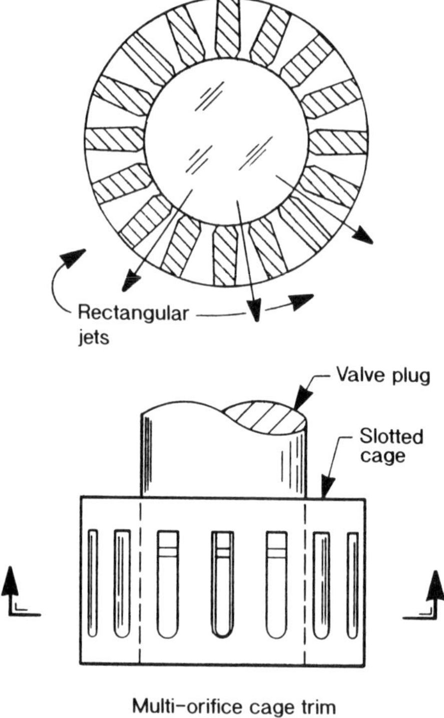

Fig. 14.23 Slotted valve cage subdivides single-step flow path in 16 separate openings ($F_d = 0.25$) to increase peak frequency and thereby transmission loss.

The beneficial effects of a reduction in velocity combined with higher transmission loss caused by increased frequency are obtained with the elaborate arrangement shown in Fig. 14.24. This combines the multistep and multipath approach by use of a layer of disks having individually cast or etched channels.

A more economical solution is to couple a static downstream pressure-reducing device, such as a multihole restrictor, with a throttling valve with a fluted disk as indicated in Fig. 14.25. The fluted disk generates multiple jets with increased noise frequencies, although most of the energy is converted by the static pressure plate. At maximum design flow the valve is sized to have a low pressure ratio of 1.1 or 1.15 and the remainder of the pressure reduction occurs across the static plate having inherent low-noise throttling due to multistage, multihole design.

Placing a silencer downstream of a valve is not cost effective because a good portion of the sound power travels upstream or radiates through the valve body and actuator, yielding noise reductions that typically do not exceed about 10 dB. Similar remarks apply to acoustic insulation of the pipe wall, where attenuations are limited to about 15 dB. The effectiveness of the various abatement procedures

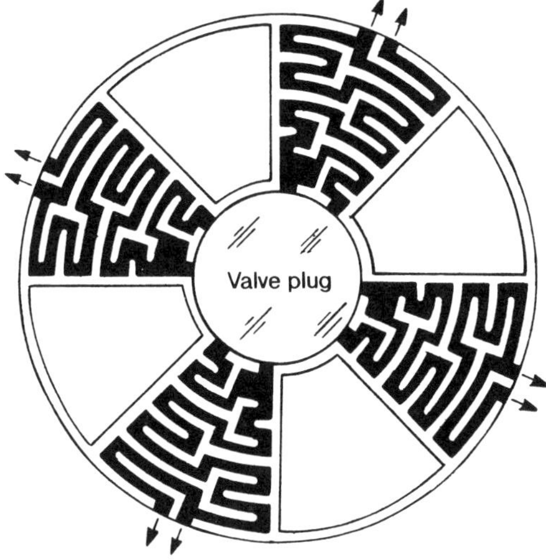

Fig. 14.24 Labyrinth-type flow path cast or machined into each of a stack of metal plates surrounding a valve plug to provide a combination of multistep and multichannel flow paths in order to reduce throttling velocity and increase peak frequency.

is summarized below:

1. One-inch-thick pipe insulation: 5–10-dB reduction
2. Doubling thickness of pipe wall: 6-dB reduction
3. Silencer downstream: 10-dB reduction
4. Silencers upstream and downstream: 20-dB reduction

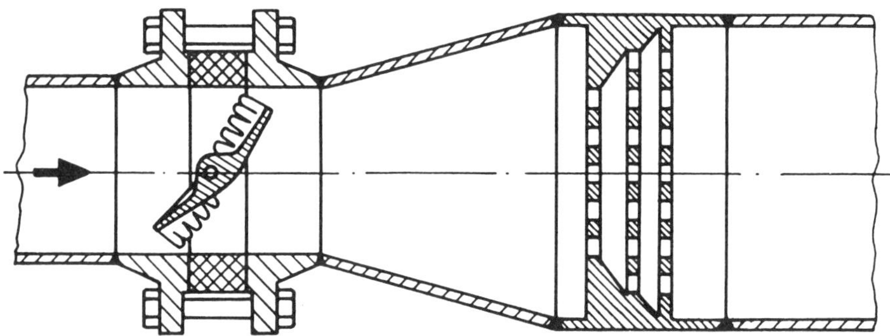

Fig. 14.25 Modulating fluted butterfly valve cooperates with triple-stage static resistance plate for purposes of pressure reduction. *Note*: Static plates see 90% of pressure drop at maximum design flow. Balance is handled by the valve.

5. Multiport resistance plate downstream of valve: 15–20-dB reduction (Use only 5–10% of valve inlet pressure as valve ΔP at maximum flow.)
6. Special low-noise valve: 15–30-dB reduction

REFERENCES

1. M. J. Lighthill, "On Sound Generated Aerodynamically. Part I: General Theory," *Proc. Roy. Soc. Lond.* **A211**, 564–587 (1952).
2. M. S. Howe, "Contributions to the Theory of Aerodynamic Sound, with Application to Excess Jet Noise and the Theory of the Flute," *J. Fluid Mech.* **71**, 625–673 (1975).
3. Society of Automotive Engineers, "Gas Turbine Jet Exhaust Noise Prediction," Report No. SAE ARP 876C, Society of Automotive Engineers, Warrendale, PA, 1985.
4. J. R. Stone, "Prediction of In-Flight Exhaust Noise for Turbojet and Turbofan Engines," *Noise Control Eng.* **10**, 40–46 (1977).
5. R. K. Amiet, "Refraction of Sound by a Shear Layer," *J. Sound Vib.* **58**, 467–482 (1978).
6. C. L. Morfey and B. J. Tester, "Noise Measurements in a Free Jet Flight Simulation Facility: Shear Layer Refraction and Facility-to-Flight Corrections," *J. Sound Vib.* **54**, 83–106 (1977).
7. A. L. Laganelli and H. Wolfe, "Prediction of fluctuating pressure in attached and separated turbulent boundary layer flow," American Institute of Aeronautics and Astronautics, Washington, D.C., Paper 89, 1989, p. 1064.
8. M. S. Howe, "The Role of Surface Shear Stress Fluctuations in the Generation of Boundary Layer Noise," *J. Sound Vib.* **65**, 159–164 (1979).
9. G. Schewe, "On the Structure and Resolution of Wall Pressure Fluctuations Associated with Turbulent Boundary Layer Flow," *J. Fluid Mech.* **134**, 311–328 (1979).
10. M. S. Howe, "Surface Pressures and Sound Produced by Low Mach Number Turbulent Flow over Smooth and Rough Walls," *J. Acoust. Soc. Am.* **90**, 1041–1047 (1991).
11. M. M. Sevik, "Topics in Hydroacoustics," in *Proceedings of the IUTAM Symposium on Aero- and Hydroacoustics*, Berlin: Springer-Verlag, Berlin, 1986.
12. D. M. Chase, "The Character of the Turbulent Wall Pressure Spectrum at Subconvective Wavenumbers and a Suggested Comprehensive Model," *J. Sound Vib.* **112**, 125–147 (1987).
13. H. H. Heller and S. E. Widnall, "Sound Radiation from Rigid Flow Spoilers Correlated with Fluctuating Forces," *J. Acoust. Soc. Am.* **47**, 924–936 (1970).
14. P. A. Nelson and C. L. Morfey, "Aerodynamic Sound Production in Low Speed Flow Ducts," *J. Sound Vib.* **79**, 263–289 (1981).
15. M. Hubert, "Untersuchungen ueber Geraeusche durchstroemter Gitter," Ph.D. Thesis, Technical University of Berlin, 1970.
16. M. S. Howe, "The Influence of Vortex Shedding on the Generation of Sound by Convected Turbulence," *J. Fluid Mech.* **76**, 711–740 (1976).
17. W. K. Blake, *Mechanics of Flow-Induced Sound and Vibration*, Vol. 2, *Complex Flow-Structure Interactions*, Academic, New York, 1986.
18. T. F. Brooks, D. S. Pope, and M. A. Marcolini, "Airfoil Self-Noise and Prediction," NASA Reference Publication No. 1218, 1989.

19. V. A. Carucci and R. T. Mueller, "Acoustically Induced Piping Vibration in High Capacity Pressure Reducing Systems," Paper No. 82-WA/PVP-8, American Society of Mechanical Engineers, New York, 1982.

20. H. D. Baumann, "On the Predicition of Aerodynamically Created Sound Pressure Level of Control Valves," Paper No. WA/FE-28, American Society of Mechanical Engineers, New York, 1970.

21. Instrument Society of America, "Flow Equations for Sizing Control Valves," Standard No. ANSI/ISA S75.01-1985, 1985.

22. H. D. Baumann, "Coefficients and Factors Relating to the Aerodynamic Sound Level Generated by Throttling Valves," *Noise Control Eng. J.* **22**, January/February 1984, pp. 1–6.

23. Instrument Society of America, "Control Valve Aerodynamic Valve Noise Prediction," Standard No. ANSI/ISA S75.17, 1989.

24. H. D. Baumann, "A Method for Predicting Aerodynamic Valve Noise Based on Modified Free Jet Noise Theories," Paper No. 87-WA/NCA-7, American Society of Mechanical Engineers, New York, 1987.

25. C. Reed, "Optimizing Valve Jet Size and Spacing Reduces Valve Noise," *Control Eng.* **9**, 63–64, 1976.

CHAPTER 15

Active Noise Control

L. J. ERIKSSON

Nelson Industries, Inc.
Stoughton, Wisconsin

15.1 INTRODUCTION

Active noise control is an old concept that has generated increased interest over the past 10–15 years. Using the principle of destructive interference of waves, an inverse pressure wave is generated to attenuate an undesired noise. In order to achieve substantial cancellation, the canceling source must produce, with great precision, an equal amplitude but inverted replica of the signal to be canceled. Only with the development of adaptive digital signal-processing theory and hardware has it become possible to maintain these relationships automatically to the desired precision without continued intervention by a human operator.

In this chapter, the key developments in active noise control are reviewed to enable the reader to understand better the current state of knowledge in this field and to apply this technology to solve noise problems. Although most of the comments may be generalized to apply to a wide variety of applications, the discussion will emphasize the use of active noise control for duct silencing because of the extensive work done in this area. In addition to the substantial number of references that are provided, more complete listings are available in several reviews[1-3] and in a very complete bibliography compiled by Guicking.[4] The topic of active vibration control is not treated in this chapter. The reader is referred to the literature.[1-4]

Noise and Vibration Control Engineering: Principles and Applications, Edited by Leo L. Beranek and István L. Vér.
ISBN 0-471-61751-2 © 1992 John Wiley & Sons, Inc.

15.2 EARLY HISTORY

The concept of actively generating a sound wave to cancel undesired noise in a duct was described in an early patent by Lueg.[5] A microphone detects the undesired noise and provides an input signal to an electronic system that drives a loudspeaker, as shown in Fig. 15.1. The microphone-to-loudspeaker spacing and electronics are adjusted so that the sound wave generated will destructively interfere with the noise propagating in the duct between the canceling loudspeaker and the receiver. The entire process depends upon the relatively slow propagation speed of the sound wave compared to the rapid processing of the electrical signal, enabling the generation of an inverted waveform before the undesired noise at the sensing microphone reaches the canceling loudspeaker. The amplitude and phase response of the electronics must be adjusted with great accuracy to obtain good cancellation.

Following Lueg's basic patent, interest in active noise control has fluctuated many times over the years. During work at the Harvard University Electro-Acoustic Laboratory in 1941, Wallace[6] proposed an active system for noise cancellation at the ear for possible use with communication systems. In the early 1950s, Olson worked on a variety of applications of active concepts for noise and vibration control.[7] Later in the same decade, Conover described an active system for use in reducing noise from power transformers.[8] All of these early projects used an analog electronic system that was manually adjusted to produce optimum results. Instability because of drift in the system parameters as well as feedback from the loudspeaker to the input microphone resulted in performance problems.

In the late 1960s, Onoda and Kido[9] published work on active noise control systems that were automatically controlled to maintain optimum performance. This work anticipated the direction of more modern schemes using adaptive digital signal-processing techniques.

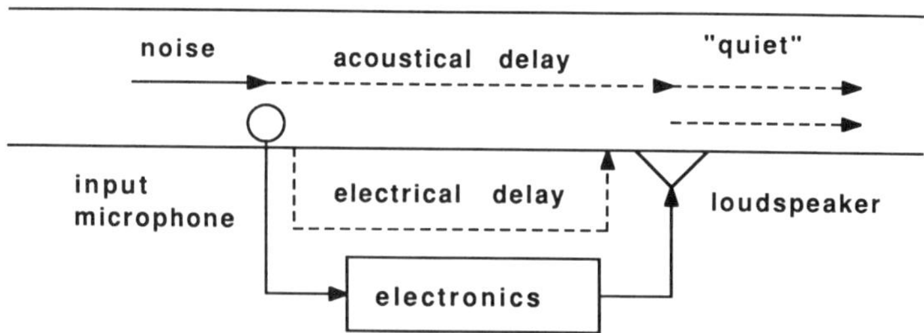

Fig. 15.1 Schematic diagram of active noise control concept in a duct. The electrical delay is much less than the acoustical delay, enabling the microphone-to-loudspeaker spacing and electronics to be adjusted such that the electrically generated sound destructively interferes with the undesired noise.

15.3 ACOUSTICS

It is interesting to note that despite the shortcomings of various nonadaptive approaches to active noise control and the early suggestions by Conover[8] as well as Onoda and Kido[9] of the need for automatic control, most early detailed theoretical studies of active noise control did not deal with adaptation but focused instead on the complex acoustical issues involved in the active cancellation of three-dimensional sound fields.[10] It was not until the 1980s that comprehensive analyses of simpler configurations such as a duct[11–13] or enclosed sound field were published.[14, 15]

In Lueg's basic configuration for silencing in a duct, an input microphone is used to detect the undesired noise, which is then processed electronically and used to drive a loudspeaker to create the sound-canceling waveform. There are two significant problems with this scheme. First, the electronics must account for the nonuniform amplitude and phase response of the transducers, filters, and amplifiers used in the system. Second, the effect of acoustical feedback from the output loudspeaker to the input microphone must be either minimized or compensated to avoid a potentially unstable response. Many of the systems to be described utilize a combination of special transducer arrangements or unique signal-processing concepts to address either or both of these problems.

A persistent question for many active noise control systems has been the identification of the mechanism of energy loss in the system. In monopole systems involving only a single canceling sound source, the energy in the sound waves is reflected back to the source by the secondary source acting as an acoustical short circuit.[12] The coupled sources see a new effective load impedance. This results in a modified power output that is brought into balance with various loss mechanisms in the acoustical system to produce a standing-wave pattern.[16] Thus, the effect of the active noise control system is to reflect energy in such a way that the primary source is unloaded. In this process, a standing wave is established between the secondary and primary sources, and the acoustical cancellation takes place beyond the secondary source. A dipole secondary source incorporating two loudspeakers to form a directional source has the additional ability to absorb acoustical energy without reflecting energy back to the primary source.[16]

15.4 FIXED-FILTER SYSTEMS

A wide variety of fixed-filter active noise control systems has been devised for use in ducts. The basic monopole system described by Lueg is shown in Fig. 15.2*a*.[17, 18] The Chelsea monopole,[19] shown in Fig. 15.2*b*, uses electrical feedback to cancel the effect of acoustical feedback received by the input microphone by subtracting a delayed electrical signal obtained at the speaker input from the input microphone signal. A special case of this configuration, known as the tight-coupled monopole,

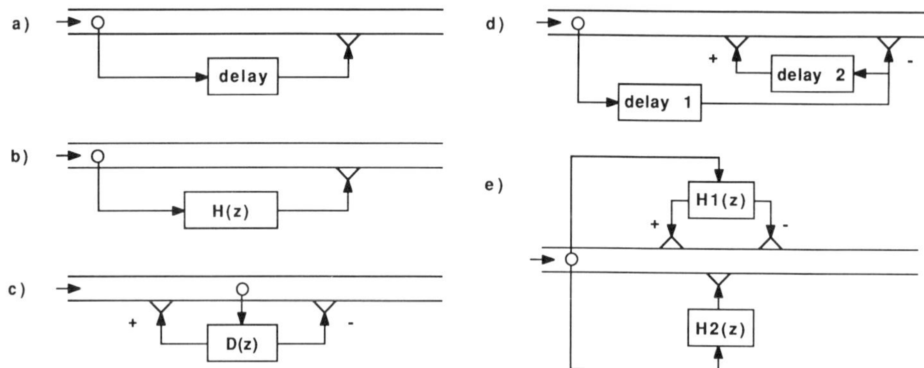

Fig. 15.2 Fixed filter noise control systems for ducts: (*a*) basic concept of Lueg with delay *d*.[5] (*b*) Chelsea monopole uses transfer function, $H(z)$, that electrically compensates for acoustical feedback.[19] Three approaches to using source configurations to reduce acoustical feedback: (*c*) Chelsea dipole.[20] (*d*) Swinbanks system.[21,22] (*e*) Jessel–Mangiante tripole.[23-25].

places the input microphone close to the loudspeaker and uses a high-gain amplifier. This configuration approximates an effective transfer function of -1 from the input microphone to the loudspeaker for the combination of the high-gain amplifier and acoustical feedback.[19]

Other techniques have used special loudspeaker configurations to cancel the feedback at the input microphone. These have included the Chelsea dipole[20] shown in Fig. 15.2*c*, the Swinbanks unidirectional loudspeaker or microphone system[21,22] shown in Fig. 15.2*d*, and the Jessel–Mangiante tripole[23-25] shown in Fig. 15.2*e*. The Chelsea dipole uses a pair of output loudspeakers driven out of phase and spaced one-half wavelength apart. The output of the loudspeakers cancels at the location of the centrally located sensing microphone while providing a noise-attenuating signal in the region away from the noise source. The Swinbanks system is arranged such that feedback from the pair of loudspeakers to the sensing microphone is reduced. This requires the electrical delay between the two out-of-phase loudspeakers to be equal to the propagation time between the speakers. The Jessel–Mangiante tripole system uses a monopole source to cancel the upstream propagation from a dipole source to eliminate feedback and produce a unidirectional source. However, all of these fixed-filter systems have basic limitations. Systems that use a source spacing based on wavelength are limited to a specific frequency range and sound speed, while those based on time delay are limited to a specific sound speed.

15.5 SIGNAL PROCESSING

Many of the recent developments in active noise control have utilized results obtained in signal-processing research over the past 20 years. Linear systems can be characterized by either their transfer function or impulse response, as shown in Fig. 15.3. In analog systems using continuous signals, this means the output of the linear system may be obtained in the time domain as the convolution of the input signal, $x(t)$, and impulse response of the system, $h(t)$, or in the frequency domain as the product of the Laplace transform of the input, $X(s)$, with the transfer function of the system, $H(s)$. In digital systems using sampled signals, the output is similarly obtained as the convolution of the discrete input signal, $x(k)$, and the discrete impulse response of the system, $h(k)$, or as the product of the z transform of the input, $X(z)$, with the discrete transfer function of the system, $H(z)$.[26,27] The digital approach has often come to be preferred due to the availability of powerful, low-cost hardware in the form of digital signal processors to perform these computations.

One of the most common methods for implementing a discrete system response, $h(k)$, is with the tapped delay line structure shown in Fig. 15.4. The scalar output of this structure, $y(k)$, is the sum of a series of products formed by multiplying a numerical coefficient or weight, $W_n(k)$, with a suitably delayed value of the scalar input, $x(k)$. Through proper selection of these coefficients, a wide variety of filter responses can be implemented with this digital filter. This filter is often referred to as a transversal filter, moving average (MA) filter, or finite impulse response (FIR) filter. This structure is easy to implement and is used in the least mean squares (LMS) adaptive filter algorithm developed by Widrow in the 1960s.[27]

In Widrow's algorithm, illustrated in Fig. 15.5, the coefficients or weights of the filter are adapted using a simple algorithm such that the filter output, $y(k)$, attempts to match a desired signal, $d(k)$, and minimize the mean-squared error between them. For the specific case of active noise control in a duct, $x(k)$ would correspond to the acoustical pressure measured by the input microphone, $y(k)$ to the acoustical pressure generated by the loudspeaker, $d(k)$ to the acoustical pressure to be reduced, and $e(k)$ to the acoustical pressure measured by the error mi-

Continuous Systems

IN
x(t)
h(t)
H(s)
OUT
y(t)
y(t) = h(t) * x(t)
Y(s) = H(s) X(s)

Discrete Systems

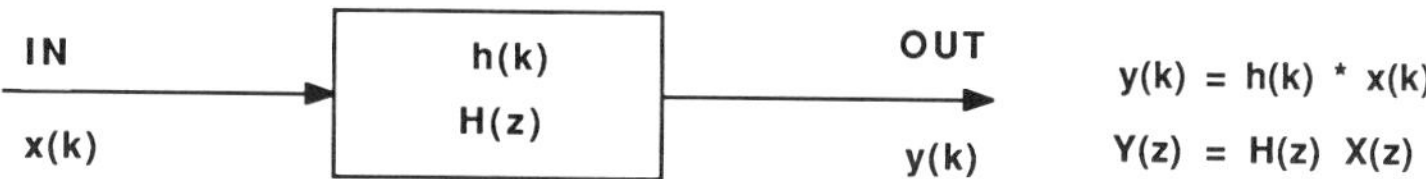

Fig. 15.3 Continuous and discrete signal, linear system analyses are related by digital-to-analog (D/A) and analog-to-digital (A/D) conversion[26]: $h(t)$ and $h(k)$ are impulse responses; $H(s)$ and $H(z)$ are transfer functions.

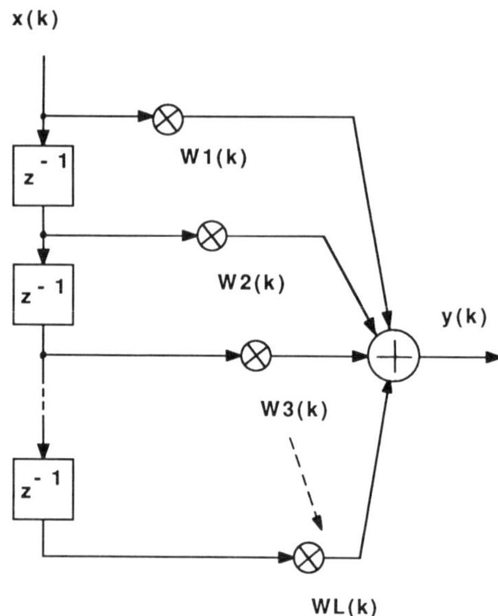

Fig. 15.4 Tapped delay line filter structure suitable for implementing transfer function [$H(z)$ of Fig. 15.3], also known as FIR, MA, or transversal filter[27]: $x(k)$, discrete input; $y(k)$, discrete output; $W_n(k)$, filter coefficient or weight; z^{-1}, delay element.

crophone located on the receiver side of the loudspeaker. The filter output and weight update equations may be written (using vector notation similar to Widrow[27]) as

$$y(k) = \mathbf{W}^T(k)\mathbf{X}(k) \tag{15.1}$$

$$\mathbf{W}(k+1) = \mathbf{W}(k) + 2M\mathbf{X}(k)e(k) \tag{15.2}$$

where $y(k)$ = scalar output at discrete time k
$\mathbf{W}(k)$ = weight vector for transversal filter
$\mathbf{W}^T(k)$ = transpose of $\mathbf{W}(k)$
$\mathbf{X}(k)$ = input vector formed by tapped delay line from scalar input $x(k)$
M = step size parameter (controls size of coefficient adjustment)
$e(k)$ = scalar error signal, $= d(k) - y(k)$
$d(k)$ = scalar desired signal

Thus, the coefficients of the filter are updated by a term proportional to the product of the input signal associated with each coefficient times the error signal. The updated coefficient, $W_n(k+1)$, is used in the FIR filter on the next iteration as indicated by the delay element leading to each coefficient.

The use of this adaptive filter for active noise control is complicated by the fact that the electrical input signal must be obtained from the acoustical pressure using the input microphone, the electrical error signal must be obtained from the acous-

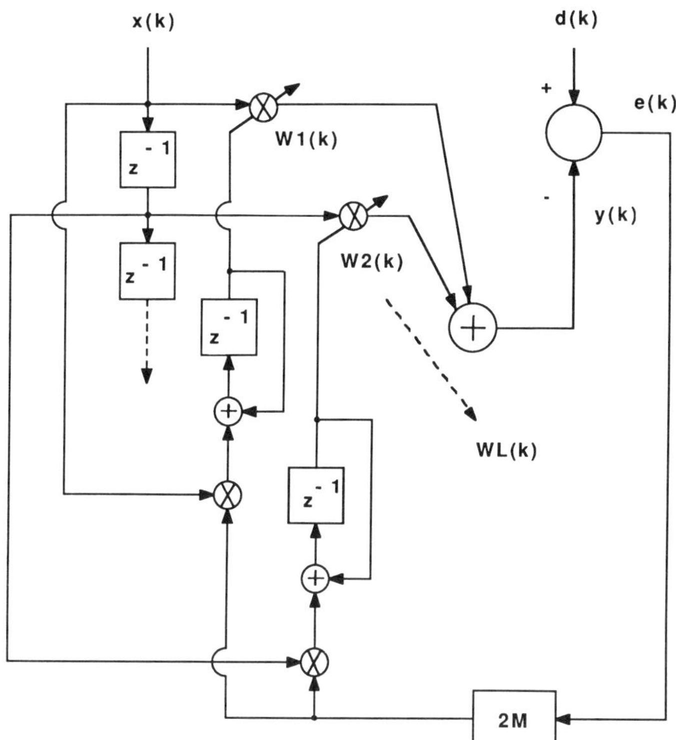

Fig. 15.5 Basic structure of LMS adaptive filter algorithm[27] described by Eqs. (15.1) and (15.2): $x(k)$, discrete input; $y(k)$, discrete output; $W_n(k)$, filter coefficient or weight; z^{-1}, delay element; $d(k)$, desired signal; $e(k)$, error signal.

tical pressure using the error microphone, and the sound-canceling acoustical pressure must be generated from the electrical output signal using the loudspeaker. This results in three acoustical/electrical transfer functions, as shown in Fig. 15.6. In addition, anti-aliasing and anti-imaging filters must usually be used on the signals from the microphones and to the loudspeaker, respectively, resulting in additional electrical transfer functions. The adaptive filter coefficients must converge to the proper value to minimize the error signal despite the presence of the transfer functions associated with these transducers and filters.

It has been shown previously that transfer functions placed in the input path of the adaptive filter (associated with the input microphone and filter for active noise control systems) do not affect the convergence of the LMS algorithm.[27] For this reason, they will not be considered in the following discussion. However, as shown by Widrow[27] and Burgess,[28] transfer functions placed in the output or error path of the adaptive filter (associated with the output loudspeaker and filters, acoustical path from the loudspeaker to the error microphone, and error microphone and filters for the active noise control system) will prevent the adaptive filter from converging properly unless these transfer functions are also placed in the input

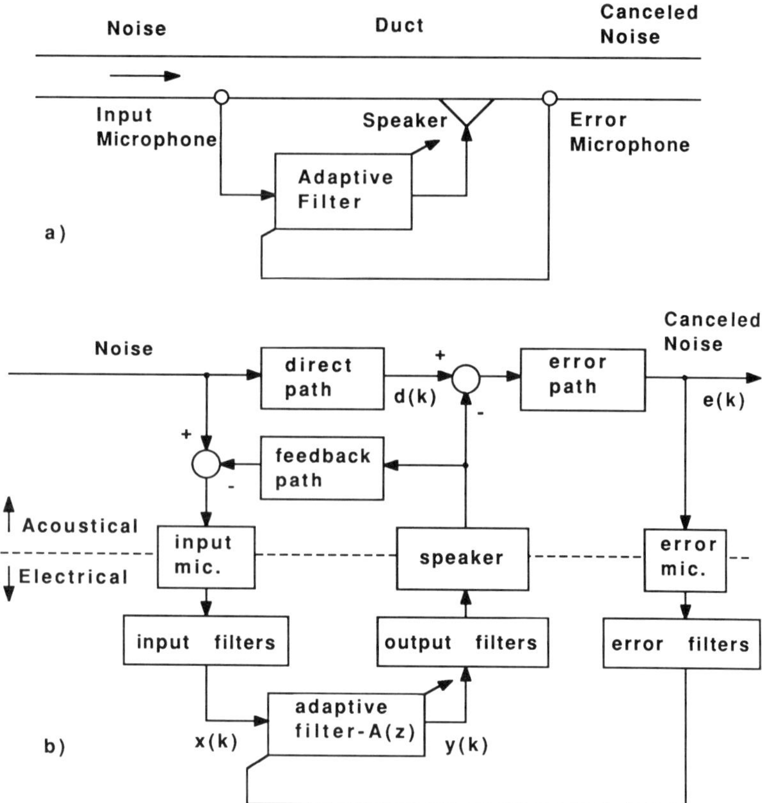

Fig. 15.6 Active noise control system in a duct including adaptive filter and transducers: (*a*) schematic diagram; (*b*) block diagram.

lines leading to the multipliers with the error signal. Widrow[27] has referred to the LMS algorithm used with a filter in these input lines to compensate for transfer functions following the adaptive filter as the *filtered-X algorithm* because of the need for the filtering of the input *X*.

An active noise control system using this approach has been discussed by Burgess[28] and is shown in Fig. 15.7. The transfer functions associated with the loudspeaker (and, e.g., its filters), $S(z)$, and error path (including, e.g., the error microphone and its filters), $E(z)$, must be determined and placed in the input lines to the multipliers with the error signal as shown.

Eriksson et al.[3,29,30] have recently discussed the use of a recursive adaptive filter that uses an infinite impulse response (IIR) filter structure with a recursive least mean squares (RLMS) algorithm to cancel the undesired noise and compensate for acoustical feedback. The IIR filter structure combines two tapped delay lines of the form shown in Fig. 15.4 into a single filter composed of a direct element, A, and recursive element, B, as shown in Fig. 15.8. Such an IIR filter

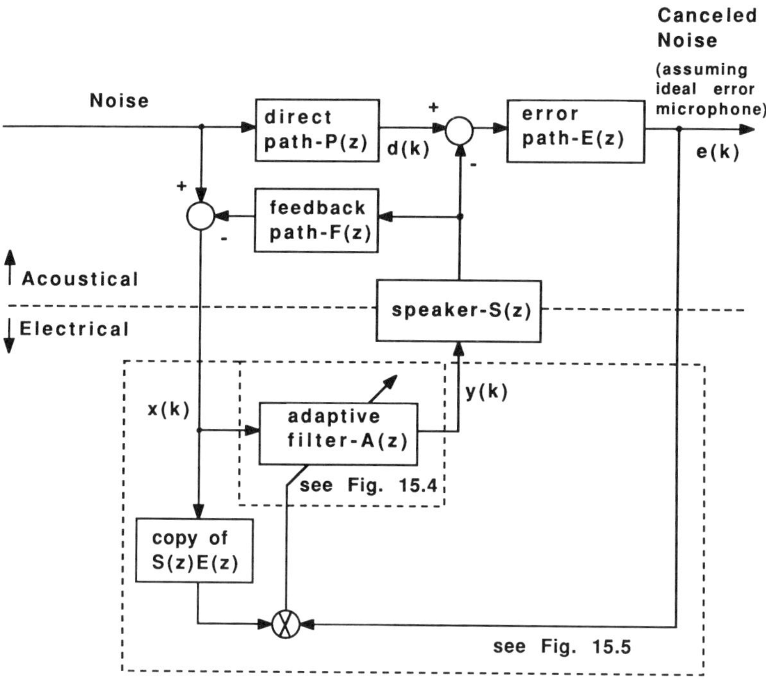

Fig. 15.7 Active noise control system described by Burgess[28] using LMS algorithm in presence of transfer functions associated with loudspeaker, $S(z)$, and error path acoustics, $E(z)$: $P(z)$, the direct path, $F(z)$, feedback path; $A(z)$, adaptive LMS filter element; $S(z)E(z)$, transfer function equal to the product of the transfer functions $S(z)$ and $E(z)$.[28]

structure is more suitable than the FIR filter structure for use in systems that include transfer functions containing poles and zeros because of the ability of the recursive element to provide the infinite time response associated with the poles in the transfer functions using a short delay line in a recursive arrangement. A comparable FIR filter structure would require a very large number of coefficients in its tapped delay line to approximate the same result. The ability to represent poles easily is especially useful for active noise control systems since acoustical feedback can be shown to introduce poles into the response of the active noise control system.[3,29,30] This can be seen in Fig. 15.8, where the poles introduced by the acoustical feedback path, F, are compensated by the poles associated with the recursive element, B, of the IIR filter structure. The zeros of the IIR filter structure are provided by the direct element, A. Both elements of the filter are adapted together using an adaptive algorithm similar to that described in Fig. 15.5 for the LMS algorithm. The coefficients of each element are updated by a term proportional to the product of the input signal associated with the coefficient times the error signal. The combined algorithm is known as a RLMS algorithm.[3,29,30]

In order to compensate for the effects of transfer functions following the adaptive filter, the ideas of Burgess and Widrow have been extended by Eriksson

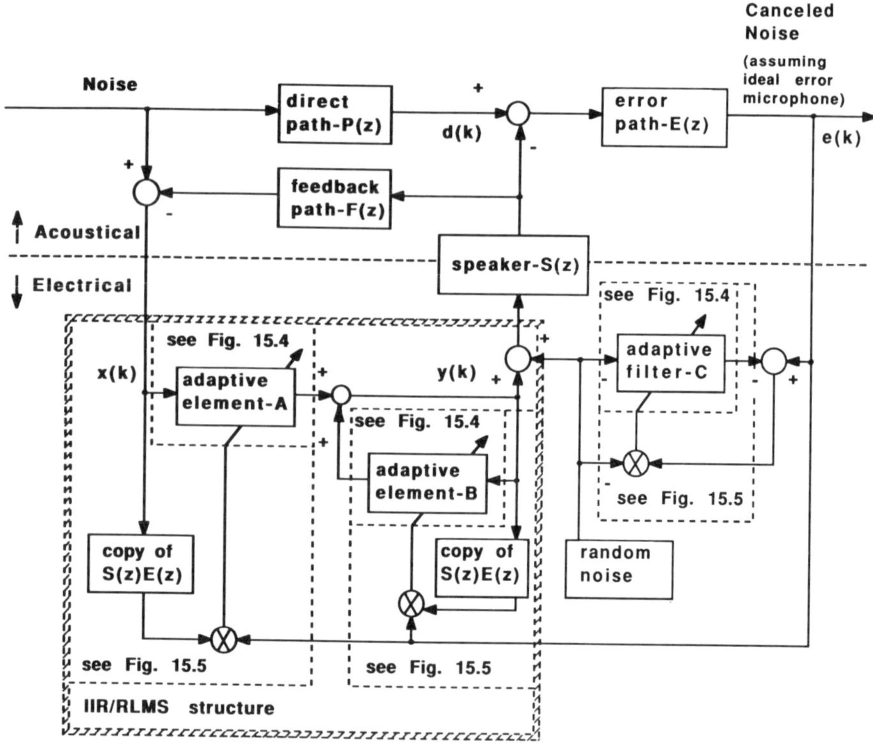

Fig. 15.8 Active noise control system described by Eriksson et al. using on-line modeling of loudspeaker, $S(z)$, and error path, $E(z)$, in RLMS model with acoustical feedback to form fully adaptive active attenuation system[29-32]: $P(z)$, direct path; $F(z)$, feedback path; $A(z)$, $B(z)$, RLMS filter elements; $C(z)$, LMS filter element.

et al.[31,32] to form the *filtered-U algorithm*.[65] The adaptive filter, C, determines the response of the loudspeakers and error path transfer functions during cancellation using a low-level, independent, random noise supplied to the loudspeaker by a random-noise generator.[31,32] A copy of the transfer function represented by the coefficients of the adaptive filter, C, is placed in the inputs to the multipliers with the error signal for both the direct and recursive elements. This supplementary adaptive process continues during cancellation to compensate for changes in loudspeaker characteristics or error path response and to ensure continued convergence of the RLMS algorithm as these characteristics change. This approach will be discussed further later in connection with Fig. 15.9*f*.

15.6 ADAPTIVE FILTER SYSTEMS

As discussed in the previous section, fixed-filter systems are dependent upon an accurate knowledge of the transfer functions present in the acoustical system. Because of changes in temperature, flow velocity, and other system parameters, these quantities are time varying and difficult to determine precisely. In addition, the nonideal, time-varying characteristics of the actual microphones, amplifiers, and loudspeakers due to changes in temperature and other physical conditions also require continual adjustment of fixed-filter systems to maintain good performance. A better solution for many applications is to use an adaptive system that has the ability to determine its parameters automatically and to update these parameters continually so that optimal performance is achieved for all system conditions.

As mentioned previously, Onoda and Kido developed an adaptive system for the control of transformer noise in the late 1960s.[9] Each harmonic of the transformer noise was filtered to modify its phase and amplitude in response to an error signal detected downstream of the noise-canceling loudspeakers. A similar arrangement for the cancellation of periodic noise in a duct using a filtered signal from the primary source is shown in Fig. 15.9*a*. Kido also described an approach for the cancellation of nonperiodic noise through the use of an input microphone in a duct[33] that is shown as an alternate input in Fig. 15.9*a*. This input signal is then passed through a filter that contains either the transfer function $H(z)$ or impulse response $h(k)$ of the acoustical system. Chaplin et al. have described a method for determining this impulse response through the use of impulse excitation of the system.[34]

Chaplin also developed an approach for the cancellation of periodic noise where

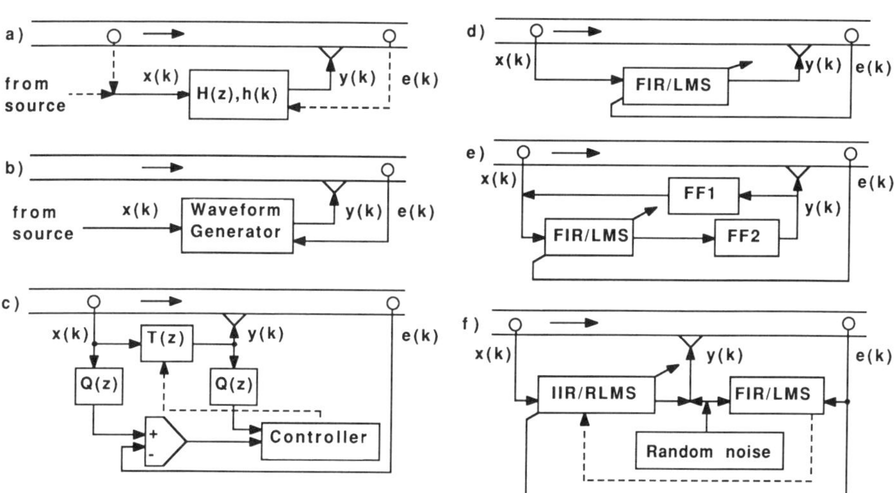

Fig. 15.9 Adaptive active noise control systems for ducts: (*a*) the adaptive input may be either from duct microphone or directly from source[33]; (*b*) or by using a waveform generator triggered from source[35]; (*c*) or by using controller with error signal and model input and output filtered by $Q(z)$ to determine coefficients of $T(z)$[36,37]; (*d*) or by using LMS algorithm to determine coefficients of FIR adaptive filter structure[28]; (*e*) or by using fixed filters (FF) to compensate for feedback (FF_1) and loudspeaker response (FF_2)[39]; (*f*) or, finally, by using IIR adaptive filter with RLMS algorithm and on-line speaker/error path modeling as shown in Fig. 15.8.[29-32]

the digital filter described above was replaced with a waveform generator.[35] This waveform generator synthesizes a periodic waveform in response to a synchronizing signal from the noise source that is optimized using an error signal from the duct, as shown in Fig. 15.9*b*.

Ross developed an approach to noise cancellation, shown in Fig. 15.9*c*, in which the coefficients of a digital filter are determined using an independent controller that responds to the error signal combined with a filtered form of the model input and output. The two controller inputs are processed using correlation techniques to obtain a least-squares estimate of the desired digital filter response.[36,37]

In contrast to the Ross method, which requires a matrix inversion, Burgess described the use of the Widrow LMS adaptive algorithm with an FIR digital filter structure for active noise control in a duct, as shown in Fig. 15.9*d*.[28] As discussed previously, this method requires knowledge of the transfer functions $S(z)$ and $E(z)$, shown in Fig. 15.7, and does not explicitly account for acoustical feedback from the loudspeaker to the input microphone.

Davidson et al. have described a system that uses adaptive control in the forward path to cancel the undesired noise and a fixed electrical path to cancel the acoustical feedback.[38]

Warnaka et al.[39] have described a system in which a fixed filter (FF_1), determined using the LMS algorithm before cancellation, is used to compensate for acoustical feedback, as shown in Fig. 15.9*e*. A second fixed filter (FF_2), also determined using the LMS algorithm before cancellation, is used to convert the loudspeaker/error path transfer functions (i.e., the transfer functions between the adaptive filter output and the error microphone, including the output filters, power amplifier, loudspeaker, and error microphone) following the adaptive filter to a pure delay. This delay must then be placed in series with the inputs to the multipliers in the weight update procedure shown in Fig. 15.7.

Takahashi et al.[40] and Hamada et al.[41] have described an adaptive system in which the feedback is compensated through use of input and error microphones that are spaced an equal propagation time from the canceling loudspeaker. Assuming that the sound from the loudspeaker is measured equally by both microphones, subtraction of these signals provides an input signal to the adaptive filter that is free from acoustical feedback. Loudspeaker and error path transfer functions are obtained prior to system operation and fixed during cancellation.

The approach described in the previous section by Eriksson et al.[29-32] utilizes a new method for dealing with acoustical feedback and loudspeaker/error path modeling, as shown in Fig. 15.9*f*. Instead of an FIR filter structure, an IIR structure is used with the RLMS adaptive algorithm to simultaneously cancel the undesired noise and compensate for acoustical feedback.[29,30] A low-level, independent, random noise is used with a second adaptive filter to continuously model the transfer functions of the loudspeaker and error path.[31,32] This is comparable to the determination on a continuous basis of the $S(z)$ and $E(z)$ that are required in the FIR approach using an LMS algorithm that was described by Burgess. The IIR adaptive filter combined with continuous modeling of the loudspeaker and error path transfer functions results in a system capable of rapidly responding to changes in the source, acoustical system, or transducer characteristics.

15.7 APPLICATION CONSIDERATIONS

Active noise control is most advantageous at low frequencies because it is not dependent on large structures to create impedance changes or long lengths to create destructive interference as are passive silencers. The requirement for the sample frequency of a digital system to be at least twice the highest frequency of interest is also more easily met for low-frequency noise. In addition, most work on active noise control in ducts to date has been at low frequency to avoid problems with higher order modes. Attenuation of both low- and high-frequency sound may be best accomplished through use of a hybrid silencer that consists of an active low-frequency section and a passive high-frequency section.[42,43]

Proper selection and application of the secondary loudspeaker source is essential for use in an active noise control system.[44] The loudspeaker must be capable of producing the acoustical output required to maintain proper cancellation of the undesired noise source. Due to the high levels of low-frequency noise produced by sources such as large industrial fans, this can require very large, high-power amplifiers. High-pass filters may be required to prevent very low frequency energy from overdriving the loudspeaker and causing premature failures. These filters complicate the overall control process for the electronic controller.[45]

Microphone placement is also a critical factor. The input microphone should not be placed at a node of any standing wave that may be present before or during cancellation. The error microphone should also avoid nodal locations before cancellation. In addition, causality considerations require that the input microphone be placed far enough upstream from the sound-canceling source to allow sufficient time for proper electronic processing of the input signal. For applications involving high-velocity gas flow, turbulent pressure fluctuations at the microphones may be a problem. These may be reduced through the use of antiturbulence probe tubes[46,47] or through the use of signal processing.[48]

Although most work on active noise control systems in ducts has focused on plane-wave propagation, higher order modes can become a problem for large duct dimensions or higher frequencies. This problem may be solved either through the use of a partitioned duct or through the use of multiple-channel systems.[49] Under certain circumstances, single-channel controllers can be used to cancel complex modes in enclosed spaces through the use of appropriate transducer locations and signal processing.[50–52]

Active noise control offers a method for duct silencing that is often considerably easier to install and does not result in the flow restriction normally associated with a passive silencer. This can result in significant energy savings[53] and may avoid the need to modify or replace the fan. These considerations can be especially significant for retrofit applications.

15.8 CASE HISTORIES

The use of active noise control for silencing discharge noise from large industrial fans is one application that has recently received considerable attention.[54–56] For example, a typical installation was on an 8400-ft^3/min centrifugal fan that discharged through a 23-in.-diameter duct. Two 18-in. diameter loudspeakers were used with an input microphone located 12 ft upstream of the loudspeakers and error microphone located near the outlet of the duct. Antiturbulence probe tubes were used on both microphones. The controller used the adaptive system described by Eriksson et al.[29–32] and featured on-line feedback compensation using an IIR adaptive filter and loudspeaker and error path compensation using an independent random noise source, as shown in Figs. 15.8 and 15.9*f*. Typical results before and after cancellation are shown in Fig. 15.10*a*. The noise reduction is shown in Fig. 15.10*b*. The natural tendency of these systems to smooth the spectrum and to eliminate sharp peaks is clearly evident. Such results mean quieter and less annoying spectra. Changes in fan speed or output are taken into account on a continuous basis since the adaptive filter and acoustical system receive the same input. In addition, the adaptive nature of the system allows it to respond automatically to changes in temperature, loudspeaker characteristics, and other system parameters.

The silencing of intake or discharge noise from fans used in heating, ventilating, and air conditioning systems is another common application for active noise control technology. Active noise control enables low-frequency noise to be reduced using compact systems that are easy to install and add no restriction to flow. Some typical results before and after cancellation with the resulting noise reduction are shown in Fig. 15.11.

Although fan silencing is an application that is receiving considerable attention, there are a number of other applications that have been successfully demonstrated. These include the silencing of gas turbine discharge noise,[57] electric power transformer noise,[58] diesel engine noise inside passenger compartments,[59] and aircraft cabin noise.[60–63]

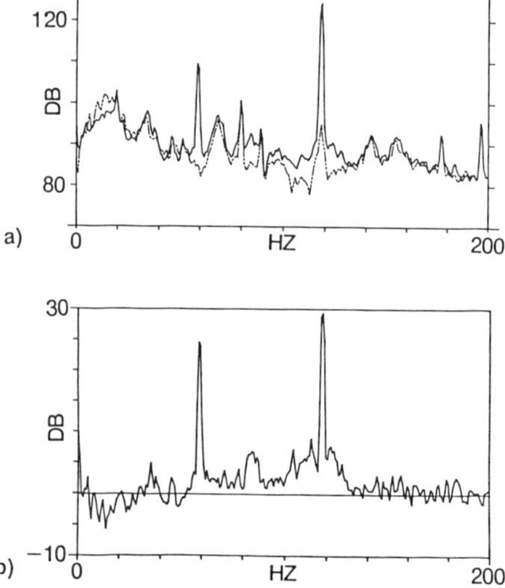

Fig. 15.10 (*a*) Sound pressure spectra at discharge of a centrifugal fan of a dust collector. Discharge duct equipped with active attenuation system of Fig. 15.8. Solid line: active cancellation off. Dotted line: active cancellation on. Mach number M_a = 0.05.[54,55] (*b*) Noise reduction with active attenuation on for system in (*a*) (M_a = 0.05).[54,55] (With permission from *Sound and Vibration* magazine and Digisonix Division, Nelson Industries, Inc.)

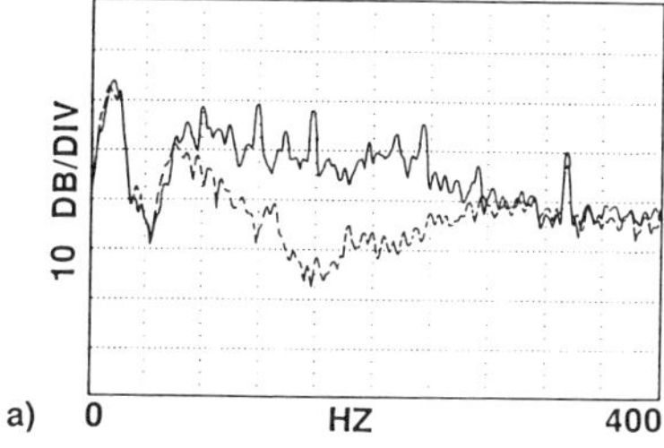

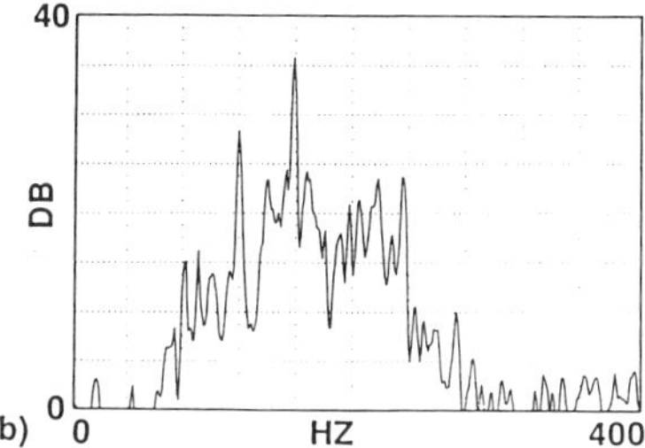

Fig. 15.11 (*a*) Sound pressure spectra in discharge duct of centrifugal fan with active attenuation system of Fig. 15.8. Solid line: active cancellation off. Dotted line: active cancellation on. (*b*) Noise reduction with active attenuation system on for data of (*a*). (Courtesy of Digisonix Division, Nelson Industries, Inc.)

15.9 FUTURE RESEARCH DIRECTIONS

The future of active noise control revolves around the development of improved algorithms, transducers, and processors.[3,64] At this time, relatively limited use has been made of more sophisticated adaptive filter algorithms that offer the potential of faster convergence, greater noise rejection capabilities, and multichannel control. For many applications, transducer performance is a serious limiting factor. In particular, there is a need for high-output, low-cost sound sources that are resistant to a variety of environmental problems, including temperature, flow, dirt, corrosion, and water. Finally, improved processor capabilities will allow current algorithms to be used more effectively and improved algorithms to be implemented.[65] Faster cycle times, more on-chip memory, and improved input–output hardware would all contribute to improved system performance.

15.10 SUMMARY

After many decades of promise, active noise control is beginning to achieve significant breakthroughs to practical applications. This progress has resulted from the development of powerful digital signal-processing concepts and processors. Current applications are utilizing active noise control for reducing low-frequency noise in a cost-effective manner while offering desirable features such as low-flow restriction in duct silencing. Widespread use is dependent upon continued development of lower cost systems with improved performance through the use of better algorithms, transducers, and processors.[66,67]

ACKNOWLEDGMENTS

The contributions of Mark Allie, Cary Bremigan, Richard Hoops, Jay Warner, and Patricia Steaffens of the Corporate Research Department of Nelson Industries and Dr. Richard A. Greiner of the University of Wisconsin—Madison are gratefully acknowledged.

REFERENCES

1. H. G. Leventhall, "Historical Review and Recent Development of Active Attenuators," paper presented at the 104th Meeting of the Acoustical Society of America, Orlando, FL, November 1982.
2. G. E. Warnaka, "Active Attenuation of Noise—The State of the Art," *Noise Control Eng.* **18**, May/June 1982, pp. 100–110.
3. L. J. Eriksson, "Active Sound Attenuation Using Adaptive Digital Signal Processing Techniques," Ph.D. Thesis, University of Wisconsin-Madison, 1985.
4. D. Guicking, *Active Noise and Vibration Control Reference Bibliography*, 3rd ed., Drittes Physikalisches Institut, University of Göttingen, 1988 (also 1991 suppl.).
5. P. Lueg, "Process of Silencing Sound Oscillations," U.S. Patent 2,043,416, June 19, 1936.
6. R. L. Wallace, Jr., "An Electro-Acoustic 'Short Circuit' for the Ear," Electro-Acoustic Laboratory, Harvard University, 1941, private communication from L. L. Beranek. See also "Active Noise and Vibration Control—1990," Report No. NCA-8, American Society of Mechanical Engineers, New York, 1990.
7. H. F. Olson, "Electronic Control of Noise, Vibration, and Reverberation," *J. Acoust. Soc. Am.* **28**, 966–972 (1956).
8. W. B. Conover and W. F. M. Gray, "Noise Reducing System for Transformers," U.S. Patent 2,776,020, January 1, 1957.
9. S. Onoda and K. Kido, "Automatic Control of Stationary Noise by Means of Directivity Synthesis," paper presented at the Sixth International Congress on Acoustics, 1968.
10. G. A. Mangiante, "Active Sound Absorption," *J. Acoust. Soc. Am.* **61**, 1516–1523 (1977).
11. S. J. Elliott and P. A. Nelson, "Models for Describing Active Noise Control in Ducts," *I.S.V.R. Technical Report No. 127*, NTIS PB85-189777, April 1984.
12. M. L. Munjal and L. J. Eriksson, "Analytical, One-Dimensional, Standing-Wave Model of a Linear Active Noise Control System in a Duct," *J. Acoust. Soc. Am.* **84**, 1086–1093, (1988).
13. M. L. Munjal and L. J. Eriksson, "Analysis of a Linear One-Dimensional Active-Noise Control System by Means of Block Diagrams and Transfer Functions," *J. Sound Vib.* **129**, 443–455 (1989).
14. P. A. Nelson, A. R. D. Curtis, S. J. Elliott, and A. J. Bullmore, "The Active Minimization of Harmonic Enclosed Sound Fields, Part I: Theory;" "Part II: A Computer Simulation," "Part III: Experimental Verification;" *J. Sound Vib.* **117**: (I) 1–13, (II) 15–33, (III) 35–58 (1987).

15. C. G. Mollo and R. J. Bernhard, ''Numerical Evaluation of the Performance of Active Noise Control Systems,'' Paper 88-WA/NCA-5 presented at the ASME Winter Annual Meeting, Chicago, November 27–December 2, 1988.
16. S. D. Snyder and C. H. Hansen, ''Active Noise Control in Ducts: Some Physical Insights,'' *J. Acoust. Soc. Am.* **86**, 184–194 (1989).
17. Kh. Eghtesadi and H. G. Leventhall, ''Comparison of Active Attenuation of Noise in Ducts,'' *Acoust. Lett.* **4**(10), 204–209 (1981).
18. M. Jessel and S. Yamada, ''Active Noise Control,'' *J. Acoust. Soc. Jpn.* (*E*) **8**(4), 151–154 (1987).
19. Kh. Eghtesadi, W. K. W. Hong, and H. G. Leventhall, ''The Tight-coupled Monopole Active Attenuator in a Duct,'' *Noise Control Eng. J.* **20**(1), 16–20 (1983).
20. Kh. Eghtesadi and H. G. Leventhall, ''Active Attenuation of Noise: The Chelsea Dipole,'' *J. Sound Vib.* **75**, 127–134 (1981).
21. M. A. Swinbanks, ''The Active Control of Sound Propagation in Long Ducts,'' *J. Sound Vib.* **27**, 411–436 (1973).
22. R. F. LaFontaine and I. C. Shepherd, ''An Experimental Study of a Broadband Active Attenuator for Cancellation of Random Noise in Ducts,'' *J. Sound Vib.* **91**, 351–362 (1983).
23. M. J. M. Jessel and G. A. Mangiante, ''Active Sound Absorbers in an Air Duct,'' *J. Sound Vib.* **23**, 383–390 (1972).
24. G. Canevet, ''Active Sound Absorption in an Air-conditioning Duct,'' *J. Sound Vib.* **58**, 333–345 (1978).
25. M. Berengier and A. Roure, ''Broad-Band Active Sound Absorption in a Duct Carrying Uniformly Flowing Fluid,'' *J. Sound Vib.* **68**, 437–449 (1980).
26. K. Ogata, *Modern Control Engineering*, Prentice-Hall, Englewood Cliffs, NJ, 1970.
27. B. Widrow and S. D. Stearns, *Adaptive Signal Processing*, Prentice-Hall, Englewood Cliffs, NJ, 1985.
28. J. C. Burgess, ''Active Adaptive Sound Control in a Duct: A Computer Simulation,'' *J. Acoust. Soc. Am.* **70**, 715–726 (1981).
29. L. J. Eriksson, M. C. Allie, and R. A. Greiner, ''The Selection and Application of an IIR Adaptive Filter for Use in Active Sound Attenuation,'' *IEEE Trans. ASSP* **ASSP-35** (4), 433–437 (1987).
30. L. J. Eriksson, ''Active Sound Attenuation System with On-Line Adaptive Feedback Cancellation,'' U.S. Patent 4,677,677, June 30, 1987.
31. L. J. Eriksson and M. C. Allie, ''Use of Random Noise for On-Line Transducer Modelling in an Adaptive Active Attenuation System,'' *J. Acoust. Soc. Am.* **85**, 797–802 (1989).
32. L. J. Eriksson, ''Active Attenuation System with On-Line Modeling of Speaker, Error Path and Feedback Path,'' U.S. Patent 4,677,676, June 30, 1987.
33. K. Kido, ''Reduction of Noise by Use of Additional Sound Sources,'' *Proc. INTER-NOISE 75*, 1975, pp. 647–650.
34. G. B. B. Chaplin and R. A. Smith, ''Improvements in and Relating to Active Sound Attenuation,'' U.S. Patent 4,122,303, October 24, 1978.
35. G. B. B. Chaplin and R. A. Smith, ''Waveform Synthesis—The Essex Solution to Repetitive Noise and Vibration,'' *Proc. INTER-NOISE 83*, 1983, pp. 399–402.

36. C. F. Ross, "An Adaptive Digital Filter for Broadband Active Sound Control," *J. Sound Vib.* **80**, 381–388 (1982).

37. C. F. Ross, "Method and Apparatus for Active Sound Control," U.S. Patent 4,480,333, October 30, 1984.

38. A. R. Davidson, Jr. and T. G. F. Robinson, "Noise Cancellation Apparatus," U.S. Patent 4,025,724, May 24, 1977.

39. G. E. Warnaka, L. A. Poole, and J. Tichy, "Active Acoustic Attenuator," U.S. Patent 4,473,906, September 25, 1984.

40. M. Takahashi, T. Kuribayashi, K. Asami, H. Hamada, T. Enokida, and T. Miura, "Electric Sound Cancellation System for Air-conditioning Ducts," Paper C5-3, *Proc. Intl. Congress Acoust. 12*, Toronto, 1986. (Also, *Proc. INTER-NOISE 86*, 1986, pp. 607–610.)

41. H. Hamada, T. Enokida, T. Miura, M. Takahashi, T. Kuribayashi, K. Asami, and Y. Oguri, "Electronic Noise Attenuation System," U.S. Patent 4,783,817, November 8, 1988.

42. L. J. Eriksson, M. C. Allie, C. D. Bremigan, and J. A. Gilbert, "Active Noise Control and Specifications for Fan Noise Problems," *Proc. NOISE-CON 88*, West Lafayette, IN, June 20–22, 1988, pp. 273–278.

43. L. J. Eriksson, M. C. Allie, and R. H. Hoops, "Hybrid Active Silencer," U.S. Patent 4,665,549, May 12, 1987.

44. I. C. Shepherd, A. Cabelli, and R. F. LaFontaine, "Characteristics of Loudspeakers Operating in an Active Noise Attenuator," *J. Sound Vib.* **110**, 471–481 (1986).

45. L. J. Eriksson and M. C. Allie, "System Considerations for Adaptive Modelling Applied to Active Noise Control," paper presented at ISCAS 88, Espoo, Finland, June 7–9, 1988.

46. I. C. Shepherd, R. F. LaFontaine, and A. Cabelli, "Active Attenuation in Turbulent Flow Ducts," *Proc. INTER-NOISE 84*, 1984, pp. 497–502.

47. M. L. Munjal and L. J. Eriksson, "An Exact One-Dimensional Analysis of the Acoustic Sensitivity of the Anti-turbulence Probe Tube in a Duct," *J. Acoust. Soc. Am.* **85**, 582–587 (1989).

48. R. Bouc and D. Felix, "A Real-Time Procedure for Acoustic-Turbulence Filtering Ducts," *J. Sound Vib.* **118**, 1–10 (1987).

49. W. V. Harrington, "The Design and Development of an Automatic Control System for the In-Duct Cancellation of Spinning Modes of Sound," M.S. Thesis, The Pennsylvania State University, June 1973.

50. L. J. Eriksson, M. C. Allie, and R. H. Hoops, "Active Acoustic Attenuation System for Higher Order Mode Non-Uniform Sound Field in a Duct," U.S. Patent 4,815,139, March 21, 1989.

51. L. J. Eriksson, M. C. Allie, R. H. Hoops, and J. V. Warner, "Higher Order Mode Cancellation in Ducts Using Active Noise Control," *Proc. INTER-NOISE 89*, 1989.

52. J. V. Warner, D. E. Waters, and R. J. Bernhard, "Digital Filter Implementation of Local Active Noise Control in a Three Dimensional Enclosure," Paper No. 88-WA/NCA-6, presented at the ASME Winter Annual Meeting, Chicago, November 27–December 2, 1988.

53. Kh. Eghtesadi, W. K. W. Hong, and H. G. Leventhall, "Energy Conservation by Active Noise Attenuation in Ducts," *Noise Control Eng. J.* **27**(3), 90–94 (1986).

54. L. J. Eriksson and M. C. Allie, "A Practical System for Active Attenuation in Ducts," *Sound Vib.* **22**(2), 30–34 (1988).

55. L. J. Eriksson, M. C. Allie, C. D. Bremigan, and J. A. Gilbert, "The Use of Active Noise Control for Industrial Fan Noise," Paper No. 88-WA/NCA-4, presented at the ASME Winter Annual Meeting, Chicago, November 27–December 2, 1988.

56. G. H. Koopmann, D. J. Fox, and W. Neise, "Active Source Cancellation of the Blade Tone Fundamental and Harmonics in Centrifugal Fans," *J. Sound Vib.* **126**, 209–220 (1988).

57. M. A. Swinbanks, "The Active Control of Low Frequency Sound in a Gas Turbine Compressor Installation," *Proc. INTER-NOISE 82*, San Francisco, 1982, pp. 423–426.

58. O. L. Angevine, "Active Acoustic Attenuation of Electric Transformer Noise," *Proc. INTER-NOISE 81*, 1981, pp. 303–306.

59. L. J. Oswald, "Reduction of Diesel Engine Noise Inside Passenger Compartments Using Active, Adaptive Noise Control," *Proc. INTER-NOISE 84*, 1984, pp. 483–488.

60. C. M. Dorling, G. P. Eatwell, S. M. Hutchins, C. F. Ross, and S. G. C. Sutcliffe, "A Demonstration of Active Noise Reduction in an Aircraft Cabin," *J. Sound Vib.* **128**, 358–360 (1989).

61. S. J. Elliott, P. A. Nelson, I. M. Stothers, and C. C. Boucher, "Preliminary Results of In-Flight Experiments on the Active Control of Propeller-induced Cabin Noise," *J. Sound Vib.* **128**, 355–357 (1989).

62. S. J. Elliott, I. M. Stothers, and P. A. Nelson, "A Multiple Error LMS Algorithm and Its Application to the Active Control of Sound and Vibration," *IEEE Trans. ASSP* **ASSP-35**, 1423–1434 (1987).

63. M. Simpson, T. Luong, C. R. Fuller, and J. D. Jones, "Full Scale Demonstration of Cabin Noise Reduction Using Active Vibration Control," AIAA Paper No. 89-1074, 1989.

64. I. C. Shepherd, R. F. LaFontaine, and A. Cabelli, "Active Attenuation in Flow Ducts: Assessment of Prospective Applications," Paper No. 86-WA/NCA-26, ASME Meeting, Anaheim, 1986.

65. L. J. Eriksson, "Development of the Filtered-U Algorithm for Active Noise Control," *J. Acoust. Soc. Am.* **89**(1), 257–265 (1991).

66. L. J. Eriksson, "Computer-aided Silencing: An Emerging Technology," *Sound Vib.* **24**(7), 42–45 (1990).

67. P. A. Nelson and S. J. Elliott, *Active Control of Sound*, Academic Press, London, U.K., 1991.

CHAPTER 16

Damage Risk Criteria for Hearing and Human Body Vibration

HENNING E. VON GIERKE and CHARLES W. NIXON
Harry G. Armstrong Aerospace Medical Research Laboratory
Wright-Patterson AFB, Ohio

16.1 INTRODUCTION

Noise and vibration are closely related biodynamic environments in terms of their origins, manifestations, and effects on people. Those effects on people that are undesirable are often threatening, induce fatigue, reduce proficiency, modify physiological responses, and harm human systems. At the present time, avoidance of excessive noise and vibration exposure is the only assured way to prevent these major hazardous effects.

The practical alternative to complete avoidance is to limit exposures in these environments to those defined as acceptable by appropriate guidelines, standards, and damage risk criteria (DRC). Scientific exposure criteria and guidelines established to curtail these effects are vital parts of comprehensive protection programs of concern to governments, industry, and affected personnel. Most adverse effects of these commonly encountered mechanical forces on human systems can be minimized and controlled through engineering and design efforts, conservation programs, personal protective equipment, and administrative actions. Central to these programs and actions are criteria and guidelines that establish acceptable exposure limits.

Noise and Vibration Control Engineering: Principles and Applications, Edited by Leo L. Beranek and István L. Vér.
ISBN 0-471-61751-2 © 1992 John Wiley & Sons, Inc.

Criteria and limits define conditions above which the risk of damage due to an exposure is considered unacceptable. Noise and vibration criteria describe exposure characteristics and corresponding undesirable effects such as noise-induced hearing loss and vibration-induced fatigue that cause decreased proficiency. Damage risk criteria have been developed and implemented worldwide. The basic exposure–effects relationships that underly these criteria are reasonably well understood and are derived from observations and experience as well as good laboratory and field studies. The various criteria contain different limiting values because of variations in interpretations of the basic data and in the rationale used to establish them. The rationale may include practical, legal, and economic considerations as well as humanitarian concerns. It is very important that the rationale underlying DRC is fully understood by the user to ensure that its application is justified and accurate.

Estimates of noise and vibration exposures and of their probable effects on populations are usually expressed in terms of population distribution statistics. These group population effects are not precise descriptors nor are they appropriate for evaluating an individual. Nevertheless, they are adopted by nations and incorporated into national regulations and laws. Many become mandatory requirements included in governmental and industrial activities involving exposure of people to noise and vibration environments.

This chapter presents contemporary regulatory and voluntary noise and vibration exposure standards and criteria along with background information that will facilitate their understanding and application in engineering control and design.

16.2 DAMAGE RISK CRITERIA FOR THE AUDITORY RANGE

Noise Factor

Permanent hearing loss and its associated problems are clearly the most critical and widespread of the various consequences of excessive noise exposure. The extent of damage to the hearing mechanism caused by noise is related to the amount of acoustical energy reaching the hearing mechanism. Such damage cannot be estimated accurately for an individual because of the variability of the noise and the susceptibility of the exposed ears. The primary factors in noise-induced hearing loss are the level of the noise, the frequency content or spectrum of the noise, the duration or time course of the noise exposure, and the susceptibility of the ear.

Exposure limits are defined in terms of level, spectrum, and duration of the noise. A-weighted sound energy of an exposure is directly related to noise-induced hearing loss. No other measure of noise exposure provides a better *cause–effect* relationship with hearing loss.[1] Impulse noise is also included in this measure for many criteria as was concluded by a special workshop on impulse noise.[2] It was agreed that there is no convincing evidence against acceptance of A-weighting measurement of all noises from 20 to 20,000 Hz in determining their permanent threshold shift (noise-induced hearing loss that does not recover to preexposure levels after cessation of exposures) hazard except when their unweighted, instantaneous, peak-sound-pressure levels exceed approximately 145 dB. Consequently, exposures are typically described in terms of the average A-weighted levels, or the equivalent continuous A-weighted sound pressure level (L_{eq}), over an average workday.

Practical measures for the prevention of noise-induced hearing loss are centered in hearing conservation programs. These programs involve definitions of acceptable noise exposure, personal hearing protection, monitoring the hearing of the affected personnel, and appropriate administrative actions to minimize and eliminate identified temporary hearing problems before they become permanent. The basis of a hearing conservation program is the definition of acceptable noise exposure or exposure criteria that specify the acceptable exposure limits and the proportion of the population to be protected. The criteria of various hearing conservation programs and applications differ in their limiting values.

The parameters of particular exposure criteria are selected to satisfy the needs of the user. Factors that may influence these selections are various interpretations of available data, policies or requirements of organizations, and the remaining uncertainty in the noise exposure–hearing loss data bases. Consequently, criteria may differ in such features as estimates of beginning hearing loss, corrections for nonnoise effects such as aging, percentage of the population to be protected, and the extent of protection to be provided.

The most obvious differences among noise exposure criteria are the sound level at which the implementation occurs and how the duration of the exposure and the sound level of the noise are combined.[3] The duration-sound level relationships are referred to as time-intensity trading rules, which assume that damage to hearing is related to total A-weighted sound level and the duration of exposure time. The equal energy relationship between these parameters results in the 3-dB rule. This and other relationships are shown in Table 16.1, which displays permissible noise exposures in A-weighted sound pressure levels for the cited criteria. The permissible A-weighted sound level for an 8-h exposure ranges from 75 dB(A) for the Environmental Protection Agency (EPA) to 90 dB(A) for the Occupational Safety and Health Administration (OSHA).

Most criteria utilize the 3-, 4-, or 5-dB rule. The 3-dB rule is based on the equal-energy concept and is the most conservative or protective of the three rules. The 4- and 5-dB rules assume that intermittency and interruptions of exposures reduce the risk to less than that expected from the total energy. Consequently, a 50% increase in exposure duration corresponds to sound level decreases of 3 and 5 dB for the respective 3- and 5-dB rules (Table 16.1). Intermittency of exposure is discussed later in the chapter.

Hearing Sensitivity

The human ear is sensitive to a much wider range of sounds than the generally cited 20 Hz–20 kHz audio frequency range. A compilation of independent measurement studies by several investigators using a wide variety of instrumentation and methodologies shows very good agreement and provides confidence in the data

TABLE 16.1 Equal Energy and Other Trading Rules Used to Define Permissible Noise Exposures in A-weighted Sound Pressure Level (dB) and Exposure Time (h)

Duration of Exposure (h)	Equal Energy[a]	OSHA	U.S. Army	U.S. Air Force	EPA	Air Force Music
24						
16				80		
8	90	90	85	84	75[b]	
4	93	95	89	88		
2	96	100	93	92		94[c]
1	99	105	97	96		
0.5	102	110	101	100		
0.25	105	115[d]	105	104		

[a]Equal energy rule of 3 dB decrease for doubling of exposure applied to a basic 8-h criterion of 90 dB(A).

[b]Threshold for detectable noise-induced permanent threshold shift (NIPTS) at 4,000 Hz; exposures exceeding 75 dB(A) may cause NIPTS exceeding 5 dB in 100% of the population after cumulative noise exposure of 10 years.

[c]Time-averaged A-weighted sound level, in dB, over a 2-h period once a week.

[d]Ceiling on exposure level and duration.

summarized in Fig. 16.1.[4] Infrasound (<20 Hz) and ultrasound (traditionally >20,000 Hz but practically above about 12,000 Hz) are normally detected by the ear only at very high sound pressure levels. The traditional audio frequency region (20 Hz–20 kHz) is well defined for stimuli including discrete tones, bands of noise, speech materials, loudness, comfort, and acceptibility. These data bases provide the information necessary to support the development of noise exposure criteria.

The sensitivity of human hearing for high-frequency sounds (3000–4000 Hz and above) gradually decreases with advancing age. This process is called presbyacusis. Auditory system components are affected in both the peripheral and the central nervous system. Although individual patterns of *presbyacusis* vary widely, normative data describing hearing sensitivity as a function of age have been compiled for various segments of society (some are reported in ref. 5). Loss of sensitivity due to accident, disease, or substances toxic to the auditory system is called *nosoacusis* while that attributed to the noises of everyday living is *sociacusis*. The major high-level noises to which people are exposed are the occupational environments.

Environmental noise occurs over the full spectrum to which the human auditory system is sensitive. Exposure to various segments of this sensory continuum produces differential effects on man. Limiting levels and durations of acoustical exposure are defined for a number of specific portions of this spectrum, which include infrasound (0.5–20 Hz), audio frequencies (20 to about 12,000 Hz), ultrasound (12,000 to about 40,000 Hz), and impulsive sounds (characterized by rapid onset and durations of less than 1 s) described in terms of peak sound pressure level and duration. Some of these limits are well substantiated by experience

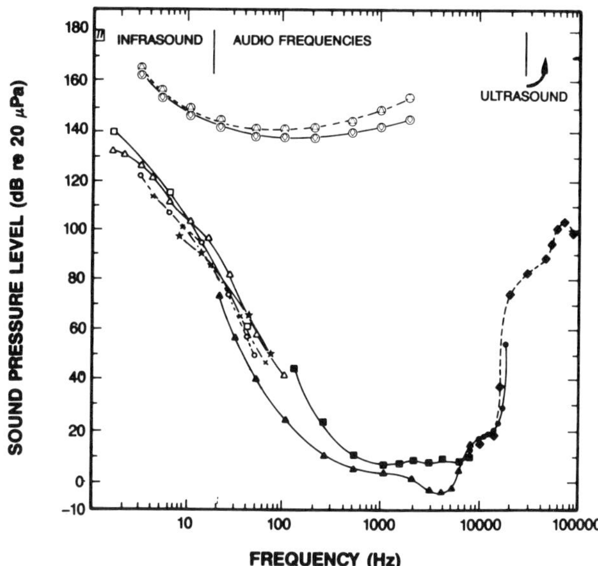

Fig. 16.1 Human auditory sensitivity and pain thresholds for pure tones, octave bands of noise, and static pressure: (△) BENOX (1953), pain MAP tones; (□) pain static pressure; (□) Bekesy (1960), MAP tones; (◇) tickle, pain tones; (▲) ISO R226 (1961), MAF tones; (◆) Corso (1963), bone conduction minus 40 dB tones; (○) Yeowart, Bryan, and Tempest (1960), MAP tones; (×) MAP octave bands of noise; (■) standard reference threshold values (American National Standard on Specifications for Audiometers) (1969), MAP tones; (●) Northern et al. (1972), MAP tones; (△) Whittle, Collins, and Robinson (1972), MAP tones; (★) Yamada et al. (1986), MAF tones. (All data adapted from ref. 4.) Minimum audible pressure (MAP) indicates that the sound was presented to the ears through earphones and the sound pressure levels were measured in an earphone–microphone coupler that approximated the cavity created by the earphone/pinna. Minimum audible field (MAF) indicates that the sound was presented to the listeners facing a loudspeaker and located in anechoic space. The sound pressure levels for MAF were measured at the location of the center of the head without the listener present. For the same listeners, MAP thresholds are generally several decibels higher than MAF thresholds.

and experimental evidence while others remain tentative until more evidence is available.

16.3 AUDIO FREQUENCY REGION

CHABA

Noise exposure criteria for the audio frequency region (20–12,000 Hz) were developed by the National Academy of Sciences—National Research Council, Committee on Hearing, Bioacoustics and Biomechanics (CHABA) in 1965.[6] This method described noise exposure in terms of pure tones, third octave, and octave bands of noise, and it includes the audio frequencies of 100–7000 Hz. Acceptable exposures to noise can be determined from 11 sets of curves. An environmental noise is considered acceptable if it produces, on average, a noise-induced permanent threshold shift (NIPTS) after 10 years or more of near daily exposure of no more than 10 dB at 1000 Hz and below, 15 dB at 2000 Hz, and no more than 20 dB at 3000 Hz and above. These criteria are based on the assumption that noise exposures producing temporary threshold shifts (TTSs) would eventually produce permanent threshold shifts (PTSs). The possible relationship that temporary threshold shift is a precursor to permanent threshold shift is still an open question. The CHABA criteria has been widely used and is an excellent tool. However, it is not simple to use or to relate to current standards, regulations, and guidelines and it is less preferred than criteria employing A-weighted sound level.

OSHA

The Occupational Safety and Health Administration (OSHA) has adopted a noise exposure limit of 90 dBA with a 5-dB trading relationship to control excessive noise exposure in industry (Table 16.1). When employees are exposed to noise at different levels during the day, ratios of the actual to the allowed duration for that level are computed and the fractions summed for the day. Total daily exposure calculated from these fractions or ratios must not exceed unity. No other corrections or adjustments are applied to these criteria.

The OSHA noise exposure criteria were verified in 1983 with the publication of "Occupational Noise Exposure; Hearing Conservation Amendment; Final Rule."[7] The basic conditions of the original OSHA noise exposure regulation remain the same with a few exceptions. Continuous A-weighted sound levels are not permitted above 115 dB(A) regardless of duration. A *permissible exposure level* (PEL) is defined as that noise dose that would result from a continuous 8-h exposure to a sound level of 90 dB(A). The limit of 90 dB(A) is a dose of 100%, which is the basic criterion level. A *time-weighted average* (TWA) is the sound level that would produce a given noise dose when the employee is exposed to that sound level continuously over an 8-h workday regardless of the length of the work shift. Workday exposures of 4 h at 90 dB, 8 h at 85 dB, or 12 h at 82 dB all correspond to a TWA of 85 dB(A) and a noise dose of 50%. The Hearing Conservation Amendment includes computational formulas and tables showing the time–intensity relationships for the 5-dB rule and the conversions of dose to time-weighted averges. Guidance is given on calculations of age corrections to audiograms. However, the use of age correction procedures is not required for compliance.

A noise dose of 50% or a TWA of 85 dB is the "action level" at which hearing conservation measures must be implemented. All workers receiving noise doses at or above the action level must be included in a hearing conservation program that requires noise monitoring, audiometric testing, hearing protection, employee training, and record keeping. A baseline audiogram is one taken within six months of the employee's first exposure above the action level, against which subsequent audiograms can be compared. An annual audiogram must be taken for each employee exposed at or above the action level. A *standard threshold shift* (STS) is a change in hearing sensitivity from the baseline audiogram that exceeds an average of 10 dB or more at 2000, 3000, and 4000 Hz in either ear. Appropriate action by the employer must be taken in response to the STS to ensure the continued protection of the hearing of the employee.

EPA

The Environmental Protection Agency (EPA) published "Information on Levels of Environmental Noise Requisite to Protect Public Health and Welfare with an Adequate Margin of Safety"[1] in 1974 in response to the Noise Control Act of 1972. The objective was to identify levels of environmental noise required to protect the public from adverse health and welfare effects. The levels for noise-induced hearing loss described in this document were based upon reviews and analyses of scientific materials as well as consultations and interpretations of experts. It was concluded that an L_{eq} of 70 dB over a 24-h day (over a 40-year working life) would protect virtually the entire population (96th percentile) for hearing conservation purposes. An $L_{eq(8)}$ limit of 75 dB was considered appropriate protection for the typical 8-h daily work period. This criterion is considered to be very restrictive for most applications, and it has not been incorporated into any DRC for occupational noise exposures.

Air Force

The U.S. Air Force (USAF) hearing conservation program[14] is based on a criterion of 84 dBA as the maximum allowable 8-h daily exposure. A trading relationship of 4 dB is utilized that allows 16-h exposures at 80 dBA and 24-h exposures at 78 dBA. Higher level continuous exposures for shorter durations, such as 96 dBA for 1 h, are limited to a maximum allowable level of 115 dBA. Daily exposures that include different levels of noise are not to exceed unity when the ratios of the actual to the allowable exposure times for each level are summated. The Air Force criteria also include limiting exposure conditions for infrasound, ultrasound, and impulsive noises.

Army

The U.S. Army (USA) hearing conservation program identifies a criterion of 85 dBA as the maximum allowable exposure regardless of duration. Personnel experiencing these exposures must be enrolled in the hearing conservation program. In training and noncombat scenarios single hearing protection will be worn in steady noises of 85–107 dB(A) and double protection from 108 to 118 dB(A). Protection requirements for military unique noise sources are determined individually. In combat, soldiers should wear hearing protection except when the functional hearing required for the operation underway is impaired by their use. The Army criteria also include limiting exposure conditions for impulsive noises.

ISO 1999.2

The International Standardization Organization (ISO) Standard 1999.2, Acoustics—Determination of Occupational Noise Exposure and Estimation of Noise-induced Hearing Impairment, 1990,[5] is a landmark document that establishes practical procedures for estimating noise-induced hearing loss in populations. Standard ISO 1999.2 does not provide a specific formula for assessing risk of hearing handicap, but it specifies uniform methods for the prediction of hearing impairment that can be used for assessment of handicap according to the formula stipulated in a specific nation. The procedures are based on the equal-energy, 3-dB rule (adopted by the ISO and member nations as a conservative criterion) and deal with the measurement and description of noise exposure, the prediction of effects of noise on hearing threshold, and the assessment of risk of noise-induced hearing impairment and handicap. Annexes, which are not part of the standard, provide calculation procedures, examples, tabular data used in the calculations, and a method for relating this information to that of the preceding standard, ISO 1999, 1975. These procedures allow agencies, industries, and governments to select parameters and establish criterion values according to their respective needs. This document will form the basis for legislation in many countries.

The ISO method describes all exposures during an average work day in terms of the A-weighted sound exposure or energy average. The integration period is taken as a working day or a working week. All noises are included in the exposure, ranging from steady state to impulses. Exposures that contain steady tonal noise or impulsive/impact noise are considered about as harmful as the same exposure without these components but about 5 dB higher in level. Exposure can be measured with personal noise dosimeters or an integrating-averaging sound-level meter. Direct and indirect methods for the determination of exposure level are discussed as well as sampling methods.

The only measure of environmental noise needed to calculate hearing impairment or risk of hearing handicap under the following conditions is the energy-averaged daily noise exposure. The maximum instantaneous sound pressure level must be less than 140 dB, the average 8-h daily exposure must not exceed 100 dB(A), and the maximum individual daily exposure must not be more than 10 dB above the average of all daily exposures to permit this determination of energy-averaged daily exposure.

Implementation of this standard follows a well-defined series of operations. The first involves determination of the age-related hearing levels of the target population for all test frequencies (e.g., population of 50-year-old males, 90th percentile at 500–6000 Hz). The long-standing difficulties of defining a "normal" population for this purpose were overcome with the utilization of two data bases. Data base A contains standardized distributions of hearing threshold of an ideal "highly screened" population free from all signs of ear disease, obstructions of wax, and without undue history of noise exposure. Data base B can be any carefully collected data base covering an *occupationally non-noise-exposed population* considered to be a valid control for the noise-exposed population under consideration. Each user of the standard can select the subpopulation most appropriate for its analysis. As an example for data base B the standard provides the data from the U.S. Public Health Service Surveys reported in 1965.[8]

Next, the predicted NIPTS of the population is calculated for all test frequencies by considering both the number of years of exposure and the average daily noise exposure levels. Data in the standard for calculating NIPTS are valid for frequencies from 500 to 6000 Hz, exposure times of 0 to 40 years, and average daily noise exposure levels between 75 and 100 dB.

The hearing handicap or risk of hearing handicap may be calculated using the appropriate NIPTS values and a formula selected by the user or a member nation. The document contains nine formulas that are proposed or commonly used among nations for assessing hearing handicap by averaging hearing threshold levels at selected audiometric test frequencies. In the United States, hearing handicap for conversational speech is assessed using the average of the hearing levels at 500, 1000, 2000, and 4000 Hz. Other procedures are available for determining overall percentage of hearing loss for purposes of compensation.

Long-Duration Noise Exposure

Noise exposure durations that exceed 8 h may occur in some work assignments and when substantial nonoccupational noise is added to that received at work. These longer duration exposures are extended by noises from daily living activities, recreation, transportation, proximity to industry, and even other vocational activities. Although the baseline of most criteria is the allowable exposure for an 8-h day, many do extend their time–intensity trading relationship to 16 or 24 h as additional guidance.

Although the 8-h day, 5-day work week is considered standard, numerous variations are employed for many occupations. Some of these are four 10-h days with three days off, three 12-h days with three and four days off, and 12 h on and 12 h off. The noise exposure criteria do not cover these exposures with the same degree of accuracy as with the standard work week. However, it is reasonable to consider the exposure per work week as a basis for calculating noise exposure (i.e., the 3-dB rule).

An important discovery in studies of effects of long-duration exposure to continuous (nonimpulsive) noise of 24 h and longer on human hearing was the phenomenon represented in Fig. 16.2 and called asymptotic threshold shift (ATS).[9] Hearing threshold levels progressively increased with time until the exposure durations reached 8–16 h. Hearing threshold levels reached a plateau or asymptote between 8 and 16 h and did not increase further with continuation of the same exposure levels for 24 and 48 h. Recovery from these asymptotic levels to the preexposure threshold levels was related to exposure time. Even though the asymptotic levels were the same for 24- and 48-h exposures to a particular stimulus, the

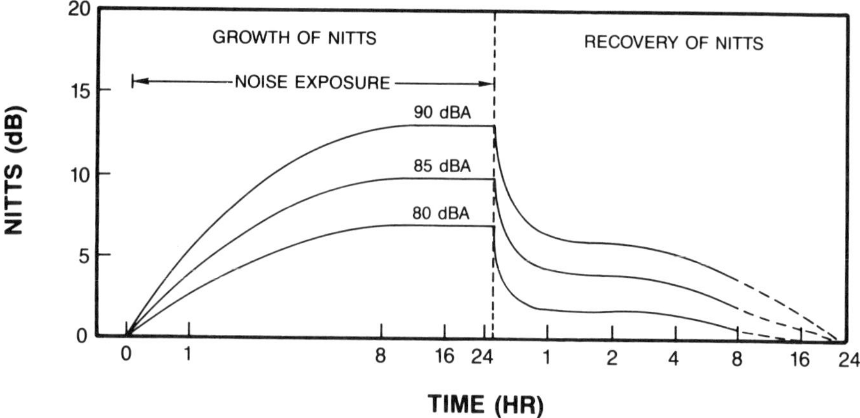

Fig. 16.2 Growth and recovery of noise-induced temporary threshold shift (NITTS) measured during and following the noise exposure at the times marked on the abscissa. The stimulus was a third-octave band of random noise centered on 1000 Hz presented at 80, 85, and 90 dBA. The curves represent the averages of the hearing levels (HL) for the 1000-, 1500-, and 2000-Hz test frequencies $\frac{1}{3}$ (HL_{1000} + HL_{1500} + HL_{2000}).

time needed to recover was significantly longer for the 48- than for the 24-h exposures. The longer period of time to recover is interpreted by some as an indication of greater risk to hearing from the same stimulus (same asymptotic threshold level) for 48 than for 24 h. On the basis of analyses of the actual recovery times the guideline was established that the period of time in effective quiet required for recovery should be at least as long as the duration of the exposure.

Audiometric data on the crew members and information on the internal noise levels of the *Voyager* lightweight aircraft (two 110-hp piston engines) during a practice (5 days) and a round-the-world flight (9 days) have provided additional important data points. The overall noise enviroments at the crew locations ranged between 99 and 103 dBA. Crew members wore communications headset and earplug equipments that allowed estimations of the exposures at the ears to be between 84 and 95 dB for the octave band at 500 Hz. Comparisons of preflight and postflight audiograms for both flights revealed substantial shifts of hearing levels. The threshold shifts from the nine-day flight were no greater than those for the five-day flight. One week following the nine-day flight the hearing thresholds of both crew members had returned to their preflight levels.

The hearing level data on the five- and nine-day exposures is consistent with laboratory data derived from both human studies of shorter duration and animal studies of similar duration. It is considered acceptable by some criteria to extend the 8-h limits to as much as 24 h using appropriate trading relationships. Noise exposures of atypical work schedules such as four 10-h days with three days off and 12 h on, 12 h off might be calculated on the bases of work week. Although recovery of the hearing thresholds occurred prior to nine days for these crew members, it remains reasonable to have personnel remain in effective quiet for a period at least as long as the exposure prior to reentering the noise.

Music Exposure Criteria

The U.S. Air Force has adopted music exposure criteria that consider customers or clients of military "clubs" to be "recreationally exposed" and employees to be "occupationally exposed." The occupationally exposed persons are governed by the same provisions as those of workers exposed in any other occupational noise. A separate set of criteria are used to control or limit the recreational exposures.

An average A-weighted sound level of 94 dB is considered acceptable when it does not exceed two h duration once a week. It is important to recognize that the 94 dB(A) guideline is not a peak or a maximum level value but is the average sound level. The average sound level concept does not specify a fixed maximum level or eliminate crescendos and special effects or even some selections. It does permit these intermittent high levels of entertainment music to be averaged in such a way that the overall performance is acceptable.

16.4 IMPULSE NOISE

Impulse or impact noise is a very brief sound or short burst of acoustical energy with a sound pressure rise of 40 dB in 0.5 s or faster that may occur singly or as a series of events. The noise may be treated as steady state when the repetition rate of a series of impulses exceeds 10 per second and the decay from the individual peaks to minima does not exceed 6 dB.

The effects on the auditory system have been examined for such characteristics of the impulsive stimulus as frequency spectrum, duration, peak pressure level, total energy, type of impulse, and rise time. Although work continues with some of these characteristics, present exposure criteria use only peak pressure level and duration and type of impulse to describe safe impulse exposures.

In 1968, CHABA developed exposure criteria for impulse noise[10] based on extensive work in the United Kingdom on firing small arms.[11] The limiting noise exposure values for the impulsive stimuli are summarized in Fig. 16.3. These criteria define exposures that should produce, on average, no more NITTS than 10 dB at 1000 Hz, 15 dB at 2000 Hz, and 20 dB at 3000 Hz and above in 95% of the exposed ears. The criteria provide for adjustments or corrections for exposure situations that vary from the basic condition. The criteria provide for a daily exposure of 100 impulses during any time period ranging from about 4 min to several hours. The values are increased for fewer and decreased for more than 100 impulses per day by a factor of 1.5 dB for each doubling or halving of the number of impulses. The allowable level must be decreased by 5 dB for impulses that strike the ear at perpendicular incidence.

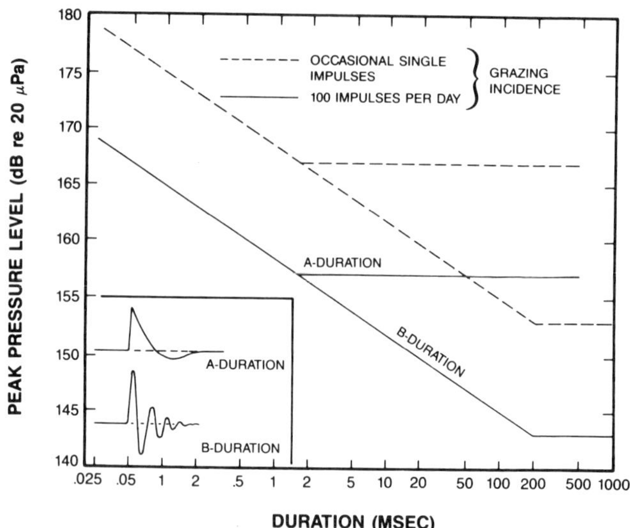

Fig. 16.3 Proposed damage risk criteria for impulse noise arriving from the front or from behind (grazing incidence) exposed persons. The *A*-duration curve reflects the simple, nonfluctuating impulse that occurs in open areas. The *B*-duration curve reflects the pressure fluctuations of impulses that occur under various reverberant conditions.

Simple, nonreverberating impulses that occur in open spaces are evaluated using the A-duration or pressure wave duration, which is the time required for the initial or principal wave to reach peak pressure level and return momentarily to zero (Fig. 16.3). The B-duration or pressure envelope duration is used for impulses that occur under various reverberant conditions and is the total time that the envelope of the pressure fluctuations (positive and negative) is within 20 dB of the peak pressure level, including reflected waves. The 143-dB floor of the B-duration curve represents the reduction of the energy entering the ear after 200 ms by action of the acoustic reflex (reflex contraction of the middle-ear muscles that reduces the transmission of energy to the inner ear). This comprehensive DRC for impulses continues to represent the present data and scientific understanding of impulse noise effects on hearing.

The OSHA amendment, the Air Force, and the Army, all limit exposure to impulse or impact noises to 140 dB peak sound pressure level. The Army has unique equipment that generates high-level impulse noises that are measured and treated individually.

Sonic booms do not constitute a threat of noise-induced hearing loss for human beings. Most of the energy in sonic booms generated by aircraft in supersonic flight is in the low and infrasonic frequency ranges and contributes little to the A-weighted sound level. Field and laboratory investigations with human exposures have revealed no significant effects on hearing from sonic booms at levels typically experienced in the community. One field study involving human exposures to extremely intense sonic booms ranging in level from about 50 to 144 lb/ft^2 observed no changes in the hearing levels of the participants.[12] Naturally, such intense sonic booms may generate higher A-weighted levels indoors than outdoors due to rattling of windows and doors.

Air bag systems are designed to provide forward crash protection of occupants during forward impacts of a motor vehicle. These systems generate a loud, impulse noise inside the vehicle upon inflation of the air cushion. In an early study[13] of a prototype system, 91 volunteers experienced this air bag deployment inside a small automobile at a median peak pressure level of 168 dB. Some temporary threshold shift was experienced by about 50% of the subjects. About 95% of those with TTS recovered preexposure hearing levels on the same day. About 5% required longer times for recovery with one subject showing a gradually returning shift at one frequency that persisted for several months.

16.5 INFRASOUND

Relationships among human exposures to infrasound (0.5–20 Hz) and resulting hearing loss are presently not sufficiently understood for the establishment of national (U.S.) or international standards on exposure limits. Few investigations have been conducted because of difficulties in measuring hearing thresholds for infrasound and in producing infrasound stimuli free from audible overtones that are required for exposure studies. Tentative criteria have been established on the basis of laboratory investigations and field experiences with noises containing intense infrasound components. These criteria have been incorporated in some Department of Defense regulations on hazardous noise exposure.[14]

Human whole-body exposures to intense levels of infrasound that exceeded 150 dB sound pressure level were reported in a classic study.[15] The sample size was small. However, the subjects were highly experienced professionals. Relationships were observed between exposure levels and human tolerance as a function of subjective "symptoms" such as localized pain, dizziness, headache, coughing, and difficulty with respiration. Some of these symptoms were reported as unpleasant but tolerable. Exposures where symptoms were unpleasant and also judged to be very close to tolerance limits were voluntarily terminated by the subjects. Hearing levels of these subjects were measured 3 min following termination of these very intense exposures, and no changes were found in hearing sensitivity.

The absence of hearing sensitivity effects immediately following the intense infrasound exposures verifies laboratory findings that infrasound is not a major threat to hearing. The exposure criteria in Fig. 16.4 have been developed on the basis of data such as that in the cited report and that presented in Fig. 16.5.[16] Numerous experimental subjects experienced exposures to 10 Hz at 144 dB for 8 min with no adverse effects. This set of safe exposure conditions was accepted as a baseline exposure from which 8-h limits (136 dB at 1 Hz and 123 dB at 20 Hz) and 24-h limits (130 dB at 1 Hz and 118 dB at 20 Hz) were extrapolated. (Development of this sound pressure level formulation is described in ref. 16).

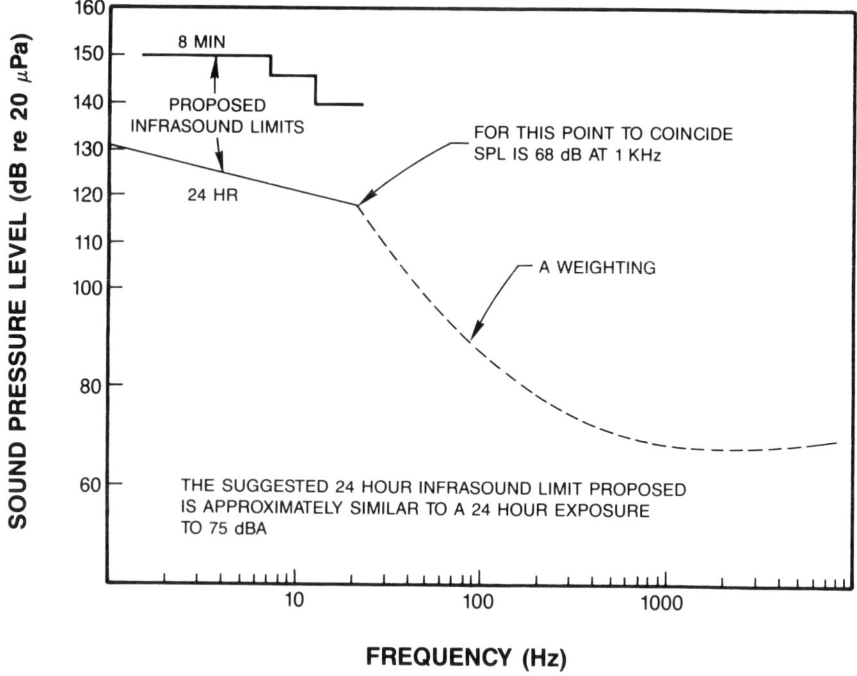

Fig. 16.4 Infrasound 8-min and 24-h exposure limits.

16.6 ULTRASOUND

Ultrasound (~ 16–40 kHz) is widespread in our society due to sources such as ultrasonic cleaners, measuring devices, drilling and welding processes, animal repellants, alarm systems, and communications control applications as well as a wide range of medical applications. Although the number of people exposed to ultrasound is large, it poses no threat to human hearing because neither temporary nor permanent hearing loss caused by ultrasound has been reported, with one exception. Temporary threshold shift was reported following experimental exposures to discrete tones in the region of 17–37 kHz at levels of 148–154 dB. The TTS occurred at subharmonics of the stimulus frequency and was likely caused by nonlinear distortion of the eardrum.

Nevertheless, ultrasound continues to be viewed as a threat to hearing as well as a cause of other subjective symptoms. Ultrasound is readily absorbed by air and its intensity diminishes rapidly with increasing distance from the source. The impedance match with human flesh is poor, and much energy is reflected away from the surface of the body. Consequently, the ear is the primary channel for transmitting airborne ultrasound to the internal systems.

Ultrasonic energy at frequencies above about 17 kHz and at levels in excess of about 70 dB may produce adverse subjective effects experienced as fullness in the

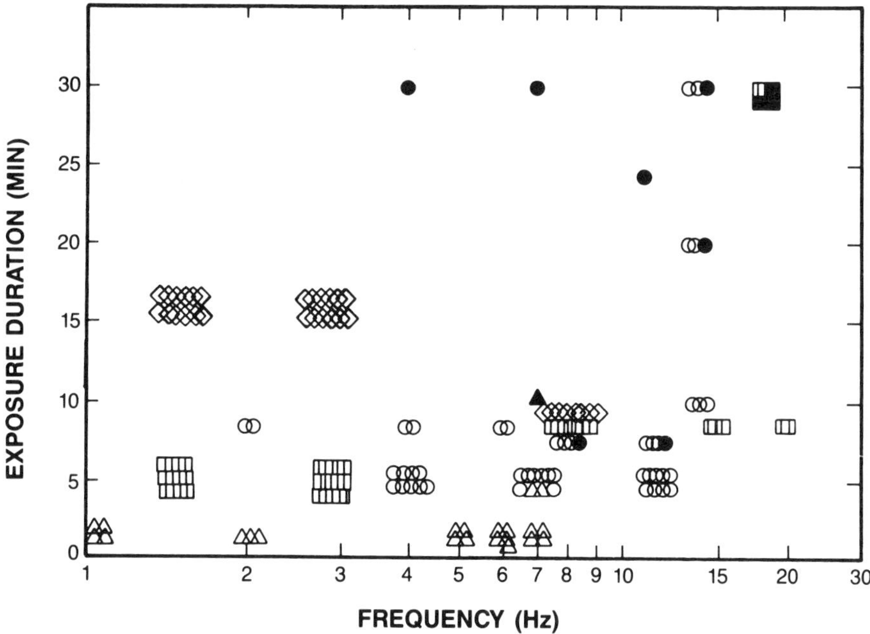

Fig. 16.5 Infrasound exposure effects on hearing. Exposure levels (filled symbols indicate that some TTS occurred): (△) >150 dB; (○) 140–149 dB; (□) 130–139 dB; (◇) 120–129 dB.

ear, tinnitus, fatigue, headache, and malaise. These subjective effects are mediated through the hearing mechanism and are related to hearing ability. Persons who do not hear in this frequency region do not experience these subjective symptoms. Women experience the symptoms more often than men, and younger individuals report them more often than older ones. This reporting is consistent with the relative hearing abilities of the three groups. Neither disorientation nor loss of balance have been attributed to ordinary exposures to ultrasound.

Ultrasound exposures usually contain various amounts of high audio frequency energy (5–20 kHz). Subjective effects attributed to the ultrasound are usually caused by the audio frequency energy. Reducing the level of the audio frequency energy in the exposure usually results in the disappearance of the auditory and subjective symptoms.

Criteria for limiting the levels of ultrasound to control auditory and subjective effects are summarized in Figs. 16.6 and 16.7. The data refer to levels at the head of the exposed person. The limiting levels at 20 kHz and above apply to protection from the subjective symptoms described earlier. Agreement is very good among the criteria developed independently by the three scientists cited in Fig. 16.6. The representative international and national criteria for airborne ultrasound are also quite similar. The World Health Organization (WHO) criterion was adopted directly from Acton.[17] Parrack's criteria were adopted by Sweden and the U.S. Con-

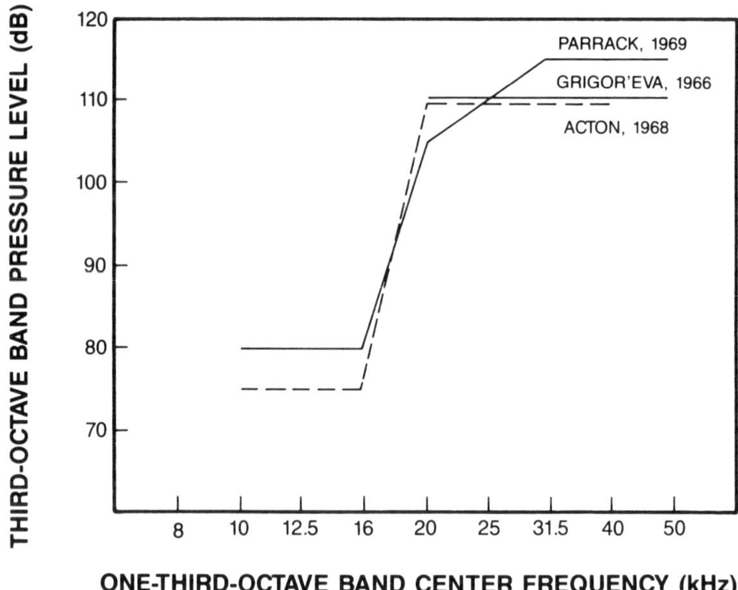

Fig. 16.6 Proposed criteria for exposure to airborne ultrasound.

ference of Governmental Industrial Hygienists. Norway accepts a level of 120 dB for frequencies higher than 22 kHz.

16.7 HEARING PROTECTION

The damage risk or noise exposure criteria presented above define allowable exposures for unprotected ears. These criteria are also used to identify exposure conditions above which personnel should be included in hearing conservation programs and where hearing protection should be worn. Hearing protection extends the limits of allowable exposures by reducing the levels at the ear, allowing personnel to experience more intense and longer duration exposures than with unprotected ears and still remain within established whole-body exposure criteria.

Hearing protector performance is influenced by characteristics of the wearer, of the protective device, and of the noise exposure. Protection is best with devices that provide good attenuation, are comfortable, fit properly, are easy to use, are in a good state of repair, and are worn. Hearing protector effectivenss is reduced by air leaks, transmission properties of the materials, and noise-induced motion of the device that produces sound under the protector. A limit on performance of ideal devices is imposed by the sound conduction properties of the tissue and bone of the head. In high-level acoustical fields, sound travels "around" the protector to the inner ear through the tissue and bone of the head. The level of the sound reaching the ear by such means is about 50 dB below that of air conduction reach-

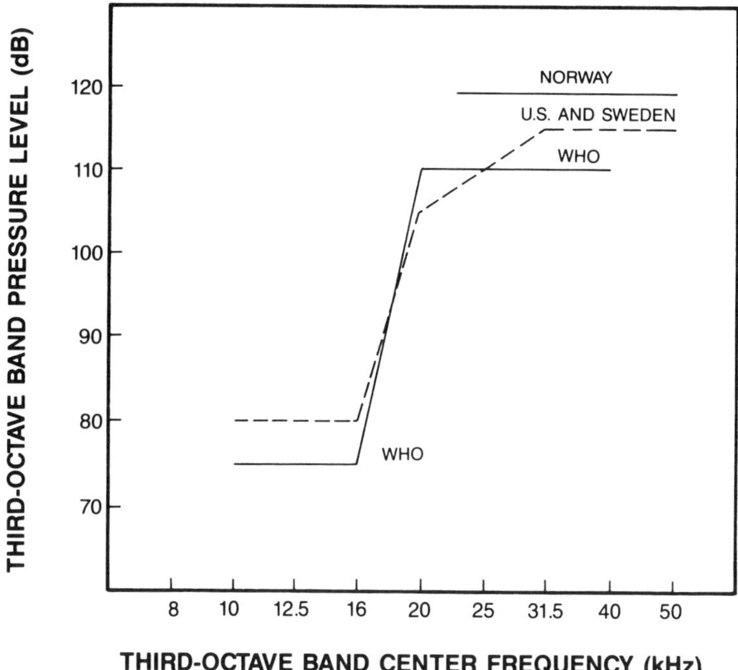

Fig. 16.7 International and national criteria for exposure to airborne ultrasound.

ing the ear through an open ear canal. Bone-conducted sound is not a major concern for most noise enviroments. Total-head enclosures extend the tissue and bone limits by about 10 dB.

Infrasound

Good insert-type hearing protection devices should provide attenuation of infrasound approaching that observed at the 125-Hz third-octave band. Earmuff hearing protectors provide very little protection and may even amplify the sound at some of these infrasonic frequencies. Exposures to levels of infrasound above 150 dB should be avoided even with maximal hearing protection because of possible adverse nonauditory effects.

Audio Frequency Region

Hearing protector performance varies as a function of the frequency spectrum of the noise. In general, conventional earplugs and earmuffs provide good sound protection at higher frequencies (above 1000 Hz). Earplugs also provide good protection at low frequencies with very good protection observed across all frequencies for certain foam inserts. Earmuffs provide poor attenuation and protection at the low frequencies.

A combination of earplug and earmuff is required when an individual protector is unable to reduce a noise to an acceptable level. The resulting attenuation is not the sum of the two protectors but an amount determined by the particular combination of devices. The attenuation of the combined units often reaches the bone conduction limits at the high frequencies. At the mid- and lower frequencies the amount of attenuation is determined primarily by the earplug. Selection of a good earplug for use with a muff will provide good double hearing protection at all frequencies.

Special earmuff-type hearing protectors, called nonlinear devices, allow face-to-face speech communication at low levels of noise and provide the typical amounts of earmuff protection at the high levels. Another type, called active noise reduction (ANR) device, utilizes electronic and acoustical means of reducing the low-frequency noise at the ear using noise cancellation techniques. The ANR device increases intelligibility and comfort and reportedly reduces fatigue when used with speech communication systems in noise.

Hearing protection devices have the potential capability to reduce and eliminate most noise-induced hearing loss. However, issues such as comfort, selection, fit, training, use, motivation, and others limit the full-performance capability of the devices to be realized in the work place.

Long-Duration Exposures

Long-duration noise exposures are typically steady state, continuous, and generally within the audio frequency range. The information on hearing protection for audio frequencies is also appropriate for long-duration noises.

Ultrasound

Conventional hearing protection devices, earplugs and earmuffs, provide good protection against airborne ultrasound at frequencies above about 20,000 Hz. Attenuation exceeding 30 dB is generally provided for frequencies from about 10,000 to 20,000 Hz. Hearing protection is most effective in eliminating subjective symptoms that occur when protectors are not worn. Reduction and elimination of subjective symptoms is also a good indication of the effectiveness of the hearing protector.

Impulses

Earmuff and earplug protectors should provide adequate attenuation for impulses comprised primarily of high-frequency energy, such as small-arms fire. Earmuff attenuation decreases as the concentration of energy in the impulses moves to the lower frequencies, as with large-caliber weapons (the attenuation vs. frequency of the protectors is not changing, only the spectral distribution of the impulse is different). The reduction of the peak sound pressure level of a particular earmuff is about 30 dB for pistol fire, about 18 dB for rifle fire, and as little as 5 dB for cannon fire. The peak level of the impulses of most pistol and standard rifle shots is reduced to less than 140 dB by good earmuffs. A combination of earplugs and earmuffs should be used for impulses requiring good low-frequency attenuation.

The USAF and the USA require single hearing protection when impulses reach 140 dB peak positive pressure and double hearing protection when the levels reach 160 dB (USAF) and 165 dB (USA).

16.8 HUMAN VIBRATION EXPOSURE

Vibratory energy can be transmitted to the human body via air to be received by the ear as sound or via solid structures to enter the body through contact areas with these structures such as feet, buttocks, hand, or arm. Although the ear is the most sensitive body organ for the reception of vibratory energy in the auditory frequency range for hearing of speech and other acoustical signals, structure-borne vibration at lower frequencies is easily transmitted through the contact areas and stimulates skin, bones, muscle, and joints and can be transmitted to all internal organs. Vibrations are part of our natural environment caused by body motion, running, and all types of muscle activity. However, their biological and psychological effects are usually negligible and harmless.

Previously, only a small part of the population might have been exposed to potentially dangerous levels of vibration while riding horseback or on horse-drawn vehicles. Today, practically the whole population is occasionally exposed to moderate levels of vibration through all means of land, sea, and air transportation. Normal living conditions and recreational activities are associated with vibration from a number of sources including automobiles, trains, aircraft, motorbikes, and boats.

Occupational exposures to vibration are more severe than nonoccupational exposures in both level of vibration and duration of the exposures as experienced by operators of agricultural and forestry tractors, earth-moving and construction equipment, all types of trucks, sea-going ships, airplanes and helicopters, as well as mining and factory workers on vibrating platforms. Most of these exposures are whole-body exposures transmitted through the feet or buttocks.[18] Hand–arm vibration caused by vibrating hand tools such as jackhammers, chipping hammers, chain saws, grinders, and other types of pneumatic or electric hand tools are of increasing occupational importance since they can result after years of exposure to a very specific vibration-induced disease, "white finger syndrome."[19]

Extensive research over the last four decades has led to a reasonable understanding of the potential pathological and physiological effects of whole-body as well as hand–arm vibration, although quantitative dose–response relationships are available only for hand–arm vibration. The potential manifestations of excessive whole-body vibration are not so unequivocally diagnosed and uniformly acknowledged. However, national and international standards have been agreed upon[20–23] that specify safe exposure levels and are generally relied upon as guidelines for the design of vehicles, equipment, structures, and for protection efforts.

The vibration environment is measured at the point of entry into the body. The vibration pick-up is fastened to the rigid structure transmitting the vibration to the body. When the vibration is transmitted by a resilient structure, such as a seat cushion, a rigid transducer support of suitable shape is interposed between the person and the structure. The linear vibrations of primary interest with respect to safety limits are measured in the appropriate directions of an orthogonal coordinate system: when standing or sitting, the z axis is used to describe whole-body vibrations in the foot- (or buttocks-) to-head directions, the x axis for the back-to-chest direction, and the y axis for the side-to-side direction. The origin of this internationally standardized biodynamic coordinate system[24] is at the heart. It is at the third metacarpal of the hand for hand–arm vibration with the x axis in the direction of the tool handle and the z axis in the direction of the arm.

Criteria are usually presented in terms of acceleration (m/s^2) or in gravity units ($1g = 9.8\ m/s^2$). The frequency range of primary interest for whole-body vibration is 1–80 Hz. Frequencies below 1 Hz are analyzed when motion sickness is of concern. The frequency range from 5 to 1500 Hz is usually monitored for hand–arm vibrations. Data are presented as rms values in third-octave bands. The crest factor, that is, the ratio of peak to rms value, must be reported.

The absorption of vibratory energy by the body is determined by the body's characteristics as a mechanical system.[25] Lumped-parameter models as well as finite-element models of the whole body as well as of its subsystems (hand–arm) give insight into the energy transmission through the body and explain the effects on specific target organs and on the overall sensitivity of the human body systems. Curves of equal sensitivity to vibration as a function of frequency have been derived for whole-body and for hand–arm vibration based on extensive studies. These curves, similar to the equal-loudness contours for hearing, change somewhat with the vibration magnitude and are not necessarily identical when the criterion to be described is equal subjective sensitivity or is equal short-term or long-term damage risk. However, since all these criteria basically reflect the mechanical response of the complex biological system and its individual variations, the curves have been simplified and standardized to reflect a practical average used to weight vibration environments with respect to their effects on humans.

The weighting functions in measuring instruments for whole-body vibrations in Table 16.2 are used in the ANSI[20,26] and ISO[21,27] standards to describe the whole-body sensitivity for the z axis and for the x and y directions. So far, the standards and the vibration exposure meter[28] use this set of curves for frequency weighting broadband whole-body vibration exposure for health risk, annoyance, interference with various activities, and fatigue. It is obvious that these weighting functions are simplifications averaging the complex variations introduced by body position and support and by the different types of response considered, for example, annoyance versus discomfort, visual acuity, or manual dexterity. Such additional weighting functions presently considered by standardizing organizations can be obtained from the literature[29,30] for particularly critical evaluations. The weighting function of Table 16.3 is used for the assessment of hand–arm vibration. (Tables 16.2 and 16.3 are the inverse of the curves of equal sensitivity presented in Figs. 16.8 and 16.9).

TABLE 16.2 Weighting Factors in Measuring Equipment Relative to Frequency Range of Maximum Acceleration Sensitivity[a] for the Response Curves of Fig. 16.8

Frequency, Hz	Weighting Factor	
	Longitudinal Vibrations[b] (z-axis)	Transverse Vibrations[c] (x, y-axes)
1.0	0.50 (−6 dB)	1.00 (0 dB)
1.25	0.56 (−5 dB)	1.00 (0 dB)
1.6	0.63 (−4 dB)	1.00 (0 dB)
2.0	0.71 (−3 dB)	1.00 (0 dB)
2.5	0.80 (−2 dB)	0.80 (−2 dB)
3.15	0.90 (−1 dB)	0.63 (−4 dB)
4.0	1.00 (0 dB)	0.5 (−6 dB)
5.0	1.00 (0 dB)	0.4 (−8 dB)
6.3	1.00 (0 dB)	0.315 (−10 dB)
8.0	1.00 (0 dB)	0.25 (−12 dB)
10.0	0.80 (−2 dB)	0.2 (−14 dB)
12.5	0.63 (−4 dB)	0.16 (−16 dB)
16.0	0.50 (−6 dB)	0.125 (−18 dB)
20.0	0.40 (−8 dB)	0.1 (−20 dB)
25.0	0.315 (−10 dB)	0.08 (−22 dB)
31.5	0.25 (−12 dB)	0.063 (−24 dB)
40.0	0.20 (−14 dB)	0.05 (−26 dB)
50.0	0.16 (−16 dB)	0.04 (−28 dB)
63.0	0.125 (−18 dB)	0.0315 (−30 dB)
80.0	0.10 (−20 dB)	0.025 (−32 dB)

[a] 4 to 8 Hz in case of z-axis vibration 1 to 2 Hz in case of y or x-axis vibration.

[b] Figure 16.8*a*.

[c] Figure 16.8*b*.

Source: Reprinted from ref. 20 by permission of the Acoustical Society of America.

In summary, broadband vibration exposures are characterized by the overall weighted acceleration values a_{z_w}, a_{x_w}, a_{y_w} for whole-body or hand–arm weighting, respectively.

16.9 HUMAN WHOLE-BODY VIBRATION GUIDELINES

No clear-cut whole-body vibration disease is generally recognized or has been described in spite of a large number of investigations in many countries.[31,32] Safe limits for instantaneous and/or short-term exposure are reasonably well established on the basis of accidental or voluntary experimental exposures. The mechanical

TABLE 16.3 Frequency Weighting in Measuring Instruments for Hand-transmitted Vibration[a]

Third-Octave-Band Center Frequency, Hz	Weighting Factor, W_f	Tolerance, % +	Tolerance, % −
6.3	1.000	12	37
8.0	1.000	12	21
10.0	1.000	12	12
12.5	1.000	12	12
16.0	1.000	12	12
20.0	0.794	12	12
25.0	0.630	12	12
31.5	0.500	12	12
40	0.397	12	12
50	0.315	12	12
63	0.250	12	12
80	0.198	12	12
100	0.157	12	12
125	0.125	12	12
160	0.099	12	12
200	0.079	12	12
250	0.063	12	12
315	0.050	12	12
400	0.039	12	12
500	0.031	12	12
630	0.025	12	12
800	0.020	12	12
1000	0.016	26	26
1250	0.012	26	37
Frequencies Outside Bands Defined Above, Hz			
0.315	0.003	26	∞
5.6	0.790	26	∞
1400	0.0088	26	∞
6300	0.0001	26	∞

[a]The tolerances and the weighting factors for frequencies outside of the third-octave bands are for frequency-weighting filters.

Source: Reprinted from ref. 26 by permission of the Acoustical Society of America.

stresses on the spinal column and the abdominal viscera (abdomen, kidney) lead to severe complaints and injuries. Chronic occupational exposures as they occur in drivers of tractors and earth-moving equipment are implicated by many investigations in the development of spinal column and other joint disorders and pathologies and of stomach and duodenal diseases.[31,32] However, their causal relationship with vibration stress has not been clearly proven, and certainly no dose–response relationship was ever proposed. Consequently, whole-body vibration exposure is not recognized as an occupational disease in most countries. The International Labor Organization (ILO) does list it with such diseases, and the regulations in some countries allow the consideration of vibration as an additional contributory risk factor in the development of some of the disorders mentioned.

Simple and practical exposure limits with respect to health and safety of the general population were first proposed in 1967 and then incorporated into the ANSI[20] and ISO[21] standards. The standards propose that these limits presented in Fig. 16.8 should not be exceeded without special analysis, justification, or precautions to protect the individual. The limits, described in rms acceleration, are considered valid up to a crest factor of approximately 6. These limits have been used for vibration control and health research and monitoring for over two decades. In many countries, no reports on health risks or injuries resulting from exposures below these limits have been reported. All studies on suspected health impairment were concerned with exposures above these limits.[32] While it might be argued that the limits are overprotective, particularly with respect to the much debated time dependency issue, there is growing evidence that they are clearly safe for the general population in "normal health."

The limits shown in Fig. 16.8 are widely used in the evaluation and design of "ride quality" of transportation vehicles and in the design of tractor seats and suspensions.

Task performance under vibration conditions depends on the nature and difficulty of the task and the reliability required. Since vibration can interfere with vision, speech, and manual dexterity and such interference can be reduced by appropriate ergonomic design of the specific tasks and the controls and displays involved, it is difficult to give general guidance with respect to acceptability. Nevertheless, the standards[20,21] present limits beyond which exposure can be regarded as carrying a significant risk of impaired working efficiency and potential fatigue known to worsen performance. The limits shown in Fig. 16.8 are for approximate general guidance. They are based on studies of aircraft pilots and vehicle drivers. For critical performance requirements, detailed analysis of the task or study of the specific references, or both, is indispensable.

Human reaction to vibration exposure in buildings is slightly modified because of the wide and different range of occupations and expectations of no environmental interference. Vibration may also be more troublesome to people at night than during the daytime. When lying down, the z-axis is parallel to the floor and the x-axis is vertical. Although the same frequency weighting curves for z axis and x, y axis exposure apply for standing, sitting, or lying, as shown in Table 16.2, a combined axis weighting is introduced to be used in cases when people use the same area in the standing, sitting, as well as reclining positon (sleeping). This combined weighting curve uses the z axis response from 1 to 2 Hz and the x, y axis response from 8 to 80 Hz. It is approximated by a low-pass filter with the attenuation $\text{dB} = 10 \log[1 + (f/5.6)^2]$. As stated in ref. 22, "building vibration frequencies range from approximately 10 Hz for low-rise structures to below 1 Hz for structures 10 stories and higher. Midfloor vibration frequencies are typically

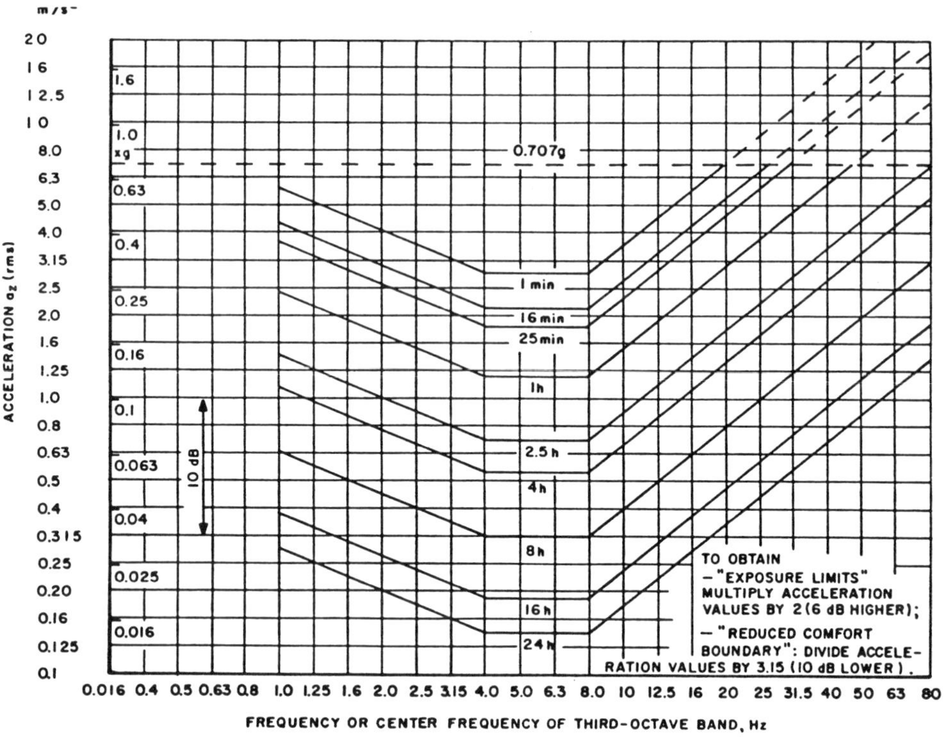

m/s²

ACCELERATION a_z (rms)

20, 16, 12.5, 10, 8.0, 6.3, 5.0, 4.0, 3.15, 2.5, 2.0, 1.6, 1.25, 1.0, 0.8, 0.63, 0.5, 0.4, 0.315, 0.25, 0.20, 0.16, 0.125, 0.1

xg: 1.6, 1.0, 0.63, 0.4, 0.25, 0.16, 0.1, 0.063, 0.04, 0.025, 0.016

0.707g

1 min, 16 min, 25 min, 1h, 2.5 h, 4h, 8h, 16h, 24h

10 dB

TO OBTAIN
– "EXPOSURE LIMITS" MULTIPLY ACCELERATION VALUES BY 2 (6 dB HIGHER);
– "REDUCED COMFORT BOUNDARY": DIVIDE ACCELERATION VALUES BY 3.15 (10 dB LOWER).

0.016 0.4 0.5 0.63 0.8 1.0 1.25 1.6 2.0 2.5 3.15 4.0 5.0 6.3 8.0 10 12.5 16 20 25 31.5 40 50 63 80

FREQUENCY OR CENTER FREQUENCY OF THIRD-OCTAVE BAND, Hz

(*a*)

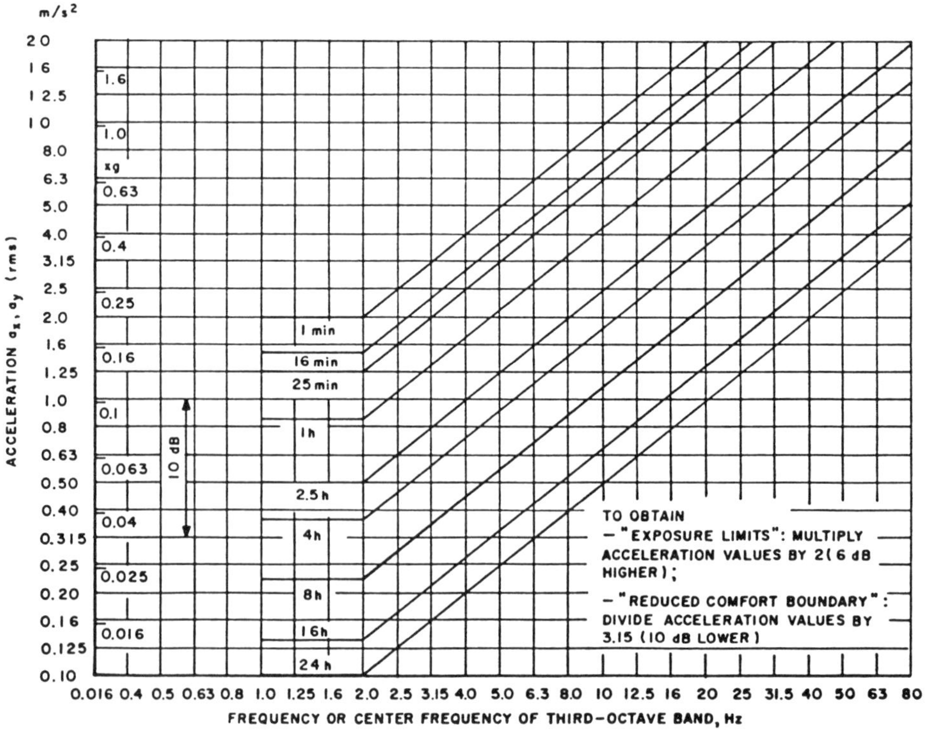

m/s²

ACCELERATION a_x, a_y (rms)

20, 16, 12.5, 10, 8.0, 6.3, 5.0, 4.0, 3.15, 2.5, 2.0, 1.6, 1.25, 1.0, 0.8, 0.63, 0.50, 0.40, 0.315, 0.25, 0.20, 0.16, 0.125, 0.10

xg: 1.6, 1.0, 0.63, 0.4, 0.25, 0.16, 0.1, 0.063, 0.04, 0.025, 0.016

1 min, 16 min, 25 min, 1h, 2.5 h, 4h, 8h, 16h, 24h

10 dB

TO OBTAIN
– "EXPOSURE LIMITS": MULTIPLY ACCELERATION VALUES BY 2 (6 dB HIGHER);
– "REDUCED COMFORT BOUNDARY": DIVIDE ACCELERATION VALUES BY 3.15 (10 dB LOWER)

0.016 0.4 0.5 0.63 0.8 1.0 1.25 1.6 2.0 2.5 3.15 4.0 5.0 6.3 8.0 10 12.5 16 20 25 31.5 40 50 63 80

FREQUENCY OR CENTER FREQUENCY OF THIRD-OCTAVE BAND, Hz

(*b*)

between 8 Hz and 16 Hz, although some massive large spans could have resonance frequencies as low as 4 Hz. Midwall frequencies are between 10 Hz and 25 Hz for light walls (e.g., wood framed residential structures) and possibly up to 50 Hz for masonry.''

Acceptability of vibration in buildings is to be evaluated relative to the ''base response curves'' which correspond to the approximate threshold of perception of the most sensitive humans. To these base response curves multiplication factors are applied to characterize higher vibration exposures. Although experience in many countries has shown that vibrations far in excess of the perception threshold are not acceptable for residential living, higher levels are usually in offices, workshops, and so on. The base response curves for rms vibration and velocity are given in Table 16.4 for the individual exposure axis as well as for the combined-axis rating[21,22]. Site multiplying factors that have been found satisfactory with respect to human response in many buildings are presented in Table 16.5. The multiplication factors for hospitals and critical working spaces are intended for use only when critical activities are taking place. When the vibrations are impulsive, the increased multiplication factors shown should not be used in the offices and workshop spaces without giving due consideration to significant disruption of working activity.

A formula for obtaining an approximate trade-off between the number of events per day and the amplitude of vibration is given by Eq. (16.1).[22] This multiplying factor F_n has been used in addition to the multiplying factors of Table 16.4 whenever $N > 3$:

$$F_n = 1.7N^{-0.5} \tag{16.1}$$

where N is the number of events per day ($N > 3$). Further formulas are given in Eq. (16.2) for taking into account the effect of discrete events with even durations, T, exceeding 1 s.[22] The factor F_d is used in addition to Table 16.4 and Eq. (16.1):

$$F_d = T^{-1.2} \quad \text{for concrete floors} \tag{16.2a}$$

$$= T^{-0.32} \quad \text{for wooden floors} \tag{16.2b}$$

Fig. 16.8 (*a*) Longitudinal (z axis) acceleration limits as a function of frequency and exposure time. The curves presented are the fatigue-decreased proficiency boundary. The exposure limits with respect to health and safety are reached by raising the curves by a factor of 2 (6 dB higher). The reduced comfort boundary (primarily in transportation vehicles) is obtained by dividing the acceleration values by 3.15 (lowering the acceleration by 10 dB). (*b*) Transverse (x and y axis) acceleration limits as a function of frequency and exposure time; curves are fatigue-decreased proficiency boundary. To obtain the exposure limits with respect to health and safety and the reduced comfort boundary, the curves should be raised or lowered as indicated. (Reprinted from ref. 20 by permission of the Acoustical Society of America.)

TABLE 16.4 Base Response Acceleration and Velocity Ratings for *z* Axis and *x*, *y* Axis Vibrations[a]

Frequency, Hz	Acceleration, m/s^2			Velocity, m/s		
	z Axis	*x*, *y* Axis	Combined Axis	*z* Axis	*x*, *y* Axis	Combined Axis
1	1.0×10^{-2}	3.6×10^{-3}	3.6×10^{-3}	1.6×10^{-3}	5.7×10^{-4}	5.7×10^{-4}
1.25	8.9×10^{-3}	3.6	3.6	1.1	4.6	4.6
1.6	8.0	3.6	3.6	8.0×10^{-4}	3.6	3.6
2	7.0	3.6	3.6	5.6	2.9	2.9
2.5	6.3	4.5	3.8	4.0	2.9	2.4
3.15	5.7	5.7	4.0	2.9	2.9	2.1
4	5.0	7.2	4.2	2.0	2.9	1.7
5	5.0	9.0	4.5	1.6	2.9	1.4
6.3	5.0	1.1×10^{-2}	4.7	1.3	2.9	1.2
8	5.0	1.4	5.0	1.0	2.9	1.0
10	8.3	1.8	8.3	1.0	2.9	1.0
12.5	7.8	2.3	7.8	1.0	2.9	1.0
16	1.0×10^{-2}	2.9	1.0×10^{-2}	1.0	2.9	1.0
20	1.3	3.6	1.3	1.0	2.9	1.0
25	1.6	4.5	1.6	1.0	2.9	1.0
31.5	2.0	5.7	2.0	1.0	2.9	1.0
40	2.5	7.2	2.5	1.0	2.9	1.0
50	3.1	9.0	3.1	1.0	2.9	1.0
63	3.9	0.11	3.9	1.0	2.9	1.0
80	5.0	0.14	5.0	1.0	2.9	1.0

[a]Root-mean-square (rms) accelerations and velocities are given. Base response ratings are intended for the most demanding conditions.

Source: Reprinted from ref. 22 by permission of the Acoustical Society of America.

16.10 HAND–ARM VIBRATION EXPOSURE CRITERIA

Excessive hand–arm vibration exposure can result in the *vibration-induced white finger disease* (VWF), or Raynaud's phenomenon, of occupational origin. This impairment of the blood circulation of the hand progresses with exposure time from intermittent numbness and tingling in selected fingers to extensive blanching of most fingers, first only in combination with cold and finally at all environmental temperatures. In the final stages the disease interferes severely with social activities and continuation of the occupation. According to Taylor and Pelmear[19], the clinical manifestations of the finger-blanching attacks progress in four stages with stage 1 signaling the onset of the disease, mainly outdoors in winter. While symptoms of the first stages might still be reversible if vibration exposure is discontinued, the later stages are considered irreversible by most researchers. In later stages, the vascular manifestations are reported to be accompanied by nerve, bone, joint,

TABLE 16.5 Site Multiplying Factors[a]

Site	Time, h	Continuous and Intermittent Vibration and Repeated Impulsive Shock	Impulsive Shock Excitation With Few Occurrences per Day ($N < 3$)
Hospital operating room and critical working areas	All	0.7–1	1
Residential (good environmental standard)	07:00–22:00	1.4–4	90
	22:00–24:00	1–1.4	1.4
	00:00–07:00	1–1.4	1.4
Office	All	4	128
Workshop	All	8	128

[a]Weighting factors for satisfactory magnitudes of building vibration in regard to human response. See text for additional details.

Source: Reprinted from ref. 22 by permission of the Acoustical Society of America.

and muscle involvement and the VWF is frequently replaced by the broader term *vibration syndrome*.[33]

The analysis of approximately 40 studies in various countries on workers exposed to hand tool vibrations allowed the derivation of the frequency-weighting curves indicating equal risk of developing VWF (Fig. 16.9)[26,34]. These weighting curves in turn allowed the different vibration exposure spectra observed in the different studies to be condensed into the single quantity called *weighted acceleration* characterizing the exposure condition. Combined with the epidemiological data on the exposure time and the time of onset of stage 1 of VWF, the dose–response relationship of Fig. 16.10 was obtained. These data allow the selection of the weighted acceleration estimated to prevent the onset of VWF in a selected percentile of a population exposed to this hand vibration for 4 h per day. The American Conference of Governmental Industrial Hygienists (ACGIH)[35] developed hand–arm vibration threshold limit values (TLVs) using protection from stage 3 upward as the limit criterion based on the same data that are incorporated in the ANSI and ISO standards. It arrives at the following TLVs:

$$a_{w_h} = 4\ \text{m/s}^2 \text{ for} \qquad 4\ \text{h} < \text{Exposure time} < 8\ \text{h}$$
$$a_{w_h} = 6\ \text{m/s}^2 \text{ for} \qquad 2\ \text{h} < \text{Exposure time} < 4\ \text{h}$$
$$a_{w_h} = 8\ \text{m/s}^2 \text{ for} \qquad 1\ \text{h} < \text{Exposure time} < 2\ \text{h}$$
$$a_{w_h} = 12\ \text{m/s}^2 \text{ for} \qquad \text{Exposure time} < 1\ \text{h}$$

These hand–arm vibration exposure guidelines and their background material, although still subject to revision based on new research results,[33] are gaining increasing recognition and use in the development of antivibration handles, gloves, and protective work rules.[36]

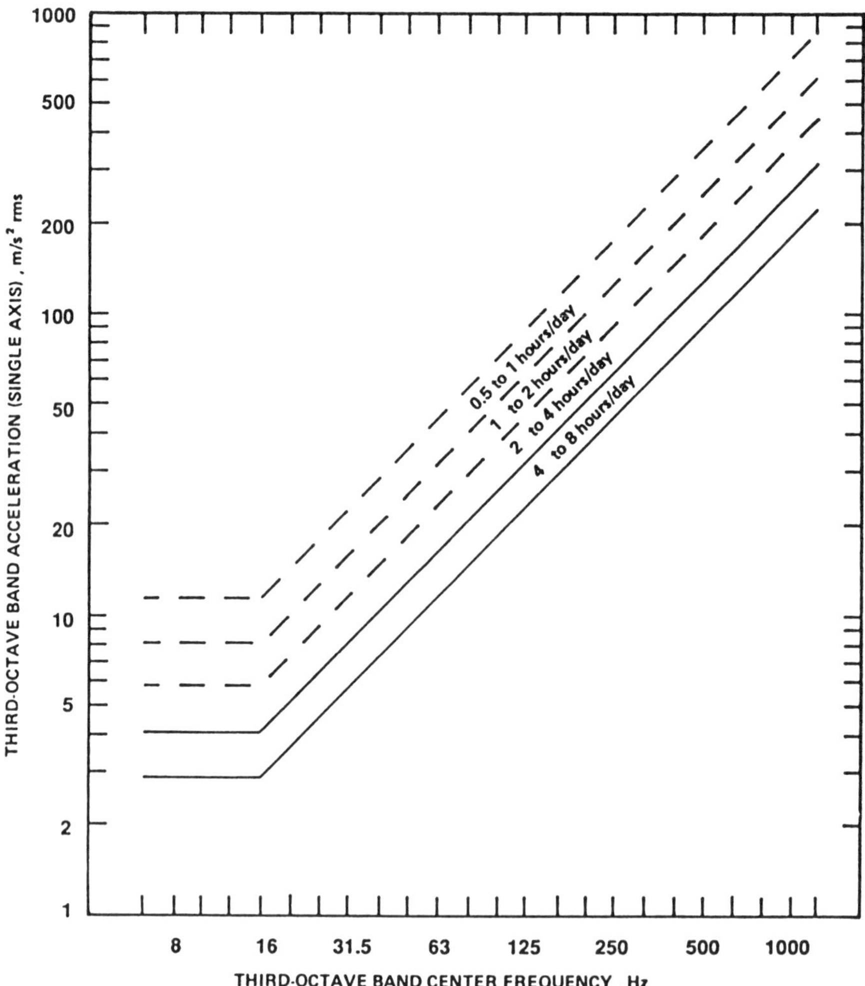

Fig. 16.9 Vibration exposure zones for the assessment of hand-transmitted vibration. The zones of daily exposure time are for rms accelerations of discrete-frequency vibration and for narrow-band or broadband vibration analyzed as third-octave-band rms acceleration. The values are for the dominant single-axis vibration generating compression of the flesh of the hand. The values are for regular daily exposure and for good coupling of the hand to the vibration source. (Reprinted from ref. 26 by permission of the Acoustical Society of America.)

16.11 COMBINED NOISE AND VIBRATION EXPOSURE

Although the combination of noise and vibration environments have a marked effect on the subject reaction to the environment and its annoyance rating, the interaction of these stimuli in affecting the health and safety limits to the separate stim-

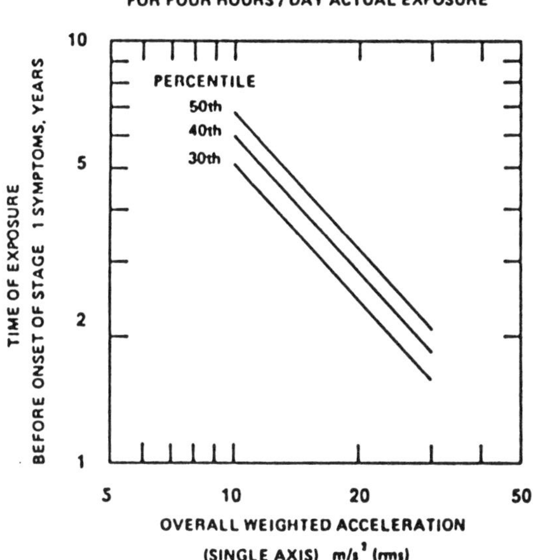

Fig. 16.10 Latent period for stage 1 symptoms of vibration syndrome for various percentiles of the population in terms of the overall weighted acceleration of the dominant single-axis vibration generating compression of the flesh of the hand. The data shown are for approximately 4 h per day of actual exposure to vibration. For other daily durations see references 26 and 34, on which this graph is based. (Reprinted from ref. 26 by permission of the Acoustical Society of America.)

uli appears minimal. The most frequently studied potential interaction is the effect of whole-body or arm-transmitted vibration on noise-induced temporary or permanent threshold shift (TTS or PTS). Human observations on such interaction are sparse and their results are conflicting. Although it is obvious that vibratory energy will reach the ear by bone and tissue conduction, this energy is, in most practical situations, small compared to the airborne stimulation required to produce temporary and permanent threshold shifts. The few studies that have found that NITTS is increased by simultaneous exposure to whole-body or hand–arm vibration utilized large vibration amplitudes. In spite of this, the change in TTS was only on the order of 0–5 dB. Available data are not sufficient, quantitative, or convincing enough to warrant at this point any modification of the health criteria recommended for exposure to the separate stimuli.[37]

REFERENCES

1. Environmental Protection Agency, "Information on Levels of Environmental Noise Requisite to Protect Public Health and Welfare with an Adequate Margin of Safety," U.S. EPA, Report No. 550/9-74004, Washington, DC, March 1974.

2. H. E. Von Gierke, D. Robinson, and S. J. Karmy, ''Results of the Workshop on Impulse Noise and Auditory Hazard,'' Institute of Sound and Vibration Research, Southampton, UK, ISVR Memorandum 618, October 1981. Also *J. Sound Vib.* **83**, 579–584 (1982).

3. J. Tonndorf, H. E. Von Gierke, and W. D. Ward, ''Criteria for Noise and Vibration Exposure,'' in C. M. Harris, (ed.), *Handbook of Noise Control*, 2nd ed., McGraw-Hill, New York, 1979.

4. C. W. Nixon, ''Excessive Noise Exposure,'' in S. Singh (ed.), *Measurement Procedures in Speech, Hearing, and Language*, University Park, Baltimore, MD, 1975.

5. International Standardization Organization, ''Acoustics-Determination of Occupational Noise Exposure and Estimation of Noise-induced Hearing Impairment,'' ISO 1999.2, Geneva 20, Switzerland, 1990.

6. K. D. Kryter, et al., ''Hazardous Exposure to Intermittent and Steady-State Noise,'' NAS-NRC Committee on Hearing, Bioacoustics, and Biomechanics, WG 46, Washington, DC, 1965.

7. Occupational Safety and Health Administration, ''Occupational Noise Exposure; Hearing Conservation Amendment,'' *Federal Register* **48**(46), 9738–9785 (1983).

8. National Center for Health Statistics, ''Hearing Levels of Adults by Age and Sex,'' Vital Statistics, Public Health Service Publication No. 1000, Series-11-No. 11, U.S. Government Printing Office, Washington, DC, 1965.

9. M. R. Stephenson, C. W. Nixon, and D. L. Johnson, ''Long Duration (24–48 Hours) Exposure to Continuous and Intermittent Noise,'' Joint EPA/USAF Report, AFAMRL Technical Report No. 82-92, Wright Patterson AFB, OH, 1982.

10. W. D. Ward et al., ''Proposed Damage Risk Criteria for Impulsive Noise (Gunfire),'' NAS-NRC Committee on Hearing, Bioacoustics, and Biomechanics, WG 46, Washington, DC, 1968.

11. R. R. A. Coles, G. R. Garinther, D. C. Hodge, and C. G. Rice, ''Hazardous Exposure to Impulse Noise,'' *J. Acoust. Soc. Am.* **43**, 336–343 (1968).

12. C. W. Nixon, H. H. Hille, H. C. Sommer, and E. Guild, ''Sonic Booms Resulting from Extremely Low Altitude Supersonic Flight: Measurements and Observations on Houses, Livestock, and People,'' AAMRL Technical Report No. 68-52, Wright Patterson AFB, OH, 1968.

13. C. W. Nixon, ''Human Auditory Response to an Air Bag Inflation Noise,'' Clearinghouse for Federal Scientific and Technical Information. PB-184-837, Arlington, VA, 1969.

14. Air Force Regulation 161-35, ''Hazardous Noise Exposure,'' 1982.

15. G. C. Mohr, J. N. Cole, E. Guild, and H. E. Von Gierke, ''Effects of Low Frequency and Infrasonic Noise on Man,'' *Aerospace Med.* **36**, 817–824 (1965).

16. C. W. Nixon and D. L. Johnson, ''Infrasound and Hearing,'' in W. D. Ward (ed.), *International Congress on Noise As A Public Health Problem*, U.S. Environmental Protection Agency, 550/9-73-008, Washington, DC, May 1973.

17. W. I. Acton, ''Exposure to Industrial Ultrasound: Hazards, Appraisal, and Control,'' *J. Soc. Occup. Med.* **33**, 107–113 (1983).

18. H. E. Von Gierke and D. E. Goldman, ''Effects of Shock and Vibration on Man,'' in C. M. Harris (ed.), *Shock and Vibration Handbook*, 3rd ed., McGraw-Hill, New York, 1988.

19. W. Taylor and P. L. Pelmear (eds.), ''Vibration White Fingers in Industry,'' Academic, London, 1975.

20. American National Standards Institute, ''Guide for the Evaluation of Human Exposure to Whole-Body Vibration,'' ANSI S3.18-1979, Acoustical Society of America, 1979.

21. International Standardization Organization (ISO), ''Evaluation of Human Exposure to Whole-Body Vibration,'' Part 1: ''General Requirements,'' ISO 2631/1-1985; Part 2: ''Continuous and Shock Induced Vibrations in Buildings (1–80 Hz),'' ISO 2631/2-1989; Part 3: ''Evaluation of Exposure to Whole-Body Vibration: z-Axis Veritcal Vibration in Frequency Range 0.1 to 0.63 Hz,'' Geneva, Switzerland.

22. American National Standard Guide to the Evaluation of Human Exposure to Vibration in Buildings, ANSI S3.29-1983, Acoustical Society of America, 1983.

23. International Standardization Organization (ISO), ''Guidelines for the Evaluation of the Response of Occupants of Fixed Structures, Especially Buildings and Off-Shore Structures, to Low-Frequency Horizontal Motion (0.063 to 1 Hz),'' ISO DP 6897, 1984.

24. International Standardization Organization (ISO), ''Mechanical Vibration and Shock—Standard Biodynamic Coordinate Systems,'' ISO DP 8727, 1983. ''American National Standard Mechanical Vibration and Shock Affecting Man—Vocabulary,'' S3.32-1982, Acoustical Society of America, 1982. International Standardization Organization (ISO), ''Mechanical Vibration and Shock Affecting Man—Vocabulary,'' ISO 5805, 1981.

25. H. E. Von Gierke, ''To Predict the Body's Strength,'' *Aviat. Space Environ. Med.*, **59**, A107–A115 (1988).

26. American National Standards Guide for the Measurement and Evaluation of Human Exposure to Vibration Transmitted to the Hand, ANSI S3.34-1986, Acoustical Society of America, 1986.

27. International Standardization Organization (ISO), ''Guidelines for the Measurement and the Assessment of Human Exposure to Hand-transmitted Vibration,'' ISO 5349-1985.

28. International Standardization Organization (ISO), ''Human Response to Vibration—Measurement Instrumentation,'' ISO DIS 8041, 1982.

29. M. J. Griffin and C. H. Lewis, ''A Review of the Effects of Vibration on Visual Acuity and Continuous Manual Control,'' Part I: ''Visual Acuity,'' Part II: ''Continuous Manual Control,'' *J. Sound Vib.* **56**(3), 383 (1978).

30. M. J. Griffin and E. M. Whitham, ''Time Dependence of Whole-Body Vibration Discomfort,'' *J. Acoust. Soc. Am.* **68**(5), 1522 (1980).

31. H. Dupuis and G. Zerlett, *The Effects of Whole-Body Vibration*, Springer-Verlag, Berlin, Heidelberg, New York, Tokyo, 1986.

32. H. Seidel and R. Heide, ''Long-Term Effects of Whole-Body Vibration: A Critical Survey of the Literature,'' *Int. Arch. Occup. Environ. Health* **58**, 1–26 1986.

33. U.S. Department of Health and Human Services, National Institute for Occupational Safety and Health (NIOSH), ''Occupational Exposure to Hand–Arm Vibration,'' DHHS (NIOSH), Publication No. 89-106, 1989.

34. A. J. Brammer and W. Taylor (eds.), *Vibration Effects on the Hand and Arm in Industry*, Wiley, New York, 1982.

35. ACGIH, "TLVs: Threshold Limit Values and Biological Exposure Indices for 1988–1989," American Conference of Governmental Industrial Hygienists, Cincinnati, OH, 1988, pp. 83–88.

36. American National Standard Guide for the Measurement and Evaluation of Gloves which are Used to Reduce Exposure to Vibration Transmitted to The Hand, ANSI S3.40-1989, Acoustical Society of America, 1989.

37. H. E. Von Gierke, "Exposure to Combined Noise and Vibration Environments," Proceedings of the Third International Congress on Noise as a Public Health Problem, ASHA Report No. 10, American Speech and Hearing Association, Rockville, MD, 1980.

CHAPTER 17

Criteria for Noise and Vibration in Communities, Buildings, and Vehicles

LEO L. BERANEK
Consultant
Cambridge, Massachusetts

Human beings are exposed to noise at work, leisure, and home and during transportation. Engineering criteria for specifying acceptable noise and vibration levels and for rating them in those locations have been developed over the past half century. The criteria presented in this chapter give guidance to the control of particular noise environments so as to accomplish one or more of the following: (a) permit satisfactory speech communication; (b) ameliorate annoyance; (c) improve sleeping conditions; and (d) improve public health and welfare over life spans. Criteria for noise levels to reduce the risk of damage to hearing were covered in the previous chapter. A portion of the material in this chapter was taken from an earlier text.[1]

17.1 SOUND LEVELS: DEFINITIONS

Different types of measurements and ratings are used in the control and evaluation of noise depending on purpose. The definitions needed in this chapter are

Noise and Vibration Control Engineering: Principles and Applications, Edited by Leo L. Beranek and István L. Vér.
ISBN 0-471-61751-2 © 1992 John Wiley & Sons, Inc.

given in Chapter 1, as taken from U.S. and international standards. They include the following:

Sound pressure level (SPL) L_p

A-weighted sound pressure level L_A

Average sound level $L_{\mathrm{av},T}$

Average A-weighted sound level L_{AT} (also called equivalent continuous A-weighted sound pressure level L_{eq})

Day–night sound level L_{dn}

A-weighted sound exposure E_{AT}

A-weighted noise exposure level L_{EAT}

17.2 CRITERIA FOR FACE-TO-FACE SPEECH COMMUNICATION

Engineering noise control most often seeks to achieve noise levels low enough for satisfactory speech communication. Except while working alone, or exercising, or in quiet meditation or asleep, people wish to communicate with each other or listen to speech or music from radio or television or in public gathering places. Music articulation and speech articulation are very similar. The tonal range of the violin, the most highly articulated instrument, is from about 200 to 8000 Hz, like that of the human voice.

Articulation Index (AI)

A procedure that calculates an articulation index for estimating the intelligibility of speech as heard by a listener in a continuous-noise background was first proposed by French and Steinberg[2] and developed further by Beranek.[3] The procedure by which the articulation index is derived is valid for speech in a noise that contains no prominent pure tones or whose one-third-octave-band spectrum has no prominent peaks or valleys. The elements of the concept are illustrated in Fig. 17.1.[4]

The middle solid curve is a plot of the average speech level produced at the position of a listener standing 1 m in front of a male talker with a raised voice. The shaded region encompasses 99% of the syllabic levels in each frequency band—the peaks rise 12 dB above and the minimums fall 18 dB below the average spectrum level. *Spectrum level* is the sound pressure or sound intensity level for $p(t)$ or $I(t)$ in Eq. (1.18) or (1.17) measured in or reduced to a bandwidth 1 Hz wide. It is also called the *power spectral density level* (see Section 1.2). When the spectrum level of a background noise is plotted on the same graph, the percentage of the middle shaded area that lies above the noise spectrum is called the *articulation index* (AI).

The frequency scale is divided into 20 bands, each of which is equally important to the intelligibility of speech syllables. For example, it is seen that four frequency regions of equal importance are approximately 380–1000 Hz, 1000–1740 Hz, 1740–2800 Hz, and 2800–5600 Hz.

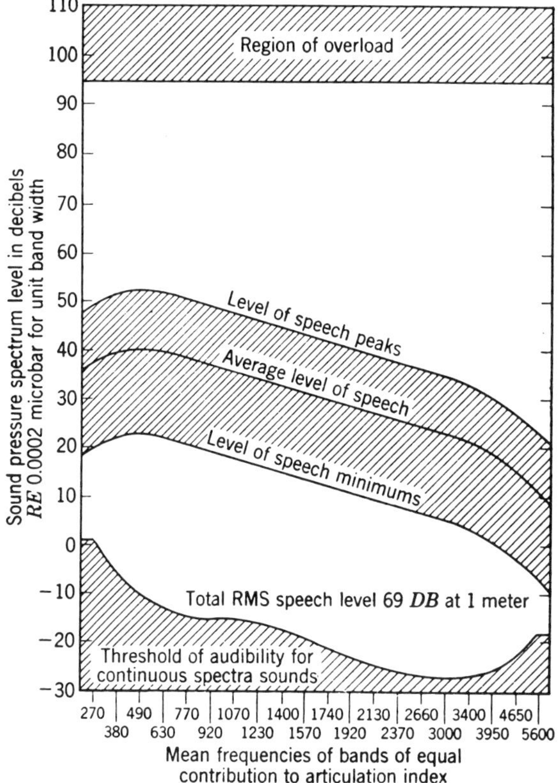

Fig. 17.1 Chart used for calculation of articulation index.[3]

The AI also becomes smaller if the speech is weak enough that part of it falls below the threshold of hearing at the bottom of the graph or is high enough (because of electronic amplification) to fall into the region of hearing overload at the top.[4]

Although many factors can influence the relation between the AI and word intelligibility, one can safely say that for an AI of 0.6 or more, conversation will be satisfactory. For an AI of 0.3 or less, speech communication will be unsatisfactory.

Speech Interference Level

The American National Standard definition for speech interference level (SIL) is as follows[5]:

> Speech-interference level in decibels is the arithmetic average of the sound pressure levels of the interfering noise *re* 20 μPa, in the four octave bands centered on the frequencies 500, 1000, 2000 and 4000 Hz.

The limiting frequencies for the four octave bands are 355–710, 710–1420, 1420–2840 and 2840–5680 Hz, respectively. These bands are not identical to but are sufficiently similar to the four equal-intelligibility regions listed above to be considered adequate substitutes.

Speech Communication Criterion Curves

Criterion curves for speech communication in noise, assuming that a talker and a listener are located in a free field (anechoic space), are given in Figs. 17.2 and 17.3.[6,7] Assuming an average female talker, Fig. 17.2 gives the permissible speech interference level, L_{SIL}, at a listener's position as a function of talker–listener separation r and the talker voice level. By "permissible" is meant "permitting just-reliable voice communication." Figure 17.3 is the same for an average male talker.

The charts are developed for an AI of approximately 0.5, which corresponds to a monosyllabic word intelligibility of at least 85%. The speech-level categories were originally developed by Clark et al.[8] Women's voices are about 4 dB weaker than men's. Figure 17.2 is applicable to mixed men–women conversations. The shaded regions indicate typical distances apart and corresponding voice levels of persons standing in noise levels, L_{SIL}.

The voice levels indicated must be adjusted upward for electronic amplification, focus, reflection from canopy, ground, or floor and downward for a barrier. Reflections from surfaces must occur within about 30 ms of the direct sound in order to merge. After a time delay of about 40 ms, reflections must be included as part of the noise or reverberation.[4]

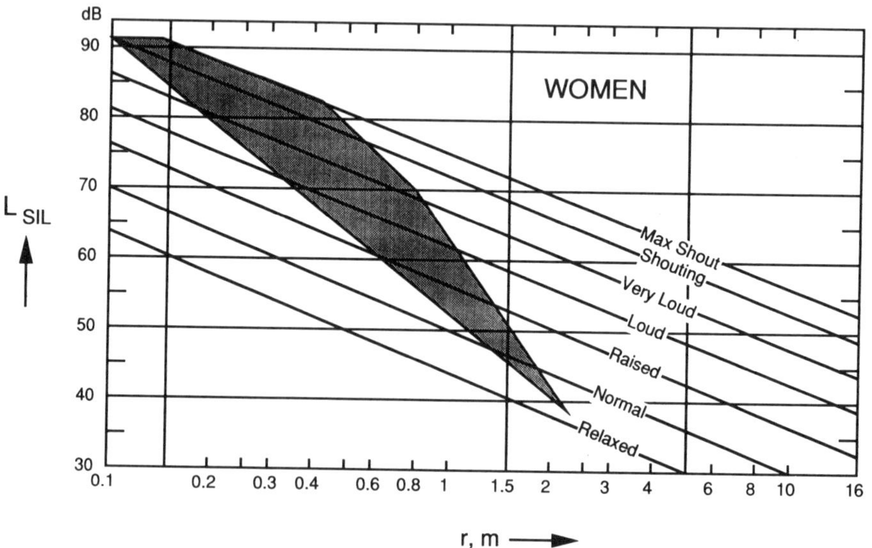

Fig. 17.2 Speech communication criterion curves, women's voices.[7]

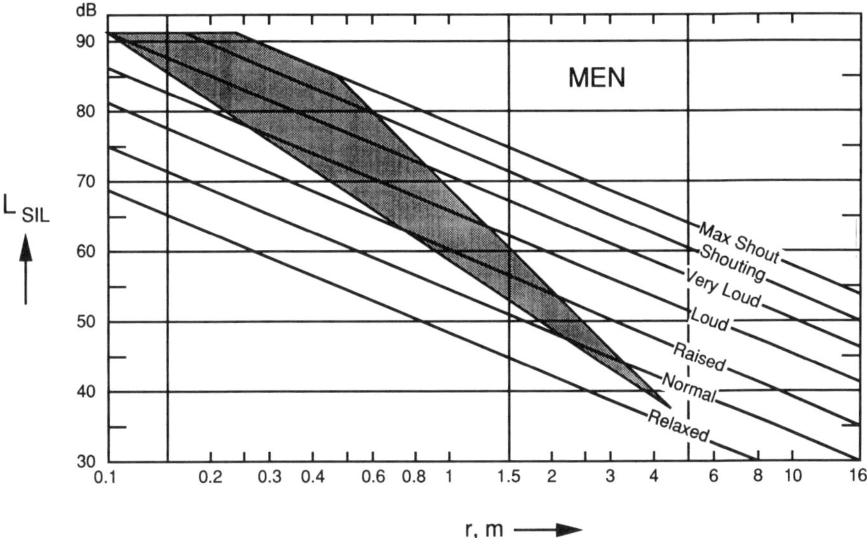

Fig. 17.3 Speech communication criterion curves, men's voices.[6]

As an example, assume a man and woman are seated in two adjacent seats in an airplane. They would be about 0.7 m apart and the SIL of the noise typically could be about 65 dB. From Fig. 17.2, the woman would probably speak with a voice level "loud." From Fig. 17.3, the man's voice level would probably be "raised." Because of reflection from the baggage rack, they might both be heard in the seats immediately ahead even though the distance would be greater.

From these graphs, the aircraft designer can choose a suitable level of conversational comfort and then specify the appropriate SIL in the seating areas.

17.3 LOUDNESS AND LOUDNESS LEVEL

General Overview

Several acoustical papers, including those by Zwicker,[9] Hellman and Zwicker,[10] Viebrock et al.,[11] and van Wyk[12] have recommended the adoption of *loudness* as a means of rating the psychological effects of noise on human beings. Loudness is a scale from which one can determine, at a given frequency, a relation between the SPL of a tone or a critical band of noise and its subjective loudness. By extension, the loudness of a noise can be calculated from its octave- or one-third-octave-band spectrum.[13,14]

By definition, a sound with a SPL of 40 dB at 1000 Hz has a loudness of 1.0 *sone*. A sound at this same frequency that sounds twice as loud (2 sones) has a SPL that is 9–10 dB greater. Each increase of 9–10 dB results in a further doubling of loudness. Similarly, for each halving of loudness the SPL decreases 9–10 dB.

The loudness of a noise can be converted to a logarithmic quantity called *loudness level*. The loudness level of a noise, with units *phons*, is defined as having a magnitude equal to the SPL in decibels of a 1000-Hz tone that sounds equally loud. A graph for converting between loudness and loudness level is given in Fig. 17.4.

The argument of those advocating loudness, in sones, for rating the noise at a particular location, centers around the logic that persons purchasing machinery, vehicles, or devices should be told that, for example, "the noise of *A* is twice as loud as that of B," rather than being told that "the noise level of *A* is 10 dB higher than that of *B*." As Zwicker says,[9]

> The citizen assumes that 50 dB(A) is half as loud as 100 dB(A) and that the reduction of the noise of (e.g.) an aircraft, from 90 dB(A) to 81 dB(A) is a reduction of only 10 percent.

Loudness, however, has low correlation with subjectively perceived annoyance, as Zwicker also seems to believe:

> From our point of view, annoyance, in contrast to loudness, has the advantage of being an unbalanced value which depends on the active or passive "use" of the noise in question. A motorcyclist riding his new machine is not at all annoyed by the noise he produces. On the other hand, the pedestrian, pushing a pram containing a sleeping child is very annoyed and so is the child! The point of view from which one listens to the noise influences the annoyance.

Broner[15] has reviewed the literature on the effects of low-frequency noise on people. He concludes:

> (a) ... annoyance due to low frequency noise which is experienced by sensitive people is more common than previously believed.
>
> (b) ... for noise containing high levels of low frequency noise, the common assumption that loudness and annoyance are equivalent breaks down. ...
>
> (c) ... annoyance did not arise from any particular sensitivity of individuals exposed but was due to the type of noise producing the annoyance [... often found in "transportation interiors, work and home ...] ... suggested that it was the slope and "turnover point" of the noise spectra of these low frequency noises rather than their absolute levels which could be important in determining annoyance. ...
>
> ... there is a wide range in the latter, from 120 dB inside motor vehicles down to 60 dB inside houses ... yet the same type of disturbance is produced at all levels.

Beranek has shown[16–19] that at work, in transportation, or in the home, while listening to radio or TV or conversing, annoyance may come from SIL that is too high for that type of space or levels in the low-frequency bands that are too high *relative* to the SIL or both. These types of annoyance occur at all overall levels, in agreement with Broner's findings and Zwicker's comment.

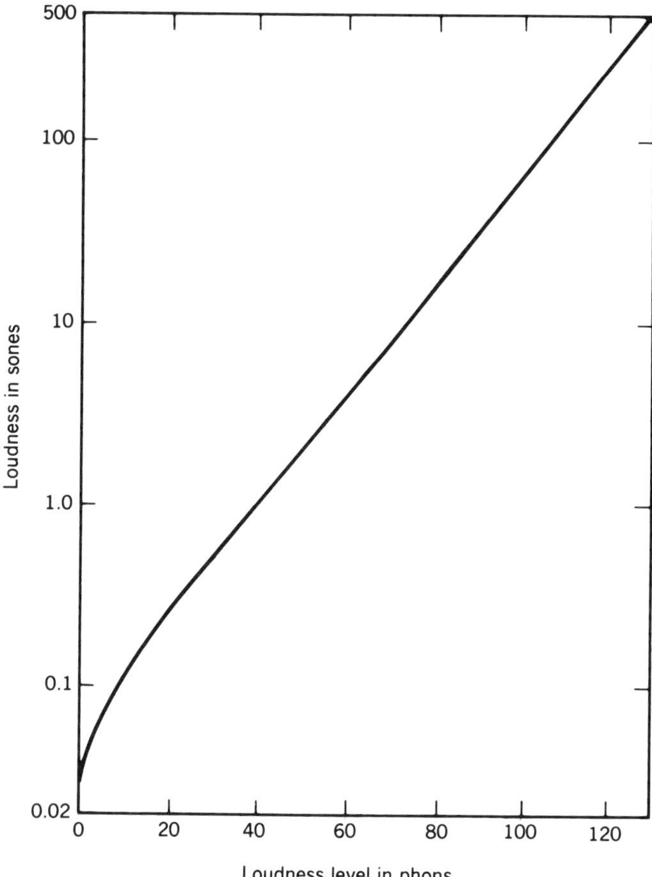

Fig. 17.4 Graph for converting from loudness to loudness level.[13]

Because the SIL is so important, another argument against a linear quality like loudness, is that the AI is determined from an area equal to the product of a logarithmic quantity (SPL) times increments along a frequency scale (Fig. 17.1). Also, if one examines the speech data of Clark et al.,[8] one finds that the distribution of successive SPLs in decibels in any frequency band versus time is Gaussian (normal distribution function). If sound pressure, a linear quantity, is substituted for the SPL, the distribution is badly skewed, and the AI that would be determined from such an area would not correlate well with the word intelligibility. Thus, a logarithmic quantity is required.

There may be examples where speech communication is not a need, such as while using a household appliance, and thus where loudness may be a better measure of comparing the relative annoyance of two *similar* products (probably not valid for comparing the relative annoyance of a food blender and a power lawn mower because one is used indoors and the other outdoors).

Calculation of Loudness

Loudness can be calculated from octave- or one-third-octave-band levels using the internationally standardized methods of Stevens, $L(S)$, or Zwicker, $L(Z)$.[9, 13, 14]

With the Stevens method, the loudnesses of the individual bands of noise are determined from their SPLs and are combined by a simple formula to give the overall loudness. In the formula, the band with the greatest loudness is weighted about three times as high as the loudness of each of the other bands.

Under the Zwicker method, the loudness of any one-third-octave band may include a contribution from the adjacent lower frequency band if the level of the lower frequency band is substantially greater. In other words, the Zwicker method takes into account the fact that a tone or a band of noise will mask weaker sounds in bands that are higher in frequency. This masking feature becomes important when the spectrum contains strong pure tone components or when one or more bands have levels that are significantly greater than the adjacent band(s) that lie above (in frequency).

Unfortunately, the two methods do not give the same overall loudness. It is commonly reported that Zwicker's method yields loudness levels that are 3.5–5.5 phons greater than the Stevens method. No standardizing body has resolved these differences, although per se they should not rule out the use of loudness or loudness level as a rating method where justified.

Approximations to Loudness and Loudness Level

It has become common practice throughout the world to measure the A-weighted SPL, L_A, of a noise and to assume that it is highly correlated with loudness level. This practice has arisen because of the universality of the internationally standardized sound-level meter and because several published papers have encouraged its use.[20–24] Those papers generally have based their acceptance of A-weighted sound level on subjective evaluations for which the test stimuli were continuous, random-like noises. Those test stimuli did not contain pure tones, nor were the levels in one or a few frequency bands considerably higher than those in bands of higher frequency.

Several papers have extended the comparisons just named to include prominent pure tones in the spectrum.[9, 10, 25] Those investigations have shown conclusively that loudness is better correlated with the Zwicker method of calculation, $L(Z)$, than with L_A. In one carefully chosen example, Hellman and Zwicker[10] show that a pink random-type noise extending from 100 to 7000 Hz that has superimposed on it a pure tone at 1000 Hz can measure (A-weighted) 6 dB less but be twice as loud. (A pink noise is one that when measured with octave or one-third-octave bands has a spectrum that is flat with band midfrequency.) If A-weighted, the band levels will differ according to Table 1.4. The example is shown in Fig. 17.5. The levels of the pure tone and the overall A-weighted level of the pink noise are given beside the datum points.

Suzuki et al.[25] have carried an even more elaborate experiment comparing L_A and LL(Z) with subjective experiments involving two types of randomlike noise

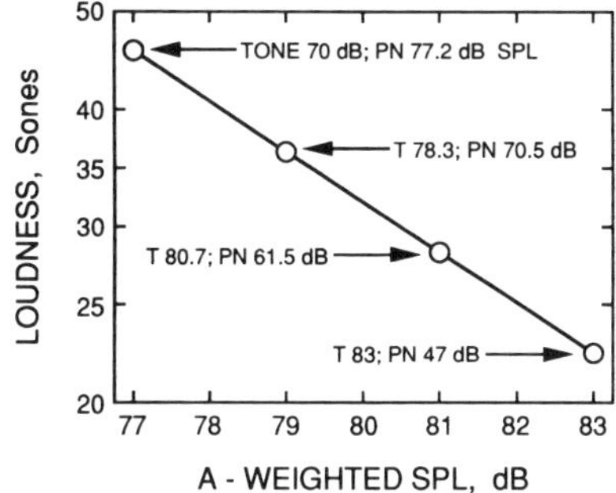

Fig. 17.5 Examples of loudness of combined "pink" noise and 1000-Hz pure tones. The calculated loudness of a pink noise–tone combination (ordinate) is plotted as a function of the A-weighted SPL of the complex noise. The pink noise levels are A-weighted.[10]

containing pure tones at different frequencies and levels. One randomlike noise was pink and the other was a third-octave-band spectrum that was flat between the 160- and 500-Hz bands and fell off −8 dB per octave below and −6 dB per octave from 500 to 2000 Hz and −12 dB above 2000 Hz. As the level of the pure tone increased with the random-noise spectrum held constant, the L_A levels were different compared to the subjective measurements by as much as 6–7 dB (pure-tone frequencies of 1000 and 500 Hz, respectively). By comparison the LL(Z) computations in phons were different a maximum of 1.5 dB at any frequency of the pure tone. Suzuki concludes that the Zwicker loudness LL(Z) is a satisfactory measure of the loudness of the two spectra studied but added, "there must be room for improvement in evaluating loudness of a complex noise including a tonal component."

Tachibana et al.[26] have selected a series of randomlike noises to present to a group of listeners who were asked to judge their loudnesses in comparison with a standard noise. All noises were presented to the listeners in an anechoic room from a full-range loudspeaker and amplifier that were compensated to have a flat frequency response. The standard (comparison) noise had a one-third-octave-band spectrum that decreased at a rate of −5 dB per octave over a range of midband frequencies from 31.5 Hz to 4 kHz. The A-weighted overall SPL of the standard noise was 35 dB in one series of judgments and 45 dB in another. No pure tones were present in the spectra. In all, 11 shapes of noise spectra were compared to the standard noise. Each spectrum had a slope that was −3 dB per octave, −5 dB per octave, −8 dB per octave, or a combination of the three.

A portion of the results are shown in Fig. 17.6. The five graphs are for (1) calculated Zwicker loudness level in phons, LL(Z) (uppermost graph); (2) calculated Stevens Mk. VI loudness level in phons, LL(S); (3) A-weighted sound level in decibels, L_A; (4) arithmetic average of all SPLs in octave bands from 63 Hz to 4 kHz, L(63–4k); and (5) the same from 31.5–4 kHz, L(31.5–4k). In each graph, the same method was used to calculate the "rating" of both the standard and the comparison noise. Thus, the straight line is for the standard noise because the comparison spectrum was the same in all cases.

The graphs clearly show that the arithmetic average of the levels in the bands from 31.5 Hz to 4 kHz or 63 Hz to 4 kHz and the Zwicker LL(Z) method give

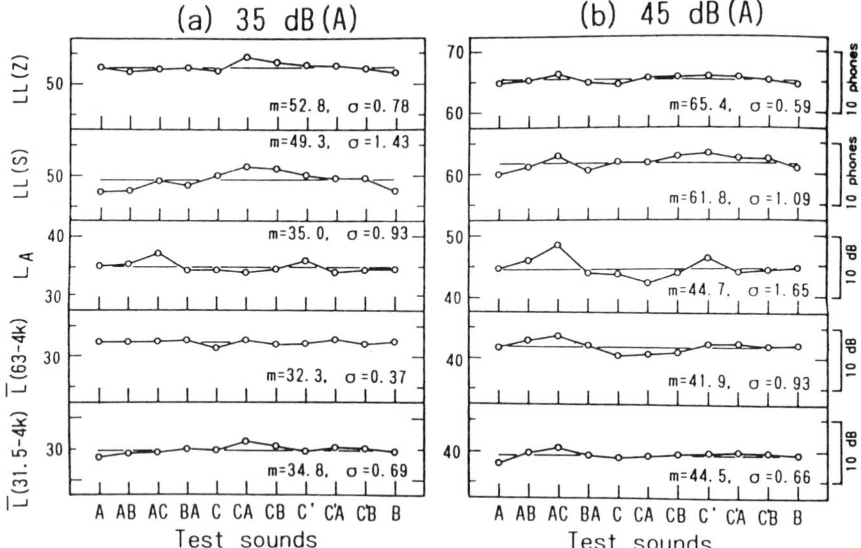

Fig. 17.6 Loudness level comparisons of 10 shapes of noise spectra with a standard noise. Two levels of noise were used. Five methods are shown for estimating loudness level.[26]

the best results and there is no advantage to the Zwicker method over band averaging when no prominent pure tones are present.

Two further tests were made using noises that had been transmitted through eight simulated walls. The source noises on the primary side of the walls were typical of indoor noise, M, and of rock music. The spectrum of the noise M was flat for the 250–1000-Hz bands and fell off in the frequency bands both above and below. Simulated transmission losses of 11 types of wall were chosen.

The results of the comparisons are shown in Fig. 17.7. Here, the arithmetic average of the levels in the bands from 63 to 4 kHz gave slightly better results than Zwicker LL(Z).

An important conclusion is that in both Figs. 17.6 and 17.7, A-weighted sound level L_A is not as satisfactory an estimate of loudness level as the arithmetic averages of band levels when no pure tones or prominent bands are present.

17.4 NOISE CRITERION CURVES FOR ENCLOSED SPACES

Noise criterion curves have two applications to assist in the writing of specifications for acceptable noise levels in a space and to rate the acceptability of a noise in an existing space. The noise criterion (NC) curves that follow are based on the premises that satisfactory conditions for speech communication or for listening to music, television, or radio are desired and that the levels in the octave bands above or below 1000 Hz compared to the level at 1000 Hz should not be so high as to produce an annoying spectrum.[16–19]

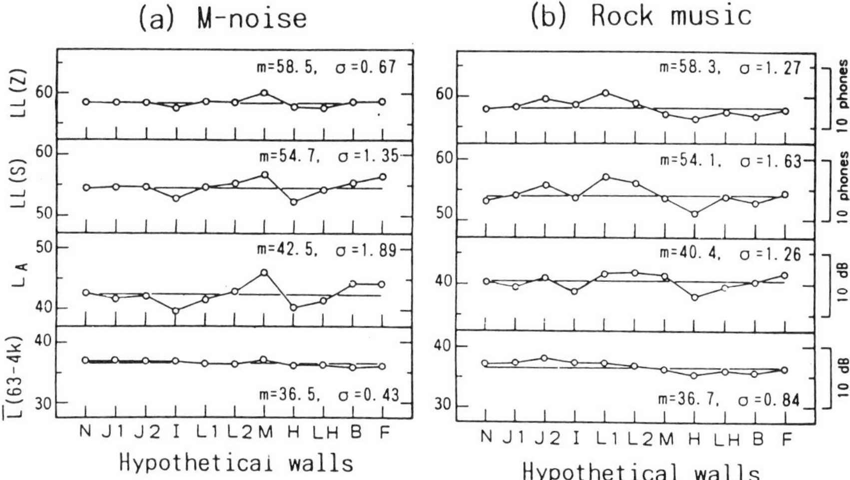

Fig. 17.7 Loudness level comparisons of noise through 11 walls having various transmission losses. Two noise sources were used: typical indoor noise (M) and rock music.[26]

Noise Criterion Curves (1957)[16,17]

The NC curves, shown in Fig. 17.8, were derived from extensive interviews of persons at work in office, industrial, and public spaces while, simultaneously, octave-band noise levels were being measured. The studies showed that the greatest contribution of noise to annoyance is interference with speech communication or with listening to music.

The NC-XX designation numbers on the curves are the average of the SPLs in the 600–1200-, 1200–2400-, and 2400–4800-ocatve frequency bands, now obsolete.[5]

The studies also showed that whenever calculated loudness levels, in phons, for a given curve (then using the method of Stevens Mk. II) were more than about 22 units above the levels, in decibels, in the three speech bands above, the persons interviewed expressed ''annoyance'' with the noise even though they were able to converse satisfactorily.

At that time, the author recommended that when determining compliance with a specified NC curve after the noise control work was completed, in no band should the measured level exceed the NC curve specified. That is to say, the measured data should be plotted in Fig. 17.8 and the NC curve that was tangent to the highest band level would establish the rating. The ''tangency'' method neither determines the adequacy of the space for speech communication nor measures spectrum imbalance. This shortcoming and the desire to extend the NC curves to lower frequency bands has led to the balanced noise criterion (NCB) curves that are described shortly.

Activity Noise The NC (and the NCB) curves apply to the total noise in a room, with normal office activities in progress and all mechanical systems operating. In

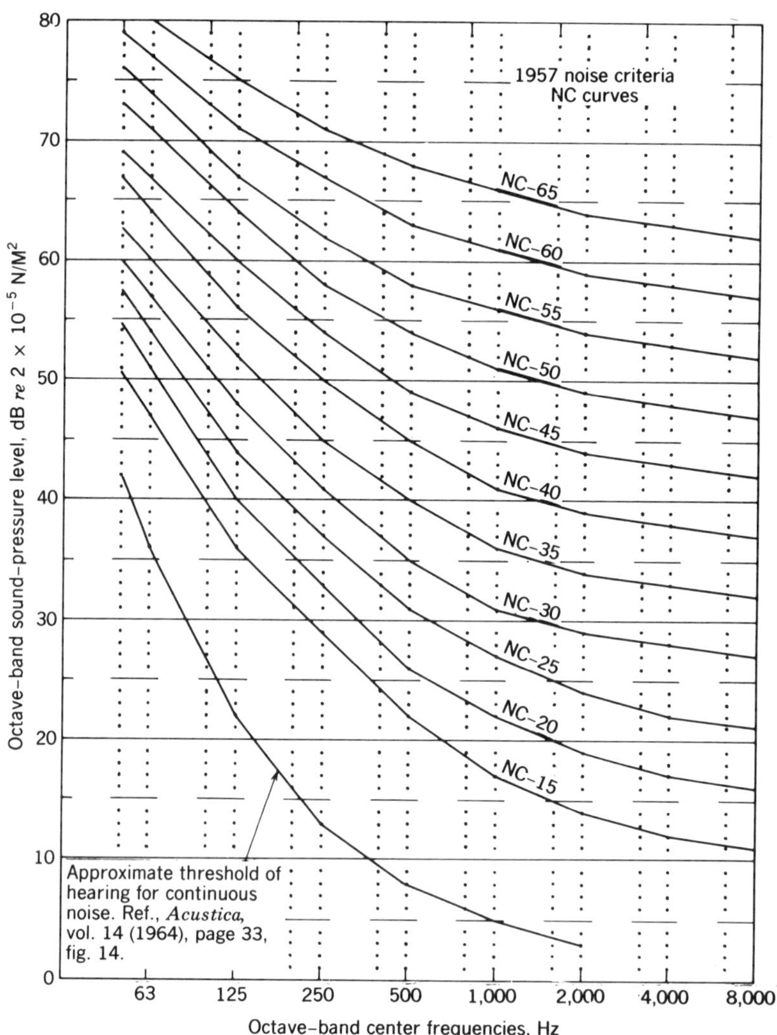

Fig. 17.8 Noise criterion (NC) curves derived from interviews and simultaneous noise measurements (1957).[16,17]

writing the specification for the air-handling system for a space, the expected activity noise for the space must be "subtracted" (on an energy basis) from the desired NC-XX (or NCB) curve to give the octave-band levels to be met later by the air-handling system when measured before the tenants move in.

As an example, curve *A* of Fig. 17.9 shows a typical noise spectrum measured in a large secretarial office with normal activity and curve *B* in the same room unoccupied with only the air-handling system operating. "Subtraction" of *B* from *A* on an energy basis yields the *activity noise* (curve *C*) of the space. If possible,

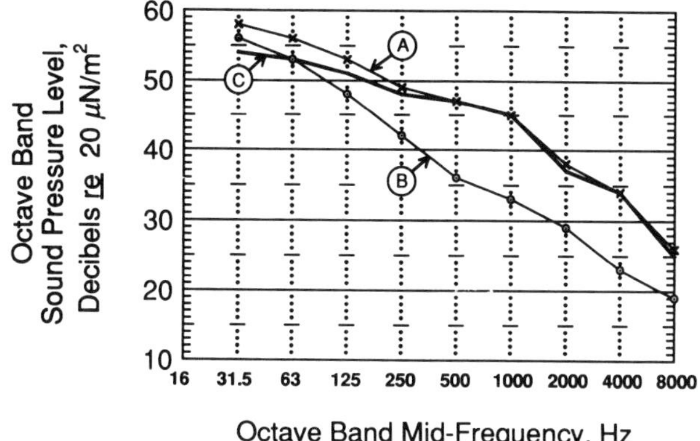

Fig. 17.9 Noise measured in large secretarial offices: *A*, with normal activity; *B*, unoccupied (HVAC noise); *C*, Activity noise per se.[19]

when writing a new specification, the acoustical engineer should measure *A* and *B* in an existing space where the similar activity takes place and determine the activity noise *C* expected for the new space.

Rumble and Hiss Occupants of a space speak of spectral imbalance using words like *rumbly*, *hissy*, *tiring*, and *annoying*. *Rumble* is used here as a generic term meaning excessive noise in the octave bands with midfrequencies below 1000 Hz, which generally arise from improperly designed air-handling systems. *Hiss* is used to mean excessive noise in the bands above 1000 Hz, which may arise from air flow noise or audible rattles induced by structural vibrations.

Room Criterion Curves (1981)[27]

Blazier derived the set of room criterion (RC) curves from a data base obtained during an ASHRAE-sponsored research project that yielded an average spectrum of noise measured in 68 unoccupied offices in which the noise levels were rated by occupants as "acceptable." The A-weighted sound levels for these spaces were in the range of 40–50 dB. The single curve that Blazier[27] derived from the ASHRAE data was so close to having a linearly decreasing spectrum at −5 dB per octave that Blazier made the anchor RC curve be a straight line with that slope and then drew the other criterion curves parallel to it. The RC curves so derived are plotted in Fig. 17.10. The rating number on each RC curve is the arithmetic average of the levels in the 500-, 1000-, and 2000-Hz bands. Because the only sources of noise in the ASHRAE studies were the air-handling systems, the curves may be used for specifying good current design of such systems. They do not take into account the hearing characteristics of persons.

An important contribution by Blazier are the two regions marked *A* and *B* in Fig. 17.10. Blazier states that in contemporary buildings with lightweight ceiling

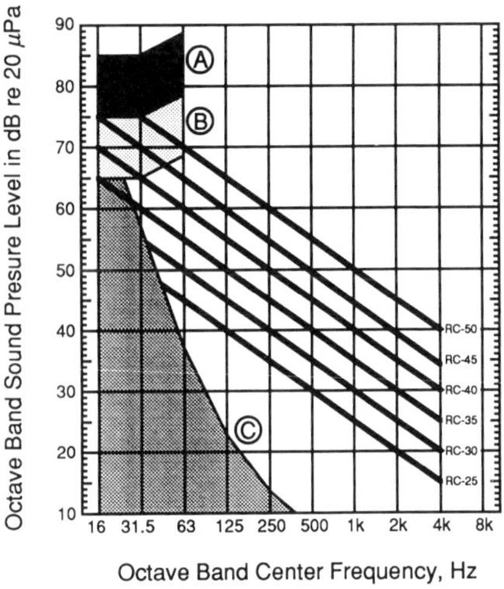

Fig. 17.10 Room noise criterion (RC) curves developed from measurements of HVAC noises in unoccupied offices: high (*A*) and low (*B*) probabilities of mechanical vibrations in rooms of lightweight construction; *C*, threshold of hearing for octave bands of noise. The linear shape of these curves represents good mechanical system design.[27]

and sidewall constructions (e.g., thin plaster or gypsum board on metal framing), mechanical vibrations may be induced by high noise levels in the octave bands below 125 Hz. In these constructions, noise levels that extend into or above region *A* are likely to produce noticeable vibrations (and often clearly annoying metallic rattles). Those that extend into region *B*, particulary the lower half, are less likely to cause noticeable vibrations.

The RC curves are more demanding than are the NC (and the NCB curves that follow) at low frequencies, certainly in those structures that are better built than assumed by Blazier. More important, the RC curves do not lead to a demonstrated means for determining the amount of noise reduction that is necessary to eliminate rumble or hiss in a postcompletion space. Finally, the RC curves were derived only for use in office spaces. A more complete set of curves is needed for other spaces such as concert halls, recording studios, auditoriums, factory spaces, industrial control rooms, and so on.

Balanced Noise Criterion Curves

The derivation of the NCB curves (see Fig. 17.11)[18,19] is based on two premises. First, the shape of the curves is determined by making equal the calculated octave-band loudnesses for those octave bands that contain the same number of critical bands.[28] The loudnesses of those octave bands containing fewer critical bands are weighted downward in proportion. Second, the rating number on an NCB curve is the average of the levels, in decibels, in the octave bands with midfrequencies at 500, 1000, 2000, and 4000 Hz, as specified in American National Standard ANSI S3.14-1977 (R-1986).[5]

The shape of the curves derived in this manner is modified slightly at the higher frequencies, yielding a slope of -3.33 dB per octave from 500 to 8000 Hz. The NCB curves incorporate regions *A* and *B* from Blazier, with the same meanings.

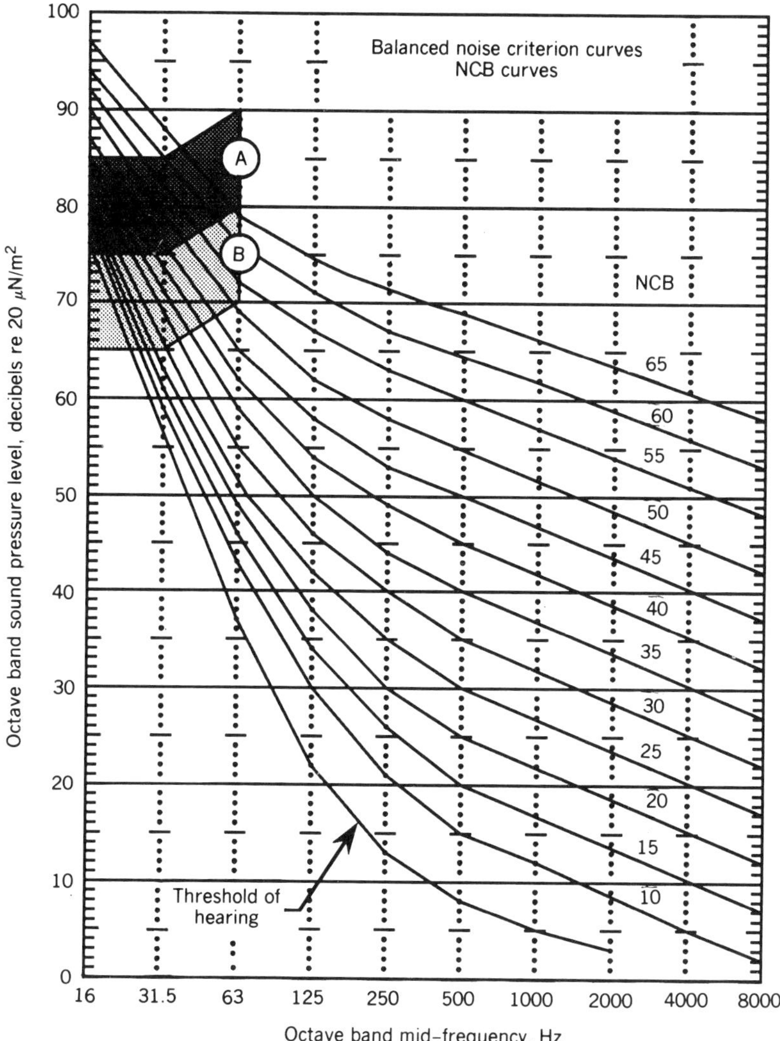

Fig. 17.11 Balanced noise criterion curves (NCB) (1989): Revised versions of NC curves, applicable to occupied rooms with HVAC systems operating. Levels in region *A* or above indicate high probability of clearly noticeable vibrations in gypsum board structures. Region *B* indicates low probability. The threshold of hearing is for octave bands.[18,19]

17.5 RECOMMENDED NCB CURVES FOR VARIOUS OCCUPIED, FUNCTIONAL, ACTIVITY AREAS

When writing specifications or judging the acceptibility of a space, a criterion curve must be chosen. Recommended NCB curves for a variety of occupied spaces are listed in Table 17.1. For almost every application a range of criterion values is shown. It is intended that the lower end of the range be selected if highly reliable speech communication or satisfaction with listening to music is desired. The upper end of the range will provide just satisfactory speech communication and should only be used if it is expected that the personnel in the space will not be overly critical.

17.6 USE OF NCB CURVES IN THE RATINGS OF NOISE OR DETERMINING CONFORMANCE WITH SPECIFICATIONS

When determining the conformance of the noise condition in a room with that specified, the octave-band spectrum of the noise in the space is usually measured before occupancy with all mechanical systems operating.[29] If the space is a private office or a conference room, this measured spectrum may be used directly for the rating. If the space is a general office or drafting room where there is appreciable activity noise, the noise of the expected activities must be "added," on an energy basis, to the noise measured in the unoccupied room (see Fig. 1.8). Then the following steps in rating the noise are followed:

Determination of Acceptability for Speech Communication

Step 1. The SIL is determined by averaging the noise levels in the 500-, 1000-, 2000-, and 4000-Hz frequency bands. The SIL is rounded to give an integral number. If the SIL so determined is equal to or less than the NCB-XX value specified, the speech communication requirement in the specification has been satisfied. If no specification exists, for example, in an existing space, the SIL is compared with the NCB values in Table 17.1 and a statement is possible as to the acceptability of the noise for speech communication in that type of space.

Determination of Whether Rumble Exists

Step 2. The measured data are plotted as in Fig. 17.12.

Step 3. To determine whether the spectrum shape is imbalanced, first add 3 dB to the SIL just determined to obtain an NCB-YY value. The curve for NCB-YY, from Fig. 17.11, interpolated to the nearest decibel is plotted on the same graph as the measured noise spectrum. If the level in any octave band with a midfrequency below 1000 Hz lies above the NCB-YY curve, the noise has rumble and the specification has not been met. In an existing space, the statement would be made, "The room is satisfactory for speech communication, but the levels in bands *T*, *U*, and *V* must be lowered by *X*, *Y*, and *Z* decibels, respectively, to prevent annoyance to room occupants."

Step 4. If the noise levels in any one of the three lowest octave bands fall within or above region *A* or within the upper half of region *B*, the room should be examined for feelable vibrations and for any resulting audible rattles.

Determination of Whether Hiss Exists

Step 5. If the SIL under step 1 and the rumble test were satisfactory, go to the set of NCB curves in Fig. 17.11 and determine, interpolating to the nearest decibel,

TABLE 17.1 Recommended Category Classification and Suggested Noise Criteria Range for Steady Background Noise as Heard in Various Indoor Occupied Functional Activity Areas

Type of Space (and Acoustical Requirements)	NCB Curve[a]	Approximate L_A, dB
Broadcast and recording studios (distant microphone pickup used)	10	18
Concert halls, opera houses, and recital halls (for listening to faint musical sounds)	10–15	18–23
Large auditoriums, large drama theaters, and large churches (for very good listening conditions)	Not to exceed 20	28
Broadcast, television, and recording studios (close microphone pickup used only)	Not to exceed 25	33
Small auditoriums, small theaters, small churches, music rehearsal rooms, large meeting and conference rooms (for very good listening), or executive offices and conference rooms for 50 people (no amplification)	Not to exceed 30	38
Bedrooms, sleeping quarters, hospitals, residences, apartments, hotels, motels, etc. (for sleeping, resting, relaxing)	25–40	38–48
Private or semiprivate offices, small conference rooms, classrooms, libraries, etc. (for good listening conditions)	30–40	38–48
Living rooms and drawing rooms in dwellings (for conversing or listening to radio and television)	30–40	38–48
Large offices, reception areas, retail shops and stores, cafeterias, restaurants, etc. (for moderately good listening conditions)	35–45	43–53
Lobbies, laboratory work spaces, drafting and engineering rooms, general secretarial areas (for fair listening conditions)	40–50	48–58
Light maintenance shops, industrial plant control rooms, office and computer equipment rooms, kitchens and laundries (for moderately fair listening conditions)	45–55	53–63
Shops, garages, etc. (for just acceptable speech and telephone communication). Levels above NC or NCB 60 are not recommended for any office or communication situation	50–60	58–68
For work spaces where speech or telephone communication is not required, but where there must be *no risk* of hearing damage	55–70	63–78

[a]See Fig. 17.11.

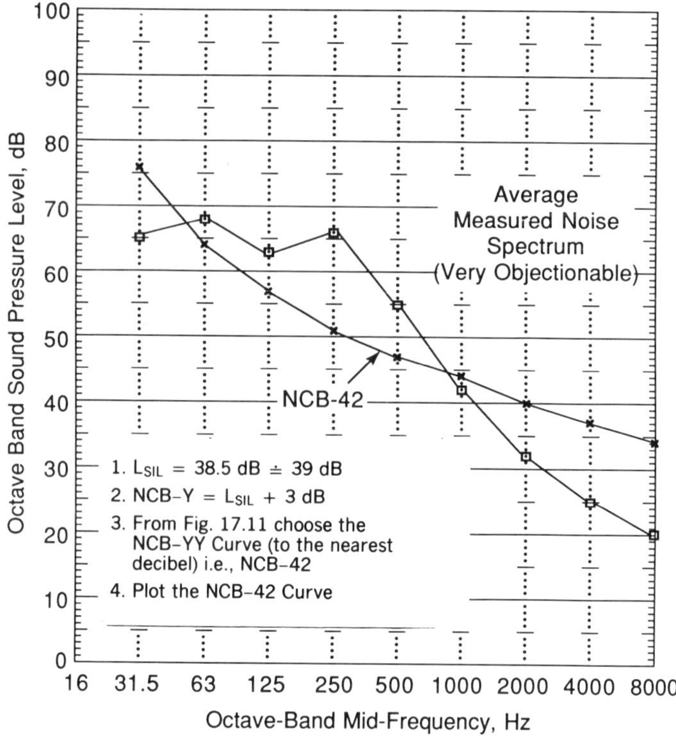

Fig. 17.12 Example of procedure for determining whether "rumble" exists using NCB curves.[19]

the NCB-ZZ curve that yields the best fit to the three bands from 125 to 500 Hz. If the levels in any of the bands with frequencies of 1000 Hz or greater lie above the NCB-ZZ curve, the spectrum is judged hissy and the levels should be reduced by the amounts indicated.

Can a single-number rating be given to a noise spectrum? In the author's opinion, a single number rating should not be given to a noise spectrum. An A-averaged noise level can be assigned, but it will not adequately measure the degree of spectrum imbalance. We have already argued that loudness in sones or loudness in phons states neither the adequacy of the space for speech commuication nor the degree of rumble or hiss. Similarly, NC, RC, or NCB ratings with modifiers like R for rumble or H for hiss are not adequate.

17.7 CRITERIA FOR NOISE IN COMMUNITIES: U.S. GOVERNMENT AND MUNICIPAL REGULATIONS[1]

The *U.S. Department of Housing and Urban Development (HUD)* has established regulations for noise abatement and control affecting the approval of new buildings and rehabilitated buildings constructed with financial assistance from the U.S. government.[30] The degree of acceptability of the noise environment at the proposed building site is determined from outdoor, day–night, average A-weighted sound level, L_{dn} [see Eq. (1.22)]. This level is obtained by adding together, on an energy basis, the mean-square A-weighted SPLs measured or taken from charts for each source including the general background noise level of nearby streets and highways and noise levels caused by aircraft flyovers and rail vehicle traffic.

The calculated (or measured) L_{dn} for a proposed site is placed in one of three categories: "normally acceptable." "normally unacceptable," and (absolutely) "unacceptable." The upper threshold of the normally acceptable L_{dn} noise level for the site is 65 dB provided there are no loud impulsive noises. If L_{dn} exceeds 65 dB, the site is normally unacceptable. If the noise level is between 65 and 70 dB, 5 dB of sound attenuation in addition to that of normal construction must be provided; and 10 dB additional for 70–75 dB. Above L_{dn} = 75 dB, the site is unacceptable.

The *U.S. Environmental Protection Agency (EPA)*, concerned only with protection of public health and welfare over life spans, has identified limiting L_{dn}

TABLE 17.2 Yearly Average[a] Sound Levels (dB) Identified as Requisite to Protect Public Health and Welfare With Adequate Margin of Safety[31]

		Indoor		Outdoor	
	Measure	Activity Interference	Hearing Loss Consideration[b]	Activity Interference	Hearing Loss Consideration[b]
Residential with outside space and farm residences	L_{dn}	45	—	55	
	L_{A24h}	—	70	—	70
Residential with no outside living space	L_{dn}	45	—	—	
	L_{A24h}	—	70		—
Commercial	L_{A24h}	[c]	70	[c]	70
Inside transportation	L_{A24h}	[c]	70	—	
Industrial	L_{A24h}[d]	[c]	70	[c]	70
Hospitals	L_{dn}	45	—	55	
	L_{A24h}	—	70	—	70
Educational	L_{A24h}	45	—	55	
Recreational areas	L_{A24h}	[c]	70	[c]	70
Farm and general unpopulated land	L_{A24h}	—	—	[c]	70

[a]Refers to average of time-mean-square sound pressures and not average of sound levels.

[b]Explanation: The exposure period that results in hearing loss at the identified average sound level is a period of 40 years.

[c]Since different types of activities appear to be associated with different sound levels, identification of a maximum sound level for activity interference may be difficult except in those circumstances where speech communication is a critical activity. See Section 17.2.

[d]An L_{A8h} of 75 dB may be identified in these situations so long as the exposure over the remaining 16 h per day is low enough to result in a negligible contribution to the 24-h average, i.e., no sound level greater than 60 dB.

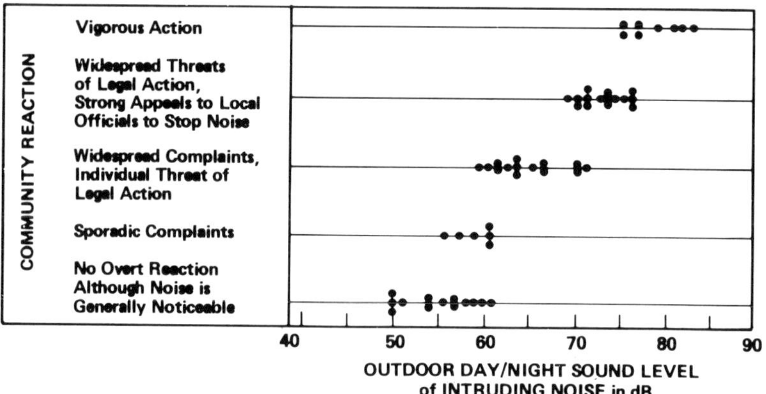

Fig. 17.13 Community responses to various outdoor day–night noise levels in numerous communities.[31]

levels in both indoor and outdoor spaces[31] (see Table 17.2). These L_{dn}'s are 45 dB for "indoor activity interference" and 55 dB for "outdoor activity interference." The EPA regards these levels as long-term goals, not standards. They are about 10 dB lower than HUD's levels "normally acceptable," which were designated for building sites.

The basis for EPA's choice of outdoor limiting levels appears to be Fig. 17.13. In living rooms in dwellings, where speech communication is most important, EPA's sound levels are several decibels too high for good intelligibility (for listening to music in concert halls, 25 dB too high). In Table 17.1, the A-weighted sound-level criterion for living rooms is given as 38–48 dB, compared to EPA's 45 dB.

Yaniv, Danner, and Bauer[32] found that for people exposed to recorded traffic noise in a simulated living room, under controlled laboratory conditions, the dividing line between average, L_A, sound levels that were "consistently acceptable" and "consistently unacceptable" by a majority was 50 dB. This number is only slightly higher than the 48 dB from Table 17.1 just quoted. The lower edge of the range in Table 17.1, 38 dB, is for a more stringent definition of acceptability.

Schultz[33] plots the relation between the percentage of people in residential areas who are "highly annoyed" for a range of outdoor L_{dn}'s in their neighborhoods. His finding is that very few are "highly annoyed" at 50-dB noise measured outdoors; 33% at 74 dB, 50% at 79 dB, and 70% at 85 dB. Thus EPA's 55 dB would cause, according to Schultz, 3% to be highly annoyed and HUD's 65 dB would cause 15% to be highly annoyed.

Typical Urban Noise

A summary of A-weighted noise levels measured in U.S. city areas, daytime and nighttime, is given in Fig. 17.14. It is seen that trucks and buses are principal sources of noise. Parenthetically, the levels in this figure are not L_{dn}, but rather the A-weighted levels in a residential area that, for example, vehicles produce

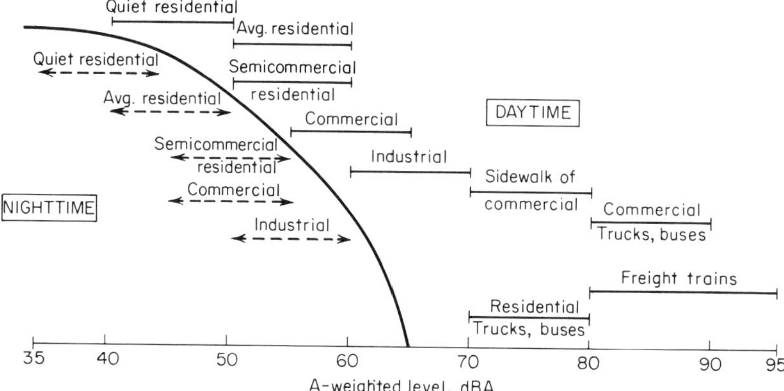

Fig. 17.14 Schematic display of typical A-weighted noise levels by day and by night in different U.S. urban areas. Noise around industrial areas is frequently caused by trucks.[1]

when they travel on the streets surrounding that neighborhood. In urban residential areas, the noise is frequently the roar of traffic on thoroughfares that may be in excess of a kilometer away. Traffic noise is likely to be relatively steady and go on all night at a reduced level. If thoroughfares exist on two or more sides of an area, the noise is often nearly uniform over the area.

U.S. City Noise Statutes Governing External Noise: External Industrial Noise. Very few cities have industrial noise statutes. Those that do generally express the limiting levels at the boundaries of residential areas. Either A-weighted sound levels, with "slow" meter setting, or octave-band spectra shaped similar to the NCB curves are specified. Nighttime limits, usually 22:00 to 07:00 hours, are likely to lie in the range of NCB-35 to NCB-50, with the median at NCB-43. This median is often obtained from a simple graphic-level recorder plots using "slow" response to avoid sharp peaks. This median corresponds to an A-weighted sound level of about 50 dB. Where daytime limits are given, they are usually about 5 to 10 dB higher.

City of San Diego's Noise Abatement and Control Ordinance. Section 59.5.0401 of San Diego's noise ordinance applies to operations in areas zoned "Industrial" that are adjacent to areas zoned "R-1 Residential." Paraphrased[34], it reads, "it is unlawful for an industrial operation to cause in an adjacent R-1 residential zone, a one-hour average sound level [L_{A1hr}] greater than the arithmetic mean of 75 and 50 decibels (62.5 decibels) between 7 a.m. and 7 p.m. (07:00–19:00); or greater than 60 dB between 7 p.m. and 10 p.m. (19:00–22:00); or greater than 57.5 decibels between 10 p.m. and 7 a.m. (22:00–07:00) the next morning. The noise subject to these limits is that part of the total noise at a given location due solely to the industrial operation."

City of Chicago's Industrial Noise Code (1984). This code is one of the most detailed.[35] It specifies that a plant, operation, or combined operations in a

TABLE 17.3 Maximum A-weighted Sound Levels (dB) Permitted by Chicago's Noise Code for Three Categories of Manufacturing[a]

Octave-Band Center Frequency	Type A, Restricted Manufacturing Zoning District	Type B, General Manufacturing Zoning District	Type C, Heavy Manufacturing Zoning District
31.5	72	72	75
63	71	71	74
125	65	66	69
250	57	60	64
500	51	54	58
1000	45	49	52
2000	39	44	47
4000	34	40	43
8000	32	37	40
A-scale level (for monitoring purposes)	55	58	61

[a]See text for measurement locations.

district zoned for restricted manufacturing (see Type A, Table 17.3) shall not caused SPLs along the district boundaries of an adjoining residence district that exceed the values given for Type A. For industrial operations, Types B and C, the levels specified must be met either at the nearest residential boundary or at 20 m from the nearest property line of the plant or operation, whichever is the greater distance.

For activities carried out in districts zoned for business and commercial (non-manufacturing) purposes, average sound levels produced by sound sources in those districts must not exceed those for Type A at the nearest residential district.

If there is no adjoining residence or business–commercial district, districts B and C shall meet the maximum SPLs for Type A at the nearest residence or business–commercial boundary line.

For all three types of manufacturing, the noise levels produced by their operations at the boundary of districts zoned for business and commercial purposes may not exceed those for Types A and B by more than 7 dB or Type C by more than 5 dB.

External Truck, Bus, Motorcycle, and Automobile Noise[36]

Regulations in the United States governing noise made by automobiles and trucks on highways are based on measurements of maximum, fast, A-weighted sound levels made at a position 50 ft from the centerline of the traffic line. The permissible fast sound-level limits for speeds above 35 mph (56 km/h) generally fall between 82 and 92 dB, with the median at 87 dB.

California's motor vehicle code[37] specifies 82 dB for automobiles, 90 dB for heavy trucks and buses, and 82 dB for motorcycles. New vehicles must not produce noise levels higher than 84 dB for automobiles and 86 for trucks, buses, and motorcycles under a specified list of tests.

Detailed wording of Chicago's ordinance is given in ref. 1. In brief, for new vehicle operating over 35 mph (56 km/h), Chicago's vehicle code[35] specifies 75 dB for automobiles, 78 dB for motorcycles, 84 dB for trucks, and 77 dB for buses. For existing vehicles, the numbers become 79 dB for automobiles, 82 dB for motorcycles, and 90 dB for trucks.

With good mufflers, modern automobiles traveling on smooth road surfaces at speeds of 65 mph (105 km/h) can easily meet a 78-dB requirement at 50 ft from the centerline of the traffic lane.[36] Trucks with good mufflers can easily meet an 86-dB requirement at 50 mph (80 km/h) up a 4% grade. Trucks with straight stacks (no muffler) produce a level of approximately 95 dB under the same conditions.

REFERENCES

1. L. L. Beranek (ed.), *Noise and Vibration Control*, rev. ed., Institute of Noise Control Engineering, Poughkeepsie, NY, 1988 Chapter 18. (Original edition by McGraw-Hill, 1971.)
2. N. R. French and J. C. Steinberg, "Factors Governing the Intelligibility of Speech Sounds," *J. Acoust. Soc. Am.* **19**, 90–119 (1947).
3. L. L. Beranek, "The Design of Speech Communication Systems," *Proc. Int. Radio Eng. (IEEE)* **35**, 880–890 (1947).
4. L. L. Beranek, *Acoustics*, McGraw-Hill, New York, 1954. Reprinted with modifications by Acoustical Society of America, New York, 1986, pp. 406–416.
5. American National Standard, S3.14-1977 (R-1986), "Rating Noise With Respect to Speech Interference," Acoustical Society of America, New York, 1977, 1986.
6. L. L. Beranek (ed.), *Noise and Vibration Control*, rev. ed., Institute of Noise Control Engineering, Poughkeepsie, NY, 1988, pp. 558–560.
7. H. Lazarus, "Prediction of Verbal Communication in Noise—A Development of Generalized SIL Curves and the Quality of Communication," *Appl. Acoust.* **20**, 245–261, **21**, 325 (1987).
8. K. C. Clark, H. W. Rudmose, J. C. Eisenstein, F. D. Carlson, and R. A. Walker, "The Effects of High Altitude on Speech," *J. Acoust. Soc. Am.* **20**, 767–786 (1948).
9. E. Zwicker, "A Means for Calculating Loudness" (in German), *Acustica* **10**, 304–308 (1960); "Meaningful Noise Measurement and Effective Noise Reduction," *Noise Control Eng. J.*, **29**, 66–76 (1987).
10. R. Hellman and E. Zwicker, "Why Can a Decrease in dB(A) Produce an Increase in Loudness?" *J. Acoust. Soc. Am.* **82**, 1700–1705 (1987).
11. V. M. Viebrock, M. J. Crocker, and W. A. Cooper, "Loudness Evaluations of Electric Clock Noise," *Appl. Acoust.* **8**, 193–201 (1975).
12. A. J. Van Wyk, "A Comparison of Measurement Methods for Assessing Human Perception of Loudness: An International Survey," *Acustica* **49**, 33–46 (1981).
13. International Standard, ISO-532-1975, "Acoustics—Methods for Calculating Loudness Level," American National Standards Institute, New York, 1975.

14. American National Standard, ANSI S3.4-1980 (R1986), "Procedure for the Computation of the Loudness of Noise," Acoustical Society of America, New York, 1980 (1986).

15. N. Broner, "Effects of Low Frequency Noise on People—A Review," *J. Sound Vib.*, **58**, 483–500 (1978).

16. L. L. Beranek, "Criteria for Office Quieting Based on Questionaire Ratings," *J. Acoust. Soc. Am.* **28**, 833–852 (1956).

17. L. L. Beranek, "Revised Criteria for Noise Control in Buildings," *Noise Control* **3**, 19–27 (1957).

18. L. L. Beranek, "Balanced Noise Criterion (NCB) Curves," *J. Acoust. Soc. Am.* **86**, 650–664 (1989).

19. L. L. Beranek, "Applications of NCB Noise Criterion Curves," *Noise Control Eng.* **33**, 45–56 (1989).

20. R. W. Young, "Don't Forget the Simple Sound Level Meter," *Noise Control*, **4**, 42–43 (1958). See also J. H. Botsford, "Using Sound Levels to Gauge the Human Response to Noise," *Sound Vib.* **3**, 16–28 (1969).

21. R. W. Young, "Single Number Criteria for Room Noise," *J. Acoust. Soc. Am.* **36**, 289–295 (1964).

22. R. W. Young, "Measurements of Noise Level and Exposure," in J. D. Chalupnik (ed.), *Transportation Noises*, A symposium on Acceptability Criteria, University of Washington, Seattle, WA, 1970.

23. R. W. Young and A. Peterson, "On Estimating Noisiness of Aricraft Sounds," *J. Acoust. Soc. Am.* **45**, 834–838 (1969).

24. R. W. Young, "Re-vision of the Speech-Privacy Calculations," *J. Acoust. Soc. Am.* **38**, 524–530 (1965).

25. Y. Suzuki, T. Sone, H. Sato, and S. Kono, "An Experimental Consideration on the Tone Correction (or Adjustment) for Environmental Noise Evaluation," *Proc. INTER-NOISE-86*, 1986, pp. 849–854.

26. H. Tachibana, H. Yano, and Y. Sonoda, "Subjective Assessment of Indoor Noises—Basic Experiments With Artificial Sounds," Proceedings of International Symposium on Environmental Acoustics," Kobe University, Kobe, Japan, May 15–16, 1989.

27. W. E. Blazier, Jr., "Revised Noise Criteria for Application in the Acoustical Design and Rating of HVAC Systems," *Noise Control Eng.* **16**, 64–73 (1981).

28. S. S. Stevens, "Perceived Level of Noise by Mark VII and Decibels (E)," *J. Acoust. Soc. Am.* **51**, 575–600 (1972).

29. L. L. Beranek, *Acoustical Measurements*, 2nd ed. American Institute of Physics, Woodbury, NY, 1988, Chapter 8. (Original edition, Wiley, 1949.)

30. "Environmental Criteria and Standards," *U.S. Federal Register* **44**(135, Pt 51), 40860–40866, (1979); also W. J. Galloway and T. J. Schultz, "Noise Assessment Guidelines—1979," Report No. HUD-CPD-586, U.S. Department of Housing and Urban Development, Washington, DC.

31. "Information on Levels of Environmental Noise Requisite to Protect Health and Welfare With an Adequate Margin of Safety," EPA Report No. 550/9-74-004, U.S. Environmental Protection Agency, Washington, DC, March 1974.

32. S. L. Yaniv, W. F. Danner, and J. W. Bauer, "Measurement and Prediction of Annoyance Caused by Time Varying Highway Noise," *J. Acoust. Soc. Am.* **72**, 200–207 (1982).

33. T. J. Schultz, *Community Noise Rating*, 2nd ed., Applied Science Publishers, New York, 1982.

34. R. W. Young, "Optimum Levels for Reporting Community Noise," private communication, January 1988.

35. Chicago, IL, Municipal Code, *Article 4—Noise and Vibration Control*, Paragraphs 17-4.1 to 17-4.30 (1984).

36. Anonymous, "Objective Limits for Motor Vehicle Noise," BBN Report No. 824, Bolt Beranek and Newman, December 1962.

37. "Vehicle Noise Measurement Regulations," and "Vehicle Code, Section 23, 130, as Amended," California Department of Highway Patrol, Sacramento, CA, (January 1968 and November 1970).

CHAPTER 18

Prediction of Machinery Noise

ERIC W. WOOD
Acentech Incorporated
Cambridge, Massachusetts

Engineering information and prediction procedures are presented describing the sound power emission characteristics of various industrial machinery. The information has been extracted from consulting project files and the results of field measurement programs. The machinery noise prediction procedures are based on studies of empirical field data sponsored by numerous clients, including the Edison Electric Institute and the Empire State Electric Energy Research Corporation, for which references 1–3 were prepared with the author's colleagues: James D. Barnes, Robert M. Hoover, Laymon N. Miller, Susan L. Patterson, Anthony R. Thompson, and Istvan L. Ver.

Characteristics identified and described for the noise produced by the machines discussed in this chapter include the estimated overall, A-weighted, and octave-band sound power levels (L_w) in dB *re* 1 pW, general directivity and tonal characteristics observed in the far field of the source, and temporal characteristics associated with the source operation. In addition, noise abatement concepts that have proven to be useful to the author during previous consulting projects are described briefly. Successful noise control treatments often require a detailed understanding of site-specific operating and maintenance requirements as well as pertinent acoustical conditions at the site. It is suggested that acoustical design handbooks or

Noise and Vibration Control Engineering: Principles and Applications, Edited by Leo L. Beranek and István L. Vér.
ISBN 0-471-61751-2 © 1992 John Wiley & Sons, Inc.

experienced acoustical engineering professionals be consulted when additional guidance is needed for specific design applications.

The information and procedures presented here can be used for numerous engineering applications requiring predictions of the approximate acoustical characteristics of new machinery installations and estimates of the noise attenuation that may be required to meet local requirements. The prediction procedures are expected to provide A-weighted sound-power-level estimates that are generally accurate to within about ± 3 dB. Octave-band sound-power-level estimates will necessarily be somewhat less accurate. The prediction procedures estimate the equivalent (energy average) L_{eq} sound power level during operation of one machine. For equipment with intermittent operating cycles, the estimated sound power levels may be reduced by 10 log(operating cycle) if long-term equivalent values are required. When several identical machines are operating simultaneously, the sound power levels can be increased by 10 log(number of machines). When equipment is located within a large plant, other nearby large equipment may provide some shielding that will reduce the noise radiated to distant neighbors in shielded directions (see Chapter 7). When equipment is located within an enclosed plant, a portion of the acoustical energy will be dissipated inside the plant, leading to further reductions in the far-field noise. It should be noted that doors, windows, and ventilation openings can substantially diminish the acoustical performance (composite sound transmission loss) of building exterior walls (see Chapter 9).

The reader should recognize that actual field data and/or manufacturers' information should be obtained when possible for machines being studied. Detailed information about the noise produced by internal-combustion reciprocating engines, electric machinery, and gears is provided in Chapters 19, 20, and 21 respectively. Information about the noise produced by gas flows and fans is presented in Chapter 14. Noise abatement information is provided in several previous chapters. Indoor sound field characteristics are described in Chapter 7, and outdoor sound propagation is discussed in Chapter 5. The various types of equipment addressed in this chapter include air compressors, boilers, coal-handling equipment, cooling towers, diesel engine powered equipment, fans, feed pumps, gas turbines, steam turbines, steam vents, and transformers. The noise emission information provided is representative of equipment that is presently available in the United States. Equipment in use in other countries may produce noise levels that are somewhat higher or lower than are presented here. Prediction formulas for the overall sound power level are provided in the text of this chapter. Adjustments to predict the A-weighted and octave band sound power levels are provided in Table 18.1.

18.1 AIR COMPRESSORS

The noise produced by air compressors is radiated from the filtered air inlet, the compressor casing, the interconnected piping and interstage cooler, as well as the motor or engine used to drive the compressor. The noise produced by air compressors can exhibit a distinctive duty cycle based on the demand for compressed air

TABLE 18.1 Adjustments Used to Estimate A-weighted and Octave-Band L_w[a]

Line	Equipment[b]	A-weighted	31	63	125	250	500	1000	2000	4000	8000
1	RR Compressor	2	11	15	10	11	13	10	5	8	15
2	C Comp Case	2	10	10	11	13	13	11	7	8	12
3	C Comp Inlet	0	18	16	14	10	8	6	5	10	16
4	S Boilers	9	6	6	7	9	12	15	18	21	24
5	L Boilers	12	4	5	10	16	17	19	21	21	21
6	CC Shakers	9	5	6	7	9	12	15	18	21	24
7	R Car Open	12	4	5	10	13	17	19	21	21	21
8	R Car Encd.	10	3	5	11	14	14	14	18	19	19
9	BL Unloaders	9	8	5	8	10	12	13	18	23	27
10	CS Unloaders	11	3	8	12	14	14	16	19	21	25
11	Coal Crushers	9	6	6	6	10	12	15	17	21	30
12	T Towers	7	7	7	7	9	11	12	14	20	27
13	C Mill (13–36)	5	—	—	4	6	8	11	13	15	—
14	C Mill (37–55)	5	—	—	6	7	4	11	14	17	—
15	ND C Towers	0	—	—	12	13	11	9	7	5	7
16	MD C Towers	10	9	6	6	9	12	16	19	22	30
17	MD C Towers 1/2S	5	9	6	6	10	10	11	11	14	20
18	D Engine	5	—	11	6	3	8	10	13	19	25
19	C Fan I + 0	5	11	9	7	8	9	9	13	17	24
20	C Fan Case	13	3	6	7	11	16	18	22	26	33
21	R Fan Case	12	10	7	4	7	18	20	25	27	31
22	Axial Fan	3	11	10	9	8	8	8	10	14	15
23	S Fd Pumps	4	11	5	7	8	9	10	11	12	16
24	L Fd Pumps	1	19	13	15	11	5	5	7	19	23
25	S St Turbines	5	11	7	6	9	10	10	12	13	17
26	L St Turbines	12	9	3	5	10	14	18	21	29	35
27	SL Blow-out	11	—	2	6	14	17	17	21	18	18
28	Xformers	—	—	−3	−5	0	0	6	11	16	23

[a]Subtract values shown from estimated overall L_w.

[b]RR: rotary and reciprocating; C: casing; S: small, L: large, CC: coal car; R: rotary car dumpers; BL: bucket ladder; CS: clamshell; T: transfer; C: coal; ND C: natural draft cooling; MD C: mechanical draft cooling; D: diesel.

and is generally considered to be omnidirectional and broadband without prominent discrete tones. The air inlet noise of certain vacuum pumps sometimes includes high levels of low-frequency pulsating noise.

Air compressors are often available from their manufacturer with special noise abatement packages that include various types of inlet mufflers, laggings, insulations, and enclosures that provide varying degrees of noise reduction. Acoustical data for these low-noise compressors should be obtained directly from the manufacturer's literature. Noise-estimating procedures presented in this section are applicable to conventional industrial-, utility-, and construction-type air compressors without special noise abatement packages.

The overall sound power level radiated by the casing and air inlet of rotary and reciprocating air compressors can be estimated using Eq. (18.1) (see also line 1 of Table 18.1). These relations for rotary and reciprocating compressors assume that the air inlet is equipped with a filter and small muffler as normally provided by the manufacturer.

$$\text{Rotary and reciprocating compressor, overall } L_w = 90 + 10 \log(\text{kW}) \quad (\text{dB}) \tag{18.1}$$

The casing and unmuffled air inlet overall sound power levels for centrifugal air compressors in the power range of 1100–3700 kW can be estimated using Eqs. (18.2) and (18.3) (see also lines 2 and 3 of Table 18.1). The insertion loss of any muffler installed at the air inlet should be deducted from the calculated inlet power levels:

$$\text{Centrifugal compressor casing, overall } L_w = 79 + 10 \log(\text{kW}) \quad (\text{dB}) \tag{18.2}$$

$$\text{Centrifugal compressor inlet, overall } L_w = 80 + 10 \log(\text{kW}) \quad (\text{dB}) \tag{18.3}$$

The term $\log(x)$ represents the common logarithm to base 10 of the value of x.

Equations (18.2) and (18.3) do not include the noise radiated by the motor or engine used to drive the compressor. Sound power levels calculated for the driving equipment should be added to the calculated sound power levels for the compressor.

Mufflers, laggings, barriers, and enclosures have all been used with varying degrees of success to control the noise radiated from compressors to nearby work spaces or neighborhoods. However, it is suggested that in many cases it is probably most practical to specify and purchase a reduced noise compressor from the manufacturer.

18.2 BOILERS

Two procedures are provided to predict the noise from boilers; one is for relatively small boilers in the range of about 50–2000 boiler horsepower (one boiler horsepower equals 15 kg of steam per hour). Another procedure is provided for large boilers of the type that serve electric-power-generating stations in the range of 100–1000 megawatts (MWe).

Small Boilers

The sound power output of small boilers is only weakly related to the thermal rating of the boiler. The combustion air fans and the burners probably radiate more noise than do the insulated side walls of small boilers. The overall sound power level of small boilers can be estimated using Eq. (18.4) (also see line 4 of Table 18.1):

$$\text{Small-boiler, overall } L_w = 95 + 4\log(\text{bhp}) \quad \text{(dB)} \tag{18.4}$$

Large Boilers

Noise levels measured in work spaces adjacent to large central-station boilers that are not enclosed in a building (open boilers) are often in the range of 80–85 dBA at the lower half and 70–80 dBA at the upper half of the boiler. Noise levels measured in the vicinity of enclosed boilers (generally found in areas with cold climates) are often about 5 dBA higher. The overall sound power output of large central-station boilers can be estimated by the relation given in Eq. (18.5) (also see line 5 of Table 18.1):

$$\text{Large-boiler, overall } L_w = 84 + 15\log(\text{MWe}) \quad \text{(dB)} \tag{18.5}$$

The noise from boilers is essentially omnidirectional and broadband in character. However, the combustion air or forced-draft fans, induced-draft fans, gas recirculation fans, overfire air fans, and drive motors sometimes produce tonal noise and may increase the noise levels on their side of the boiler. A small number of boilers has also been found to exhibit a strong discrete tone, usually at a frequency between about 20 and 100 Hz during operation at particular boiler loads. This is caused by vortex shedding from heat exchanger tubes that excite an acoustical resonance within the boiler that in turn causes coherent (rather than random) vortex shedding to occur at a boiler resonance frequency. The low-frequency tonal noise radiates from the boiler side walls and has caused strong adverse reactions from residential neighbors. Boiler operators have also expressed concern about possible fatigue failure of vibrating tube sheets. Installation of one or more large metal plates to subdivide the interior of the boiler volume and thereby change its resonance frequency has successfully corrected this problem. Additional information about this resonant condition and its control is provided in reference 4.

The typical broadband noise radiated by boiler side walls has been successfully reduced with the use of a well-insulated exterior enclosure, which also serves as weather protection for equipment and workers. The noise produced by fans serving the boiler can also be controlled as described in Section 18.6.

18.3 COAL-HANDLING EQUIPMENT

A wide variety of noise-producing equipment, such as car, ship and barge unloaders, transfer towers, conveyors, crushers, and mills, is used to unload, transport, and condition coal for use at large industrial and utility boilers. The results of numerous noise-level surveys have been studied and condensed to prepare the following general noise prediction procedures for this equipment.

Coal Car Shakers

Coal car shakers are used to vibrate coal cars during bottom unloading. The unloading operation of each car often occurs for about 2–5 min when the coal is not frozen. When frozen coal is located in the car, the shake-out operation can last for up to 10 min or longer. Car shakers are often located inside a sheet metal or masonry building with little or no interior insulation for sound absorption. Shaker buildings include large openings at both ends and sometimes smaller openings and windows along the side walls. They also include dust collection and ventilation systems with fans that produce noise. The overall sound power produced during a coal car shake-out can be estimated using the following relation (also see line 6 of Table 18.1):

$$\text{Coal car shake-out, overall } L_w = 141 \text{ dB} \qquad (18.6)$$

Noise produced by the dust collection and ventilation fans can be estimated using the relations provided in Section 18.6.

The far-field noise produced by car shake-outs can be reduced by enclosing the operation inside a metal or masonry building or by reducing the openings in any existing building. In addition, modern car-shaking equipment is sometimes available that produces measured A-weighted sound levels that are about 10 dB lower than are estimated with Eq. (18.6). Careful train movements can reduce the impact noise associated with indexing the train through the shaker operation. Limiting the shake-out operation to daytime hours will also reduce the community noise impacts.

Rotary Car Dumpers

Rotary car dumpers are often used to unload long unit trains equipped with special rotary couplings. Two or 3 min are typically required to unload unfrozen coal from each car. A 100-car unit train can be unloaded in about 3–5 h unless delays are encountered. The overall sound power produced during rotary car unloading cycles can be estimated using Eqs. (18.7) and (18.8) for open and enclosed facilities (also see lines 7 and 8 of Table 18.1). However, short-duration impact sounds can be as much as 10–25 dB greater than the equivalent values given by these relations:

$$\text{Rotary car open unloading, overall } L_w = 131 \text{ dB} \qquad (18.7)$$

$$\text{Rotary car enclosed unloading, overall } L_w = 121 \text{ dB} \qquad (18.8)$$

Noise abatement methods for rotary car dumpers are similar to those discussed previously for car shakers.

Bucket Ladder and Clamshell Bucket Unloaders

Bucket ladder and clamshell bucket unloaders are often used to unload coal delivered by ship or barge. This equipment is located outdoors along the dock line with the electric motors and speed reduction gears often located within metal enclosures for weather protection. Acoustical data studied to prepare the prediction relations provided here were obtained at bucket ladder unloaders with free-digging rates in the range of 1800–4500 metric tons per hour and clamshell unloaders with free-digging rates in the range of 1600–1800 metric tons per hour. The energy-averaged overall sound power level produced during operation of bucket ladder and clamshell bucket unloaders can be estimated using Eqs. (18.9) and (18.10) (see also lines 9 and 10 of Table 18.1):

$$\text{Bucket ladder unloader, overall } L_w = 123 \text{ dB} \tag{18.9}$$

$$\text{Clamshell bucket unloader, overall } L_w = 131 \text{ dB} \tag{18.10}$$

Noise control treatments for bucket ladder and clamshell bucket unloaders generally employ insulated enclosures for the drive equipment and unloading operations are limited to daytime hours.

Coal Crushers

Coal is sometimes conveyed to a crusher building where the chunks are reduced in size in preparation for firing in a power plant boiler. Noise produced during crushing operations is a composite of the crusher, metal chutes, conveyors, drive motors, and speed reducers. Coal crushing is usually an intermittent operation and the noise is often omnidirectional without major tonal components. The overall sound power level produced during operation of the crusher can be estimated in accordance with the following relation (also see line 11 of Table 18.1):

$$\text{Coal crusher, overall } L_w = 127 \text{ dB} \tag{18.11}$$

The far-field noise associated with coal-crushing operations can be reduced with the use of a well-insulated and well-ventilated building that encloses the operation.

Coal Transfer Towers

Transfer towers reload coal from one conveyor to another conveyor as the coal is relocated within the coal yard on its way to the plant. Transfer tower noise is a composite of coal impacts, local conveyors, drive motors, and ventilation systems. Coal transfer is usually an intermittent operation, and the noise is often omnidirectional without major tonal components. The overall sound power level produced during coal transfer in open buildings can be estimated in accordance with the following relation (also see line 12 of Table 18.1):

$$\text{Transfer tower, overall } L_w = 123 \text{ dB} \tag{18.12}$$

Coal transfer operations can be enclosed inside a well-ventilated building to reduce the far-field community noise. In this case, the estimated far-field noise can be reduced by the composite sound transmission loss of the exterior building walls. Fan noise associated with the ventilation system should be estimated in accordance with the relations given in Section 18.6.

Coal Mills or Pulverizers

Coal mills and pulverizers crush and size coal in preparation for burning in a boiler. The resulting noise is associated with internal impacts, drive motors, speed reducers, and fans. The noise is essentially broadband, omnidirectional, reasonably steady, and continuous while the plant is operating. The overall sound power level produced by a coal mill located in a building with large openings can be estimated in accordance with the following relations for mills in the size ranges of 13–36 metric tons per hour and 37–55 metric tons per hour (also see lines 13 and 14 of Table 18.1). When coal mills are located within a closed building, the estimated far-field noise can be reduced by the composite sound transmission loss of the exterior building walls:

$$\text{Coal mill, overall } L_w = 110 \text{ dB} \qquad \text{13–36 metric tons per hour} \tag{18.13}$$

$$\text{Coal mill, overall } L_w = 112 \text{ dB} \qquad \text{37–55 metric tons per hour} \tag{18.14}$$

18.4 COOLING TOWERS

The sound power output of cooling towers is caused primarily by the water splash in the fill and in the basin as well as by the fans, motors, and gears used to provide draft in mechanical towers. The noise is usually continuous and somewhat directive for rectangular towers. The noise produced by at least one pair of large natural-draft hyperbolic cooling towers included a low-frequency discrete tone associated with aerodynamic vortex shedding at the tower base.

Methods that can be used to estimate the sound power output of natural-draft and mechanical induced-draft cooling towers are provided below. Major tower manufacturers can provide additional useful information regarding noise estimates and noise abatement.

Natural-Draft Cooling Towers

The overall sound power output radiated from the rim of large hyperbolic natural-draft cooling towers can be estimated using Eq. (18.15) (also see line 15 in Table

18.1). Noise radiated by the top of the tower is usually not significant at ground elevations compared to the rim-radiated noise:

$$\text{Natural-draft cooling tower rim, overall } L_w = 86 + 10 \log Q \qquad (18.15)$$

where Q is the water flow rate in gallons per minute (one cubic meter per minute is equal to 265 U.S. gallons per hour).

Inlet mufflers and fan assistance have been installed to control rim-radiated noise for at least one hyperbolic cooling tower in Europe located near residential neighbors. However, noise abatement is expensive and not common for large hyperbolic towers.

Mechanical Induced-Draft Cooling Towers

The sound power produced by mechanical induced-draft cooling towers operating at full fan speed and at half fan speed can be estimated using the following relations (also see lines 16 and 17 of Table 18.1):

$$\text{Mechanical-draft tower full-speed, overall } L_w = 96 + 10 \log(\text{fan kW}) \qquad (18.16)$$

$$\text{Mechanical-draft tower half-speed, overall } L_w = 88 + 10 \log(\text{fan kW}) \qquad (18.17)$$

where kW is the full-speed fan power rating in kilowatts for both relations.

The above relations apply in all horizontal directions from round towers and most directions away from the inlet face of rectangular towers. For directions away from the enclosed ends of rectangular towers, the far-field noise is several decibels less than estimated above due to the effects of shielding by the solid closed ends. For a line of as many as 6–10 rectangular towers, the far-field noise from the enclosed ends could be as much as 5–6 dB less than estimated, based on omnidirectional radiation using Eqs. (18.16) and (18.17) above.

Reduced fan speed is a common method used to reduce the noise from mechanical-draft cooling towers during evening and nighttime hours when excess cooling capacity may be available due to reduced ambient air temperatures. In multiple-tower installations, it is preferable to reduce the fan speed for all towers rather than to shut down unneeded towers. Another common noise control method is to install wide-chord high-efficiency fan blades that can provide the necessary fan performance at reduced speeds. In addition, mechanical-draft towers with centrifugal fans may be selected because they can sometimes be designed to produce less noise than propeller fans. Mufflers have also been installed at the air inlet and discharge to reduce both the fan and water noise. However, the mufflers must be protected from the wet environment, can become coated with ice during freezing weather, and introduce an additional aerodynamic restriction that must be overcome by the fan. Barrier walls and partial enclosures have been successfully used to shield neighbors from cooling-tower noise; however, barrier walls that avoid excessive restrictions to the air flow often limit their usefulness.

18.5 DIESEL-ENGINE-POWERED EQUIPMENT

When machinery such as compressors, generators, pumps, and construction equipment is powered by a diesel engine, it is usually the diesel engine that is the dominant source of noise. Detailed information about the acoustical characteristics of internal combustion engines is provided in Chapter 19.

Mobile construction and coal yard equipment powered by diesel engines, such as dozers, loaders, and scrapers, often produces in-cab noise levels as high as 95–105 dBA. Methods to retrofit and reduce in-cab noise levels for several loaders and dozers have been developed, field tested, and documented in references 5 and 6. Well-designed cabs and noise control features are now often available from major manufacturers when purchasing new diesel-engine-powered mobile equipment.

A simple relation for estimating the maximum exterior overall sound power level for naturally aspirated and turbocharged diesel engines used to power equipment is provided in Eq. (18.18) (also see line 18 in Table 18.1):

$$\text{Diesel engine equipment, overall } L_w = 99 + 10\log(\text{kW}) \qquad (18.18)$$

The above relation assumes that the engine is equipped with a conventional exhaust muffler in good working condition as typically provided by the engine manufacturer and further assumes that the engine is operating at rated speed and power. Noise associated with material impacts during equipment operation is not included.

Equipment used on construction sites often operates at part power. Measurements obtained at the operating equipment indicate that work-shift-long equivalent L_{eq} sound levels are therefore typically about 2–15 dB less than the maximum values provided by Eq. (18.18). It is assumed that the following values could be subtracted from the Eq. (18.18) maximum levels to obtain work-shift-long equivalent L_{eq} levels. When project-long equivalent L_{eq} levels are required, the estimated values can be further reduced to account for the percentage of time that the equipment will actually operate on the construction site [10 log(operating time/project time)]:

3–4 dB	Backhoes, rollers
5–6 dB	Dozers, graders, haulers, loaders, scrapers
7–8 dB	Air compressors, concrete batch plants, mobile cranes, trucks
12–13 dB	Derrick cranes, paging systems

When stationary diesel-engine-driven equipment is located inside of masonry or metal buildings, it is necessary to consider the sound power radiated by the engine inlet, the engine casing, the engine exhaust, and the engine cooling fan as well as the driven equipment. It is also necessary to account for the attenuation expected from the building walls and openings, the inlet filter/muffler, the exhaust muffler, and mufflers to be installed at building openings that serve ventilation and cooling systems. Additional information about the propagation of noise from equipment located inside buildings and about muffler systems is provided in Chapters 7 and 10.

18.6 INDUSTRIAL FANS

The noise produced by industrial fans is caused by the dynamic interaction of the gas flow with rotating and stationary surfaces of the fan. Noise produced by shear flow is usually not considered to be important. Broadband fan noise results from the random aerodynamic interactions between the fan and the gas flow. The prominent discrete tones produced by centrifugal fans results from the periodic interaction of the outlet flow and the cutoff located directly downstream of the blade trailing edges. Tonal noise produced by axial flow fans results from periodic interactions between distorted inflow and the rotor blades as well as the rotor wakes and nearby downstream surfaces including struts and guidevanes. This tonal noise is usually most prominent at the harmonics of the passing frequency of the fan blade (number of blades times the rotation rate in revolutions per second). Detailed information about the noise associated with gas flows and fans is provided in Chapter 14.

The noise produced by new forced-draft fans can be controlled most easily with the use of ducted and muffled inlets. Sometimes the ducted inlet is extended to a location in the plant where warm air is available and additional ventilation is needed. To provide further noise reduction, combined thermal and acoustical insulation lagging should be considered for the inlet and outlet ducts and the fan housing. Open-inlet forced-draft fans can be located within an acoustically treated fan room that contains the fan and its noise. However, workers inside the fan room during inspections and maintenance will be exposed to high levels of fan noise and should wear ear protection.

High-performance thermal-acoustical insulation lagging is available and can be used to control noise radiated from inlet and outlet ducts and the housings of industrial fans with ducted inlets and outlets.

Discharge noise of large induced-draft fans radiated from the top of the stack to residential neighbors has caused serious community noise problems. The fan and duct system design should include provisions to control this noise if residential areas are located within $\frac{1}{2}$ mile to 1 mile of the plant. Mufflers used for induced-draft fan service at boilers and furnaces should be of the dissipative, parallel-baffle, open-cavity type that is tuned to the harmonics of fan blade passing frequency. These mufflers are built to avoid fly-ash clogging and to include erosion protection that ensures adequate long-term performance. They can usually be designed to introduce a pressure loss in the range of about 10–40 mm water gauge. Consideration should also be given to the noise attenuation that will be provided by any flue gas scrubbers, filters, or precipitators that are to be located between the induced-draft fan and its stack. Alternatively, variable-speed fan drives can be installed to reduce nighttime noise at cycling plants that shed load during nighttime hours. This has the added benefit of also reducing the power consumed by the fans as well as reducing the erosion rates and stresses of rotating components. When multiple fans are installed, they should be operated at the same speed to avoid acoustical beats between the fan tones.[3,7]

Well-designed and sized inlet and outlet ducts that properly manage the flow are essential to avoid excessive fan noise. Inlet swirl, distorted inflow, and excessive turbulence can result in high noise levels and reduced system efficiency. Useful design guidelines have been established by the Air Movement and Control Association (AMCA) for large industrial fan installations.

Procedures are provided in this section to estimate the overall sound power output of large industrial fans operating at peak efficiency conditions with undistorted inflow. Experience indicates that fans operating with highly distorted inflow or at off-peak efficiency conditions often produce sound power levels that may be 5–10 dB higher than are indicated below. Centrifugal and axial flow fans produce less sound power when operating at low speeds and reduced working points than during full-load, high-speed operation. For part-speed operation, the sound power levels estimated below can be reduced by about 55 log(speed ratio).

Centrifugal-Type Forced-Draft and Induced-Draft Fans with Single-Thickness, Backward-curved or Backward-inclined Blades or Airfoil Blades

The overall sound power level radiated from the inlet of forced-draft centrifugal fans and the outlet of induced-draft centrifugal fans (with single-thickness, backward-curved or backward-inclined blades or airfoil blades) can be estimated with the relationship provided in Eq. (18.19) (also see line 19 of Table 18.1):

$$\text{Centrifugal fan, overall } L_w = 10 + 10 \log Q + 20 \log(\text{TP}) \qquad (18.19)$$

where Q is fan flow rate in cubic meters of gas per minute and TP is fan total pressure rise in newtons per square meter at rated conditions.

To account for the tonal noise, 10 dB should be added to the octave bands containing the fan blade passing frequency and its second harmonic. For multiple-fan installations, the estimated sound power levels should be increased by 10 log N, where N is the number of identical fans.

The overall sound power level radiated from the uninsulated casing of centrifugal fans can be estimated with the relationship provided in Eq. (18.20) (also see line 20 of Table 18.1). To account for the tonal noise, 5 dB should be added to the octave band containing the fan blade passing frequency:

$$\text{Centrifugal fan casing, overall } L_w = 1 + 10 \log Q + 20 \log(\text{TP}) \qquad (18.20)$$

where Q and TP are as defined for Eq. (18.19).

The sound power radiated by the uninsulated discharge breaching is about 5 dB less than the sound power estimated using Eq. (18.20) for the fan casing.

The overall sound power level radiated from the casings of radial-blade centrifugal fans with ducted inlets and outlets, such as are sometimes used for gas recirculation service, can be estimated with the relationship provided in Eq. (18.21) (also see line 21 of Table 8.1):

$$\text{Radial fan casing, overall } L_w = 13 + 10 \log Q + 20 \log(\text{TP}) \qquad (18.21)$$

where again Q and TP are as defined for Eq. (18.19).

To account for the tonal noise, 10 dB should be added to the octave bands containing the fan blade passing frequency and its second harmonic.

Axial Flow Forced-Draft and Induced-Draft Fans

The overall sound power level radiated from the inlet of forced-draft axial flow fans and the outlet of induced-draft axial flow fans can be estimated with the following relation:

$$\text{Axial flow fan, overall } L_w = 24 + 10 \log Q + 20 \log(\text{TP}) \qquad (18.22)$$

where Q and TP have been defined in Eq. (18.19).

To account for the tonal noise, 6 dB should be added to the octave band containing the fan blade passing frequency and 3 dB should be added to the octave band containing its second harmonic. For multiple-fan installations, the above estimated values should be increased by 10 log N, where N is the number of identical fans.

Ventilation Fans

The sound power output of industrial ventilation fans can be estimated using Eq. (18.19) for centrifugal ventilation fans or Eq. (18.22) for axial flow ventilation fans.

18.7 FEED PUMPS

It is common for the noise radiated by a pump–motor set to be dominated by the motor noise. One exception is the relatively high-flow, high-head pumps used for boiler and reactor feedwater service in large modern power generating stations that operate in the United States. These feed pumps are usually driven by an electric motor, an auxiliary steam turbine, or the main turbine–generator shaft. Midsize pumps generally produce broadband noise without strong tonal components. The larger pumps, however, commonly produce both broadband noise and strong mid-frequency tonal noise. Feed pump noise is omnidirectional and continuous during plant operation.

A high performance thermal-acoustical blanket insulation, described in reference 8, has been developed and evaluated at field installations specifically for use with equipment such as noisy feedwater pumps, valves, and turbines. This thermal-acoustical insulation has been found to provide far better noise attenuation than conventional blanket insulations. It is also easier to remove and reinstall during maintenance and inspections than most rigid insulations. It has been successfully used as a retrofit insulation at installed pumps, turbines, valves, and lines. It is also being considered for use by equipment manufacturers at new installations requiring reduced noise levels. The use of flexible thermal-acoustical insulation to control noise avoids the mechanical, structural, and safety problems that can be associated with the large rigid enclosures that have sometimes been used at boiler feedwater pump installations. Note that experience in the Federal Republic of Germany indicates that it is possible to design and operate boiler feedwater pumps that produce relatively low noise levels without the use of external means for controlling the noise.

The overall sound power level radiated by boiler and reactor feedwater pumps can be estimated using the information provided below for pumps in the power range of 1–18 MWe.

Rated Power (MWe)	Overall Sound Power (dB)
1	108
2	110
4	112
6	113
9	115
12	115
15	119
18	123

Octave-band and A-weighted sound power levels for boiler and reactor feedwater pumps can be estimated by subtracting the values provided in line numbers 23 and 24 of Table 18.1 from the overall power level estimated above.

18.8 INDUSTRIAL GAS TURBINES

Industrial gas turbines are often used to provide reliable, economic power to drive large electric generators, gas compressors, pumps, or ships. Flight-worthy gas turbines used in aircraft operations are not addressed in this chapter.

The noise produced by industrial gas turbine installations is radiated primarily from four general source areas: the inlet of the compressor, the outlet of the power turbine, the casing and/or enclosure of the rotating components, and various auxiliary equipment including cooling and ventilating fans, the exhaust-heat recovery steam generator, poorly gasketed or worn access openings, and electric transformers. Each of the above areas should receive considerable attention if an industrial gas turbine is to be located successfully near a quiet residential area.

Operation of unmuffled industrial gas turbines of moderate size would produce sound power levels of 150–160 dB or greater. Because the resulting noise levels would be unacceptable, essentially all industrial gas turbine installations include at least a moderate degree of noise abatement provided by the manufacturer. Noise abatement performance requirements may call for reducing the sound power output to levels as low as 100 dB for installations located near sensitive residential areas.

Noise radiated from the compressor inlet includes high levels of both broadband and tonal noise, primarily at frequencies above 250–500 Hz. This mid- and high-frequency noise is controlled with the use of (a) a conventional parallel-baffle or tubular muffler comprised of closely spaced thin baffles and (b) inlet ducting, plenum walls, expansion joints, and access hatches that are designed to contain the noise and avoid flanking radiation.

Noise radiated by the casings of rotating components is typically contained with the use of thermal insulation laggings and close-fitting steel enclosures that include interior sound absorption and well-gasketed access doors. It is not uncommon for the enclosure to be mounted directly on the structural steel base that supports the rotating components. This encourages the transmission of structure-borne noise to the enclosure and can limit its effective acoustical performance at installations requiring high degrees of noise reduction. Off-base mounting of the enclosure can be considered for installations where its noise abatement performance must be improved. In addition, many large industrial gas turbines are being installed within a turbine–generator building that provides weather protection and additional reduction of noise radiated through the enclosure from the rotating components.

Noise radiated from the power turbine outlet includes high levels of both broadband and tonal noise at low, mid, and high frequencies, with the low-frequency noise being the most difficult and expensive to control at simple-cycle installations (without heat recovery boilers) located near residential areas. The wavelength of sound at about 30 Hz is on the order of 15–20 m in the gas stream at the back end of the power turbine, thereby increasing the required size of the discharge mufflers. Furthermore, low-frequency exhaust noise has caused significant community noise problems when the sound level in the 31-Hz octave band exceeded about 70 dB at sensitive residential areas. This low-frequency exhaust noise might be best controlled by the manufacturer with improved gas management designs behind the power turbine. The outlet noise is typically controlled with the use of (a) conventional parallel-baffle or tubular mufflers that include thick acoustical treatments; (b) additional dissipative or reactive muffler elements tuned to further attenuate low-frequency noise; and (c) outlet ducts, plenum chamber, expansion joints, and access hatches that include adequate acoustical treatments to avoid flanking along these paths and their radiation of excessive noise. Heat recovery steam generators, installed downstream of the power turbine to absorb waste exhaust heat, also serve to attenuate the exhaust noise and can reduce the performance requirements of the exhaust muffler.

Noise produced by auxiliary cooling and ventilation fans can be controlled effectively with the use of high-performance low-speed fan blades, mufflers, or partial enclosures depending on specific site and installation requirements. The control of transformer noise, if required for a specific installation, is discussed in Section 18.11 of this chapter. The control of noise radiated from the side walls of heat recovery steam generators can be controlled with the use of spaced-out insulation laggings designed for thermal performance with enhanced acoustical performance. When additional noise attenuation is needed, the heat recovery steam generator is located within the plant building.

Sound-power-level prediction procedures are not provided here for industrial gas turbines because of the rather wide range of noise control treatments that are available within the industry. Many manufacturers offer their machines with a variety of optional noise abatement treatments, ranging from modest to highly effective. For example, 50–100 MWe installations, with or without heat recovery boilers, are readily available with noise abatement treatments resulting in noise levels in the range of 50–60 dBA at 120 m. Installations with lower noise levels are available from certain manufacturers and can be designed by several independent professionals who have specialized in the control of noise from industrial gas turbines.

One early but important step in the design of a new gas turbine installation is the preparation of a reasonable and well-founded technical specification by the buyer or buyer's acoustical consultant that fully describes the site-specific noise requirements. Methods and procedures for preparing gas turbine procurement specifications that describe expected noise-level limits are included in ANSI B133.8-1977. If residential neighbors are located in the vicinity of the site, the reader is cautioned to consider fully the site-specific needs when planning control of the low-frequency exhaust noise. Several independent consultants with many years of gas turbine experience have suggested that the noise-level threshold of complaints resulting from low-frequency noise in the 31-Hz octave band is about 65–70 dB, measured at residential wood-frame homes. Higher levels of low-frequency noise can sometimes vibrate the walls and rattle the windows and doors of wood-frame homes and result in varying degrees of annoyance. Gas turbine manufacturers, however, correctly indicate that low-frequency noise control treatments are expensive and that many industrial gas turbines are operating that produce higher levels of low-frequency noise in residential areas without causing complaints.

18.9 STEAM TURBINES

Two procedures are provided to predict the sound power radiation from steam turbines; one is for the relatively small turbines in the power range of 400–8000 kW that operate at about 3600–6000 rpm and are often used to drive auxiliary equipment at a plant where steam is readily available. Another procedure is provided for the large steam turbine generators in the range of about 200–1100 MWe used at central electricity generating stations. Thermal-acoustical blanket insulation that has been developed to control noise radiated from equipment including steam turbines is discussed in Section 18.7 and reference 8.

Auxiliary Steam Turbines

The overall sound power output of auxiliary steam turbines, with common thermal insulation installed, can be estimated using Eq. (18.23) (see also line 25 of Table

18.1). This noise is considered to be omnidirectional, generally nontonal, and continuous when the driven equipment is operating:

$$\text{Auxiliary steam turbine, overall } L_w = 93 + 4 \log(\text{kW}) \qquad (18.23)$$

Large Steam Turbine–Generators

The overall sound power output of large steam turbine–generators can be estimated using Eq. (18.24). This includes the noise radiated by the low-pressure, intermediate-pressure, and high-pressure turbines as well as the generator and shaft-driven exciter. The turbine–generator produces both tonal and broadband noise. The tonal components produced by the generator are typically most noticeable at 60 and 120 Hz for 3600-rpm machines and 30, 60, 90, and 120 Hz for 1800-rpm machines:

$$\text{Large steam turbine–generator, overall } L_w = 113 + 4 \log(\text{MW}) \qquad (18.24)$$

Octave band and A-weighted sound power levels for large steam turbine–generators can be calculated by subtracting the values provided in line number 26 of Table 18.1 from the overall sound power level estimated above using Eq. (18.24).

Most major manufacturers of large steam turbine–generators will provide their equipment with additional noise abatement features that reduce the noise by about 5–10 dBA, when required by a well-written purchase specification. The reverberant noise inside a turbine building can also be reduced through the proper selection of the building siding. The use of a building siding with a fibrous insulation sandwiched between a perforated metal inner surface and a solid exterior surface will provide improved midfrequency sound absorption and thereby reduce the reverberant buildup of noise within the turbine building and can also reduce the noise radiated to the outdoors.

18.10 STEAM VENTS

Atmospheric venting of large volumes of high-pressure steam is probably one of the loudest noise sources found at industrial sites. The overall sound power produced during steam line blow-outs at large central stations prior to starting a new boiler can be estimated using Eq. (18.25) (see also line 27 of Table 18.1). This noise is broadband and only occurs for a few minutes during each blow-out for the first few weeks of boiler operation:

$$\text{Steam line blow-out, overall } L_w = 177 \text{ dB} \qquad (18.25)$$

The actual sound power level produced during the venting of high-pressure gas is, of course, related to various variables including the conditions of the flowing gas and the geometry of the valve and pipe exit. However, the above relationship will provide a reasonable estimate of the noise associated with the blow-out of steam lines at large boilers used for utility service.

Large heavy-duty mufflers are sometimes purchased or rented that reduce the noise by about 15–30 dB during steam line blow-outs. The noise produced by the more common atmospheric vents and commonly encountered valves can be estimated using prediction procedures available from many valve manufacturers and their representatives. Many manufacturers offer valves with special low-noise trims, orifice plates, and inline mufflers that effectively reduce noise generation and radiation. The low noise trim can also reduce the vibration and maintenance that is sometimes associated with valves used in high-pressure-drop service.

18.11 TRANSFORMERS

The noise radiated by electrical transformers is composed primarily of discrete tones at even harmonics of line frequency, that is, 120, 240, 360, ... Hz when the line frequency is 60 Hz, and 100, 200, 300, ... Hz when the line frequency is 50 Hz. This tonal noise is produced by magnetostrictive forces that cause the core to vibrate at twice the electrical line frequency. The cooling fans and oil pumps at large transformers produce broadband noise when they operate; however, this noise is usually less noticeable and therefore less annoying to nearby neighbors. The tonal core noise should be considered omnidirectional and continuous while the transformer is operating. The broadband fan and pump noise occurs only during times when additional cooling is required.

The technical literature includes numerous relations and guidelines for the prediction of noise produced by transformers. Reference 10 reports the results of measurements obtained at 60 transformer banks and indicates that the space-averaged A-weighted sound level produced by the core of the average transformer (without built-in noise abatement) at an unobstructed distance of about 150 m is well represented by the relationship given in Eq. (18.26). Ninety-five percent of the A-weighted noise data reported in reference 10 lies within about ±7 dB of this relation for transformers with maximum ratings in the range of 6–1100 MVA:

$$\text{Average A-weighted core sound level at 150 m, } L_p = 26 + 8.5 \log(\text{MVA}) \tag{18.26}$$

where MVA is the maximum rating of the transformer in million volt-amperes.

The space-averaged sound pressure levels of the transformer core tones at 120, 240, 360, and 480 Hz at 150 m can be estimated by adding the following values of the A-weighted sound level of Eq. (18.26):

120	240	360	480
17	5	−4	−8

Another relation for transformer A-weighted sound levels versus distance, extracted from reference 11, is

$$\text{Space-averaged far-field sound level } L_p = L_n - 20 \log \frac{d}{\sqrt{S}} - 8 \tag{18.27}$$

where L_n = circumferential average sound pressure levels measured at National Electrical Manufacturers Association (NEMA) close-in measurement positions (A-weighted or tonal)

S = total surface area of four side walls of transformer tank

d = distance from transformer tank (in units that are compatible with tank side wall area), must be greater than $\sqrt{S}$

Equation (18.27) can be used if L_p and L_n represent the A-weighted sound levels or the sound pressure levels of the discrete tones produced by the transformer tank. The octave-band sound-pressure levels of the transformer core noise can be obtained directly from the sound pressure levels of the discrete tones estimated above. The octave-band sound pressure levels of the total transformer noise with the cooling fans operating can be estimated by subtracting the values provided in line number 28 of Table 18.1 from the A-weighted sound level estimated with Eq. (18.27). This applies to conventional cooling-fan systems with motors in the power range of about $\frac{1}{6}$–1 hp operating at about 1000–1700 rpm with two- or four-bladed propeller fans. Further information about special low-noise cooling systems should be obtained from the manufacturer. Excess attenuation should be considered when estimating sound levels at distances beyond about 150 m.

The National Electrical Manufacturers Association has published standard tables of close-in noise levels that can be expected from transformers. Experience indicates that the noise close to most operating transformers is often equal to or somewhat less than the NEMA standard values. However, the noise produced by converter transformers operating at a.c.-to-d.c. converter terminals can include discrete tones at frequencies up to about 2000 Hz and can be about 5–10 dB greater than the NEMA standard values. This additional high-frequency noise has been found to be unusually noticeable and disturbing to residential neighbors when the converter transformer is located in quiet rural–suburban areas.

Two basic methods are available for reducing the far-field noise produced by transformers. First, manufacturers are able to produce transformers with reduced flux density that generate noise levels as much as 10–20 dB lower than the NEMA standard values. New high-efficiency transformers with low electrical losses usually produce core noise levels that are less than the NEMA standard values. High-efficiency low-speed cooling fans and cooling-fan mufflers are also available from some manufacturers when needed for special siting applications. In some cases, the cooling fans are eliminated from the transformer tank and replaced with oil-to-water heat exchangers.

Second, barrier walls, partial enclosures, and full enclosures can be provided to shield or contain the transformer noise. They are usually fabricated with masonry blocks or metal panels. The interior surfaces of the barrier or enclosure walls should usually include sound absorption that is effective at the prominent transformer tones. Care must be used to ensure adequate strike distances and space for cooling-air flows. If the fin-fan oil coolers are to be located outside of the enclosure, some attention should be given to the structure-borne and oil-borne noise from the core that can be radiated by the coolers. Space and provisions must be provided for inspections, maintenance, and transformer removal. It is also important to ensure that the walls are structurally isolated from the transformer to avoid the radiation of structure-borne noise resulting from transformer-induced vibration. Further information about the control of transformer noise is available from most major transformer manufacturers as well as from reference 9.

Instead of, or in addition to, reducing the noise radiated from the transformer substation, it is sometimes possible to site a transformer far from residential neighbors or within an existing noisy area, such as close to a well-traveled highway, where the ambient sounds mask the transformer noise partially.

18.12 SUMMARY

The machinery–noise prediction procedures and relations presented in this chapter are based primarily on extensive field measurement data collected by the author and his colleagues during many years of consulting projects. The results obtained when using these relations should be useful for many engineering applications. The reader is cautioned, however, that site-specific installation conditions and individual equipment characteristics can cause noise levels to be somewhat higher or lower than predicted, and detailed knowledge of these exceptions can be important for critical applications. Also, many items of equipment can be purchased with reduced noise, can be installed so as to reduce noise, or can be fitted with effective noise control treatments.

The author continues to add to and update his library of equipment noise emission data used in the preparation of this chapter. Readers with access to new or useful data or information on equipment noise characteristics or control, and wanting to share their information, are encouraged to send copies to the author.

REFERENCES

1. L. N. Miller, E. W. Wood, R. M. Hoover, A. R. Thompson, and S. L. Patterson, *Electric Power Plant Environmental Noise Guide*, rev. ed., Edison Electric Institute, Washington, DC, 1984.
2. J. D. Barnes, L. N. Miller, and E. W. Wood, "Power Plant Construction Noise Guide," BBN Report No. 3321, Empire State Electric Energy Research Corporation, New York, 1977.
3. I. L. Vér and E. W. Wood, "Induced Draft Fan Noise Control: Technical Report and Design Guide," BBN Report Nos. 5291 and 5367, Empire State Electric Energy Research Corporation, New York, 1984.
4. I. L. Vér, "Perforated Baffles Prevent Flow-Induced Resonances in Heat Exchangers," in *Proceedings of DAGA 1982*, Deutschen Arbeitsgemeinschaft für Akustik, Göttingen, Germany, 1982, pp. 531–534.
5. *Bulldozer Noise Control*, manual prepared by Bolt Beranek and Newman for the U.S. Bureau of Mines Pittsburgh Research Center, Pittsburgh, PA, May 1980.
6. *Front-End Loader Noise Control*, manual prepared by Bolt Beranek and Newman for the U.S. Bureau of Mines Pittsburgh Research Center, Pittsburgh, PA, May 1980.

7. R. M. Hoover and E. W. Wood, "The Prediction, Measurement, and Control of Power Plant Draft Fan Noise," Electric Power Research Institute Symposium on Power Plant Fans—The State of the Art, October 1981.

8. C. C. Thornton, C. B. Lehman, and E. W. Wood, "Flexible-Blanket Noise Control Insulation—Field Test Results and Evaluation," *Proc. INTER-NOISE 84*, June 1984.

9. C. L. Moore, A. E. Hribar, T. R. Specht, D. W. Allen, I. L. Vér, and C. G. Gordon, "Power Transformer Noise Abatement," Report No. EP 9-14, Empire State Electric Energy Research Corporation, October 1981.

10. R. J. Sawley, C. G. Gordon, and M. A. Porter, "Bonneville Power Administration Substation Noise Study," BBN Report No. 3296, Bolt Beranek and Newman, Cambridge, MA, September 1976.

11. I. L. Vér, D. W. Anderson, and M. M. Miles, "Characterization of Transformer Noise Emissions," Vols. 1 and 2, BBN Report No. 3305, Empire State Electric Energy Research Corporation, New York, July 1977.

CHAPTER 19

Noise and Vibration Control of the Internal Combustion Reciprocating Engine

THEO PRIEDE
University of Cape Town
Cape Town, South Africa

19.1 INTRODUCTION

This chapter considers the origin and control of vibrations and noise radiated by the structure of the internal combustion reciprocating engine. The two predominant sources of engine noise, namely the inlet and exhaust, which can be effectively reduced by acoustical design of silencers, are not considered here.

The noise radiated by engine surfaces is determined by judicious choices of basic engine design factors that are primarily dictated by considerations of performance, stresses, emissions, weight, and reliability. Fortunately, proper choices among these alternatives in design can both reduce the noise and often improve the performance of the product.

Today the internal combustion reciprocating engine has an overall monopoly in road transportation. There are only two distinct combustion systems used, both conceived some hundred years ago by two German engineers.

In 1876 Nicholas Otto demonstrated the spark ignition gasoline engine and in

Noise and Vibration Control Engineering: Principles and Applications, Edited by Leo L. Beranek and István L. Vér.
ISBN 0-471-61751-2 © 1992 John Wiley & Sons, Inc.

1892 Rudolf Diesel demonstrated the compression ignition engine, now known as a diesel engine.

Initially the spark ignition engine covered a wide range of size and power but was gradually displaced in most applications by the more efficient diesel engine. The spark ignition engine, however, has become the unchallenged power unit for passenger cars.

Apart from its use in stationary plants, the diesel engine is used for powering ships, locomotives, and construction equipment and by now almost exclusively in commercial road transportation.

The reason why internal combustion engines have exclusive use in road transportation are their high specific power coupled with the fact that they use liquid hydrocarbon fuels, which have high energy density per unit mass and volume. For an operating radius of 500 miles, the energy storage on the vehicle amounts to only 1–3% of the total vehicle weight.

The material presented in this chapter is based on investigations carried out by the writer and his colleagues at the British Company CAV Lucas, at the Institute of Sound and Vibration Research (ISVR) University of Southampton, and recently, at the Energy Research Institute, University of Cape Town, South Africa. The sound pressure levels (SPLs) presented in the figures were obtained at a distance of 1 m from the side of the engine, according to SAE measurement procedure J-1074, February 1987, and are referenced to 20 μPa. Vibration acceleration levels are referenced to $1g = 9.81 \text{ m/s}^2$ if not noted otherwise.

19.2 COMBUSTION PROCESSES, COMBUSTION CHAMBERS, AND CHARACTERISTICS OF GAS LOADS

Gas loads, meaning the pressure in the combustion chamber, in both gasoline and diesel engines are impulsive. Figure 19.1 illustrates this for a four-stroke-cycle engine. As can be seen, the duration of the rapid rise and fall of a gas load is about 120° crank angle, that is, of the total period of 720°. The peak pressures, which depend on the combustion system used, are between 30 and 200 bars. When the diagrams are enlarged (as illustrated in Fig. 19.2) along the time scale, the fundamental differences between the cylinder pressure developments of gasoline and diesel engines become apparent and also reveal their effects on emitted noise.

In the gasoline engine a premixed fuel–air mixture is compressed to about one-eighth to one-tenth of its original volume. The mixture is ignited by a small heat source generated by a spark. Propagation of the turbulent flame is slow at first followed by an increasing rate. For that reason the pressure rise owing to combustion blends smoothly with the compression curve at first followed by exponential pressure rise (see Fig. 19.2). These characteristics of pressure development result in quiet operation of the engine.

In the diesel engine only air is compressed, but to a very much higher degree (about $\frac{1}{16}$ to $\frac{1}{22}$ of its original volume) than in gasoline engines. The compression increases the air temperature to 600–800°C. Into this compressed hot air, liquid

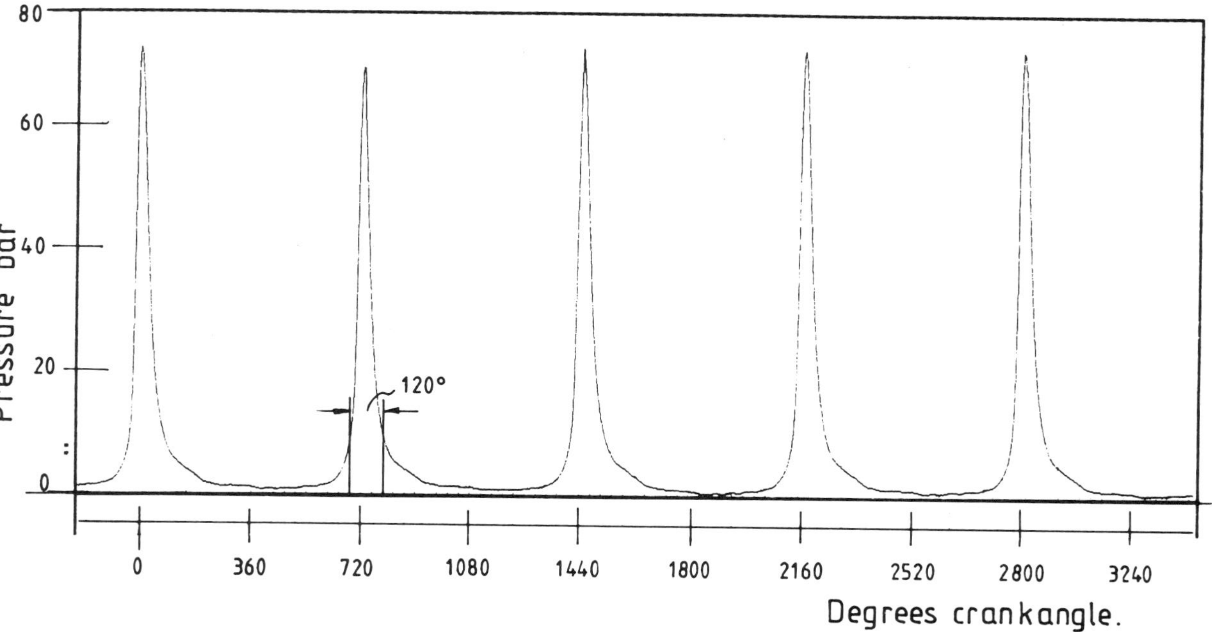

Fig. 19.1 Characteristic of shape cylinder pressure pulse in internal combustion engines.

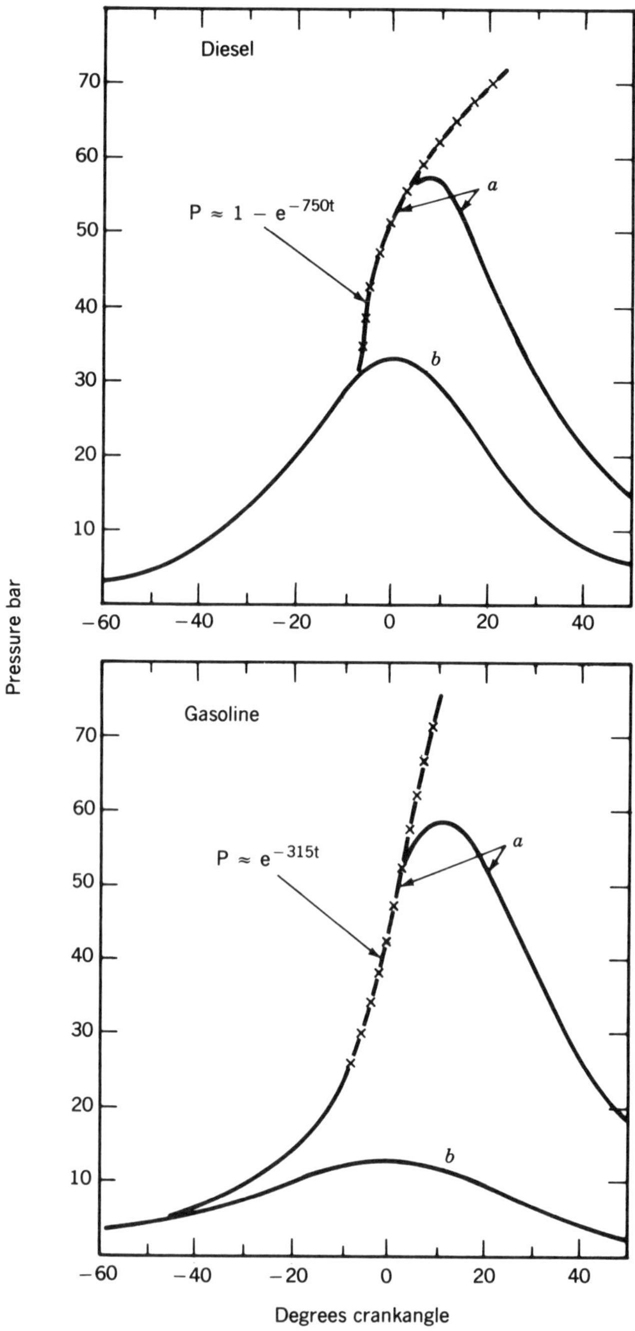

Fig. 19.2 Cylinder pressure pulses in diesel and gasoline engines: *a*, combustion; *b* compression.

fuel is injected in the form of atomized spray. The fuel droplets evaporate and heat up until the fuel–air mixture attains a condition where spontaneous ignition occurs in a significant volume of the mixture. This is accompanied by a rapid or almost instantaneous pressure rise, which results in explosive noise, followed by a diffusion flame that is associated with a reduction of the rate of pressure rise.

There are varying opinions about the optimum combustion chamber design, with the result that there is a substantial number of radically different types in use.

Figure 19.3 illustrates schematically some of the combustion chamber shapes for gasoline and diesel engines. For gasoline car engines the commonly used shapes are wedge shape, bath tub, and bowl-in-piston chambers. High-performance engines use hemispherical chambers. More recently, high-turbulency, lean-burn high-compression chambers have been introduced for low emissions.

Fuel-injected diesel engines can be divided into two distinct groups, namely the direct-injection engines (DI engines) and the indirect-injection engines (IDI engines). In DI engines the combustion chamber is contained in the piston, its shape being determined by the mechanism of fuel–air mixing employed. The quiescent chamber is used for engines of 2 liters per cylinder capacity and above, the mixing is achieved by the momentum of the spray alone. A large number of sprays (seven to nine) are used with injection pressures of 1000 bars and more. In this system combustion occurs at the instant of highest premixing, which results in the noisiest combustion. The quiescent chamber is used for large truck engines and most stationary, locomotive, and marine engines.

In engines below 2 liters per cylinder capacity, the introduction of swirl is accomplished by tangential inlet ports with injection pressures of 450 bars using four-hole nozzles. The premixed part of the sequence is significantly reduced giving quieter combustion. This combustion system is used in the majority of commercial vehicles.

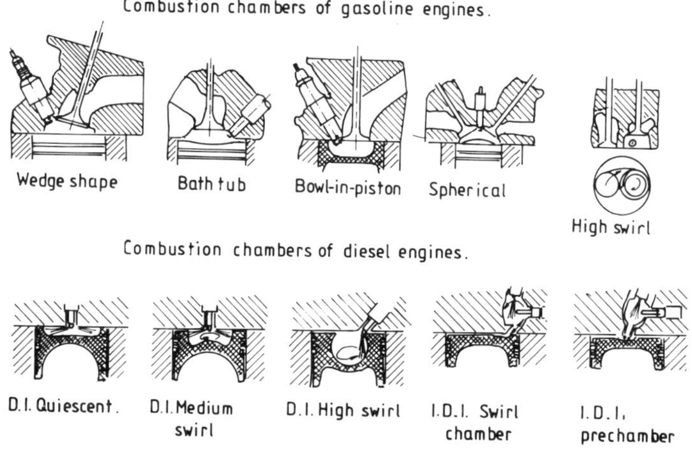

Fig. 19.3 Types of gasoline and diesel engine combustion chambers.

The Multiple Annular Nozzle (M.A.N.) type M systems offer the quietest form of diesel combustion. A very high swirl is obtained by helical inlet ports in a spherical combustion chamber. Two sprays are directed on the combustion chamber wall, thus reducing the premixed part of combustion to the minimum. For cylinder capacities below 0.5 liter, the DI multihole injectors present sizable difficulties because of the small size of holes required in the injectors. To overcome this problem, IDI engines were developed for cars, of which there are two types: the swirl chamber system developed by Ricardo and the prechamber system. Each uses a single-spray-pintle annular orifice nozzle with an injection pressure of about 250 bars. The majority of car diesel engines use the Ricardo swirl chamber, which has quiet combustion over a wide speed range of up to 5000 rpm.

19.3 BASIC PRINCIPLES OF ENGINE VIBRATION AND NOISE

An engine consists of two basic structure elements:

(a) the internal load-carrying structure, which includes the piston connecting rod and the crankshaft system, and
(b) the outer load-carrying cylinder block and crankcase.

The internal load-carrying structure is separated from the main outer load-carrying structure by running clearances shown in the upper part of Fig. 19.4*b*. Excitation forces can be classified in two categories: the unidirectional (P) and reversible forces (F). Unidirectional (P) forces, caused by combustion, induce noise in the vicinity of top dead center (TDC). These combustion induced forces are caused by the rapid pressure rise resulting from the high rate of heat release at the onset of combustion. As shown in Figure 19.4*a*, this pressure rise also induces significant pressure oscillations within the combustion chamber. The response of the engine structure to these pressure fluctuations, $P(t)$, can be modeled as a linear spring-mass system with mass M, damping coefficient C, and dynamic stiffness K:

$$M\ddot{x} + C\dot{x} + Kx = P(t) \tag{19.1}$$

Where x is the displacement, $\dot{x} = j\omega x$ is the velocity, and $\ddot{x} = -\omega^2 x$ is the acceleration.

Reversible forces F are the cause of mechanically induced noise. Because of the crank mechanism and inertia, these forces change direction. They accelerate various elements of the internal load-carrying structure across the clearances and thus cause impacts that induce the engine structure into vibration.

The equation that describes system behavior for these reversible forces accounts for the fact that transverse motion is governed by the rate of change of acceleration, $\dddot{x}$.

$$M\dddot{x} = \frac{dF}{dt} \text{ const} \tag{19.2}$$

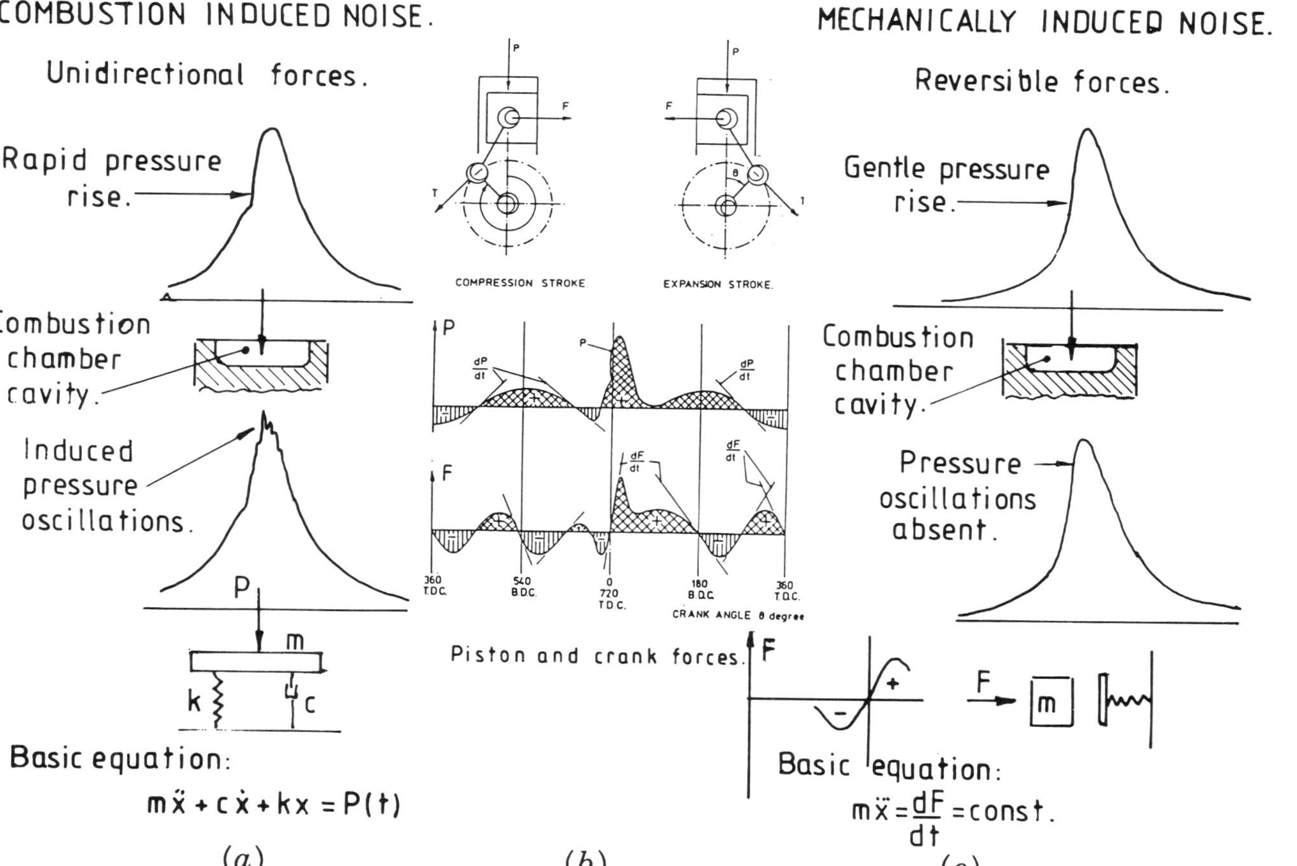

Fig. 19.4 Mechanism of combustion and mechanically induced engine noise.

For rotating parts the equation is similar, except it contains the rate of change of angular acceleration $\dddot{\theta}$ and moment of inertia J:

$$J\dddot{\theta} = \frac{dT}{d\theta} = \text{const} \tag{19.3}$$

where T is torque.

In engines where the cylinder pressure development is smooth, as shown in Fig. 19.4*c*, there are no pressure oscillations in the combustion chamber and the mechanically induced noise predominates.

In all engines both sources of noise are present. In normally aspirated DI diesel engines the combustion-induced noise is predominant while in gasoline engines the mechanically induced noise is predominant.

Combustion-induced Noise: The Concept of Cylinder Pressure Spectrum

During the World War I Sir Harry Ricardo[1,2] recognized the qualitative relation between rapidity of combustion and noise, stating that "explosion strikes the walls of the cylinder with a hammer blow."He also recognized that rapid rise in gas pressure has the identical effect on engine structure noise as would be produced by metallic impact.

Because the structure response decays rapidly during the time interval between successive combustions, engine vibration and noise are basically transient phenomena whose causes are divided into two parts:

(a) a part representing a forced motion of engine structure determined by the form of the gas force and controlled by the stiffness of structure and

(b) a part representing the "ringing" vibration of the structure at its natural frequencies whose amplitude depends on the shape and time of the applied gas force and the dynamic characteristics of the engine structure and its damping.

Hinze[3] arrived at very important conclusions in regard to the effect of the particular force shape on structural response. His study revealed that the important factor in the force–time diagram is the ratio of the period of rapid force rise (Δt) and the natural period of the vibratory system, T_0:

$$\tau = \frac{\Delta t}{T_0} \tag{19.4}$$

Only when the ratio τ exceeds a value of about 0.25 does the force rise become a controlling factor in determining the amplitude of the resultant vibration. This means that at high engine speeds the actual shape of the gas force, either gasoline or diesel, has negligible effect.

Figure 19.5*a* illustrates a force diagram for the main bearing of a six-cylinder diesel engine. The bearing force is the sum of the forces from two adjacent cylinders acting on their respective crankpins. Figure 19.5*b* is the resultant bearing cap vibration in the vertical direction, measured with an accelerometer. The onset of the rapid change of the magnitude of the bearing force (caused by the abrupt pressure rise in engine cylinder) gives rise to a transient, heavily damped vibration. During a full cycle there are only two vibration transients that are particularly predominant. These are caused by the firing of the adjacent cylinders. By integrating twice, the vibration acceleration shown in Fig. 19.5*b* produces the waveform of the displacement of the bearing cap, as shown in Fig. 19.5*c*. Not surprisingly, it follows closely the form of the applied bearing force.

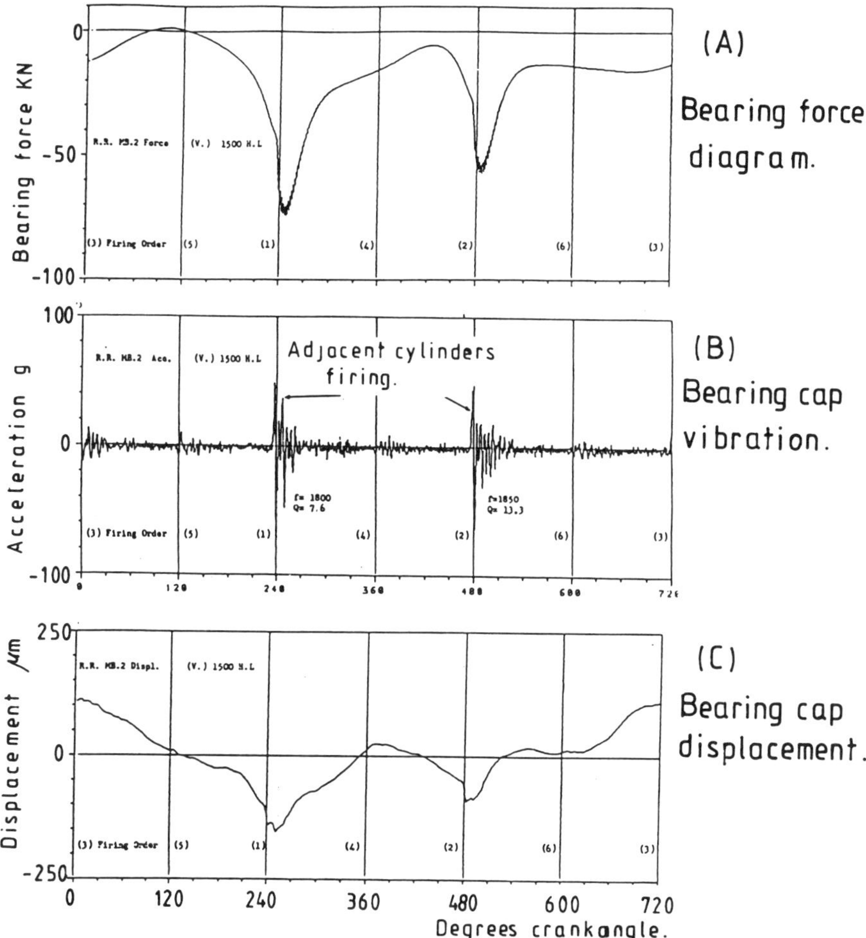

Fig. 19.5 Transient vibration of engine structure resulting from rapid rise in cylinder pressure.

The high level of diesel engine noise has been a concern of engine manufacturers for many decades. In particular, manufacturers of fuel injection equipment, being to some extent responsible for the form of cylinder pressure development, need a technique to assess the noise-making properties of cylinder pressure irrespective of the engine structure.

This is done by frequency analysis of the pressure development in the cylinder,[4] that is, a Fourier series analysis that provides a full description of the exciting propensities of the cylinder pressure.

When an electronic frequency analyzer is used, it is essential to check its accuracy by applying a waveform for which the amplitude of harmonics can be defined by a mathematical formula. A simple method of checking is to subject the analyzer to an electronically generated sawtooth wave, for which the amplitudes of the harmonics decrease by 20 dB per 10-fold increase of harmonic number. The computer programs used for harmonic analyses of digital cylinder pressure diagrams require similar checking.

Cylinder pressure and spectra are shown in the upper and lower part of Fig. 19.6 for a DI diesel engine with two injection timings (A and B). With a narrow-band analyzer, as seen in *A* and *B* of the left lower part of Fig. 19.6, the harmonics up to about the 20th harmonic can clearly be distinguished, but above the 20th harmonic the curve becomes continuous owing to the higher harmonic density.

It can be seen that the more abrupt pressure development of timing A has a significantly higher level of harmonics from about 200 Hz and above. A more or less similar increase is found in the emitted noise as measured in one-third-octave bands, as shown by the two (*A* and *B*) curves inserted in the right lower graph. It is concluded that the level of emitted noise is proportional to the level of cylinder pressure spectrum and correspondingly also that the engine noise is controlled by the rapid rise in cylinder pressure. By application of cylinder pressure spectra as input information in studies of engine noise, many aspects of the noise characteristics can be explained.

Mechanically Induced Noise

From the numerous mechanically induced sources of noise the most readily identified is the piston slap which has received major attention. Meier[5] published a detailed theoretical and experimental analysis of piston sideways movement and introduced piston pin offset for controlling piston slap noise. Ross and Ungar[6] developed an impact energy method for assessment of parameters that control the intensity of noise. More recently, computer programs supported by advanced experimental techniques[7–10] have led to quiet piston designs. At present, piston slap noise is seldom a predominant source of engine noise.

Here it is intended only to discuss the parametric analysis of the mechanism of piston slap, which is also relevant to other impacts in the engine, for example, bearing impact and impacts in timing gears. Figure 19.7 illustrates the side force diagram of a 2-liter cylinder and the resultant transient linear vibration of the liner.

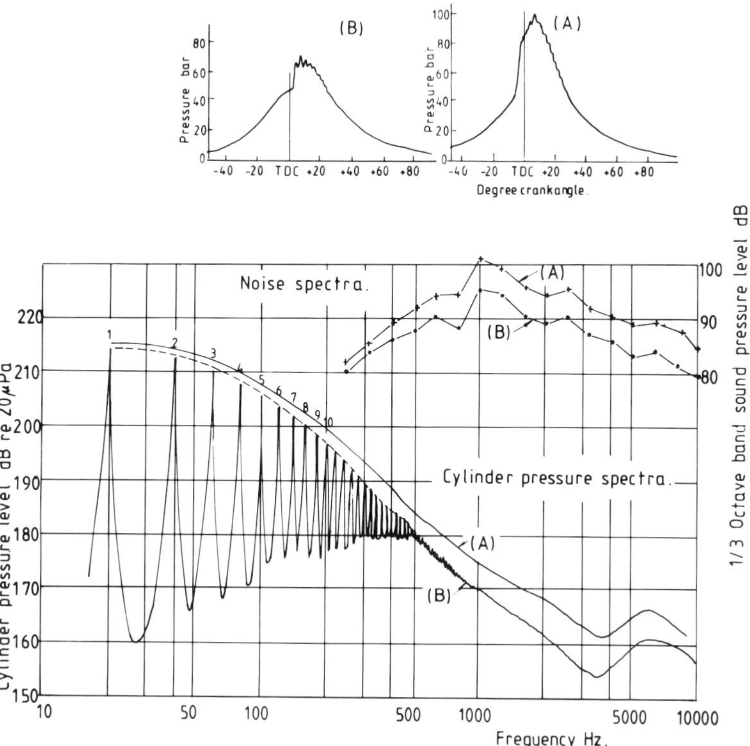

Fig. 19.6 Relationship between cylinder pressure spectra and emitted noise in a DI normally-aspirated diesel engine for two injection timings A and B.

A simplified piston slap model is shown in Figure 19.8. The rate of growth of the side force is constant and reaches its maximum around the TDC of the working cycle, where the force on the piston, mainly due to combustion, is highest. The rate of the side force is only controlled by the peak value of the force, and therefore the form of the gas force due to combustion is irrelevant.

The equation of lateral motion for the piston is given by

$$M\dddot{x} = \text{const} = \frac{dF}{dt} = \sigma_0 \tag{19.5}$$

where σ_0 is the force growth rate.

Integrating the equation three times yields t_i, the time required for the piston to cross the clearance:

$$t_i = \left(\frac{6M\delta}{\sigma_0}\right)^{1/3} \tag{19.6}$$

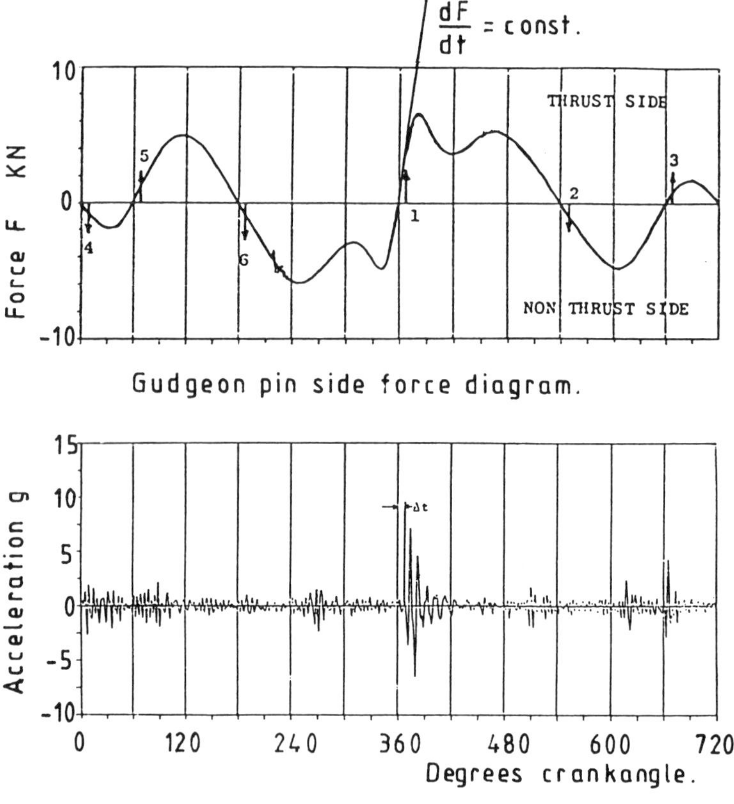

Fig. 19.7 Piston-slap-induced cylinder liner vibration.

where δ is the piston clearance and M is the piston mass. V_i, the piston impact velocity against the liner, is:

$$V_i = \left(\frac{9\sigma_0\delta^2}{2M}\right)^{1/3} \tag{19.7}$$

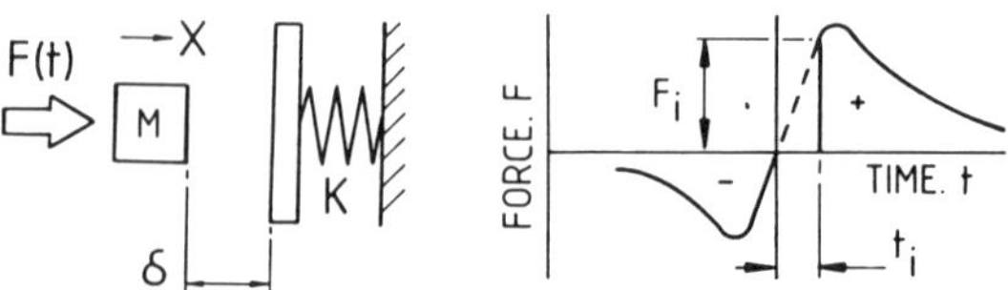

Fig. 19.8 Piston slap model.

Kinetic energy on impact is $\frac{1}{2}MV_i^2$, and by equating to the maximum energy absorbed by the spring, the resulting force amplitude can be shown to be

$$|F_0|_{\mathrm{KE}} = (\tfrac{81}{4}\sigma_0^2\delta^4 MK^3)^{1/6} \tag{19.8}$$

where K is the dynamic stiffness of the liner.

Although the side force on the piston is continuous, the liner reaction force is zero while the piston is traveling across the clearance. Thus even if the piston's vertical movement were zero, when it reaches the other side of the bore, the liner would suddenly experience a piston side force appropriate to the crankangle. This force will be termed the potential energy force $|F_0|_{\mathrm{PE}}$ to distinguish it from that produced by the kinetic energy of the impact. Its magnitude at impact is directly proportional to the time taken for the piston to cross the clearance.

Therefore,

$$|F_0|_{\mathrm{PE}} = \sigma_0 t_i = (6M\delta\sigma_0^2)^{1/3} \tag{19.9}$$

The spectral amplitude of this suddenly applied force will decay by 20 dB per 10-fold increase of speed. In addition, its magnitude will also increase by an additional factor of speed.

Therefore,

$$|S_0|_{\mathrm{PE}} = (6M\delta\sigma_0^2)^{1/3}N \tag{19.10}$$

where N is the repetition rate and S_0 is the spectral amplitude.

For a given engine, the intensity of the radiated noise, I, is proportional both to the square of the force amplitude and to the speed. Hence, from Eqs. (19.8) and (19.10).

$$I_{\mathrm{KE}} \sim N^{2/3}\delta^{4/3} \tag{19.11}$$

and

$$I_{\mathrm{PE}} \sim N^{10/3}\delta^{2/3} \tag{19.12}$$

The kinetic energy component of noise is relatively insensitive to speed but is much more sensitive to clearance, while for the potential energy component of noise the reverse is true. In addition to the parameters contained in Eqs. (19.11) and (19.12), the kinetic and potential energy components of noise depend on engine size. This size effect can be approximately related to engine bore B as follows:

$$\sigma \sim B^2 \qquad \delta \sim B \qquad M \sim B^3 \qquad K \sim B \tag{19.13}$$

In addition, the radiating area is proportional to B. Accordingly, the following simple relationship can be obtained:

$$I_{\mathrm{KE}} \sim B^7 \quad \text{and} \quad I_{\mathrm{PE}} \sim B^5 \tag{19.14}$$

The relations in Eq. (19.14) indicate that for small engines piston slap is dominated by potential energy components and for large engines the kinetic energy component predominates. It is the potential energy component that predominates crankshaft bearing impact, and for that reason the bore relation tends to control all automotive engine noise irrespective of the combustion system used.

19.4 ORIGINS AND CHARACTERISTICS OF NOISE OF VARIOUS ENGINE TYPES

In general, the engine structures are very similar for different combustion systems. The type of combustion system, however, determines to a great extent the level and spectral characteristics of emitted noise. Therefore, in the following sections engines are classified according to the type of combustion system used.

Direct-Injection Normally Aspirated Diesel Engine

It is generally found for a normally aspirated DI diesel engine that the cylinder pressure diagrams remain similar irrespective of the engine speed if the horizontal axis represents crank angle instead of time. The amplitudes of harmonics of cylinder pressure are also the same (i.e., independent of engine speed) except that on a frequency scale the cylinder pressure spectra are geometrically similar but have a shift parallel to the frequency axis, corresponding to the change in engine speed.

Noise–Speed Relationship

Figure 19.9 shows the shape of the cylinder pressure pulse (lower left), its frequency composite (center), and the one-third-octave-band spectrum of the radiated noise (upper right) of a 118-mm-bore DI engine over its speed range from 1000 to 2600 rpm. The level of cylinder pressure harmonics decreases at almost a constant rate above the 20th harmonic by some 32 dB/decade. Since the predominant radiation of the engine, as shown by the noise spectrum, lies in the high-frequency range from 800 to 2000 Hz, the increase of cylinder pressure excitation in that frequency range depends on the slope of the cylinder pressure spectrum. The 32-dB/decade slope of the cylinder pressure spectrum would give a 9.7-dB increase per doubling of engine speed, or 32 dB per 10-fold increase of speed. A straight line relation between noise level (dBA) and speed depicted in the upper left part of Figure 19.9 also shows the same (32 dBA) per decade rate of increase of noise with speed.

Provided that the engine noise is controlled by rapid pressure rise, it can be

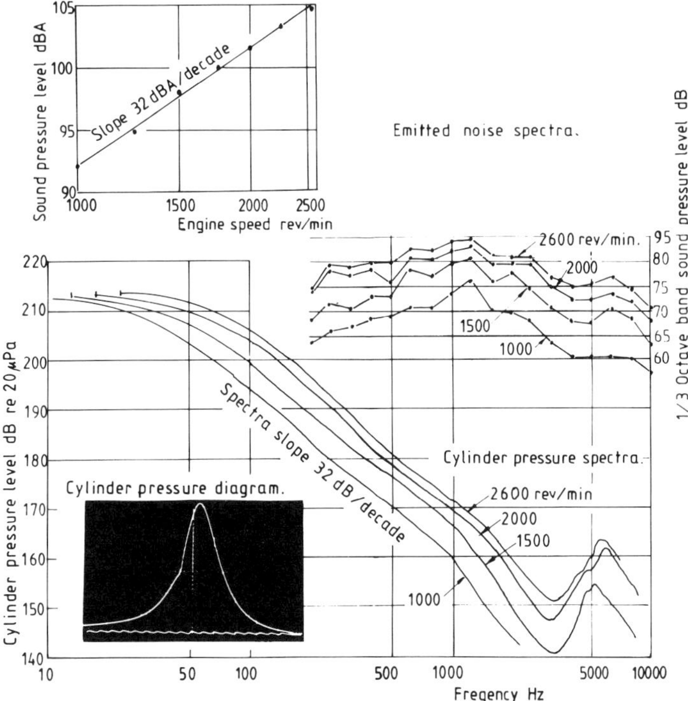

Fig. 19.9 Effect of engine speed on characteristics of cylinder pressure spectra and emitted noise in a DI normally aspirated engine.

concluded that the rate of increase in noise is controlled by the form of cylinder pressure pulse, which in turn, determines the slope of the cylinder pressure spectrum.

The intensity (sound power flow per unit area) of radiated engine noise thus can be expressed in the form

$$I \sim N^n \tag{19.15}$$

The exponent n may be called a *combustion index*, which in diesel engines can vary between 2.5 and 4.5. The smoother the cylinder pressure development, the higher is the index n. Note here that the radiated noise is less for smooth combustion pulse; only its rate of growth with speed is higher. For a quiescent chamber, which has more abrupt cylinder pressure development, the index n is 2.5–3.0. For medium-swirl engines n is 3.0–4.0, while for a high-swirl MAN type M engine index n is of the order of 4.5.

Some engine-noise-versus-speed curves are shown in Fig. 19.10a for DI engines of different bore diameters. They all show similar characteristics of about 30 dBA per 10-fold increase in speed. The marked effect of bore diameter is noted.

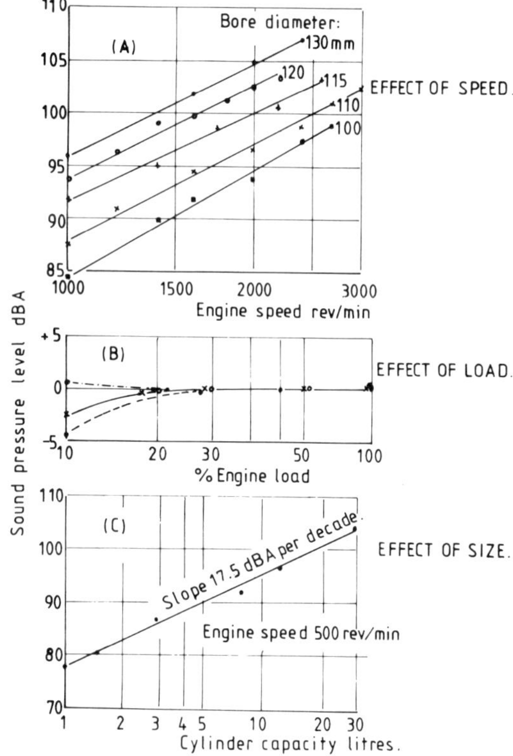

Fig. 19.10 Effect of speed, load, and engine size on the noise of a DI normally aspirated diesel engine.

Effect of Engine Load

Reduction of engine load is associated with reduction of peak cylinder pressures. However, the initial pressure rise at light loads is just as abrupt as at full-load conditions. This results in some reduction of cylinder pressure level in the low-frequency range, while the level over the high-frequency range remains basically the same. For that reason the effect of load on engine noise is minimal and of the order of 1–4 dBA, as shown in Fig. 19.10*b*. In some combustion systems where the combustion chambers are more susceptible to cooling effects, the noise at light loads may even be greater. For all practical purposes it can be assumed that the effect of load on noise can be neglected.

Effect of Engine Size

The effect of engine size on noise was determined[11] by measurement of the noise of engines with a wide range of cylinder capacities ranging from 0.35 to 30 liters per cylinder. It was found that engine noise increased by 17.5 dBA per 10-fold increase of cylinder capacity, as shown in Fig. 19.10*c*. Thus a relationship with size and intensity of noise was established:

$$I \sim V^{1.75} \tag{19.16}$$

for a given crankshaft speed, where V is the total cylinder volume.

Dependence of Engine Noise on Engine Parameters

An empirically derived expression on the relation between engine noise and speed, load, and size can be summarized as follows:

$$I = N^n V^{1.75} \tag{19.17}$$

where the exponent n is determined by the combustion system.

Equation (19.17) indicates that engine noise is not a function of engine power and that the two controlling parameters are speed and size. It also confirms earlier observations by practicing engineers that on a power-for-power basis a large engine running slowly is quieter than a smaller one running faster. The reason becomes obvious from Eq. (19.17), which indicates that the rate of increase of noise with speed is at least twice that of the increase of noise with size ($n \geq 3.5$).

Since the cylinder capacity $V \sim B^3$, where B is the cylinder bore, the intensity of noise can be expressed as

$$I \sim N^n (B^3)^{1.75} \simeq N^n B^5 \tag{19.18}$$

which leads to a formula derived at the Institute of Sound and Vibration Research (ISVR) predicting the parametric dependence of engine noise,

$$\text{SPL(dBA)} = 10n \log_{10}[N(\text{rpm})] + 50 \log_{10} B + \text{const} \tag{19.19}$$

where bore B is in inches and the average value of the constant is 31.5.

The constant in Eq. (19.19) depends on the response of engine structure and its radiation efficiency.

The formula was checked by carefully controlled experiments using two V8 engines of the same cylinder capacity but of different stroke–bore ratio. The noise spectra of the two engines running at the same speed are shown in Fig. 19.11. The cylinder pressure pulse shape and the spectra were adjusted to be identical. As can be seen, a considerably greater overall level of noise is measured from the larger bore engine. The higher noise level corresponds directly to the larger total gas force applied to the pistons.

By using the larger bore engine, the designer is usually attempting to achieve the maximum possible engine power output per bulk volume (kilowatts per cubic meter) at the expense of noise. Both the larger gas force applied to the pistons and the higher rated speed result in increased noise.

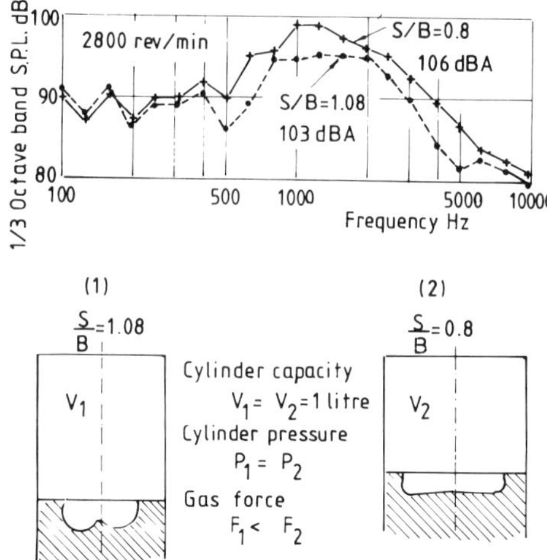

Fig. 19.11 Effect of bore-to-stroke ratio on emitted noise.

In another experiment[12] the bore diameter on the same engine was increased. This experiment also confirmed the B^5 relationship given in Eqs. (19.18) and (19.19).

The bore relationship indicates that the power output of an engine can be best increased by increasing the number of cylinders. This way the engine noise is affected only very slightly. Experimental data have shown that in-line 4, in-line 6, and V8 engines of the same bore have the same level of noise.[13] If reduction of noise is required, a quieter engine will be obtained if the total needed engine power is produced by a larger number of cylinders.

Cylinder Pressure Oscillations

The gas volume in the combustion chamber has natural modes of vibration, and therefore pressure changes occurring locally with a time scale that is short compared with the time period of the first natural frequency will give rise to pressure oscillations in the gas. Pressure oscillations commonly seen on diesel pressure diagrams occur in this way. Hickling et al.[14,15] provided the most comprehensive study of this phenomenon. The fundamental circumferential mode tends to be predominant. The frequency of this mode is determined by the cylinder bore and temperature.

In automotive DI diesel engines (bore diameter 100–130 mm) the frequency of pressure oscillations is between 4000 and 6000 Hz and their contribution to engine noise is minimal. In larger DI diesel engines the pressure oscillations are of larger amplitude, and these can become predominant sources of noise in the frequency range of 800–2000 Hz.

The frequency of pressure oscillations are also lower (1500–2500 Hz) in small IDI diesel engines because for these engines the frequency of the first resonant mode is determined by a Helmholtz-type resonator formed by the prechamber and main chamber above the piston. Due to smooth cylinder pressure development, it is seldom a serious source of noise.

Gasoline Engine Noise

The measured cylinder pressure spectra of gasoline engines show low levels, particularly in the high-frequency range from 800 Hz upwards. However, the noise spectra of gasoline engines have characteristics similar to those of equivalent size high-speed diesel engines used in cars. As shown in Fig. 19.12, the noise spectra peak in both cases lies in the same 800–2000-Hz frequency range. However, for a given speed the level of noise of the gasoline engines is somewhat lower. In both engine types the noise is due to vibration of the engine structure at its natural frequencies. The major difference is that excitation of the structure lies in a different origin.

Raff and Grover[16] identified the exciting forces in the gasoline engines as being caused by impulsive oil pressures (i.e., hydraulic forces) that develop in the main

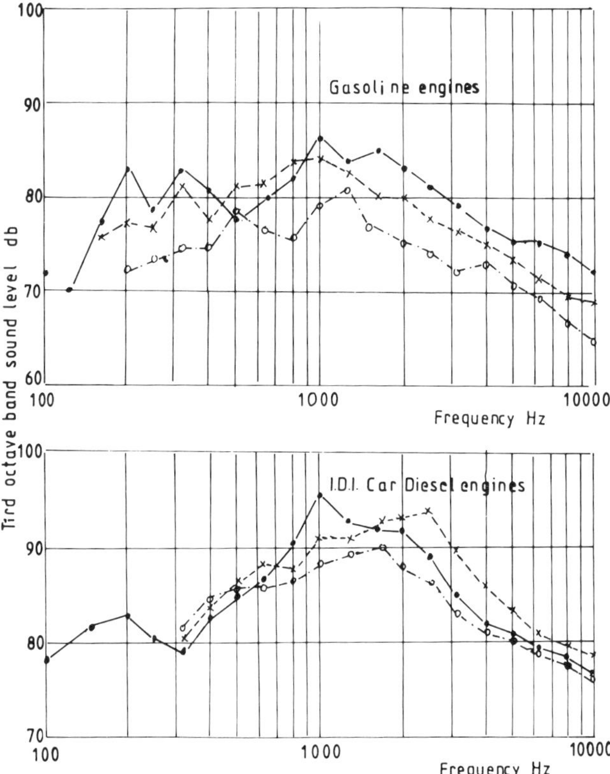

Fig. 19.12 Typical one-third-octave-band noise spectra of car diesel and gasoline engines.

crankshaft bearings owing to rapid movements of the crankshaft across its clearances. They found that the peak cylinder pressure and not the rapid rate of pressure rise is the controlling factor of the noise of the gasoline engine.

Figure 19.13 shows curves of the A-weighted sound pressure level versus speed of some selected gasoline engines running at full load. The engines ranged from low- to high-performance designs. Also shown in Fig. 19.13 are the noise–speed characteristics of small IDI diesel engines. These engines have similar noise-versus-speed dependence as gasoline engines but higher noise levels. The difference in the noise level between the various engines is of the order of 15 dBA. In this sample, the engines with higher peak cylinder pressures produced higher noise levels. However, the engine structure is also an important parameter, since some engines do have higher vibratory respone than others.

The noise–speed characteristics of most engines also show two distinct slopes. In the low-speed range of up to 2500–3500 rpm, the rate of increase of noise is between 25 and 35 dBA per 10-fold increase of speed or noise intensity proportional to

$$I \sim N^{2.5-3.5} \tag{19.20}$$

In the high-speed range the increase of noise is considerably greater, about 50–70 dBA per 10-fold increase of speed. The noise intensity is therefore proportional to

$$I \sim N^{5-7} \tag{19.21}$$

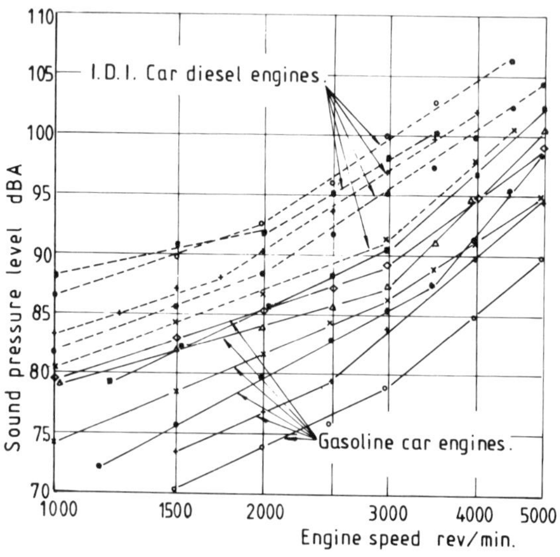

Fig. 19.13 Noise-versus-speed relationships of gasoline and IDI diesel engines.

The effect of peak cylinder pressure on noise was investigated by motoring tests on an engine where combustion chamber volume was reduced to yield a compression ratio of about 40:1. This enabled the tests to be carried out at compression pressures of up to 90 bars (i.e., pressures equal to a running engine). The air inlet manifold was fitted with a butterfly throttle valve, so that compression pressures could be decreased to 4 bars.

It was found that at the low motoring speeds noise was mainly affected by the compression pressure. With no throttling of air intake (cylinder pressure 90 bars) the engine sounded like a typical diesel engine with apparent diesel knock very much in evidence. By throttling the air intake (reduction of compression pressure) the diesel knock could be made to disappear, leaving the engine sounding like a typical gasoline engine at no load.

The noise–speed relation of the motored engine is shown in Fig. 19.14. The characteristics are identical to those of a gasoline engine shown in Figure 19.13. Cylinder pressure affects engine noise most markedly in the low-speed range, where the rate of increase is $N^{2.4-3.8}$, and considerably less at high speeds, where the rate of increase of noise with speed is N^5. The large increase of noise due to higher compression pressures at low speeds reduces the slope of the noise–speed curve from 35 to 25 dB/decade.

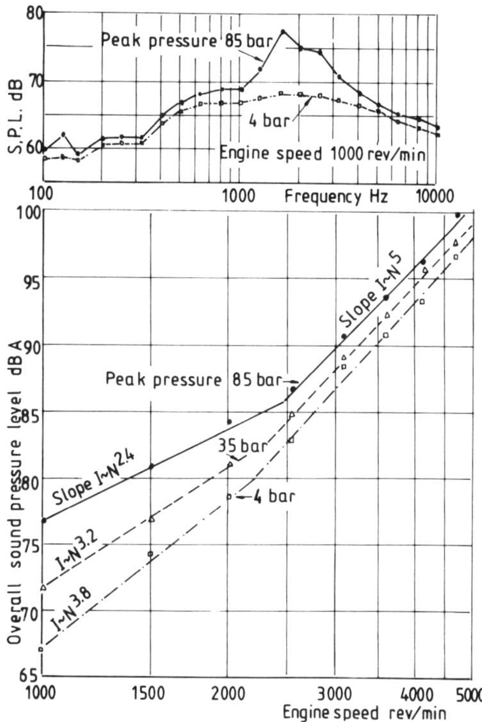

Fig. 19.14 Effect of compression pressure on the noise produced by a motored engine.

Knock-induced Noise in Gasoline Engines

Knock in gasoline engines is due to spontaneous ignition of the fuel–air mixture prior to the arrival of the flame front. The pressure diagram of combustion with severe knock is shown in Fig. 19.15*a*, which has a certain similarity with the normally aspirated diesel engine, except that rapid pressure rise occurs at the end of combustion, after TDC. The pressure rise is almost instantaneous and is often of much greater magnitude than in diesel engines. In the example shown in Fig. 19.15*a* the pressure rise is of the order of 50 bars and the initial peak-to-peak pressure oscillations of the order of 40 bars.

Figure 19.15*b* shows that the curve of noise level (dBA) versus speed exhibits a very marked change with the onset of knock, particularly in the low-speed range. The third-octave-band noise spectra depicted in the upper curves in Fig. 19.15*c* was obtained at 1500 rpm. The rapid pressure rise from knock excites the engine structure to vibration both in the frequency range from 1000 to 3000 Hz and to a far greater extent above 5000 Hz. At higher engine running speeds, for example, at 4000 rpm, the effect of knock on engine structural components is almost negligible in the frequency range from 1000 to 4000 Hz. This suggests that at higher speeds the forces that develop in the engine main bearings are predominant. However, in the high-frequency range, from 5000 Hz upward, the effect of knock-induced noise remains strong.

IDI Car Diesel Engine Noise

In the last decades there has been a considerable increase of application of high-speed diesel engines in passenger cars, mainly to achieve good economy. Most of the engines are 0.35–0.5 liters cylinder capacity with speed ranges of up to 4500–5000 rpm employing Ricardo swirl chambers.

Reasonably smooth combustion is obtained by using high compression ratios of 20 : 1–22 : 1. The cylinder pressures are therefore higher than in the gasoline engine and also the levels of noise are significantly higher. The noise-versus-speed characteristics, as shown in the upper chart of Fig. 19.13, are very similar in shape to engine characteristics with a very rapid increase of noise with speeds above a certain transition point around 2000 rpm.

Noise of Turbocharged Diesel Engines

There has been a very rapid increase in the application of turbocharged diesel engines, particularly in heavy-goods vehicles. Turbocharging provides a high engine output, better fuel consumption, greater flexibility in control of smoke and emissions, and reduction of engine weight but only moderate noise reductions.[17]

Figure 19.16 shows an example of pressure diagrams measured on a typical turbocharged engine over a large speed and load range. The curves shown dashed in Fig. 19.16 (included for the purpose of comparison) were obtained on a normally aspirated diesel engine at full load.

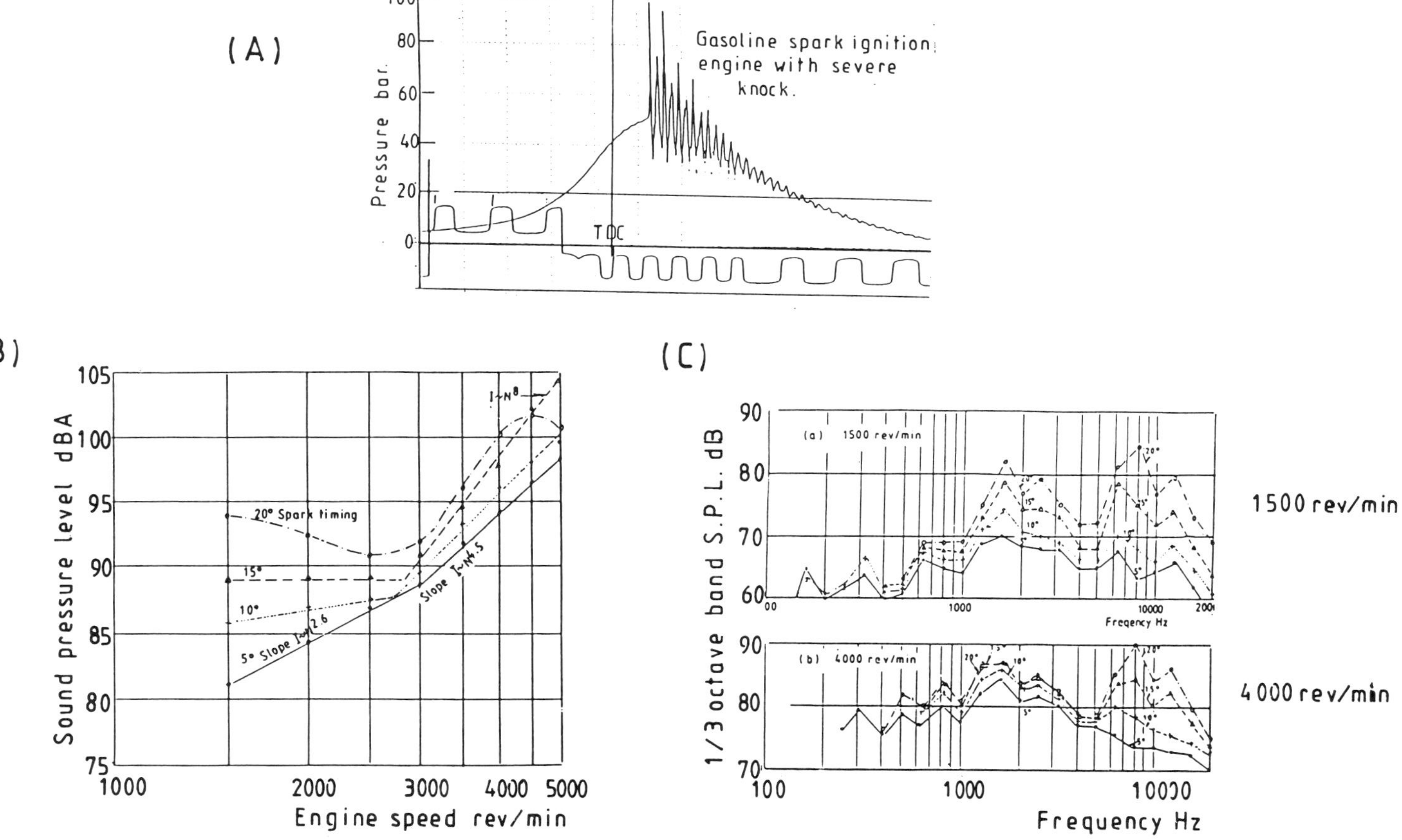

Fig. 19.15 Knock-induced noise in gasoline engines: (*a*) cylinder pulse with knock; (*b*) engine noise versus engine speed; (*c*) one-third-octave band spectrum of the engine noise with knock.

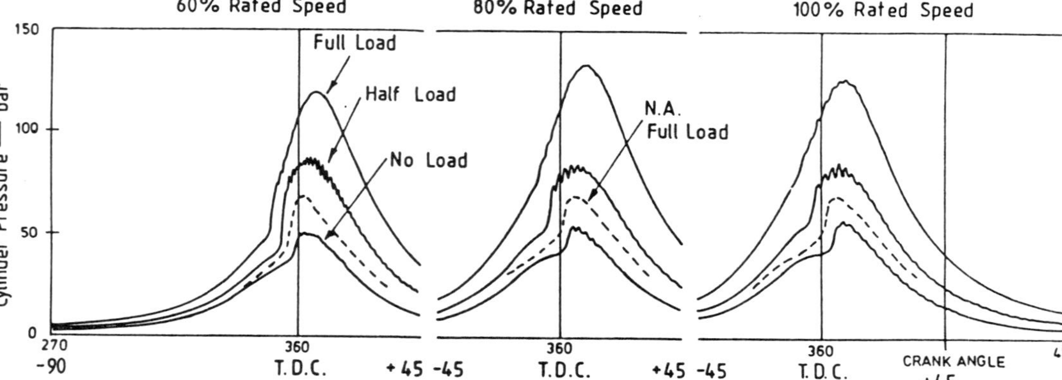

Fig. 19.16 Pressure diagrams of turbocharged DI diesel engines; dependence on engine load and speed. Dashed curves: aspirated engines at full load.

Apart from very light loads the peak cylinder pressures are of the order of 30–60% higher for the turbocharged diesels than those obtained for normally aspirated diesels. These higher peak cylinder pressures accentuate considerably the mechanical noise of the engine. The smooth cylinder pressure development only occurs over a small range of operating conditions of the engine, namely, in the small area close to rated speed and load.

With reduction of speed and load from the rated values the cylinder pressure pulse progressively becomes abrupt in nature, resembling that of a normally aspirated DI diesel engine.

Because the pressure rise in turbocharged engines is of far greater magnitude, the effect of the abrupt nature of the pressure diagram on noise is even greater than in normally aspirated engines. Figure 19.17*a* shows noise, speed, and load relationships, which are characteristic for turbocharged engines.

The effect of smoother combustion on engine noise is disappointing for the following reasons:

(a) The smoother cylinder pressure development, which amounts to some 15 dB reduction of combustion-induced noise, only occurs over a small region when the turbocharger is working close to rated speed and load (region 3 in Fig. 19.17*a*).

(b) The smoother cylinder pressure pulse as shown in Fig. 19.16 is achieved with a factor of 2 increase in peak pressure compared to the normally aspirated engine. This high peak cylinder pressure increases the mechanical impact component of the engine noise, which controls the emitted noise at rated conditions.

In region 1 of Fig. 19.17*a*, which covers the lighter load range, the cylinder pressure pulse becomes abrupt, producing high rates of pressure rise exceeding that of normally aspirated engines.

In region 3 of Fig. 19.17*a*, which is the most important part of the automotive engine operating range, both sources, namely combustion and mechanical, are of equal importance in producing engine noise. At all speeds the maximum engine noise levels fall in this range. This is also illustrated in Fig. 19.17*b* by the vibration acceleration oscillograms.

Due to the combination of the two sources of noise (combustion and mechanical), the noise-versus-speed relationships as shown in Fig. 19.17*c* for the turbocharged engines have on the average the following functional dependence:

$$\begin{aligned} &\text{Full load:} && I \sim N^{1.6-2} \\ &\text{Half load:} && I \sim N^{1.5-1.8} \\ &\text{No Load:} && I \sim N^{2.6-3.5} \end{aligned} \qquad (19.22)$$

The no-load noise relationship is the same as that of normally aspirated DI engines.

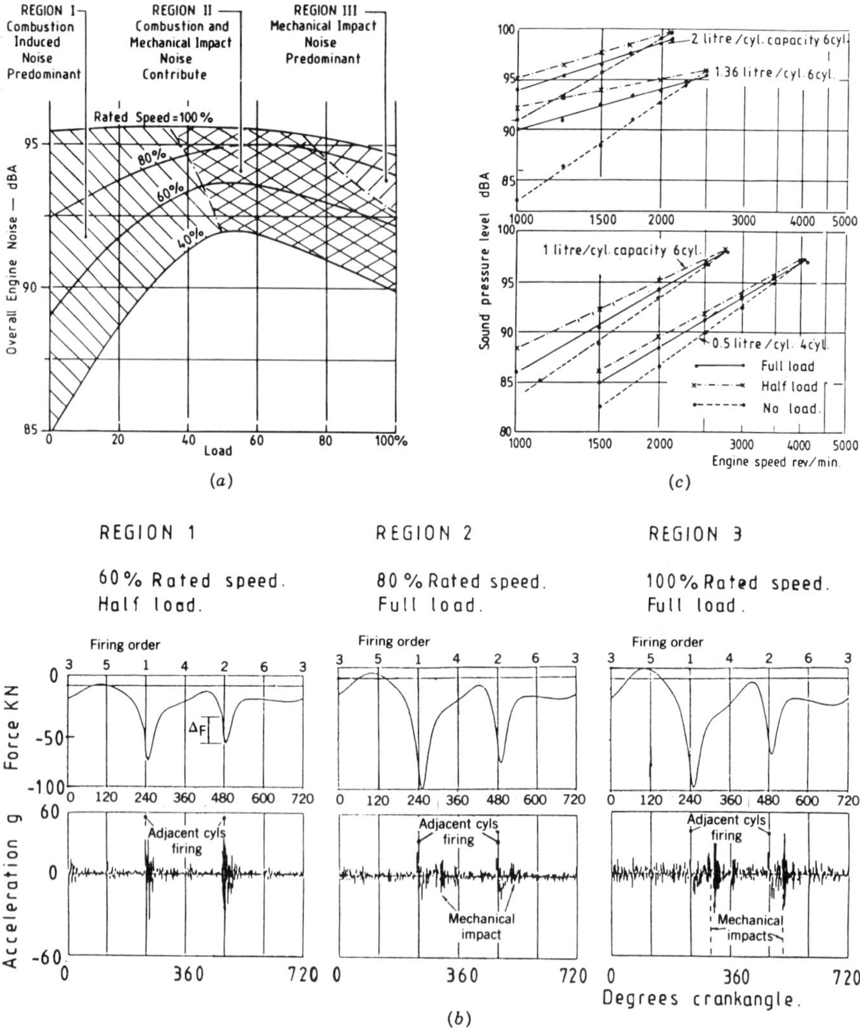

Fig. 19.17 Dependence of engine noise and vibration of a turbocharged DI diesel engine on speed and load.

Noise of Two-Stroke Cycle Diesel Engine

When considering noise of two- and four-stroke engines, a very basic question arises regarding the basis on which the two engines should be compared. One comparison can be on the same cycle repetition rate. The two-stroke cycle engine would thus be running at half the speed of the four-stroke cycle engine. Accepting this method of comparison, there is still a fundamental difference between the pressure diagrams of the two engines that considerably affects the exiting propensities as expressed by their aspects. This difference is illustrated in Fig. 19.18.

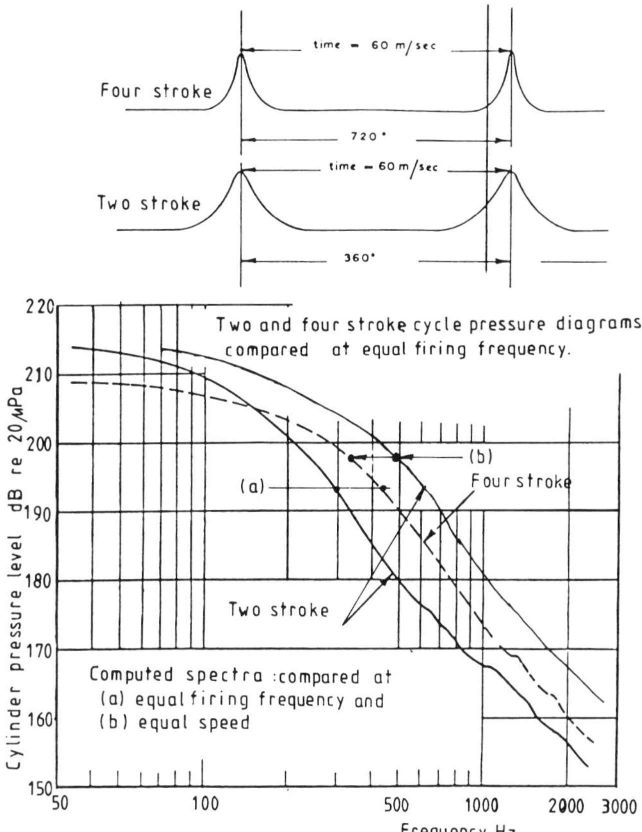

Fig. 19.18 Comparison of cylinder pressure spectra of two- and four-stroke-cycle engines.

The pulse shape of the two-stroke cycle engine is wider than that of the four-stroke cycle engine. This difference has a very profound effect on the level of harmonic components. For identical pressure diagrams on a degree basis and comparing them on the basis of repetition rate, which roughly mean the same power, the following observations can be made:

(a) The low frequency of the two-stroke cycle components engine are higher, up to the fifth harmonic.

(b) At frequencies above the crossover point, the noise levels of the two-stroke cycle cylinder pressure spectrum becomes 6 dB lower than that of the four-stroke engine.

If the cylinder pressure spectra are compared on the same-speed basis (i.e., rpm), the spectrum of the two-stroke cycle engine is 6 dB higher than that of the four-stroke cycle engine.

In practice, at equivalent engine speeds the two-stroke cycle engine has a considerably higher level of noise. At the same power output (i.e., two-stroke cycle engine running at a lower speed) the emitted noise is comparable.

19.5 COMPARISON OF VARIOUS ENGINE TYPES

Since the cylinder pressure spectrum describes fully the exiting propensities of the gas force over the whole frequency range of interest, it can be compared with the external noise spectrum and thus leads to a full description of the structural acoustical characteristics of the engine as follows:

$$\text{Sound attenuation (dB)} = \text{SPL}_{\text{cylinder}} - \text{SPL}_{\text{engine}} \tag{19.23}$$

When the tests are carried out, it is essential to ensure that noise due to gas loads is predominant, that is, in the range where there is an increasing linear relationship between cylinder pressure level and level of noise. Shown in Fig. 19.19 is the case when cylinder pressure exceeds the critical level.

Under this circumstance the attenuation is generally found to be independent of operating conditions such as speed and load and therefore gives a measure of the

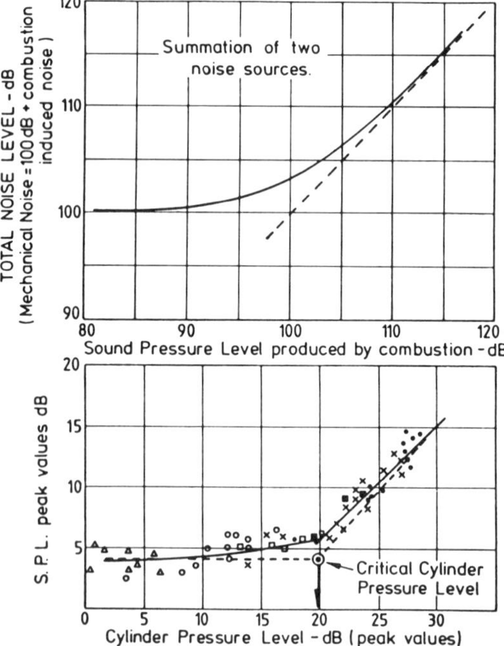

Fig. 19.19 Experimental determination of critical cylinder pressure level, above which engine noise is controlled by cylinder pressure pulse and below which by mechanical noise.

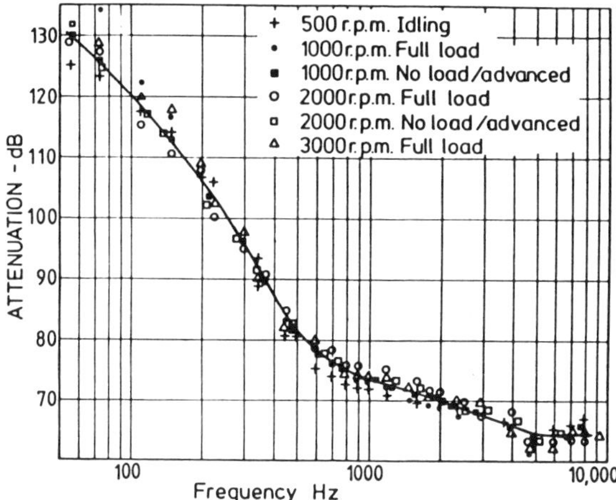

Fig. 19.20 Acoustical attenuation of engine structure (SPL cylinder–SPL engine).

noisiness of the engine structure as shown in Fig. 19.20. The attenuation is very high in the low-frequency range (around 130 dB), decreasing with increasing frequency to about 70 dB at 1000 Hz and above. The attenuation curves enable the designer to make comparisons between engine structures to be considered for series production and to predict the combustion induced noise for a given cylinder pressure pulse.

Over the years the ISVR has compiled a substantial library of engine test data that includes cylinder pressure, engine structure vibration, and noise analyses. These results permit the examination of the relations that exist in practice between the combustion-induced noise of various engine designs under a variety of operating parameters.

Figure 19.21*a* summarizes the measured A-weighted sound-pressure-level reults obtained with 60 different engines running at 2000 rpm. The data are plotted as a function of the bore diameter. Though the data points exhibit substantial scatter, it becomes obvious that engines with certain features of design and combustion system fall within certain classified groups. Each group appears to fall within a band of 3 dBA and each band follows a distinct slope corresponding to a fifth-power dependence on engine bore diameter.

The engine classification, from most noisy to least noisy, are as follows:

1. Opposed piston two-stroke engines.
2. Two-stroke DI engines.
3. Direct-injection normally aspirated engines.
4. Turbocharged diesel engines.
5. Normally aspirated IDI diesel engines.
6. Gasoline car engines.

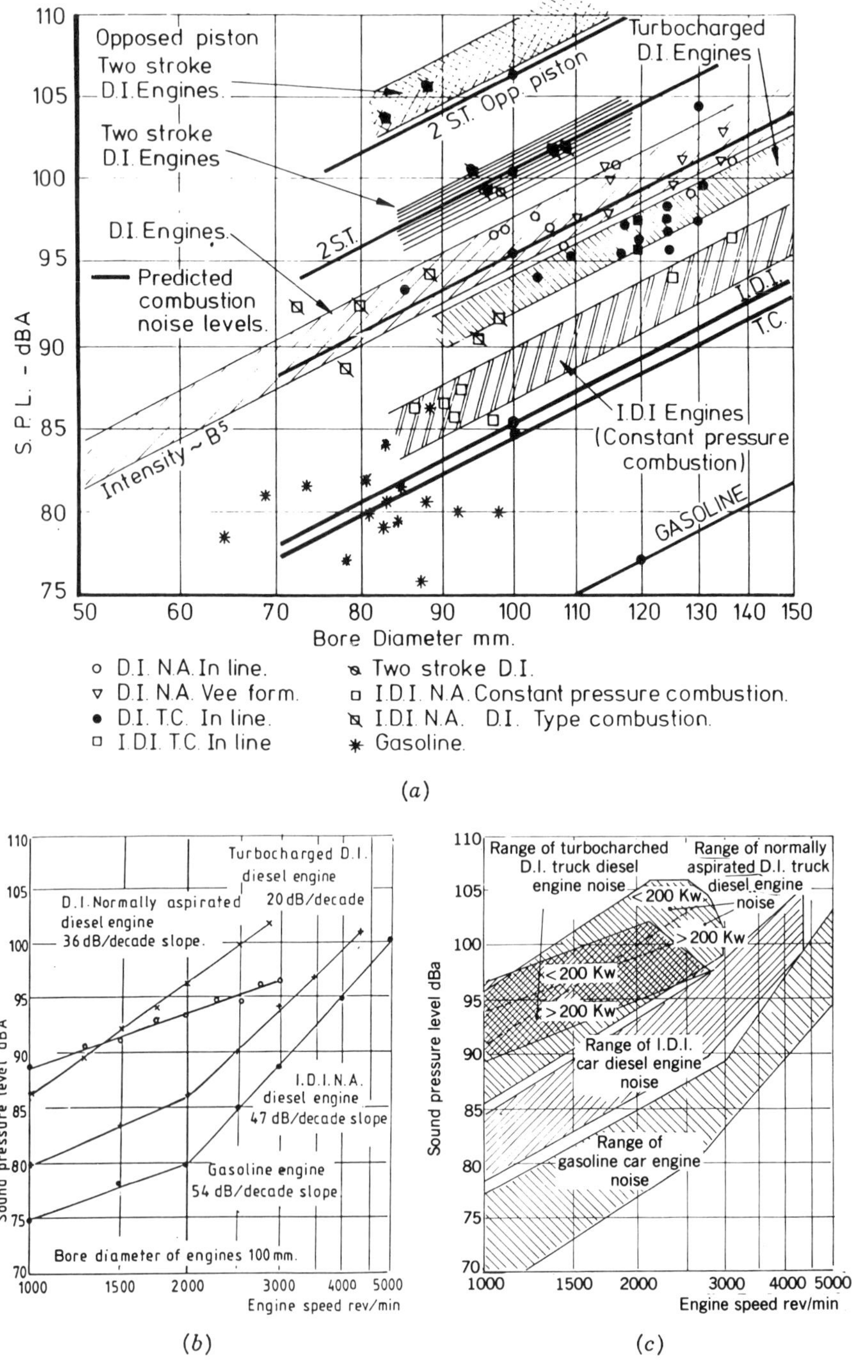

Fig. 19.21 Comparison of the noise of engines with different combustion systems as a function of: (*a*) bore diameter, (*b*) engine speed, and (*c*) engine types.

Superimposed on the graphs are predicted combustion-induced levels of noise for each of the groups, calculated by using the engine structure acoustical attenuation and cylinder pressure pulse exciting properties.

Only the normally aspirated DI engines and two-stroke cycle engines give a reasonable agreement with predicted noise. Even the predicted values fall in the lower part of the band, indicating that other noise sources of the engine could be of nearly comparable level to that of combustion-induced noise.

Turbocharged diesel engines, which in practice are only some 3–4 dBA quieter than normally aspirated engines from the combustion point of view, should be at least 10 dBA quieter.

In the IDI car diesel engines the predicted combustion-induced noise lies some 5 dBA lower than the measured values. In gasoline car engines the predicted combustion-induced noise is also well below the actual measured values. Figure 19.21*b* shows the noise-versus-speed relationship of engines of the same bore diameter but with different combustion systems and Fig. 19.21*c* shows ranges of various automotive engine noise.

For a given engine bore diameter, the range of combustion-induced noise can differ by as much as 25 dBA depending on the combustion system used. This can also be confirmed when comparing cylinder pressure spectra of various combustion systems used as shown in Fig. 19.22. The lowest cylinder pressure levels are obtained with only a pulse resulting from compression pressure, where the level of high-frequency harmonic components decrease by some 100 dB per 10-fold increase of frequency. Gasoline engine spectra decay by some 50 dB/decade, while diesel engine spectra decay by 20 to 40 dB/decade, the former representing

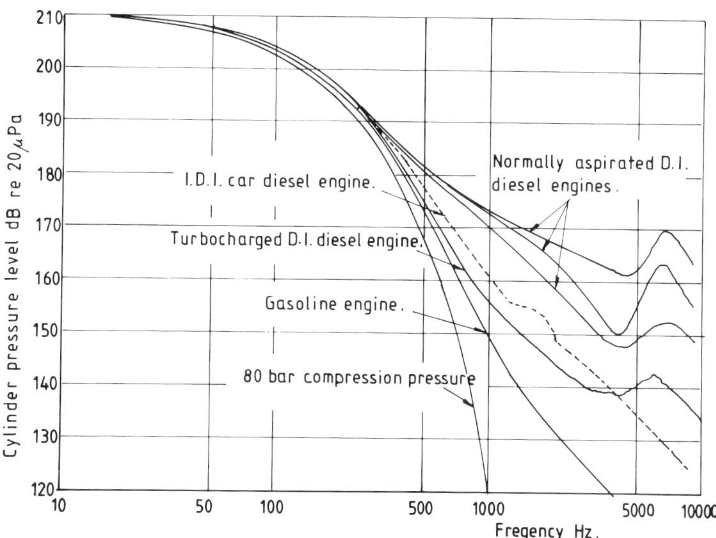

Fig. 19.22 Comparison of cylinder pressure spectra of various combustion systems at 2000 rpm.

almost instantaneous pressure rise. The range of cylinder pressure levels of different combustion systems at 1000 Hz differ from each other by about 25 dB.

Russel and Haworth[18] were the first to introduce a so-called combustion noise meter that was based on comprehensive studies of the relationships between diesel engine structure and characteristics of cylinder pressure developments.

The meter consists of a cylinder pressure transducer and signal conditioning amplifier and includes a structure response filter in the circuit. It is used to measure combustion-induced noise in the development of diesel combustion systems.

19.6 VIBRATION CHARACTERISTICS OF ENGINE STRUCTURE

Based on detailed experimental and analytical investigations of engine structures, Lalor and Petyt[19] classified the structural response into two distinct groups. These are beam-bending modes where both sides of the engine structure move in phase, as illustrated in Fig. 19.23*a*, and the panel modes, where both sides vibrate independently, as illustrated in Fig. 19.23*b* for a six-cylinder in-line engine.

Depending on engine size, beam modes occur typically from about 100 to 1500 Hz. In this mode, the engine casing behaves as if it were an elastic beam. This mode is also found with V-form engines although, due to the greatly increased lateral bending stiffness, the natural frequencies are correspondingly higher. In

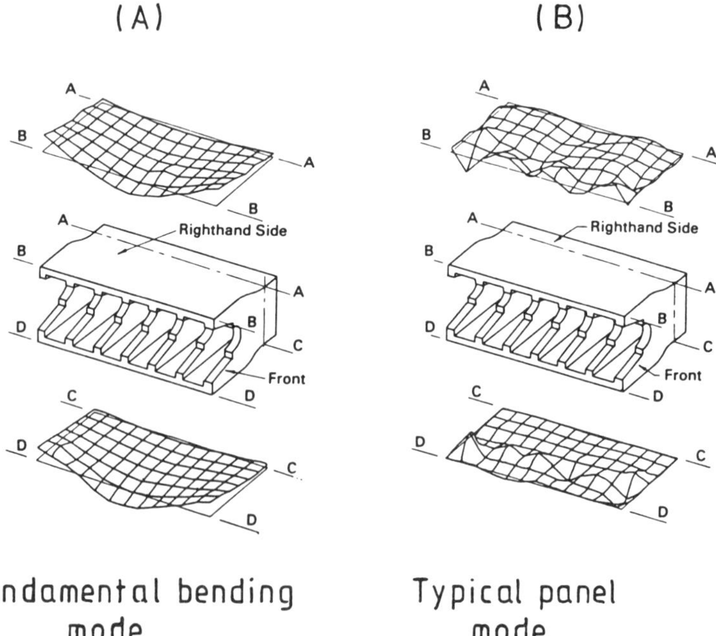

Fig. 19.23 Beam and panel modes of engine structure vibration.

addition, V engines display a bank-to-bank mode, where the cylinder banks vibrate like prongs of a tuning fork, which also belongs to this group.

An empirical formula can be used for estimating the fundamental bending mode for in-line engines:

$$f = \frac{225}{l^{1.5}} \text{ Hz} \tag{19.24}$$

where l is block length in meters.

For V-form engines the bank-to-bank frequency is as follows:

$$f = \left(\frac{2.79}{d}\right)^3 \quad \text{Hz} \tag{19.25}$$

where d is the perpendicular distance from the crankshaft axis of the cylinder block flame face in meters.

Although the block-bending and the V-engine bank-to-bank modes are fundamentals, the structure will also exhibit higher order modes, which extend the influence of this group well up the frequency range. However, when the distance between nodal lines reduces to the span of a typical cast panel (e.g., between the adjacent stiff main bearing sections of the crankcase wall), the block starts to exhibit panel modes. In this second mode of vibration, each cast panel of the block structure vibrates independently at frequencies well above the frequency of the fundamental beam mode, as shown in Fig. 19.23*b*. The natural frequencies of these panels will depend on their size and shape. Consequently, each engine will have a range of frequencies where its different panels vibrate in their fundamental mode.

In practice, there is usually an overlap frequency range between the two groups, where both modal types can be identified. This is particularly true with deep-skirted crankcase designs, where the natural frequencies of the crankcase walls below the axis can be quite low.

Lalor and Petyt[19] were able to obtain accurate correlation between experimental and calculated modal characteristics up to 2600 Hz with a model that is inherently simple.

For large in-line diesel engines from 10 to 14 liters capacity, the fundamental bending frequency is around 200–250 Hz, which in practice does not present a noise problem. For 6–8-liter-capacity engines the fundamental bending mode is about 300–350 Hz and this can be significant in controlling the overall (dBA) value of engine noise. As has been shown,[13] the bending mode for small four-cylinder car diesel engines is around 600–800 Hz, which in some instances can be the predominant source of noise. It has also been shown[13] that by the installation of an integral bearing beam, an 8-dB reduction of the noise radiated by this bending mode was possible without affecting the high-frequency noise above 1000 Hz.

Ochiai and Nakano[20] investigated the relationship between the vibration behavior of the crankcase bending and torsional vibration of an 8-liter in-line six-cylinder engine. By theoretical analysis they found that with an additional inertia mass at the front of the crankshaft, the bending mode could be changed into a torsional mode. The practical application of this work resulted in a 13-dB decrease in noise in the 315-Hz one-third-octave band.

Third-octave-band analysis of the engine block vibration over a closely marked grid yields a reasonable assessment of the vibration characteristics of an engine. Figure 19.24 shows average vibration acceleration level (dB *re* 1*g*) of a four-cylinder 2-liter diesel engine measured in the 1600 Hz center frequency third octave band, which was the frequency of the predominant engine noise. There is an increase of 10 dB of vibration from the top of the cylinder block down to the crankcase lower oil pan flange. Crankcase vibration suggests the presence of a second bending mode. From these results Chan and Anderton,[21] developed a method to calculate the emitted noise.

In assessing the structure vibration, the deformation of the cylinder block to static loads has been informative. Using a high-pressure oil pump and replacing the normal piston rings with rubber sealing rings, the engine is loaded to pressures of 70–80 bars (pressures equal to combustion pressures in a running engine) and the deformation of the structure is measured using a noncontacting capacitive displacement probe.

Figure 19.25 shows the deflected shapes measured at a horizontal section through the crankshaft axis and the deflection of the bearing caps when individual cylinder cavities were loaded statically. The deflection is localized to the vicinity of the particular loaded cylinder, with little deflection occurring beyond the next cylinder centerline. It is usually observed that the bearing caps deflect outward in the axial direction and downward, while side walls deflect inward.

Tests with axial static load applied across the bearing caps confirmed the bulkhead-to-side-wall coupling. Vibration measurements of the engine along the crankshaft axis were performed on a running engine. The results are shown in Fig. 19.26 indicating that transient vibrations induced by rapid pressure rise are localized in

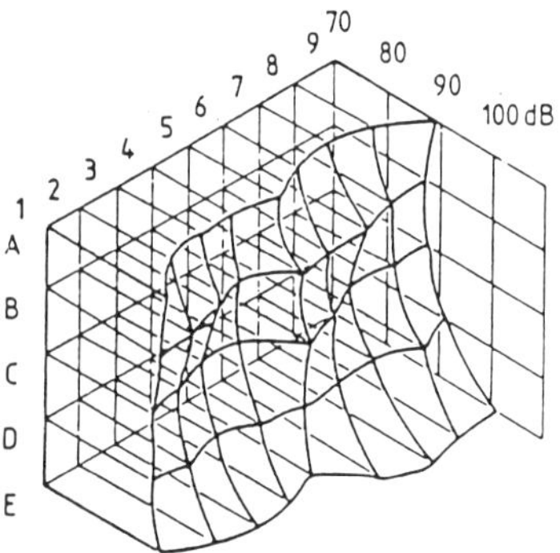

Fig. 19.24 Spatial variation of engine block vibration; 1 g = 84 dB.

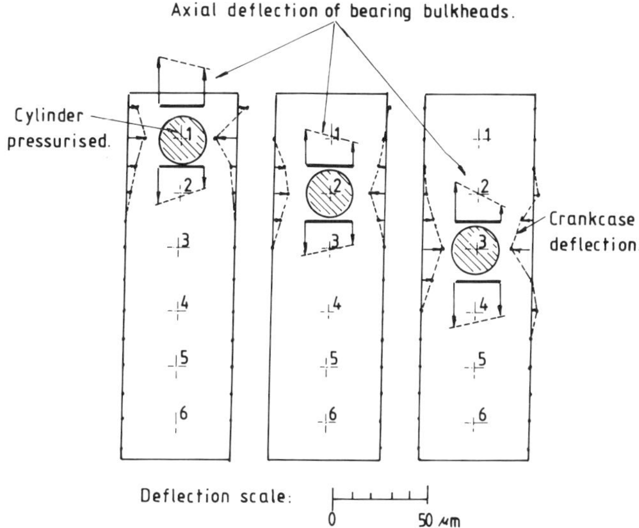

Fig. 19.25 Deformation of engine structure due to static cylinder loads.

the vicinity of the firing cylinder. The time domain response shows a decrease of the vibration amplitude by some 10–15 dB one cylinder away from the firing cylinder. The radiation of noise from the engine is therefore controlled by localized sources at any time instant, and it may therefore be assumed that the rest of the engine structure acts as a shielding baffle.

The axial vibration of the crankshaft and bulkheads generally produce high levels of noise from the front of the engine and the spectrum contains strong tonal components. Noise spectra from the sides are generally broader. Torsional vibration dampers specifically designed to incorporate axial damping provide worthwhile reduction of noise radiated from the front. Also effective are rubber-isolated pulleys in both axial and horizontal directions for smaller four-cylinder engines.

Covers

The engine design also requires unstressed lightweight covers for retention of oil and water. The principal covers of the engine are for the valve gear, oil pan, and front timing. There are also other minor covers depending on the engine design requirements.

The covers are attached to the load-carrying structure and are exited by vibratory forces. The covers are driven by the vibration of the load-carrying structure, although their own natural frequency, to some extent, amplifies vibration in particular frequency regions.

In Fig. 19.27 the range of one-third-octave-band acceleration levels of various covers is compared with that of the basic structure (cylinder block and crankcase) for a four-cylinder 4-liter diesel engine. On this engine the timing cover and oil pan are aluminum castings and the valve cover is a sheet metal stamping.

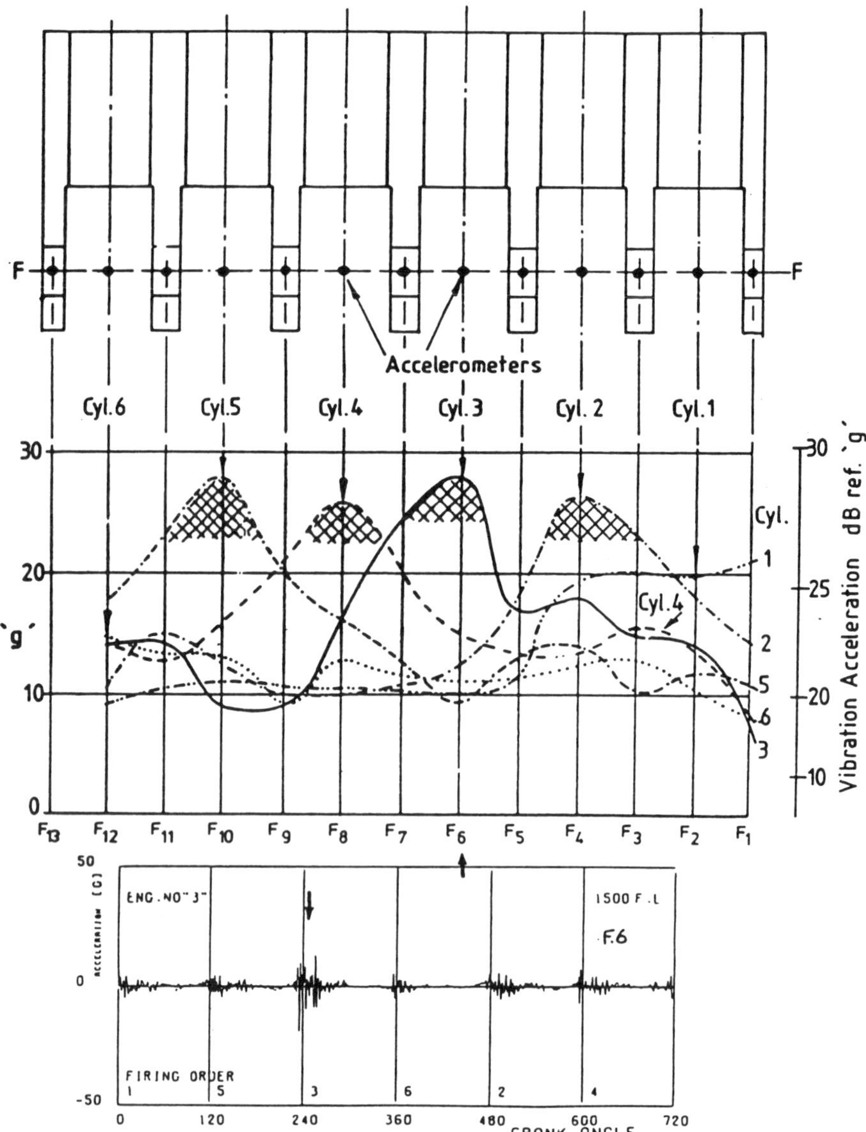

Fig. 19.26 Typical time domain response of engine structure vibration.

The maximum vibration levels measured on all the covers exceed the vibration levels of the basic engine load-carrying structures. For example, on the oil pan, which is attached to the lower part of the crankcase, the vibration levels are up to 8 dB higher in the frequency range from 800 to 4000 Hz.

It can also be noted that on the cast aluminum timing cover and oil pan, the vibration levels are augmented in the higher frequency range (higher natural frequency of their structure) than with the pressed steel cover, where the levels are

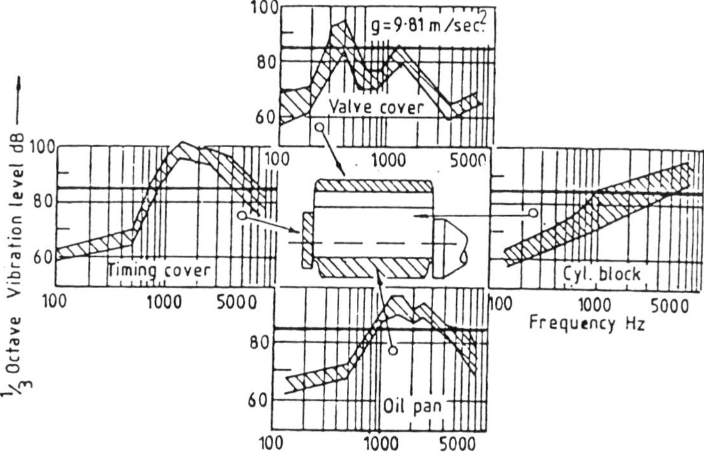

Fig. 19.27 Comparative vibration levels of various surfaces of a 4-liter, four-cylinder diesel engine; 1 g = 84 dB.

augmented in the lower frequency range, around 300 Hz. This indicates that sheet metal covers offer a worthwhile reduction of high-frequency vibration.

19.7 REDUCTION OF ENGINE NOISE BY STRUCTURAL DESIGN

The contribution of individual sources to the overall level of engine noise can be determined by a number of methods.

A very simple method is the shielding technique using a lead shield with a Fiberglass blanket on the engine side. Either the entire engine or a particular part of the engine is covered to assess the contributions to the overall noise. Fahy[22] and Chung et al.[23] proposed an acoustical intensity method that can be used to identify the sources of noise directly. Figure 19.28 illustrates a typical example of the sources of noise from a V-form engine that provides valuable information to the engine designer by giving the relative reductions in overall (dBA) that could be achieved if the noise from each individual source (accumulative noise reduction) were to be practically eliminated. In the example shown in Fig. 19.28, the oil pan contributes some 36% of the total acoustical power radiated from the side of the engine, whereas the valve cover and heads contribute 22%. Reducing these sources to a minimum should reduce the overall noise by about 3.5 dBA. On other engines the cylinder block and crankcase can be major contributors to overall noise.

Over the past 30 years numerous designs have been explored to achieve significant reductions of diesel engine noise. Some of the examples are shown in Fig. 19.29. Grover and Priede[24] built 12 engine prototypes with engine powers from 30 to 250 kW. These were skeleton, crankframe, and bedplate designs with the outer walls made from highly damped sandwich material. With these designs around 10 dBA reduction of noise has been achieved. The bedplate design, shown

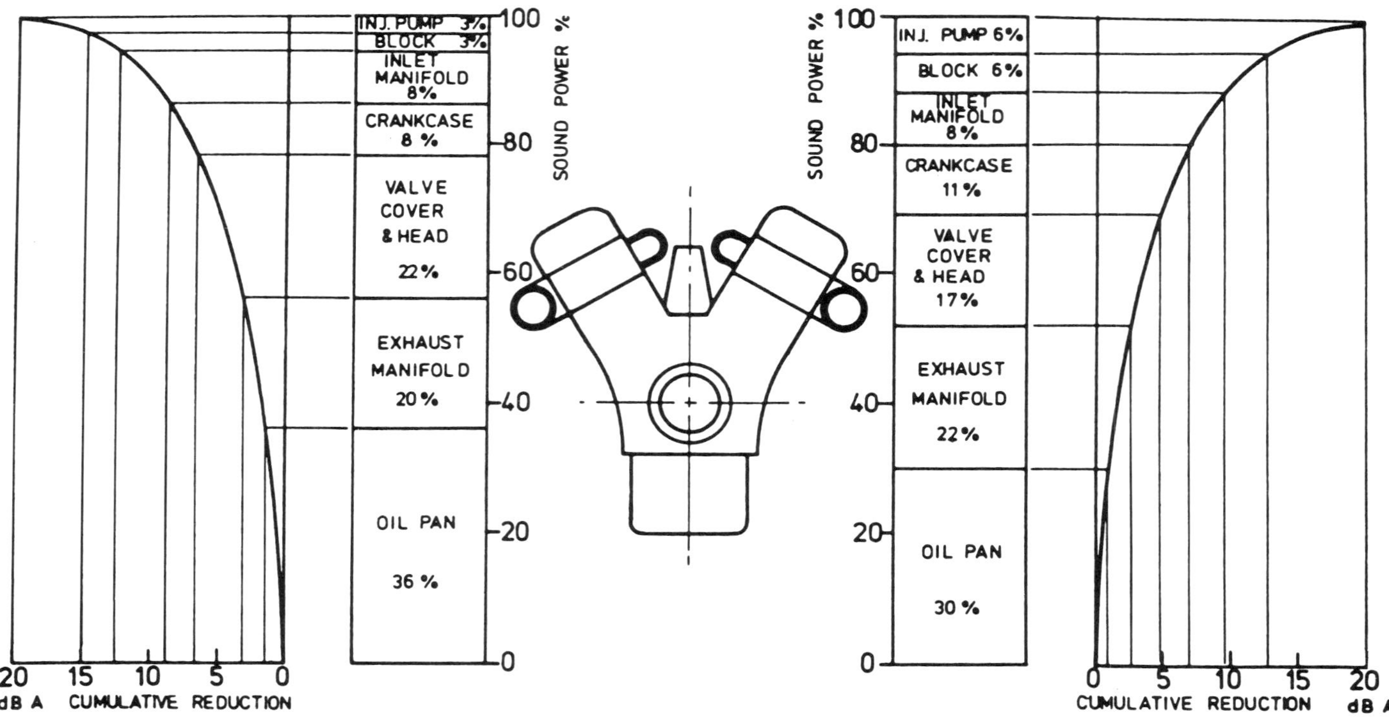

Fig. 19.28 Relative contribution of various sources to engine noise.

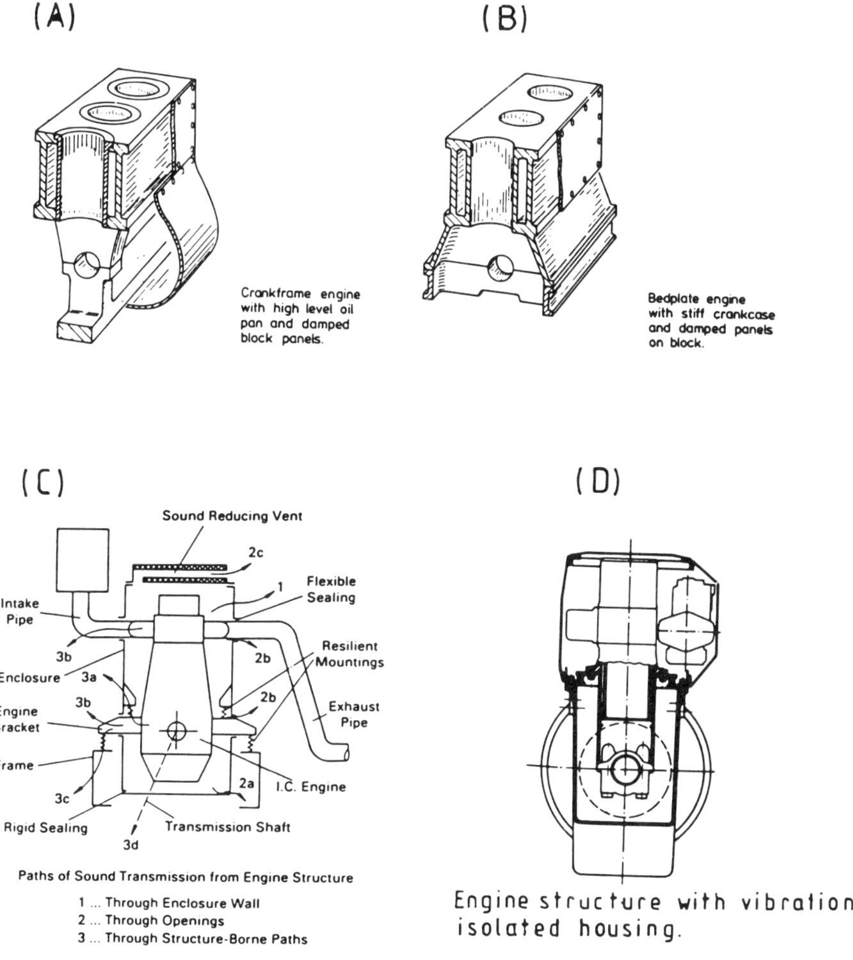

Fig. 19.29 Examples of experimental engine prototypes.

in Fig. 19.29*b*,was used for the low-noise truck project of the United Kingdom in 1979. Thien[25] perfected noise enclosures for existing engines based on detailed theoretical and experimental investigations. He achieved reductions of radiated noise by 18–20 dBA. Thien[25] developed an interesting design of a car diesel engine using the cylinder block integral with bearing supports, with a large vibration isolated ribbed aluminum oil pan and head cover. This engine gave 14 dBA noise reduction with engines of conventional design.

Although none of the designs have direct application in production engines, some of the key design features, such as rubber-isolated valve covers and inlet manifolds and covers made from sandwich material, are in series production.

Figure 19.30 shows the noise levels of a large number of diesel engines of different types and sizes plotted against specific weight, that is, total block weight

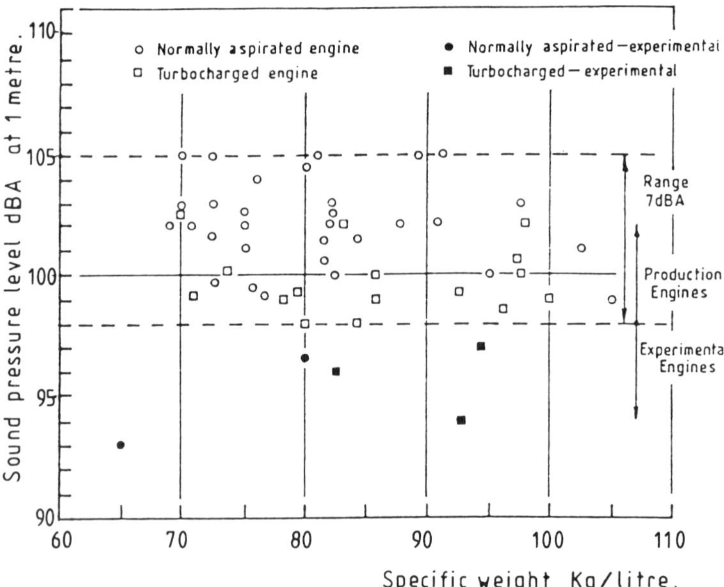

Fig. 19.30 Relation between specific weight and overall noise of various engine types at rated conditions.

against total engine capacity (kilograms per liter). As can be seen, there is no direct relationship between the weight of an engine and its noise for any class. The important interpretation of these data is that for quiet engine design the engine need not be heavier than normal and can in fact be significantly lighter. Furthermore, it indicates that the weight distribution in the engine structure may not be optimum. Lalor[26] developed a weight optimization technique using a three-dimensional finite-element model to reduce the static deflection of an engine under static hydraulic load. He proved that engine deflections can be significantly reduced by redistribution of the casting thicknesses; which were associated with some reductions of engine noise. The gradual development of the finite-element technique by the Automotive Design Advisory Unit of ISVR, University of Southampton, led to dynamic optimization procedures that are now successfully applied in practice.

Unlike engine stress calculations where finely meshed complex block elements are used because of the need to evaluate local stress concentrations, meshes for vibration prediction may be much broader and generally the finite-element model comprises thin plates, solid element, and beams.

Present-day computer power makes it practical to calculate the first 70 vibration modes usually necessary for noise prediction. This requires a representative model with about 10,000 degrees of freedom.

Such models, as illustrated in Fig. 19.31*a* for a six-cylinder in-line engine, are sufficiently detailed for all vibration predictions. Details of the construction of one bay of the crankcase of this typical model is shown in the Fig. 19.31*b*.

Although these relatively simple models can be used to obtain the natural fre-

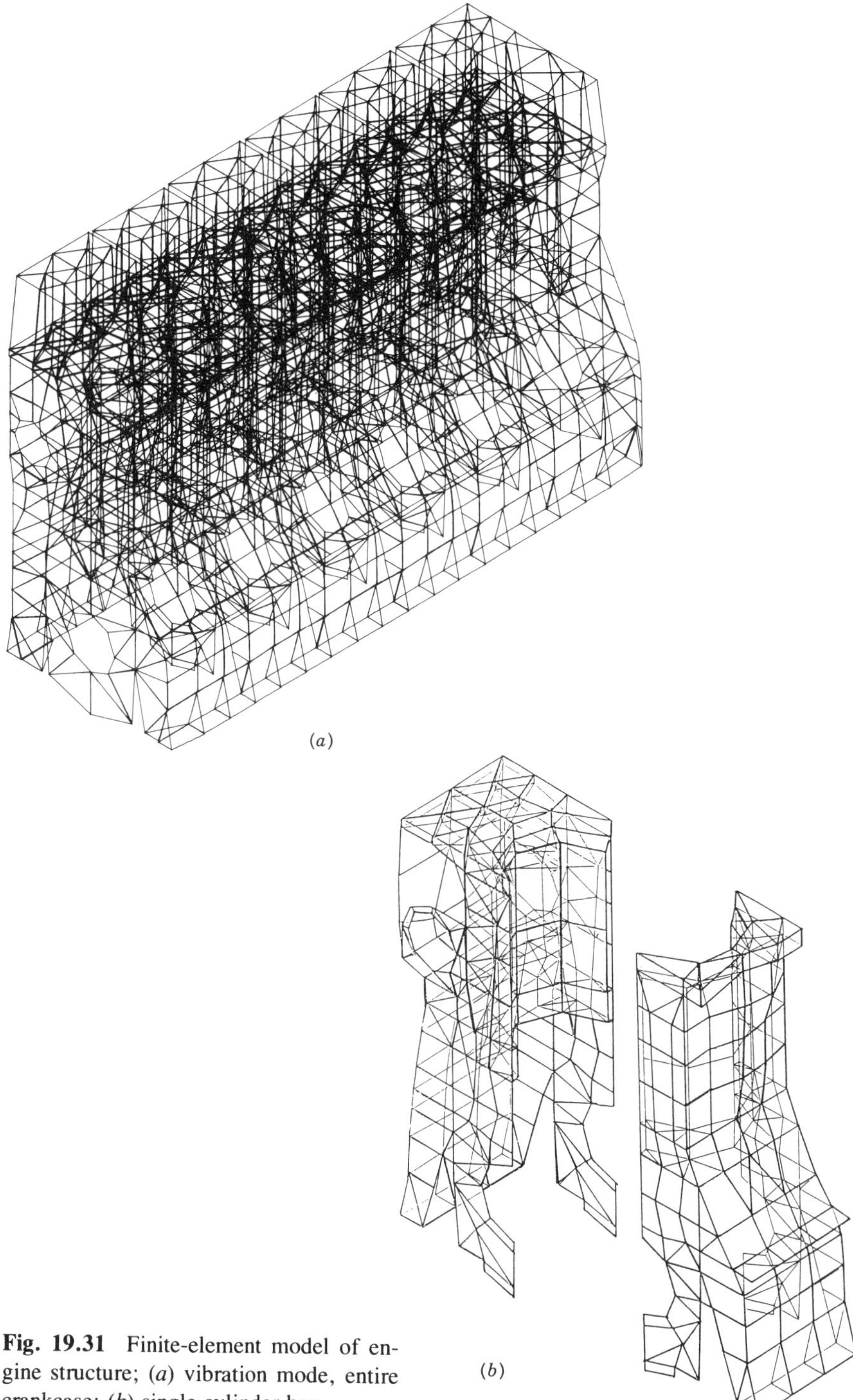

Fig. 19.31 Finite-element model of engine structure; (*a*) vibration mode, entire crankcase; (*b*) single cylinder bay.

quencies of the structure, to ascertain the forced response correctly, accurate values of loadings must be used such as gas and fuel injection forces and piston and bearing impacts as well as realistic damping characteristics. These values can be directly measured, calculated, or estimated from previous experience.

Engine covers, such as the oil pan, can be modeled in a similar way as the engine block, but it is usually considered convenient to run each part separately and to combine the results using substructuring techniques. Alternatively, the cover response to inputs of imposed block displacements can be studied.

A major advantage of using the finite-element technique as a development tool is that the effect of changes to the structure can be predicted with a high degree of confidence after the basic model has been established and experimentally verified. Recently this concept has been expanded by linking it with optimization algorithms, which will automatically search out the best combination of design parameters for minimum-noise radiation within prescribed limits of weight, thickness, and shape.

19.8 FUTURE TRENDS AND DIRECTIONS

It is to be expected that legislation will become more stringent, increasing the pressure on manufacturers to produce quieter power units. At the same time driver expectations for a more refined environment will continue to increase. Also influential will be the details of the engine combustion system, which will be dictated by numerous influences such as the quality of the fuels available and emissions restrictions, which will result in a reassessment of noise control measures.

The general aim will be for a low-noise engine of lower weight that will require greater precision in casting technology and the extensive use of light alloys.

The choice of design parameters will be strongly influenced by noise control considerations. These will be assisted by numerical analyses, and the final design will be the result of detailed modeling of the engine structure and its component parts with more accurate definition of the complex exiting forces involved. This will be made possible by advances in electronic and transducer technologies.

REFERENCES

1. H. R. Ricardo, *The High Speed Internal Combustion Engine*, 1st ed., Blackie & Sons, Glasgow, 1931.
2. Sir Harry Ricardo, F. R. S., *Memories and Machines. The Pattern of my Life*, Constable & Co., London 1968.
3. J. O. Hince, "Effects of Cylinder Pressure Rise on Engine Vibrations," ASME Paper No. 49-OG p-3, American Society of Mechanical Engineers, New York, April 1949.

4. T. Priede, "Relation between Form of Cylinder Pressure Diagram and Noise in Diesel Engines," *Proc. Mech. Eng. (A.D)*, 63 (1960–61).
5. A. Meier, "Zur Kinematic der Colbergerausche," *Automobitechnische Zeitscrift* **54**(6), 123 (1952).
6. D. Ross and E. E. Ungar, "On Piston Slap as a Source of Engine Noise," ASME. Paper No. 65 OGP-10, American Society of Mechanical Engineers, New York, 1965.
7. M. D. Rohrle, "Affecting Diesel Engine Noise by the Piston," paper presented at the SAE Diesel Engine Noise Conference, SP-397, Paper No. 750799, 1975, p. 51.
8. A. M. Laws, D. A. Parker, and B. Turner, "Piston Movement in the Diesel Engine," paper presented at the 10th International Congress on Combustion Engines, CIMAC, Washington DC, 1973.
9. R. Munro and A. Parker, "Transverse Movement Analysis and Its Influence on Diesel Piston Design," paper presented at the SAE Diesel Engine Noise Conference SP-357, Paper No. 750800, 1975, p. 68.
10. S. D. Haddad and H. L. Pullen, "Piston Slap as a Source of Noise and Vibration in Diesel Engines," *J. Sound Vib.* **34**(2), 249–260 (1974).
11. T. Priede and E. C. Grover, "Noise of Industrial Diesel Engines," paper presented at the Institute of Mechanical Engineers Symposium on Noise from Power Plant Equipment, Paper No. 7a, Southampton, September 1966.
12. P. Chan and D. Anderton, "The Effect of Engine Bore and Engine Noise, Surface Vibration and Combustion for a Six-Cylinder Engine," ISVR Report No. 74/2, 1974.
13. T. Priede, E. C. Grover, and N. Lalor, "Relation between Noise and Basic Structural Vibration of Diesel Engines," SAE Paper No. 690450, 1969.
14. R. Hickling, F. H. K. Chen, and D. A. Feldmaier, "Pressure Pulsations in Engine Cylinders," *Engine Noise*, Plenum, New York, 1982, pp. 3–33.
15. R. Hickling, D. A. Felmaier, and S. H. Sung, "Knock-induced Cavity Resonance in Open Chamber Diesel Engines," *J. Acoust. Soc. Am.* **65**, 6 (1979).
16. J. A. Raff and E. C. Grover, "A Primary Noise Generation Mechanism in Petrol Engines," paper presented at the Institute for Mechanical Engineers Conference on Passenger Car Engines, Paper No. C.320, 1973.
17. T. Priede, J. M. Baker, E. C. Grover, and R. Ghazi, "Characteristics of Exciting Forces and Structural Response of Turbocharged Diesel Engines." SAE Paper No. 850972, 1985.
18. M. F. Russel and R. Haworth, "Combustion Noise from High Speed Direct Injection Diesel Engines," SAE Paper No. 850973, 1985.
19. N. Lalor and M. Petyt, "Noise Assessment of Engine Structure Designs by Finite Element Techniques," in *Engine Noise*, Plenum, New York, 1982.
20. K. Ochiai and M. Nakano, "Relation between Crankshaft Tortional Vibration and Engine Noise," S.A.E. Paper No. 790365, 1979, p. 185 SP 80.
21. C. M. P. Chan and D. Anderton, "The Correlation of Machine Structure Surface Vibration and Radiated Noise," *Proc. INTER-NOISE 72*, Washington, 1972.
22. F. J. Fahy, "A Technique for Measuring Sound Intensity with a Sound Level Meter," *Noise Control Eng.* **2**(3), 155 (1977).
23. J. Y. Chung, J. Pope, and D. A. Feldmaier, "Application of Acoustic Intensity Measurement to Engine Noise Evaluation," SAE Paper No. 790502, 1979, p. 80.

24. E. C. Grover and T. Priede, "Current Research Leading to Low Noise Engine Development," paper presented at 11th ICA Congress, Lyon, 1983.

25. G. E. Thien, "The Use of Enclosures for Reducing Engine Noise," in *Engine Noise*, Plenum, New York, 1982, pp. 345–385.

26. N. Lalor, "Computer Design of Engine Structures for Low Noise," SAE Paper No. 790364, 1979.

CHAPTER 20

Noise and Vibration of Electrical Machinery

LÁSZLÓ TÍMÁR-PEREGRIN
Technical University of Budapest
Budapest, Hungary

20.1 INTRODUCTION

Noise and vibration are undesirable side effects of the energy conversion in electrical machines. Compared to other types of machines, electrical machines can be generally classified as quiet. This is due to systematic noise control research and development work.

Electrical drives using inverters were first introduced in the 1970s and their use is increasing. Inverters produce a supply voltage (current) that is rich in harmonics. In many cases, these harmonics considerably enhance the electromagnetic noise and vibration level. Today, engineers encounter a new challenge with implusive noise caused by the transient operation of electrical machines. Measuring transient noise and vibration requires special transducers and signal-processing methods.

The causes of noise and the vibration of electrical machines, motors, and generators are electromagnetic, mechanical, or aerodynamic. Usually each of these contributes to the generation of noise and vibration with different frequency spectra depending on the type of electrical machine. The electromagnetic field needed to fulfill the basic function of an electrical machine, energy conversion, also causes

Noise and Vibration Control Engineering: Principles and Applications, Edited by Leo L. Beranek and István L. Vér.
ISBN 0-471-61751-2 © 1992 John Wiley & Sons, Inc.

vibration and noise. Factors that contribute mainly to mechanical deformation and vibration of electromagnetic origin and give rise to sound include winding formation, fluctuation in the air gap permeance due to slotting, eccentricity, and local saturation (see Fig. 20.1).

All the other elements of the electrical machine (in addition to the winding and iron body) can be thought of as passive machine elements assisting in the energy conversion process. Also the bearing, the sliding contacts, the fan, and the ventilation system are examples belonging to this group. Some of these auxiliary elements, for instance, the bearings, can give rise to substantial mechanical vibrations of the machine due to their inherent geometric defects of random nature. Other types of small mechanical components (e.g., brushes) do not produce any noticeable vibration of the machine body or housing but may generate considerable noise. The sliding contacts and aerodynamic processes, which are not detectable in the vibration of the electrical machine, belong to this category.

The vibrational energy components of the electrical machine with frequencies within the audible frequency range result in airborne sound radiation. The electrical motor is connected mechanically to the loading engine and to the foundation through the base plate or frame. A considerable part of the vibrational energy in the audible frequency range is transmitted to the connected structures in the form of structure-borne sound. The lighter elements of the connected structure with large surface area can act as efficient secondary sound radiators. Their contribution adds to the airborne sound directly emitted by the machine. The low-frequency part of the vibrational energy of the electrical machine, which is not audible (frequency below 30 Hz) or is audible but with considerable damping by the ear (frequency below 1000 Hz), is called the vibration of the machine.

Which of the three major noise-generating mechanisms dominates depends on the type of the machine and its operating range. In low-power a.c. machines, electromagnetic noises are generally most important. With the increasing power ratings of a.c. machines (at a given rpm) the importance of electromagnetic noises decreases, while the contribution of mechanical noises (particularly aerodynamic noise) increases. For high-powered electrical machines, the noise of aerodynamic origin dominates. If, on the other hand, the power is held constant, similar changes in the dominant noise-generating mechanism are found as the speed is increased.

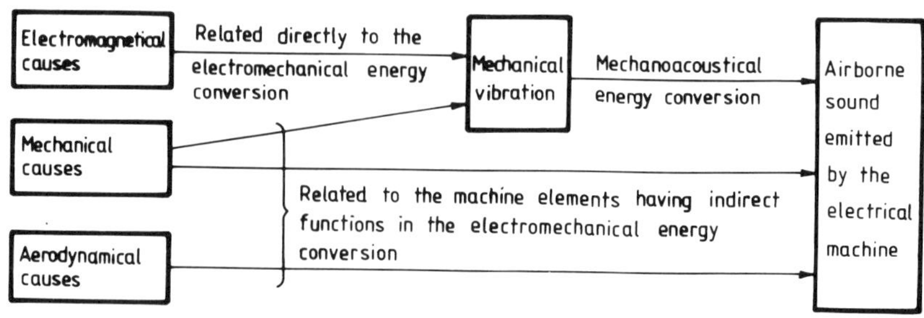

Fig. 20.1 Airborne sound generation.

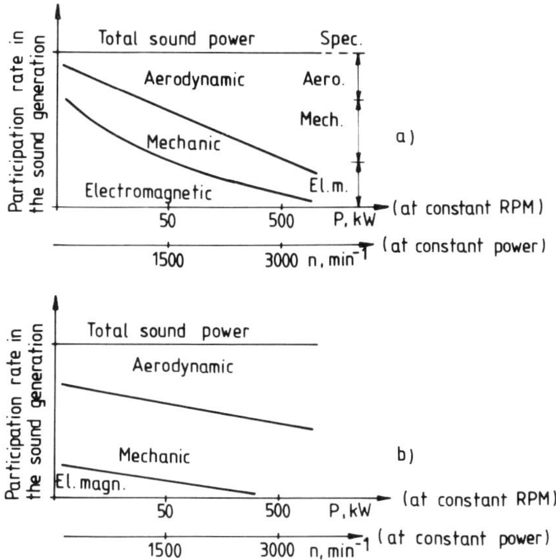

Fig. 20.2 Percential participation rate of different sound sources in the generation of sound power as function of rated power and speed; (*a*) a.c. machines; (*b*) d.c. machines.

Figure 20.2*a* gives a qualitative picture of the relative contributions of the different noise-generating mechanisms.

The relative contribution of different noise components is somewhat different for d.c. machines. At low speeds or low power ratings, the noise from sliding contacts will dominate (see Fig. 20.2*b*).

Manufacturers of motors can design quieter machines by controlling the dominating noise source. For high-powered or high-speed a.c. machines, one first reduces the aerodynamic noise. Once the aerodynamic noise is reduced, the contribution of noise of mechanical and electromagnetic origins can be addressed.

For fractional horsepower motors, the shaded pole motor is the least expensive but has a low starting torque, large slip, and low efficiency. The split-phase motor has higher starting torque, reduced slip, and increased efficiency but needs a starting switch. In both types of motors, one can expect tonal components in the radiated noise (hum) and torque pulses at two times the line frequency and its harmonics. Synchronous motors with solid rotors produce smaller torque pulses, and hence less electromagnetic noise, than the shaded pole or split-phase motors of comparable speed and power rating.

20.2 NOISE AND VIBRATION OF ELECTROMAGNETIC ORIGIN

Theoretical Analysis

To achieve efficient conversion of electrical energy into mechanical energy, electrical machines must have a suitably distributed magnetic field in the air gap. This field is sinusoidally distributed in a.c. machines and is kept constant under the poles in d.c. machines.

The magnetic field in the air gap is generated by windings in slots (the space between two poles in the d.c. machines may be considered as large slots). As a result of slotting, the current-carrying conductors develop a stepped magnetomotive force (MMF) along the periphery of the stator bore. Since the field is periodical in space, it can be expanded into spatial Fourier series. In addition to the fundamental component with p number of pole pairs, space harmonic MMF waves also develop. Another result of slotting is that the magnetic conductivity of the air gap becomes nonuniform. The air gap, which constitutes a greater part of the magnetic resistivity of the magnetic circuit, fluctuates periodically due to the relative movement of the rotor to the stator. The greatest fluctuation of the conductivity among all the rotating electrical machine types is observed in induction machines. This is because the induction machine takes all its magnetizing current from the main (also the rotor field is generated by the stator currents). To keep magnetizing current low, the air gap is made as small as possible. This is the principal reason why vibration and noise of electromagnetic nature usually dominates other sources in induction machines. Magnetic conductivity harmonics are caused either by eccentricity of the rotor or by local iron saturation in the magnetic circuit.

The product of the MMF waves and the conductivity waves, gives one the flux density space harmonics in the air gap. The space harmonic waves rotate either in the direction of the fundamental harmonic wave or in the opposite direction.

Caused by radial flux density waves, tensile stresses develop that act on both the stator and the rotor. The tensile stress is proportional to the square of the flux density. The radial tensile force is the product of the tensile stress and the surface area of the stator bore.

The unbalanced magnetic pull, which has an angular frequency equal to the mechanical rotational frequency of the rotor, is created by the product of the first field harmonic, originating from eccentricity, and the fundamental wave of the stator flux density. Because of its low frequency, it is a direct, significant source of vibration, and the noise will be influenced by it only indirectly.

The mixed product of the space harmonics of the stator and rotor has an important role in generating vibrations and noise. A component of the tensile stress wave developed has the form

$$P_r(x, t) = P_r \cos(rx - \omega_r t - \phi_r) \tag{20.1}$$

where P_r is the amplitude, r is the mode number, ω_r is the angular frequency, and ϕ_r is the initial phase angle of the tensile stress wave. The radial force waves in the rotating electrical machine are waves of different mode numbers characterized by different spatial distributions in the air gap around the periphery. These stress wave components rotate with different angular velocities, either in the same or in the opposite direction as the rotor. They act on both the stator and rotor.

With the exception of the bending force wave of mode number $r = 1$ (e.g., unbalanced magnetic pull) for which the rotor is more flexible, all other modes result in a more significant deformation of the stator than the rotor. Bending forces of mode number $r = 1$ present the greatest problem in high-powered rotating machines with a flexible rotor and large bearing span. A rotor is called flexible if its first critical speed (defined as the rotation speed that corresponds to the first bending resonance of the shaft-rotor assembly) is below or just above the rotational speed of the rotor.

Any force wave of frequency f_r and mode number r gives rise to a set of vibrations (the mode number of the deformations will be j, $0 \leq j \leq \infty$, and frequency f_r) in the rotating electrical machine. The instantaneous values of the stator deformation patterns (i.e., resonant vibration modes) of a stator ring are shown in Fig. 20.3 for different vibrational mode numbers. The magnitudes of the vibration response components depend on the geometric dimension of the machine, the magnitude of the exciting force, the difference between the frequency of the exciting force and the resonance frequencies of the machine, and damping. If the excitation frequency f_r is close to or equal to any of the mechanical resonance frequencies of the machine, then resonance amplification occurs. The result may be dangerously high vibrations, increased noise, or both. The rms value of the vibration velocity produced by a force wave with amplitude P_r, frequency f_r, and mode number r is found as

$$v_{fr} = \sqrt{2}\,\pi f_r \sum_j H_j P_r \tag{20.2}$$

where H_j is the mechanical transfer function of the electrical machine. The transfer function H_j can be split into two components ($H_j = H_{\mathrm{stat}(j)} H_{\mathrm{dyn}(j)}$). One of them, $H_{\mathrm{stat}(j)}$, contains the geometric dimension of the machine, the material properties and the mode number j. For $f_r = 0$, this component yields the static deformation. The second component, $H_{\mathrm{dyn}(j)}$, depends on the difference between the frequency of the exciting electromagnetic force f_r and the resonance frequency $f_{\mathrm{res}(j)}$ char-

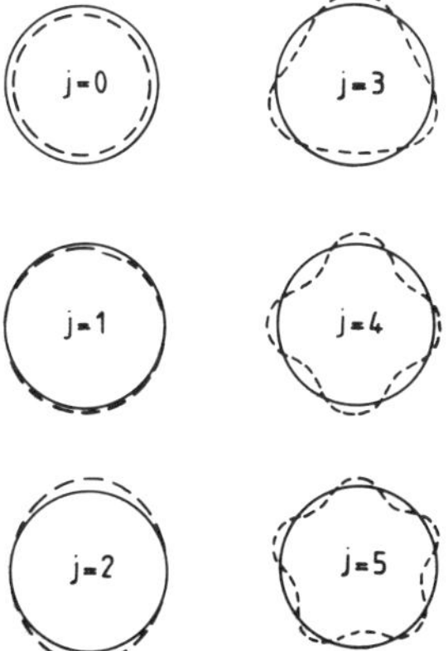

Fig. 20.3 Deformational waves of various mode numbers in a cylindrical electrical machine.

acteristic of the vibration mode with mode number, and on the internal damping. The $H_{dyn(j)}$ component is called the magnification factor. Note that the mechanical resonance frequencies become lower with increasing size of the machine. Consequently, more resonance frequencies appear within the audible range as the machine size increases. The type of machine mounting influences the lower order resonance frequencies. The main source of the damping in electric motors is the interface between the winding and the laminated core.

Based on practical experience, the force waves classified as serious in terms of vibrations or noise can be selected on the basis of their mode number and frequency. Low- and medium-powered electrical machines can be considered extremely rigid for force waves with a mode number higher than 6. Consequently, the effect of force waves with mode numbers over 6 can usually be neglected. For airborne noise, the frequency range from 200 Hz to 6 kHz is important. Vibration problems usually occur from 10 Hz to 2 kHz.

The sound power radiated by an electrical motor with the frequency f_r is given by

$$W_{fr} = \rho c \sigma v_{fr}^2 S_{\text{rad}} \tag{20.3}$$

where ρ is the density of medium, c is the speed of sound in the medium (the value of ρc for air at room temperature is 415 N $\cdot$ s/m^3). σ is the radiation efficiency (also called the radiation factor), v_{fr} is the rms value of the mechanical vibration velocity, measured on the surface of the machine, and S_{rad} is the area of the machine surface with which v_{fr} is associated. Determining the values of S_{rad} and σ in

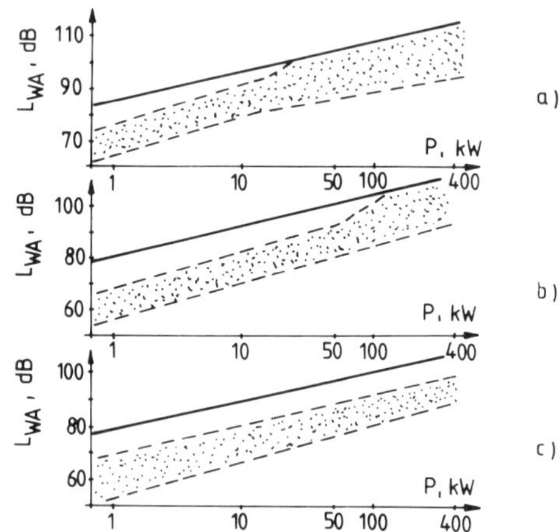

Fig. 20.4 Maximum permissible A-weighted sound power level values (solid lines) for low-voltage, totally enclosed rotating electrical machines (according to International Standard IEC 34-9) and range of realized sound power levels (shaded region) of induction machines as function of rated power for: (*a*) two-pole induction motors; (*b*) four-pole induction motors; (*c*) six-pole induction motors.

the case of complex-shaped electrical machines is quite difficult. Procedures to estimate these parameters are given in references 1 and 2. The spectra of the electromagnetic vibration and noise of electrical machines are typical pure-tone spectra with frequency components ranging from twice the supply or line frequency up to many kilohertz. Figure 20.4 depicts curves showing the maximum permissible sound power level values of low-voltage electrical motors[3] as a function of the rated power and expected range of noise level for commercial motors.

Special Problems Related to Electromagnetic Noise

Change in Speed. When deriving the expression for the frequency of the exciting force, one gets the following equation for induction motors:

$$f_r = An + Bf_l \tag{20.4}$$

where A and B are machine constant, f_l is the frequency of the supply voltage, and n is the shaft rotational speed per second. A resonance may occur in the steady-state mode of operation as well as in the transient mode of operation. The value of f_r changes linearly with the shaft rotation speed. Therefore, resonance may occur in the acceleration period in cases where the mechanical resonance frequency of the machine in steady-state operations is much lower than the frequency of the exciting force. In other words, the smoothly running machine might "roar up" during the acceleration period. This transient resonance may show up under other types of transient operating conditions, like reversing, or in the generator braking mode of multispeed induction motors.

If the motor will be used at other speeds or over a wider speed range than was assumed at its design, a large increase in noise and vibration may occur due to resonance.

Change in Load. The change in the loading condition of an induction motor may have a large influence on the vibration and noise because with a change of load, the speed of the motor changes and with it the frequency of the exciting force in the motor. If the exciting frequency at no-load condition, f_{r0}, approaches the resonance frequencies of the stator or rotor, $f_{\text{res}(j)}$, a large increase or decrease in vibration and noise may occur when the load of the induction motor varies. In extreme cases, the tonal component of electromagnetic origin may increase by 30–40 dB or decrease by 20–30 dB upon loading. Whether the noise increases or decreases with load depends on the relative position of f_{r0} to $f_{\text{res}(j)}$. This does not mean, however, that the A-weighted sound power level would change by the same amount. The change in L_{WA} can be computed as the result of the A-weighted changes of the different tonal components. Without detailed analysis or experimental data it is not possible to predict how loading will affect the A-weighted sound level.

Harmonics Content of Supply Voltage. The frequency of the exciting force f_r depends on the frequency of the supplying voltage system. In the case of variable-speed drives, the motor speed is varied by varying the frequency of a frequency converter (inverter) connected between the mains and the electrical motor. The frequency of the output voltage of the converter varies over a wide range (e.g., from 5 up to 120 Hz).

Correspondingly, the spectral components of the electromagnetic exciting force occupy the frequency domain from 0 Hz up to the kilohertz region. Consequently, machines that are quiet without a frequency converter may become noisy when driven by a converter, especially whenever the frequency of excitation coincides with a structural resonance. It is not uncommon when that happens for the vibration to increase 16 times its previous value and the noise level to increase by 25 dB.

Frequency converters have another, undesirable influence on the motor as well. The output voltage of the inverter is rich in harmonics. The phenomenon can be modeled by assuming that the motor is driven by a different supply voltage at each of the frequencies and superimposing the resulting vibration. For converter-supplied motors, exciting forces have many more frequency components so there is a greater likelihood that one of them will coincide with one of the mechanical resonances than if the motor were driven by a pure sine wave.

Control of Electromagnetic Noise and Vibration

Both noise and vibration control should start at the design stage. Because the causes of electromagnetic noise and vibration are the same, measures that reduce the vibrations also reduce the radiated noise.

Vibration of electromagnetic origin is proportional to the square of the fluctuation of the flux density. The airborne sound power, which is proportional to the square of the vibration velocity, varies with the fourth power of the flux density. The change in noise level with flux density is

$$\Delta L = 40 \log_{10} \frac{B^*}{B} \qquad (20.5)$$

where, B^*/B is the ratio of a flux density to a reference flux density.

The flux density in the air gap can be reduced either by increasing the number of windings at the design stage or by reducing the supply voltage to the machine. The first is done by the manufacturer, the second by the user. Reducing the supply voltage might seem to be an easy way of decreasing the vibration or noise, but it is very rarely applied in practice. The reason is that the machine becomes more expensive, and the output power and torque of the machine decrease with the decreasing flux density. If a reduction in torque is tolerable, then the air gap can be increased by reducing the rotor diameter. This is because the flux density is inversely proportional to the size of the air gap. The magnitude of the slot harmonics will be decreased as well.

In order to reduce the vibration and noise of electromagnetic origin in induction machines, it is important to choose a proper combination of rotor slot number and stator slot number. In choosing the correct slot numbers, one has to avoid, above all, the generation of force waves with small mode numbers. The mode number of the force waves is given by the sum/difference of the stator and rotor space harmonic orders, so one can determine the wanted slot number combination.[4] During the long history of electrical machine design, many investigators[5] tried to find the optimum slot number combinations. However, there are other important considerations when choosing slot numbers, such as avoiding parasitic torques during run up and keeping the parasitic losses at a reasonable value. Usually, one can find a proper slot combination for the steady-state mode of operation that will avoid coincidence of the exciting force frequency with the mechanical resonance frequency. The situation is more difficult when either the load changes or the supply frequency varies (e.g., for variable-speed a.c. drives). With variable-speed drives, it is practically impossible to avoid resonance, so one has to devote even more attention to finding the least offending slot number.

By using special frequency converters, there exists a simple technical way to avoid the resonances caused by a change in load. Assume that the induction motor has been designed to be quiet at the rated speed and at the rated output power. The actual value of the output power is usually less than the rated one. The machine speed can be kept constant by appropriately reducing the magnitude of the supply voltage to compensate for the reduced load. Thus the resonance can be avoided. In addition, with the flux reduced, the noise and vibration of electromagnetic origin will be reduced as well. At the same time losses are decreased.

Of course, one can avoid the resonance by modifying the dynamic properties of the machine, that is, by detuning the vibrating system. These considerations can be carried out at the design stage only. Usually for this solution the built-in iron volume is increased, which makes the machine heavier and more expensive.

Vibration and noise of electromagnetic origin can be influenced by proper design of the slot inlet shape as well. Pitch shortening (in a.c. windings) mainly influences the no-load noise favorably.

Another favorable measure that contributes to noise reduction of induction motors is to skew slots, which can be implemented in the design stage only. This action reduces the amplitude of the torque pulses. Proper skewing of the slots can eliminate the most troublesome flux density space harmonics. Skewing modifies the vibration by the skewing factor ξ_s, and the expected reduction in sound power level is[6]

$$\Delta L = 20 \log_{10} \xi_s \tag{20.6}$$

For medium and large induction machines with long rotors, skewing results in torsional excitation. Then the motor vibration is the result of both radial and torsional excitation. Therefore, skewing may result in an increase or a decrease in vibration/noise depending on whether the reduction in radial excitation is smaller or larger than the increase in torsional excitation.

20.3 VIBRATION AND NOISE OF MECHANICAL ORIGIN

Bearings

Both ball and sleeve bearings are extensively used in electrical machines. In terms of vibration and noise, sleeve bearings are better, but because they are produced in small quantities, they are usually more expensive. Their installation and maintenance require special care. Therefore, they are rarely used in small or medium-sized machines except when extremely quiet operation is needed. The frequency spectra of sleeve bearings are typical pure-tone spectra, containing the frequency of the shaft rotation and its harmonics.

The manufacturers of electrical machines mostly use roller or ball bearings. The frequency of the exciting forces induced by roller bearings can fall anywhere within a wide frequency range of from 10 to 5000 Hz due to the variety of possible causes of noise and vibration in these components. Roller bearings emit more noise than ball bearings, and their frequency spectrum is also wider.[7]

Based on broad experience,[8] the A-weighted sound pressure level of ball bearing noise (at 1 m measuring distance) can be approximated by the curvs in Fig. 20.5.

The shaft height on the horizontal axis has been standardized internationally. The uncertainty of the statistical approximation is about ± 2 dB. The spectral distribution of a rolling bearing has a character of a white noise with a weak maximum value at approximately 40% of the shaft rotation speed.[7]

Frequency analysis of bearing vibration and noise provides diagnostic information on bearing defects, fabrication inaccuracies, and installation errors.[7] In order to diminish the noise of bearings, it is customary to install a preloading spring parallel to the bearing to maintain the axial position of the bearing rotor. Because a weak spring is useless and a spring that is too strong would shorten the lifetime of the bearing, the optimal level of axial preload is determined experi-

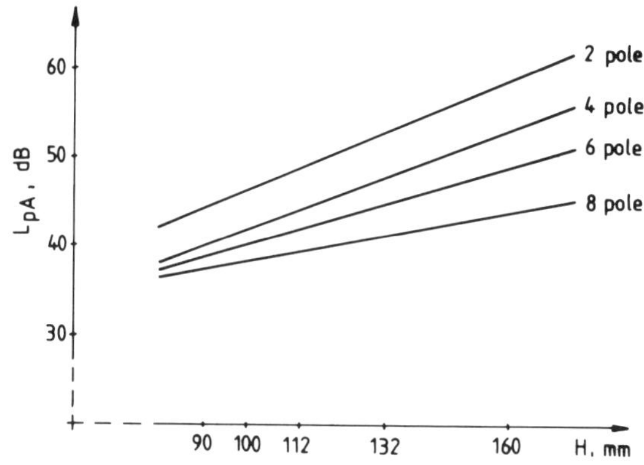

Fig. 20.5 The A-weighted sound pressure level of ball bearing in a.c. machines at 1 m measuring distance as function of shaft height.

mentally. By placing elastic inserts between the bearing and the end shield, the vibration transfer to the end shields and to the housing can be reduced. If the inserts are good electrical insulators, they also hinder the generation of unipolar flux and, through this, the emergence of unbalanced magnetic pull. Special bearings made with plastic cages produce less noise than those with metal cages.

Sliding Contacts

In slip-ring induction machines and in synchronous machines, sliding contacts (brushes) are required to establish current transfer to the rotating parts. A sliding contact is a source of both vibration and noise. The brush–slip-ring system is excited by forces, surface irregularities, sliding friction, and load changes. Uniform radial forces make an annular slip ring expand, lifting it from its seat so that the sliding surface becomes eccentric. To avoid this, the slip rings are fitted on a prestressed shaft. If the prestress is larger than the centrifugal force, the ring does not become deformed, and no vibration will be produced. The intensity of noise generated by the friction between the sliding surfaces is usually low compared with other machine noises. The frequency of the noise generated by eccentric slip rings coincides with the rotational frequency of the rotor.

The brush–commutator system is more complicated than the slip ring, and is the main source of noise in d.c. machines. The construction of a brush–commutator system, the manufacturing technology, the tolerances, and the deflections under load are the parameters that control the generation of the noise. The effect of these parameters cannot be separated due to their interrelations. High-frequency brush noise is produced by the friction between the brush and sliding surface. The friction itself is a function of the condition of the film, the patina (oxidated surface layer) formed on the sliding surface that consists of copper oxide, graphite dust, and moisture. Patina is formed during operation under current. When the color of the sliding surface becomes bronze, both the friction and noise reach a minimum. The discrete lines in the brush noise frequency spectrum correspond to the product of the rotational frequency of the rotor, the number of commutator segments, and its harmonics. An accurate procedure to calculate the sound power level of brush noise does not exist. The magnitude of noise can be controlled by robust cast brush holders and stiff mounting brackets. In order to produce a quiet brush, it is most important to build up and maintain the patina layer on the commutator. The use of a noise enclosure over the sliding system is impractical because of ventilation requirements.

Rotor Imbalance

Rotor imbalance is one of the most important causes of mechanical vibration. If the rotor mass distribution is not balanced, a rotating centrifugal force is created which acts on the bearings, exciting them into vibration at the shaft rotational frequency. This force increases with the square of the rotation speed. Simple static balancing of the rotor is applicable only to rotating-disk-shaped parts. The exact location of the rotational plane of the center of gravity of cylindrical rotors is difficult to determine, and it is also very difficult and sometimes impossible to place the balancing weights at the proper position. In such a case one has to attach split-balance weights near the bearings, in the so-called balancing planes.[4] The location of the balancing planes can be different depending on the rotor construction. For squirrel-cage induction motors equipped with a metal fan, the balancing planes are the drive-side face of the rotor core or the plane of the fan blades on the opposite side. This process is called dynamic balancing and is done in addition to static balancing.[4]

The main causes of rotor imbalance are manufacturing inaccuracies. Imbalance is a result of asymmetry in shape, material nonuniformities, nonobservance of working tolerances, misalignment of bearings, asymmetric elastic deformations due to centrifugal forces, asymmetrical thermal deflections, and aerodynamic imbalance.

Rotor imbalance affects the vibroacoustical parameters of the electrical motor both directly and indirectly. The direct effect is purely kinematic. The imbalance rotating mass gives rise to a centrifugal force at the shaft rotation frequency that acts upon the elastically coupled vibrating system comprising the rotor and stator. The most important indirect effect of rotor imbalance is the fluctuation of air gap permeance due to dynamic eccentricity (which is the result of the dynamic force bending the shaft) and the resultant flux density harmonics.

The precise balancing of the rotor can reduce the vibration of rotating electrical machines up to a certain limit. Balancing is favored both for technical and economical reasons. The perfectness of the balancing is limited by the sensitivity of the balancing rig and by the associated cost. Large rotating machines are usually balanced in situ after they have been installed.

20.4 NOISE OF AERODYNAMIC ORIGIN

The major part of the sound power emitted by low-power, high-speed electrical machines (like two-pole induction motors) and medium- and high-powered electrical machines is of aerodynamic origin. The cooling air transfers the heat to water or ambient air. Air flow is driven by a fan or a group of fans. The capacity of the fan is determined by the amount of heat to be transported and the pressure boost required to maintain the cooling-air flow. Based on the spectral distribution of the fan noise, one can talk about broadband noise and tonal (siren) noise. Methods to predict and measures to control fan noise are given in Chapters 14 and 18. Only the aspects of fan noise that are specific to electric motor cooling systems are discussed below.

Broadband Noise

Broadband noise, which is continuously distributed over the frequency range of 100 Hz–10 kHz, is caused by turbulence and flow separation. It is inherent in the operation of the fan and cannot be eliminated but at best can be reduced to a reasonable and acceptable level.

There is no procedure for the exact calculation of fan noise, but the following empirical formula is suitable for estimating octave-band power levels.[9]

$$L_W(\text{oct}) = (40 \pm 4) + 10 \log_{10} \frac{\dot{V}}{\dot{V}_0} + 20 \log_{10} \frac{\Delta p_t}{\Delta p_{t0}} - \Delta L_{\text{oct}} \qquad (20.7)$$

where $L_W(\text{oct})$ is the octave band sound power level in dB *re* 10^{-12} W, $\dot{V}$ is the volumetric flow in m^3/s, $V_0 = 1\ \text{m}^3/\text{s}$, Δp_t is the total pressure increment, $\Delta p_{t0} = 1$ Pa, and ΔL_{oct} is the spectral correction factor given in Table 20.1. The $\pm$4dB in the constant term indicates the range of typical deviations. The important point of the formula is that the radiated sound power depends on the volume flow rate (delivered volume) and, in particular, on the total pressure increment produced. Experimental results[10] indicate that if the actual volume flow rate differs by $\pm 50\%$ from the design value, L_W increases by 3–5 dB. If strong pure-tone components are present in the spectrum, Eq. (20.7) may underestimate the real sound power level in the octave band where the tone may dominate by 8–10 dB. In sum, the formula gives only the minimum level for the produced noise and highlights the fact that one can reduce the noise considerably by reducing the losses in the flow circuit.

The spectral distribution of the fan noise depends on the fan type. For preliminary calculation purposes one can get acceptable results if the octave-band sound power level is determined by using different correction factors (given in Table 20.1) in Eq. (20.7) for each octave band. If the diameter of the opening is increased, the low-frequency component of the radiated noise grows, and if the diameter is decreased, the low-frequency components diminish.

Almost all the low-power electrical motors are equipped with radial fans with straight blades. This allows operation in both rotational directions. Straight blades are undesirable from the aerodynamic and acoustical point of view and it is very difficult to reduce their noise, there being no generally accepted method to cure the problem.

The sound power emitted by medium-size and large electrical machines depends strongly on the design of the ventilation system. The internal ventilation system is located on one side in short machines and on both sides in long machines. Two-

TABLE 20.1 Spectral Correction Factor ΔL_{oct} for Broadband Fan Noise For Suction-Opening Diameter 400 mm

Fan and Blade Type	Octave-Band Center Frequency, Hz							
	63	125	250	500	1000	2000	4000	8000
Radial flow fan, straight blades	13	10	13	13.5	15	20	23	26
Radial flow fan, backward-curved blades	14	11	11	11.5	13	16	19	22
Axial flow fan	17	14	9	7.5	8	11	16	18

sided ventilation is better because the temperature distribution within the active part of the machine is more uniform. The volume of air fed to each fan and the total pressure drop are both reduced by almost 50%. The acoustical consequence is a 4–5-dB reduction in noise, as obtained from Eq. (20.7).[10]

The flow of the internal cooling air is usually maintained by radial flow fans in small machines and by axial flow fans in large machines.

Figures 20.6*a*, *b* and 20.7*a* show examples of the three most common arrangements for closed machines. Figure 20.7*b* presents a special solution where the tube separates the air passing through the vent section from the air cooling the winding head.

Control of Ventilation Noise

A straightforward way of reducing motor ventilation noise by design is to choose cooling systems and configurations that are inherently quiet. The quietest machines are totally enclosed and equipped with a water–air heat exchanger. The noise of

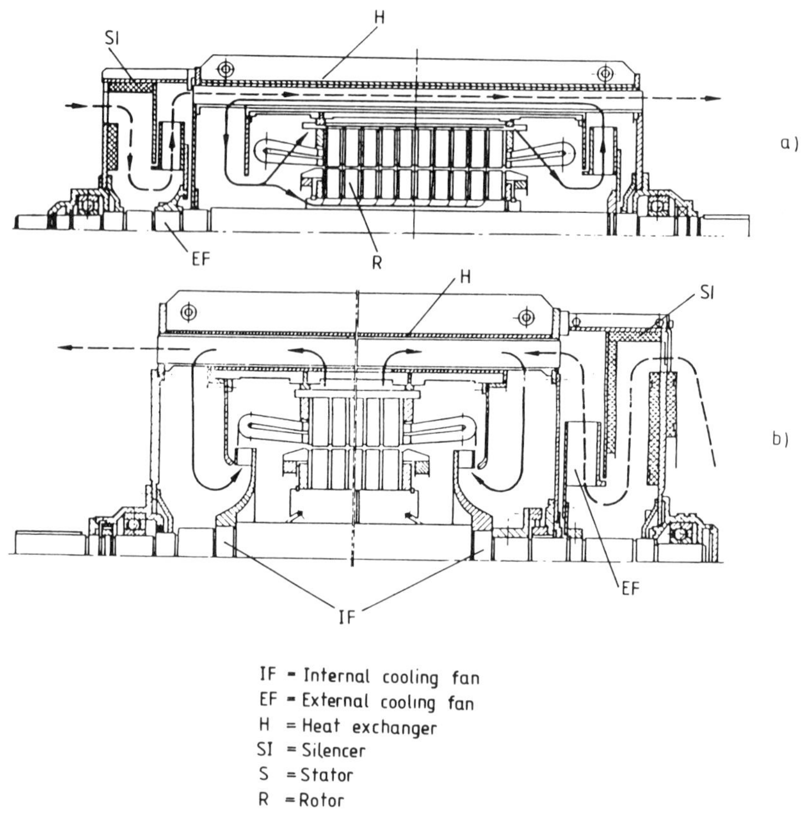

Fig. 20.6 Cooling of totally enclosed induction machines equipped with air-to-air heat exchanger and radial blade internal fan: (*a*) one-sided cooling; (*b*) double-sided cooling.

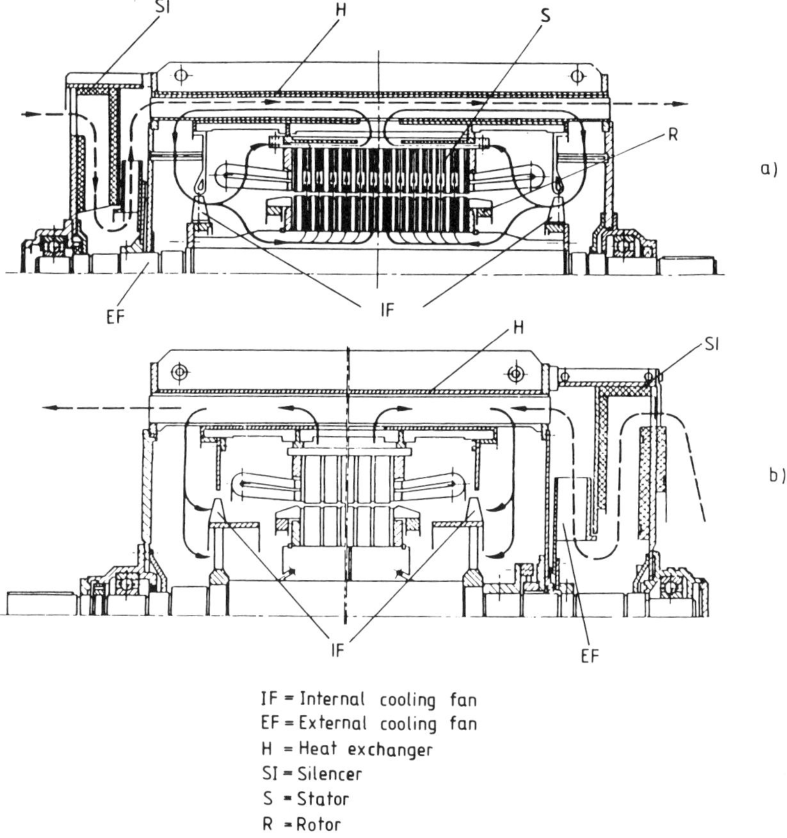

Fig. 20.7 Cooling of totally enclosed induction machines equipped with air-to-air heat exchanger and two-sided axial flow internal fan: (*a*) normal axial fan; (*b*) axial blades on tubular spokes.

totally enclosed high-powered inducation motors can be reduced by as much as 17 dB by replacing the air-to-air heat exchanger with a water-to-air heat exchanger. Another effective way to reduce ventilation noise is to connect the ventilation air circuit of the machine to the closed external ventilation duct. In this case, the fan of the electrical machine becomes unnecessary. The direction of rotation of medium-size and large machines is usually fixed. Therefore, they can be equipped with backward-curved fan blades. The noise level of backward-curved blades is lower, but the pressure produced by them is also lower. According to experience, an 8–10-dB noise reduction is achievable in the case of two-pole machines by replacing straight fan blades with backward-curved blades.

The change from radial flow to axial flow fans can result in a noise reduction of 4 dB. However, this change in design is feasible only if the lower pressure produced by the axial flow fan is sufficient to produce the required air flow. In many cases there is ample thermal reserve in the machine. In such a case the decrease in sound power owing to volume flow rate, D, and the total pressure increments can be approximated by ΔL_1:

$$\Delta L_1 = k \log_{10} \frac{D_{\text{new}}}{D_{\text{old}}} \tag{20.8a}$$

If the speed of the motor, n, is changed (e.g., in the case of motors working in variable-speed drives), the sound power level of the ventilation system will decrease approximately by

$$\Delta L_2 = k \log_{10} \frac{n_{\text{new}}}{n_{\text{old}}} \tag{20.8b}$$

where $k = 50$ for small and medium-sized machines and $k = 60$ for large machines. One should bear in mind, however, the thermal consequences of such a noise reduction. A machine may have thermal reserve because it has been oversized, it works in a low-temperature environment, or the needed output power is permanently less than the rated power.

Mufflers

If the external cooling air in an open circuit cooling or in a totally enclosed machine equipped with an air-to-air heat exchanger is routed through properly designed mufflers or lined ducts located at the inlet and outlet, the noise can be reduced effectively by as much as 15–20 dB and the costs involved are generally much lower than those of building an enclosure for the machine.

The most common mufflers are either the parallel-baffle or concentric cylinder types. The acoustical design of duct mufflers is treated in Chapter 10.

Drive-side installation of mufflers frequently gives rise to problems owing to the presence of a clutch and the driven machine. In such cases a 90° and or 180° change of flow direction is almost inevitable. It is good design to split the air flow and route the two branches separately.

Silencers cause pressure drop. Following their installation the originally required total pressure must be increased. If fan pressure is not increased, the volume flow of cooling air will decrease and may result in unacceptable heat buildup inside the machine.

In machines with forced cooling (the ventilation system is driven by a separate ventilator motor), the volume of the cooling air, can perhaps be reduced, say, by throttling the air flow or by the reduction of speed of the ventilator motor. The latter solution is an energy-saving one. In the case of electrical machines with their own shaft-mounted cooling fan, the noise can be reduced by throttling. In special cases even the fan can be removed or its diameter can be reduced. However, removal may cause mechanical imbalance of the rotor and a machine without a fan cannot be loaded to its rated output power because thermal stress may result.

Tonal Noise

A pure tone is produced by a fan if a periodic disturbance of the inflow acts on it or if the flow downstream of the fan blades interacts with a stationary obstacle such as vanes or mounting struts. Owing to the tonal character of the sound produced, the phenomenon is called a siren effect. The lowest frequency of the pure tone is given by the product of the shaft rotational speed (in reciprocal seconds) and the number of blades. If there is more than one stationary obstacle downstream of the rotor, each one behaves like an individual sound source. The magnitude of the pure-tone sound produced is strongly affected by the distance δ of the stationary obstacle from the rotor and the angle of attack of the flow with respect to the boundary surfaces of the stationary obstacle. The appearance of pure-tone components can be avoided by obeying the empirical inequality $\delta \leq \frac{1}{30}v^2$, where δ is the distance of the stationary obstacle from the rotating fan in millimeters and v is the peripheral speed of the blade in meters per second. Another method of pure-tone ventilation noise reduction is use of unevenly spaced blades.[4]

20.5 VIBROACOUSTICAL MEASUREMENTS OF ELECTRICAL MACHINES

Vibration Measurements on Rotating Electrical Machines

The qualification of an electrical machine is based on a prescribed measurement of the vibration velocities at stipulated measuring points.

Such vibration measurements are usually made by contact sensors such as piezoelectric accelerometers. A signal proportional to the vibration velocity is obtained by integrating their rms output within the range of 10–1000 Hz using a bandpass filter.

The standards strive to mechanically isolate the machine from its environment. In the case of small and medium-sized machines, flexible mounting receives priority, isolating the machine from its surroundings by means of undertuned springs. For large machines, a completely rigid mounting is better. Rigid mounting should be applied, of course, for small machines if the vibration of a machine aggregate is being measured. In the case of flexible mounting, the proper undertuning must be checked. The spring-mounted machine or the machine placed on a flexible baseplate forming a vibrating system should have a resonance frequency lower than one-quarter of the rotational frequency of the rotating machine. This requirement will be met if the static deformation of the elastic element remains within the range

$$\tfrac{1}{2}L \geq \delta \geq 4.2/n^2 \tag{20.9}$$

where L is the unloaded length of the elastic element in millimeters, δ is the change of the length of the elastic element in millimeters due to the weight of the machine, and n is the rotational frequency of the machine in reciprocal seconds.

The relevant standards strictly specify the operating conditions during vibration measurements; for example, for induction motors, the vibration measurements are normally made under no-load condition at rated voltage.

The number and location of measuring points depend on the type and design of the machine. Important measuring points are the bearing planes and the points where the machine is coupled to its surroundings. Vibration is measured in three mutually orthogonal X, Y, and Z directions where the Z axis always corresponds to the rotational axis of the machine.

The vibration classification is based on the vibration severity measured. For small and medium-size rotating electrical machines ISO 2373 gives the maximum permissible vibration severity values as presented in Table 20.2.[11]

In the case of large machines the direct measurement of the vibration of the shaft by means of a noncontact transducer is important in analytical, diagnostic, and safety testings. Noncontact transducers measure displacement, so the characteristic quantity in the vibration of rotating shafts will always be the vibrational displacement. The output signal of a noncontact transducer is a relative quantity. For additional information and instructions see the international standard ISO/DIS 7919/1.

Noise Measurements on Rotating Electrical Machines

The sound power level emitted by an electrical machine has been accepted internationally as the quantity characteristic to the machine as a noise source. Determination of sound power is reduced to the measurement of sound pressure level. The basic difference between sound pressure measurements conducted in the direct (anechoic) field and those in the diffuse (reverberant) field is that in the former the intensity and in the latter the energy density are determined from the sound pressure.

The energy density of a diffuse sound field may be determined by measuring the sound pressure level at various points in the surrounding space. The sound power is calculated from the energy density. If the sound pressure level is averaged

TABLE 20.2 Vibration Quality Grades of Small- and Medium-Sized Rotating Electrical Machines[a]

Vibration Quality Grade	Nominal Speed n, s^{-1}		Maximum Permissible Vibration Severity Values, mm/s		
	Over	Up to	$56 \leq h \leq 132$ mm	$132 < h \leq 225$ mm	$225 < h \leq 400$ mm
Normal, N	10	60	1.8	2.8	4.5
Reduced, R	10	30	0.71	1.12	1.8
	30	60	1.12	1.8	2.8
Special, S	10	30	0.45	0.71	1.12
	30	60	0.71	1.12	1.8

[a]From ISO 2373 (ref. 11).

for the whole of the diffuse field, the sound power level is

$$L_W = \overline{L}_p + 10 \log_{10} \frac{V}{V_0} - 10 \log_{10} \frac{T}{T_0} - 14 \tag{20.10}$$

where $\overline{L}_p$ is the space-averaged sound pressure level, V is the volume of the test room, $V_0 = 1\ \text{m}^3$, T is the reverberation time, and $T_0 = 1$ s.

In the direct field it can generally be assumed that the intensity vector and the normal vector of the measuring surface have the same direction and that the pressure wave and the particle velocity wave are in phase. The points for measuring the noise (representing a unit of the measuring surface) are placed along a measuring surface that differs in shape according to the machine type and the standards used. The measuring surface can be a hemisphere, parallelepiped, or a conformal surface resting on a rigid floor, according to relevant standards, ISO 1680/1,2.[12] (For the parallelepiped measuring surface see Fig. 20.8.) The smallest parallel-

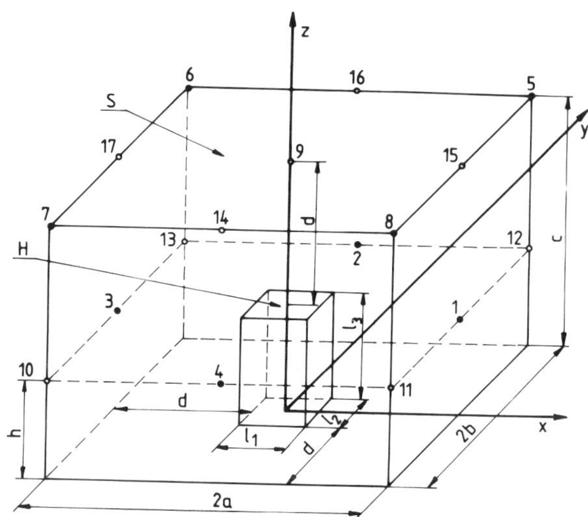

The co-ordinates of the key measuring points:

Serial No.	X	Y	Z
1	a	0	h
2	0	b	h
3	-a	0	h
4	0	-b	h
5	-a	b	c
6	-a	b	c
7	a	-b	c
8	a	-b	c
9	0	0	c

Fig. 20.8 Rectangular parallelepiped measuring surface and measuring points: H, reference surface; S, measuring surface; (1–9) key measuring points; (10–17) supplementary measuring points; l_1, l_2, and l_3: dimensions of reference surface d is measuring distance.

epiped containing the measured electrical machine is called a reference surface. The measuring surface follows the reference surface usually in the measuring distance of 1 m.

Small rotating electrical machines can usually be considered as pointlike sound sources. The most suitable in this case is the hemispheric measuring surface. For medium and large electrical machines, the conformal or the parallelepiped surface is used.

The sound power level radiated by the electrical machine should be determined at rated operating conditions, that is, at rated supply voltage, frequency, speed, and load. Most of the electrical rotating machines tested are electric motors, which can run alone under no-load condition. Many standards accept the measurement of noise in no-load operation, but in many cases it is not enough to know the sound power level radiated by the machine in no-load operation.

When measuring the sound pressure level, the A-weighting filter is used. If it is important to know the components that make up the noise of the tested machine, an octave or one-third-octave analysis is carried out.

In practice, when measuring the noise of a machine, the sound pressure level L_p', is the result of the background noise L_{pb} and the sound pressure level L_p emitted by the machine. If $L_p' - L_{pb} > 10$ dB, the maximum error will be 0.4 dB if the calculation is made with $L_p' = L_p$. The measured sound pressure levels must be corrected by a factor K_1 taking into account the background noise:

$$L_p = L_p' - K_1 \tag{20.11a}$$

where

$$K_1 = 10 \log_{10} \frac{10^{0.1\Delta}}{10^{0.1\Delta} - 1} \tag{20.11b}$$

and

$$\Delta = L_p' - L_{pb} \tag{20.11c}$$

The increase of sound pressure level due to the reflected wave should be considered together with the K_2 environmental correction to account for the influence of reflected sound:

$$K_2 = 10 \log_{10} \left(1 + \frac{4/S_0}{A/S}\right) \tag{20.12}$$

where A is the sound absorption area in square meters. The absorption area may be computed by the product of the mean absorption coefficient of the internal wall surfaces and the total surface area of the testing room (see Chapter 7). In practice, the absorption area may be approximated by measuring reverberation time and applying the formula. $A = 0.164/T$ (where V is the volume of the testing room in cubic meters and T is the reverberation time).

The sound power level emitted by the electrical machine can be computed by the equation

$$L_W = \overline{L}_p - K_2 + 10 \log_{10} \frac{S}{S_0} \tag{20.13}$$

where S is the measuring surface and $\overline{L}_r$ is the average of the sound pressure level corrected by K_1. Based on the A-weighted sound power level, rotating electrical machines can be classified in five quality grades, numbered from 0 to 4. According to the IEC Publication 34-9, the A-weighted sound pressure level of machines (with IP 44 protection grade) with noise Class 1 must not be higher than the values presented in Table 20.3. The maximum permissible L_{WA} values of Class 2, 3, and 4 machines are at least 5, 10, and 15 dB less than the values permitted for Class 1 machines, respectively.

20.6 SPECIAL NOISE MEASUREMENTS

Large Immobile Machines

The acoustical field existing in a real workshop is complicated and does not correspond to the requirements of noise measurement standards. In this case the use of a reference sound source can give a good solution. The sound pressure level measurement on the operating machine and the determination of the correction factor K_1 are to be carried out as above. The machine is then turned off and a reference sound source with a verified sound power $L_{W,\text{ref}}$ must be placed along the sides or on the top of the investigated machine in m different positions. The sound pressure level must be measured in the same measuring points as for the operating machine and the correction factor K_1 again applied to get $L'_{W,\text{ref}}$, the sound power level of the reference sound source under the actual acoustical circumstances of measuring space. The difference of $L_{W,\text{ref}}$ and $L'_{W,\text{ref}}$ is the correction factor K_2. By means of K_2 the sound power level emitted by the investigated machine is determined.

The accuracy of this method depends on whether the spectral characteristics of the investigated machine and the reference sound source are similar. If, however, the spectral characteristics are different, the above procedure should be applied to the octave-band sound pressure levels, and the A-weighted sound power level then calculated from the octave-band sound power levels.

Noise Measurement under Load

The protection of a human being's hearing from noise requires knowledge of the noise characteristics of electrical machines under operating condition. The no-load noise level can differ substantially from the level under load. In order to decide in advance whether the no-load noise measurement is sufficient, without carrying out

TABLE 20.3 Maximum Permissible A-weighted Sound Power Level Values (dB) of Class 1 Machines of IP 44 Protection Grade[a]

Rated Output Power P, kW (kVA)	Rated Rotational Frequency n, s^{-1}					
	$n \leq 16$	$16 < n \leq 22$	$22 < n \leq 31.7$	$31.7 < n \leq 39.3$	$39.3 < n \leq 52.5$	$52.5 < n \leq$
$P \leq 1.1$	71	74	78	81	84	88
$1.1 < P \leq 2.2$	74	78	82	85	88	91
$2.2 < P \leq 5.5$	78	82	86	90	93	95
$5.5 < P \leq 11$	82	85	90	93	97	98
$11 < P \leq 22$	86	88	94	97	100	100
$22 < P \leq 37$	90	91	98	100	102	102
$37 < P \leq 55$	93	94	100	102	104	104
$55 < P \leq 110$	96	98	103	104	106	106
$110 < P \leq 220$	99	102	106	107	109	110
$220 < P \leq 630$	102	105	108	109	111	113
$630 < P \leq 1100$	105	108	111	111	112	116
$1100 < P \leq 2500$	107	110	113	113	113	118
$2500 < P \leq 6300$	109	112	115	115	115	120

[a] From IEC 34-9 (ref. 3).

the noise measurement under load, one has to separate the total sound power L_W by origin. First one has to separate the sound power of the ventilation system of the machine investigated, $L_{W,v}$, by reducing the volume of the external forced cooling air to zero (it gives L'_W). So one gets

$$L_{W,v} = 10 \log_{10} (10^{0.1L_W} - 10^{0.1L'_W}) \tag{20.14}$$

Extrapolating the curve L'_W versus U_{main} (measured by decreasing voltage) to zero voltage, one gets the sound power level of mechanical origin, $L_{W,m}$. So the sound power level of electromagnetic origin can be computed:

$$L_{W,e} = 10 \log_{10} (10^{0.1L'_W} - 10^{0.1L_{W,m}}) \tag{20.15}$$

If relation (20.16) holds, then the result of the no-load noise test can be accepted as characteristic of the electrical machine and the noise qualification must be done on this basis:

$$10 \log_{10} (10^{0.1L_{W,m}} + 10^{0.1L_{W,v}}) - L_{W,e} > 8 \text{ dB} \tag{20.16}$$

If, however, the above inequality does not hold, the noise test under load must be made as well. If one does not have the possibility to carry out a perfect noise test under load, then one has to determine the sound power level in the no-load condition of the idling electrical machine L_{W0} and approximate the change in the noise level ΔL_l due to the loading. The ΔL_l can be determined in the acoustical near field of the machine. Then the investigated electrical machine and the loading engine have to be coupled together. One has to measure the sound pressure level in the near field in no-load (the averaged sound pressure level is $\overline{L}_0$) and in the same measuring points at rated loading condition (the averaged sound pressure level is $\overline{L}_l$). The loading correction factor can be defined as $\Delta L_l = \overline{L}_l - \overline{L}_0$. Adding this factor to L_{W0}, one can approximate the sound power level of the electrical machine emitted under load,

$$L_{Wl} = L_{W0} + \Delta L_l \tag{20.17}$$

since the subtraction while computing ΔL_l eliminates the near-field error. If $\Delta L_l < +1$, then L_{W0} should be accepted as the characteristic acoustical parameter of the machine, while in the case of $\Delta L_l \geq +1$, L_{Wl} should be used to determine the noise quality grade.

20.7 EXPERIMENTAL IDENTIFICATION OF NOISE SOURCES IN ELECTRICAL MACHINES

In practice, one frequently encounters a noise complaint but has no idea at first which element is the direct cause. The first step toward reducing the noise is a

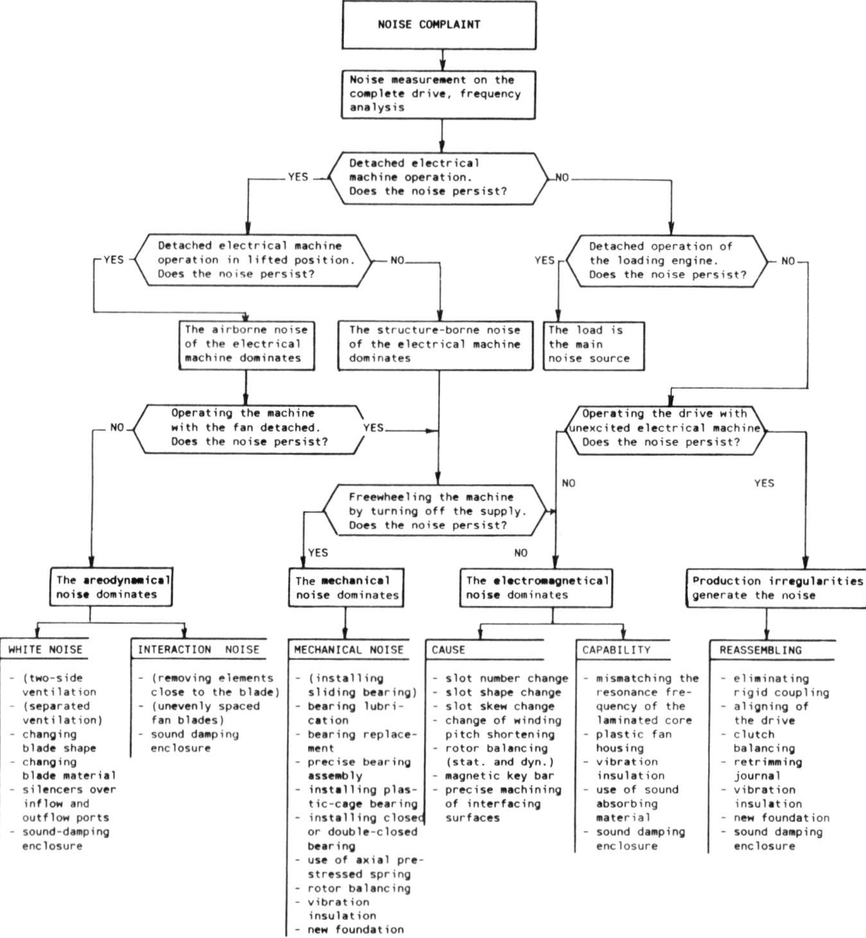

Fig. 20.9 Noise source identification process at electrical drives.

general survey of the possible sources. Figure 20.9 helps in this respect, illustrating the logical steps of noise analysis. The two main categories shown in the figure are statements and decisions. Each statement is preceded by a measurement. The measurement, is the indispensable tool of probing reality and raising and solving problems. The flowchart of the expert system shown in the figure clearly reveals three fundamental cases. The noise source can be the electrical machine itself or another machine coupled to it, although it could happen that each component is quiet with low vibration levels but the noise is introduced by improper coupling. Through relatively simple experiments one can quickly identify the sources of noise. The possible methods of noise reduction are listed briefly.

REFERENCES

1. P. L. Timár and J. Palatinszky, "Modelling the Sound Radiating Electrical Machine" (in Hungarian), *Elektrotechnika*, **66**(7–8), 313–319 (1973).
2. S. J. Yang, "Effect of Length–Diameter Ratio on Noise Radiation for Electrical Machines," *Acustica* **32**(4), 255–261 (1975).
3. IEC Publications 34-9, "Rotating Electrical Machines: Noise Limits," American National Standards Institute, New York.
4. P. L. Timár, *Noise and Vibration of Electrical Machines*, Elsevier, Amsterdam, 1989.
5. H. Sequenz, "Die Wahl der Nutenzahlen bei Käfigankermotoren," *Elektrotech. Maschinenbau* **50**(8), 428–434 (1932).
6. H. Jordan and M. Weis, "Die Nutenschrägung und ihre Wirkungen," *ETZ-Arch.* **88**(10), 528–533 (1967).
7. S. J. Yang, *Low-Noise Electrical Motors*, Clarendon, Oxford, 1981.
8. "Research of Noise Reduction Possibilities for Squirrel-Cage Asynchronous Machines" (in Hungarian), sponsored by EVIG—United Electric Machine Works, Technical University of Budapest, Department of Electrical Machines, 1977.
9. I. Kurutz, P. L. Timár, and E. Tóth, "Calculation of Aerodynamic Noise High-powered Rotating Electrical Machines," *Elektrotechnika* **78**(11), 413–415 (1985).
10. W. Amey and H. Pohl, "Massnahmen zur Geräuschminderung bei Hochspannungsmotoren mittlerer Leistung," *Siemens-Zeitschrift* **47**(6), 466–469 (1973).
11. ISO 2373, "Mechanical Vibration of Certain Rotating Electrical Machinery with Shaft Heights between 80 and 400 mm—Measurement and Evaluation of Vibration Severity," American National Standards Institute, New York.
12. ISO 1680/1.2, "Test Code for the Measurement of Airborne Noise Emitted by Rotating Electrical Machinery. Engineering Method, Survey Method," American National Standards Institute, New York.

BIBLIOGRAPHY

Ellison, A. J. and C. J. Moore, "Acoustic Noise and Vibration of Rotating Electric Machines," *Proc. IEE* **115**(11), 1633–1640 (1968).

Erdélyi, E, "Predetermination of Sound Pressure Levels of Magnetic Noise of Polyphase Induction Motors," *JAIEE* **94**(12, Pt. III), 1269–1280 (1955).

Frohne, H., "Über die primären Bestimmungsgrössen der Lautstärke bei Asynchronmotoren," Dissertation, T. H. Hannover, 1959.

Heller, B. and V. Hamata, *Harmonic Field Effects in Induction Machines*, Elsevier, Amsterdam, 1977.

Kleinrath, H., *Stromrichtergespeiste Drehfeldmaschinen*, Springer, Vienna, 1980.

Lachonius, L., "The Water-cooled Electrical Motor Solves Noise Problems" (in German), *Antriebstechnik* **9**, 533–535 (1974).

Narolski, B., "Beiträge zur Berechnung des magnetischen Geräusches von Asynchronmotoren," *Acta Tech. CSAV* **2**, 156–171 (1965).

Timár, P. L. and M. Poloujadoff, "Method of Comparative Measurement as a Survey Process to Determine the A-weighted Sound Power Level of Rotating Electrical Machines," *IEEE Trans. PAS* **103**(7), 1816–1821 (1984).

Tsivitse, P. J. and P. R. Weihsmann, "Polyphase Induction Motor Noise," *IEEE Trans. IGA* **7**(3), 339–358 (1971).

Yang, S. J. and A. J. Elison, *Machinery Noise Measurement*, Clarendon, Oxford, 1985.

Chapter 21

Elements of Gear Noise Prediction

WILLIAM D. MARK*

Bolt Beranek and Newman, Inc.
Cambridge, Massachusetts

Each pair of meshing gears in a gearbox is a source of vibratory excitation. Each such source gives rise to vibratory motions of the gearing elements (gears and pinions), which excite the shafting, bearings, and supporting structure and eventually cause the radiation of sound by panels, beams, and so on. In this chapter, a general methodology is described for computing the frequency spectra of the vibratory responses of the supporting structures of gear element bearings, where these vibrations are caused by the meshing action of parallel-axis helical or spur gears of nominal involute design. The influences of tooth elastic deformations, intentional and unintentional deviations of tooth running surfaces from equispaced perfect involute surfaces, gear design parameters, and gear element structural transfer functions are included in the methodology. Effects of friction between the teeth are neglected. The principles described in Chapter 9 can be applied to estimate the sound radiated from gear housings using vibration spectra computed by the methods described in this chapter.

Rigid gear elements with equispaced perfect involute teeth transmit exactly uniform angular motion.[1,2] The lack of uniformity in the transmission of angular motion by real gear elements is the principal source of vibratory excitation arising

**Current affiliation:* The Pennsylvania State University, Applied Research Laboratory and Engineering Science and Mechanics Department, University Park, Pennsylvania.

Noise and Vibration Control Engineering: Principles and Applications, Edited by Leo L. Beranek and István L. Vér.
ISBN 0-471-61751-2 © 1992 John Wiley & Sons, Inc.

from meshing gear pairs. This displacement form of excitation is called the transmission error. It is caused by (1) the temporal variation in the total mesh stiffness that is a consequence of the alternating number of (elastic) teeth simultaneously in contact and the changing mesh geometry and (2) geometric deviations in the running surfaces of the teeth from equispaced perfect involute[1,2] surfaces. In Section 21.1, an analytical model of the transmission error and its use are described. Expressions are given for the Fourier series coefficients of the transmission error components attributable to elastic deformations of the teeth and geometric deviations of the tooth running surfaces in Sections 21.2 and 21.3, respectively. A simple, useful model of mesh-to-pinion-bearing structural transfer functions is developed in Section 21.4. In Section 21.5, examples are provided of the various components of transmission error spectra and the mesh-to-pinion-bearing structural transfer function. The effects of key gear system parameters on vibration response and noise spectra are discussed in Section 21.6.

Figure 21.1 illustrates a meshing pair of helical[1,2] gears with rigid, equispaced perfect involute teeth. The plane of the paper in the lower portion of the figure is

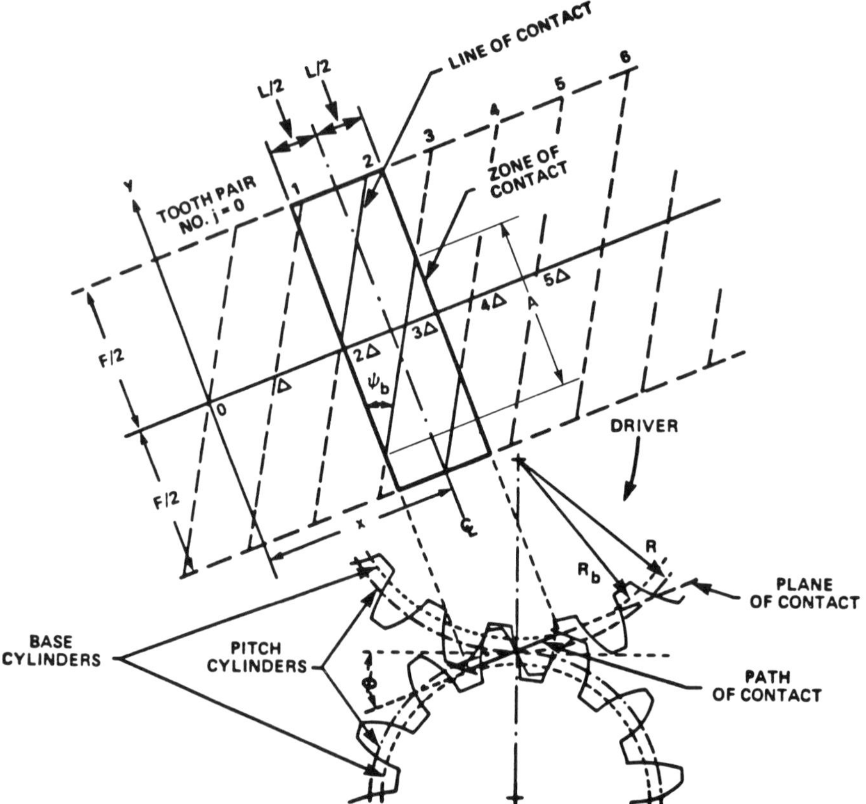

Fig. 21.1 Lines of contact and zone of contact in plane of contact for a helical gear pair with rigid, equispaced, perfect involute teeth.

normal to the gear axes, whereas the plane of the paper in the upper portion of the figure coincides with the plane of contact that is tangent to the base cylinders[1,2] of the two meshing gears, as shown in the lower portion of the figure. A rectangular zone of tooth contact is shown in the upper portion of the figure. The width F of this zone of contact is determined by the facewidth F of the running surfaces of the teeth, and its length L is determined by the points of intersection of the tips of the running surfaces with the plane of contact as illustrated in the lower portion of the figure. For the case of spur gears where the base helix angle ψ_b shown in the figure is zero, the lines of tooth contact in the upper portion of the figure become parallel with the zone of contact centerline, but the zone of contact remains unchanged.

In the material that follows, it is assumed that all tooth contact takes place in the theoretical plane of contact shown in Fig. 21.1 and that the actual zone of tooth contact in the plane of contact is a known rectangular region of width F and length L as shown in the upper portion of the figure. This latter assumption allows us to separate the attenuating effects of multiple tooth contact from the effects of elastic deformations and intentional and unintentional geometric deviations of the running surfaces of the teeth from equispaced perfect involute surfaces.[3-5] It will be seen that this separation permits one to assess how the gear design parameters influence the excitation independently of the values of tooth elastic deformations and geometric deviations from equispaced perfect involute surfaces.

When the torque transmitted by a gear pair is assumed to be constant, the vibratory excitation arising from the gear pair is periodic. When the zone of tooth contact is assumed to be the same for all meshing pairs of teeth as indicated above, the vibratory excitation has two basic periods that are the rotational periods of the two gears of the pair. It is convenient, then, to describe this periodic excitation in the frequency domain. Using a frequency domain description of the excitation, the attenuating effects of multiple tooth contact will be shown to take the form of transfer functions that describe these attenuating effects as a function of frequency. These frequency domain descriptions of the excitation may then be multiplied by the structural transfer functions of the gear system elements to compute the vibration spectra of bearing and gear casing responses, which in turn can be used to compute radiated noise spectra. Derivation of the main results presented herein can be found in references 3–5.

21.1 TRANSMISSION ERROR

Tooth Contact Geometry

As indicated in Fig. 21.1, all contact between each mating pair of teeth on rigid perfect involute gears occurs along a straight line in the theoretical plane of contact shown in the upper portion of the figure. We may imagine these lines of tooth contact to be drawn on a fictitious belt riding on the base cylinders of the two meshing gears as illustrated in the figure. The spacing between the running surfaces of adjacent teeth, measured in the plane of contact in a direction normal to the gear axes (base pitch), is Δ. As the two gears shown in Fig. 21.1 rotate, the dashed lines of contact drawn on this fictitious belt pass through the fixed zone of contact, where they are shown solid.

Coordinate x shown in the upper portion of Fig. 21.1 measures the position of the belt relative to the fixed center of the zone of tooth contact. From Fig. 21.1, it is readily seen that x can be defined as

$$x \triangleq R_b^{(1)}\theta^{(1)} = R_b^{(2)}\theta^{(2)} \tag{21.1}$$

where $R_b^{(1)}$ and $R_b^{(2)}$ are the base cylinder radii of the two meshing gears and $\theta^{(1)}$ and $\theta^{(2)}$ are their rotational positions illustrated in Fig. 21.2. From the involute tooth construction (ref. 2, pp. 163–167) and the above discussion, it follows that rigid perfect involute gears transmit exactly uniform angular velocities. The vibratory excitation arising from a gear pair is a function of the independent variable x defined by Eq. (21.1).

Transmission Error Definition

The vibratory excitation caused by elastic deformations of the teeth and gear bodies and by geometric deviations of the tooth running surfaces from equispaced perfect involute surfaces is described by the transmission error.[6–8] In defining the traditional transmission error $\zeta(x)$ considered in this chapter, it is assumed that the two shafts of the generic pinion–gear pair under consideration can move relative to each other but that they always remain parallel. In order to provide a precise definition of the transmission error, we must consider the instantaneous positions of the actual gears under consideration relative to the positions of their rigid perfect involute counterparts illustrated in Fig. 21.1. The *transmission error* $\zeta(x)$ describes the amount that mating teeth come together relative to the teeth of their rigid per-

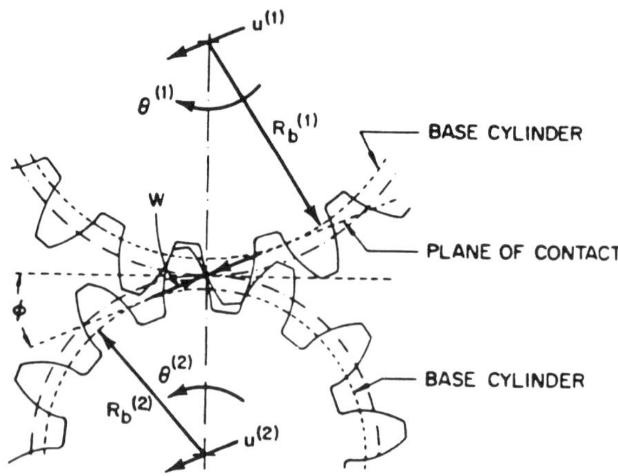

Fig. 21.2 Pair of meshing parallel-axis gears. Plane of paper is transverse plane.[9] Upper gear is gear element 1 and lower gear is gear element 2.

fect involute counterparts. The transmission error is "measured" in the plane of the paper of Fig. 21.2 (transverse plane[9]) in the direction parallel with the plane of contact. However, the shafting supports of a loaded real pair of gears also deform under loading. Let $u^{(1)}(x)$ and $u^{(2)}(x)$ denote positive shaft centerline *displacements* in directions parallel with the plane of contact as shown in Fig. 21.2, where these displacements are measured from the shaft centerline positions (base circle centers) of the rigid perfect involute counterparts to the actual gears under consideration. Denoting rotational deviations in the positions of the two gears shown in Fig. 21.2 from the positions of their rigid perfect involute counterparts by $\delta\theta^{(1)}$ and $\delta\theta^{(2)}$, positive in the directions of $\theta^{(1)}$ and $\theta^{(2)}$, respectively, we may define[3] the transmission error $\zeta(x)$ as

$$\zeta(x) \triangleq R_b^{(1)}\,\delta\theta^{(1)}(x) + u^{(1)}(x) - R_b^{(2)}\,\delta\theta^{(2)}(x) - u^{(2)}(x) \tag{21.2}$$

which may be seen from Fig. 21.2 to express in terms of the variables $\delta\theta^{(1)}$, $u^{(1)}$, $\delta\theta^{(2)}$, and $u^{(2)}$ the amount that the teeth have come together relative to their rigid perfect involute counterparts. These variables are used in writing the equations of motion of a gear system.

The transmission error defined by Eq. (21.2) must be compensated for by elastic deformations of the teeth and gear bodies and by geometric deviations of the tooth running surfaces from equispaced perfect involute surfaces. The assumption that the zone of tooth contact illustrated in Fig. 21.1 is known and fixed allows[3] one to decompose $\zeta(x)$ into a contribution $\zeta_W(x)$ arising from elastic deformations of the teeth and gear bodies and contributions $\zeta^{(1)}(x)$ and $\zeta^{(2)}(x)$ arising from geometric deviations of the tooth running surfaces of gears 1 and 2, respectively, of the meshing pair from equispaced perfect involute surfaces; that is,

$$\zeta(x) = \zeta_W(x) + \zeta^{(1)}(x) + \zeta^{(2)}(x) \tag{21.3}$$

Use of Transmission Error in Equations of Motion

Let $C(x)$ denote the total mesh compliance and $W(x)$ the total force transmitted by the mesh as illustrated in Fig. 21.2. Then

$$\begin{aligned}\zeta_W(x) &= C(x)\,W(x)\\ &= [\overline{C} + \delta C(x)]\,[\overline{W} + \delta W(x)]\\ &= \overline{C}\,W(x) + \delta C(x)\,[\overline{W} + \delta W(x)]\end{aligned} \tag{21.4}$$

where $\overline{C}$ and $\overline{W}$ denote, respectively, the average values of $C(x)$ and $W(x)$ with respect to x, and

$$\delta C(x) \triangleq C(x) - \overline{C} \tag{21.5}$$

$$\delta W(x) \triangleq W(x) - \overline{W} \tag{21.6}$$

Dividing both sides of Eq. (21.4) by $\overline{C}\overline{W}$ gives

$$\frac{\zeta_W(x)}{\overline{C}\overline{W}} = \frac{W(x)}{\overline{W}} + \frac{\delta C(x)}{\overline{C}}\left[1 + \frac{\delta W(x)}{\overline{W}}\right] \tag{21.7}$$

Hence, whenever $\delta W(x)/\overline{W} << 1$, one has

$$\zeta_W(x) \approx \overline{C}W(x) + \overline{W}\,\delta C(x) \qquad \frac{\delta W(x)}{\overline{W}} << 1 \tag{21.8}$$

Let W_0 denote the value of the force W shown in Fig. 21.2 transmitted by the mesh, which we assume here to be constant. Define

$$\zeta_0(x) \triangleq W_0\,\delta C(x) + \zeta^{(1)}(x) + \zeta^{(2)}(x) \tag{21.9}$$

Then, by combining Eqs. (21.2), (21.3), (21.8), and (21.9) and using $\overline{W} = W_0$, we have, whenever $\delta W(x)/\overline{W} << 1$,

$$R_b^{(1)}\,\delta\theta^{(1)}(x) + u^{(1)}(x) - R_b^{(2)}\,\delta\theta^{(2)}(x) - u^{(2)}(x) - \overline{C}W(x) \approx \zeta_0(x) \tag{21.10}$$

When the equations of motion of a gear system are written, temporal variations in the displacements $\delta\theta^{(1)}$, $u^{(1)}$, $\delta\theta^{(2)}$, $u^{(2)}$ and the mesh force W for each gear pair in the system, as in Fig. 21.2, are unknowns that must be determined by solving the equations of motion of the system. Because the temporal fluctuations in the mesh forces W are unknowns, an equation in addition to the equations of motion of the system is required for each gear mesh in the system. Equation (21.10) is that equation. The left-hand side of Eq. (21.10) contains the unknown displacements and the unknown mesh force $W(x)$, and the right-hand side of Eq. (21.10) is $\zeta_0(x)$, which may be regarded as known.

Except for the additive *constant* $W_0\overline{C}$, it follows from Eqs. (21.3) and (21.4) that the quantity $\zeta_0(x)$ defined by Eq. (21.9) is the transmission error $\zeta(x)$ defined using the *static* mesh force W_0. Hence, except for the additive constant $W_0\overline{C}$, $\zeta_0(x)$ is the *static transmission error*. It follows from Eq. (21.10) that for cases where fractional fluctuations in the mesh force $W(x)$ are small, the unsteady components $\zeta_0(x)$ of the static transmission error may be regarded as the principal source of vibratory excitation arising from meshing gear pairs. The static transmission error is a *displacement* form of excitation (not a force).

When writing the equations of motion of a gear system, the independent variable is time t. From Eq. (21.1), it follows that the independent variable x used in Eq. (21.10) can be expressed in terms of time t by

$$x = R_b^{(1)}[\omega_1 t + \theta_0^{(1)}] = R_b^{(2)}[\omega_2 t + \theta_0^{(2)}] \tag{21.11}$$

where ω_1 and ω_2 are the rotational speeds of gears 1 and 2, respectively, in radians per unit time, and $\theta_0^{(1)}$ and $\theta_0^{(2)}$ are the rotational positions of gears 1 and 2, respectively, at $t = 0$.

Since the quantities $\delta\theta^{(1)}(x)$, $u^{(1)}(x)$, $\delta\theta^{(2)}(x)$, $u^{(2)}(x)$, and $W(x)$ in the left-hand side of Eq. (21.10) all are unknowns or dependent variables describing the state of the gear pair, whereas the quantity $\zeta_0(x)$ defined by Eq. (21.9) may be regarded as known, Eq. (21.10) can be considered to be a "*mesh constraint equation*"[10] relating the unknown dynamic variables of the gear pair. The above-described concepts have been generalized in the literature to cases where gear shafts are not assumed to remain parallel[10–14] and fractional variations in the mesh loading are not assumed to be small.[3, 10–14]

Harmonic Representation of Static Transmission Error

The contribution $\zeta^{(i)}(x)$, $i = 1, 2$ as appropriate, to the transmission error, Eq. (21.3) or (21.9), arising from geometric deviations of the tooth running surfaces on gear i from equispaced perfect involute surfaces is periodic in x with period equal to the base circle circumference $N^{(i)}\Delta$ of gear i, where $N^{(i)}$ is the number of teeth on gear i and Δ is the base pitch. Thus, $\zeta^{(i)}(x)$ can be represented by the complex Fourier series[3,5]

$$\zeta^{(i)}(x) = \sum_{n=-\infty}^{\infty} \alpha_n^{(i)} \exp\left(\frac{i2\pi nx}{N^{(i)}\Delta}\right) \tag{21.12}$$

where index n counts the *rotational* harmonic contributions from each gear i of the meshing pair. Note that $i = \sqrt{-1}$ is not to be confused with index i, which designates the gear under consideration and in the mathematics always appears within parentheses. The expansion coefficients $\alpha_n^{(i)}$ are given by

$$\alpha_n^{(i)} = \frac{1}{N^{(i)}\Delta} \int_{-N^{(i)}\Delta/2}^{N^{(i)}\Delta/2} \zeta^{(i)}(x) \exp\left(\frac{-i2\pi nx}{N^{(i)}\Delta}\right) dx \tag{21.13}$$

The stiffnesses of all mating *pairs* of teeth of a gear pair generally may be regarded to be governed by the same function. When the force $W(x)$ transmitted by the mesh (Fig. 21.2) is assumed to be a constant value $W(x) = W_0$, the elastic deformation component, Eq. (21.4), of the transmission error is periodic in x with period equal to the base pitch Δ. The complex Fourier series representation of this elastic deformation component is[3,5]

$$\zeta_{W_0}(x) = \sum_{p=-\infty}^{\infty} \alpha_{W_p} \exp\left(\frac{i2\pi px}{\Delta}\right) \tag{21.14}$$

where index p counts the *tooth meshing* harmonics, and

$$\alpha_{W_p} = \frac{1}{\Delta} \int_{-\Delta/2}^{\Delta/2} \zeta_{W_0}(x) \exp\left(-\frac{i2\pi px}{\Delta}\right) dx \tag{21.15}$$

When a gear pair is running at uniform speed, it follows by comparing the exponents in Eqs. (21.12) and (21.14) that pairs of rotational and tooth meshing harmonics, n and p, respectively, contributing to the same frequency are related (for integer p) by

$$\frac{n}{N^{(i)}\Delta} = \frac{p}{\Delta} \tag{21.16}$$

or

$$n = pN^{(i)} \qquad p = 0, \pm 1, \pm 2, \ldots \tag{21.17}$$

which applies to either gear, $i = 1, 2$, of the meshing pair, where $N^{(i)}$ is the number of teeth on the gear under consideration. Equation (21.17) states that the frequency of the $N^{(i)}$th *rotational* harmonic of each gear i coincides with the tooth meshing fundamental frequency $p = 1$ of the gear pair, the frequency of the $2N^{(i)}$th rotational harmonic of each gear i coincides with tooth meshing harmonic $p = 2$, and so on. For different numbers of teeth $N^{(1)}$ and $N^{(2)}$, the frequencies of the rotational harmonics n arising from the two gears generally will differ, but their tooth meshing harmonic frequencies always conicide.

It follows from the additive relationship, Eq. (21.3), and the above comments that the complex Fourier series expansion coefficients α_{mp} of the *tooth meshing* harmonic components of the static transmission error are[3,5]

$$\alpha_{mp} = \alpha_{Wp} + \alpha^{(1)}_{pN^{(1)}} + \alpha^{(2)}_{pN^{(2)}} \qquad p = 0, \pm 1, \pm 2, \ldots \tag{21.18}$$

where α_{Wp} is the elastic deformation contribution from the *gear pair* determined by Eq. (21.15) and $\alpha^{(1)}_{pN^{(1)}}$ and $\alpha^{(2)}_{pN^{(2)}}$ are tooth running surface geometric deviation contributions determined from the deviations of the teeth on gears 1 and 2 for $n = pN^{(1)}$ and $n = pN^{(2)}$, respectively, by Eq. (21.13). *It can be shown that the tooth meshing harmonic contributions $\alpha^{(1)}_{pN^{(1)}}$ and $\alpha^{(2)}_{pN^{(2)}}$ arise from the component of the tooth running surface deviations that is common to all teeth on the gear or pinion under consideration. This component is obtained from the deviation surface that is constructed by forming the average of the deviation surfaces of all teeth on the gear element under consideration.*[3,5] Formulas for the various Fourier series expansion coefficients are given next.

21.2 FOURIER SERIES COEFFICIENTS OF ELASTIC DEFORMATION COMPONENT OF STATIC TRANSMISSION ERROR

Let us define

$$Q_a \triangleq \frac{FL}{A\Delta} \qquad Q_t \triangleq \frac{L}{\Delta} \tag{21.19}$$

where the parameters F, L, A, and Δ all are defined[3] in Fig. 21.1. Quantities Q_a and Q_t are the axial and transverse contact ratios, respectively; that is, Q_a is the average number of teeth simultaneously in contact in the axial direction, and Q_t is the average number of teeth simultaneously in contact in a plane cut normal to the gear axes (transverse plane). Let

$$K_T(x) = \frac{1}{C(x)} \tag{21.20}$$

be the total mesh stiffness, that is, the total stiffness of all tooth pairs simultaneously in contact at the rotational positions of the gears designated by the variable x (Fig. 21.1), and let $\overline{K}_T$ denote the average value with respect to x of the periodic function $K_T(x)$. Then, assuming that the tooth pair stiffness per unit length of line of contact (Fig. 21.1) is a constant, one can show[3-5] that the Fourier series coefficients, Eq. (21.15), of the elastic deformation component $\zeta_{W_0}(x)$ of the static transmission error are approximated *for helical gears* by

$$\alpha_{Wp} \approx -\frac{W_0}{\overline{K}_T} \frac{\sin(p\pi Q_a)}{p\pi Q_a} \frac{\sin(p\pi Q_t)}{p\pi Q_t} \qquad p = \pm 1, \pm 2, \ldots \tag{21.21}$$

and *for spur gears* by

$$\alpha_{Wp} \approx -\frac{W_0}{\overline{K}_T} \frac{\sin(p\pi Q_t)}{p\pi Q_t} \qquad p = \pm 1, \pm 2, \ldots \tag{21.22}$$

The nonfluctuating component $\alpha_{W_0} \approx W_0/\overline{K}_T$ of $\zeta_{W_0}(x)$ is absent from the static transmission error contribution $W_0\ \delta C(x)$ in Eq. (21.9); however, its remaining Fourier series coefficients are given by Eq. (21.21) or (21.22), as appropriate. For double helical or herringbone gears,[1,2] the appropriate value of Q_a for use in Eq. (21.21) is the value computed from one of the two helices (using one-half of the *total* active facewidth of the gear). For spur gears with adequate tip relief, Eq. (21.22) tends to overestimate the magnitudes of the high-order tooth meshing harmonics.

The assumption that the tooth pair stiffness per unit length of line of contact is a constant is quite reasonable in the case of helical gears[15-18]; however, it should be regarded only as a first approximation in the case of spur gears.[19-21] The mean total length of line of contact can be expressed as[3,4]

$$\bar{l} = Q_a Q_t \Delta \csc \psi_b \tag{21.23}$$

where Δ is the base pitch, ψ_b is the base helix angle (Fig. 21.1), and $\csc \psi_b \triangleq (\sin \psi_b)^{-1}$.

From Eq. (21.2) and Fig. 21.2, it follows that the transmission error $\zeta(x)$ and mesh loading W both are defined in the direction determined by the intersection of the plane of contact and the transverse plane; hence, the elastic deformation component $\zeta_{W_0}(x)$ of the static transmission error also is defined in this direction. Tooth pair stiffness values K_N for helical gears[17] obtained from deflections and loadings both measured *normal* to tooth surfaces must be adjusted by

$$K_T = K_N \cos^2 \psi_b \tag{21.24}$$

to obtain tooth pair stiffness values K_T defined for deflections and loadings both measured in the *transverse* plane.

21.3 FOURIER SERIES COEFFICIENTS OF GEOMETRIC DEVIATION COMPONENTS OF TRANSMISSION ERROR

Decomposition of Tooth Running Surface Deviations into Elementary "Errors"

We turn now to prediction of the Fourier series coefficients, Eq. (21.13), of the transmission error contributions $\zeta^{(1)}(x)$ and $\zeta^{(2)}(x)$ in Eqs. (21.3) and (21.9) arising from deviations of the tooth running surfaces of gears 1 and 2, respectively, of a meshing pair from equispaced perfect involute surfaces.

Let $\eta_{Cj}^{(i)}(y, z)$ denote the deviation of the running surface of tooth j of gear i, $i = 1, 2$ as appropriate, measured as a function of the *Cartesian* coordinates y, z illustrated in Fig. 21.3. Coordinate axis y is parallel to the gear axis; its origin is placed at the exact midpoint of the range F of y where tooth contact is assumed to take place. Coordinate z is defined by

$$z = R_b \epsilon \sin \phi + c \tag{21.25}$$

where R_b is the base cylinder radius of the gear being considered, ϵ is the involute roll angle[9] measured in radians, and ϕ is the pressure angle defined as the angle between the plane of contact and the pitch plane (Fig. 21.1). The constant c is chosen so that the origin of the coordinate z is placed at the exact midpoint of the range D of z where tooth contact is assumed to take place. From Figs. 21.1 and 21.3 one can see that distances D and L are related by[3,5] $D = L \sin \phi$. Deviations $\eta_{Cj}^{(i)}(y, z)$ are "measured" in the direction defined by the intersection of the plane of contact and a plane normal to the gear axes.

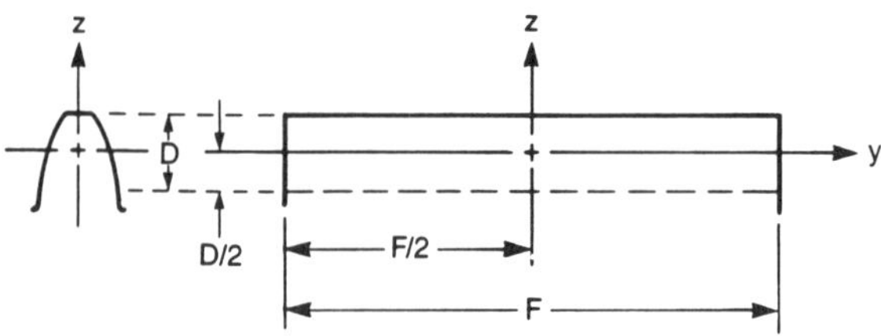

Fig. 21.3 Coordinate system used for description of tooth surface deviations $\eta_{Cj}^{(i)}(y, z)$. Coordinate y is parallel to gear axis and coordinate z is defined by Eq. (21.25).

The analysis and understanding of the effects of such deviations on the vibratory excitation are greatly enhanced by decomposing the deviations into elementary "errors."[3-5] A generic deviation surface $\eta_{Cj}^{(i)}(y, z)$ can be expanded[4,5] in a complete set of normalized two-dimensional Legendre polynomials

$$\eta_{Cj}^{(i)}(y, z) = \sum_{k=0}^{\infty} \sum_{l=0}^{\infty} c_{j,kl}^{(i)} \psi_{yk}(y) \psi_{zl}(z) \tag{21.26}$$

where the normalized polynomials $\psi_{yk}(y)$ and $\psi_{zl}(z)$ are defined[4,5] by

$$\psi_{yk}(y) \triangleq (2k + 1)^{1/2} P_k\left(\frac{2y}{F}\right) \qquad -\tfrac{1}{2}F < y < \tfrac{1}{2}F \tag{21.27}$$

and

$$\psi_{zl}(z) \triangleq (2l + 1)^{1/2} P_l\left(\frac{2z}{D}\right) \qquad -\tfrac{1}{2}D < z < \tfrac{1}{2}D \tag{21.28}$$

where the functions $P_n(x)$ are the usual Legendre polynomials[22]

$$\begin{aligned} P_0(x) &= 1 \\ P_1(x) &= x \\ P_2(x) &= \tfrac{1}{2}(3x^2 - 1) \\ P_3(x) &= \tfrac{1}{2}(5x^3 - 3x) \\ &\vdots \end{aligned} \tag{21.29}$$

and where the expansion coefficients $c_{j,kl}^{(i)}$ in Eq. (21.26) are defined from the surface deviations $\eta_{Cj}^{(i)}(y, z)$ by[4,5]

$$c_{j,kl}^{(i)} = \frac{1}{FD} \int_{-D/2}^{D/2} \int_{-F/2}^{F/2} \eta_{Cj}^{(i)}(y, z) \psi_{yk}(y) \psi_{zl}(z)\, dy\, dz \tag{21.30}$$

Equation (21.26) can be interpreted as an expansion of the tooth surface deviation $\eta_{Cj}^{(i)}(y, z)$ in a canonical set[4,5] of the elementary errors

$$\psi_{yk}(y)\psi_{zl}(z) = [(2k + 1)(2l + 1)]^{1/2} P_k\left(\frac{2y}{F}\right) P_l\left(\frac{2z}{D}\right) \tag{21.31}$$

Each elementary error is designated by a pair of nonnegative integer indices k, l, where index k is associated with the lead (axial) direction and index l is associated with the profile (radial) direction. From the definition, Eq. (21.29), of the Legendre polynomials,[22] we see that the low-order elementary errors possess the simple interpretations described in Table 21.1. Figure 21.4 illustrates the terms $k = 0$, $l = 0, 1, 2, 3$, and $l = 0$, $k = 0, 1, 2$. Thus, the normalized Legendre polynomials provide us with a convenient mathematical tool for representing tooth errors in a form closely allied with gearing industry practice.

The normalization factors $(2k + 1)^{1/2}$ and $(2l + 1)^{1/2}$ in the definitions, Eqs. (21.27) and (21.28), are not normally associated with the Legendre polynomials.[22] These factors have been introduced[4,5] so that the expansion coefficients, Eq. (21.30), satisfy the relation

$$\frac{1}{FD}\int_{-D/2}^{D/2}\int_{-F/2}^{F/2}[\eta_{Cj}^{(i)}(y, z)]^2\,dy\,dz = \sum_{k=0}^{\infty}\sum_{l=0}^{\infty}[c_{j,kl}^{(i)}]^2 \qquad (21.32)$$

that is, $[c_{j,kl}^{(i)}]^2$ directly measures the contribution of the term $c_{j,kl}^{(i)}\psi_{yk}(y)\,\psi_{zl}(z)$ in Eq. (21.26) to the mean-square value of the tooth surface deviation $\eta_{Cj}^{(i)}(y, z)$ described by the left-hand side of Eq. (21.32). From Eq. (21.32) it follows that the expansion coefficients $c_{j,kl}^{(i)}$ have the same dimension as the deviations $\eta_{Cj}^{(i)}(y, z)$ and that $c_{j,kl}^{(i)}$ can be regarded as a direct measure of the elementary error, Eq. (21.31), contribution to the tooth surface deviation $\eta_{Cj}^{(i)}(y, z)$. *These properties greatly facilitate interpretation of numerical values of the coefficients* $c_{j,kl}^{(i)}$ *obtained from measurements of tooth running surfaces*. Systematic methods for generating the coefficients $c_{j,kl}^{(i)}$ using standard gearing metrology equipment are described in references 23 and 24.

Harmonic Number Spectra of Elementary Error Contributions

We continue to assume that the tooth pair stiffness per unit length of line of contact is a constant. All of the information describing deviations of the running surfaces of the $N^{(i)}$ teeth on gear i from equispaced perfect involute surfaces is contained in the Legendre expansion coefficients $c_{j,kl}^{(i)}$, $j = 0, 1, \ldots, N^{(i)} - 1$. To obtain an expression for the Fourier series expansion coefficients, Eq. (21.13), it is necessary[3-5] to first generate the finite discrete Fourier transform[25] $B_{kl}^{(i)}(n)'$ with respect to j of the sequence $c_{j,kl}^{(i)}$, $j = 0, 1, \ldots, N^{(i)} - 1$, for each pair of indices

TABLE 21.1 Elementary Error Classifications Using Normalized Two-Dimensional Legendre Polynomials.[a]

$k = 0, l = 0$	Tooth spacing deviations
$k = 0, l = 1$	Pure involute slope deviations
$k = 1, l = 0$	Pure linear lead deviations
$k = 0, l = 2$	Pure involute fullness deviations
$k = 1, l = 1$	Combined linear lead involute slope deviations
$k = 2, l = 0$	Pure lead crowning deviations

[a]From refs. 4 and 5.

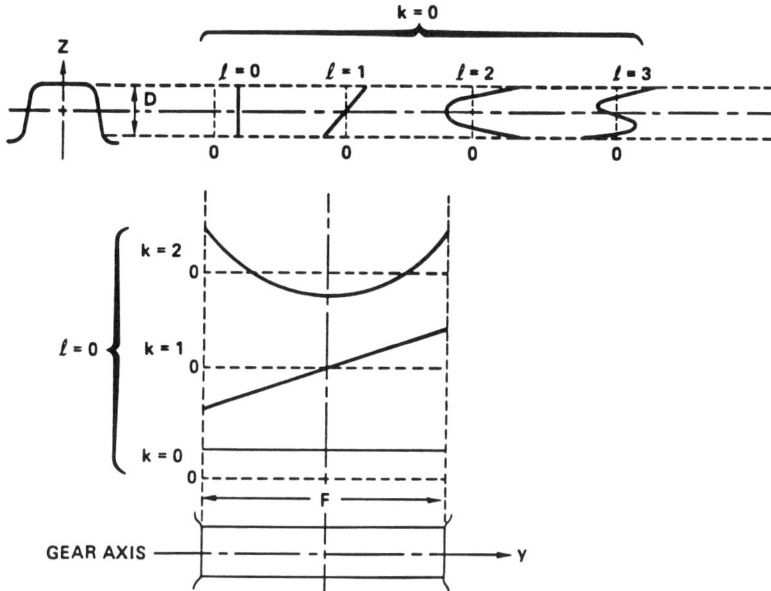

Fig. 21.4 Interpretation of elementary errors derived from Legendre polynomials $P_n(x)$, $n = 0, 1, 2, \ldots$. See Eq. (21.31) and Table 21.1.

$k = 0, 1, 2, \ldots, l = 0, 1, 2, \ldots$, in the case of helical gears, and $k = 0, l = 0, 1, 2, \ldots$, in the case of spur gears, where

$$B_{kl}^{(i)}(n)' \triangleq \frac{1}{N^{(i)}} \sum_{j=0}^{N^{(i)}-1} c_{j,kl}^{(i)} \exp\left(-\frac{i2\pi nj}{N^{(i)}}\right), \qquad n = 0, \pm 1, \pm 2, \ldots \tag{21.33}$$

The discrete Fourier transform $B_{kl}^{(i)}(n)'$ is periodic in harmonic number n with period $N^{(i)}$; that is,

$$B_{kl}^{(i)}(n + pN^{(i)})' = B_{kl}^{(i)}(n)' \qquad p = 0, \pm 1, \pm 2, \ldots \tag{21.34}$$

since $\exp(-i2\pi pj) \equiv 1$ for $p = 0, \pm 1, \pm 2, \ldots$ because j always is an integer. Thus, although $B_{kl}^{(i)}(n)'$ is of interest for all integer harmonic numbers n, all values of $B_{kl}^{(i)}(n)'$ outside of the range $0 \leq n \leq (N^{(i)} - 1)$ can be obtained from its values within this range using Eq. (21.34). *The quantity* $B_{kl}^{(i)}(n)'$ *can be regarded as the harmonic number spectrum of the elementary error contribution* $c_{j,kl}^{(i)}\psi_{yk}(y)\psi_{zl}(z)$ *in Eq. (21.26), whose strength for each tooth* j *is characterized by* $c_{j,kl}^{(i)}$, *as one can see from Eq. (21.32).*

Expressions for Fourier Series Coefficients

The Fourier series expansion coefficients, Eq. (21.13), of the tooth surface deviation contributions $\zeta^{(i)}(x)$ to the transmission error can be expressed in terms of the elementary error spectra $B_{kl}^{(i)}(n)'$ *for helical gears* by[4,5]

$$\alpha_n^{(i)} = \sum_{k=0}^{\infty} \sum_{l=0}^{\infty} B_{kl}^{(i)}(n)' \, \hat{\phi}_{kl}\left(\frac{n}{N^{(i)}\Delta}\right) \qquad n = 0, \pm 1, \pm 2, \ldots \tag{21.35}$$

and *for spur gears* by

$$\alpha_n^{(i)} = \sum_{l=0}^{\infty} B_{0l}^{(i)}(n)' \hat{\phi}_{0l}\left(\frac{n}{N^{(i)}\Delta}\right) \qquad n = 0, \pm 1, \pm 2, \ldots \tag{21.36}$$

where the $B_{kl}^{(i)}(n)'$ are to be determined by Eqs. (21.33) and (21.34), and the *mesh transfer functions* $\hat{\phi}_{kl}(n/N^{(i)}\Delta)$ are described in the next section.

According to Eq. (21.18), the values of $\alpha_n^{(i)}$ at the *tooth meshing* harmonics $n = pN^{(i)}$, Eq. (21.17), are of particular interest. At the tooth meshing harmonics $n = pN^{(i)}$, the values of the elementary error spectra $B_{kl}^{(i)}(n)'$ are, according to Eq. (21.33),

$$\begin{aligned} B_{kl}^{(i)}(pN^{(i)})' &= \frac{1}{N^{(i)}} \sum_{j=0}^{N^{(i)}-1} c_{j,kl}^{(i)} \exp(-i2\pi pj) \\ &= \frac{1}{N^{(i)}} \sum_{j=0}^{N^{(i)}-1} c_{j,kl}^{(i)} \equiv B_{kl}^{(i)}(0)' \end{aligned} \tag{21.37}$$

according to Eq. (21.33) evaluated at $n = 0$. In going from the first to the second line in Eq. (21.37), we have used the fact that $\exp(-i2\pi pj) \equiv 1$ since p and j always are integers. Let us define the average of the expansion coefficients $c_{j,kl}^{(i)}$ with respect to tooth number j over all $N^{(i)}$ teeth on a gear element as

$$a_{kl}^{(i)} \triangleq \frac{1}{N^{(i)}} \sum_{j=0}^{N^{(i)}-1} c_{j,kl}^{(i)} \tag{21.38}$$

Combining Eqs. (21.37) and (21.38) with Eqs. (21.35) and (21.36) evaluated at $n = pN^{(i)}$, we obtain *for helical gears*

$$\alpha_{pN^{(i)}}^{(i)} = \sum_{k=0}^{\infty} \sum_{l=0}^{\infty} a_{kl}^{(i)} \hat{\phi}_{kl}\left(\frac{p}{\Delta}\right) \qquad p = 0, \pm 1, \pm 2, \ldots \tag{21.39}$$

and *for spur gears*

$$\alpha_{pN^{(i)}}^{(i)} = \sum_{l=0}^{\infty} a_{0l}^{(i)} \hat{\phi}_{0l}\left(\frac{p}{\Delta}\right) \qquad p = 0, \pm 1, \pm 2, \ldots \tag{21.40}$$

which are expressions for the *tooth meshing* harmonic Fourier series coefficients of the contributions to the transmission error arising from deviations of the tooth running surfaces of gear element i from equispaced perfect involute surfaces. From Eqs. (21.38)–(21.40), it follows that these tooth meshing harmonic Fourier series coefficients depend only on the average value, Eq. (21.38), of the Legendre expansion coefficients $c_{j,kl}^{(i)}$. Equivalently, then, these Fourier series coefficients $\alpha_{pN^{(i)}}^{(i)}$ depend only on the *average deviation surfaces*, which is the average with respect to tooth number j of the individual deviation surfaces $\eta_{Cj}^{(i)}(y, z)$ used in generating the individual expansion coefficients $c_{j,kl}^{(i)}$ by Eq. (21.30).

The remaining Fourier series coefficients $\alpha_n^{(i)}$, $n \neq pN^{(i)}$ represent the *rotational* harmonic contributions from gear element i to the transmission error. These contributions arise from the differences $\epsilon_{Cj}^{(i)}(y, z)$ of the individual tooth running surfaces from the mean surface; that is, from the individual deviation surfaces defined by[3–5]

$$\epsilon_{Cj}^{(i)}(y, z) \triangleq \eta_{Cj}^{(i)}(y, z) - m_C^{(i)}(y, z) \tag{21.41}$$

where

$$m_C^{(i)}(y, z) \triangleq \frac{1}{N^{(i)}} \sum_{j=0}^{N^{(i)}-1} \eta_{Cj}^{(i)}(y, z) \tag{21.42}$$

is the mean deviation surface defined as the average of all of the deviation surfaces $\eta_{Cj}^{(i)}(y, z)$. Thus, if the center of rotation of gear element i is the base cylinder center and if there are no machining errors on the teeth, then the rotational harmonic contributions $n \neq pN^{(i)}$ to the transmission error from gear element i should be negligible.

Formulas for Mesh Transfer Functions

Using the notation[26]

$$\text{rect}(x) \triangleq \begin{cases} 1 & |x| < \frac{1}{2} \\ 0 & |x| > \frac{1}{2} \end{cases} \tag{21.43}$$

let us define[4]

$$w_{00}(g\Delta - n') \triangleq \tfrac{1}{2} \,\text{rect}\,[\tfrac{1}{2}(g\Delta - n')]\,\{1 + \cos[\pi(g\Delta - n')]\} \tag{21.44}$$

and

$$w(g\Delta) \triangleq \begin{cases} \frac{1}{2}[1 - \cos(\pi g\Delta)] & |g\Delta| < 1 \\ 1 & |g\Delta| \geq 1 \end{cases} \tag{21.45}$$

If we now define the general weighting function $w_{kl}(g\Delta, n')$ from Eqs. (21.44) and (21.45) as[4]

$$w_{kl}(g\Delta, n') \triangleq \begin{cases} w_{00}(g\Delta - n') & k = 0, \quad l = 0 \\ w(g\Delta) & \text{otherwise} \end{cases} \tag{21.46}$$

we have[4] for the mesh transfer functions used in Eqs. (21.35), (21.36), (21.39), and (21.40),

$$\begin{aligned} \hat{\phi}_{kl}(g) \approx{}& (-i)^{k+l}[(2k+1)(2l+1)]^{1/2} \\ &\times \Bigg(j_k(\pi Q_a g\Delta) j_l(\pi Q_t g\Delta) - \sum_{\substack{n'=-\infty \\ \text{except} \\ n'=0}}^{\infty} \\ &\{w_{kl}(g\Delta, n') j_0(n'\pi Q_a) j_0(n'\pi Q_t) \\ &\times j_k[(g\Delta - n')\pi Q_a] j_l[(g\Delta - n')\pi Q_t]\} \Bigg) \end{aligned} \tag{21.47}$$

where $i = \sqrt{-1}$ and $j_n(x)$ is the spherical Bessel function[27] of the first kind of order n which can be defined in terms of the Bessel functions $J_{n+1/2}(x)$ of the first kind of order $n + \frac{1}{2}$ by

$$j_n(x) \triangleq \left(\frac{\pi}{2x}\right)^{1/2} J_{n+1/2}(x) \tag{21.48}$$

In the above formulas, Δ is the base pitch illustrated in Fig. 21.1, and Q_a and Q_t are, respectively, the axial and transverse contact ratios defined by Eq. (21.19) and Fig. 21.1. The spherical Bessel functions of order $n = 0$ in Eq. (21.47) have the particularly simple form[27]

$$j_0(x) = \frac{\sin x}{x} \tag{21.49}$$

For the case of *spur gears*, the version of $\hat{\phi}_{kl}(g)$ required in Eqs. (21.36) and (21.40) can be expressed as

$$\begin{aligned} \hat{\phi}_{0l}(g) \approx{}& (-i)^l(2l+1)^{1/2} \Bigg\{ j_l(\pi Q_t g\Delta) - \sum_{\substack{n'=-\infty \\ \text{except} \\ n'=0}}^{\infty} \\ & w_{0l}(g\Delta, n') j_0(n'\pi Q_t) j_l[(g\Delta - n')\pi Q_t] \Bigg\} \end{aligned} \tag{21.50}$$

since, for spur gears, we have $A = \infty$ from Fig. 21.1; hence, $Q_a = 0$ from Eq. (21.19) and $j_0(0) = 1$ from Eq. (21.49).

A physical interpretation of the roles of the various functions on the right-hand sides of Eqs. (21.35), (21.36), (21.39), and (21.40) will be provided later in this chapter. Simple approximations to the envelopes of the mesh transfer functions, Eqs. (21.47) and (21.50), can be found on pp. 1780, 1781, and 1786 of reference 4.

21.4 MESH-TO-PINION-BEARING STRUCTURAL TRANSFER FUNCTIONS

Vibratory energy must be transmitted from the gear meshes through the gearing elements in order to excite the gear casing walls, for example. Since the pinion of a meshing pinion–gear pair ordinarily has less inertia than the gear, at the tooth meshing fundamental and higher frequencies most of this vibratory energy usually is transmitted to the casing by the pinion. A simple two-degree-of-freedom pinion model is described next that can be used to compute the passage of vibratory energy from a pinion–gear mesh to the pinion bearings.

Two-Degree-of-Freedom Pinion Model

Let gear element 2 shown in Fig. 21.2 be the pinion of the meshing pinion–gear pair shown there. Assume that gear element 1 has sufficient inertia so that the displacements $u^{(1)}$ and $\delta\theta^{(1)}$ caused by the tooth meshing fundamental and higher frequencies of the mesh force W are negligible in comparison with the comparable pinion displacements $u^{(2)}$ and $\delta\theta^{(2)}$. Assume further that the shaft of pinion element 2 is elastic and that it is connected to a massive rotational element, so that at the tooth meshing fundamental and higher frequencies the rotational vibratory motions $\delta\theta^{(2)}$ of the pinion are taken up entirely by elastic deformations of its shaft of torsional stiffness K rather than by rotational motions of the massive element to which it is connected. Let u denote the pinion shaft displacement $u^{(2)}$ in Fig. 21.2 in the direction normal to the shaft centerline and parallel with the plane of contact. Let θ denote the vibratory component $\delta\theta^{(2)}$ of the pinion rotation $\theta^{(2)}$ shown in Fig. 21.2, where positive directions of u and θ coincide with positive directions of $u^{(2)}$ and $\theta^{(2)}$, respectively. If the pinion is modeled as a rigid body except for elastic deformations of its teeth, then its equations of motion are

$$m\ddot{u} + r\dot{u} + ku = W \tag{21.51}$$

and

$$I\ddot{\theta} + K\theta = RW \tag{21.52}$$

where m and I are its mass and rotational moment of inertia, respectively, R is its base cylinder radius $R_b^{(2)}$, W is the transmitted force shown in Fig. 21.2, r is the real part of the bearing–bearing support impedance, and k is the bearing–bearing support stiffness; hence, $-k/\omega$ is the imaginary part of the bearing-bearing support impedance. According to Eqs. (21.51) and (21.52), the only vibratory energy dissipation taking place in the above-described pinion model is energy transmitted to the casing, which is governed parametrically by the real part r of the pinion bearing–bearing support impedance.

The mesh constraint equation, Eq. (21.10), is to be combined with the pinion equations of motion, Eqs. (21.51) and (21.52). Setting $u^{(1)} = 0$ and $\delta\theta^{(1)} = 0$ in Eq. (21.10) as indicated above and $R_b^{(2)} = R$, $\delta\theta^{(2)} = \theta$, and $u^{(2)} = u$ and dropping the overbar from the total mesh compliance $\overline{C}$ in Eq. (21.10), we have

$$-R\theta - u - CW = \zeta_0 \tag{21.53}$$

or

$$W = -C^{-1}(\zeta_0 + R\theta + u) \tag{21.54}$$

Eliminating W by substituting Eq. (21.54) into Eqs. (21.51) and (21.52) gives

$$m\ddot{u} + r\dot{u} + (k + C^{-1})u + RC^{-1}\theta = -C^{-1}\zeta_0 \tag{21.55}$$

and

$$RC^{-1}u + I\ddot{\theta} + (K + R^2C^{-1})\theta = -RC^{-1}\zeta_0 \tag{21.56}$$

which are a pair of simultaneous differential equations for the pinion displacements $u(t)$ and $\theta(t)$ with the static transmission error Eq. (21.9) as the driving function.

Pinion Model Structural Transfer Function

To predict the noise radiated by the gear casing, one requires the pinion bearing–bearing support displacement $u(t)$. This quantity is most conveniently determined in the frequency domain by utilizing the complex frequency response function $\hat{u}/\hat{\zeta}_0$ to be obtained from Eqs. (21.55) and (21.56), where $\hat{u}$ and $\hat{\zeta}_0$ are the complex sinusoidal amplitudes defined by

$$u = \hat{u}\exp(i\omega t) \tag{21.57}$$

and

$$\zeta_0 = \hat{\zeta}_0\exp(i\omega t) \tag{21.58}$$

Hence,

$$\dot{u} = i\omega\hat{u}\exp(i\omega t) \tag{21.59}$$

and

$$\ddot{u} = -\omega^2 \hat{u} \exp(i\omega t) \tag{21.60}$$

with comparable expressions holding for the variable θ. Eliminating the temporal derivatives in Eqs. (21.55) and (21.56) using such expressions, one obtains

$$(-m\omega^2 + i\omega r + k + C^{-1})\hat{u} + RC^{-1}\hat{\theta} = -C^{-1}\hat{\zeta}_0 \tag{21.61}$$

and

$$RC^{-1}\hat{u} + (-I\omega^2 + K + R^2C^{-1})\hat{\theta} = -\ RC^{-1}\hat{\zeta}_0 \tag{21.62}$$

which is a pair of simultaneous algebraic equations for the complex sinusoidal amplitudes $\hat{u}$ and $\hat{\theta}$. Dividing both sides of Eqs. (21.61) and (21.62) by $\hat{\zeta}_0$, one can solve algebraically for $\hat{u}/\hat{\zeta}_0$ using Cramer's rule. After minor algebraic manipulations, the results can be expressed as

$$\frac{\hat{u}}{\hat{\zeta}_0} = -\frac{N(\omega)}{D_r(\omega) + iD_i(\omega)} \tag{21.63}$$

$$= -\frac{N(\omega)\,[D_r(\omega) - iD_i(\omega)]}{D_r^2(\omega) + D_i^2(\omega)} \tag{21.64}$$

where

$$N(\omega) \triangleq -I\omega^2 + K \tag{21.65}$$

$$D_r(\omega) \triangleq CN(\omega)\,[-m\omega^2 + k + C^{-1}] + R^2[-m\omega^2 + k] \tag{21.66}$$

and

$$D_i(\omega) \triangleq r\omega[CN(\omega) + R^2] \tag{21.67}$$

From Eq. (21.64), it follows that

$$\left|\frac{\hat{u}}{\hat{\zeta}_0}\right|^2 = \frac{N^2(\omega)}{D_r^2(\omega) + D_i^2(\omega)} \tag{21.68}$$

Discussion

Several general features of the *mesh-to-pinion-bearing* structural transfer function given by Eqs. (21.63)–(21.67) are readily apparent. First, since the pinion response $u(t)$ and the excitation $\zeta_0(t)$ both are *displacements*, the structural transfer functions $\hat{u}/\hat{\zeta}_0$ are dimensionless. This property makes them very easy to interpret. At frequencies where $|\hat{u}/\hat{\zeta}_0|$ is less than unity, static transmission error components are *attenuated* by the structural transfer function $\hat{u}/\hat{\zeta}_0$, whereas at frequencies where $|\hat{u}/\hat{\zeta}_0|$ is greater than unity, static transmission error components are *amplified* by the structural transfer function $\hat{u}/\hat{\zeta}_0$.

The dynamical system modeled by Eqs. (21.51) and (21.52) or Eqs. (21.55) and (21.56) has two degrees of freedom, the pinion displacement u parallel with the plane of contact and the pinion rotation θ. Hence, the system possesses two resonant frequencies. The values of the two undamped ($r = 0$) resonant frequencies are the solutions of $D_r(\omega) = 0$, as one cas see from Eqs. (21.63) and (21.67). From Eqs. (21.65) and (21.66) one can observe that $D_r(\omega)$ is a quadratic expression in ω^2.

The value of the structural transfer function $\hat{u}/\hat{\zeta}_0$ is zero at the frequency $\omega = \omega_\theta \triangleq (K/I)^{1/2}$ where $N(\omega) = 0$, as one can see from Eqs. (21.63) and (21.65). This value of $\omega = \omega_\theta$ is the natural frequency of the torsional system described by Eq. (21.52).

It is instructive to evaluate the transfer function $\hat{\theta}/\hat{\zeta}_0$ at the torsional natural frequency $\omega = \omega_\theta$ where $(\hat{u}/\hat{\zeta}_0) = 0$. From Eqs. (21.61) and (21.62), one can readily show, using Cramer's rule, that

$$\frac{\hat{\theta}}{\hat{\zeta}_0} = -\frac{R[-m\omega^2 + k] + iRr\omega}{D_r(\omega) + iD_i(\omega)} \tag{21.69}$$

where $D_r(\omega)$ and $D_i(\omega)$ are defined by Eqs. (21.66) and (21.67), respectively. Evaluating Eq. (21.69) at the torsional natural frequency $\omega = \omega_\theta$ where $N(\omega) = 0$ gives $(\hat{\theta}/\hat{\zeta}_0) = -1/R$ or

$$\hat{\zeta}_0 = -R\hat{\theta} \quad \text{at } \omega = \omega_\theta \tag{21.70}$$

Thus, at the frequency $\omega = \omega_\theta$ where $\hat{u} = 0$, the transmission error ζ_0 is taken up entirely by the rotational motion θ, as one can see from Eq. (21.70) and Fig. 21.2.

The above-described behavior of the two-degree-of-freedom system, Eqs. (21.55) and (21.56), is essentialy equivalent to that of the dynamic vibration absorber,[28] except that the excitation ζ_0 of our system is a displacement rather than a force. In the present application, the torsional degree of freedom plays the role of the "absorber system." This concept is potentially useful only in gearing systems designed to operate at a constant speed, so that $\omega_\theta \triangleq (K/I)^{1/2}$ and the tooth meshing fundamental frequency can be set equal to each other, thereby attempting to eliminate the gear casing excitation u at this frequency. In attempting to utilize this principle, the most accurate dynamic model possible should be used to calculate the frequency that sets $\hat{u}/\hat{\zeta}_0$ equal to zero.

At $\omega = 0$, one can see from Eqs. (21.63) and (21.65)–(21.67) that the value of $\hat{u}/\hat{\zeta}_0$ is

$$\frac{\hat{u}}{\hat{\zeta}_0} = -\frac{1}{1 + k[C + R^2/K]} \qquad \omega = 0 \tag{21.71}$$

which describes the roles of the bearing–bearing support stiffness k, mesh compliance C, and torsional stiffness K in determining the fraction of the static transmission error taken up by the pinion displacement u at frequencies near $\omega = 0$ for the simplified model described by Eqs. (21.51) and (21.52).

Finally, the asymptotic behavior of $\hat{u}/\hat{\zeta}_0$ as $\omega \to \infty$ can be obtained from Eq. (21.63) by retaining the terms containing the highest powers in ω in the numerator and denominator, that is,

$$\frac{\hat{u}}{\hat{\zeta}_0} \sim \frac{1}{Cm\omega^2} \qquad \omega \to \infty \tag{21.72}$$

which, in particular, is independent of both the bearing–bearing support stiffness k and the real part of the bearing–bearing support impedance r.

21.5 EXAMPLES

Examples of the various components of the static transmission error spectra and the mesh-to-pinion-bearing structural transfer function are provided below. A more detailed discussion of transmission error spectra may be found in references 4 and 5.

Transmission Error Tooth Meshing Harmonic Spectra

The Fourier series expansion coefficients of the *tooth meshing* harmonic components $p = 0, \pm 1, \pm 2, \ldots$ of the static transmission error from a meshing gear pair are given by the sum of the three terms on the right-hand side of Eq. (21.18). The first term arising from elastic deformations of the teeth is given for helical and spur gear pairs, respectively, by Eqs. (21.21) and (21.22) when the tooth pair stiffness per unit length of line of contact is assumed to be constant. Figure 21.5 is a plot of the normalized rms harmonic amplitudes $\sqrt{2}\,|\alpha_{W_p}|/(W_0/\overline{K}_T)$ computed using Eq. (21.22) for a spur gear pair with transverse contact ratio $Q_t = 1.819$.

The individual second and third terms on the right-hand side of Eq. (21.18) each represents the contribution of one gear of the meshing pair arising from deviations of its mean tooth surface from a perfect involute surface. Each of these two terms is given by (21.39) for helical gears or Eq. (21.40) for spur gears.

Evaluation of the Fourier series coefficients using Eq. (21.40) was carried out[4] for a spur gear with 50 teeth and a transverse contact ratio of $Q_t = 1.819$. Measured deviations of four of the tooth profiles from perfect involute profiles are shown in Fig. 21.6. Since each of the four deviation curves shown in Fig. 21.6 is virtually identical with the others, it was assumed that all tooth profiles on the gear are identical except for tooth spacing errors. Hence, for all $j = 0, 1, \ldots, 49$ it was assumed in this case that $a_{0l}^{(i)} = c_{j,0l}^{(i)}$, $l \neq 0$. See Eq. (21.38). The Legendre expansion coefficients $a_{0l}^{(i)} = c_{j,0l}^{(i)}$, for use in Eq. (21.40) were computed using

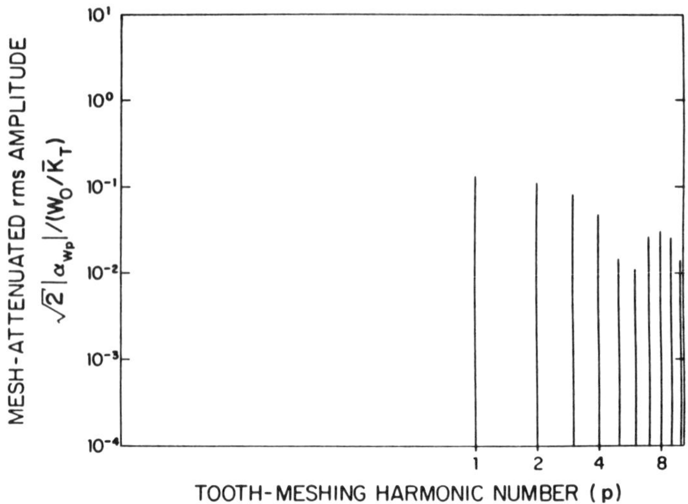

Fig. 21.5 Normalized contribution to static transmission error spectrum from elastic tooth deformations of a spur gear pair with transverse contact ratio $Q_t = 1.819$.

Eq. (21.30) from the second profile from the top displayed in Fig. 21.6. In carrying out this computation, the surface deviation $\eta_{Cj}^{(i)}(y, z)$ was assumed to be independent of axial position y. Using the expansion coefficients $a_{0l}^{(i)}$ through $l = 20$, the profile deviation was reconstructed using the summation over the index l ($k = 0$) in Eq. (21.26). Excellent agreement with the second profile deviation in Fig. 21.6 was obtained using this reconstruction. Using these same 21 expansion coefficients $a_{0l}^{(i)}$, the Fourier series coefficients $\alpha_{pN^{(i)}}^{(i)}$ were computed using Eq. (21.40); the corresponding rms amplitudes $\sqrt{2}\,|\alpha_{pN^{(i)}}^{(i)}|$ are displayed in Fig. 21.7. After computing the comparable Fourier series coefficients for the mating gear, the Fourier series coefficients α_{mp} of the tooth meshing harmonics would be obtained by forming the sum of the three complex terms on the right-hand side of Eq. (21.18).

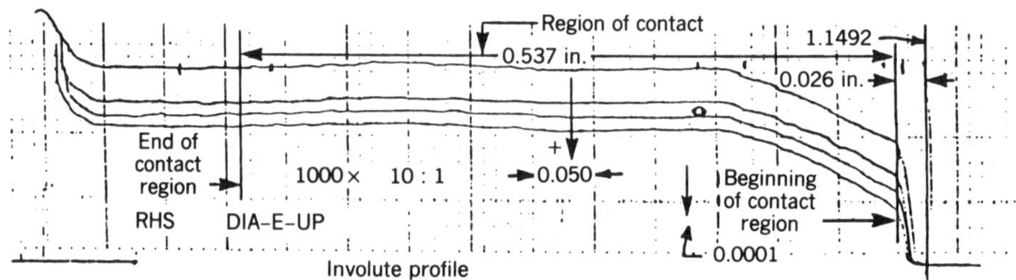

Fig. 21.6 Measured deviations from involute profiles of four teeth spaced approximately 90° apart on a spur gear with 50 teeth. Tooth tips are to the right. (Measurements were obtained from Bell Helicopter Textron.)

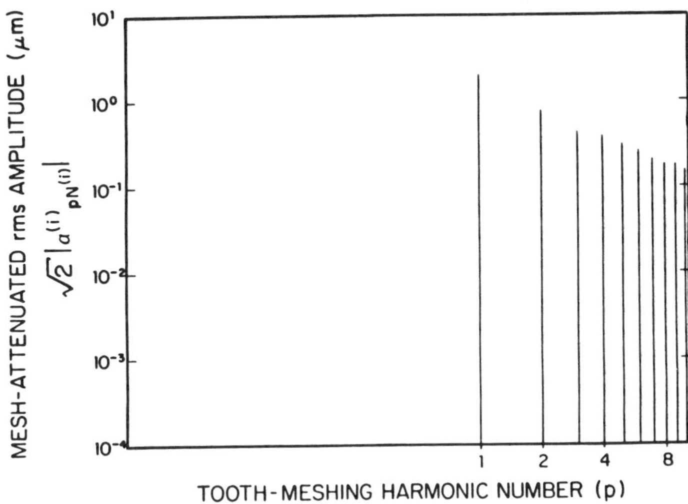

Fig. 21.7 Mean profile deviation rms harmonic contributions to the transmission error spectrum of 50-tooth spur gear whose profile deviations are shown in Fig. 21.6; Q_t = 1.819.

The assumption that the tooth pair stiffness per unit length of line of contact is a constant can be avoided by utilizing Eqs. 68, 69, and 94 of reference 3 to calculate the Fourier series coefficients of the individual contributions instead of Eqs. (21.21) or (21.22) and (21.39) or (21.40).

Transmission Error Rotational Harmonic Spectra

Expressions for the Fourier series expansion coefficients of the *rotational* harmonic contributions to the transmission error from an individual gear of a meshing pair are given by Eqs. (21.35) and (21.36) for helical and spur gears, respectively. If all tooth running surfaces are assumed to be identical except for tooth spacing errors, then only the single term $k = 0$, $l = 0$ will contribute to the summations in Eqs. (21.35) and (21.36). Furthermore, it follows from Eqs. (21.26)–(21.31) that when all running surfaces are identical except for tooth spacing errors, the expansion coefficient $c^{(i)}_{j,00}$ is identical with the *corrected accumulated* tooth spacing deviation[24] (error) of tooth j as measured by a typical pitch error measurement apparatus.[29]

The accumulated tooth spacing errors are displayed in Fig. 21.8 for the 50-tooth gear whose profile deviations are shown in Fig. 21.6. Notice that the accumulated errors of tooth number 50 and tooth number 0 are the same since they are the same tooth.

Figure 21.9 displays the one-sided spectrum $\sqrt{2}\,|B^{(i)}_{00}(n)'|$ of the accumulated tooth spacing errors shown in Fig. 21.8, where $B^{(i)}_{00}(n)'$ is computed from the accumulated errors $c^{(i)}_{j,00}$ by Eq. (21.33). Notice that the tooth spacing error spectrum

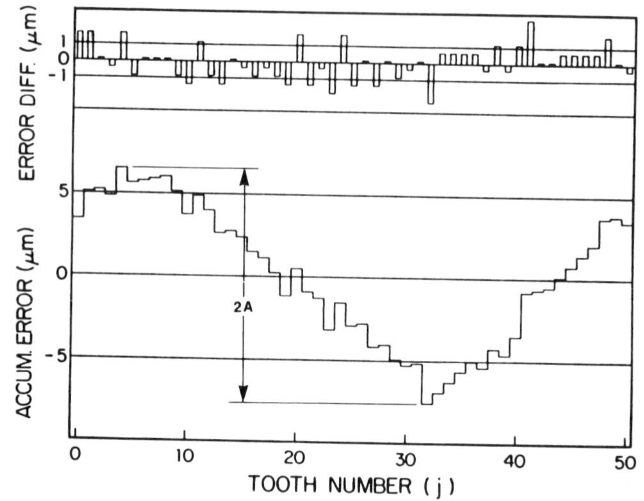

Fig. 21.8 Tooth spacing errors of 50-tooth spur gear whose profile deviations are shown in Fig. 21.6. (Original data obtained from Bell Helicopter Textron.)

$\sqrt{2}\,|B_{00}^{(i)}(n)'|$ shown in Fig. 21.9 is periodic with period equal to the number of teeth $N^{(i)} = 50$, as is indicated by Eq. (21.34).

According to Eq. (21.35) or (21.36), the tooth spacing error spectrum $B_{00}^{(i)}(n)'$ must be multiplied by the mesh transfer function $\hat{\phi}_{00}(n/N^{(i)}\Delta)$ for tooth spacing errors to obtain the contribution from tooth spacing errors to the rotational harmonic spectrum of the transmission error. The magnitude of the tooth spacing error mesh transfer function $|\hat{\phi}_{00}(n/N^{(i)}\Delta)|$ computed by Eq. (21.50) for a 50-tooth spur

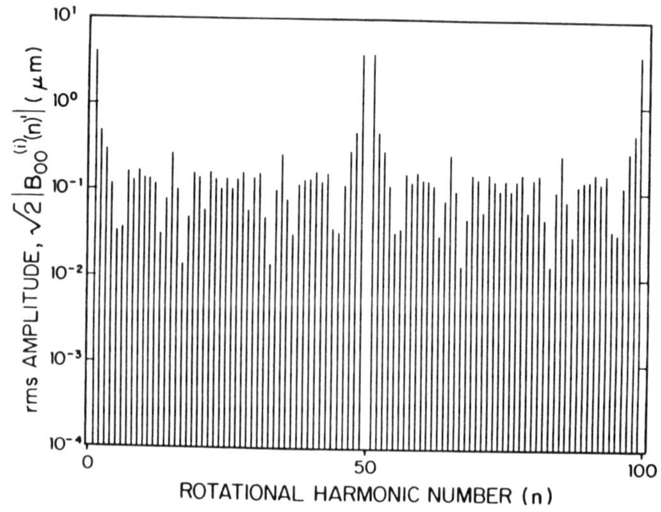

Fig. 21.9 One-sided spectrum of the accumulated tooth spacing errors shown in Fig. 21.8. Computation carried out using Eq. (21.33).

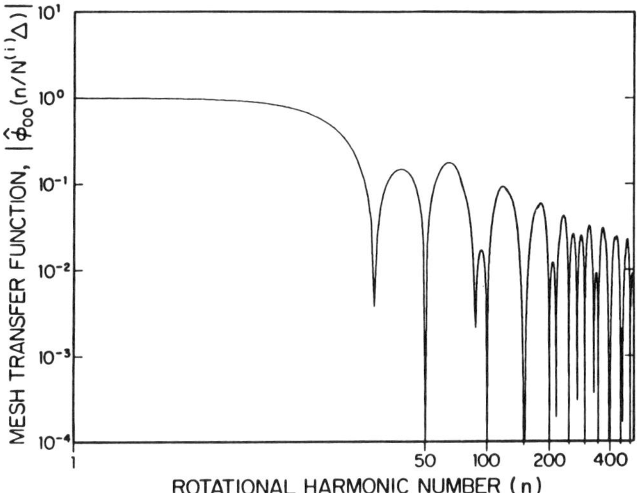

Fig. 21.10 Tooth spacing error mesh transfer function magnitude for 50-tooth spur gear with transverse contact ratio $Q_t = 1.819$.

gear with transverse contact ratio of $Q_t = 1.819$ is shown in Fig. 21.10. Notice that Fig. 21.10 has a logarithmic harmonic number axis, whereas Fig. 21.9 has a linear harmonic number axis.

According to Eq. (21.36), the contribution from tooth spacing errors to the rotational harmonic transmission error spectrum of a single gear is obtained by forming the product of the line spectrum shown in Fig. 21.9 and the mesh transfer function shown in Fig. 21.10. The magnitude of this product is shown in Fig. 21.11, which displays the *envelope* of the contributions from the tooth spacing errors shown in Fig. 21.8 to the rms values of the rotational harmonics $\sqrt{2}\,|\alpha_n|$. Figure 21.11 also is plotted on a logarithmic harmonic number axis. A rotational harmonic contribution occurs at each discontinuity in slope of the data shown in Fig. 21.11.

It is of interest to compare the spur gear mesh transfer function for tooth spacing errors shown in Fig. 21.10 with its helical gear counterpart shown in Fig. 21.12, which was computed by Eq. (21.47) using $k = 0$, $l = 0$ for a 50-tooth helical gear with a transverse contact ratio of $Q_t = 1.819$ and an axial contact ratio of $Q_a = 3.19$. Thus, the two gears for which the mesh transfer functions shown in Figs. 21.10 and 21.12 were computed are the same except that Fig. 21.10 is for a spur gear and Fig. 21.12 is for a helical gear with $Q_a = 3.19$. At the larger harmonic numbers, the helical gear mesh transfer function attenuates tooth spacing error spectra significantly more than the spur gear mesh transfer function.

Mesh-to-Pinion-Bearing Structural Transfer Function

The absolute value $|\hat{u}/\hat{\zeta}_0|$ of the mesh-to-pinion-bearing structural transfer function given by the square root of Eq. (21.68) is plotted on logarithmic coordinates

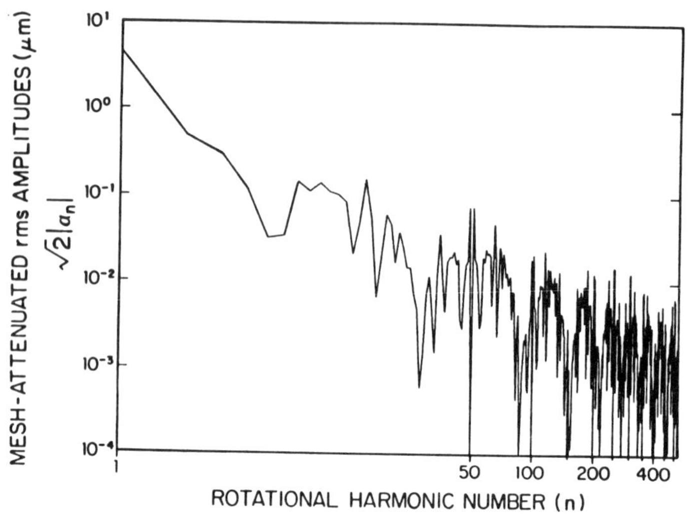

Fig. 21.11 Envelope of contribution to transmission error spectrum rms harmonic amplitudes arising from tooth spacing errors shown in Fig. 21.8 for 50-tooth spur gear with transverse contact ratio $Q_t = 1.819$. A discrete harmonic occurs at each discontinuity in slope.

in Fig. 21.13. The set of parameters used in computing Fig. 21.13 is listed in Table 21.2.

The values of the two resonant frequencies shown in Fig. 21.13 are obtained from the solution of $D_r(\omega) = 0$, where $D_r(\omega)$ is given by Eq. (21.66). Using the parameters listed in Table 21.2, we find the values of these resonant frequencies

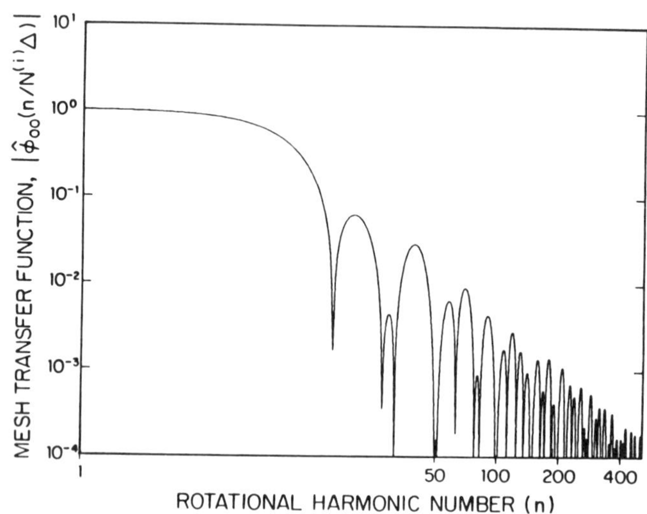

Fig. 21.12 Tooth spacing error mesh transfer function magnitude for 50-tooth helical gear with transverse contact ratio $Q_t = 1.819$ and axial contact ratio $Q_a = 3.19$. Compare with Fig. 21.10.

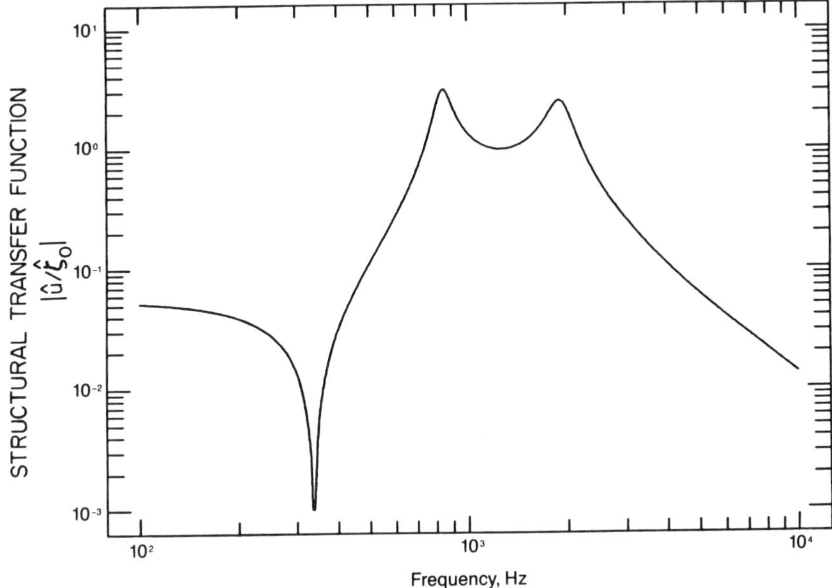

Fig. 21.13 Magnitude of mesh-to-pinion-bearing structural transfer function given by square root of Eq. (21.68) for parameters listed in Table 21.2.

to be 833 and 1931 Hz, which are in good agreement with Fig. 21.13. The value of the null frequency is the solution of $N(\omega) = 0$, which from Eq. (21.65) is $\omega_\theta = (K/I)^{1/2}$. Using the parameters in Table 21.2, we find the value of this null frequency to be 337 Hz. The value of $|\hat{u}/\hat{\zeta}_0|$ at $\omega = 0$ is found from Eq. (21.71) and Table 21.2 to be $|\hat{u}/\hat{\zeta}_0| = 5.60 \times 10^{-2}$. Using the asymptotic formula, Eq. (21.72), the value of $|\hat{u}/\hat{\zeta}_0|$ at $\omega = 2\pi \times 10^4$ is found to be $|\hat{u}/\hat{\zeta}_0| \approx 1.3 \times 10^{-2}$. These values also are in good agreement with Fig. 21.13. Finally, from Fig. 21.13 we see that for frequencies within the range of about 740–2200 Hz ($|\hat{u}/\hat{\zeta}_0| > 1$) the structural transfer function $|\hat{u}/\hat{\zeta}_0|$ *amplifies* or leaves unchanged the transmission error displacements, whereas for frequencies below and above this range, $|\hat{u}/\hat{\zeta}_0| < 1$, $|\hat{u}/\hat{\zeta}_0|$ *attenuates* the transmission error displacements.

The structural transfer function, Eq. (21.63), is the ratio of response to excitation *displacements*. To compute the sound power radiated from a structure, one

TABLE 21.2 Parameters Used in Calculating Fig. 21.13 from the Square Root of Eq. (21.68)

$I = 1.40\ \text{N} \cdot \text{m} \cdot \text{s}^2$
$K = 6.27 \times 10^6\ \text{N} \cdot \text{m}$
$C = 2.61 \times 10^{-10}\ \text{m/N}$
$m = 76.0\ \text{N} \cdot \text{s}^2/\text{m}$
$k = 5.25 \times 10^9\ \text{N/m}$
$R = 1.36 \times 10^{-1}\ \text{m}$
$r = 2.00 \times 10^5\ \text{N} \cdot \text{s/m}$

requires the *velocity* spectrum of the vibration. By multiplying the transfer function $\hat{u}/\hat{\zeta}_0$ by $i\omega$, one obtains the ratio of velocity response to displacement excitation.

21.6 NOISE CONTROL RAMIFICATIONS

Transfer Function Method

A transfer function method[3–5] has been outlined above for estimating the velocity spectrum Fourier series coefficients $\alpha_{\text{BV}}(\omega)$ of the bearing–bearing support motions normal to a pinion or gear shaft centerline, where these motions are caused by the transmission error of the pinion–gear mesh. The bearing–bearing support velocity history $v_B(t)$ can be expressed by the Fourier synthesis

$$v_B(t) = \sum \alpha_{\text{BV}}(\omega)\exp(i\omega t) \tag{21.73}$$

where $\alpha_{\text{BV}}(\omega)$ denotes the complex *bearing velocity* amplitude arising from a generic Fourier series coefficient of one of the transmission error components. If we denote by α_ζ one of these Fourier series coefficients, Eqs. (21.13), (21.15), or the superposition, Eq. (21.18), then $\alpha_{\text{BV}}(\omega)$ can be expressed as

$$\alpha_{\text{BV}}(\omega) = i\omega \frac{\hat{u}}{\hat{\zeta}_0} \alpha_\zeta \tag{21.74}$$

where $\hat{u}/\hat{\zeta}_0$ denotes the dimensionless ratio of complex sinusoidal bearing displacement response divided by complex sinusoidal transmission error mesh displacement excitation, which is the mesh-to-bearing structural transfer function. For the simple pinion model described in Section 21.4, this frequency-dependent structural transfer function is given by Eq. (21.63); its absolute value is illustrated in Fig. 21.13. The term $i\omega$ on the right-hand side of Eq. (21.74) converts the bearing displacement response amplitude $\hat{u}$ to velocity. For each of the cases where α_ζ denotes the Fourier series coefficient, Eq. (21.13), of a transmission error *rotational* harmonic component of order n from gear element i, frequency ω in Eqs. (21.73) and (21.74) is taken to be $\omega = \omega_n \triangleq n\omega^{(i)}$, $n = 0, \pm 1, \pm 2, \ldots$, where $\omega^{(i)}$ denotes the shaft rotation frequency of gear element i in radians per second. When α_ζ denotes the Fourier series coefficient, Eq. (21.15) or (21.18), of a transmission error *tooth meshing* harmonic component of order p, frequency ω in Eqs. (21.73) and (21.74) is taken to be $\omega = \omega_p \triangleq pN^{(i)}\omega^{(i)}$, $p = 0, \pm 1, \pm 2, \ldots$. See Eq. (21.17). The summation in Eq. (21.73) is a summation over all rotational harmonic components, Eq. (21.13), $n = 0, \pm 1, \pm 2, \ldots, n \neq pN^{(i)}$, of *each* of the two gear elements of the meshing pair and over all tooth meshing harmonic components, Eq. (21.18), $p = 0, \pm 1, \pm 2, \ldots$, of the gear pair.

Sound radiated from gear housings can be estimated from the bearing–bearing support velocity amplitude coefficients $\alpha_{\text{BV}}(\omega)$ using methods outlined in Chapter 9. Housing resonant frequencies should be avoided in the frequency ranges where the tooth meshing fundamental frequency and its harmonics occur.

Three transmission error sources contribute to the vibration excitation arising from each gear mesh: elastic deformations of the teeth whose Fourier series coefficients $\alpha_\zeta = \alpha_{W_p}$ are given by Eqs. (21.21) or (21.22) and deviations of the tooth running surfaces of each of the two meshing gear elements from equispaced perfect involute surfaces whose Fourier series coefficients $\alpha_\zeta = \alpha_n^{(i)}$ or $\alpha_\zeta = \alpha_{pN^{(i)}}^{(i)}$ are given by Eq. (21.35) or (21.36) or (21.39) or (21.40). Each of these six expressions for the transmission error component Fourier series coefficients α_ζ consists of products of terms $W_0/\overline{K}_T$, $B_{kl}^{(i)}(n)'$ or $B_{0l}^{(i)}(n)'$, or $a_{kl}^{(i)}$ or $a_{0l}^{(i)}$ that characterize elastic deformations or tooth deviations independently of the principal mesh design parameters, and terms $\sin(p\pi Q_a)\sin(p\pi Q_t)/[(p\pi Q_a)(p\pi Q_t)]$ or $\sin(p\pi Q_t)/(p\pi Q_t)$, or $\hat{\phi}_{kl}$ or $\hat{\phi}_{0l}$ that, for given harmonics p or harmonic number relative positions $n/N^{(i)}$, depend on the axial and transverse contact ratios Q_a and Q_t, or Q_t alone. *This transfer function method allows one to assess, independently, the effects of structural transfer functions $\hat{u}/\hat{\zeta}_0$, elastic deformations $W_0/\overline{K}_T$, mean tooth deviations characterized by $a_{kl}^{(i)}$ or $a_{0l}^{(i)}$, tooth errors characterized by $B_{kl}^{(i)}(n)$ or $B_{0l}^{(i)}(n)$, and the principal gear mesh design parameters that affect noise which are the axial and transverse contact ratios Q_a and Q_t.*

Effects of Shaft Speed, Numbers of Teeth, and Dynamic System Parameters

Normally, the strongest vibratory excitation components arising from a gear mesh are the tooth meshing harmonic components that occur at the tooth meshing fundamental frequency and its harmonics, $\omega_p = pN^{(i)}\omega^{(i)}$ radians per second, or $f_p = pN^{(i)}f^{(i)}$ Hz, $p = 1, 2, \ldots$, where $\omega^{(i)}$ denotes the shaft rotational speed of gear element i in radians per second, $f^{(i)} = \omega^{(i)}/2\pi$, and $N^{(i)}$ is the number of teeth on gear element i. Ordinarily, the shaft speeds and their ratios are controlled by considerations other than noise. *Nevertheless, the designer still has some control over the ranges of tooth meshing harmonic excitation frequencies $f_p = pN^{(i)}f^{(i)}$ through choice of the number of teeth $N^{(i)}$ on the pinion or gear, even though the tooth ratio $N^{(1)}/N^{(2)}$ of the gear pair may be governed by other considerations.*

If the number of teeth $N^{(i)}$ can be chosen jointly with the gearing element dynamic parameters so that the low-order tooth meshing harmonic frequencies $f_p = pN^{(i)}f^{(i)}$ avoid the resonant ranges of the mesh-to-bearing structural transfer functions where the excitation harmonics are amplified, then quieter operation should be achieved. This frequency range would be the range from about 740 to 2200 Hz for the particular mesh-to-pinion-bearing structural transfer function illustrated in Fig. 21.13. In addition to affecting the tooth meshing harmonic excitation frequencies, the numbers of teeth $N^{(i)}$ also affect the axial and transverse contact ratios Q_a and Q_t that control the attenuating effects of multiple tooth contact as discussed later in this section.

The low-order rotational harmonic components at frequencies $\omega_n = n\omega^{(i)}$ radians per second or $f_n = nf^{(i)}$ hertz, $n = 1, 2, \ldots$, also can be quite strong. The designer may have some control over the mesh-to-bearing structural transfer functions in this low-frequency region through potential adjustments in the system stiffnesses and compliances, as is illustrated by Eq. (21.71) for the pinion structural transfer function shown in Fig. 21.13.

Finally, apart from the effects of the mesh-to-bearing structural transfer function represented in Eq. (21.74) by $\hat{u}/\hat{\zeta}_0$, the complex bearing velocity amplitudes $\alpha_{BV}(\omega)$ are increased in proportion to shaft speed $\omega^{(i)}$ because of the term ω on the right-hand side of that equation.

Effects of Gear Mesh Design Parameters

All of the functions $[\sin(p\pi Q_a)\sin(p\pi Q_t)/(p\pi Q_a p\pi Q_t)]$, $[\sin(p\pi Q_t)/(p\pi Q_t)]$, $\hat{\phi}_{kl}(n/N^{(i)}\Delta)$, $\hat{\phi}_{0l}(n/N^{(i)}\Delta)$, $\hat{\phi}_{kl}(p/\Delta)$, and $\hat{\phi}_{0l}(p/\Delta)$ in Eqs. (21.21), (21.22), (21.35), (21.36), (21.39), and (21.40), respectively, play similar roles. These functions represent the attenuating effects of multiple tooth contact on elastic deformations of the teeth, machining errors, and mean deviations of the running surfaces of the teeth from equispaced perfect involute surfaces. It follows from Eqs. (21.44)–(21.50) that for a given tooth meshing harmonic number p or a given ratio $n/N^{(i)}$ of rotational harmonic number n to number of teeth $N^{(i)}$, the above-mentioned $\sin x/x$ and $\hat{\phi}$ functions are dependent only on the axial and transverse contact ratios Q_a and Q_t for helical gears or on Q_t alone for spur gears. However, from Eq. (21.17) one can see that the ratio $n/N^{(i)}$ describes the positions of the rotational harmonics n relative to the tooth meshing harmonics p. *Hence, for given tooth meshing harmonics p or rotational harmonic number relative positons $n/N^{(i)}$, the attenuating effects of multiple tooth contact are controlled by the axial and transverse ratios Q_a and Q_t for helical gears or the transverse contact ratio Q_t alone for the spur gears.*[4] Expressions for Q_a and Q_t in terms of common mesh design parameters are[4,5]

$$Q_a = \frac{FN^{(i)}\cos\phi\tan\psi}{\pi D_b^{(i)}} \qquad Q_t = \frac{DN^{(i)}\csc\phi}{\pi D_b^{(i)}} \tag{21.75}$$

where
- F = active facewidth measured parallel to gear axes
- $N^{(i)}$ = number of teeth on gear i
- ϕ = transverse pressure angle
- ψ = pitch cylinder helix angle
- $D_b^{(i)}$ = base cylinder diameter of gear i
- D = active tooth height

The active tooth height D is measured in the coordinate z define by Eq. (21.25). The active tooth dimensions F and D are illustrated in Fig. 21.3.

Figures 21.10 and 21.12 show that in the regions of the tooth meshing fundamental frequency $p = 1$ ($n = 50$) and beyond, the envelope of each mesh transfer function decays in a linear fashion on the log-log coordinates of these figures. This behavior is predicted by examining the asymptotic (large-argument) behavior[4] of the envelopes of the dominant terms in Eqs. (21.47) and (21.50) at $g = p/\Delta$:

$$\left|\hat{\phi}_{kl}\left(\frac{p}{\Delta}\right)\right|_{\text{env}} \sim \frac{[(2k+1)(2l+1)]^{1/2}}{p^2\pi^2 Q_a Q_t} \quad \text{for helical gears} \tag{21.76}$$

$$\left|\hat{\phi}_{0l}\left(\frac{p}{\Delta}\right)\right|_{\text{env}} \sim \frac{(2l+1)^{1/2}}{p\pi Q_t} \quad \text{for spur gears} \tag{21.77}$$

The dependence of the above asymptotic forms, Eqs. (21.76) and (21.77), on p, Q_a, and Q_t is the same as the dependence of the *envelopes* of Eqs. (21.21) and (21.22), respectively, on these same parameters. In particular, *for helical gears* it follows from Eq. (21.21) and the asymptotic form, Eq. (21.76), that, *to a first approximation, a single mesh design parameter $Q_a Q_t$ governs the overall behavior of the attenuating effects of multiple tooth contact on elastic deformations of the teeth, machining errors, and mean deviations of the running surfaces of the teeth from equispaced perfect involute surfaces.*[4] This parameter, which is the product $Q_a Q_t$ of the axial and transverse contact ratios, is called the *aggregate contact ratio*.[5] An intuitive understanding of this conclusion is aided by the fact that the Fourier series coefficients α_p of the instantaneous total length of line of contact within the zone of contact shown in Fig. 21.1 divided by the mean total length of line of contact α_0 are

$$\frac{\alpha_p}{\alpha_0} = \frac{\sin(p\pi Q_a)}{p\pi Q_a} \frac{\sin(p\pi Q_t)}{p\pi Q_t} \qquad p = 0, \pm 1, \pm 2, \ldots \tag{21.78}$$

whose *envelope* depends on Q_a and Q_t only through their product $Q_a Q_t$, the aggregate contact ratio. For spur gears ($Q_a = 0$), the result, Eq. (21.78) reduces to $\alpha_p/\alpha_0 = \sin(p\pi Q_t)/(p\pi Q_t)$. Equation (21.21) and the asymptotic form, Eq. (21.76), suggest that, *to a first approximation*, the mesh design parameters affect the decibel value of the vibratory excitation arising from a meshing pair of helical gears as $-20 \log Q_a Q_t$. *Thus, to minimize the vibratory excitation, the product $Q_a Q_t$ of the axial and transverse contact ratios should be maximized. In the case of spur gears, the transverse contact ratio should be maximized. To minimize noise, every effort should be made to hold the transverse contact ratio Q_t, computed from the active tooth height, above the value of 2 for both helical and spur gears.*

Finally, it follows from Eqs. (21.21), (21.22), (21.35), (21.36), (21.39), (21.40), (21.47), and (21.50) that, *to a first approximation, the decibel reduction in the vibratory excitation and noise in changing from a spur gear to a helical gear of the same active transverse contact ratio while holding the specific mesh loading $W_0/\overline{K}_T$ and machining accuracies unchanged is, on the average, about* $20 \log(\sqrt{2} p\pi Q_a)$, where $p = 1, 2, \ldots$ denotes the tooth meshing harmonic number of the frequency region for which the comparison is being made. This rule of thumb, *which should be regarded only as very crude first approximation*, illustrates *the enormous advantage, from a noise and vibration viewpoint, of large* *axial contact ratio helical gears in comparison with spur gears for frequencies in the regions of the tooth meshing harmonics*. When comparing helical gears to spur gears with adequate tip relief, this rule of thumb tends to overestimate the advantage of helical gears in the regions of the larger harmonic numbers p. However, helical gears offer no significant advantage over spur gears in reducing the low-order *rotational* harmonics $n = 1, 2, \ldots$, that arise, principally, from tooth-spacing errors.

Effects of Specific Mesh Loading and Tip/Root/End Relief Modifications

In spur gears, sufficient tip and/or root relief should be applied to compensate for the elastic deformations and spacing errors so that smooth tooth contact initiation is ensured.[30] In helical gears, end relief, tip/root relief, or a combination of end relief and tip/root relief, accomplishes this same purpose.[17,30] Thus, deviations of the mean running surfaces of the teeth from equispaced perfect involute surfaces characterized by the Legendre expansion coefficients $a_{kl}^{(i)}$ and $a_{0l}^{(i)}$ in Eqs. (21.39) and (21.40), respectively, are of the same order of magnitude as the deviations arising from elastic deformations of the teeth under full loading conditions. It follows then from Eqs. (21.21) and (21.22) that, *to a first approximation*, decibel changes in the tooth meshing harmonics of the vibratory excitation and response vary with the specific mesh loading $W_0/\overline{K}_T$ as 20 log $(W_0/\overline{K}_T)$, where we note that $\overline{K}_T$ is the mean *total* mesh stiffness.

Effects of Tooth Spacing Errors

Usually, the tooth spacing error term $k = 0$, $l = 0$ in Eqs. (21.35) and (21.36) is the largest machining error contributor to the *rotational* harmonics $n \neq pN^{(i)}$ of the vibration excitation and response. From Fig. 21.8 one can see that a typical measured accumulated tooth spacing error chart has a stronge once-per-revolution component. This component shows up in the accumulated tooth spacing error spectrum illustrated in Fig. 21.9 as a strong rotational fundamental harmonic $n = 1$ and also as strong rotational harmonics *immediately adjacent* to the positions of the tooth meshing harmonics that occur at integer multiples of the number of $N^{(i)} = 50$ teeth on the rotational harmonic number axis shown in Fig. 21.9.

Thus, we consider separately the lowest order rotational harmonics $n = 1, 2, \ldots$ and the rotational harmonics immediately adjacent to the tooth meshing harmonics that often are referred to as sidebands of the tooth meshing harmonics. It can be seen from Figs. 21.10 and 21.12 that the attenuating effects of multiple tooth contact contribute no attenuation to the low-order tooth spacing error rotational harmonics since the mesh transfer functions $|\hat{\phi}_{00}(n/N^{(i)}\Delta)|$ are unity in the regions of the first few rotational harmonics $n = 1, 2, \ldots$ for both spur and helical gears. Thus, the first few harmonics, $n = 1, 2, \ldots$ in Fig. 21.9 of the accumulated tooth spacing error chart shown in Fig. 21.8 will contribute directly to the static transmission error spectrum with no attenuation provided by multiple tooth contact. This result is illustrated by the first mesh attenuated rotational harmonic shown in Fig. 21.11, which has the same amplitude as that harmonic shown in Fig. 21.9.

The behavior of the mesh-to-bearing structural transfer functions in the neighborhood of $\omega = 0$ is important for determining the magnification or attenuation of these low-order rotational harmonics in the structural path between the mesh and the gear housing. This behavior is illustrated by Eq. (21.71) and Fig. 21.13 for our simple mesh-to-pinion-bearing structural transfer function model.

The rotational harmonics in the neighborhoods of the tooth meshing harmonics also are of considerable interest. From Figs. 21.9–21.12 one can see that these rotational harmonics are significantly attenuated by the smoothing action of multiple tooth contact for both spur and helical gears. Both the envelopes of the mesh transfer functions shown in Figs. 21.10 and 21.12 and the narrow deep nulls in these mesh transfer functions in the immediate neighborhoods of the locations of the tooth meshing harmonics at $n = 50, 100, 150, \ldots$ contribute to attenuation of the strong rotational harmonics shown in Fig. 21.9 in the neighborhoods of the tooth meshing harmonics.

The deep nulls shown in Figs. 21.10 and 21.12 in the neighborhoods of these tooth meshing harmonic locations systematically occur only in the mesh transfer functions for tooth spacing errors designated by the index pair $k = 0$, $l = 0$ in Eqs. (21.47) and (21.50). The mesh transfer functions for the other classes of errors (Table 21.1 and Fig. 21.4), k and l not both zero, do not systematically possess such deep nulls in the neighborhoods of the tooth meshing harmonics.[4,5]

It is of interest to estimate when the rotational harmonics immediately adjacent to the locations $n = 50, 100, 150, \ldots$ of the tooth meshing harmonics in Figs. 21.9 and 21.11 become of comparable strength to the tooth meshing harmonic strengths in the transmission error spectrum, in which case the tooth meshing harmonics no longer would be dominant in the vibration excitation and response. Using a number of reasonable simplifying approximations, one can obtain a relation that will ensure with a reasonable degree of confidence that the tooth meshing harmonic components are stronger than the tooth spacing error contributions to their neighboring rotational harmonic components for gear pairs transmitting moderate to high torque. The three main simplifying approximations are as follows:

1. The tooth meshing harmonic components, Eq. (21.18), of the transmission error spectrum can be estimated from the elastic deformation contributions alone, which are given by Eqs. (21.21) or (21.22).
2. The accumulated tooth spacing error chart, as illustrated by Fig. 21.8, is dominated by a single sinusoidal component with period equal to the number of teeth, so that the maximum range $2A$ of accumulated error is approximately equal to twice the amplitude of this once-per-revolution sinusoidal component.
3. The attenuating effects of the above-mentioned narrow deep nulls in the tooth spacing error mesh transfer functions in the immediate neighborhoods of the tooth meshing harmonics located at $n = 50, 100, 150, \ldots$ in Figs. 21.10 and 21.12 can be ignored.

Using these approximations, one can show from Eqs. (21.18), (21.21)–(21.36), (21.47), (21.49), and (21.50) for $k = 0$, $l = 0$ that the tooth meshing harmonic contributions to the transmission error should be larger than the tooth spacing error contributions to their neighboring rotational harmonics if one has

$$A < \frac{W_0}{\overline{K}_T} \tag{21.79}$$

where A is the half-range of the accumulated tooth spacing error chart, as illustrated in Fig. 21.8, and $W_0/\overline{K}_T$ is the nonfluctuating (d.c.) component of the elastic deformation of the mesh that occurs in Eqs. (21.21) and (21.22). Because of the attenuating effects of the narrow deep nulls in the tooth spacing error mesh transfer functions in the immediate neighorhoods of the tooth meshing harmonics (compare approximation 3 above), satisfaction of the relation, Eq. (21.79), ordinarily should ensure in practice that the rotational harmonic contributions from tooth spacing errors are smaller than the tooth meshing harmonic contributions.

The relation, Eq. (21.79), can be used to specify the maximum allowable range $2A$ of accumulated tooth spacing errors illustrated in Fig. 21.8. For example, for a moderately heavily loaded gear pair one has, approximately, $(W_0/\overline{K}_T) \approx 25$ μm (1000 μin.) For this loading, the sideband contributions from tooth spacing errors should be smaller than the tooth meshing harmonic contributions provided that the maximum accumulated tooth spacing error range $2A$ is not larger than about 50 μm (2000 μin.).

ACKNOWLEDGMENT

Support for the writing of this chapter was provided by the Science Development Program of Bolt Beranek and Newman Inc.

REFERENCES

1. V. M. Faires and R. M. Keown, *Mechanism*, 5th ed., McGraw-Hill, New York, 1960; republished by Robert E. Krieger Publishing Co., Huntington, NY, 1980, Chapters 7 and 8.
2. A. Sloane, *Engineering Kinematics*, Macmillan, New York, 1941; republished by Dover, New York, 1966, pp. 163–197.
3. W. D. Mark, "Analysis of the Vibratory Excitation of Gear Systems: Basic Theory," *J. Acoust. Soc. Am.* **63**(5), 1409–1430 (1978).
4. W. D. Mark, "Analysis of the Vibratory Excitation of Gear Systems. II: Tooth Error Representations, Approximations, and Application," *J. Acoust. Soc. Am.* **66**(6), 1758–1787 (1979).

5. W. D. Mark, "Gear Noise Excitation," in R. Hickling and M. M. Kamal (eds.), *Engine Noise: Excitation, Vibration, and Radiation*, Plenum, New York, 1982, pp. 55–89.
6. H. Walker, "Gear Tooth Deflection and Profile Modification," *The Engineer*, October 14, 1938, pp. 409–412; October 21, 1938, pp. 434–436; August 16, 1940, pp. 102–104.
7. S. L. Harris, "Dynamic Loads on the Teeth of Spur Gears," *Proc. Inst. Mech. Eng.* **172**, 87–100 (1958).
8. R. W. Gregory, S. L. Harris, and R. G. Munro, "Dynamic Behavior of Spur Gears," *Proc. Inst. Mech. Eng.* **178**, 207–218 (1963–1964).
9. M. L. Baxter, Jr., "Basic Theory of Gear-Tooth Action and Generation," in D. W. Dudley (ed.), *Gear Handbook*, McGraw-Hill, New York, 1962, pp. 1-1 to 1-13.
10. W. D. Mark, "Use of the Generalized Transmission Error in the Equations of Motion of Gear Systems," *ASME J. Mechan. Transmiss. Automat. Design*, **109**(2), 283–291 (1987).
11. W. D. Mark, "The Generalized Transmission Error of Spiral Bevel Gears," *ASME J. Mechan. Transmiss. Automat. Design*, **109**(2), 275–282 (1987).
12 W. D. Mark, "Effects of Bearing Offset and Flexibility on the Mesh Force Distribution of Spiral Bevel Gears," *ASME J. Mechan. Transmiss. Automat. Design* **110**(2), 203–210 (1988).
13. W. D. Mark, "An Extremum Principle for Computation of the Zone of Tooth Contact and Generalized Transmission Error of Spiral Bevel Gears," *ASME J. Mechan. Transmiss. Automat. Design* **110**(2), 211–220 (1988).
14. W. D. Mark, "The Generalized Transmission Error of Parallel-Axis Gears," *ASME J. Mechan. Transmiss. Automat. Design* **111**(3), 414–423 (1989).
15. K. Hayashi, "Load Distribution on the Contact Line of Helical Gear Teeth," *Bull. JSME*, **6**(22) (Part 1, "Fundamental Concept"), pp. 336–343; (Part 2, "Gears of Large Tooth Width"), pp. 344–353, 1963.
16. M. D. Trbojevic, "Load Distribution on Helical Gear Teeth," *The Engineer*, August 9, 1957, pp. 187–190; August 16, 1957, pp. 222–224.
17. A. W. Davis, "Marine Reduction Gearing," *Proc. Inst. Mech. Eng.* **170**(16), 477–498 (1956).
18. H. Walker, "Helical Gears," *The Engineer*, July 12, 1946, pp. 24–26; July 19, 1946, pp. 46–48; July 26, 1946, pp. 70–71.
19. R. W. Cornell and W. W. Westervelt, "Dynamic Tooth Loads and Stressing for High Contract Ratio Spur Gears," *ASME J. Mechan. Design* **100**(1), 69–76 (1978).
20. R. W. Cornell, "Compliance and Stess Sensitivity of Spur Gear Teeth," *ASME J. Mechan. Des.* **103**(2), 447–459 (1981).
21. C. Weber, "The Deformation of Loaded Gears and the Effect of Their Load-carrying Capacity," Sponsored Research (Germany), British Department of Scientific and Industrial Research, Report Nos. 3, 4, and 5, 1949.
22. E. D. Rainville, *Special Functions*, Macmillan, New York, 1960, Chapter 10.
23. W. D. Mark, "Analytical Reconstruction of the Running Surfaces of Gear Teeth Using Standard Profile and Lead Measurements," *ASME J. Mechan. Transmiss. Automat. Design* **105**(4), 725–735 (1983); **106**(1), 22 (1984).

24. W. D. Mark, "Analytical Reconstruction of the Running Surfaces of Gear Teeth. Part 2: Combining Tooth Spacing Measurements with Profile and Lead Measurements," *ASME J. Mechan. Transmiss. Automat. Design* **109**(2), 268–274 (1987).

25. G. D. Bergland, "A Guided Tour of the Fast Fourier Transform," *IEEE Spectrum*, **6**, 41–52 (1969). Reprinted in L. R. Rabiner and C. M. Rader (eds.), *Digital Signal Processing*, IEEE Press, New York, 1972, pp. 228–239.

26. P. M. Woodward, *Probability and Information Theory With Applications to Radar*, 2nd ed., Pergamon, Oxford, 1964, p. 29.

27. M. Abramowitz and I. A. Stegun, *Handbook of Mathematical Functions with Formulas, Graphs, and Mathematical Tables*, U.S. Government Printing Office, Washington, DC, 1964, pp. 437, 438.

28. L. Meirovitch, *Elements of Vibration Analysis*, 2nd ed., McGraw-Hill, New York, 1986, pp. 131–134.

29. H. E. Merritt, *Gear Engineering*, Halsted, Wiley, New York, 1971, pp. 76–78.

30. H. Sigg, "Profile and Longitudinal Corrections on Involute Gears," AGMA Paper No. 109.16, the American Gear Manufacturers Association, Alexandria, VA, 1965.

BIBLIOGRAPHY

Kohler, H. K, A. Pratt, and A. M. Thompson, "Dynamics and Noise of Parallel-Axis Gearing," *Proc. Inst. Mech. Eng.* **184**, 111–121 (1969–1970).

Mark, W. D., "The Role of the Discrete Fourier Transform in the Contribution to Gear Transmission Error Spectra from Tooth-spacing Errors," *Proc. Inst. Mech. Eng.* **201**(C3), 227–229 (1987).

Mark, W. D., "Gear Noise," in C. M. Harris (ed.), *Handbook of Acoustical Measurements and Noise Control*, 3rd ed., McGraw-Hill, New York, 1991, Chapter 36.

Niemann, G. and J. Baethge, "Drehwegfehler, Zahnfederhärte und Geräusch bei Stirnrädern" *VDI-Z* **112**, 205–214 (1970).

Opitz, H., "Noise of Gears," *Philos. Trans. Roy. Soc. Lond. Ser. A* **263**, 369–380 (1968–1969).

Remmers, E. P., "Analytical Gear Tooth Profile Design," ASME Paper No. 72-PTG-47, American Society of Mechanical Engineers, New York, 1972.

Remmers, E. P., "Gear Mesh Excitation Spectra for Arbitrary Tooth Spacing Errors, Load and Design Contact Ratio," *ASME J. Mechan. Design* **100**(4), 715–722 (1978).

Smith, J. D., *Gears and Their Vibration*, Marcel Dekker, New York, 1983.

Welbourn, D. B., "Fundamental Knowledge of Gear Noise—A Survey," *Conference on Noise and Vibrations of Engines and Transmissions*, Cranfield Institute of Technology, The Institution of Mechanical Engineers, London, 1979.

APPENDIX A

General References

D. A. Bies and C. H. Hansen, *Engineering Noise Control*, Unwin Hyman, London (1988).

C. M. Harris, Ed., *Shock and Vibration Handbook. 3rd Edition*, McGraw-Hill, New York (1988).

L. L. Beranek. *Acoustical Measurements*, Revised Edition, Acoustical Society of America, New York (1988).

R. B. Randall, *Frequency Analysis*, 3rd Edition, Bruel & Kjaer. Naerum, Denmark (1987).

R. N. Bracewell, *The Fast Fourier Transform and Its Applications*, 2nd Edition, McGraw-Hill, New York (1986).

M. C. Junger and D. Feit, *Sound, Structures, and Their Interaction, 2nd Edition*, MIT Press, Cambridge, MA (1986).

J. S. Bendat and A. G. Piersol, *Random Data, Analysis and Measurement Procedures*, 2nd Edition, Wiley, New York (1986).

L. L. Beranek, *Acoustics*. Reprinted with changes by the Acoustical Society of America, New York (1986).

M. Belanger, *Digital Processing of Signals: Theory and Practice*, Wiley, New York (1985).

F. Fahy, *Sound and Structural Vibration—Radiation, Transmission and Response*, Academic, New York (1985).

M. J. Crocker, *Noise Control*, Van Nostrand Reinhold, New York (1984).

L. E. Kinsler, A. R. Frey, A. B. Coppens and J. V. Sanders, *Fundamentals of Acoustics*, 3rd Edition, Wiley, New York (1982).

L. Cremer and H. A. Mueller, *Principals and Applications of Room Acoustics*, Vols. 1 and 2, translated by T. J. Schultz Applied Science, London and New York (1982).

J. D. Erwin and E. R. Graf. *Industrial Noise and Vibration Control*, Prentice-Hall, Englewood Cliffs, NJ (1979).

Noise and Vibration Control Engineering: Principles and Applications, Edited by Leo L. Beranek and István L. Vér.
ISBN 0-471-61751-2

© 1992 John Wiley & Sons, Inc.

A. V. Oppenheim and R. W. Schafer, *Digital Signal Processing*, Prentice-Hall, Englewood Cliffs, NJ (1975).

R. L. Lyon, *Statistical Energy Analysis of Dynamical Systems*, MIT Press, Cambridge, MA (1975).

H. Kuttruff, *Room Acoustics*, Applied Science Publishers, London (1973).

M. P. Blake and W. S. Mitchell, *Vibration and Acoustic Measurement Handbook*, Spartan Books, New York (1972).

APPENDIX B

English System of Units

The English system of units is inherently confusing. In everyday American life, the *pound* (abbreviated lb) is used either as a force or as a weight, both having the units of force. To further complicate matters, some technical writers seek to provide a system parallel to the metric system by using either of two quantities for force and either of two corresponding quantities for mass. Thus, some adopt the pound as the unit of force and define a *slug* as the unit of mass. Others define a *poundal* as the unit of force and adopt the pound as the unit of mass. Neither the slug nor the poundal has found general acceptance in the literature, although we use the former in this text.

As one well-traveled acoustician said, "I have determined that 1 kg of butter bought in Zurich is exactly the same amount as 2.2 lb bought in New York. Whenever I wish to solve a technical problem in America without confusion, I immediately divide the number of pounds by 2.2 to obtain the equivalent number of kilograms. Then I work in the mks system, where force and mass are clearly distinguished." Let us take a moment to distinguish further between mass and force.

The mass of a body is defined as $m = F/\ddot{x}$, where F is the vector sum of all forces acting on the center of gravity of the unrestrained body and $\ddot{x}$ is the acceleration produced in the direction of the force F.

The weight of a body is defined as $w = \text{mg}$, where g is the acceleration of gravity (9.81 m/sec^2 on earth) and w is the force that must act on the otherwise unrestrained body to keep it stationary when exposed to the gravitational field. Consequently, a body has the same mass but different weight on the moon than on earth. On earth the weight of a 1 kg mass, which is designated as one kilopond (1 kp), is 1 kp weight = 1 kg mass $\times$ 9.81 m/sec^2 = 9.81 Newtons.

Noise and Vibration Control Engineering: Principles and Applications, Edited by Leo L. Beranek and István L. Vér.
ISBN 0-471-61751-2 © 1992 John Wiley & Sons, Inc.

CONSISTENT SYSTEMS OF UNITS USED IN THIS TEXT

Two consistent systems of units are used in this text, the *mks* and the *fss* systems. To describe them, let us start with Newton's second law.

$$\text{Force} = \text{mass} \times \text{acceleration} \tag{B.1}$$

In the *meter-kilogram-second* system

$$\text{No. of newtons} = \text{no. of kg} \times \text{no. of m/sec}^2 \tag{B.2}$$

In the *foot-slug-second* system

$$\text{No. of pounds force (lb)} = \text{no. of slugs} \times \text{no. of ft/sec}^2 \tag{B.3}$$

The relations among the magnitudes of the units are

$$\begin{aligned} 1 \text{ kp} &= 2.205 \text{ lb weight} \\ 1 \text{ kg} &= 0.0685 \text{ slug} \\ 1 \text{ slug} &= 14.59 \text{ kg} \\ 1 \text{ newton} &= 0.225 \text{ lb force} \\ 1 \text{ lb force} &= 4.448 \text{ newtons} \\ 1 \text{ slug} &= 32.17 \text{ lb weight} \\ 1 \text{ lb weight} &= 0.03108 \text{ slug} = 0.454 \text{ kg} \end{aligned}$$

Example B.1: If 1 kg mass is to be accelerated 1 m/sec^2, we see, by Eq. (B.2), that a force of 1 newton is required. How many pounds (force) is required for the same result?

Solution: 1 kg equals (2.205/32.17) slug and 1 m/sec^2 = 3.28 ft/sec^2. Thus, by Eq. (B.3), 0.225 lb (force) is required.

INCONSISTENT SYSTEMS OF ENGLISH UNITS USED IN THE LITERATURE

Two inconsistent systems of English units are commonly encountered, the *fps* and the *ips* systems.

In the *foot-pound-second* system (inconsistent system)

$$\text{No. of pounds force (lb)} = \frac{\text{no. of pounds weight (lb)}}{g} \times \text{no. of ft/sec}^2 \tag{B.4}$$

where g is the acceleration due to gravity in units of ft/sec^2, that is, 32.17 ft/sec^2.

In the *inch-pound-second system* (inconsistent system)

$$\text{No. of pounds force (lb)} = \frac{\text{no. of pounds weight (lb)}}{g} \times \text{no. of in./sec}^2 \quad \text{(B.5)}$$

where g is the acceleration due to gravity in units of in./sec^2, that is, 3.86 in./sec^2. Mechanical engineers often use the *in.-lb-sec* system in the field of shock and vibration.

Example B.2: One kilogram mass is accelerated five meters per second^2. Find the force necessary to do this in newtons and pounds (force).

Solution:

$$\begin{aligned} 1 \text{ kg} &= 2.2 \text{ lb weight} = 0.0685 \text{ slug} \\ 5 \text{ m/sec}^2 &= 16.4 \text{ ft/sec}^2 \\ \text{F (newtons)} &= 1 \times 5 = 5 \text{ newtons} \\ \text{F (lb)} &= 0.0685 \times 16.4 = 1.124 \text{ lb (force)} \end{aligned}$$

APPENDIX C

Conversion Factors

The following values for the fundamental constants were used in the preparation of the factors:

$$
\begin{aligned}
1 \text{ m} &= 39.37 \text{ in.} = 3.281 \text{ ft} \\
1 \text{ lb (weight)} &= 0.4536 \text{ kp} = 0.03108 \text{ slug} \\
1 \text{ slug} &= 14.594 \text{ kg} \\
1 \text{ lb (force)} &= 4.448 \text{ newtons} \\
\text{Acceleration due to gravity} &= 9.807 \text{ m/sec}^2 \\
&= 32.174 \text{ ft/sec}^2 \\
\text{Density of } H_2O \text{ at } 4^\circ\text{C} &= 10^3 \text{ kg/m}^3 \\
\text{Density of Hg at } 0^\circ\text{C} &= 1.3595 \times 10^4 \text{ kg/m}^3 \\
1 \text{ U.S. lb} &= 1 \text{ British lb} \\
1 \text{ U.S. gallon} &= 0.83267 \text{ British gallon}
\end{aligned}
$$

Noise and Vibration Control Engineering: Principles and Applications, Edited by Leo L. Beranek and István L. Vér.
ISBN 0-471-61751-2 © 1992 John Wiley & Sons, Inc.

TABLE C.1 Conversion Factors

To convert	Into	Multiply by	Conversely, multiply by
acres	ft^2	4.356×10^4	2.296×10^{-5}
	$miles^2$ (statute)	1.562×10^{-3}	640
	m^2	4,047	2.471×10^{-4}
	hectare ($10^4\ m^2$)	0.4047	2.471
atm	in. H_2O at 4°C	406.80	2.458×10^{-3}
	in. Hg at 0°C	29.92	3.342×10^{-2}
	ft H_2O at 4°C	33.90	2.950×10^{-2}
	mm Hg at 0°C	760	1.316×10^{-3}
	lb/in.^{2}	14.70	6.805×10^{-2}
	newtons/m^2	1.0132×10^5	9.872×10^{-6}
	kg/m^2	1.033×10^4	9.681×10^{-5}
°C	°F	(°C x 9/5) + 32	(°F − 32) x 5/9
cm	in.	0.3937	2.540
	ft	3.281×10^{-2}	30.48
	m	10^{-2}	10^2
circular mils	in.^{2}	7.85×10^{-7}	1.274×10^6
	cm^2	5.067×10^{-6}	1.974×10^5
cm^2	in.^{2}	0.1550	6.452
	ft^2	1.0764×10^{-3}	929
	m^2	10^{-4}	10^4
cm^3	in.^{3}	0.06102	16.387
	ft^3	3.531×10^{-5}	2.832×10^4
	m^3	10^{-6}	10^6
deg (angle)	radians	1.745×10^{-2}	57.30
dynes	lb (force)	2.248×10^{-6}	4.448×10^5
	newtons	10^{-5}	10^5
dynes/cm^2	lb/ft^2 (force)	2.090×10^{-3}	478.5
	newtons/m^2	10^{-1}	10
ergs	ft-lb (force)	7.376×10^{-8}	1.356×10^7
	joules	10^{-7}	10^7
ergs/cm^3	ft-lb/ft^3	2.089×10^{-3}	478.7
ergs/sec	watts	10^{-7}	10^7
	ft-lb/sec	7.376×10^{-8}	1.356×10^7
ergs/sec-cm^2	ft-lb/sec-ft^2	6.847×10^{-5}	1.4605×10^4
fathoms	ft	6	0.16667
ft	in.	12	0.08333
	cm	30.48	3.281×10^{-2}
	m	0.3048	3.281
ft^2	in.^{2}	144	6.945×10^{-3}
	cm^2	9.290×10^2	0.010764
	m^2	9.290×10^{-2}	10.764
ft^3	in.^{3}	1728	5.787×10^{-4}
	cm^3	2.832×10^4	3.531×10^{-5}
	m^3	2.832×10^{-2}	35.31
	liters	28.32	3.531×10^{-2}

TABLE C.1 Conversion Factors—*Continued*

To convert	Into	Multiply by	Conversely, multiply by
ft H_2O at 4°C	in. Hg at 0°C	0.8826	1.133
	lb/in.^{2}	0.4335	2.307
	lb/ft^2	62.43	1.602×10^{-2}
	newtons/m^2	2989	3.345×10^{-4}
gal (liquid U.S.)	gal (liquid Brit. Imp.)	0.8327	1.2010
	liters	3.785	0.2642
	m^3	3.785×10^{-3}	264.2
gm	oz (weight)	3.527×10^{-2}	28.35
	lb (weight)	2.205×10^{-3}	453.6
hp (550 ft-lb/sec)	ft-lb/min	3.3×10^4	3.030×10^{-5}
	watts	745.7	1.341×10^{-3}
	kw	0.7457	1.341
in.	ft	0.0833	12
	cm	2.540	0.3937
	m	0.0254	39.37
in.^{2}	ft^2	0.006945	144
	cm^2	6.452	0.1550
	m^2	6.452×10^{-4}	1550
in.^{3}	ft^3	5.787×10^{-4}	1.728×10^3
	cm^3	16.387	6.102×10^{-2}
	m^3	1.639×10^{-5}	6.102×10^4
kg	lb (weight)	2.2046	0.4536
	slug	0.06852	14.594
	gm	10^3	10^{-3}
kg/m^2	lb/in.^{2} (weight)	0.001422	703.0
	lb/ft^2 (weight)	0.2048	4.882
	gm/cm^2	10^{-1}	10
kg/m^3	lb/in.^{3} (weight)	3.613×10^{-5}	2.768×10^4
	lb/ft^3 (weight)	6.243×10^{-2}	16.02
liters	in.^{3}	61.03	1.639×10^{-2}
	ft^3	0.03532	28.32
	pints (liquid U.S.)	2.1134	0.47318
	quarts (liquid U.S.)	1.0567	0.94636
	gal (liquid U.S.)	0.2642	3.785
	cm^3	1000	0.001
	m^3	0.001	1000
$\log_e n$, or $\ln n$	$\log_{10} n$	0.4343	2.303
m	in.	39.371	0.02540
	ft	3.2808	0.30481
	yd	1.0936	0.9144
	cm	10^2	10^{-2}
m^2	in.^{2}	1550	6.452×10^{-4}
	ft^2	10.764	9.290×10^{-2}
	yd^2	1.196	0.8362
	cm^2	10^4	10^{-4}
m^3	in.^{3}	6.102×10^4	1.639×10^{-5}
	ft^3	35.31	2.832×10^{-2}

TABLE C.1 Conversion Factors—*Continued*

To convert	Into	Multiply by	Conversely, multiply by
m^3 *(Cont.)*	yd^3	1.3080	0.7646
	cm^3	10^6	10^{-6}
microbars (dynes/cm^2)	lb/in.^{2}	1.4513×10^{-5}	6.890×10^4
	lb/ft^2	2.090×10^{-3}	478.5
	newtons/m^2	10^{-1}	10
miles (nautical)	ft	6080	1.645×10^{-4}
	km	1.852	0.5400
miles (statute)	ft	5280	1.894×01^{-4}
	km	1.6093	0.6214
miles^2 (statute)	ft^2	2.788×10^7	3.587×10^{-8}
	km^2	2.590	0.3861
	acres	640	1.5625×10^{-3}
mph	ft/min	88	1.136×10^{-2}
	km/min	2.682×10^{-2}	37.28
	km/hr	1.6093	0.6214
nepers	db	8.686	0.1151
newtons	lb (force)	0.2248	4.448
	dynes	10^5	10^{-5}
newtons/m^2	lb/in.^{2} (force)	1.4513×10^{-4}	6.890×10^3
	lb/ft^2 (force)	2.090×10^{-2}	47.85
	dynes/cm^2	10	10^{-1}
lb (force)	newtons	4.448	0.2248
lb (weight)	slugs	0.03108	32.17
	kg	0.4536	2.2046
lb H_2O (distilled)	ft^3	1.602×10^{-2}	62.43
	gal (liquid U.S.)	0.1198	8.346
lb/in.^{2} (weight)	lb/ft^2 (weight)	144	6.945×10^{-3}
	kg/m^2	703	1.422×10^{-3}
lb/in.^{2} (force)	lb/ft^2 (force)	144	6.945×10^{-3}
	N/m^2	6894	1.4506×10^{-4}
lb/ft^2 (weight)	lb/in.^{2} (weight)	6.945×10^{-3}	144
	gm/cm^2	0.4882	2.0482
	kg/m^2	4.882	0.2048
lb/ft^2 (force)	lb/in.^{2} (force)	6.945×10^{-3}	144
	N/m^2	47.85	2.090×10^{-2}
lb/ft^3 (weight)	lb/in.^{3} (weight)	5.787×10^{-4}	1728
	kg/m^3	16.02	6.243×10^{-2}
poundals	lb (force)	3.108×10^{-2}	32.17
	dynes	1.383×10^4	7.233×10^{-5}
	newtons	0.1382	7.232
slugs	lb (weight)	32.17	3.108×10^{-2}
	kg	14.594	0.06852
slugs/ft^2	kg/m^2	157.2	6.361×10^{-3}
tons, short (2,000 lb)	tonnes (1,000 kg)	0.9075	1.102
watts	ergs/sec	10^7	10^{-7}
	hp (550 ft-lb/sec)	1.341×10^{-3}	745.7

INDEX